Plate Tectonics and Geomagnetic Reversals

A SERIES OF BOOKS IN GEOLOGY

Editors: James Gilluly
A. O. Woodford

Plate Tectonics and Geomagnetic Reversals

Readings, selected, edited, and with Introductions by

ALLAN COX
STANFORD UNIVERSITY

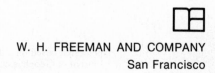

W. H. FREEMAN AND COMPANY
San Francisco

Library of Congress Cataloging in Publication Data

Cox, Allan, 1926–   comp.
    Plate tectonics and geomagnetic reversals

    Bibliography: p.
    1. Plate tectonics—Addresses, essays, lectures.
2. Paleomagnetism—Addresses, essays, lectures.
3. Heat budget (Geophysics)—Addresses, essays, lectures. I. Title.
QE511.4.C68   551.1′3   73-4323
ISBN 0–7167–0259–2
ISBN 0–7167–0258–4 (pbk)

Printed in the United States of America

Cover design based on a drawing from
the Journal of Geophysical Research.   4 5 6 7 8 9 10

Contents

Description of Introductory Plates

Frontispiece. San Andreas fault cutting across the Carrizo Plains in California. Photo courtesy of Robert E. Wallace, U.S. Geological Survey. ii

Section I. Gemini XI photograph of Ethiopia, Somali, the Red Sea, the Gulf of Aden, and Saudi Arabia. Photo courtesy of NASA. 1

Section II. Surf and headlands along the coast of California at Big Sur at the site of a former subduction zone. Photography by Cole Weston, reprinted with permission of the publisher, Sierra Club Books, from *Not Man Apart*. 9

Section III. San Andreas fault in the Carrizo Plains cutting and off-setting a stream. The nearly horizontal fault line marks the boundary between the nearer North America plate and the farther Pacific plate, which is moving to the right or northward. 39

Section IV. Eruption of Mayan Volcano, Phillipines, on May 1, 1968. Large blocks are being ejected from the vent. Small tongues of molten lava are speeding down the steep sides of the volcano. Photo by U.S. Air Force. 137

Section V. Young pillow basalt lava flow on the sea floor east of Hawaii at a water depth of 1,650 meters. Photo by James G. Moore and Richard S. Fiske, U.S. Geological Survey. 221

Section VI. View of the campus of Stanford University immediately after the 1906 earthquake. It is said that when David Starr Jordan, who was then president of the university, saw this statue, which had toppled from a second-story ledge, he remarked, "My dear Aggasiz, many is the time that I have contemplated you in the abstract, but this is the first time I have seen you in the concrete." Photo courtesy of Stanford Archives. 283

Section VII. Research vessel Argo of Scripps Institution of Oceanography on the Nova Expedition in the Tasman Sea, August 1967. Photo by C. E. Abranson, courtesy of William Menard. 407

Section VIII. Himalayas at dawn from Apollo 9. Photo by NASA. 439

Section IX. San Andreas fault cutting through the Mecca Hills in southern California. Photo by Robert E. Wallace, U.S. Geological Survey. 529

Preface

The impetus for publishing this set of articles was the realization that, year after year, I and my students at Stanford were making copies (with varying degrees of legibility and legality) of the same classic articles on plate tectonics and geomagnetic reversals. At first we did this because there was so little material on the subject available, but even after the appearance of several excellent textbooks on plate tectonics and paleomagnetism, we still found ourselves using many of the original articles as supplementary reading. Students somehow seem to sense more acutely the excitement of discovery when they are given a sense of participation by having the scientist himself describe it to them.

As scientists, most of us, when we write about our research, almost completely depersonalize it. Although there may be good reasons for this, it is a rather odd thing to do because we almost always feel a deep sense of personal involvement in our research. As some anomalous data fall into place or as a new idea begins to stir, the individual scientist, working perhaps late at night, experiences a feeling that surely is similar to the elation of the composer as he first hears new music in his mind, or to the exhilaration of the athlete as he breaks a record. Yet our science, as we write about it, tends to be science with the scientists left out. The effect is that we inadvertently may leave students with the impression that science is a routine, matter-of-fact business, which it surely isn't.

In the present book I have tried to put the scientists back into the scientific story. Wherever possible I have done this using the scientists' own words, quoting from statements given to me in the spring of 1972 by many of the authors whose articles are included in this book. No pretense is made of presenting a complete history of research in plate tectonics and geomagnetic reversals. My intent, rather, is to convey something of the feeling of intense involvement and excitement experienced by earth scientists as they contributed to important new ideas.

I acknowledge with gratitude the photographs and personal statements supplied by the several dozen scientists who are the true authors of this book. I also thank Kathleen Hart and R. Brian Hart for their help with editing, and Heather Sandretto for help with the bibliography.

November 1972   *Allan Cox*

PARADIGM OF PLATE TECTONICS

1

Introduction

The earth sciences are currently in an intellectual ferment as a result of recent advances in the study of magnetic reversals, sea-floor spreading, and plate tectonics. The last decade, in which most of these advances were made, was a most unusual one in the history of earth science. Previously, during the century following the great works of Charles Darwin, the earth sciences were characterized by increasing specialization and divergence. Paleontologists, seismologists, geomagnetists, geologists, and marine geophysicists became better and better at what they were doing, but at the same time they seemed to have less and less to say to each other. Then, in a remarkable series of articles written between 1962 and 1968, the trend toward divergence was reversed and many of the main threads of geologic research were brought together to form the fabric of plate tectonics.

Four main lines of independent experiments and measurements gave rise to this remarkable synthesis: (1) mapping of the topography of the sea floor using echo depth sounders; (2) measuring the magnetic field above the sea floor using proton-precession magnetometers; (3) timing the north-south flips of the earth's magnetic field using the magnetic memory of rocks from the continents and their radiometric ages; (4) determining very accurately the location of earthquakes using the worldwide net of seismometers originally developed to detect nuclear blasts. Each of these lines of research had been pursued for a decade or more for separate reasons. Their coming together to form the observational basis for plate tectonics must surely constitute one of the classic examples of serendipity in the history of science. It is difficult to imagine a central committee responsible for planning research in tectonics that would have had the imagination to foresee the unlikely path of development of this major scientific advance.

The central idea of plate tectonics is comparable to the theory of the Bohr atom in its simplicity, its elegance, and its ability to explain a wide range of observations about our planet. It explains why earthquakes and volcanoes are concentrated in remarkably narrow belts. It explains why some of these belts, such as the San Andreas fault zone, have only shallow earthquakes, whereas others, such as the Aleutian arc, have deep earthquakes. It explains why the sea floor is much younger than the continents. Plate tectonics links all of these phenomena by postulating the formation of moving crustal plates. Where two plates are moving apart, new oceanic floor is formed by the solidification of molten rock in the opening crack. Where two plates converge, one plate usually thrusts to depths of 700 km beneath the other, forming deep ocean trenches, deep earthquakes and, as the descending plate melts, volcanoes.

The earth's magnetic field acts as a chronometer, accurately timing these events. The heart of the timing system is located in the earth's liquid core, where a magnetic field is generated by electrical currents. This magnetic chronometer is binary in the sense that it has two stable states: a "normal" state in which the magnetic field is directed toward the north, and a "reversed" state in which the field is directed toward the south. The field switches back and forth between these two states at irregular intervals that may be as short as 30 thousand years or as long as several million years. The magnetic chronometer in the earth's core leaves a magnetic mark on rocks as they form at the earth's surface. As a result, rocks contain information about their age stored in a magnetic memory physically similar to the magnetic memory of a computer. Sediments that form on the sea floor contain such a memory, as do lava flows. Deciphering the magnetically coded information in rocks has permitted the element of time to be introduced into the description of plate tectonics, thus providing fundamental information about the *rate* of tectonic processes. This quantitative element, in turn, allowed plate tectonics to advance beyond earlier qualitative tectonic theories.

The main ideas of plate tectonics were put forward in several dozen articles by a handful of scientists, many of them still young and most of them interested in oceanography. Oceanography had undergone a great expansion during the late 1950's. As more and more geologic and geophysical information became available, the oceanographers noticed that the crests of certain submarine mountain ranges were places where a lot was happening. These mountain ranges are major physiographic features of our planet. They rise about two kilometers above the adjacent sea floor, they are several thousand kilometers wide at the base, and some of them extend more than one-quarter the way around the world. A good example is the mid-Atlantic ridge, which splits the North and South Atlantic oceans and then extends around the southern tip of Africa into the Indian Ocean. Terming these great mountain ranges "rises" or "ridges," the oceanographers noticed from echo bottom soundings that the ridges commonly had a well-developed rift or median valley running along their summits. The geomagnetists then noticed that the earth's magnetic field was unusually strong over the summit of the mountain ranges. Seismologists noticed that earthquakes were concentrated in a narrow band along the crests of the rises. Marine geologists found that the blanket of sediments on the sea floor was extremely thin or missing on the summits of the rises and that it thickened laterally on the flanks of the mountain ranges. Finally, marine geophysicists noticed that the flow of heat through the floor of the ocean was much greater near the crests of the mountain ranges.

All of these observations were soon explained in terms of plate tectonics. The crest of a submarine mountain range is the superficial manifestation of a crack 100 kilometers deep separating two "plates," a plate being a segment of the earth's crust mechanically uncoupled from other such segments along deep cracks. As two plates move away from each other, molten material rises from below to fill the incipient crack. High heat flow and shallow earthquakes accompany the formation of new crust as molten lava is forced through cracks in the older crust and is then extruded onto the sea floor. On contact with the sea water, the lava is quickly quenched and during quenching becomes strongly magnetized parallel to the earth's magnetic field. The magnetization of the lava flows is responsible for the strong magnetic anomalies observed over the ridges. Near the crest of a submarine mountain range the rocks that make up the 100-kilometer-thick crustal plate are hotter and therefore less dense than at greater distances from the crest. Because of this, they are more buoyant and float higher on the semifluid mantle. In fact, the reason for the very existence of the submarine mountain ranges is because they have not yet cooled and subsided. Plate tectonics was able to offer a simple explanation for a remarkably wide range of diverse observations. Conversely, it was the interest of specialists from different disciplines in a common phenomenon — the mid-oceanic ridges — that gave the initial impetus to the development of modern plate tectonics.

A Scientific Revolution

The term "scientific revolution" has been used in a technical rather than promotional sense by the historian of science Thomas Kuhn (1962). A scientific revolution is defined by Kuhn as a new pattern of generalizations about nature that emerges suddenly to guide the future direction of research. Kuhn's viewpoint contrasts with the traditional one that science grows in a continuous manner by a process of steady accretion. Terming the process by which a scientific revolution occurs a "paradigm change," Kuhn argues convincingly that this process is a discontinuous one in the sense that social revolutions are discontinuous: there is no reconciliation of opposing views within the context of some higher authority. The new viewpoint embodying the paradigm change simply wins by gaining the consent of the relevant community. After the scientific revolution, that is, after the paradigm change has been generally accepted, important new lines of research are guided by the new generalizations.

Many facets of plate tectonics and sea-floor spreading were anticipated by earlier theories. This was true also of such major paradigm changes as the discovery of oxygen and the development of the fluid theory of electricity, both of which had strong precedents in earlier theories. What occurred at the times of paradigm change was a sudden acceptance by a sizable segment of the scientific community because of the presentation of new experimental data and compelling theoretical arguments in favor of the theories. Once accepted, each of the new paradigms served to shape future research.

The history of our present-day ideas about plate tectonics and sea-floor spreading is similar to the above examples. They clearly had their antecedents in earlier theories of continental drift, mantle convection, and sea-floor renewal. Yet prior to the mid-1960's, research in tectonics could scarcely be regarded as being in a state of scientific health. Experimentation and observation, not having a firm theoretical basis, were poorly focused and lacked specific objectives. With the introduction of the concept of plate tectonics, many new avenues for well-focused theoretical and experimental research suddenly became apparent in many of the subdisciplines of earth science. Many lines of productive research closely related to plate tectonics continue up to the present time.

In the introductory chapters of this book we have followed Kuhn in using the terms "scientific revolution" and "paradigm" to describe plate tectonics and sea-floor spreading, rather than the more traditional "hypothesis" or "theory." In this way a sterile argument was avoided about whether plate tectonics should be described as a hypothesis or a theory. Moreover the development of plate tectonics, although describable in terms of several theories of the history of science, fits the pattern of Kuhn's scientific revolutions surprisingly well. It appears reasonable, therefore, to regard developments in the earth sciences during the past decade as the emergence of a major new scientific paradigm.

Plan of the Book

This collection of readings is divided into sections corresponding to the main lines of research that eventually contributed to plate tectonics. Each section begins with an introduction in which the articles that follow are tied to earlier work and the important contributions of the articles are spotlighted. The introductory chapters also describe how many of the contributing scientists became interested in research that was related to plate tectonics. For the most part, this historical material is based on statements from the individual scientists that were kindly provided at the time this book was assembled in 1972. Omissions of important contributions are unfortunately inevitable in any anthology of finite size. In the introductory chapters some of the important articles that are not included are described, and a supplemental reading list is given at the end of these chapters. Although emphasis is placed on articles written since 1962, a few key articles written earlier are included to remind us that in tectonics, as in other areas of human endeavor, original thinking did not begin with the present generation.

The articles are all reprinted in their entirety, even though some of them cover the same ground or contain material that is not of general interest. The reason for doing this is to allow the student to have a view of science in the rough as it is actually being done, before it is refined and polished on the way to the textbook. The student will note that the aim of the scientific investigator may be slightly off from what turns out to have been the important scientific target. He will also note that some discoveries are made before their time in the sense that there is a long delay from the time of discovery to the time when it is generally accepted and used. A good example is Matuyama's discovery in 1929 that the most recent reversal of the magnetic field took place at the beginning of the Pleistocene. On the other hand, other discoveries are not made until after the time is ripe for them, as evidenced by their simultaneous discovery by scientists working independently and by their rapid acceptance and application in subsequent research. A good example is the independent work by Dan McKenzie, Robert Parker, Jason Morgan, and Xavier Le Pichon on the geometry of plate movements. It is hoped that the section introductions and the articles themselves will convey something of the excitement of discovery felt by the scientists participating in this fruitful decade of research.

REFERENCES AND READING LISTS

References from the individual articles are combined in a bibliography at the end of the book. Where the original article used an abbreviated form of citation, complete titles and inclusive page numbers have been supplied insofar as was possible. For those articles originally cited as being "in press" that have now been published, we have completed the references. It is hoped that the listing of complete titles and the provision of reading lists will prove useful to students who may wish to pursue additional reading. To help the student bring himself up-to-date in this rapidly expanding field, several hundred recent references that are not cited in the individual articles have been added to the reading lists and to the general bibliography. Reading lists for the following subjects begin on the pages given.

THE BEGINNING: MARINE GEOLOGY

2

Introduction and Reading List

Guyots and Rises

The key figure in marine geologic research leading to plate tectonics was Harry H. Hess. This is said with the realization that the broad fabric of geologic research has many threads stretching far back into the past. Yet if a single seminal observation were to be singled out in the development of modern plate tectonics, it would be Hess's discovery in 1946 of flat-topped submarine volcanoes, which he termed "guyots" (Hess, 1946). (Contemporary students may be interested in knowing that Hess's marine research began in a happier and more innocent era when even the young did not question doing basic research under military auspices. Much of his research was done during World War II when he was navigator and then commanding officer of the U.S.S. *Cape Johnson* on her cruises in the central Pacific.) Hess's interpretation was that the guyots (1) form as volcanic islands near rise crests, (2) are truncated by wave erosion, and (3) then are submerged to depths of several kilometers as they migrate down the flanks of the rises. This idea was exciting to marine geologists because it demonstrated that the sea floor is very mobile and is still a convincing piece of evidence that sea-floor spreading has occurred.

The 1950's were a time of major geographic exploration over the world's oceans. This exploration yielded many results that were crucial to the development of plate tectonics. Although the shape of the ocean bottom had been known in a general way since the pioneering voyages of the H.M.S. *Challenger* and the

Figure 2-1
Harry Hess

U.S.S. *Tuscarora* in the 1870's, the research of the 1950's served to focus attention on two important geomorphic features: fracture zones and oceanic rises (or ridges—the two terms are equivalent).

Maurice Ewing and Bruce Heezen, sailing from Lamont Geologic Observatory at the Palisades on the Hudson River, concentrated on exploring an unusual set of submarine mountains in the Antarctic Ocean, in the Indian Ocean and running down the center of the North and South Atlantic oceans. These were termed ridges, although this name is somewhat misleading in view of the huge dimensions of these mountain ranges: typically 2 km high, 2000 km wide, and 10,000 km long.

Ewing and Heezen made the important discovery that at the very crest of the ridges there is usually a narrow trough or rift valley (Ewing and Heezen, 1956). This central rift valley was interpreted by Carey (1958b) and Heezen (1960) to be a narrow block sinking under tension as the sea floor on either side of the valley moves apart, that is, as the sea floor spreads. In the late 1950's the interesting idea was advanced that these rifts were evidence that the earth was undergoing steady expansion, the rifts being places where the earth was bursting at the seams (Egyed, 1956, 1957; Carey, 1958b; Heezen, 1960). However paleomagnetic studies have now pretty well ruled out expansion large enough to account for the formation of all the ocean basins (Cox and Doell, 1961; Van Andel and Hospers, 1968).

Sailing from Scripps Institution of Oceanography at La Jolla in Southern California, H. W. Menard explored the submarine mountain ranges of the Pacific

Figure 2-2
Maurice Ewing

Ocean. He carefully explored the crest of the longest of the Pacific mountain ranges, the East Pacific rise, which curves across the South Pacific from Australia to Chile and then northward to the Gulf of California. To his surprise, he was unable to find rift valleys at the summit like those found on the mid-Atlantic ridge. For a while it seemed that the East Pacific rise and the ridges of the Atlantic, Indian, and Antarctic oceans were different kinds of mountains produced by different tectonic processes. We now know that both are produced by sea-floor spreading and the terms rise and ridge, which were used for awhile to make distinct the difference in geomorphology, have come to be used interchangeably. For reasons still not clear, central rift valleys do not tend to form on those rises where, as in the South Pacific, the spreading rate is high.

The careful mapping of the ocean basins done during the 1950's helped set the scene for plate tectonics by demonstrating that the rises are not simply isolated submarine mountain ranges, but rather that they constitute a simple, continuous,

Figure 2-3
H. William Menard (left)

world-encircling system. And simple patterns in data always prove tempting to theoreticians.

The Darwin Proto-rise versus Hot Spots

Present-day rises are "living" geologic features in the sense that they owe their elevation to a dynamic process taking place today. Is there any evidence in the geologic record for the existence of ancient oceanic rises that are now "dead?" Menard (1958, 1964) and Hess (Chapter 4) suggested that there is, in the form of a zone of guyots with tops submerged 1–2 km below sea level that extends northwestward across the central Pacific. The interpretation of this well-studied set of guyots has challenged earth scientists up to the present time. Menard's explanation (1964) was that (1) the ocean floor bulged upward due to rising mantle convection which did not, however, produce ocean-floor spreading; (2) volcanic islands formed and were truncated; and (3) the bulge subsided as convection stopped. Menard termed this ancient ridge system the Darwin rise. Hess's explanation (Chapter 4) was similar but included sea-floor spreading in a direction perpendicular to the northwest trend of the rise. A new interpretation, and one that is stimulating much contemporary research, is that the band of guyots marks "hot spots" in the floor of the Pacific Ocean, that is, places where plumes of molten magma rise from the mantle (Chapter 49).

Fracture Zones

One of the most important scientific results to come from oceanographic exploration in the Pacific was the discovery of fracture zones in 1952 by H. W. Menard and R. S. Dietz (1952). Fracture zones have now been found in all ocean basins. Initially these zones attracted attention simply as unusual geomorphic features on the ocean floor. The fracture zones are long, thin bands of submarine mountains. They typically comprise several ridges and troughs with a relief of a few kilometers that trend parallel to the general trend of the zone, which in turn is in many instances roughly perpendicular to the trend of adjacent oceanic rises. A typical fracture zone is 50–100 km wide and of variable length up to several thousand kilometers.

Fracture zones usually separate regions of different depth. They trace paths that are segments of circles. Those in the Pacific are among the longest and most nearly perfect natural geometric features on the surface of the earth (Menard, 1964). From the time of the discovery of fracture zones, it was recognized that they are somehow associated with lateral faulting. Their full significance, however, was appreciated only after the development of plate tectonic theory, which uses fracture zones to determine the direction of movement of plates, as will be described in Section III. The difference in depth on the two sides of a fracture zone provides useful information about the thermal history of the ocean floor (Chapter 38).

Sea-floor Spreading versus Continental Drift

The difference between sea-floor spreading as it is now understood and the older idea of continental drift is concerned mainly with how the continents move relative to the oceanic crust. In his classic study Alfred Wegener (1912) envisaged the continents as moving like shallow rafts through a sea of basalt and upper mantle material. This idea met with opposition from geophysicists including Harold Jeffreys (1929) on the grounds that the oceanic basaltic layer was too strong to permit continents to plow through it. A second objection arose as geophysicists came to realize that the structure of continents is deep (MacDonald, 1963). For example, the seismic zone is known to extend 700 km down into the mantle beneath South America. If it was difficult for the geophysicist to imagine the drift of shallow continents such as Wegener described, it was even more difficult for him to imagine movement through the earth's upper mantle of a continent with a deep structure. For these and other reasons of a more geologic nature, many earth scientists were reluctant for many decades to accept Wegener's hypothesis of continental drift.

Interest in continental drift was revived with the publication in 1956 by E. Irving (1956) and S. K. Runcorn (1956) of paleomagnetic data substantiating that drift between North America and Europe had occurred since the Paleozoic Era. During the subsequent decade as the paleomagnetic evidence for continental drift became stronger and stronger, it became increasingly important to find a model consistent with the geophysical and geologic evidence.

The needed model was provided in Harry Hess's classic paper (Chapter 4). *History of the Ocean Basins.* Hess suggested that the continents do not plow through the oceanic crust but are carried passively on a mantle that is overturning due to thermal convection. The mid-oceanic rises are places where hot mantle material rises to form new crust, the topographic elevation of the rises being due to the lower density of the hotter rocks. Evidence for the so-called deep structure of continents such as the seismic zone beneath the west coast of South America actually points more toward a *process* than a structure. The process is downward movement on the descending limb of the convection cell. R. Dietz (1961) coined the term "sea-floor spreading" to describe the overall process.

The idea of thermal convection in the mantle is an old one. It was probably introduced by W. Hopkins in 1839 (Hopkins, 1839). It was further developed by A. Holmes, F. A. Vening Meinesz, and D. Griggs in the 1930's. Discussions of some of the earliest references are given by Meyerhoff (1968), Dietz (1968), and Hess (1968).

Arthur Holmes presented some of the concepts of ocean-floor spreading in an article in 1931 and in his classic textbook *Principles of Physical Geology* (Chapter 3). He saw the basaltic layer as a conveyor belt on top of which a continent is carried along until it comes to rest at a place where the belt turns downward and sinks into the earth. He also realized that new oceanic crust is generated by the intrusion of basaltic magma into the oceanic crust and the extrusion of lava on the sea floor, and that a continent moves *with* the adjacent ocean floor, and not through it. However Holmes saw the formation of new oceanic crust as a rather diffuse process rather than one occurring at the mid-oceanic rises. The rises themselves he thought were sialic fragments of the original continents left behind over the stagnant zone of the rising convection cell. In the crucial matter of explaining the rises, Holmes was wrong. However he was closer to modern thinking than was Hess in proposing that eclogite rather than Hess's serpentine played an important part in the convective process.

In developing the concept of ocean-floor spreading, Hess was particularly influenced by the work of Griggs (1939), which even today has a very modern ring. Griggs discussed "the attractive possibility of a convection cell covering the whole Pacific basin, comprising sinking peripheral currents localizing the circum-Pacific mountains and rising currents in the center. Such an interpretation would partially explain the sweeping of the Pacific basin clear of continental material. The seismologists all agree that the foci of deep earthquakes in the circum-Pacific region seem to be on planes inclined about 45° toward the continents. It might be possible that these quakes were caused by slipping along the convection-current surfaces."

Hess's great contribution was in using geologic and geophysical data to show that some of the earlier speculative ideas about convection in the mantle were wrong and that others were right. It was Hess who, for the first time, argued convincingly from data that new oceanic crust was being generated at the rises. Hess's classic paper marks the emergence of the new paradigm of sea-floor spreading and plate tectonics.

READING LIST

Historical Development of the Paradigm

Amstutz, A., 1955, Structures alpines, subductions successives dans l'Ossola: Acad. Sci. Paris, C. R., ser. D, v. 241, p. 967–969.

Carey, S. W., 1955, The orocline concept in geotectonics, part 1: Roy. Soc. Tasmania, Pap. Proc., v. 89, p. 255–288.

Carey, S. W., ed., 1958a, Continental drift; a symposium: Hobart, Univ. of Tasmania, Geol. Dept., 375 p.

Carey, S. W., 1958b, The tectonic approach to continental drift, in Carey, S. W., ed., Continental drift; a symposium: Hobart, Univ. of Tasmania, Geol. Dept., p. 177–358.

Dietz, R. S., 1961, Continent and ocean basin evolution by spreading of the sea floor: Nature, v. 190, p. 854–857.

Dietz, R. S., 1968, Reply: J. Geophys. Res., v. 73, p. 6567.

Du Toit, A. L., 1937, Our wandering continents: Edinburgh, Oliver and Boyd, 366 p.

Egyed, L., 1956, The change of the earth's dimensions determined from paleogeographical data: Pure Appl. Geophys., v. 33, p. 42–48.

Griggs, D. T., 1939, A theory of mountain building: Amer. J. Sci., v. 237, p. 611–650.

Gunn, R., 1947, Quantitative aspects of juxtaposed ocean deeps, mountain chains, and volcanic ranges: Geophysics, v. 12, p. 238–255.

Hess, H. H., 1946, Drowned ancient islands of the Pacific basin: Amer. J. Sci., v. 244, p. 772–791.

Hess, H. H., 1948, Major structural features of the western North Pacific: an interpretation of H. O. 5485, Bathymetric Chart, Korea to New Guinea: Geol. Soc. Amer., Bull., v. 59, p. 417–446.

Hess, H. H., 1955, The oceanic crust: J. Mar. Res., v. 14, p. 423–439.

Hess, H. H., 1968, Reply: J. Geophys. Res., v. 73, p. 6569.

Holmes, A., 1931, Radioactivity and earth movements: Geol. Soc. Glasgow, Trans., v. 18, p. 559–606.

Holmes, A., 1933, The thermal history of the earth: Wash. Acad. Sci., J., v. 23, p. 169–195.

Hopkins, W., 1839, Researches in physical geology: Roy. Soc. London, Phil. Trans., v. 129, p. 381–385.

Kennedy, G. C., 1959, The origin of continents, mountain ranges and ocean basins: Amer. Scientist, v. 47, p. 491–504.

Menard, H. W., 1965, Sea floor relief and mantle conviction: Phys. Chem. Earth, v. 6, p. 315–364.

Meyerhoff, A. A., 1968, Arthur Holmes: Originator of spreading ocean floor hypothesis: J. Geophys. Res., v. 73, p. 6563–6565.

Vening Meinesz, F. A., 1952, The origin of continents and oceans: Geol. Mijnbouw, v. 31, p. 373–384.

Vening Meinesz, F. A., 1962, Thermal convection in the earth's mantle, *in* Runcorn, S. K., ed., Continental drift: New York, Academic Press, p. 145–176.

Wegener, A., 1912, Die Entstehung der Kontinente: Geol. Rundschau, v. 3, p. 276–292.

Rises

Drake, C. L., Campbell, N. J., Sander, G., and Nafe, J. E., 1963, A mid-Labrador sea ridge: Nature, v. 200, p. 1085–1086.

Ewing, M., and Heezen, B. C., 1956, Some problems of antarctic submarine geology, *in* Crary, A., Gould, L. M., Hurlburt, E. O., Odishaw, H., and Smith, W. E., eds. Antarctica in the International Geophysical Year: Amer. Geophys. Union, Geophys. Monogr. 1, p. 75–81.

Ewing, M., Le Pichon, X., and Ewing, J., 1966, Crustal structure of the mid-Atlantic ridge: J. Geophys. Res., v. 71, p. 1611–1636.

Heezen, B. C., 1960, The rift in the ocean floor: Sci. Amer., v. 203, p. 98–110.

Heezen, B. C., and Ewing, M., 1961, The mid-oceanic ridge and its extension through the Arctic basin, *in* Raasch, G. O., ed., Geology of the Arctic, v. 1: Toronto, Univ. of Toronto Press, p. 622–642.

Heezen, B. C., and Ewing, M., 1963, The mid-oceanic ridge, *in* Hill, M. N., ed., The sea, v. 3: New York, Wiley-Interscience, p. 388–410.

Heezen, B. C., Ewing, M., and Miller, E. T., 1953, Trans-Atlantic profile of total magnetic intensity and topography, Dakar to Barbados: Deep-Sea Res., v. 1, p. 25–33.

Heezen, B. C., and Tharp, M., 1965, Tectonic fabric of Atlantic and Indian oceans and continental drift, *in* Blackett, P. M. S., Bullard, E. C., and Runcorn, S. K., eds., Symposium on continental drift: Roy. Soc. London, Phil. Trans., ser. A, v. 258, p. 90–108.

Heezen, B. C., Tharp, M., and Ewing, M., 1959, The floors of the oceans, 1, The North Atlantic: Geol. Soc. Amer., Spec. Pap. 65, 122 p.

Hess, H. H., 1959b, Nature of the great oceanic ridges, *in* Sears, M., ed., International Oceanographic Congress preprints: Washington, D. C., American Association for the Advancement of Science, p. 33–34.

Hess, H. H., 1965, Mid-oceanic ridges and tectonics of the sea floor, *in* Whittard, W. F., and Bradshaw, R., eds., Submarine geology and geophysics: London, Butterworth, p. 317–332.

Menard, H. W., 1958, Development of median elevations in the ocean basins: Geol. Soc. Amer., Bull., v. 69, p. 1179–1186.

Menard, H. W., 1967b, Sea-floor spreading, topography, and the second layer: Science, v. 157, p. 923–924.

Talwani, M., Le Pichon, X., and Heirtzler, J. R., 1965b, East Pacific rise: the magnetic pattern and the fracture zones: Science, v. 150, p. 1109–1115.

Fracture Zones

Heezen, B. C., Bunce, E. T., Hersey, J. B., and Tharp, M., 1964a, Chain and Romanche fracture zones: Deep-Sea Res., v. 11, p. 11–33.

Heezen, B. C., Gerard, R. D., and Tharp, M., 1964b, The Vema fracture zone in the equatorial Atlantic: J. Geophys. Res., v. 69, p. 733–739.

Matthews, D. H., 1963. A major fault scarp under the Arabian Sea displacing the Carlsberg ridge near Socotra: Nature, v. 198, p. 950–952.

Matthews, D. H., 1966, The Owen fracture zone and the northern end of the Carlsberg ridge: Roy. Soc. London, Phil. Trans., ser. A, v. 259, p. 227–239.

Menard, H. W., 1966, Fracture zones and offsets of the east Pacific rise: J. Geophys. Res., v. 71, p. 682–685.

Menard, H. W., 1967a, Extension of northeastern Pacific fracture zone: Science, v. 155, p. 72–74.

Menard, H. W., and Dietz, R. S., 1952, Mendocino submarine escarpment: J. Geol., v. 60, p. 266–278.

Vacquier, V., 1959, Measurement of horizontal displacement along faults in the ocean floor: Nature, v. 183, p. 452–453.

Vacquier, V., 1962, Magnetic evidence for horizontal displacements in the floor of the Pacific Ocean, *in* Runcorn, S. K., ed., Continental drift: New York, Academic Press, p. 135–144.

Vacquier, V., 1965, Transcurrent faulting in the ocean floor, *in* Blackett, P. M. S., Bullard, E., and Runcorn, S. K., eds., Symposium on continental drift: Roy. Soc. London, Phil. Trans., ser. A, v. 258, p. 77–81.

Vacquier, V. Raff, A. D., and Warren, R. E., 1961, Horizontal displacements in the floor of the northeastern Pacific Ocean: Geol. Soc. Amer., Bull., v. 72, p. 1251–1258.

3

The Machinery of Continental Drift: the Search for a Mechanism

ARTHUR HOLMES
1944

From *Principles of Physical Geology*, p. 505–509, 1944. Reprinted with permission of Thomas Nelson and Sons, Ltd., and The Ronald Press.

It has been shown that in looking for a possible means of "engineering" continental drift we must confine ourselves to processes operating within the earth. To be appropriate, the process must be capable (*a*) of disrupting the ancestral Gondwanaland into gigantic fragments, and of carrying the latter radially outwards: Africa and India towards the Tethys; Australasia, Antarctica, and South America out into the Pacific; (*b*) of disrupting Laurasia, though much less drastically, and again with radially outward movements towards the Tethys and the Pacific. We have already seen that the peripheral orogenic belts probably mark the regions where opposing systems of sub-crustal currents came together and turned downwards. The movements required to account for the mountain structures are in the same directions as those required for continental drift, and it thus appears that sub-crustal convection currents may provide the sort of mechanism for which we are looking (Fig. 3–1).

To explain the peripheral orogenic belts three systems of convection currents are called for (or three co-ordinated groups of systems), with their ascending centres situated beneath Gondwanaland, Laurasia, and the Pacific respectively. Incidentally, it should be noticed that the coalescence of the usual chaotic or small convective systems into three gigantic ones involves a coincidence that can rarely have happened in the earth's history, and one that is just as likely to have come about during the Mesozoic era as at any other time. The often-asked question: How is it that Pangæa did not begin to break up and unfold until Mesozoic time? thus ceases to have any significance. If continental drift could have been caused by the gravitational forces invoked by Wegener, then it should have occurred once and for all very early in the earth's history, since those forces have always been in operation. If convection currents are necessary,

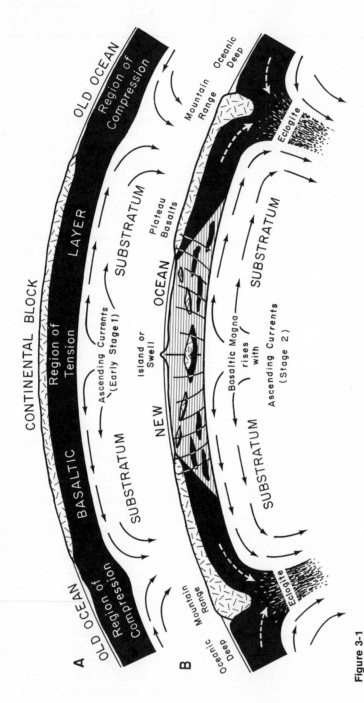

Figure 3-1
Diagrams to illustrate a purely hypothetical mechanism for "engineering" continental drift. In (A) sub-crustal currents are in the early part of the convection cycle. In (B) the currents have become sufficiently vigorous to drag the two halves of the original continent apart, with consequent mountain building in front where the currents are descending, and ocean floor development on the site of the gap, where the currents are ascending.

continental drift may have accompanied all the greater paroxysms of mountain building in former ages but, if so, it would usually have been on no more than a limited scale. That there was a quite exceptional integration of effort in Mesozoic and Tertiary times is forcibly suggested by eruptions of plateau basalts and building of mountains on a scale for which it would be hard to find a parallel in any earlier age.

There are, therefore, good reasons for supposing that at this critical period of the earth's history the convective circulations became unusually powerful and well organised. Currents flowing horizontally beneath the crust would inevitably carry the continents along with them, provided that the enormous frontal resistance could be overcome. The obstruction that stands in the way of continental advance is the basaltic layer, and obviously for advance to be possible the basaltic rocks must be continuously moved out of the way. In other words, they must founder into the depths, since there can be nowhere else for them to go (Fig. 3–1).

Now this is precisely what would be most likely to happen when two opposing currents come together and turn downwards beneath a cover of basaltic composition. The latter then suffers intense compression, and like the sial in similar circumstances it is eventually drawn in to form roots. On the ocean floor the expression of such a down-turning of the basaltic layer would be an oceanic deep. The great deeps bordering the island festoons of Asia and the Australasian arc (Tonga and Kermadec) probably represent the case where the sialic edge of a continent has turned down to form the inner flanks of a root, while the oceanic floor contributes the outer flanks.

It is not difficult to see that a purely basaltic root must have a very different history from one composed of sial. The density of sial is not significantly increased by compression. Consequently, when a sialic root is no longer being forcibly held down, it begins to rise in response to isostasy, heaving up a mountain range as it does so. But when rocks like basalt or gabbro (density 2.9 or 3.0) are subjected to intense dynamic metamorphism they are transformed into schists and granulites and finally into a highly compressed type of rock called *eclogite,* the density of which is about 3.4. Since this change is known to have happened to certain masses of basaltic rocks that have been involved in the stresses of mountain building, it may safely be inferred that basaltic roots would undergo a similar metamorphism into eclogite. Such roots could not, of course, exert any buoyancy, and for this reason it is impossible that tectonic mountains could ever arise from the ocean floor. On the contrary, a heavy root formed of eclogite would continue to develop downwards until it merged into and became part of the descending current, so gradually sinking out of the way, and providing room for the crust on either side to be drawn inwards by the horizontal currents beneath them (Fig. 3–1).

The eclogite that founders into the depths will gradually be heated up as it shares in the convective circulation. By the time it reaches the bottom of the substratum it will have begun to fuse, so forming pockets of magma which, being of low density, must sooner or later rise to the top. Thus an adequate source is provided for the unprecedented flows of plateau basalt that broke through the continents during Jurassic and Tertiary times. Most of the basaltic magma, however, would naturally rise with the ascending currents of the main convectional systems until it reached the torn and outstretched crust of the disruptive basins left behind the advancing continents or in the heart of the Pacific. There it would escape through innumerable fissures, spreading out as sheet-like intrusions within the crust, and as submarine lava flows over its surface. Thus, in a general way, it is possible to understand how the gaps rent in the crust come to be healed again; and healed, moreover, with exactly the right sort of material to restore the basaltic layer. To sum up: during large-scale convective circulation the basaltic layer becomes a kind of endless travelling belt on the top of which a continent can be carried along, until it comes to rest (relative to the belt) when its advancing front reaches the place where the belt turns downwards and disappears into the earth.

To go beyond the above indication that a mechanism for continental drift is by no means inconceivable would at present be unwise. Many serious difficulties still remain unsolved.

In particular, it must not be overlooked that a successful process must also provide for a general drift of the crust over the interior: a drift with a northerly component on the African side sufficient to carry Africa over the Equator, and Britain from the late Carboniferous tropics to its present position. The northward push of Africa and India, of which the Alpine system and the high plateau of Tibet are spectacular witnesses, could not have been sufficient by itself to shove Europe and Asia so far to the north. To achieve this the aid of exceptionally powerful sub-Laurasian currents directed towards the Pacific is required. The total northward components might then overbalance the southward components, and a general drift of the crust would be superimposed on the normal radial directions of drift.

It must be clearly realised, however, that purely speculative ideas of this kind, specially invented to match the requirements, can have no scientific value until they acquire support from independent evidence. The detailed complexity of convection systems, and the endless variety of their interactions and kaleidoscopic transformations, are so incalculable that many generations of work, geological, experimental, and mathematical, may well be necessary before the hypothesis can be adequately tested. Meanwhile it would be futile to indulge in the early expectation of an all-embracing theory which would satisfactorily correlate all the varied phenomena for which the earth's internal behaviour is responsible. The words of John Woodward, written in 1695 about ore deposits, are equally applicable to-day in relation to continental drift and convection currents: "Here," he declared, "is such a vast variety of phenomena and these many of them so delusive, that 'tis very hard to escape imposition and mistake."

4

History of Ocean Basins

HARRY H. HESS

1962

The birth of the oceans is a matter of conjecture, the subsequent history is obscure, and the present structure is just beginning to be understood. Fascinating speculation on these subjects has been plentiful, but not much of it predating the last decade holds water. Little of Umbgrove's (1947) brilliant summary remains pertinent when confronted by the relatively small but crucial amount of factual information collected in the intervening years. Like Umbgrove, I shall consider this paper an essay in geopoetry. In order not to travel any further into the realm of fantasy than is absolutely necessary I shall hold as closely as possible to a uniformitarian approach; even so, at least one great catastrophe will be required early in the Earth's history.

PREMISES ON INITIAL CONDITIONS

Assuming that the ages obtained from radioactive disintegrations in samples of meteorites approximate the age of the solar system, then the age of the Earth is close to 4.5 aeons.[1] The Earth, it is further assumed, was formed by accumulation of particles (of here unspecified character) which initially had solar composition. If this is true, then before condensation to a solid planet the Earth lost, during a great evaporation, a hundred times as much matter as it now contains. Most of this loss was hydrogen. An unknown but much smaller amount of heavier elements was lost to space as well. The deficiency of the atmosphere in the inert gases points clearly to their loss. Urey (1957) suggests loss of nitrogen, carbon, and water, and perhaps a considerable proportion of original silicate material. He also points out that the lack of concentration of certain very

From Petrological Studies: A Volume in Honor of A. F. Buddington, ed. by A. E. J. Engel, H. L. James, and B. F. Leonard, p. 599–620, 1962. Reprinted with permission of the Geological Society of America.

---

[1] Aeon = 10^9 years (H. C. Urey).

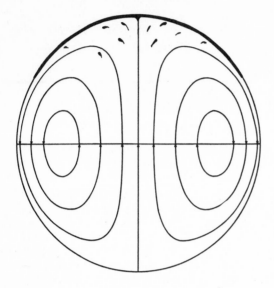

Figure 4-1
Single-cell (toroidal) convective overturn of Earth's interior. After Vening Meinesz. Continental material extruded over rising limb but would divide and move to descending limb if convection continued beyond a half cycle

volatile substances at the Earth's surface indicates that it never had a high surface temperature. This low temperature more or less precluded escape of large amounts of material after the Earth condensed and suggests that the loss occurred when the material forming the Earth was very much more dispersed so that the escape velocity from its outer portion was comparatively low. The condensation was rapid, and some light elements and volatile compounds were trapped within the accumulated solid material of the primordial Earth. I will assume for convenience and without too much justification that at this stage the Earth had no oceans and perhaps very little atmosphere. It is postulated that volatile constituents trapped within its interior have during the past and are today leaking to the surface, and that by such means the present oceans and atmosphere have evolved.

THE GREAT CATASTROPHE

Immediately after formation of the solid Earth, it may have contained within it many short-lived radioactive elements; how many and how much depends on the time interval between nuclear genesis and condensation. The bricketted particles from which it was made might be expected to have a low thermal conductivity at least near its surface as suggested by Kuiper (1954). The temperature rose, lowering the strength and perhaps starting partial fusion. The stage was thus set for the *great catastrophe* which it is assumed happened forthwith. A single-cell (toroidal) convective overturn took place (Fig. 4–1) (Vening Meinesz, 1952), resulting in the formation of a nickel-iron core, and at the same time the low-melting silicates were extruded over the rising limbs of the current to form the primordial single continent (Fig. 4–1). The single-cell overturn also converted gravitational energy into thermal energy (Urey, 1953). It is postulated that this heat and a probably much larger amount of heat resulting from the energy involved in the accumulation of the Earth were not sufficient to produce a molten Earth. The great quantitative uncertainties in this assumption can be gauged from MacDonald's analysis (1959).

The proposed single-cell overturn brought about the bilateral asymmetry of the Earth, now possibly much modified but still evident in its land and water hemispheres. After this event, which segregated the core from the mantle, single-cell convection was no longer possible in the Earth as a whole (Chandrasekhar, 1953).

The critical question now facing us is what percentage of the continental crustal material and of the water of the oceans reached the surface in the *great catastrophe*. On the basis that continental material is still coming to the surface of the Earth from the mantle at the rate of

1 km³/year* accepting Sapper's (1927, p. 424) figure on the contribution of volcanoes over the past 4 centuries, and assuming uniformitarianism, this means 4×10^9 km³ in 4 aeons or approximately 50 percent of the continents. So we shall assume that the other half was extruded during the catastrophe. The percentage of water is much harder to estimate. Rapid convective overturn might be much less efficient in freeing the water as compared to the low-melting silicates. The water might be expected to be present as a monomolecular film on grain surfaces. The low-melting silicate droplets could coagulate into sizable masses as a result of strong shearing during the overturn. On the other hand, shearing that would break down solid crystals to smaller size might increase their surface areas and actually inhibit freeing of water films. The best guess that I can make is that up to one-third of the oceans appeared on the surface at this time.

It may be noted that a molten Earth hypothesis would tend toward the initial formation of a thin continental or sialic layer uniformly over the Earth with a very thin uniform world-encircling water layer above it. Later it would require breaking up of this continental layer to form the observed bilateral asymmetry. With the present set of postulates this seems to be a superfluous step. Bilateral asymmetry was attained at the start, and it would be impossible ever to attain it once a core had formed, unless George H. Darwin's hypothesis that the moon came out of the Earth were accepted.

We have now set the stage to proceed with the subject at hand. Dozens of assumptions and hypotheses have been introduced in the paragraphs above to establish a framework for consideration of the problem. I have attempted to chose reasonably among a myriad of possible alternatives, but no competent reader with an ounce of imagination is likely to be willing to accept all of the choices made. Unless some such set of confining assumptions is made, however, speculation spreads out into limit-

---

*This figure includes felsic volcanic material probably derived from partial melting within the continental crust but does not include magmas that formed intrusions which did not reach the surface.

less variations, and the resulting geopoetry has neither rhyme nor reason.

TOPOGRAPHY AND CRUSTAL COLUMNS

If the water were removed from the Earth, two distinct topographic levels would be apparent: (1) the deep-sea floor about 5 km below sea level, and (2) the continental surface a few hundred meters above sea level. In other words, the continents stand up abruptly as plateaus or mesas above the general level of the sea floor. Seismic evidence shows that the so-called crustal thickness—depth to the M discontinuity—is 6 km under oceans and 34 km under continents on the average. Gravity data prove that these two types of crustal columns have the same mass—the pressure at some arbitrary level beneath them, such as 40 km, would be the same. They are in hydrostatic equilibrium. It is evident that one cannot consider the gross features of ocean basins independent of the continental plateaus; the two are truly complementary.

Whereas 29 per cent of the Earth's surface is land, it would be more appropriate here to include the continental shelves and the slopes to the 1000-m isobath with the continents, leaving the remainder as oceanic. This results in 40 per cent continental and 60 per cent oceanic crust. In 1955 I discussed the nature of the two crustal columns, which is here modified slightly to adjust the layer thicknesses to the more recent seismic work at sea (Raitt, 1956; Ewing and Ewing 1959) (Fig. 4–2). A drastic change, however, has been made in layer 3 of the oceanic column, substituting partially serpentinized peridotite for the basalt of the main crustal layer under the oceans as proposed elsewhere (Hess, 1959a). Let us look briefly into the facts that seemed to necessitate this change.

That the mantle material is peridotitic is a fairly common assumption (Harris and Rowell, 1960; Ross, Foster, and Myers, 1954; Hess, 1955). In looking at the now-numerous seismic profiles at sea the uniformity in thickness of layer 3 is striking. More than 80 per cent of the profiles show it to be 4.7 ± 0.7 km thick.

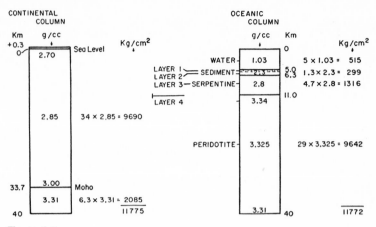

Figure 4-2
Balance of oceanic and continental crustal columns

Considering the probable error in the seismic data to be about ± 0.5 km, the uniformity may be even greater than the figures indicate. It is inconceivable that basalt flows poured out on the ocean floor could be so uniform in thickness. Rather, one would expect them to be thick near the fissures or vents from which they were erupted and thin or absent at great distance from the vents. The only likely manner in which a layer of uniform thickness could be formed would be if its bottom represented a present or past isotherm, at which temperature and pressure a reaction occurred. Two such reactions can be suggested: (1) the basalt to eclogite inversion (Sumner, 1954; Kennedy, 1959), and (2) the hydration of olivine to serpentine at about 500°C (Hess, 1954). The common occurrence of peridotitic inclusions in oceanic basaltic volcanic rocks (Ross, Foster, and Myers, 1954) and absence of eclogite inclusions lead the writer to accept postulate (2). Furthermore, the dredging of serpentinized peridotites from fault scarps in the oceans (Shand, 1949)[2], where the displacement on the faults may have been sufficient to expose layer 3, adds credence to this supposition. This choice of postulates is made here and will control much of the subsequent reasoning. The seismic velocity of layer 3 is highly variable; it ranges from 6.0 to 6.9

km/sec and averages near 6.7 km/sec, which would represent peridotite 70 per cent serpentinized (Fig. 4–3).

MID-OCEAN RIDGES

The Mid-Ocean Ridges are the largest topographic features on the surface of the Earth. Menard (1958) has shown that their crests closely correspond to median lines in the oceans and suggests (1959) that they may be ephemeral features. Bullard, Maxwell, and Revelle (1956) and Von Herzen (1959) show that they have unusually high heat flow along their crests. Heezen (1960) has demonstrated that a median graben exists along the crests of the Atlantic, Arctic, and Indian Ocean ridges and that shallow-depth earthquake foci are concentrated under the graben. This leads him to postulate extension of the crust at right angles to the trend of the ridges. Hess (1959b) also emphasizes the ephemeral character of the ridges and points to a trans-Pacific ridge that has almost disappeared since middle Cretaceous time, leaving a belt of atolls and guyots that has subsided 1–2 km. Its width is 3000 km and its length about 14,000 km (Fig. 4–4). The present active mid-ocean ridges have an average width of 1300 km, crest height of about 2½ km, and total length of perhaps 25,000 km.

The most significant information on the structural and petrologic character of the

---

[2]J. B. Hersey reports dredging serpentinized peridotite from the northern slope of the Puerto Rico Trench (Personal communication, 1961)

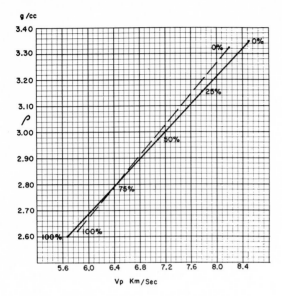

Figure 4-3
Relationship between seismic velocity, density, and per cent serpentinization. Solid curve for room temperature and pressure. Dashed curve estimated for T and P at 15 km depth. Curves based on measurements in laboratory by J. Green at the California Research Laboratory, La Habra, with variable temperatures up to 200°C and pressures up to 1 kilobar. The 100 per cent serpentinized sample measured by F. Birch at Harvard at pressures from 0 to 10 kilobars at room temperature (Data from Hess, 1959a)

ridges comes from refraction seismic information of Ewing and Ewing (1959) (Fig. 4–5) on the Mid-Atlantic Ridge, and Raitt's (1956) refraction profiles on the East Pacific Rise. The sediment cover on the Mid-Atlantic Ridge appears to be thin and perhaps restricted to material ponded in depressions of the topography. On the ridge crest, layer 3 has a seismic velocity of from 4 to 5.5 km/sec instead of the normal 6 to 6.9 km/sec. The M discontinuity is not found or is represented by a transition from layer 3 to velocities near 7.4 km/sec. Normal velocities and layer thicknesses, however, appear on the flanks of ridges.

Earlier I (1955, 1959b) attributed the lower velocities (*ca.* 7.4 km/sec) in what should be mantle material to serpentinization, caused by olivine reacting with water released from below. The elevation of the ridge itself was thought to result from the change in density (olivine 3.3 g/cc to serpentine 2.6 g/cc). A 2-km rise of the ridge would require 8 km of complete serpentinization below, but a velocity of 7.4 km/sec is equivalent to only 40 per cent of the rock serpentinized. This serpentinization would have to extend to 20-km depth to produce the required elevation of the ridge. This reaction, however, cannot take place at a temperature much above 500° C, which, considering the heat flow, probably exists at the bottom of layer 3, about 5 km be-

low the sea floor, and cannot reasonably be 20 km deep. Layer 3 is thought to be peridotite 70 per cent serpentinized. It would appear that the highest elevation that the 500° C isotherm can reach is approximately 5 km below the sea floor, and this supplies the reason for uniform thickness of layer 3 (Fig. 4–6).

CONVECTION CURRENTS IN THE MANTLE AND MID-OCEAN RIDGES

Long ago Holmes suggested convection currents in the mantle to account for deformation of the Earth's crust (Vening Meinesz, 1952; Griggs, 1939; 1954; Verhoogen, 1954; and many others). Nevertheless, mantle convection is considered a radical hypothesis not widely accepted by geologists and geophysicists. If it were accepted, a rather reasonable story could be constructed to describe the evolution of ocean basins and the waters within them. Whole realms of previously unrelated facts fall into a regular pattern, which suggests that close approach to satisfactory theory is being attained.

As mentioned earlier a single-cell convective overturn of the material within the Earth could have produced its bilateral asymmetry, segregating the iron core and primordial continents in the process. Since this event only

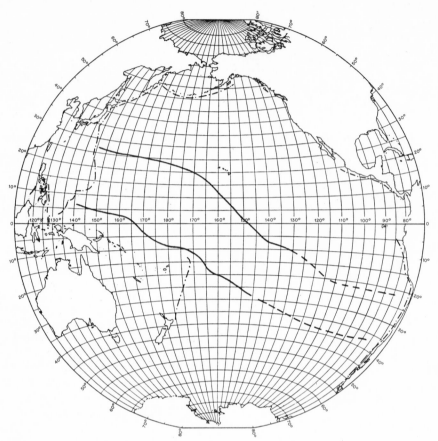

Figure 4-4
Former location of a Mid-Pacific Mesozoic ridge

multicell convection in the mantle has been possible. Vening Meinesz (1959) analyzed the spherical harmonics of the Earth's topography up to the thirty-first order. The peak shown in the values from the third to fifth harmonic would correlate very nicely with mantle-size convection currents; cells would have the approximate diameter of 3000 to 6000 km in cross section (the other horizontal dimension might be 10,000–20,000 km, giving them a banana-like shape).

The lower-order spherical harmonics of the topography show quite unexpected regularities. This means that the topography of a size smaller than continents and ocean basins has a greater regularity in distribution than previously recognized.

Paleomagnetic data presented by Runcorn (1959), Irving (1959), and others strongly sug-

gest that the continents have moved by large amounts in geologically comparatively recent times. One may quibble over the details, but the general picture on paleomagnetism is sufficiently compelling that it is much more reasonable to accept it than to disregard it. The reasoning is that the Earth has always had a dipole magnetic field and that the magnetic poles have always been close to the axis of the Earth's rotation, which necessarily must remain fixed in space. Remanent magnetism of old rocks shows that position of the magnetic poles has changed in a rather regular manner with time, but this migration of the poles as measured in Europe, North America, Australia, India, etc., has not been the same for each of these land masses. This strongly indicates independent movement in direction and amount of large portions of the Earth's surface

29

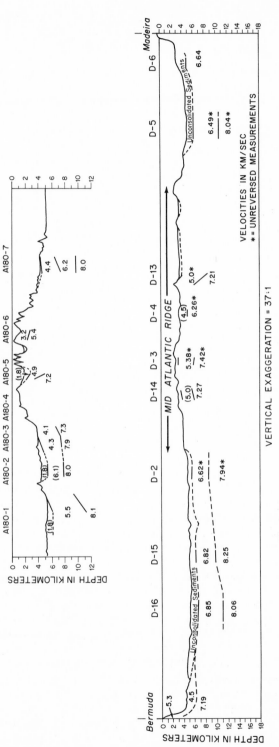

Figure 4-5
Seismic profiles on the Mid-Atlantic Ridge, by Ewing and Ewing (1959)

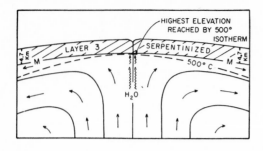

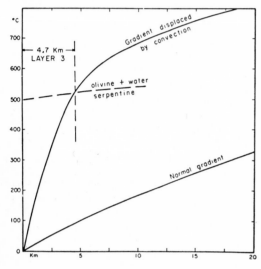

Figure 4-6
Diagram to portray highest elevation that 500°C isotherm can reach over the rising limb of a mantle convection cell, and expulsion of water from mantle which produces serpentinization above the 500°C isotherm

with respect to the rotational axis. This could be most easily accomplished by a convecting mantle system which involves actual movement of the Earth's surface passively riding on the upper part of the convecting cell. In this case at any given time continents over one cell would not move in the same direction as continents on another cell. The rate of motion suggested by paleomagnetic measurements lies between a fraction of a cm/yr to as much as 10 cm/yr. If one were to accept the old evidence, which was the strongest argument for continental drift, namely the separation of South America from Africa since the end of the Paleozoic, and apply uniformitarianism, a rate of 1 cm/yr results. This rate will be accepted in subsequent discussion. Heezen (1960) mentions a fracture zone crossing Iceland on the extension of the Mid-Atlantic rift zone which has been widening at a rate of 3.5 m/1000 yrs/km of width.

The unexpected regularities in the spherical harmonics of the Earth's topography might be attributed to a dynamic situation in the present Earth whereby the continents move to positions dictated by a fairly regular system of convection cells in the mantle. Menard's theorem that mid-ocean ridge crests correspond to median lines now takes on new meaning. The mid-ocean ridges could represent the traces of the rising limbs of convection cells, while the circum-Pacific belt of deformation and volcanism represents descending limbs. The Mid-Atlantic Ridge is median because the continental areas on each side of it have moved away from it at the same rate — 1 cm/yr. This is not exactly the same as continental drift. The continents do not plow through oceanic crust impelled by unknown forces; rather they ride passively on mantle material as it comes to the surface at the crest of the ridge and then moves laterally away from it. On this basis the crest of the ridge should have only recent sediments on it, and recent and Tertiary sediments on its flanks; the whole Atlantic Ocean and possibly all of the oceans should have little sediment

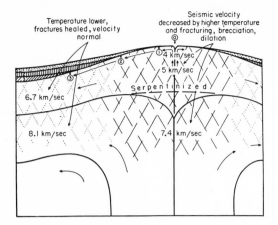

Temperature lower,
fractures healed, velocity
normal

Seismic velocity
decreased by higher temperature
and fracturing, brecciation,
dilation

Figure 4-7
Diagram to represent (1) apparent progressive
overlap of ocean sediments on a mid-ocean ridge
which would actually be the effect of the mantle
moving laterally away from ridge crest, and
(2) the postulated fracturing where convective
flow changes direction from vertical to
horizontal. Fracturing and higher temperature
could account for the lower seismic velocities
on ridge crests, and cooling and healing of the
fractures with time, the return to normal
velocities on the flanks

older than Mesozoic (Fig. 4–7). Let us look a bit further at the picture with regard to oceanic sediments.

Looking over the reported data on rates of sedimentation in the deep sea, rates somewhere between 2 cm and 5 mm/1000 yrs seem to be indicated. Writers in the last few years have tried hard to accept the lowest possible rate consistent with the data in order to make the thickness jibe with the comparatively thin cover of sediment on the ocean floor indicated by seismic data. Schott's figures for the Atlantic and Indian oceans as corrected by Kuenen (1946) and further corrected by decreasing the number of years since the Pleistocene from 20,000 years to 11,000 years indicate a rate of 2 cm/1000 yrs. Hamilton's (1960) figures suggest 5 mm/1000 yrs. A rate of 1 cm/1000 yrs would yield 40 km in 4 aeons or 17 km after compaction, using Hamilton's compaction figures. A 5-mm rate would still give 8.5 km compacted thickness instead of 1.3 km as derived from seismic data. This 1 order of magnitude discrepancy had led some to suggest that the water of the oceans may be very young, that oceans came into existence largely since the Paleozoic. This violates uniformitarianism to which the writer is dedicated and also can hardly be reconciled with Rubey's (1951) analysis of the origin of sea water. On the system here suggested any sediment upon the sea floor ultimately gets incorporated in the continents. New mantle material with no sedimentary cover on it rises and moves outward from the ridge. The cover

of young sediments it acquires in the course of time will move to the axis of a downward-moving limb of a convection current, be metamorphosed, and probably eventually be welded onto a continent.

Assuming a rate of 1 cm/1000 yrs one might ask how long, on the average, the present sea floor has been exposed to deposition if the present thickness of sediment is 1.3 km. The upper 0.2 km would not yet have been compacted and would represent 20 million years of deposition. The remaining 1.1 km now compacted would represent 240 million years of accumulation or in total an average age of the sea floor of 260 million years. Note that a clear distinction must be made between the age of the ocean floor and the age of the water in the oceans.

In order to explain the discrepancy between present rate of sedimentation in the deep sea and the relatively small thickness of sediment on the floor of the oceans, many have suggested that Pleistocene glaciation has greatly increased the rate of sedimentation. The writer is skeptical of this interpretation, as was Kuenen in his analysis (1946)[3]. Another discrepancy of the same type, the small number of volcanoes on the sea floor, also indicates the apparent youth of the floor. Menard estimates there are in all 10,000 volcanic seamounts in the oceans. If this represented 4 aeons of

---

[3]The Mohole test drilling off Guadalupe Island in 1961 suggests a rate of sedimentation in the Miocene of 1 cm/1000 yrs or a little more.

volcanism, and volcanoes appeared at a uniform rate, this would mean only one new volcano on the sea floor per 400,000 years. One new volcano in 10,000 years or less would seem like a better figure. This would suggest an average age of the floor of the ocean of perhaps 100 to 200 million years. It would account also for the fact that nothing older than late Cretaceous has ever been obtained from the deep sea or from oceanic islands.

Still another line of evidence pointing to the same conclusion relates to the ephemeral character of mid-ocean ridges and to the fact that evidence of only one old major ridge still remains on the ocean floor. The crest of this one began to subside about 100 million years ago. The question may be asked: Where are the Paleozoic and Precambrian mid-ocean ridges, or did the development of such features begin rather recently in the Earth's history?

Egyed (1957) introduced the concept of a great expansion in size of the Earth to account for apparent facts of continental drift. More recently Heezen (1960) tentatively advanced the same idea to explain paleomagnetic results coupled with an extension hypothesis for mid-ocean ridges. S. W. Carey (1958) developed an expansion hypothesis to account for many of the observed relationships of the Earth's topography and coupled this with an overall theory of the tectonics of the Earth's crust. Both Heezen and Carey require an expansion of the Earth since late Paleozoic time (*ca.* 2×10^8 years) such that the surface area has doubled. Both postulate that this expansion is largely confined to the ocean floor rather than to the continents. This means that the ocean basins have increased in area by more than 6 times and that the continents until the late Paleozoic occupied almost 80 per cent of the Earth's surface. With this greatly expanded ocean floor one could account for the present apparent deficiency of sediments, volcanoes, and old mid-ocean ridges upon it. While this would remove three of my most serious difficulties in dealing with the evolution of ocean basins, I hesitate to accept this easy way out. First of all, it is philosophically rather unsatisfying, in much the same way as were the older hypotheses of continental drift, in that there is no apparent mechanism within

the Earth to cause a sudden (and exponential according to Carey) increase in the radius of the Earth. Second, it requires the addition of an enormous amount of water to the sea in just the right amount to maintain the axiomatic relationship between sea level-land surface and depth to the M discontinuity under continents, which is discussed later.

MESOZOIC MID-PACIFIC RIDGE

In the area between Hawaii, the Marshall Islands, and the Marianas scores of guyots were found during World War II. It was supposed that large numbers of them would be found elsewhere in the oceans. This was not the case. The Emperor seamounts running north-northwest from the west end of the Hawaiian chain are guyots, a single linear group of very large ones. An area of small guyots is known in the Gulf of Alaska (Gibson, 1960). There are a limited number in the Atlantic Ocean north of Bermuda on a line between Cape Cod and the Azores, and a few east of the Mid-Atlantic Ridge; other than these only rare isolated occurrences have been reported.

Excluding the areas of erratic uplift and depression represented by the island arcs, lines can be drawn in the mid-Pacific bounding the area of abundant guyots and atolls (Fig. 4-4), marking a broad band of subsidence 3000 km wide crossing the Pacific from the Marianas to Chile. The eastern end is poorly charted and complicated by the younger East Pacific Rise. The western end terminates with striking abruptness against the eastern margin of the island-arc structures. Not a single guyot is found in the Philippine Sea west of the Marianas trench and its extensions, although to the east they are abundant right up to the trenches.

Fossils are available to date the beginning of the subsidence, but only near the axis of the old ridge. Hamilton (1965) found middle Cretaceous shallow-water fossils on guyots of the Mid-Pacific mountains, and Ladd and Schlanger (1960) reported Eocene sediments above basalt at the bottom of the Eniwetok bore hole. It should also be noted that atolls of the Caroline, Marshall, Gilbert, and Ellice islands pre-

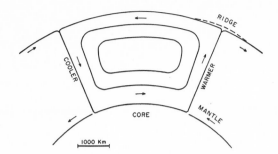

Figure 4-8
Possible geometry of a mantle convection cell

dominate on the southern side of the old ridge, whereas guyots greatly predominate on the northern side. Hess (1946) had difficulty in explaining why the guyots of the mid-Pacific mountain area did not become atolls as they subsided. He postulated a Precambrian age for their upper flat surfaces, moving the time back to an era before lime-secreting organisms appeared in the oceans. This became untenable after Hamilton found shallow-water Cretaceous fossils on them. Looking at the same problem today and considering that the North Pole in early Mesozoic time, as determined from paleomagnetic data from North America and Europe, was situated in southeastern Siberia, it seems likely that the Mid-Pacific mountain area was too far north for reef growth when it was subsiding. The boundary between reef growth and nonreef growth in late Mesozoic time is perhaps represented by the northern margins of the Marshall and Caroline islands, now a little north of 10° N, then perhaps 35° N. Paleomagnetic measurements from Mesozoic rocks, if they could be found within or close to this area, are needed to substantiate such a hypothesis.

The old Mesozoic band of subsidence is more than twice as wide as the topographic rise of present-day oceanic ridges. This has interesting implications regarding evolution of ridges which are worth considering here. Originally I attributed the rise of ridges to release of water above the upward-moving limb of a mantle convection cell and serpentinization of olivine when the water crossed the 500-degree C isotherm. As mentioned above, this hypothesis is no longer tenable because the high heat flow requires that the 500-degree C isotherm be at very shallow depth. The topographic rise

of the ridge must be attributed to the fact that a rising column of a mantle convection cell is warmed and hence less dense than normal or descending columns. The geometry of a mantle convection cell (Fig. 4–8) fits rather nicely a 1300-km width assuming that the above effects causes the rise.

Looking now at the old Mesozoic Mid-Pacific Ridge with the above situation in mind, volcanoes truncated on the ridge crest move away from the ridge axis at a rate of 1 cm/yr. Eventually they move down the ridge flank and become guyots or atolls rising from the deep-sea floor. Those 1000 km from the axis, however, were truncated 100 million years before those now near the center of the old ridge (Fig. 4–9). On this basis it would be very interesting to examine the fauna on guyots near the northern margin of the old ridge or to drill atolls near the southern margin to see if the truncated surfaces or bases have a Triassic or even Permian age. At any rate the greater width of the old ridge and its belt of subsidence compared to present topographic ridges could be explained by the above reasoning.

Turning to a reconsideration of the Mid-Atlantic Ridge it appears that layer 3, with a thin and probably discontinuous cover of sediments, forms the sea floor. The dredging of serpentinized peridotite from fault scarps at three places on the ridge (Shand, 1949) points to such a conclusion. The abnormally low seismic velocity, if this is layer 3, might be attributed to intense fracturing and dilation where the convective flow changes direction from vertical to horizontal. The underlying material, which ordinarily would have a velocity of 8 km/sec or more, has a velocity approximately 7.4 km/sec partly for the same reason

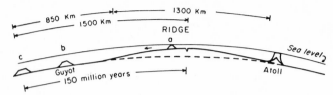

Figure 4-9
Diagram to show progressive migration of volcanic peaks, guyots, and atolls, from a ridge crest to the flanks, suggesting that the wave-cut surfaces of guyots or the bases of atolls may become older laterally away from the crest

but also because of its abnormally high temperature (Fig. 4–7). The interface between layer 3 and the 7.4 km/sec material below is thus the M discontinuity. The increase in velocity of layer 3 to about 6.7 km/sec and of the sub-Moho material to 8 km/sec as one proceeds away from the ridge crest may be attributed to cooling and healing of the fractures by slight recrystallization or by deposition from solution in an interval of tens of millions of years.

DEVELOPMENT OF THE OCEANIC CRUST (LAYER 3) AND THE EVOLUTION OF SEA WATER

Assuming that layer 3 is serpentinized peridotite, that the water necessary to serpentinize it is derived by degassing of the rising column of a mantle convection cell, and that its uniform thickness (4.7 ± 0.5 km) is controlled by the highest level the 500° C isotherm can reach under these conditions, we have a set of reasonable hypotheses which can account for the observed facts (Fig. 4–6).

The present active ridge system in the oceans is about 25,000 km long. If the mantle is convecting with a velocity of 1 cm/yr a vertical layer 1 cm thick of layer 3 on each side of the ridge axis is being formed each year. The material formed is 70 per cent serpentinized, based on an average seismic velocity of 6.7 km/sec, and this serpentine contains 25 per cent water by volume. If we multiply these various quantities, the volume of water leaving the mantle each year can be estimated at 0.4

km³. Had this process operated at this rate for 4 aeons, 1.6×10^9 km³ of water would have been extracted from the mantle, and this less 0.3×10^9 km³ of water now in layer 3 equals 1.3×10^9 km³ or approximately the present volume of water in the oceans.[4]

The production of layer 3 by a convective system and serpentinization must be reversed over the downward limbs of convection cells. That is, as layer 3 is depressed into the downward limb it will deserpentinize at 500° C and release its water upward to the sea. Thus the rate of entry of juvenile water into the ocean will equal the rate of acquisition of water from the mantle to form layer 3 over the rising limbs of convection cells.

It is not at present possible to check against the record the assumption that the process outlined went far back to the beginning of geologic history at a uniform rate. If Africa and South America moved away from each other at the rate of 2 cm a year they would have been adjacent to each other about 200 million years ago. Presumably this was the beginning of the convection cells under the present ridge. The assumption of a rate of movement for convection of 1 cm/yr was based on the above situation because the geologic record suggests splitting apart near the end of the Paleozoic Era. The convection cells under the Mesozoic

---

[4]The estimate of how much of the present Mid-Ocean Ridge system is active is uncertain. That fraction of the system with a median rift was used in this estimate. The whole system is approximately 75,000 km long. The velocity of 1 cm/yr is also uncertain. If it were 0.35 cm/yr, as Heezen mentions for widening of the Iceland rift, this coupled with a 75,000 km length of the ridge system would give the required amount of water for the sea in 4 aeons.

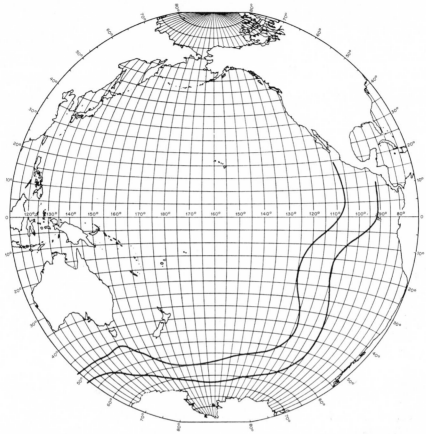

Figure 4-10
Approximate outline of East Pacific Rise, which possibly represents an oceanic ridge
so young that it has not yet developed a median rift zone and pre-Rise sediments still
cap most of its crest

Mid-Pacific Ridge ceased to function about 100 million ago inasmuch as the crest is known to have begun to subside at this time. It must have taken at least 150 million years at 1 cm/yr for the flanks of the ridge to spread to a width of 3000 km, and possibly the convection cells were in operation here for several times this long. The East Pacific Rise crosses the Mesozoic ridge at right angles and presumably did not come into existence until recent times, but certainly less than 100 million years ago. No evidence of older ridges is found in the oceans, suggesting that convection is effective in wiping the slate clean every 200 or 300 million years. This long and devious route leads to the conclusion that the present shapes and floors of ocean basins are comparatively young features.

RELATIONSHIP OF THICKNESS OF CONTINENTS TO DEPTH OF THE SEA

In Figure 4–2 the balance of oceanic and continental columns is portrayed. The layer thicknesses are derived from seismic profiles, and the densities are extrapolated from seismic velocities and petrologic deduction (Hess, 1955). Gravity measurements during the past half century have shown that the concept of isostasy is valid — in other words that a balance does exist. The oceanic column is simpler than

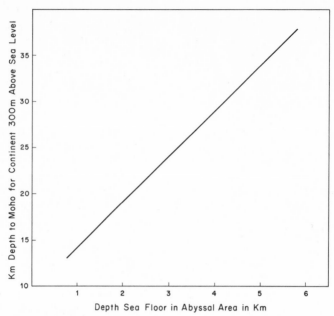

Figure 4-11
Graph portraying depth to the M discontinuity under continents
vs. depth of abyssal areas in oceans, computed from balance of
crustal columns

the continental column and less subject to conjecture with regard to layer thicknesses or densities. The main uncertainty in the continental column is its mean density. Given the thickness of the crust, this value was derived by assuming that the pressure at 40 km below sea level under the continents equalled that for the same depth under the oceans, or 11,775 kg/cm². The mean density of the continental crust then becomes 2.85 g/cc. The latitude that one has for changing the numerical values in either of the two columns is small. The error in the pressure assumed for 40 km depth is probably less than 1 per cent.

The upper surface of the continent is adjusting to equilibrium with sea level by erosion. But as material is removed from its upper surface, ultimately to be deposited along its margins in the sea, the continent rises isostatically. If undisturbed by tectonic forces or thermal changes it will approach equilibrium at a rate estimated by Gilluly (1954) as 3.3×10^7 yrs half life. It is thus evident that, if the oceans were half as deep, the continents would be eroded to come to equilibrium with the new sea level, they would rise isostatically, and a

new and much shallower depth to the M discontinuity under continents would gradually be established. A thinner continent but one of greater lateral extent would be formed inasmuch as volume would not be changed in this hypothetical procedure. The relationship between depth of the oceans, sea level, and the depth to the M discontinuity under continents is an axiomatic one and is a potent tool in reasoning about the past history of the Earth's surface and crust.

The oft-repeated statement that the amount of water in the sea could not have changed appreciably since the beginning of the Paleozoic Era (or even much further back) because the sea has repeatedly lapped over and retreated from almost all continental areas during this time interval is invalid because the axiomatic relationship stated in the last paragraph would automatically require that this be so regardless of the amount of water in the sea.

One can compute the pressure at 40 km depth for an ocean with 1, 2, 3, or 4 km of water and equate this to continental columns for the same pressure at 40 km, distributing the amount of crustal material (density 2.85

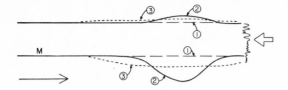

Figure 4-12
Diagram to illustrate thickening of a continent by deformation. Initially a mountain system and much larger root are formed, but both spread laterally with time and isostatic adjustment

g/cc) and mantle material (density 3.31 g/cc) in such proportion that balance is established. This computation is shown graphically in Figure 4–11). Assuming, as has been done in this chapter, that the oceans have grown gradually with time, one must suppose that the continents were much thinner in the early Precambrian. This could possibly be recognizable in the difference of tectonic pattern in very old terrains as compared to present continental structure.

If there is gradual increase of water in the sea one may ask why continents are not eventually flooded and why are there not continental-type areas now a kilometer or more below sea level. No extensive areas of this sort are found. Part of the answer might lie in the generation of new continental material at a rate equivalent to eruption of new water. An increase of depth of the sea by 1 km allows thickening of the continents by about 5 times this amount, which would be several times in excess of the estimated 1 km³ per year extraction of magma from the mantle. Even if this were an underestimate there is no reason why continents might not extend laterally rather than grow thicker. The answer seems to be that there is more than enough energy in the crustal regime of the Earth to thicken the continents to an extent that they are maintained somewhat above the equilibrium level (Fig. 4–12). A continent will ride on convecting mantle until it reaches the downward-plunging limb of the cell. Because of its much lower density it cannot be forced down, so that its leading edge is strongly deformed and thickened when this occurs. It might override the downward-flowing mantle current for a short distance, but thickening would be the result as before.

The Atlantic, Indian, and Arctic oceans are surrounded by the trailing edges of continents moving away from them, whereas the Pacific Ocean is faced by the leading edges of contin-

ents moving toward the island arcs and representing downward-flowing limbs of mantle convection cells or, as in the the case of the eastern Pacific margin, they have plunged into and in part overridden the zone of strong deformation over the downward-flowing limbs.

RECAPITULATION

The following assumptions were made, and the following conclusions reached:

(1) The mantle is convecting at a rate of 1 cm/yr.

(2) The convecting cells have rising limbs under the mid-ocean ridges.

(3) The convecting cells account for the observed high heat flow and topographic rise.

(4) Mantle material comes to the surface on the crest of these ridges.

(5) The oceanic crust is serpentinized peridotite, hydrated by release of water from the mantle over the rising limb of a current. In other words it is hydrated mantle material.

(6) The uniform thickness of the oceanic crust results from the maximum height that the 500° C isotherm can reach under the mid-ocean ridge.

(7) Seismic velocities under the crests of ridges are 10–20 per cent lower than normal for the various layers including the mantle, but become normal again on ridge flanks. This is attributed to higher temperature and intense fracturing with cooling and healing of the fractures away from the crest.

(8) Mid-ocean ridges are ephemeral features having a life of 200 to 300 million years (the life of the convecting cell).

(9) The Mid-Pacific Mesozoic Ridge is the only trace of a ridge of the last cycle of convecting cells.

(10) The whole ocean is virtually swept clean (replaced by new mantle material) every 300 to 400 million years.

(11) This accounts for the relatively thin veneer of sediments on the ocean floor, the relatively small number of volcanic seamounts, and the present absence of evidence of rocks older than Cretaceous in the oceans.

(12) The oceanic column is in isostatic equilibrium with the continental column. The upper surface of continents approaches equilibrium with sea level by erosion. It is thus axiomatic that the thickness of continents is dependent on the depth of the oceans.

(13) Rising limbs coming up under continental areas move the fragmented parts away from one another at a uniform rate so a truly median ridge forms as in the Atlantic Ocean.

(14) The continents are carried passively on the mantle with convection and do not plow through oceanic crust.

(15) Their leading edges are strongly deformed when they impinge upon the downward moving limbs of convecting mantle.

(16) The oceanic crust, buckling down into the descending limb, is heated and loses its water to the ocean.

(17) The cover of oceanic sediments and the volcanic seamounts also ride down into the jaw crusher of the descending limb, are metamorphosed, and eventually probably are welded onto continents.

(18) The ocean basins are impermanent features, and the continents are permanent although they may be torn apart or welded together and their margins deformed.

(19) The Earth is a dynamic body with its surface constantly changing. The spherical harmonics of its topography show unexpected regularities, a reflection of the regularities of its mantle convection systems and their seconary effects.

In this chapter the writer has attempted to invent an evolution for ocean basins. It is hardly likely that all of the numerous assumptions made are correct. Nevertheless it appears to be a useful framework for testing various and sundry groups of hypotheses relating to the oceans. It is hoped that the framework with necessary patching and repair may eventually form the basis for a new and sounder structure.

GEOMETRY OF PLATE TECTONICS

5

Introduction

Postulates

In view of the insight into convective processes and their geologic consequences shown in the work of A. Holmes, F. A. Vening Meinesz, D. T. Griggs, and H. H. Hess, the question naturally arises whether there is anything new in plate tectonics other than the name. The answer is yes, there is. In plate tectonics the additional step was taken of proceeding beyond the qualitative concept of ocean floor spreading to the quantitative calculation of the direction and velocity of plate movements.

This was done by making two idealized geometrical postulates about crustal deformation that are almost Euclidian in their simplicity. Before giving the postulates, several definitions are needed.

Definition 1, plates. The lithosphere, defined as the rigid outer shell of the earth (roughly 100 km thick), is divided by a network of boundaries into separate blocks which are termed "plates."

Definition 2, boundaries. Boundaries are lines separating plates. Boundaries are of three types.

a. *Ridges,* where two plates are diverging, permitting the upwelling of magma that creates new lithosphere. (The direction of relative motion of the two plates need not be perpendicular to the ridge.)

b. *Trenches* or *sinks,* where two plates are converging, with one plate moving beneath the other eventually to be absorbed into the mantle, or "destroyed." (The direction of relative motion of the two plates need not be perpendicular to the trench.)

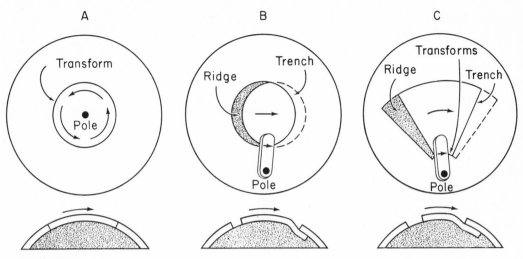

Figure 5-1
The tennis ball experiment. (A) A circle is cut from the tennis ball and pivoted to rotate about its center (solid dot), which is the pole of relative motion. (B) The cut-out circle is attached to a rigid arm which pivots about a new pole (solid dot), creating a ridge and trench but no transform faults. (C) A transform fault is created by cutting the upper boundary of the small plate along a circle concentric about the pivot point (i.e., pole)

c. *Transform faults,* where two plates are moving tangential to each other. Lithosphere is neither created nor destroyed. The direction of relative motion of the two plates is exactly parallel to the fault.

Postulate 1. The plates are internally rigid but are uncoupled from each other. At their boundaries two plates may pull apart or slip one beneath the other, but within the plates there is no deformation.

To visualize better the three kinds of boundaries, we can use a tennis ball as a model of the spherical shell of the lithosphere and a solid inner ball as a model of the earth's interior. Take a compass, draw a circle on the tennis ball, and cut along the circle (Figure 5–1a). The tennis ball is now divided into two "plates." Now rotate the circular plate keeping its center fixed relative to the rest of the tennis ball. The motion of the circular plate is everywhere tangential to the boundary so that the entire boundary is by definition a transform fault. The center of rotation of the small plate is termed its "pole of motion" relative to the large plate. This pole (and its antipodal point on the opposite side of the ball) are the only two points that remain in a fixed position relative to both plates.

To generate boundaries of the ridge type and trench type, it is necessary to rotate our circular plate about a pole that is not located at the center of the circle. To visualize this on our model, first attach an arm rigidly to the circular plate. Now attach the arm by a pivot point to the tennis ball (Figure 5–1b). This pivot point is the new pole of relative motion. As the arm and circular plate rotate about this pole, the leading edge of the circular plate is thrust under (or over) the larger plate, this boundary being by definition a trench. Along the opposite trailing edge or ridge boundary, a gap opens up. On our tennis ball model, this gap is a depression. However in the real earth the incipient crack is filled with molten magma which solidifies to form new lithosphere. Being hot, this new lithosphere is less dense than the adjacent lithosphere and rises to form a ridge in response to buoyant forces.

Is it possible for all three types of boundaries to be represented along the perimeter of a single plate? Yes, but not if the plate is a circle. Returning to our model, draw two arc segments concentric about a common center on a new tennis ball (Figure 5–1c). Now form a small plate by connecting the ends of the two arcs and cut the ball along the boundary. The last two boundaries may be drawn perpendicular to the arc segments, but they need not be. Again attach a rigid arm to the small plate and connect the arm to the rest of the ball by means of a pivot, placing the pivot exactly at the center of the original concentric circles. Now rotate the small plate about the pivot (i.e., pole). On opposite sides of the arc segments the relative motion of the two plates will be tangential to the plate bounary. Therefore these boundaries are by definition transform faults. The leading and trailing edges of the small plate are again trench and ridge boundaries. Note that two conditions must be satisfied for the existence of a boundary of the transform type. (1) Part of the plate boundary must be a segment of a circle. (2) The pole of relative motion (i.e., the pivot point) must be located at the center of the circle. Note also that the opening crack is wider at greater distances from the pole.

These geometrical relationships can be generalized in the form of the following definition and postulate.

Definition 3, pole of relative motion. The pole of relative motion between two plates is the unique point on the globe that does not move relative to either of the two plates. (Strictly speaking, each pole has an antipodal point on the opposite side of the globe.) The pole may be visualized as a pivot point about which the two plates rotate relative to each other.

Postulate 2. The pole of relative motion between a pair of plates remains fixed relative to the two plates for long periods of time.

The following theorems follow from Postulate 2.

Theorem 1. Transform faults between two plates lie along segments of concentric small circles centered on the pole of relative motion of the two plates.

Theorem 2. The pole of relative motion for two plates may be found by constructing perpendiculars to local segments of transform faults. The common intersection of the perpendiculars is the pole.

Theorem 3. The width W of new lithosphere formed adjacent to a ridge in a given interval of time decreases from a maximum width W_0 at an arc distance $A = 90°$ from the pole of relative motion to zero width at the pole itself. Quantitatively, $W = W_0 \sin A$ where A is the arc distance from the pole to the point of observation and W is the width of new lithosphere measured parallel to the direction of relative motion between the two plates.

These are the basic postulates and theorems of plate tectonics. They are discussed more fully in Chapter 8. In addition, an elegant set of geometrical theorems has been developed to describe the behavior of junctions where three boundaries come together (Chapter 10). Together with the use of magnetic stripes to determine spreading rates at the rises (Section V), it is these postulates and theorems that give plate tectonics mathematical rigor and permit quantitative calculations of the poles and rates of relative plate motion.

Of course, every geologist knows that these postulates do not provide a complete explanation for all geologic information—if the continents were rigid blocks, there would be no need to study structural geology. Yet the conclusions drawn from plate tectonics offer rational explanations for so many of the earth's major tectonic features that the basic assumption appears to be justified with only two minor modifications: *most* large-scale deformation occurs in narrow zones between plates that are *nearly* rigid. The questions of how much is "most" and how rigid is "nearly rigid" are discussed in the articles of Section IX.

The Birth of a Theory

By late 1964 the stage was set at four widely separated institutions for the birth of the new theory of plate tectonics. The four institutions were Cambridge University in England, Princeton University in New Jersey, Lamont Geological Observatory in Palisades, New York, and Scripps Institution of Oceanography in La Jolla, California.

Cambridge had long been a world center of highly creative geophysical research. The modern school of paleomagnetism with its rigorous statistical approach and its emphasis on the problems of polar wandering and continental drift began at Cambridge with the research of J. Hospers, S. K. Runcorn, K. Creer, and E. Irving, all of whom were graduate students in the early 1950's. Since the time of George Darwin most geophysicists had believed that the earth was too rigid to permit either polar wandering or continental drift. This changed in 1956 with Irving's (1956) and Runcorn's (1956) demonstrations using paleomagnetic data that both polar wandering and continental drift had occurred.

Since then the world of geophysics has not been the same. Interest in global tectonics was sustained at Cambridge through the late 1950's with Edward Bullard's pioneering studies of heat flow through the sea floor and by his general interest in marine geophysics. In 1965 the paths of an astonishingly large number of creative people crossed at Cambridge. Harry Hess visited for a few months from Princeton, as did J. Tuzo Wilson from the University of Toronto. Four of the research students then at Cambridge were later to make very important contributions to plate tectonics: Fred Vine (Chapters 23 and 24), Dan McKenzie (Chapters 7, 10, and 37), Robert Parker (Chapter 7), and John Sclater (Chapter 38). Drummond Matthews (Chapter 22) was on the staff, and the previous year Vine and Matthews had published their explanation of linear marine magnetic anomalies (Section V).

Figure 5-2
J. Tuzo Wilson

In late 1964 Bullard, Everett, and Smith (1965) had just completed an important study of continental drift. Bullard wanted to find a quantitative way to describe the fit of the coastline of South America to that of Africa and hit upon a theorem of Euler which states that the displacement of any curve on the surface of a sphere (such as a coastline) may be described as the rotation of the curve about some axis through the center of the sphere (even though the actual displacement may have been via a more complex route). The introduction of an axis of rotation to describe displacement on a sphere anticipated an important geometrical technique that was later used in plate tectonics.

J. Tuzo Wilson spent his sabbatical leave at Cambridge in early 1965. Wilson had begun his education in physics but switched to geology because he liked field work—a motivation responsible for many careers in earth science. By the early 1950's he had become interested in the origin of the ancient shield area that constitutes the stable central core of North America. He then became in-terested in the origin of island arcs, which lead to an early interest in large shear faults. By 1960 he had become a convert to the theory of continental drift and began a systematic study of oceanic islands. In 1963 he found that, in general, islands that are farther from the mid-ocean ridges are older than those that are closer (Wilson, 1963a). Noting that the Hawaiian Islands are progressively older to the west, he proposed, as early as 1963, that the islands were formed by the movement of a plate over a hot spot deep in the mantle (Wilson, 1963c). In 1964 Wilson attended the symposium organized by the Royal Society in London at which the current state of knowledge about continental drift was summarized (Blackett et al., 1965). In early 1965 all of these lines of thought came together in the course of conversations Wilson had at Cambridge with Bullard, Hess, Matthews, and Vine. In early 1965 Wilson wrote the short paper (Chapter 6) that contains the main elements of plate tectonics.

In considering Wilson's ideas, it is useful to compare them with those of S. W. Carey (1958b), whose work had considerably influenced Wilson. Carey envisaged

Figure 5-3
Dan P. McKenzie astride the
Hayward fault, a branch of the
San Andreas fault

large-scale crustal deformation that was pervasive and continuous whereas Wilson restricted large-scale deformation to narrow mobile belts that form a network between rigid plates. Wilson presented the main evidence for plate tectonics with remarkable completeness in his short, classic 1965 paper (Chapter 6). He already saw that submarine fracture zones mark the traces of transform faults; he saw that some fracture zones are active now whereas others trace zones of former activity; he recognized the origin of the fracture zones that offset the mid-Atlantic ridge; and he recognized that the Juan de Fuca ridge was a continuation of the East Pacific rise offset by the San Andreas fault, which he identified as a transform. Most important of all, his theory made specific predictions that could be tested, as described in Sections V and VI.

Putting the Plates on a Globe

Wilson's original theory was worked out for flat plates on a plane. The next important steps were to consider the consequences of plates that together form a spherical shell (i.e., the earth's lithosphere) and to give Wilson's concepts more mathematical rigor. Dan McKenzie and Robert L. Parker (Chapter 7) at Scripps Institution of Oceanography and W. Jason Morgan at Princeton (Chapter 8), working independently, took this next step. Dan McKenzie's interest in plate tectonics went back to Cambridge, where he was just starting research when Vine and Matthews published their important paper (Chapter 22). He met Hess and Wilson at Cambridge, but was not convinced of the validity of plate tectonics until he heard papers presented by Fred Vine and Lynn Sykes at an important NASA symposium in New York in November, 1966 (Phinney, 1968). In June, 1967, working at Scripps Institution of Oceanography in La Jolla, McKenzie got the idea of using rigid-body rotations to describe plate motions while rereading the paper by Bullard, Everett, and Smith (1965) on fitting the continents together. Their paper, together with Wilson's, made clear to him the importance of using rotation poles to describe plate motions on a sphere. Being a novice in seismology at the time, he did not describe the first motion of earthquakes in the usual way but used slip vectors, which show a worldwide pattern parallel to the direction of plate motion. Robert Parker had just completed a general computer program called SUPERMAP for plotting worldwide geophysical data using any conceivable projection. Parker introduced the idea of using a Mercator projection in plate tectonics, taking for the pole of the projection the same coordinates as the pole of relative motion between the two plates. This has the fortunate result that transforms and slip vectors are horizontal lines if the correct pole is used. Maps of this type have proved very useful in subsequent studies (e.g., Chapters 9 and 45).

At Princeton, W. Jason Morgan had, via a different route, developed similar techniques and arrived at the same conclusions. He became interested in plate tectonics through contacts with his fellow faculty members at Princeton Harry Hess and Walter Elsasser. By 1967 Elsasser had developed a theory to explain how stress could be transmitted over great distances by a rigid oceanic plate resting on a soft layer. Morgan became interested in the great fracture zones of the Pacific Ocean that Menard (1967) had shown could almost be fitted with a set of great circles—almost but not quite. Morgan found that the fracture zones could be fitted much more accurately by a set of small circles drawn about a point that he soon realized had the significance of being a pole of relative motion between two plates. From this he developed the geometrical technique of locating poles by finding the intersection of great circles perpendicular to transforms: either active transforms like the San Andreas fault or inactive ones like the great fracture zones of the Pacific basin. He was the first to develop a method for finding pole locations from variations in the rate of sea-floor spreading, as determined from magnetic stripes. He also produced the first map showing the major plates of the world.

The final step in this initial formulation of plate tectonics was taken by Xavier Le Pichon at Lamont Geological Observatory. Le Pichon took the methods of McKenzie, Parker, and Morgan, programmed them to handle the vast amount of geophysical information available at Lamont, and proceeded to produce his masterful analysis of all the major plates of the world. Le Pichon also carried the history of plate motions back into the past using magnetic stripes together with available geologic data.

READING LIST

Additional articles describing application of the principles of plate tectonics to geologic and geophysical problems are given in the reading lists for Sections VI, VIII, and IX.

6

A New Class of Faults and their Bearing on Continental Drift

J. TUZO WILSON
1965

From *Nature*, vol. 207, p. 343–347, 1965. Reprinted with permission of the author and Macmillan Journals, Ltd.

TRANSFORMS AND HALF-SHEARS

Many geologists (Bucher, 1933) have maintained that movements of the Earth's crust are concentrated in mobile belts, which may take the form of mountains, mid-ocean ridges or major faults with large horizontal movements. These features and the seismic activity along them often appear to end abruptly, which is puzzling. The problem has been difficult to investigate because most terminations lie in ocean basins.

This article suggests that these features are not isolated, that few come to dead ends, but that they are connected into a continuous network of mobile belts about the Earth which divide the surface into several large rigid plates (Figure 6–1). Any feature at its apparent termination may be transformed into another feature of one of the other two types. For example, a fault may be transformed into a mid-ocean ridge as illustrated in Figure 6–2, A. At the point of transformation the horizontal shear motion along the fault ends abruptly by being changed into an expanding tensional motion across the ridge or rift with a change in seismicity.

A junction where one feature changes into another is here called a transform. This type and two others illustrated in Figures 6–2,B and C may also be termed half-shears (a name suggested in conversation by Prof. J. D. Bernal). Twice as many types of half-shears involve mountains as ridges, because mountains are asymmetrical whereas ridges have bilateral symmetry. This way of abruptly ending large horizontal shear motions is offered as an explanation of what has long been recognized as a puzzling feature of large faults like the San Andreas.

Another type of transform whereby a mountain is transformed into a mid-ocean ridge was suggested by S. W. Carey (1955) when he proposed that the Pyrenees Mountains were com-

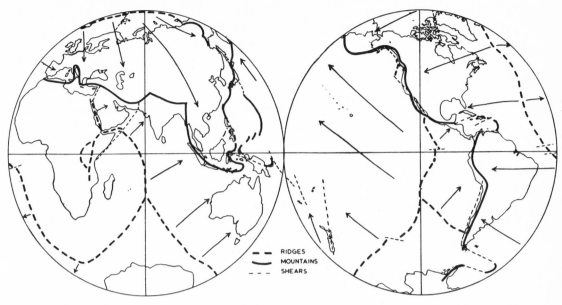

Figure 6-1
Sketch map illustrating the present network of mobile belts around the globe. Such belts comprise the active primary mountains and island arcs in compression (solid lines), active transform faults in horizontal shear (light dashed lines), and active mid-ocean ridges in tension (heavy dashed lines)

pressed because of the rifting open of the Bay of Biscay (presumably by the formation of a mid-ocean ridge along its axis). The types illustrated are all dextral, but equivalent sinistral types exist.

In this article the term 'ridge' will be used to mean mid-ocean ridge and also rise (where that term has been used meaning mid-ocean ridge, as by Menard (1964) in the Pacific basin). The terms mountains and mountain system may include island arcs. An arc is described as being convex or concave depending on which face is first reached when proceeding in the direction indicated by an arrow depicting relative motion (Figures 6–2 and 6–3). The word fault may mean a system of several closely related faults.

TRANSFORM FAULTS

Faults in which the displacement suddenly stops or changes form and direction are not true transcurrent faults. It is proposed that a separate class of horizontal shear faults exists which terminate abruptly at both ends, but which nevertheless may show great displace-

ments. Each may be thought of as a pair of half-shears joined end to end. Any combination of pairs of the three dextral half-shears may be joined giving rise to the six types illustrated in Figure 6–3. Another six sinistral forms can also exist. The name transform fault is proposed for the class, and members may be described in terms of the features which they connect (for example, dextral transform fault, ridge-convex arc type).

The distinctions between types might appear trivial until the variation in the habits of growth of the different types is considered as is shown in Figure 6–4. These distinctions are that ridges expand to produce new crust, thus leaving residual inactive traces in the topography of their former positions. On the other hand oceanic crust moves down under island arcs absorbing old crust so that they leave no traces of past positions. The convex sides of arcs thus advance. For these reasons transform faults of types A, B, and D in Figure 6–4 grow in total width, type F diminishes and the behaviour of types C and E is indeterminate. It is significant that the direction of motion on transform faults of the type shown in Figure 6–3,A is the reverse of that required to offset

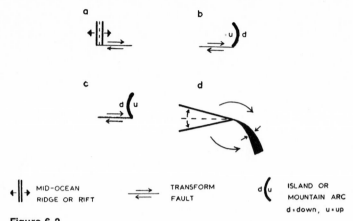

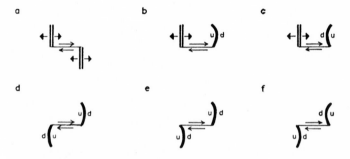

Figure 6-2
Diagram illustrating the four possible right-hand transforms: a, ridge to dextral half-shear: b, dextral half-shear to concave arc; c, dextral half-shear to convex arc; d, ridge to right-hand arc

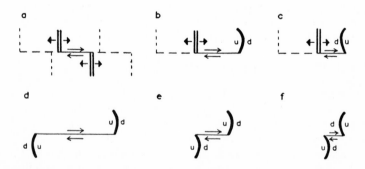

Figure 6-3
Diagram illustrating the six possible types of dextral transform faults; a, ridge to ridge type; b, ridge to concave arc; c, ridge to convex arc; d, concave arc to concave arc; e, concave arc to convex arc; f, convex arc to convex arc. Note that the direction of motion in a is the reverse of that required to offset the ridge

Figure 6-4
Diagram illustrating the appearance of the six types of dextral transform faults shown in Figure 6-3 after a period of growth. Dashed lines, traces of former positions now inactive, but still expressed in the topography

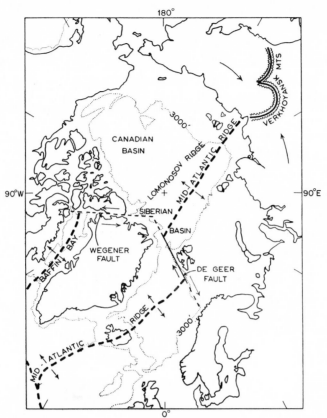

Figure 6-5
Sketch map of the northern termination of the Mid-Atlantic Ridge.
This involves two large transform faults (Wegener and De Geer
faults) and transformation into the Verkhoyansk Mountains

the ridge. This is a fundamental difference between transform and transcurrent faulting.

Many examples of these faults have been reported and their properties are known and will be shown to fit those required by the constructions above. If the class as a whole has not heretofore been recognized and defined, it is because all discussions of faulting, such as those of E. M. Anderson, have tacitly assumed that the faulted medium is continuous and conserved. If continents drift this assumption is not true. Large areas of crust must be swallowed up in front of an advancing continent and re-created in its wake. Transform faults cannot exist unless there is crustal displacement, and their existence would provide a powerful argument in favour of continental drift and a guide to the nature of the displacements involved. These proposals owe much to the ideas of S. W. Carey, but differ in that I suggest that the plates between mobile belts are not readily deformed except at their edges.

The data on which the ensuing accounts are based have largely been taken from papers in two recent symposia (Blackett et al., 1965; Hurley, 1964) and in several recent books (Menard, 1964; Hill, 1963; Runcorn, 1962) in which many additional references may be found.

NORTH ATLANTIC RIDGE TERMINATION

If Europe and North America have moved apart, an explanation is required of how so large a rift as the Atlantic Ocean can come to a relatively abrupt and complete end in the cul-de-sac of the Arctic Sea. Figure 6–5 illustrates one possible explanation.

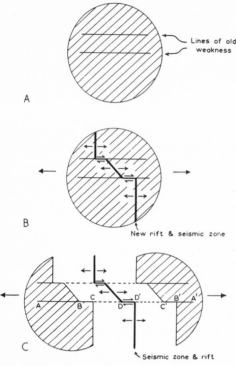

Figure 6-6
Diagram illustrating three stages in the rifting
of a continent into two parts. This could
represent South America and Africa. There will
be seismic activity along the heavy lines only

Wegener (1924) suggested that the strait between Greenland and Ellesmere Island was formed by a fault, here postulated to be a sinistral transform fault (ridge-ridge type). Wegmann (1948) named another between Norway, Spitsbergen and Greenland, the De Geer line, which is here regarded as a dextral transform fault (ridge-ridge type). The extension of the Mid-Atlantic ridge across the Siberian basin was traced by Heezen and Ewing (1961) while Wilson (1963b) proposed its transform into the Verkhoyansk Mountains by rotation about a fulcrum in the New Siberian Islands. In accordance with the expectations from Figure 6–4,A earthquakes have been reported along the full line of the De Geer fault in Figure 6–5, but not along the dashed older traces between Norway and Bear Island and to the north of Greenland. The Baffin Bay ridge and Wegener fault are at present quiescent. W. B. Harland (1961) and

Canadian geologists have commented on the similarities of Spitsbergen and Ellesmere Island.

EQUATORIAL ATLANTIC FRACTURE ZONES

If a continent in which there exist faults or lines of weakness splits into two parts (Figure 6–6), the new tension fractures may trail and be affected by the existing faults.

The dextral transform faults (ridge-ridge type) such as *AA'* which would result from such a period of rifting can be seen to have peculiar features. The parts *AB* and *B'A'* are older than the rifting. *DD'* is young and is the only part now active. The offset of the ridge which it represents is not an ordinary faulted displacement such as a transcurrent fault would produce. It is independent of the distance through which the continents have moved. It is confusing, but true, that the direction of motion along *DD'* is in the reverse direction to that required to produce the apparent offset. The offset is merely a reflexion of the shape of the initial break between the continental blocks. The sections *BD* and *D'B'* of the fault are not now active, but are intermediate in age and are represented by fracture zones showing the path of former faulting.

Figure 6–7 shows that the Mid-Atlantic ridge and the fracture zones in the equatorial Atlantic may well be a more complex example of this kind. If so the apparent offsets on the ridge are not faulted offsets, but inherited from the shape of the break that first formed between the coasts of Africa and the Americas. Figure 6–7 is traced from Heezen, Bunce, Hersey and Tharp (1964a) with additions to the north from Krause (1964). The fracture zones are here held to be right-hand transform faults and not left-hand transcurrent faults as previously stated. If the fracture zones can be traced across the Atlantic and are of the type postulated, then the points where they intersect the opposite coasts are conjugate points which would have been together before rifting.

It seems possible that the old fault in Pennsylvania and the offset of the Atlantic Coast described by Drake and Woodward (1963) are of the same nature, although it is suggested

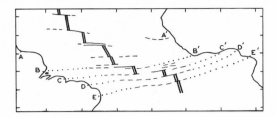

Figure 6-7
Sketch (after Krause, 1964, and Heezen et al., 1964a) showing how the Mid-Atlantic Ridge is offset to the left by active transform faults which have dextral motions if the rift is expanding (see Figure 6-4,a). Double vertical lines, mid-ocean ridge; solid horizontal lines, active fault; dashed lines, inactive fault trace; dotted lines, hypothetical extension of fault

that it is not usual for a fracture zone to follow a line of seamounts, and that the fracture zone may extend eastward, not south-east.

A POSSIBLE EXPLANATION OF THE TERMINATION OF THE CARLSBERG RIDGE

Another type of transform fault is found in the Indian Ocean (Figure 6–8). If the Indian Ocean and Arabian Gulf opened during the Mesozoic and Cenozoic eras by the northward movement of India, new ocean floor must have been generated by spreading of the Carlsberg ridge. This ends abruptly in a transcurrent fault postulated by Gregory (1920) off the east coast of Africa. A parallel fault has been found by Matthews (1963) as an offset across the

Carlsberg ridge and traced by him to the coast immediately west of Karachi. Here it joins the Ornach-Nal and other faults (Hunting Survey Corp., 1960) which extend into Afghanistan and, according to such descriptions as I can find, probably merge with the western end of the Hindu Kush. This whole fault is thus an example of a sinistral transform fault (ridge-convex arc type).

At a later date, probably about Oligocene time according to papers quoted by Drake and Girdler (1964), the ridge was extended up the Red Sea and again terminated in a sinistral transform fault (ridge-convex arc type) that forms the Jordan Valley (Quesnell, 1958) and terminates by joining a large thrust fault in south-eastern Turkey (Z. Ternek, private communication). The East African rift valleys are a still later extension formed in Upper Miocene time according to B. H. Baker (private communication).

The many offsets in the Gulf of Aden described by Laughton (1966) provide another example of transform faults adjusting a rift to the shape of the adjacent coasts.

POSSIBLE RELATIONSHIPS BETWEEN ACTIVE FAULTS OFF THE WEST COAST OF NORTH AMERICA

This tendency of mid-ocean ridges to be offset parallel to adjacent coasts is thought to be evident again in the termination of the East Pacific ridge illustrated in Figure 6–9. The San Andreas fault is here postulated to be a dextral transform fault (ridge-ridge type) and not a transcurrent fault. It connects the termination of the East Pacific ridge proper with another short length of ridge for which Menard (1964) has found evidence off Vancouver Island. His explanation of the connexion — that

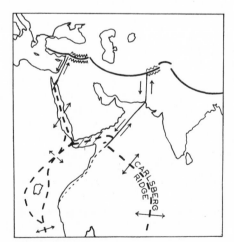

Figure 6-8
Sketch illustrating the end of the Carlsberg mid-ocean ridge by a large transform fault (ridge to convex arc type) extending to the Hindu Kush, the end of the rift up the Red Sea by a similar transform fault extending into Turkey and the still younger East African rifts

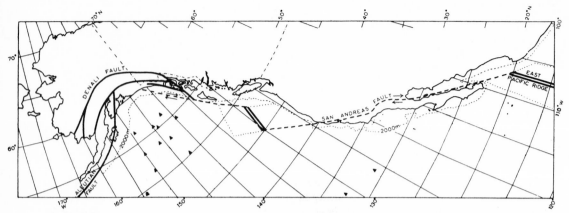

Figure 6-9
Sketch map of the west coast of North America showing major structural features. These include the approximate location of a submarine thrust fault along the Aleutian trench, the Denali faults (after St. Amand, 1957), the San Andreas and another large transform fault (after Benioff, 1962) and part of the East Pacific ridge and another mid-ocean ridge (after Menard, 1964)

the mid-ocean ridge connects across western United States—does not seem to be compatible with the view that the African rift valleys arc also incipient mid-ocean ridges. The other end of the ridge off Vancouver Island appears to end in a second great submarine fault off British Columbia described by Benioff (1962) as having dextral horizontal motion.

In Alaska are several large faults described by St. Amand (1957). Of the relations between them and those off the coast he writes: "If the two systems represent one consistent system, some interesting possibilities arise. One that the San Andreas and Alaska Complex is a gigantic tear fault, along which the Pacific Basin is being slid, relatively speaking under the Alaska Mainland, and the Bering Sea. On the other hand, if the whole system is a strike-slip fault having consistent right-lateral offset, then the whole of the western north Pacific Basin must be undergoing rotation."

St. Amand was uncertain, but preferred the latter alternative, whereas this interpretation would favour the former one. Thus the Denali system is considered to be predominantly a thrust, while the fault off British Columbia is a dextral transform fault.

At a first glance at Figure 6–9 it might be held that the transform fault off British Columbia was of ridge-concave arc type and that it connects with the Denali system of thrust faults, but if the Pacific floor is sliding under

Alaska, the submarine fault along the Aleutian arc that extends to Anchorage is more significant. In that case the Denali faults are part of a secondary arc system and the main fault is of ridge-convex arc type.

**FURTHER EXAMPLES FROM
THE EASTERN PACIFIC**

If the examples given from the North and Equatorial Atlantic Ocean, Arabian Sea, Gulf of Aden and North-west Pacific are any guide, offsets of mid-ocean ridges along fracture zones are not faulted displacements, but are an inheritance from the shape of the original fracture. The fracture zones that cross the East Pacific ridge (Sykes, 1963) are similar in that their seismicity is confined to the offset parts between ridges. An extension of this suggests that the offsets in the magnetic displacements observed in the aseismic fracture zones off California may not be fault displacements as has usually been supposed, but that they reflect the shape of a contemporary rift in the Pacific Ocean. More complex variants of the kind postulated here seem to offer a better chance of explaining the different offsets noted by Vacquier (1962) along different lengths of the Murray fracture zone than does transcurrent faulting. If the California fracture zones are of this character and are related to the

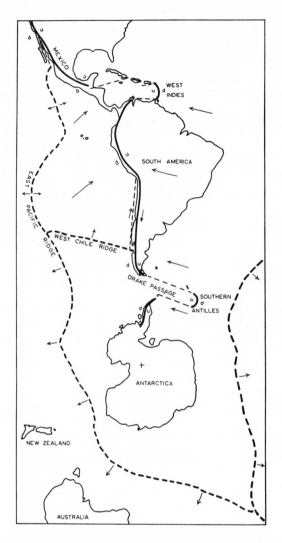

Figure 6-10
Sketch map of Mexico, South America, Antarctica, and part of the mid-ocean ridge system. This illustrates that the great loop of the ridge about Antarctica can grow only by increasing in diameter. Heavy dashed lines, mid-ocean ridges; light dashed lines, transform faults

Darwin rise as postulated by Hess, then the Darwin rise should be offset in a similar pattern.

The southern Andes appear to provide an example of compression combined with shearing. The compressional features are obvious. The existence of dextral shearing is also well known (St. Amand, 1961). It is suggested that the latter may be due to the transformation of the West Chile ridge into a dextral transform fault (ridge-convex arc type) along the Andes which terminates at the northern end by thrusting under the Peruvian Andes (Figure 6–10).

The observation that there is little seismicity and hence little movement south of the point where the West Chile ridge interests the Andes can be explained if it is realized that the ridge system forms an almost complete ring about Antarctica, from which expansion must everywhere be directed northwards. This may explain the absence of an isthmus across Drake Passage.

It would also appear that the faults at the two ends of the South Antilles and West Indies arcs are examples of dextral and sinistral pairs of transform faults (concave-concave arc types). According to Figure 6–4 both these arcs should be advancing into the Atlantic and inactive east-west faults should not be found beyond the arcs.

This article began by suggesting that some aspects of faulting well known to be anomalous according to traditional concepts of transcurrent faults could be explained by defining a new class·of transform faults of which twelve varieties were shown to be possible.

The demonstration by a few examples that at least six of the twelve types do appear to exist with the properties predicted justifies investigating the validity of this concept further.

It is particularly important to do this because transform faults can only exist if there is crustal displacement and proof of their existence would go far towards establishing the reality of continental drift and showing the nature of the displacements involved.

7

The North Pacific: An Example of Tectonics on a Sphere

DAN P. McKENZIE
ROBERT L. PARKER
1967

From *Nature*, v. 216, p. 1276–1280, 1967. Reprinted with permission of the authors and Macmillan Journals Ltd.

The linear magnetic anomalies (Vine and Matthews, 1963; Vine, 1966) which parallel all active ridges can only be produced by reversals of the Earth's magnetic field (Vine and Matthews, 1963) if the oceanic crust is formed close to the ridge axis (Hess, 1962). Models (Matthews and Bath, 1967) have shown that the anomalies cannot be observed in the North Atlantic unless most dyke intrusion, and hence crustal production, occurs within 5 km of the ridge axis. The spreading sea floor (Hess, 1962) then carries these anomalies for great horizontal distances with little if any deformation. The epicentres of earthquakes also accurately follow the axis and are offset with it by transform faults (Sykes, 1963; 1967). The structure of island arcs is less clear, though the narrow band of shallow earthquakes suggests that crust is consumed along a linear feature. These observations are explained if the sea floor spreads as a rigid plate, and interacts with other plates in seismically active regions which also show recent tectonic activity. For the purposes of this article, ridges and trenches are respectively defined as lines along which crust is produced and destroyed. They need not also be topographic features. Transform faults conserve crust and are lines of pure slip. They are always parallel, therefore, to the relative velocity vector between two plates—a most useful property. We have tested this paving stone theory of world tectonics in the North Pacific, where it works well. Less detailed studies of other regions also support the theory.

The movement of blocks on the surface of a sphere is easiest to understand in terms of rotations. Any plate can clearly be moved to a given position and orientation on a sphere by two successive rotations, one of which carries one point to its final position, a second about an axis through this point then produces the required orientation. These two rotations are equivalent to a single rotation about a different axis, and therefore any relative motion of two

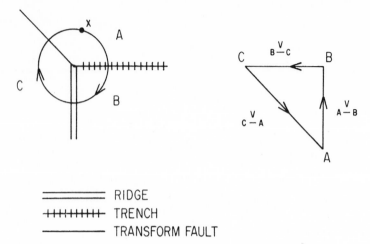

======= RIDGE
++++++++ TRENCH
———————— TRANSFORM FAULT

Figure 7-1
The circuit and its vector diagram show how a ridge and a trench can
meet to form a transform fault

plates on the surface of a sphere is a rotation about some axis. This is Euler's theorem, and has been used to fit together the continents surrounding the Atlantic (Bullard et al., 1965). If one of two plates is taken to be fixed, the movement of the other corresponds to a rotation about some pole, and all relative velocity vectors between the two plates must lie along small circles or latitudes with respect to that pole. If these small circles cross the line of contact between the two plates, the line must be either a ridge or a trench depending on the sense of rotation. Neither of these structures conserves crust. If the line of contact is itself a small circle, then it is a transform fault. This property of transform faults is very useful in finding the pole position and is a consequence of the conservation of crust across them. There is no geometric reason why ridges or trenches should lie along longitudes with respect to the rotational pole and in general they do not do so. The pole position itself has no significance, it is merely a construction point. These remarks extend Wilson's (1965c) concept of transform faults to motions on a sphere, the essential additional hypothesis being that individual aseismic areas move as rigid plates on the surface of a sphere.

There are several points on the surface of the Earth where three plates meet. At such points the relative motion of the plates is not completely arbitrary, because, given any two velocity vectors, the third can be determined. The method is easier to understand on a plane than on a sphere, and can be derived from the plane circuit in Fig. 7–1. Starting from a point x on A and moving clockwise, the relative velocity of B, $_Av_B$ is in the direction AB in the vector diagram. Similarly the relative velocities $_Bv_C$ and $_Cv_A$ are represented by BC and CA. The vector diagram must close because the circuit returns to x. Thus:

$$_Av_B + {}_Bv_C + {}_Cv_A = 0 \qquad (1)$$

The usual rules for the construction of such traingles require three parameters to be known, of which at least one must be the length of a side, or spreading rate. Transform faults on both ridges and trenches are easy to recognize, and they determine the direction, but not the magnitude, of the relative velocities. The magnetic lineations are one method of obtaining $_Bv_C$, though this value must be corrected for orientation unless the spreading is at right angles to the ridge. Then the triangle in Fig. 7–1 determines both $_Av_B$ and $_Cv_A$. This method is probably most useful to determine the rate of crustal consumption by trenches. Equation (1) must be used with care, because it only applies rigorously to an infinitesimal circuit round a point where three (or more) plates meet. If the circuit is finite, the rotation

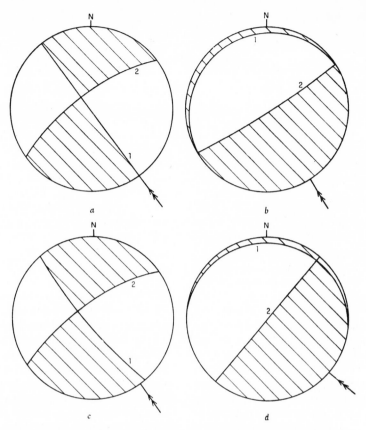

Figure 7-2
Mechanism diagrams for four circum-Pacific earthquakes. The lower
half of the focal sphere is projected stereographically on to a horizontal
surface, and the rarefraction quadrants are shaded. The horizontal
projection of the slip vector in plane 1 is marked with a double arrow.
(a) June 28, 1966, Parkfield (McEvilly, 1966a), strike slip; (b) September
4, 1964, Alaska (Stauder and Bollinger, 1966a), overthrust; (c) June 14,
1962, Near and Aleutian Islands (Stauder and Bollinger, 1964a), strike
slip; (d) October 20, 1963, Kurile Islands (Stauder and Bollinger,
1966b), overthrust

of the plates also contributes to their relative
velocity, and therefore these simple rules no
longer apply.

Equation (1) is easily extended to the corresponding problem on a spherical surface
because angular velocities behave like vectors
(Goldstein, 1950):

$$_A\omega_B + {}_B\omega_C + {}_C\omega_A = 0 \qquad (2)$$

The sign convention takes a rotation which is
clockwise when looked at from the centre of
the sphere to be a positive vector which is
pointing outward along the rotation axis. By
adding more terms, equation (2) can be ex-

tended to circuits crossing more than three
plates and applies to all possible circuits on
the surface. ω diagrams for three plates are no
more difficult to construct than those for v,
because the third vector must lie in the plane
containing the other two. This result does not
apply to diagrams for four or more plates,
which are three dimensional and therefore less
easy to draw.

These geometrical ideas can now be applied
to the North Pacific. There are many fault
plane solutions for earthquakes in the area, and
these are used in a new way in order to determine the direction of the horizontal projection

of the slip vector. Unlike the projection of the principal stress axes, that of the slip vector varies in a systematic manner over the entire region. This is clearly a consequence of spreading of the sea floor, which determines the relative motion, not the stress field.

The North Pacific was chosen for several reasons. The spreading rate from the East Pacific rise is the most rapid yet measured (Vine, 1966), and should therefore dominate any slight movements within the plate containing North America and Kamchatka (Gutenberg and Richter, 1954). The belt of earthquake epicentres which extends from the Gulf of California to Central Japan without any major branches (Gutenberg and Richter, 1954) suggests that the area contains only two principal plates. Also, the belt of seismic activity between them is one of the most active in the world and many fault plane solutions are available (Sykes, 1967; Stauder, 1960; Stauder and Udias, 1963; Udias and Stauder, 1964; McEvilly, 1966; Stauder and Bollinger, 1964a; 1966a; 1966b). It is an advantage that the trend of the belt which joins the two plates varies rapidly over short distances, because this illustrates the large variety of earthquake mechanisms which can result from a simple rotation (Fig. 7–2). It is also helpful that the outlines of the geology and topography of the sea floor are known.

Fault plane solutions which were obtained from the records of the world-wide network of standardized stations now give excellent and consistent results (Sykes, 1967; Stauder and Bollinger, 1966a). The directions of principal stress axes, however, which were determined from first motions, vary widely over short distances (Fig. 7–2) and are therefore difficult to use directly. The concept of spreading of the sea floor suggests that the horizontal projection of the slip vector is more important than that of any of the stress axes, and Fig. 7–2 shows that his is indeed the case. The examples which are illustrated are stereographic projections of the radiation field in the lower hemisphere on to a horizontal plane (Stauder and Bollinger, 1966a). The direction of the projection of the slip vector in plane one is obtained by adding or subtracting 90° from the strike of plane two if the planes one and two are orthogonal. The slip directions which are

shown give the motion of the oceanic plate relative to the plate containing North America and Kamchatka. For each case in Fig. 7–2 there are two possible slip directions, but, whereas one changes in direction slowly and systematically between Baja California and Japan, the other shows no consistency even for earthquakes in the same area. The ambiguity is therefore unimportant in this case. If all the earthquakes between the Gulf of California and Japan are produced by a rotation of the Pacific plate relative to the continental one, any pair of widely spaced slip directions can be used to determine the pole of relative rotation. The two which are used here are the strike of the San Andreas between Parkfield and San Francisco, and the average slip vector of all the aftershocks in the Kodiak Island region (Stauder and Bollinger, 1966a) of the 1964 Alaskan earthquake. A pole position of 50° N., 85° W. was obtained by construction on a sphere. If the paving stone theory applies, all slip vectors must be parallel to the latitudes which can be drawn with respect to this pole. Though this prediction could be tested by tabulations, a simpler and more obvious test is to plot the slip vectors on a map of the world in Mercator projection, taking the projection pole to be the rotation axis (Fig. 7–3). The Mercator projection has two advantages; it is conformal, which means that angles are locally preserved and slip vectors can be plotted directly, and also all small circles centred on the projection pole are parallel. Because the upper and lower boundaries of Fig. 7–3 are themselves small circles, the theory requires all slip vectors to be parallel both to each other and to the top and bottom. This prediction was tested on eighty published (Sykes, 1967; Stauder, 1960; Stauder and Udias, 1963; Udias and Stauder, 1964; McEvilly, 1966; Stauder and Bollinger, 1964a; 1966a; 1966b) fault plane solutions for shallow earthquakes during and after 1957. Of these, about 80 per cent had slip vectors with the correct sense of motion and within ±20° of the direction required by Fig. 7–3. Most of the fault plane solutions for earthquakes before this date also agreed with the sense and direction of motion. Representative slip vectors in Fig. 7–3 show the motion of the Pacific plate relative to the continental one, which is taken to be fixed. The rotation vector

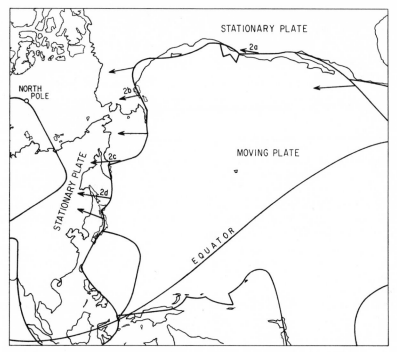

Figure 7-3
A Mercator projection of the Pacific with a pole at 50° N., 85° W. The arrows
show the direction of motion of the Pacific plate relative to that containing
North America and Kamchatka. If both plates are rigid all slip vectors must
be parallel with each other and with the upper and lower boundaries of the
figure. Possible boundaries of other plates are sketched

is therefore negative and points inward at 50°
N., 85° W. The agreement with theory is remarkable over the entire region. It shows that
the paving stone theory is essentially correct
and applies to about a quarter of the Earth's
surface.

The disadvantage of the Mercator projection
is the distortion it introduces around the poles.
It is therefore difficult to use Fig. 7–3 to estimate spreading velocities. For this purpose an
orthonormal projection is more useful (Fig.
7–4), for the spreading rate is then proportional to distance from the centre if this is
taken at the pole of rotation. In this projection,
which is simply a vertical projection on to a
plane at right angles to the rotation axis, rigid
body rotations on the plane, and all slip vectors
must be tangents to concentric circles about
the centre of projection (Fig. 7–4). This projection is useful if spreading rates, rather than

angles, are known. There are as yet few such
measurements in the North Pacific.

The large active tectonic areas of the North
Pacific are now clear from Fig. 7–3. The fault
systems of the San Andreas, Queen Charlotte
Islands and Fairweather form a dextral transform fault joining the East Pacific rise to the
Aleutian trench. The strike slip nature of these
faults is clear from field observations (Tocher,
1960; Hill and Dibblee, 1953; Allen, 1965)
and from the fault plane solutions (for example,
Fig. 7–2,a). In Alaska the epicentral belt of
earthquakes changes direction (Gutenberg
and Richter, 1954) (Fig. 7–3) and follows the
Aleutian arc. The fault solutions also change
from strike slip to overthrust (Stauder and
Bollinger, 1966a) (for example, Fig. 7–2,b),
and require that the islands and Alaska should
override the Pacific on low angle (~7°) faults.
Though the direction of slip remains the same

Figure 7-4
An orthonormal projection of the
North Pacific centred on the Mercator
Pole. Slip vectors are tangents to
concentric circles about the centre.

along the entire Aleutian arc, the change in
strike changes the fault plane solutions from
overthrusting in the east to strike slip in the
west (Fig. 7–2,*c*). A sharp bend occurs between the Aleutians and Kamchatka (Fig.
7–3). Here the fault plane solutions change
back to overthrust (Fig. 7–2,*d*). This motion
continues as far as Central Japan, where the
active belt divides (Fig. 7–3) and the present
study stops. Thus the North Pacific contains
the two types of transform faults which require
trenches (Wilson, 1965c), and clearly shows
the dependence of the fault plane solutions on
the trend of the fault concerned.

The variation of trend also controls the distribution of trenches, active andesite volcanoes, intermediate and deep focus earthquakes
(Gutenberg and Richter, 1954). All these
phenomena occur in Mexico, Alaska, the Eastern Aleutians, and from Kamchatka to Japan,
but are absent where the faults are of a strike
slip transform nature. This correlation is
particularly obvious along the Aleutian arc,
where all these features become steadily less
important as Kamchatka is approached (Gutenberg and Richter, 1954), then suddenly
reappear when the trend of the earthquake belt
changes. Though it is clear from these remarks

that the paving stone theory applies to the
North Pacific region as a whole, there are
some small areas which at first sight are
exceptions.

The most obvious of these is the complicated region of the ocean floor off the coast
between northern California and the Canadian
border (Wilson, 1965b). The difficulties begin
where the San Andreas fault turns into the
Mendocino fault. Fig. 7–5 shows that the
change in trend of the epicentres is possible
only if crust is consumed between *C* and *A* (or
created in *B*, which is unlikely). The earthquakes along the coast of Oregon (Berg and
Baker, 1963) and the presence of the volcanoes of the Cascade range, one of which has
recently been active and all of which contain
andesites, support the idea that crust is destroyed in this area. In the same area two remarkable seismic station corrections which
possess a large azimuthal variation (Bolt and
Nuttli, 1966) also suggest that there is a high
velocity region extending deep into the mantle
similar to that in the Tonga-Kermadec (Oliver
and Isacks, 1967) region. These complications
disappear when the ridge and trench structures
join again and become the Queen Charlotte
Islands fault.

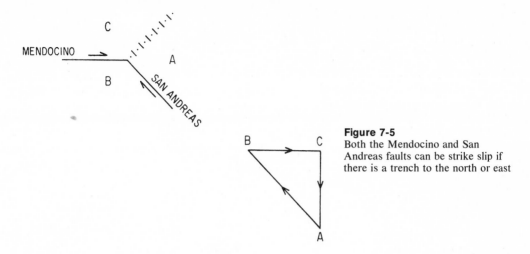

Figure 7-5
Both the Mendocino and San Andreas faults can be strike slip if there is a trench to the north or east

Another complicated area is in Alaska between 147.5° W. and the north end of the Fairweather fault (Stauder and Bollinger, 1966a). In this same area the local uplift after the 1964 earthquake suggested that several faults were active (Stauder and Bollinger, 1966a), and therefore the tectonics cannot be understood without more fault plane solutions.

The third area is in the Kurile Islands where two fault plane solutions (Fig. 7–6,*a* and *b*) require dip slip faulting and crustal extension. This motion is completely different from most of the solutions in the area, which agree well with the rest of the North Pacific. Both earthquakes occurred beneath the steep wall of the Kurile trench on the island arc side, and are consistent with gravity slides down into the trench. The terraces which would result from such slides are common features of the trenches of both Japan and the Aleutians (Ludwig et al., 1966; Gates, 1956). There is also one fault plane solution which requires that the Pacific should be overthrusting the Kurile Islands (Fig. 7–6,*c*) though the crustal shortening is consistent with the regional pattern.

The two ends of the North Pacific belt may also be discussed with the help of vector circuits. The end in Central Japan gives the trivial result that two trenches can join to give a third. The other end at the entrance to the Gulf of California is the circuit in Fig. 7–1, and shows how the East Pacific rise and the Middle America trench combine to become the San Andreas transform fault.

The North Pacific shows the remarkable success of the paving stone theory over a quarter of the Earth's surface, and it is therefore expected to apply to the other three-quarters. It is, however, only an instantaneous phenomenological theory, and also does not apply to intermediate or deep focus earthquakes. The evolution of the plates as they are created and consumed on their boundaries is not properly understood at present, though it should be possible to use the magnetic anomalies for this purpose. The other problem is the nature of the mechanism driving the spreading. It is difficult to believe that the convection cells which drive the motion are closely related to the boundaries of the plates.

One area where the evolution is apparent lies between the plate containing the Western Atlantic, North and South America (Gutenberg and Richter, 1954) and the main Pacific plate. The transform faults in the South-East Pacific are east–west; therefore the ocean floor between the rise and South America is moving almost due east relative to the main Pacific plate. The motion of the Atlantic plate relative to the Pacific is given by the San Andreas, and is towards the south-east. If the motion of the Atlantic plate is less rapid than that of the South-Eastern Pacific north of the Chile ridge, then the crust must be consumed along the Chile trench. The faults involved must have both overthrust and right-handed strike slip components. The present motion on the San Andreas is not in conflict with the east–west transform faults of the North-Eastern Pacific

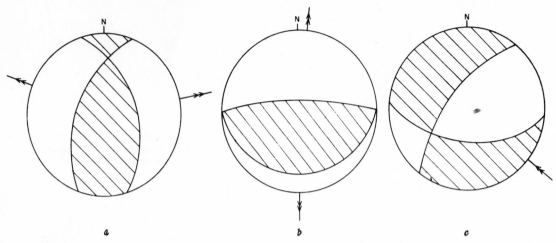

Figure 7-6
Mechanism diagrams drawn as in Figure 7-2 for three earthquakes in the Kurile Islands. (a) September 15, 1962 (Stauder and Bollinger, 1964a), extension by normal faulting. (b) November 15, 1963 (Stauder and Bollinger, 1966b), extension by normal faulting. (c) May 22, 1963 (Stauder and Bollinger, 1966b), island arc overthrust from the Pacific side

if there were originally a plate of ocean floor between North America and the main Pacific plate joined to that which still exists to the west of Chile. This piece of ocean floor has since been consumed, and therefore the direction of spreading in the Pacific appears to have changed in the north but not in the south. This explanation requires changes in the shape of the plates but not in their relative motion, and therefore differs from those previously suggested (Vine, 1966; Sykes, 1967). This study suggests that a belief in uniformity and the existence of magnetic anomalies will permit at least the younger tectonic events in the Earth's history to be understood in terms of sea floor spreading.

8

Rises, Trenches, Great Faults, and Crustal Blocks

W. JASON MORGAN
1968

From *Journal of Geophysical Research*, v. 73, p. 1959–1982, 1968. Reprinted with permission of the author and the American Geophysical Union. Copyright 1968.

A geometrical framework with which to describe present day continental drift is presented here. This presentation is an extension of the transform fault concept (Wilson, 1965c) to a spherical surface. The surface of the earth is divided into about twenty units, or blocks, as shown in Figure 8–1. Some of these blocks are of continental dimensions (the Pacific block and the African block); some are of subcontinental dimensions (the Juan de Fuca block, the Caribbean block, and the Persian block). The boundaries between blocks are of three types and are determined by present day tectonic activity. The first boundary is the rise type at which new crustal material is being formed. The second boundary is the trench type at which crustal surface is being destroyed; that is, the distance between two landmarks on opposite sides of a trench gradually decreases and at least one of the landmarks will eventually disappear into the trench floor. Other compressive systems in which the distance between two points decreases and the crust thickens, e.g., the folded mountains north of the Persian Gulf, are considered to be of this second type. The third boundary is the fault type at which crustal surface is neither created nor destroyed. Each block in Figure 8–1 is surrounded by some combination of these three types of boundaries. For example, Arabia is separated from Africa by the Aden-Red Sea rise and fracture zone system and by the Aqaba-Dead Sea fault. Arabia is separated from the Indian-Australian block by the Owen fracture zone (considered to be a transcurrent fault), and it is separated from Persia and from Europe by the compressive-type features in Iran and Turkey.

The compressive-type boundary seems to be the most difficult to delineate. The Tonga-New Zealand-Macquarie system has the well-developed Tonga trench at its northern end and the anomalous Macquarie ridge at its southern end. We suppose that this ridge is the

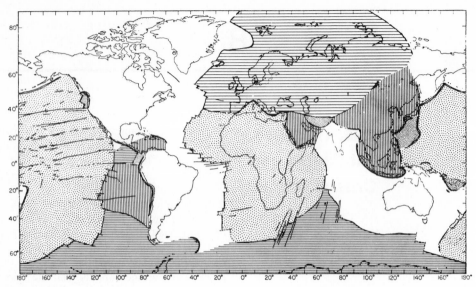

Figure 8-1
The crust is divided into units that move as rigid blocks. The boundaries between blocks are
rises, trenches (or young fold mountains). The boundaries drawn in Asia are tentative, and
additional sub-blocks may be required. (Figure is based on Sykes's, 1969, map of the ridge system
with additional features from Heezen and Tharp's, 1965, tectonic map)

result of slow compression and that fast compression leads to the trench-type structure. (In the terminology used here, the pole of rotation of the Pacific block relative to the Indian-Australian block is located near the southern end of the Macquarie ridge.) This results in a slow rate of closing along the Macquarie ridge (near the pole) and a fast closing of the Tonga trench and an equally fast slipping along the fault between New Guinea and the Fiji Islands. We have supposed that slow compressive systems are difficult to identify and have freely placed such boundaries at likely places. For example, a compressive-type boundary has been placed in the Mediterranean Sea between Europe and Africa. There might, in fact, be two almost parallel compressive belts in this region with a series of sub-blocks between them: the western Mediterranean, the Balkans, and others. The boundaries in the complex area around Central America are based on linear belts of earthquakes, and it is believed this subdivision is correct. The area east of New Guinea is less certain; it is believed that there is a fault between New Guinea and Fiji primarily accommodating the westward motion of the Pacific block and that there is a

trench just south of this fault primarily accommodating the northward motion of the Indian-Australian block. The boundaries in Siberia and Central Asia are very uncertain. There is no compelling reason to separate China from the North American block. The Ninetyeast ridge between India and Australia and the mid-Labrador Sea ridge between Greenland and North America are probably fossil boundaries.

We now make the assumption that gives this model mathematical rigor. We assume that each crustal block is perfectly rigid. If the distances between Guadalupe Island, Wake Island, and Tahiti, all within the Pacific block, were measured to the nearest centimeter and then measured again several years later, we suppose these distances would not change. The distance from Wake Island to Tokyo would, however, shorten because there is a trench between these two points, and the distance from Guadalupe Island to Mexico City would increase because there is a rise between these two points. But within the Pacific block, or any other crustal block, we shall assume there is no stretching, injection of large dikes, thickening, or any other distortion that would change distances between points. If this hy-

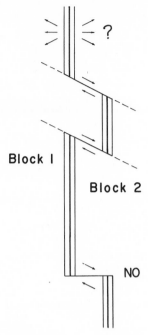

Figure 8-2
The motion of the left-hand block relative to the right-hand block cannot be determined from the strike of the ridge, but it can be determined from the strike of the transform faults. The fault at the bottom has a strike inconsistent with the other two faults and should not co-exist with them

pothesis is true, our conclusions will be in accord with observation. If this hypothesis is only partially valid, perhaps we will be able to assess the extent of such distortion by comparing observations with this model.

As will be demonstrated later (see Figure 8–4), the relative motion between two blocks may be represented by an angular velocity vector. Suppose the velocity of North America relative to Africa is ω_{Am-Af} and the velocity of the Pacific relative to North America is ω_{Pac-Am}. We may find the velocity of the Pacific relative to Africa by vector addition; $\omega_{Pac-Af} = \omega_{Pac-Am} + \omega_{Am-Af}$. We may also find the angular velocity of the Pacific relative to Africa by another route: first Africa to Antarctica and then Antarctica to the Pacific. Will the ω_{Pac-Af} so found equal that found via the other route? It is not believed the hypothesis of ri-

gidity would rigorously meet this test. Such features as the African rift system, the Cameroon trend, and the Nevada-Utah earthquake belt are most likely the type of distortion denied in the rigidity hypothesis. Nevertheless, it is of interest to see how far this simplying concept of rigidity can be applied.

We begin by considering blocks sliding on a plane. In this simple case we ignore the possibility of rotations and consider translations only. Figure 8–2 shows two rigid blocks separated by a rise and faults. From the rise alone, we cannot tell the direction of motion of one block relative to the other; the motion does not have to be perpendicular to the axis of the ridge. (There appears to be a tendency for the ridge to adjust itself to be almost perpendicular to the direction of spreading, but this is a dynamical consideration and not a requirement of geometry.) From the direction of a single transform fault, however, we can decide upon the direction of relative motion of the two blocks. The fault shown at the bottom of Figure 8–2 is incompatible with the two faults above and would not occur. The magnetic anomaly pattern (which will be parallel to the ridge crest) may now be projected along a line parallel to the direction of relative motion, and the velocity of one block relative to the other may be determined from the spacing of the anomalies.

Figure 8–3 shows three blocks separated by a trench, a rise, and two faults at three successive time intervals. The blocks have velocities relative to our coordinate system as shown in the figure for time 1. The four circles in this figure represent circular markers placed on the sea floor. At times 2 and 3, these markers have moved according to the velocity of their respective block (their original coordinates are shown by the dotted circles). We see that the strike of an offset depends on the difference between the velocities of the two sides. The active segment between the offsets of the ridge crest, and the extensions of this fracture zone, will have the same strike out to a distance that corresponds to the time interval during which the velocity difference of the two blocks has had its present azimuth. Further, we see that, if the ridge pattern remains symmetric, the axis of the ridge will have a 'drift' velocity equal to the vector average of the velocities of the two

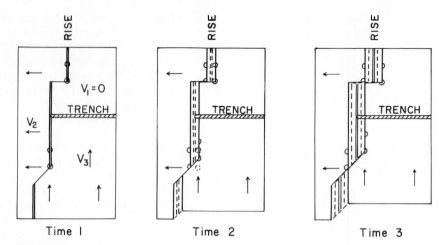

Figure 8-3
Three crustal blocks bounded by a rise, trench, and faults are shown at three successive time intervals. Note the motion of the four circular markers placed on the ridge crest at time 1: the solid segments show the motion of these circles; the dotted segments show the original coordinates of these markers. The strike of a transform fault is parallel to the difference of the velocities of the two sides; the crest of the ridge drifts with a velocity that is the average of the velocities of the two sides

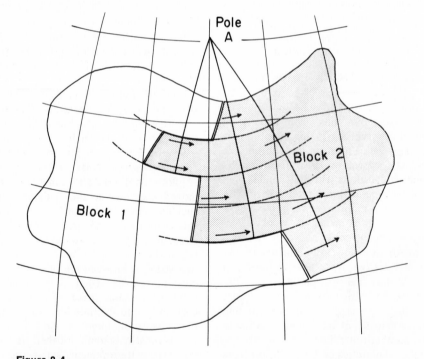

Figure 8-4
On a sphere, the motion of block 2 relative to block 1 must be a rotation about some pole. All faults on the boundary between 1 and 2 must be small circles concentric about the pole A

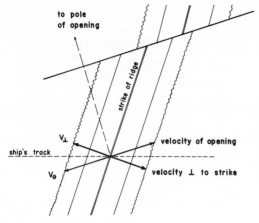

Figure 8-5
The magnetic profile measured along the ship's track must be projected parallel to the stike of the ridge

sides. Note that the two transform faults extending into block 2 on the left are not parallel. All faults north of the trench (between blocks 1 and 2) would run east to west as the one shown, and all faults south of the trench (between blocks 2 and 3) would have a 45° strike as shown. An example of where the strike of transform faults changes in this manner occurs off the coast of Mexico at the intersection of the Middle America trench, the East Pacific rise, and the Gulf of California.

We now go to a sphere. A theorem of geometry states that a block on a sphere can be moved to any other conceivable orientation by a single rotation about a properly chosen axis. We use this theorem to prove that the relative motion of two rigid blocks on a sphere may be described by an angular velocity vector by using three parameters, two to specify the location of the pole and one for the magnitude of the angular velocity. Consider the left block in Figure 8–4 to be stationary and the right block to be moving as shown. Fault lines of great displacement occur where there is no component of velocity perpendicular to their strike; the strike of the fault must be parallel to the difference in velocity of the two sides. Thus, all the faults common to these two blocks must lie on small circles concentric about the pole of relative motion.

The velocity of one block relative to another will vary along their common boundary; this velocity has a maximum at the 'equator' and vanishes at the poles of rotation. It is convenient to let the 'half-velocity perpendicular to the strike of the ridge' be the form in which the observations are placed. We choose 'half-velocity' since this is the form in which sea floor spreading rates are commonly quoted. There appears to be some self-adjusting mechanism in the rifting process that gives rise to a symmetric magnetic anomaly pattern, but

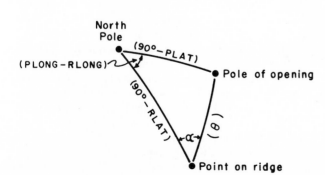

Figure 8-6
The angular relations used in deriving the formula for spreading velocity

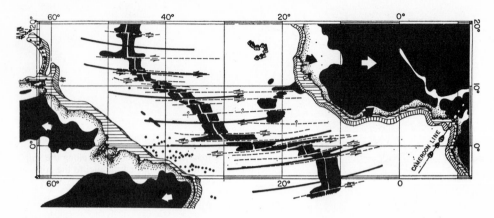

Figure 8-7
The strike of the transform faults in the equatorial Atlantic are compared with circles concentric about a pole at 62°N, 36°W. These circles indicate the present motion of Africa relative to South America. (Figure is adapted from Heezen and Tharp, 1965)

there is no geometrical requirement that spreading rates be equal on both sides. To allow for the possibility of unequal rates on the two sides of a ridge, we define half-velocity to be half the distance from a recognizable feature of the magnetic pattern to the corresponding feature on the other side of the ridge divided by the appropriate time. We choose 'perpendicular to the strike of the ridge' since this means an observed rate need be determined only once, and this value is then compared to a choice of models calculated with different pole positions and angular velocities. The angular relations used to project a pattern of magnetic anomalies from a ship's track to a line perpendicular to the strike of the ridge are shown in Figure 8–5. If we know the latitude and longitude of a point on the crest of the ridge, and if we know the strike of the ridge at this point, we calculate the velocity of spreading perpendicular to the strike of the ridge according to:

$$\theta = \arccos [\sin (RLAT) \sin (PLAT) + \cos (RLAT) \cos (PLAT) \cos (PLONG - RLONG)]$$

$$\alpha = \arcsin [\sin (PLONG - RLONG) \cos (PLAT)/\sin \theta]$$

$$V_\perp = V_{max} \sin \theta \cos (STRIKE - \alpha)$$

The quantities used in these formulas are shown in Figure 8–6.

THE MOTION OF THE AFRICAN BLOCK RELATIVE TO THE SOUTH AMERICAN BLOCK

Figure 8–7 shows the offsets of the ridge in the equatorial Atlantic Ocean. A set of circles concentric about a pole at 58°N, 36°W is plotted on a figure of Heezen and Tharp (1965). Figure 8–8,*a* shows how this pole position was obtained. Great circles were constructed perpendicular to the strike of each fracture zone offsetting the crest of the ridge listed in Table 8–1. The intersections of the great circles define the pole of rotation; the great circles are analogous to meridians and the fault lines are analogous to lines of latitude about this pole. As we see in Figure 8–8, the perpendiculars intersect at grazing angles and give good control in longitude but poor control in latitude. All of the perpendiculars except the one constructed for the fracture zone at 14.5°N pass through the circle drawn in the figure, 57.5°N (±2°), 36.5°W(±4°). Several other circles of about the same radius, which included all the perpendiculars except perhaps two or three, were also drawn. The centers of these circles ranged from 51°N to 63°N and from 35°W to 38°W. The limits on the location of the pole of rotation by this method are estimated to be 58°N(±5°), 36°W(±2°).

The strikes of transcurrent earthquakes used in Figure 8–8,*b* are listed in Table 8–2. The

Table 8–1
Strike of faults on the Mid-Atlantic ridge.

Name	Latitude	Longi-tude	Strike	Ref-erence
Atlantis	30.0°N	42.3°W	99°	a
	18.5°N	46.8°W	95°	a
	14.5°N	46.0°W	91°	a
Vema V	10.8°N	42.3°W	92°	b
Vema W	10.2°N	40.9°W	94°	b
Vema X	9.4°N	40.0°W	92°	b
Vema Y	8.8°N	38.7°W	92°	b
Vema Z	7.6°N	36.6°W	91°	b
	7.2°N	34.3°W	91°	b
	4.0°N	31.9°W	88°	c
	1.9°N	30.6°W	86°	c
St. Paul's	1.1°N	26.0°W	86°	c
	1.1°S	24.0°W	81°	c
Romanche	0.1°S	18.0°W	77°	c
Chain	1.3°S	14.5°W	75°	c
	1.9°S	12.9°W	82°	c
	2.9°S	12.5°W	73°	c
	7.5°S	12.3°W	73°	a

a. Heezen and Tharp (1965).
b. Heezen et al. (1964b).
c. Heezen et al. (1964a).

common intersection in Figure 8–8,*b* is not so good as the intersection in Figure 8–8,*a*. It is interesting that a circle of about 6° radius can be used both here and in the Pacific-North America case (see Figure 8–14,*f*) to illustrate the departure from a point intersection. The bulk of the epicenters are about 45° (5000 km) from the point of intersection in both of these cases. This suggests that the accuracy to which the fault planes are determined, or the accuracy to which the first motions represent the strike of a long fault, is the cause of the scatter in Figures 8–8,*b* and 8–14,*f*.

There is one fracture zone north of approximately 20°N listed in Table 8–1; there are two earthquake solutions north of this point listed in Table 8–2. This latitude roughly divides the ridge into a part between North America and Africa and a part between South America and Africa. There is no striking difference between these three values north of 20°N and those south of this latitude, and there is no line of earthquakes or other indication of tectonic activity entirely separating North America and South America. We shall assume that North

America and South America at present move as a single block. The Caribbean area almost entirely separates the Americas, and perhaps there is a slow relative movement with gradual distortion in the Atlantic Ocean area. If there is relative movement, the velocities and displacements involved will be very slow at this 'hinge' or 'pole' somewhere between the Lesser Antilles and the mid-Atlantic ridge. In contrast, the Azores-Gibraltar ridge is presumed to be a major transcurrent fault between the African and European blocks. All fracture zones north of the Azores are between Europe and America and have a different pole of opening.

Figure 8–9 shows several observed spreading rates in the Atlantic Ocean compared with the model. Since the ridge runs almost north-south with only a minimum of doubling back at the equator, latitude is a convenient coordinate against which to plot the rates. To use the preceding formulas for spreading rate, knowledge of the latitude, longitude, and strike of the ridge is needed at each point along the ridge. These quantities were obtained from figures in

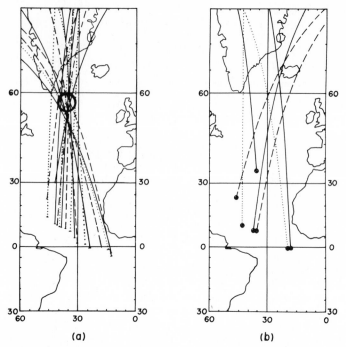

Figure 8-8
Great circles perpendicular to the strike of offsets of the mid-Atlantic
ridge are shown in (a). With one exception, all of these lines pass
within the circle centered at 58°N, 36°W. Great circles perpendicular
to the strike determined by earthquake mechanism solutions are shown
in (b)

Talwani et al. (1961) (northern region), Heezen and Tharp (1965) (equatorial region), and
Heirtzler and Le Pichon (1965) (southern region). The solid line in Figure 8–9 was calculated with these quantities and with the choice
of $PLAT = 62°N$, $PLONG = 36°W$, $V_{max} =$
1.8 cm/yr. The dashed line was calculated
without the strike correction; it does not give
the half-velocity perpendicular to the strike of
the ridge but, rather, it gives the half-velocity
parallel to the direction of spreading.

This figure was originally calculated with
$PLAT = 58°N$, the latitude of the center of the
circle in Figure 8–8,a. With this pole position
the computed curve does not satisfactorily fit
the observed points: the points south of 20°S
alone fit $V_{max} = 1.8$ cm/yr; the points north of
20°N better fit $V_{max} = 2.2$ cm/yr. If the pole is
chosen farther north, say at 62°N as shown in
the figure, a single curve apparently fits both
the northern and southern portions of the data
within the scatter of the points. The velocity
pattern is sensitive to the latitude but not the

longitude of the chosen pole, whereas the intersection of the perpendiculars in Figure
8–8 was just the opposite. A pole at 62°N
(±5°), 36°W (±2°) with a maximum velocity of
1.8 (±0.1) cm/yr satisfies both of these criteria.

The observed spreading rates were inferred
from magnetic profiles over the mid-Atlantic
ridge in the following manner. The two points
Chain 44 and Chain 61 were determined by
Phillips (1967). The magnetic profiles used for
obtaining the other points may be found in
Heirtzler and Le Pichon (1965), Talwani et al.
(1961), Vacquier and Von Herzen (1964), and
U. S. Naval Oceanographic Office (1965). The
strike of the ridge at the crossings of Vema 4,
17, and 10 was assumed to be 38°, 38°, and 30°,
respectively. Zero strike was assumed at the
crossings of Argo, Vema 18, Zapiola, Vema
12, and Project Magnet Flight 211. Simulated
magnetic anomaly profiles at locations near
each crossing of the ridge were calculated with
the normal-reversed time scale (and computer
program) of F. J. Vine (1966). The spreading

Table 8–2
Strikes determined from transcurrent earthquakes
on the Mid-Atlantic ridge.

Name	Latitude	Longitude	Strike	Reference
8	35.29°N	36.07°W	86°	a
5	23.87°N	45.96°W	103°	a
4	10.77°N	43.30°W	90°	a
3	7.80°N	37.35°W	97°	a
2	7.45°N	35.82°W	100°	a
18	0.49°S	19.95°W	84°	b
1	0.17°S	18.70°W	87°	a

a. Sykes (1967).
b. Sykes (1968a).

rates were established by matching features on the computed profiles (known time) to corresponding features on the observed profiles (known distance from the crest of the ridge). Only that portion of Vine's time scale between 0 and 5 million years was used in determining the rates; as noted by Phillips (1967) either the spreading rates were about 25% faster previous to 5 million years ago or the time scale needs adjustment.

More magnetic profiles should be analyzed to critically test this hypothesis. The scatter in spreading rate values determined from adjacent ship tracks is likely due to numerous small

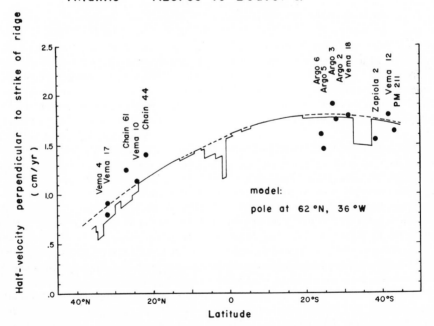

Figure 8-9
Spreading rates determined from magnetic anomaly profiles are compared with the values calculated with the model. The solid line shows the predicted rate perpendicular to the strike of the ridge; the dashed line shows the rate parallel to the direction of spreading

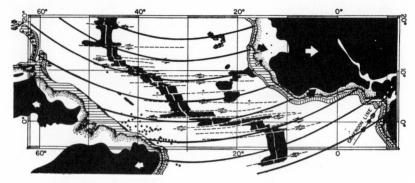

Figure 8-10
The strike of the faults in the equatorial Atlantic are compared with circles concentric about a pole at 44.0°N, 30.6°W, the pole about which South America must be rotated to make its coastline (500-fm isobath) coincide with the coastline of Africa (Bullard et al., 1965). These circles indicate the average motion since drifting began. (Figure is adapted from Heezen and Tharp, 1965)

fracture zones offsetting the anomaly pattern. An aeromagnetic survey of the scale of the U.S. Naval Oceanographic Office survey of the Reykjanes ridge (Heirtzler et al., 1966) would permit an unequivocal determination of spreading rate and would afford a stringent test to this model. The area just south of 20°N has particular significance; a uniform change or discontinuous change in spreading rate here would show whether North America and South America move as a single block or as two blocks.

We may contrast this present motion of Africa and South America with the average motion of these two continents since they first split apart. This average motion is shown in Figure 8–10 and is quite different from the present motion indicated in Figure 8–7. The total length of the transform faults in this region suggests that about half of the motion of these two continents has been about the present pole. The earlier half of this total motion would have followed lines tending more northeast to southwest than the strike of the features observed in the center of the ocean.

THE MOTION OF
THE PACIFIC BLOCK RELATIVE TO
THE NORTH AMERICAN BLOCK

Figure 8–11 shows the great fracture zones of the Pacific block. Menard's (1967) demonstra-

tion that these great fracture zones are not all great circles initiated the present investigation of crustal blocks. A set of circles concentric about a pole at 79°N, 111°E are superposed on this figure. Except for the Mendocino and Pioneer fracture zones, the concentric circles are nearly coincident with the fracture zones. The Mendocino and Pioneer fracture zones depart from the circles farther west than do the other fracture zones; this departure is likely related to North America 'overriding' and interfering with the flow of the northernmost end of the rise at an earlier date. These old fracture zones indicate that the Pacific once moved away from North America toward trenches off New Guinea and the Philippines. About 10 m.y. ago this pattern changed, and the Pacific now moves toward the Japan and Aleutian trenches.

We now consider the present boundary between the North American block and the Pacific block. This boundary is of the fault type from the Gulf of California to the Gulf of Alaska and is of the trench type along the Alaska Peninsula and Aleutian Islands. A small region, the Juan de Fuca block, is an anomalous region between these two large blocks. The boundaries of this block are shown in Figure 8–12. The trend of the Mendocino and Blanco fracture zones is not parallel to the trend of the San Andreas and Queen Charlotte faults, and we wish to justify exclusion of this region in our consideration of the motion of the Pacific block relative to the North American

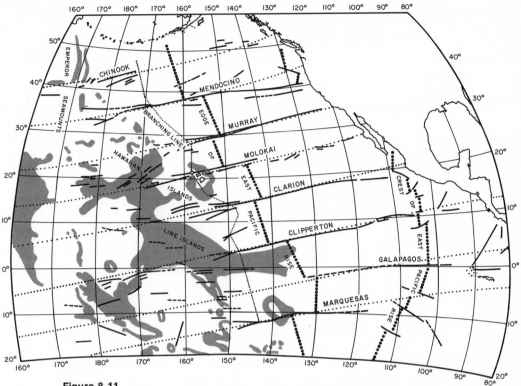

Figure 8-11
Old fracture zones in the Pacific are compared with circles concentric about a pole at
79°N, 111°E. (Figure is adapted from Menard, 1967a)

block. The belt of earthquakes between the Pacific and North America splits into two branches here (see Figure 8–12); the major branch follows the Mendocino and Blanco fracture zones, a less active branch follows what we suppose to be a compressive zone along the coast of Oregon and Washington. Figure 8–13 shows straight lines drawn on the magnetic diagram of Raff and Mason to indicate faults that offset the magnetic pattern. These faults are especially prominent in the color version of this figure appearing in Vine (1968). As noted by Raff and Mason (1961), there are many faults at angles oblique to the main trends, and these faults suggest many small blocks moving independently of one another in this region. Detailed surveys of other rises will, perhaps, show that their magnetic patterns are equally broken up by oblique faults; in this event the argument advanced here on the anomalous nature of this small re-

gion will prove false. Note in particular the triangular shaped region ABC bounded by the coast line from 47°N to 42°N, the Blanco fracture zone AB, and the fault BC, which begins at the intersection of the Blanco fracture zone and the Juan de Fuca ridge and heads at a 45° strike toward Puget Sound. The magnetic pattern is offset on the fault BC by about 70 km. This fault extended intercepts the Cascade Range near Mount Baker (49°N), and the Blanco fracture zone extended intercepts the Cascade Range at Lassen Peak (40°N). The volcanic Cascade Range lies between these two peaks. The Cascades and the sedimentary Coast Range were both formed at the close of the Pliocene and during the Pleistocene time. The Coast Ranges are probably the fill of an uplifted trench, and the Cascades are probably the volcanic counterpart of this now extinct system. The activity of this region is probably in its last stages; the fault BC is not seismically

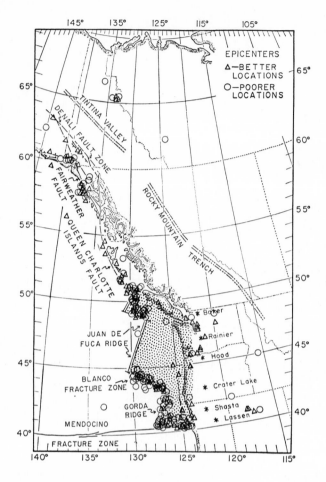

Figure 8-12
The seismic belt along the west coast splits and follows the east and west boundaries of the Juan de Fuca block. Several large volcanic cones in the Cascades are identified. (Figure is adapted from Tobin and Sykes, 1968)

active at present. It appears as if this triangular block ABC has moved eastward into North America with the crust thickening beneath the Coast Range or Cascades. The motion of this triangular block has possibly accommodated a variable spreading rate along the Juan de Fuca ridge; note the variable spreading rate along the Gorda ridge indicated by its fan-shaped magnetic anomaly pattern. On the basis of these arguments, we shall assume that the Juan de Fuca block moves independently of the Pacific and North America and shall ignore it in our discussion.

The faults used in this paper are listed in Table 8–3. The latitude and longitude of points at the northern and southern ends of a straight fault segment are listed here along with the length and strike of the segment between the two end points. A great circle was constructed which passes midway between the points at right angles to the segment joining them. As shown in Figure 8–14,*e*, these great circles perpendicular to the fault segments have a common intersection near 53°N, 53°W; we shall return to this summary figure after discussing the faults in each region.

The fault segments in the Fairweather-Queen Charlotte region generally follow the line drawn by St. Amand (1957). On July 10, 1958, a large earthquake occurred on the Fairweather fault, and the first nine faults listed in the table are from papers based on field work in this area immediately after this earthquake. The Fairweather fault has a total length of 200 km, of which 90% is covered with ice, water, or unconsolidated sediments. The only observed fault traces were observed at the northern tip of fault 2, at the southern tip of fault

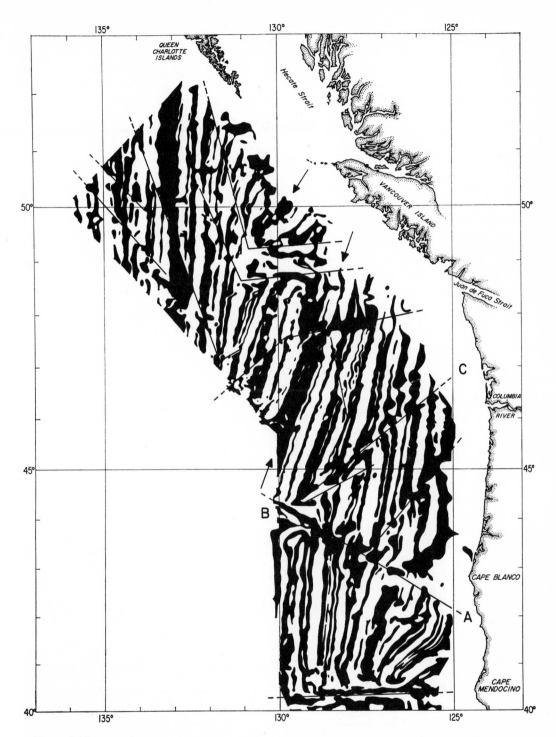

Figure 8-13
Superposed on Raff and Mason's summary diagram of the magnetic anomalies in the Juan de Fuca region are arrows that indicate the axes of the three short ridge lengths in the area and straight lines that indicate faults offsetting the anomaly pattern (Vine, 1966)

Table 8–3
Strike of major transcurrent faults on the west coast of North America.

Name	Latitude and longitude northern end	Latitude and longitude southern end	Length (approx., km)	Strike (approx., deg)	Reference
FAIRWEATHER—QUEEN CHARLOTTE					
1. Russell Fiord	60.043°N 139.512°W	59.848°N 139.274°W	26	149	a
2. SE of Nunatak Fiord	59.814°N 139.000°W	59.390°N 138.336°W	60	142	b
3. Alsek River and Glacier	59.390°N 138.336°W	59.046°N 138.000°W	45	153	b
4. Fairweather Trench Lituya Bay	59.031°N 138.000°W	58.583°N 137.361°W	61	143	b, c
5. North Dome	58.583°N 137.361°W	58.553°N 137.309°W	5	139	c
6. La Perouse Glacier	58.553°N 137.309°W	58.504°N 137.211°W	8	134	c
7. South Dome	58.504°N 137.211°W	58.483°N 137.176°W	3	138	c
8. Kaknau Creek	58.442°N 137.124°W	58.393°N 137.073°W	6	151	c
9. Crillion Lake to Palma Bay	58.583°N 137.361°W	58.393°N 137.073°W	29	142	5, 6, 7, 8
10. Chichagof Island	58.00°N 136.65°W	57.48°N 136.00°W	70	146	d
11. Baranof Island	57.00°N 135.54°W	56.00°N 134.64°W	125	154	d
12. Queen Charlotte Islands (N)	54.00°N 133.43°W	52.93°N 132.42°W	130	151	d, e
13. Queen Charlotte Islands (S)	52.93°N 132.42°W	52.00°N 131.15°W	130	140	d, e
NORTHERN CALIFORNIA					
1. Pt. Arena	39.00°N 123.69°W	38.48°N 123.19°W	72	143	f
2. Tomales Bay	38.24°N 122.98°W	37.91°N 122.68°W	45	144	f
3. San Andreas Lake	37.70°N 122.50°W	37.55°N 122.37°W	20	145	f
4. [Unnamed]	37.48°N 122.31°W	37.29°N 122.11°W	27	140	f
5. San Juan Bautista	37.12°N 121.91°W	36.91°N 121.59°W	37	130	f
6. San Juan Bautista	36.47°N 121.07°W	36.29°N 120.89°W	26	141	f
7. [Unnamed]	36.29°N 120.89°W	36.00°N 120.57°W	43	139	f
8. [Unnamed]	35.71°N 120.26°W	35.36°N 119.91°W	50	141	f
9. [Unnamed]	35.36°N 119.91°W	35.07°N 119.58°W	44	137	f

Table 8–3 (continued)

Name	Latitude and longitude northern end	Latitude and longitude southern end	Length (approx., km)	Strike (approx., deg)	Reference
	SOUTHERN CALIFORNIA				
1. Elsinor	33.82°N 117.57°W	33.63°N 117.32°W	31	133	*f*
2. Elsinor	33.27°N 116.87°N	33.00°N 116.41°W	52	125	*f*
3. San Jacinto	33.55°N 116.64°W	33.31°N 116.32°W	40	131	*f*
4. San Jacinto	32.28°N 115.14°W	32.09°N 114.93°W	29	136	*g*
5. San Andreas	33.74°N 116.22°W	33.57°N 116.00°W	28	132	*f*
6. Imperial	33.00°N 115.62°W	32.69°N 115.33°W	44	142	*f*
7. Imperial	33.00°N 115.62°W	32.50°N 115.17°W	70	143	*g*
	GULF OF CALIFORNIA				
1. Sal si Puedes	29.339°N 113.617°W	28.444°N 112.495°W	147	132	*h*
2. [Unnamed]	26.716°N 111.151°W	26.225°N 110.420°W	90	127	*h*
3. [Unnamed]	25.150°N 109.764°W	24.670°N 109.000°W	94	125	*h*
4. [Unnamed]	24.116°N 109.000°W	23.473°N 108.000°W	124	125	*h*

a. Davis and Sanders (1960), Figure 1.
b. Davis and Sanders (1960), Figure 2, and Tocher (1960), Table 1.
c. Tocher (1960), Table 1 and Plate 1.
d. St. Amand (1957), Figure 7.
e. 200-fm line on U.S.G.S. Geologic Map of North America, 1965.
f. Geologic Map of California, Plate 1 of California Div. Mines and Geology Bulletin 190, 1966.
g. Biehler et al. (1964), Chart 1.
h. Rusnak et al. (1964), Plate 3.

4, and along faults 5, 7, and 8. The strikes of the longer faults (no. 2, 60 km; no. 3, 45 km; no. 4, 61 km) are thus inferred from the topography of the glacier filled troughs. The three short observed fault traces (no. 5, 5 km; no. 7, 3 km; no. 8, 6 km) and the segment through La Perouse glacier that connects the end points of 5 and 7 (no. 6, 8 km) were used in constructing Figure 8–14,*a*. There is a 17° difference in the strikes of parts of these four segments. However, even the very straight San Andreas fault in northern California has 10° variations rather continuously along short (5 km) segments of its total length. (See for example figures in Oakeshott, 1966, and Dibblee, 1966). To reduce this scatter, an average fault was constructed from the beginning of fault 5 to the end point of fault 8. This average fault (no. 9, 26 km) was used in Figure 8–14,*e* in place of faults 5, 6, 7, and 8.

In northern California, nine segments of the San Andreas fault are tabulated; lines constructed perpendicular to these segments are shown in Figure 8–14,*b*. Only one fault, no. 5 near San Juan Bautista, has a strike notably different from the others. At this location, the San

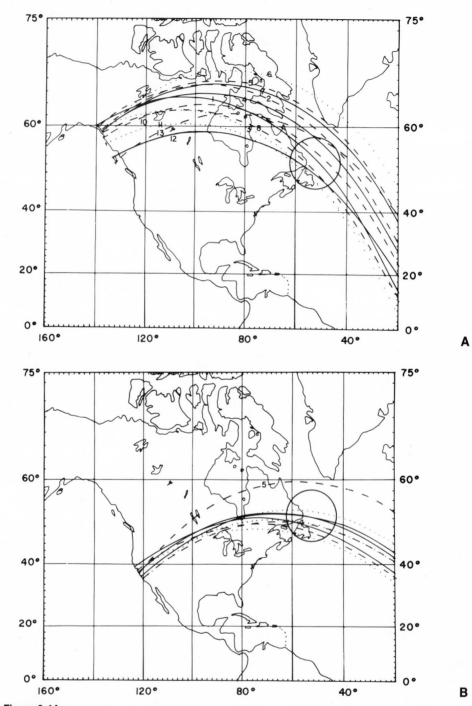

Figure 8-14
Parts (a) through (d) show great circles constructed perpendicular to the strikes of
fault segments observed in the Fairweather-Queen Charlotte, northern California,
southern California, and Gulf of California regions. Part (e) is a composite of the four
separate regions with the exceptions noted in the text. Part (f) shows great circles
constructed perpendicular to strikes determined from earthquake mechanism solutions.
The circle of intersection drawn has the coordinates 53°N (±6°), 53°W (±10°)

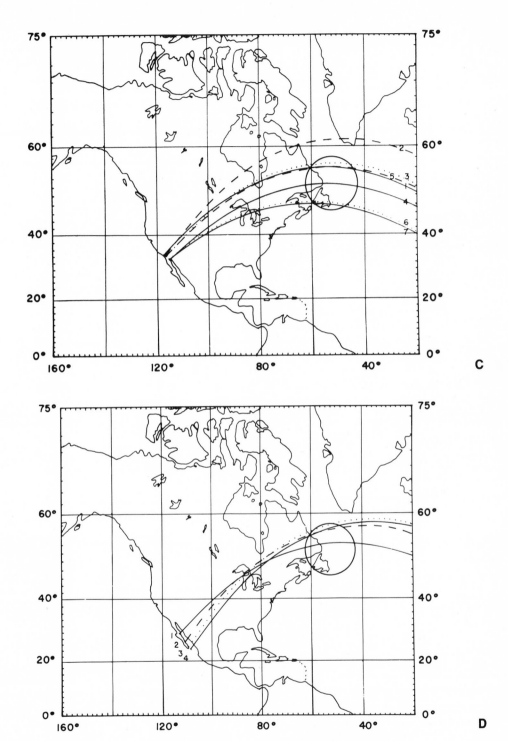

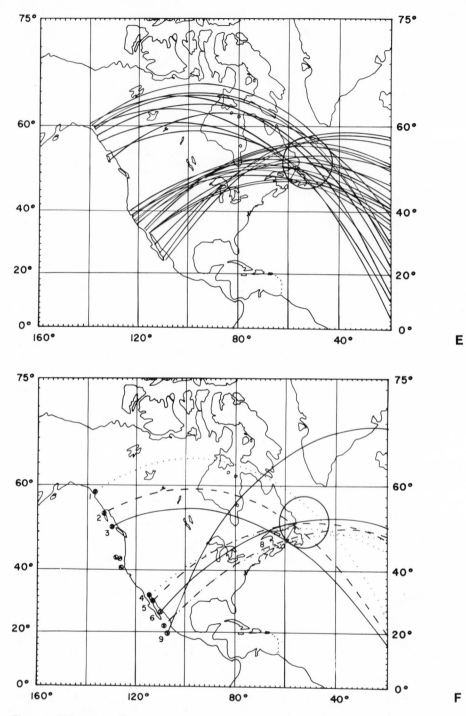

Figure 8-14 *(continued)*

Andreas fault splits into two parallel branches, the northwestern branch being the San Andreas fault and the northeastern branch being the active Hayward fault and Calaveras fault. In light of the special nature of this junction, fault 5 has been omitted from the summary diagram in Figure 8–14,e.

In southern California, the fault segments chosen are presumably the major faults of the region. This choice was based on articles by Biehler et al. (1964) and Allen et al. (1965). One segment, the Elsinor no. 2, has a strike notably different from the others, and it was omitted from Figure 8–14,e. The two entries for the Imperial fault are not independent: one entry was taken from a source that extended only to the international border; the second entry includes that half of the Imperial fault which is in Mexico. The tensional or compressional nature of individual features in southern California is qualitatively explained by the difference between the strike of a feature and the average strike of the region. The Salton trough (general strike 150°) is a depressed region; the Salton Sea is 75 meters below sea level. The observed surface faults in the trough do not run the length of the trough but are arranged en echelon with each fault having a strike of about 140°. In contrast, the Transverse Range (general strike 110°) reaches an altitude of 3 km. A major transcurrent fault, the Garlock fault, runs east-west just north of the Transverse Range. The motion on this fault is such as to move a wedge-shaped block eastward to relieve the north-south compressive stresses occurring at this bottleneck. A similar pattern occurs in Arabia. The strikes of the Gulf of Aqaba and the Dead Sea are inclined to one side, and the strike of the short mountain range in Lebanon is inclined to the other side, of a smooth small circle representing the motion of Arabia relative to the African-Mediterranean block.

The four faults in the Gulf of California were taken from the fault map of Rusnak et al. (1964). The line perpendicular to the longest and best defined of these faults, the Sal si Puedes fault, passes through the center of the circle of intersection as shown in Figure 8–14,d. With hindsight, a different choice of faults in the southern part of the Gulf could be made that would allow all the perpendiculars

to pass near the center of the circle. The apparently systematic shift in the lines 1, 2, 3, and 4 should not be considered significant.

The lines in Figures 8–14,a and 8–14,b, with the exceptions noted in the text above, are all drawn in Figure 8–14,e. All of these lines pass through the circle centered at 53°N, 53°W. The size and location of the circle, 53°N (±6°), 53°W (±10°), were chosen to fit the intersections in Figures 8–14,e and 8–14,f; the same circle has been drawn in all six figures. The scatter in the strikes of neighboring faults is larger on this boundary than was the scatter in the Atlantic Ocean. We cannot determine the distance between a region and the pole from the intersection of the great circles constructed for Alaska alone or California alone. This might be expected if the surface expression of a fault in continental regions is more irregular than in oceanic regions. Alaska and California are sufficiently far apart to compensate for the scatter in strikes in a single region. The great circles of the two regions intersect at large angles and precisely locate the best pole. We may suppose that the accuracy in determining the center of the circle of intersection is half the radius of the circle drawn in the figure; the pole is then located at 53°N (±3°), 53°W (±5°).

The strikes of fault planes of nine earthquake mechanism solutions are listed in Table 8–4. Lines drawn perpendicular to these strikes are shown in Figure 8–14,f. The epicenters of three additional earthquakes for which solutions have been determined (Tobin and Sykes, 1967) are also shown in this figure. These three earthquakes occurred on the boundary between the Pacific and Juan de Fuca blocks; therefore their perpendiculars were not constructed. Earthquake 9 on the Rivera fracture zone occurs on a tongue of the North American block that sticks out into the Pacific block and partially surrounds part of the Middle America trench. It is not surprising that the direction of this earthquake departs from the strike of the other earthquakes. Earthquakes 7 and 8, which are only 200 km away from the Rivera earthquake, however, have strikes parallel to the earthquakes farther north in the Gulf.

We may assign a rate to the motion of the Pacific block relative to North America if we assume that the motion along the San Andreas

Table 8–4
Strikes determined from transcurrent type earthquake mechanism solutions.

Name	Lati-tude (°N)	Longi-tude (°W)	Strike (deg)	Refer-ence
1. Fairweather	58.33	136.92	145	a
2. Queen Charlotte	54.10	132.58	151	b
3. Queen Charlotte	50.81	130.15	155	c
4. Gulf of California	31.72	114.42	138	d
5. Gulf of California	29.68	113.74	135	d
6. Gulf of California	26.26	110.22	132	d
7. Gulf of California	21.36	108.65	134	d
8. Gulf of California	21.26	108.75	133	d
9. Rivera Fault Zone	18.87	107.18	112	e

a. Stauder (1960).
b. Hodgson and Milne (1951).
c. Tobin and Sykes (1967).
d. Sykes (1968a).
e. Sykes (1967).

fault is 6 ± 1 cm/yr (Hamilton and Myers, 1966). Converting this value into half-velocity and taking into account that the San Andreas fault is 49° from the pole at 53°N, 53°W, we find $V_{max} = 4.0 \pm 0.6$ cm/yr. This rate is based on recent movements of the San Andreas fault. In the Atlantic, the magnetic pattern produced during the past 5 million years was used to determine the spreading rate; hence, that rate is an average rate for 5 million years. The rate for the Pacific block relative to North America might be based on the same standard if magnetic profiles oriented parallel to the transform faults in the Gulf of California were analyzed.

THE MOTION OF THE ANTARCTIC BLOCK RELATIVE TO THE PACIFIC BLOCK

A study of the motion of the Antarctic block relative to the Pacific block was made, and a pole at $71 \pm 2°S$, $118 \pm 6°E$ with an equatorial half-velocity of 5.7 ± 0.3 cm/yr was found. The data on which this study is based are given in Pitman et al. (1968) and also in Heirtzler (1968). Le Pichon (1968) has also investigated this region and has found a pole position practically identical to that listed above. A listing of the strikes of faults and spreading rates is given in Le Pichon's paper and will not be repeated here.

Six large fracture zones offsetting the Pacific-Antarctic ridge have recently been delineated by the authors listed above. Great circles were constructed perpendicular to the strike of these fracture zones, and, as shown in Figure 8–15, these great circles all pass within 2° of a pole at 71°S, 118°E. This pole position was chosen using both the constructed great circles and the spreading rate data, which will be discussed next. The great circles intersect at grazing angles and give good control only in the latitude of the pole; the spreading rates provide the control in the other direction.

Magnetic profiles of twelve crossings of this ridge have been presented by Heirtzler (1968) and Pitman et al. (1968). Ten of these profiles were analyzed in the following manner to obtain the spreading rates shown in Figure 8–16. The profile Eltanin 19N was taken to be the standard, and a spreading rate of 4.40 cm/yr was assigned to it. This rate is in agreement with the Vine (1966) time scale used here in the analysis of the Atlantic Ocean spreading rates; if a different time scale is to be used, all rates here will be scaled up or down by the same factor. The central portion of this profile, corresponding to the spreading within about the last 8 m.y. was examined and the distance from the center of the profile to each distinctive peak or valley of the profile was noted. Each profile was so examined and the ratio of its spreading rate relative to the standard was

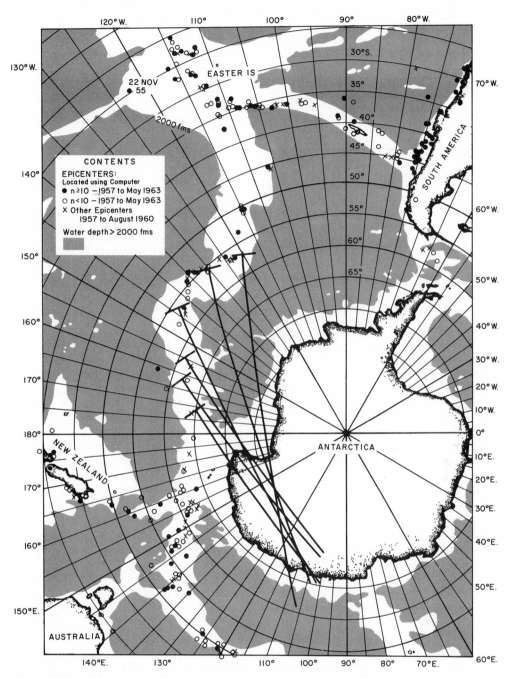

Figure 8-15
Great circles constructed perpendicular to the strike of fracture zones offsetting
the Pacific-Antarctic ridge are plotted on Syke's (1963) seismic map of this
region. The great circles all pass within 2° of the pole at 71°S, 118°E

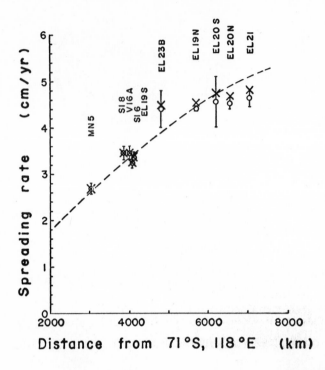

Figure 8-16

Spreading rates on the Pacific-Antarctic ridge are compared with a model with $V_{max} = 5.7$ cm/yr about a pole at 71°S, 118°E. The circles are the spreading rates measured perpendicular to the strike of the ridge; the crosses are these rates projected parallel to the direction of spreading

found. On some profiles, an apparently perfect matching of each peak of the profile to the corresponding peak of the standard was made, and the error in determining the rate is small. On other profiles the peaks within the interval corresponding to less than 8 m.y. could not be matched with certainty, or a large number of offsets broke the pattern into many short matchable segments, and the error in determining the rate is large. The open circles and error flags in Figure 8–16 show the rates and estimated errors so found. These are rates perpendicular to the strike of the ridge, as the profiles examined had been projected perpendicular to the ridge. The crosses in the figure show V_θ, the rate projected parallel to the direction of spreading as described in Figure 8–5. The dashed line computed for $V_{max} = 5.7$ cm/yr should pass through all of the crosses if there were no errors in the analysis (or in the hypothesis of rigid blocks).

No attempt was made to construct a diagram predicting the spreading rate perpendicular to the strike of the ridge as was done for the Atlantic. The crossings of the ridge are spaced about 1 every 500 km, and the strike of the ridge is simply not known. It is interesting to

note that the three profiles easiest to interpret, EL19N, SI8, and SI6, had ship's tracks inclined 7° or less from the direction of spreading inferred from a pole position at 71°S, 118°E. In general, the greater the angle between the ship track and the direction of spreading, the more often offsets in the magnetic pattern were observed. It seems likely, therefore, that the Pacific-Antarctic ridge is offset by many small fracture zones (the offsets noted here ranged from 10 km to 40 km); this ridge has, perhaps, a pattern similar to that observed in the equatorial Atlantic Ocean. If this is so, the 'local' strike of the ridge, the strike that must be used in projecting a profile, might differ significantly from the 'general' strike determined from widely spaced crossings of the ridge. The crossings SI3 and SI5 presented in the references above are not shown in Figure 8–16 because of the uncertainty in projection. Rates of 2.50 ± 0.50 cm/yr and 2.68 ± 0.10 cm/yr were determined for SI3 and SI5, respectively, along the projected profiles presented in Pitman et al. (1968). If the strike of the ridge is perpendicular to the fracture zones here, i.e., the ridge strike is 45°, the velocities parallel to the spreading direction are 2.0 cm/yr and 2.5

Table 8–5
Prediction of Antarctica–Africa Pole from closure
of Africa–North America–Pacific–Antarctica–Africa circuit.

	Latitude (°N)	Longitude (°E)	V_{max} (cm/yr)	$\frac{1}{2}a\omega_x$ (cm/yr)	$\frac{1}{2}a\omega_y$ (cm/yr)	$\frac{1}{2}a\omega_z$ (cm/yr)
ω_{Am-Af}	62 ± 5	-36 ± 2	1.8 ± 0.1	0.7 ± 0.1	$-0.5 \pm .1$	1.6 ± 0.1
ω_{Pac-Am}	53 ± 3	-53 ± 5	4.0 ± 0.6	1.4 ± 0.3	$-1.9 \pm .4$	3.2 ± 0.5
$\omega_{Ant-Pac}$	-71 ± 2	118 ± 6	5.7 ± 0.3	-0.9 ± 0.2	$1.6 \pm .2$	-5.4 ± 0.3
ω_{Ant-Af}	-25 ± 30	-35 ± 20	1.6 ± 0.5	1.2 ± 0.4	$-0.8 \pm .5$	-0.6 ± 0.6

cm/yr, respectively. If the ridge is running east to west here, i.e., the ridge strike is 90°, the velocities parallel to the spreading directions are 3.4 cm/yr and 3.3 cm/yr. Three or four closely spaced tracks are needed in this area to establish the strike of the ridge and pin down the spreading rate at this critical end of Figure 8–16.

THE MOTION OF THE ANTARCTICA BLOCK RELATIVE TO THE AFRICAN BLOCK

We are now in a position to estimate the motion of the Antarctic block relative to the African block by summing the angular velocity vectors found above to describe the motion of Antarctica relative to the Pacific, the Pacific relative to North America, and North America relative to Africa. The three pole positions and rates found above to describe these motions are listed in the first three columns of Table 8–5. The angular velocity vectors represented by these angles and rates are transformed into angular velocity in Cartesian coordinates under the headings $\frac{1}{2}a\omega_x$, $\frac{1}{2}a\omega_y$, $\frac{1}{2}a\omega_z$. The one-half emphasizes that half the spreading rate was used in the calculation, and the factor a (the radius of the earth) allows us to express our results in units of cm/yr. The first three vectors are then added to obtain the predicted ω_{Ant-Af}, and this is transformed back into a pole position and spreading rate. The errors listed in the table were calculated by adding the squares of the errors of each term contributing to the result. More precise values for the three measured angular velocity vectors could reduce this error in ω_{Ant-Af} to a negligible amount, but there would still remain possible

systematic errors. Our value for ω_{Am-Af} was determined primarily from data between South America and Africa, and, if North America is moving relative to South America, we must change this rate. If the rate between North America and Africa is more than 1.8 cm/yr, say 2.2 cm/yr as suggested by the data in Figure 8–9, the pole of ω_{Ant-Af} would be shifted northward. If North America is presently splitting apart in Nevada, we need an additional term to correct the rate we found for the Pacific relative to western North America into a rate for the Pacific ralative to eastern North America. This additional rate is probably small, and its pole is likely to be somewhere in Canada; such an additional angular velocity vector would shift the resultant pole of ω_{Ant-Af} northward and westward. In addition, any general distortion of the blocks would invalidate the rigidity hypothesis and introduce error in the resultant ω_{Ant-Af} found by summing along this path.

The sign of the ω_{Ant-Af} we have found is such that we should expect Africa and Antarctica to be moving apart. The few magnetic profiles available from the Atlantic-Indian rise do not show the characteristic symmetric pattern of the other ridge crests: this has led Vine (1966) to speculate that the Atlantic-Indian rise is an extinct rise. The results here suggest otherwise; the lack of a recognizable magnetic pattern over the crest is then supposedly the result of crossing a highly fractured region of the ridge at angles oblique to the direction of spreading. If more significance than just the sign is given to the value of ω_{Ant-Af}, we may expect Africa and Antarctica to be separating about a pole in the South Atlantic Ocean with a maximum half-rate of about 1.5 cm/yr. Great circles were constructed perpendicular to the strikes of the Malagasy fracture zone and the nearby Prince Edward fracture zone. These two great circles intersect at grazing angles, and a unique pole position could not be determined from their intersection. A pole at 15°S, 15°W is compatible with these circles and is within the error limits shown in Table 8–5; there is no contradiction between the observations of the rise and that predicted by closure. As stated above, there are no magnetic profiles across this rise with which to check the predicted spreading rate, but it is possible to estimate this rate by closure around the triple junction of rises in the center of the Indian Ocean. The mid-Indian Ocean rise between Antarctica and Australia is opening north to south at a rate of about 3.0 cm/yr (Le Pichon, 1968), and the Carlsberg ridge is opening more or less north to south at a rate of about 1.5 cm/yr. The difference between these rates agrees with the value of 1.5 cm/yr listed in Table 8–5.

If the closure principle demonstrated here is shown to have acceptable precision, we may use rates measured over rises (or across transcurrent faults on land) to predict the velocity difference between the two sides of a trench. It will be interesting to see if properties of trench systems can be correlated with the rate of closing or the angle between the velocity difference of the two sides and the axis of the trench.

CONCLUSION

The evidence presented here favors the existence of large 'rigid' blocks of crust. That continental units have this rigidity has been implicit in the concept of continental drift. That large oceanic regions should also have this rigidity is perhaps unexpected. The required strength cannot be in the crust alone; the oceanic crust is too thin for this. We instead favor a strong tectosphere, perhaps 100 km thick, sliding over a weak asthenosphere. Theoretical justification for a model of this type has been advanced by Elsasser (1967). In the simple two-dimensional picture of a rise and a trench with a continent between them, we imagine a conveyor-belt process in which the drifting continent need have no great strength. In the model considered here, we may have local hot spots on the rise and faster sinking at some places on the trenches. The crustal blocks should have the mechanical strength necessary to average out irregular driving sources into a unifrom motion; the tectosphere should be capable of transmitting even tensile stresses. The crustal block model can possibly explain the median position of most oceanic rises and the symmetry of their magnetic pattern. We assume that the location of the rises is not fixed by some deep-seated thermal source but is determined by the motion of the blocks. Suppose a crustal block is under tension and splits along some line of weakness. The forces that tore it apart continue to act, and the blocks move apart creating a void, say, 1 km wide and 100 km deep, which is filled with mantle material. As the blocks move farther apart, they split down the center of the most recently injected dike, since this is the hottest and weakest portion between the two blocks. Even if one block remains stationary with respect to the mantle and only one block moves, we will have a symmetric pattern if a new dike is always injected up the center of the most recent dike. If the initial split was entirely within a large continental block, this control of mantle convection by boundary conditions at the top surface will result in a ridge crest with a median position.

9

Sea-floor Spreading and Continental Drift

XAVIER LE PICHON
1968

From *Journal of Geophysical Research*, v. 73, p. 3661–3697, 1968. Reprinted with permission of the author and the American Geophysical Union. Copyright 1968.

It has long been recognized that if continents are being displaced on the surface of the earth, these displacements should not in general involve large-scale distortions, except along localized belts of deformation. Recent studies of the physiography of the ocean floor (Heezen, 1962) and of the distribution of sediments in the oceans (Ewing and Ewing, 1964) did not reveal widespread indications of compression or distortion of large oceanic blocks. Consequently, the displacements inferred in the spreading-floor hypothesis of Hess (1962) and Dietz (1961) should not result in large-scale deformation of the moving blocks. Morgan (1968b) has investigated the important implications of these observations on the geometry of the displacements of ocean floor and continents. In this paper we try to carry this attempt further and to test whether the more uniformly distributed data on sea-floor spreading now available are compatible with a non-expanding earth. The discussion will be confined to a preliminary investigation of the global geometry of the pattern of earth surface movements as implied by the spreading-floor hypothesis. We use Morgan's exposition of the problem as a basis. Parts of these results were previously reported by Le Pichon and Heirtzler (1968) and Heirtzler et al. (1968).

Let us assume that large blocks of the earth's surface undergo displacements and that the only modifications of the blocks occur along some or all of their boundaries, that is, the crests of the mid-ocean ridges, where crustal material may be added, and their associated transform faults, and the active trenches and regions of active folding or thrusting, where crustal material may be lost or shortened. Then the relative displacement of any block with respect to another is a rotation on the spherical surface of the earth. For example, if the Atlantic Ocean is opening along the mid-Atlantic ridge, the movement should occur in such a way as not to deform or

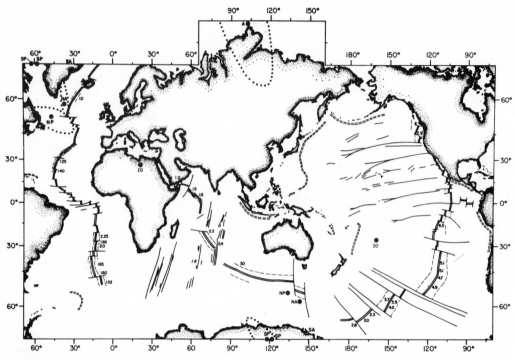

Figure 9-1

Available data on sea-floor spreading. The axes of the actively spreading mic-ocean ridges are shown by a double line; the fracture zones by a single line; anomaly 5 ($\approx$ 10 m.y. old) by a single dashed line; the active trenches by a double dashed line. The spreading rates are given in centimeters per year. The locations of the centers of rotation obtained from spreading rates are shown by ×; those obtained from the azimuths of the fracture zones by +. NA stands for North Atlantic; SA for South Atlantic; NP for North Pacific; SP for South Pacific; IO for Indian Ocean; A for Arctic. The ellipses drawn around the NA, NP, SP, and A centers of rotation obtained from the fracture zones are the approximate loci of the points at which the standard deviation equals 1.25 times the minimum standard deviation. The ellipse around the IO center of rotation is too small to be shown. These ellipses indicate how fast the least-squared determination converges

distort the large bodies of horizontally strati-fied sediments lying in its basins and at the continental margins. It should not involve large-scale distortion of the African or South American continents. Motion of the African relative to the South American block (one block including the continent and its adjacent basins) should be everywhere parallel to the transform faults (Wilson, 1965c), which should be arcs of a small circle about the center of this movement of rotation. The angular velocity of rotation should be the same everywhere. This implies that the spreading rate increases as the sine of the distance (expressed in degrees of arc) from the center of rotation and reaches a maximum at a distance of 90° from this center, along the equator of rotation.

Morgan (1968) has shown that the fracture zones in the Atlantic Ocean between 30°N and

10°S are very nearly small circles centered about a point near the southern tip of Green-land and that the spreading rates already de-termined roughly agree with the velocities required for a movement of opening of the Atlantic Ocean about this point. Thus sea-floor spreading in the Atlantic Ocean does not in-volve distortion of the oceanic and continental blocks on each side of it. Morgan has shown similarly that the fault systems along the west coast of North America (e.g., the Denali, San Andreas, and Gulf of California fault systems) were compatible with a movement of rotation of the Pacific Ocean floor away from North America about a point also situated near the southern tip of Greenland.

Recent work (Pitman et al., 1968; Dickson et al., 1968; Le Pichon and Heirtzler, 1968; Heirtzler et al., 1966, 1968; Herron and

Table 9–2
Centers of rotation obtained by least-squares fitting.

Area	Latitude	Longitude	Number	Standard deviation	Angular rate $(10^{-7}\ deg/yr)$
South Pacific (Antarctica–Pacific)					
From fracture zone	**70S**[a]	**118E**	6	4.5[b]	**10.8**
From spreading rate	68S	123E	11	0.058[c]	10.8
Atlantic (America–Africa)					
From fracture zone	**58N**	**37W**	18	2.9[b]	**3.7**
From spreading rate	69N	32W	9	0.065[c]	3.7
North Pacific (America–Pacific)					
From fracture zone	**53N**	**47W**	32	5.7[b]	**6.0**
Indian Ocean (Africa–India)					
From fracture zone	**26N**	**21E**	5	0.6[b]	**4.0**
Arctic Ocean (America–Eurasia)					
From fracture zone	**78N**	**102E**	4	9.1[b]	**2.8**

[a] Boldface values are those used in computing movements of different blocks.
[b] Deviation of measured from computed azimuths, in degrees.
[c] Deviation of measured from computed normalized spreading rates (actual spreading rate divided by maximum spreading rate).

Heirtzler, 1967; Herron, 1971; 1972) has greatly extended our knowledge of the pattern of spreading since the end of the Mesozoic. The locations and extents of the large fracture zones in parts of the North and Equatorial Atlantic and in the Indian Ocean (Heezen and Tharp, 1964, 1965), in the North and Equatorial Pacific (Menard, 1964), and in the South Pacific (Pitman et al., 1968) are now reasonably well known. These data are adequate for a preliminary examination of the global geometry of continental and oceanic drift deduced from the spreading-floor hypothesis.

We first show that the opening of the South Pacific, the Atlantic, the Arctic, the North Pacific, and the Indian oceans can each be described by a single rotation. The parameters of these rotations are obtained.

Second, we adopt a simple earth model consisting of six large rigid blocks. Using the parameters obtained in the first part, we obtain the vectors of differential motion between blocks along all the boundaries. The picture obtained is in reasonable agreement with physiographic, seismic, and geological data.

We then use the same type of analysis to study the movements of continental drift and sea-floor spreading since Mesozoic time. The Atlantic, South Indian, and South Pacific oceans are studied in greater detail. The data

suggest a history of episodic spreading directly related to the major orogenic phases.

MAIN OCEAN OPENING MOVEMENTS AS DETERMINED FROM SEA-FLOOR SPREADING

Spreading Rates

Vine and Wilson (1965), using Vine and Matthews' (1963) hypothesis, first tried to relate the magnetic pattern over the crests of the ridges to the known geomagnetic time scale in order to determine the spreading rate over the last few million years. To date, 31 determinations of spreading rate at the axis of the ridge during Plio-Pleistocene times have been published. The results are listed in Table 9–1 and shown in Figure 9–1. The numbers represent the mean spreading rate in centimeters per year on one limb, assuming the motion to be perpendicular to the axis of the ridge and symmetrical about it. The total rate of addition of new crust is equal to twice the spreading rate. The precision of the measurements is probably not better than 0.1 cm/yr. Figure 9–1 also shows the location of a magnetic anomaly presumed to be 10 m.y. old which marks the outer boundary of the axial magnetic pattern (anomaly 5 of Heirtzler et al., 1968).

The data reveal that the process of addition of new crust is now occurring in all oceans. The spreading rates vary between about 1 cm/yr (in the Arctic Ocean) and as much as 6 cm/yr in the Equatorial Pacific Ocean. Spreading rates have been obtained for all branches of the mid-ocean ridge system except the southwest mid-Indian Ocean ridge; its axial magnetic pattern could not be interpreted simply in terms of the spreading-floor hypothesis (Vine, 1966; Le Pichon and Heirtzler, 1968). The number of determinations is now sufficiently large to show that the values of the spreading rate vary rather smoothly and systematically by more than a factor of 2 within a given ocean. The maximum spreading rate is found south of the equator in the Atlantic, Indian, and Pacific oceans. If there are no regions where the earth's surface is destroyed to compensate for the creation of new earth's surface, the earth must then be expanding in an asymmetrical way: the equatorial circumference is increasing faster than any longitudinal circumference. This argument will be explored further in a later section. In any case, the data available suggest a relatively simple pattern of opening of the oceans, the Atlantic and Pacific oceans opening about approximately the same axis and being linked by two oblique openings, one in the Indian Ocean and one in the Arctic Ocean.

Directions of Motion

As indicated earlier, the movement of spreading away from the axes of the ridges should be parallel to the seismically active portions of the transform faults. Figure 9–1 shows the locations of the major fracture zones over the mid-ocean ridge system according to the sources listed above. The degree of accuracy of mapping is extremely unequal, and large errors may exist in some areas, as south of Australia, for example.

We have two independent sets of data from which to determine the center of rotation: the spreading rates and the azimuths of the transform faults at their intersections with the ridge axis. The mapping of the fracture zones away from the crests of the ridges allows us to determine whether the geometry of the spreading has been the same during the whole geological time required for the creation of these transform faults.

Determination of the Parameters of Rotation

To test the simple geometrical concept of rotation of rigid blocks, we used the following method. For each of the five principal lines of opening (Arctic, Atlantic, Indian, South Pacific, and North Pacific), if the data were adequate, we obtained by least-squares fit (1) the location of the center of rotation (or its antipode) and the angular velocity best fitting the spreading rates and (2) the location of the center of rotation best fitting the azimuths of the transform faults at their intersection with the ridge axis. The numerical method of fitting minimized the sum of the squares of the residuals of the normalized spreading rates (i.e., actual spreading rate divided by maximum spreading rate) in the first case and of the azimuths in the second case. (See in the appendix an outline of the numerical method of computation.) The data for regions in the Atlantic and South Pacific oceans near the equator of rotation are sufficient to allow a good determination of the maximum spreading rate (respectively 2.05 and 6 cm/yr). For the Indian, North Pacific, and Arctic oceans, the data are inadequate to allow a determination of the center of rotation by use of the spreading rates only.

The values of the standard deviation for each fit and the importance of the disagreement between the locations of the centers of rotation obtained by the two methods give a first indication of how well the movement of spreading can be approximated by a single rotation. In addition, a graphical test was made in which the properties of the Mercator projection were used. If the axis of rotation is the axis of projection for a Mercator map, the transform faults should be along lines of latitude, the ridge axis should in general be along lines of longitude (as spreading generally occurs perpendicularly to the ridge crest), and the distance to a given anomaly should be constant on the map (as it varies as the sine of

Table 9–1

Measured spreading rates[a]. Numbers in parentheses were computed from center of rotation determined from spreading rates by least squares.

Pacific			Atlantic			Indian		
Latitude	Longitude	Rate (cm/yr)	Latitude	Longitude	Rate (cm/yr)	Latitude	Longitude	Rate (cm/yr)
48N	127W	2.9	60N	29W	0.95	19N	40E	1.0
17S	113W	6.0 (5.9)	28N	44W	1.25 (1.3)	13N	50E	1.0
40S	112W	5.1 (5.3)	22N	45W	1.4 (1.5)	7N	60E	1.5
45S	112W	5.1 (5.1)	25S	13W	2.25 (2.0)	5N	62E	1.5
48S	113W	4.7 (5.0)	28S	13W	1.95 (2.0)	22S	69E	2.2
51S	117W	4.9 (4.8)	30S	14W	2.0 (2.0)	30S	76E	2.4
58S	149W	3.9 (3.6)	38S	17W	2.0 (1.9)	43S	93E	3.0
58S	149W	3.7 (3.6)	41S	18W	1.65 (1.9)			
60S	150W	4.0 (3.4)	47S	14W	1.60 (1.6)			
63S	167W	2.3 (2.8)	50S	8W	1.53 (1.5)			
65S	170W	2.0 (2.6)						
65S	174W	2.8 (2.4)						

[a]Arctic Ocean: ≈ 1.0 cm/yr; Norwegian Sea: ≈ 1.0 cm/yr.

the distance from the center of rotation). This test was made, with the help of a digital computer with plotter, by rotating the pole of the system of coordinates to the center of rotation determined by least squares and by replotting the map in this new coordinate system.

The results of the least-squares determinations of the centers of rotations are listed in Table 9–2 and their locations are shown in Figure 9–1. The parameters of rotation adopted for the calculation of the movements of the different blocks are underlined in the table. The rate is given in units of 10^{-7} deg/yr (1° in 10 m.y.), which is nearly equal to 1 cm/yr at the equator of rotation (0.5 cm/yr for the spreading rate). The graphical tests of the calculations of the centers of rotation are shown in Figures 9–2, 9–3, and 9–4, which should be compared with Figure 9–1. In these figures the latitude is the distance in degrees from the equator of rotation.

The South Pacific Ocean

Spreading Rates. The spreading rates used for the determination of the South Pacific rotation are listed in Table 9–1 and compared with the spreading rates computed for the center of rotation so determined. They were obtained from Pitman et al. (1968) and E. M.

Herron (1971). Note the good agreement between theoretical and measured spreading rates in Table 9–1.

Fracture Zones. The azimuths of the transform faults used for the determination of the center of rotation are compared in Table 9–3 with the theoretical azimuths. The determination depends heavily on the azimuth of the southernmost fracture zone. It is difficult to know which part of the standard deviation of 4.5° is due to errors in mapping the fracture zones and reading their strikes. However, the agreement between the two methods of determination is good, since the distance between the two points is only 200 km (see Figure 9–1).

Graphical Test. Figure 9–2 is a graphical test of this determination obtained by rotating the pole of projection from the present geographic pole to 69°S, 123°E, which was an early determination of the center of rotation. In this new projection the fracture zones in the South Pacific lie along lines of latitude. Figure 9–2 also shows the locations of three key anomalies, numbers 5, 18, and 31, which are specifically 10, 45, and 70 m.y. old according to the Heirtzler et al. (1968) provisional time scale. Anomaly 31 approximately marks the

Table 9–3
Azimuths of fracture zones used for least-squares fitting[a]. Numbers in parentheses were computed from center of rotation determined from fracture zones by least squares.

South Pacific			Indian Ocean			Arctic Ocean		
Lati-tude	Longi-tude	Azimuth	Lati-tude	Longi-tude	Azimuth	Lati-tude	Longi-tude	Azimuth
56S	122W	117 (114)	12N	46E	35 (34)	82N	12W	132 (131)
56S	124W	118 (115)	13N	50E	30 (30)	71.5N	12W	125 (116)
57S	141W	121 (121)	9N	55E	32 (32)	66.5N	20W	98 (110)
62S	152W	123 (128)	15N	60E	24 (23)	52N	35W	96 (101)
63S	160W	125 (131)	18N	61E	19 (19)			
65S	170W	141 (137)						

[a]Azimuths of fracture zones for the Atlantic Ocean and the North Pacific Ocean are as listed by Morgan (1968).

outer limit of the correlatable pattern of magnetic anomalies covering the present system of mid-ocean ridges. Figure 9–2 shows, independently of the latitude or of the side of the crest on which the anomaly is situated, that the longitudinal separation between any of these key anomalies and the axis is nearly the same. The agreement would be slightly better if the center of rotation used had been the one listed in Table 9–2 (70°S, 118°E).

We can conclude that the movement of rotation now characteristic of the South Pacific has been prevalent during the time necessary for the opening of the ocean, i.e., presumably since the end of the Mesozoic. The center of rotation is near 70°S, 118°E, and the angular rate of opening is 10.8×10^{-7} deg/yr. This rotation is not much different from the rotation now characteristic of the North Pacific, as the ridge crests and transform faults in the Gulf of California and north of the Mendocino fracture zone aligned nearly along lines of longitude and latitude, respectively, in Figure 9–2. It disagrees, however, by about 30° from the set of 'fossil' fracture zones mapped by Menard (1964) in the North and Equatorial Pacific. Note in particular that north of the intersection of the East Pacific rise with the West Chile ridge, the anomaly 5 lines (10 m.y.) are in most cases perpendicular to the 'fossil' fracture zones and oblique to the now active transform faults. Thus, apparently, the pattern of spreading in the North and Equatorial Pacific has changed in the last 10 m.y. and now agrees more closely with that prevailing in the

South Pacific. The converging movements of spreading in the North and South Pacific prior to 10 m.y. ago must have resulted in relative compression west of the ridge along a line corresponding to the Tuamotu ridge and in relative extension along a line corresponding to the Chile ridge. Consequently, there should now be little if any spreading at the axis of the Chile ridge. Herron and Heirtzler (1967) have attributed the existence of the extensional Galapagos rift zone to a similar process.

Figure 9–2 also shows that the movement of separation of America away from Africa is about approximately the same axis of rotation as the opening of the South Pacific. (See how the two anomaly 18 lines parallel each other in this projection.)

The Atlantic Ocean South of the Azores: Rotation of America Away from Africa

If there is no differential movement between South and North America, the movement of the whole American continent away from Africa should be reducible to a single rotation. On the other hand, because this movement may be different from the movement between Eurasia and the Greenland-America block, the movement between Africa and America can be deduced only from the part of the mid-Atlantic ridge south of the Azores. North of the Azores, as we will see later, the mid-Atlantic ridge belongs to the Arctic system, which also includes the Arctic ridge, the Norwegian ridge, and the

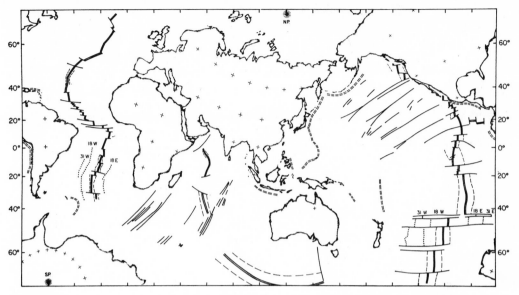

Figure 9-2
The pole of projection has been shifted to the South Pacific center of rotation (69°N, 157°W). The continent outlines, ridge crests and fracture zone locations, and the location of anomalies 5, 18, and 31 were digitized from the original of Figure 9-1, which was traced on the World Map of the U.S. Navy Oceanographic Office (scale of 1/39,000,000 at the equator). The outlines of the continents north of 80°N or south of 70°S are not as accurate, being traced by hand in each case from an atlas

Reykjanes ridge, and determines the movement of the Greenland-America block away from Eurasia.

Spreading Rates. The published spreading rates for the area south of 40°N used for the determination of the movement of rotation are listed in Table 9–1 and compared with the theoretical spreading rates. The two rates for the area north of the equator are from Phillips (1967). All the others, for the area between 25°S and 50°S, are from Dickson et al. (1968). Because of the rapid decrease in spreading rate south of 30°S, the equator of rotation has to be near 15°S–20°S as in Figure 9–2, and the center of rotation best fitting the spreading rates is near 69°N and 32°W. However, this determination is mostly based on data from the South Atlantic, where the true trend of the ridge crest and of the magnetic anomalies is poorly known.

Fracture Zones. The reliable data on the azimuths of the fracture zones at their intersection with the ridge crest are for the area between 30°N and 8°S. A list of 18 transform faults in this area given by Morgan (1968b) was used in computing the best center of rotation giving a point near 58°N, 37°W. Morgan arrived at the same location by using a slightly different method. The distance between this center of rotation determined from fracture zones mostly in the North Atlantic and the center of rotation obtained from spreading rates mostly in the South Atlantic is about 1200 km. It is probably too large to be explained by errors in determinations of spreading rates and can possibly be attributed to some small differential movement between North and South America. The question will be resolved when good mapping of the fracture zones is available for the South Atlantic.

Graphical Test. Figure 9–3 shows that, when the origin of coordinates is rotated to 58°N, 37°W, the fracture zones between 30°N and 10°S are nearly exactly along lines of latitude as required for a movement of rotation around this point. But the longitudinal distance to anomaly 18 in the South Atlantic decreases

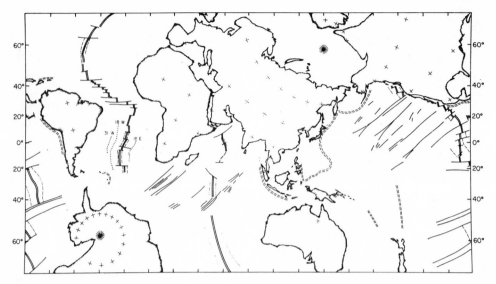

Figure 9-3
The pole of projection has been shifted to the Atlantic Ocean center of rotation (58°N, 37°W)

toward the south instead of staying constant. In Figure 9–2, on the contrary, the longitudinal distance to anomaly 18 is constant with latitude, but the fracture zones north of the equator do not follow lines of latitude. This is further evidence for some small differential movement between North and South America.

We conclude that although there is some evidence for differential movement north and south of the equator, the data suggest that a single rotation around a point near the southern tip of Greenland, at a rate of 3.7×10^{-7} deg/yr, describes the movement of America away from Africa as a good first approximation.

The Arctic Opening from the Azores to the Lena River Delta: Movement of Eurasia away from the Greenland-America Block

It is assumed that Eurasia, on one hand, and Greenland-America, on the other, are moving as rigid units. Their movement of separation should then be a single rotation produced by spreading along a system of ridges going from the shelf north of the Lena River delta (northern Siberia) to the Azores and including the

Arctic ridge. Recent, mostly still unpublished data have indicated that this system is actively spreading at a rate of about 1 cm/yr in the Arctic according to Demenitskaya and Karasik (1967) and Ostenso and Wold (1967), about 1 cm/yr in the Norwegian Sea according to Heirtzler et al. (1968), and 0.95 cm/yr over the Reykjanes ridge according to Pitman and Heirtzler (1966).

Fracture Zones. The published data on exact trends of the fracture zones are still rare: the four given in Table 9–3 come from Johnson (1967) for 53°N, Heirtzler et al. (1967) for 66° and 71°N, and Sykes (1965) for 82°N. The abrupt change of 27° between the azimuths of the faults at 71°N and 66°N implies some important distortion between 71°N and 66°N. This distortion may be related to the presumed zone of extension which includes the Bresse and Limagne in southern France, the Rhine graben (most active in Oligocene time (Cogné et al., 1966), and possibly the North Sea subsiding zone (Closs, 1967). In the absence of more adequate data, a single center of rotation was obtained from these four fracture zones for the whole Arctic system, at 78°N and 102°E, with a standard deviation of 9.1°. This preliminary fit gives a

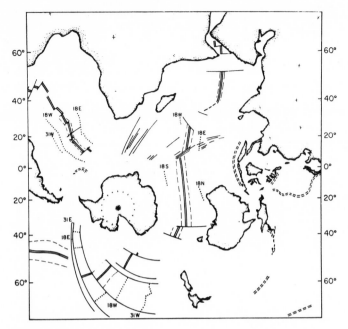

Figure 9-4
The pole of projection has been shifted to the Indian Ocean center
of rotation (26°N, 21°E)

maximum deviation of 12° between measured and computed azimuths (see Table 9–3). A rate of rotation of about 2.8×10^{-7} deg/yr was chosen on the basis of the rate of spreading over the Reykjanes ridge, account being taken of the 60° angle between the ridge axis and the transform fault at 53°N.

Graphical Test. Figure 9–5 is a world map in which the pole of the system of coordinates has been rotated to 79°N, 111°E, which is the center of rotation determined by Morgan for the 'fossil' system of fracture zones in the North Pacific. It is slightly different from the 78°N, 102°E, center of rotation determined for the Arctic system of rotation. However, Figure 9–5 is useful in evaluating the movement involved in the opening of the Arctic system. The fracture zones between 53°N and the Azores are poorly determined, and their trend might be different from the ones shown in Figures 9–1 to 9–5. A much better estimate of the movement involved in the separation of America from Eurasia will be possible when the new data are published. The present deter-

mination is reasonable as a first approximation, however, and it is interesting that it falls close to the center of rotation determined by Bullard et al. (1965) for the best fit of Greenland to Europe (73°N, 96.5°E).

The Movement of the North Pacific Ocean away from North America

Wilson (1965c) pointed out that the San Andreas fault could be understood as a ridge-ridge type of transform fault between the Gorda and Juan de Fuca ridges and the termination of the East Pacific rise in the Gulf of California. Similarly he noted that the Denali and Queen Charlotte Island fault systems could be understood as a ridge-arc type of transform fault connecting the Juan de Fuca ridge to the Aleutian arc. Thus the resulting movement should be a movement of rotation of the Pacific Ocean floor away from North America, the direction of motion being given by the azimuths of the transform faults.

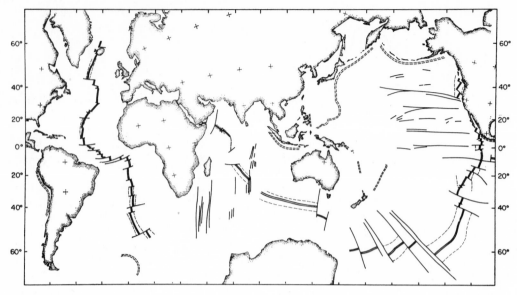

Figure 9-5
The pole of projection has been shifted to the center of rotation of the "fossil" Pacific fracture zones (79°N, 111°E)

Fracture Zones. Morgan (1968b) has compiled a list of the azimuths of this system of faults at 32 locations between 60°N and 24°N. Using this list, he found the center of rotation by least squares to be at 53°N, 47°W. The standard deviation of 5.7° is not excessively large, considering that the measured azimuths range over 29° from 154° to 125° and the computed ones from 145° to 130°. The difference between this location and the one of the center of rotation determined by Morgan (53°N, 53°W) may come from his elimination of some of the azimuths listed in his paper.

Morgan did not include the Blanco fracture zone in his list, as he considered that the Juan de Fuca and Gorda ridge blocks were moving independently of the North America block. Comparing the trends of the fracture zones at both ends of the Juan de Fuca ridge with the trends predicted by the 53°N, 47°W, center of rotation shows a systematic disagreement of 20°, which implies compression and deformation of the continent adjacent to this block.

Velocity of Rotation. The only locations where spreading rates have been determined in the North Pacific are over the Juan de Fuca-Gorda blocks, where the spreading rate at

46°N is 2.9 cm/yr (Vine, 1966). However, as there is presumably some differential movement between the ocean block and the continental block east of the ridge, the actual rate of angular rotation of the Pacific Ocean floor away from North America should be somewhat smaller than 6.5×10^{-7} deg/yr (i.e., (2.9 cm/yr $\times$ 2 $\times$ cos 20°)/(sin 50° $\times$ 1.11 $\times$ 10^7 cm), 50° being approximately the distance to the center). On the other hand, Rusnak et al. (1964) have estimated the total displacement involved in the creation of the Gulf of California as between 300 and 450 km. This displacement is supposed to have occurred since late Miocene time (which is in reasonable agreement with the date of anomaly 5, i.e., 10 m.y.) and would lead to an average rate of rotation of 3.5 to 5.2×10^{-7} deg/yr if the movement started 10 m.y. ago. For the global calculations of the displacements, a rate of 6×10^{-7} deg/yr was chosen for the rotation of the Pacific Ocean away from North America around 53°N, 47°W, which implies a rate of slippage of about 5 cm/yr along the San Andreas fault. This estimate agrees closely with the present velocity of regional shear strain, which is about 6 cm/yr (Hamilton and Myers, 1966).

The Indian Ocean

Figure 9–1 and Table 9–1 show that the data on spreading in the Indian Ocean cannot be simply explained by a single rotation of the northeast Indian Ocean away from the southwest Indian Ocean. The spreading rates increase continuously from the Red Sea-Gulf of Aden region to south of Australia while the trends of the fracture zones change abruptly south of the Rodriquez fracture zone near 20°S, being apparently influenced by the junction with the southwest branch of ridge over which no spreading rate has yet been determined. Unfortunately, the fracture zones south of Australia are poorly known, and it is not possible to determine directly a center of rotation of Australia away from Antarctica. On the other hand, the fracture zones within the Gulf of Aden, including the Owen fracture zone (Laughton, 1966), are better known and can be used to determine the movement of rotation of the northern Indian Ocean away from the Africa block (Le Pichon and Heirtzler, 1968).

Fracture Zones. The five azimuths used for the determination of the center of rotation were obtained from Figure 14 of Laughton; two were from the Gulf of Aden, three from the Owen fracture zone. They are listed in Table 9–3 and compared with the computed values for the center of rotation at 26°N, 21°E. The standard deviation between measured and computed azimuths is only 0.6° in spite of the fact that the azimuths range over 16°, from 19° to 35°. These trends are in good qualitative agreement with the trends of the fracture zones over the Carlsberg ridge as shown in the *National Geographic Magazine* diagram of the Indian Ocean (based on Heezen and Tharp's revised interpretation). On the basis of the fracture zones, it is not possible to distinguish between movements on each side of the Owen fracture zone. From Table 9–1 it can be seen that the spreading rates for the Gulf of Aden (1 cm/yr, W. Ryan, in preparation) and for the Carlsberg ridge (1.5 cm/yr) do not indicate large differential movement along the Owen fracture. On the other hand, it is not possible to reconcile the trends of the

Rodriguez and Amsterdam fracture zones with the rotation around 26°N, 21°E.

Graphical Test. In the test, illustrated in Figure 9–4, the pole of the system of coordinates has been rotated to 26°N, 21°E. Although the agreement with a single movement of rotation around this pole is quite good for the area north of the Rodriquez fracture zone, south of the point where the three branches of ridge join together the pattern of fracture zones assumes new trends. This can be explained if there is some movement of opening along the southwest branch of the ridge. The movement of rotation of Australia away from Antarctica will then be the sum of the rotation of Africa away from Antarctica and Arabia-India away from Africa. Le Pichon and Heirtzler have noted on the basis of the anomaly pattern that, if there is active spreading on the southwest mid-Indian Ocean ridge, it is unlikely to exceed 1 cm/yr. In a later section we will try to reconcile these conclusions. Le Pichon and Heirtzler have also mentioned the possibility of some small differential movement between the northwest and northeast Indian Ocean along the Ninetyeast ridge (which is slightly seismic, L. R. Sykes, personal communication). As a first approximation we conclude that the relative movement of the whole northern Indian Ocean (including Australia) away from Africa occurs around a point near 26°N, 21°E. The angular rate of opening is about 4×10^{-7} deg/yr.

GLOBAL GEOMETRY OF PRESENT EARTH SURFACE DISPLACEMENTS

In the preceding section, the system of large crustal displacements has been reduced as a first approximation to five ocean-opening movements. Each movement has been described as a single rotation of one rigid block away from the other, and the parameters of rotation, i.e., the axis of rotation and the rate of angular rotation, have been determined (see Table 9–2). Figure 9–1 summarizes the data available on sea-floor spreading, showing the two locations at which each axis of rotation pierces the earth's surface. Those movements result from the addition of new crustal material at the crests of the mid-ocean ridges, so that, if the earth is not expanding, there should be other boundaries of crustal blocks along which surface crust is shortened or destroyed. In the spreading-floor hypothesis, these boundaries are the active trenches and Tertiary mountain belt systems. However, if the earth is not expanding, what is the mechanism which results in this pattern of movements? It is difficult to imagine large-scale convection currents rising immediately below the crests of the ridges, with elongated conical shapes on the ridges, rates of motion increasing from zero near the axis of rotation to a maximum under the equator of rotation, and staggered sections offset along transform faults. Rather the pattern of opening of the oceans seems to be the response of a thick lithosphere (able to transmit stress along great distances), which breaks apart along lines of weakness, to some underlying state of stress as variously favored by Elsasser (1968), McKenzie (1967), Morgan (1968b), and Oliver and Isacks (1967). Carey (1958) and more recently Heezen and Tharp (1965) have argued that a possible cause of this state of stress is the expansion of the earth, a cause which does not require that the crust be destroyed as rapidly as it is created at the crests of the ridges.

The Expansion Hypothesis

Without entering into other considerations, we will consider first the pattern of deformation of the earth's surface, assuming that there are no zones of compression. We can then easily find, with the data in Table 9–2, how much the circumference of any great circle on the surface of the earth is increasing per year. Ideally, each great circle should expand at the same rate, so that the earth's surface maintains its nearly spherical shape. Such an idea can be tested rapidly by considering Figure 9–2 in which the axis of projection is the axis of rotation of the South Pacific; it is also close to the Atlantic and North Pacific axes of rotation. Note that in the projection of Figure 9–2 most of the spreading occurs about axes which run north-south, whereas very few ridge axes run east-west. The great circle formed by the equator of rotation in Figure 9–2 expands at a rate

of about 17 cm/yr (4 at the mid-Atlantic ridge, 12 at the East Pacific rise, and 1.5 at the Carlsberg ridge). On the other hand, the great circles of longitude of Figure 9–2 are parallel to the crests of the Atlantic and Pacific ridges and would at most intersect the Arctic and South Indian Ocean ridges. The resulting rates of expansion vary between 0 and 7 cm/yr. There is no evidence for important pole migration during the last 10 m.y., yet the equator of rotation in Figure 9–7 would have increased its circumference by 1700 km and some of the great circles of longitude would not have expanded. This implies a differential expansion of as much as 270 km between the average radius along extreme great-circle circumferences and possibly as much as 500 km between individual radii. It is unacceptable, and it becomes even more so when we recognize that the present pattern of spreading has prevailed during the whole Cenozoic era. Consequently, in the expansion hypothesis, we have to assume some compensating large-scale processes of earth's surface shortening by compression or thrust to maintain the nearly spherical surface of the earth. The expansion hypothesis then loses most of its appeal. Other strong arguments have previously been advanced against the expansion hypothesis (e.g., Runcorn, 1965), and it will not be considered further here.

This discussion has led us to recognize that the axes of rotation of spreading which have prevailed during the Cenozoic are not randomly distributed. They have maintained a systematic preferred orientation. In particular the two major openings, in the Atlantic and Pacific oceans, have their centers of rotation within 15° of each other. This observation suggests that a simple pattern of the primary state of stress has continued over the last 60 m.y. (along the lines of latitude of Figure 9–2). The axis of revolution is close to the axis of rotation of the movements of spreading ($\approx$60°N and 50°W).

Determination of the Movements Between Blocks

If we assume that the earth is spherical and that the length of its radius does not change with time, we can then proceed to the complete

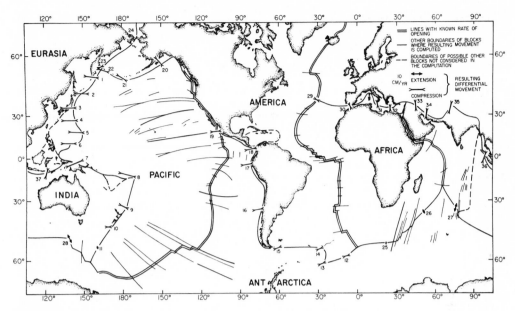

Figure 9-6
The locations of the boundaries of the six blocks used in the computations. The numbers next to the vectors of differential movement refer to Table 9-5. Note that the boundaries where the rate of shortening exceeds about 2 cm/yr account for most of the world earthquake activity

determination of the movements of the major crustal blocks relative to each other. This, of course, presupposes the determination of the boundaries of the blocks, other than ridge crests, i.e., the lines of compression or shear between blocks. It is further necessary to assume that all blocks, and consequently all ridge crests and other boundaries, may migrate over the surface of the earth (which is another difficulty involved in the hypothesis of convection currents rising immediately below the ridge crests). To make the problem entirely determinate, we divide the earth's surface into six rigid blocks (as compared with the twenty blocks used by Morgan) which stay undeformed except at their boundaries, where surface may be added or destroyed. These simplifications will lead to a mathematical solution which can be considered a first-approximation solution to the actual problem of earth's surface displacements.

Figure 9–6 shows the five boundaries (double lines) of the six blocks and the names adopted for each. Subsidiary blocks, not used in the computation, are shown by dashed line. Two of these, the Fiji and Philippines basins, can be understood as providing the means of absorbing the differential movement between

two large blocks along two nearly parallel boundaries, instead of only one. The Caribbean block presumably absorbs whatever differential movement there may be between the North and South American blocks. In the Indian Ocean, the Ninetyeast ridge may allow some differential movement between the northwest and northeast Indian blocks. Except possibly for the last, none of these subsidiary boundaries would significantly affect the general picture obtained here. The Pacific is the only entirely oceanic block. Eurasia is considered to be a single block (except for India and Arabia), and an arbitrary boundary passes between Alaska and northeastern Siberia.

The resulting instantaneous rotational vector between two blocks can be obtained from the geometrical or vector sum of the rotational vectors of the blocks. These can be obtained from the five rotational vectors previously determined (the five whose parameters are underlined in Table 9–2). For example, the America-Antarctica vector is the sum of the Pacific-Antarctica and America-Pacific vectors, and the India-Antarctica vector is the sum of the four vectors: Pacific-Antarctica, America-Pacific, Africa-America, and India-Africa. The more vectors that are involved in

Table 9–4
Instantaneous centers of rotation deduced from
Table 9–2.

Blocks	Lati-tude	Longi-tude	Rate[a] $(10^{-7}\ deg/yr)$
America–Antarctica	79.9S	40.4E	−5.44
Africa–Antarctica	42.2S	13.7W	+3.24
India–Antarctica	4.5S	18.1E	+5.96
India–Eurasia	23.0N	5.2W	−5.50
India–Pacific	52.2S	169.2E	−12.3
Eurasia–Pacific	67.5S	138.5E	−8.15
Africa–Eurasia	9.3S	46.0W	−2.46

[a]Positive value indicates extension; negative, compression.

the sum, the greater is the probable inaccuracy. Also, the sums involving the America-Pacific and America-Eurasia vectors may be less accurate than the others. Table 9–4 lists the parameters of seven resulting rotational vectors. Tables 9–2 and 9–4 were used to compute the vectors of differential movement at 37 locations along the boundaries of the six blocks. The results are shown in Figure 9–6 and listed in Table 9–5.

It is important to realize that the number of blocks is chosen in such a way that the problem can be solved. The initial data are the five known movements of opening of the oceans, shown by a double line in Figure 9–6. The two rotations corresponding to spreading at the axes of the southwest and southeast mid-Indian Ocean ridges are not known and will be determined by the computation. Note that if these two rotations were known, the problem would be overdetermined. This provides a severe test of the computation, as the computed rotations should indicate extension, not compression, over the southeast and southwest mid-Indian Ocean ridges, and directions and rates should agree with the physiographic and magnetic data. Although the axis of a spreading ridge is a well-defined block boundary, the locations of the other boundaries are not as well defined. The choice is guided by the seismic and physiographic data. Active deep-sea trenches and the Alpine-Himalayan mountain belt are obvious choices. If other boundaries than the ones indicated by single full lines in Figure 9–6 are chosen, the resulting vectors of differential movement will be different, but not the relative vectors of rota-

tion. By using Tables 9–2 and 9–4, we can consequently rapidly estimate the effect of choosing somewhat different nonspreading boundaries. Finally, as mentioned earlier, the dashed lines represent boundaries along which it appears that some differential movement occurs. But the data available do not allow a unique determination of the movement along them.

Vectors 1 to 6 represent differential movement between Eurasia and Pacific. They are perpendicular to the trench system (corresponding to pure compression or thrust) and have magnitude of 8 to 9 cm/yr. Vectors 4 to 6 would have a different direction and a smaller magnitude if part of the differential movement were absorbed on the west side of the Philippines basin.

Vectors 7 to 11 represent differential movement between the Indian and Pacific blocks. These results may be affected by a possible error in the India-Antarctica rotational vector as discussed later. However, the results obtained get some support from a recent study of deep earthquake focal mechanisms of Isacks et al. (1968). Vectors 7 and 8 indicate mostly strike-slip motion at a rate of 9 to 11 cm/yr along the east-west boundary, from New Guinea to the northern hook of the Tonga trench. Vector 8 also indicates pure compression or thrust in the main part of the Tonga trench. If part of the movement is absorbed on the west side of the Fiji basin, as it is in the Philippines basin, the direction of vector 8 will be somewhat different and its magnitude smaller; it will then be more comparable to vector 9, which is representative of the southernmost Kermadec trench. South of the Kermadec trench, one gets closer to the center of India-Pacific rotation (situated somewhat north of Macquarie Island). The magnitude of the vector of differential movement decreases and a strike-slip component appears. The rate of motion is less than 2 cm/yr south of 45°S. The small amount of expansion found near Macquarie Island (vector 11) comes from its location just south of the center of rotation. This disagrees with Sykes' (1967) finding of thrust faulting for an earthquake mechanism near Macquarie Island. However, a small change in the location of the India-Antarctica center of rotation could modify this last result.

Vectors 12 to 19 represent differential movement between Antarctica and America. Vectors 12 to 15 along the southern boundary indicate strike-slip motion at a rate of 2 to 4 cm/yr, except along the north-south part of the South Sandwich trench where there is compression or thrust. The location of the southern boundary is somewhat arbitrary but is partly justified by the existence of seismic activity along it (L. R. Sykes, personal communication). The rate of motion exceeds 5 cm/yr at the southern boundary of the active part of the Chile trench (vector 16) and stays between 5 and 6 cm/yr up to the northern end of the Middle America trench. A problem is posed by the Panama continental rise, a filled trench (Ross and Shor, 1965) for which the computed rate of differential motion is 5.9 cm/yr. A solution to this problem may have been given by Herron and Heirtzler (1967), who were able to show the existence of an east-west Galapagos rift zone which began to form about 7 m.y. ago by north-south spreading at a rate of 2 to 3 cm/yr (see Figure 9–1). Spreading along the axis of the Galapagos rift zone would change the orientation of vectors 18 and 19 from east to northeast. In this hypothesis, the Panama filled trench, like the filled trench east of the Juan de Fuca ridge (Ewing et al., 1968), is a remnant of the once continuous trench system along the west coast of North America. They were abandoned and filled at the time of the change in the pattern of spreading between 10 and 5 m-y. ago (Vine, 1966). It is also possible that some of the compression is taken up in the Lesser Antilles instead of in the west coast of Panama (Wilson, 1965c).

Vectors 20 to 22 indicate compression or thrust at a rate of 5 to 6 cm/yr on the eastern part of the Aleutian trench and strike slip on the western part. Vectors 23 and 24 indicate relatively small (1.5 cm/yr), mostly strike-slip motion between Alaska and Eurasia. This computed movement may result partly from errors in the determination of the Arctic center of rotation or may be absorbed by crustal deformation somewhere within Eurasia (Verkhoyansk Mountains) or America.

Vectors 25 to 28 are the only vectors computed for the crests of the mid-ocean ridge system. They are consequently comparable to the available data on spreading along these portions of the mid-ocean ridges. Note that the magnitude and direction of the extensional movement computed is in reasonable general agreement with the data. Vectors 25 and 26 indicate some spreading along the crest of the southwest mid-Indian Ocean ridge, the spreading rate (half of the total movement) increasing progressively from 0.8 to 1.3 cm/yr from southwest to northeast. This combination of small spreading rate and numerous large transform faults might explain the difficulty in recognizing the magnetic pattern and yet account for the well-developed physiography of this ridge (Le Pichon and Heirtzler, 1968). Vectors 27 and 28 also indicate spreading along the axis of the southeast mid-Indian Ocean ridge. The computed spreading rates (2.9 to 3.4 cm/yr) agree well with the measured spreading rate of 3 cm/yr (see Figure 9–1 and Table 9–1). The direction of vector 27 also agrees well with the trend of the Amsterdam fracture zone. However, vector 28 seems to have an asimuth disagreeing by about 15° with the apparent trend of the fracture zone. Part of this discrepancy may be due to the existence of differential movement along the Ninetyeast ridge. Part must undoubtedly be due to accumulated errors in the determination of the rotational vectors, as the India-Antarctica rotation is the sum of four different vectorial rotations.

Vectors 29 and 32 represent differential motion between Eurasia and Africa. Vectors 29 and 30 indicate compression at a rate of 1.5 to 1.9 cm/yr along the Azores-Gibraltar ridge and some strike-slip component near the western end. A small error in the location of the center

Table 9–5
Computed differential movements between blocks as given in Figure 9–6.

	Latitude	Longitude	Rate[a] (cm/yr)	Azimuth	Location
			EURASIA–PACIFIC		
1.	51N	160E	−7.9	99	Kurile Trench
2.	43N	148E	−8.5	93	Kurile Trench
3.	35N	142E	−8.8	91	Japan Trench
4.	27N	143E	−9.0	91	Japan Trench
5.	19N	148E	−9.0	93	Mariana Trench
6.	11N	142E	−8.9	91	Mariana Trench
			INDIA–PACIFIC		
7.	3S	142E	−11.0	69	New Guinea
8.	13S	172W	−9.1	107	N. Tonga Trench
9.	34S	178W	−4.7	113	S. Kermadec Trench
10.	45S	169E	−1.7	89	S. New Zealand
11.	55S	159E	+1.6	339	Macquarie Island
			AMERICA–ANTARCTICA		
12.	58S	7W	−2.6	72	Southwest Atlantic
13.	61S	26W	−2.7	68	S. South Sandwich Trench
14.	55S	29W	−3.3	72	N. South Sandwich Trench
15.	55S	60W	−3.7	73	Cape Horn
16.	35S	74W	−5.2	79	S. Chile Trench
17.	4S	82W	−6.0	81	N. Peru Trench
18.	7N	79W	−5.9	80	Panama Gulf
19.	20N	106W	−5.3	83	N. Middle America Trench
			AMERICA–PACIFIC		
20.	57N	150W	−5.3	136	E. Aleutian Trench
21.	50N	178W	−6.2	119	W. Aleutian Trench
22.	54N	162E	−6.3	108	W. Aleutian Trench
			AMERICA–EURASIA		
23.	56N	165E	−1.6	68	Aleutian-Kurile Islands
24.	66N	169W	−1.4	62	Alaska-Siberia
			AFRICA–ANTARCTICA		
25.	53S	22E	+1.5	8	S. Southwest Indian Ridge
26.	37S	52W	+2.7	331	N. Southwest Indian Ridge
			INDIA–ANTARCTICA		
27.	36S	75E	+5.8	17	W. Southeast Indian Ridge
28.	50S	138E	+6.4	333	E. Southeast Indian Ridge
			AFRICA–EURASIA		
29.	40N	31W	−1.5	117	Azores
30.	36N	6W	−1.9	153	Gibraltar
31.	38N	15E	−2.4	169	Sicily
32.	35N	25E	−2.6	176	Crete

Table 9–5 (continued)

	Latitude	Longitude	Rate[a] (cm/yr)	Azimuth	Location
			INDIA–EURASIA		
33.	37N	45E	−4.3	176	Turkey
34.	30N	53E	−4.8	186	Iran
35.	35N	72E	−5.6	193	Tibet
36.	0	97E	−6.0	23	W. Java Trench
37.	12S	120E	−4.9	19	E. Java Trench

[a]Positive value indicates extension; negative, compression.

of rotation could change this last result. Vectors 31 and 32 indicate nearly north-south compression at a rate of about 2.5 cm/yr within the Mediterranean. The choice of the boundary within the Mediterranean is somewhat arbitrary, as the Mediterranean boundary probably consists of several zones of compression. This vector, then, would be the resultant of the different partial components.

Vectors 33 to 37 correspond to differential movement between Eurasia and the Indian block. The rate of compression reaches a maximum of 6 cm/yr in the Java trench and is about 5 cm/yr in the Himalyas. However, the same remark should be made for this boundary as for the Mediterranean boundary.

Validity of the Pattern of Motion Obtained

Two serious limitations should be placed on the picture just obtained of the general pattern of motion of large crustal blocks on the surface of the earth. The first is obvious; it comes from the fact that we have been forced to make great simplifications and generalizations in order to solve the problem. The second results from the inadequacy and inaccuracy of parts of the data used to determine the pattern of spreading (e.g., preliminary charts of fracture zones, different methods of determining spreading rates). With more and better data and with a more careful consideration of the geology and the seismicity, a better picture can be obtained.

The principal weakness of this picture is that it ignores the seismic and physiographic evidence indicating differential movement around the Caribbean Sea. An examination of the seismic belt along the Cayman trough and the western part of the Puerto Rico trench shows that it follows exactly a small circle centered on 28°S, 71°W. This is consistent with the idea that this zone acts as a transform fault (Wilson, 1965a). The main differential movement, then, is parallel to this zone and perpendicular to the Lesser Antilles trend. It is possible to account for this movement by a small difference in the vectors of rotation for North and South America. Better data on the fracture zones in the Atlantic Ocean will show whether the present seismic activity of the Caribbean results from differential motion between North and South America or whether it corresponds mainly to the thrusting of a Pacific plate toward the east as hypothesized by Wilson.

We have shown, however, that the general pattern of spreading revealed in the last two years gives a geometrically consistent picture of the pattern of motion at the surface of the earth. We see that none of the ridges can be understood as an isolated feature but that each is part of an intricate pattern which transfers new earth's surface from sources, principally the axes of the Atlantic and Pacific ridges, to sinks, most of which are situated along the Pacific western margin and the Alpine-Himalayan belt. The geological history indicates that the major sinks, like the Alpine-Himalayan belt, have been localized in the same regions for periods several times longer than the life of the present mid-ocean ridges. The pattern of movement at the mid-ocean ridge crests is consequently controlled not only by the directions of the driving stress pattern and the locations of zones of weakness where crust may part but also by the locations of the major sinks. In this picture the roles of the Indian and Arctic ridges are to reconcile the geometrical incompatibility of the Atlantic source and the Alpine-Himalayan sink. Indeed, the easiest way to transform a vector of movement into another one having a different orientation is by placing an intermediary zone of spreading between the two. Thus the east-west Atlantic vector of movement is transformed by spreading along the Carlsberg ridge into a north-south North Indian Ocean vector of movement which is perpendicular to the Himalayan sink. Similarly, the east-west East Pacific rise vector of movement is transformed by north-south spreading along the Galapagos rift zone into a southwest-northeast vector of movement which is perpendicular to the Middle America trench. It is possible that similar subsidiary spreading exists in the Philippines basin (J. Ewing, personal communication), the Fiji basin, the South Sandwich basin, the Eastern Mediterranean, etc., transforming the directions of vectors of differential movement from one border to the other.

The pattern of motion obtained reveals that zones of folding and compression have differential rates of motion smaller than about 6 cm/yr, whereas active trenches (zones of thrust) have differential rates larger than 5 cm/yr. The obvious exceptions are the west part of the Puerto Rico trench and the east-west part of the South Sandwich trench, which seem to be mostly characterized by strike-slip motion as are the northern end of the Tonga trench and the western end of the Aleutian trench. It is somewhat puzzling, however, that seismic reflection and refraction do not detect significant structural differences between these two types of trenches (J. Ewing, personal communication). Another point of interest is that, when the rate exceeds 8 or 9 cm/yr, there is a tendency for the system to decouple itself into two parallel systems of trenches, each of which absorbs part of the movement.

Finally, it should be noted that along the Pacific borders the main system of trenches has been a western system since latest Miocene time. However, we will see later that during early Cenozoic time the main Pacific system of trenches probably was along the eastern border (west coast of America).

SEA-FLOOR SPREADING AND CONTINENTAL DRIFT DURING THE CENOZOIC

Relative Time Scale

Vine (1966) and Pitman (1967) were able to relate the Raff and Mason (1961) and Peter (1966) lineations to an over-all magnetic pattern which had been created at the crest of the East Pacific rise by sea-floor spreading. More recently, a series of publications summarized by Heirtzler et al. (1968) established the fact that the magnetic pattern covering the whole of the East Pacific rise can also be recognized in the South Atlantic and Indian oceans. Other works, in preparation, have tentatively recognized the pattern in the Labrador and Norwegian seas. In the areas covered by the identified magnetic pattern, which corresponds to the surface of the mid-ocean ridge system, we know the relative age of any portion of crust with respect to another. A numbering system of key anomalies was established by Pitman et al. (1968), the numbers increasing with age from 1 at the crest to 32 at the outer limit of the recognizable pattern. Three of these key anomalies are shown in Figures 9–1 to 9–4. By rotating the blocks by the amount necessary to superpose a given anomaly and by using the assumptions developed in the preceding section, it is possible to obtain the relative positions of the continents and ocean basins on each side of it at the time this anomaly was created. However, the geological time to which this reconstruction applies is not known.

Absolute Time Scale

Vine (1966) derived an absolute time scale for the magnetic pattern by assuming that the spreading rate had been constant throughout the time required to produce this pattern (about 80 m.y.). The result of this twentyfold extrapolation from the known time scale (3.5 m.y.) proved reasonable, as it agreed with many geological deductions on the age of the mid-ocean ridge system (e.g., Heezen, 1962). Furthermore, this concept of steady movement during a time of the order of 100 m.y.

agreed with the original spreading-floor concept of convection currents reaching to the surface. It was difficult to imagine such large-scale convection currents being able to stop and restart, accelerate or slow down, in times as small as a few million years.

Heirtzler et al. (1968) tried by the same approach to derive an absolute time scale. However, when the magnetic patterns in the oceans were compared, it appeared that the relative spreading rates between oceans had shown great and apparently rather abrupt variations. This is illustrated by the inset in Figure 9–11, which is a figure from Heirtzler et al. showing the distance from an anomaly to the crest in an ocean plotted versus the distance from the same anomaly to the crest in the South Atlantic Ocean. These distances were obtained from type profiles which were sufficiently far from the center of rotation to be representative of the movement of opening. From this figure we can deduce that the ratio between spreading rates in the North and South Pacific was about 1 from anomaly 31 to 24, increased rapidly to more than 3 from anomaly 24 to 18, and decreased again to 0.7 after anomaly 5. No systematic gap in the anomaly pattern was found, however, and it was concluded that if there had been a total interruption of spreading it had to have occurred simultaneously in all oceans. This was considered unlikely. Heirtzler et al. consequently derived their absolute time scale (inset, Figure 9–11) by assuming a constant spreading rate in the South Atlantic, the ocean which apparently showed the minimum amount of systematic variations of relative spreading rate. This time scale, which is not greatly different from Vine's time scale, was in reasonable agreement with dates of seismic reflectors based on core data. However, these results showed the postulate that spreading occurred at a steady rate for times of tens of millions of years to be wrong.

Episodic Spreading

On the basis of seismic reflection studies, Ewing and Ewing (1967) suggested that the sediment distribution could be explained much more easily in terms of an episodic spreading.

A discontinuity in sediment thickness near anomaly 5 (10 m.y. in the Heirtzler et al. time scale) was interpreted as indicating an interruption of spreading possibly covering all of Miocene time. Ewing et al. (1968) further suggested that the sedimentary structures in the Atlantic basins and the western Pacific implied some important reorganization of spreading, possibly accompanied by a large interruption, most probably at the Mesozoic-Cenozoic boundary. Thus the study of sediment distribution led Ewing et al. to hypothesize three main episodes of drifting: (1) Mesozoic, during which the basins formed; (2) early Cenozoic, during which most of the mid-ocean ridge area was created; and (3) latest Cenozoic, during which the crestal regions appeared. Langseth et al. (1966), to explain the distribution of heat-flow values, had also suggested that spreading was episodic.

It should be noted that the ridge-basin and crest-flank boundaries described by Ewing et al. separate three different provinces of magnetic anomalies. Anomaly 5 is the outer limit of a high-amplitude, short-wavelength axial magnetic pattern easily distinguished from the larger-wavelength flank anomalies by a zone of short-amplitude, short-wavelength anomalies (Heirtzler and Le Pichon, 1965; Talwani et al., 1965b). This important difference was described by Vine (1966) and Heirtzler et al. (1968) as perhaps reflecting a change in the average periodicity of reversal of the geomagnetic field and possibly a change of intensity, indicating two different behaviors of the geomagnetic field during early and late Cenozoic time. The outer limit of the magnetic pattern correlated to this day coincides with the ridge-basin boundaries in the Atlantic and the 'opaque layer' boundary in the Pacific, dated by Ewing et al. as latest Mesozoic. We will see later that there are evidences from the magnetic anomalies for major reorganizations in the pattern of spreading at the beginning of each episode. On these bases, we can modify the Heirtzler et al. time scale by giving to anomaly 32 an age of 60 m.y. (early Paleocene) instead of 77 m.y. and by placing an interruption of spreading about 10 m.y. long at anomaly 5, preceded by a general slowing down of the movement. This time scale would not violate any of the core or sediment distribution data. It would also explain the change in magnetic pattern without advocating some over-all change in the average reversal periodicity of the geomagnetic field between early and late Cenozoic times. This adjusted time scale does not disagree with the time scale of Heirtzler et al. by more than 17 m.y.

The objections to the hypothesis of episodic spreading made on the basis of convection current inertia and peak-to-peak correlation of magnetic patterns between oceans are not valid if spreading really corresponds to the response of a thick rigid lithosphere to some underlying stress pattern. In this case, all openings are interrelated and should persist until one or several of them become so poorly adjusted to the stress pattern that a readjustment in the pattern of spreading is necessary. Ewing et al. (1968) and Le Pichon and Heirtzler (1968) have stressed the fact that such episodic spreading can be more easily correlated with the major orogenies.

On the basis of the magnetic pattern now known, it is possible to reconstruct the earth's surface configuration at the beginning of the two latest major readjustments of spreading — late Miocene (anomaly 5) and early Paleocene (anomalies 31 and 32) — within the assumptions stated in the preceding section. We emphasize that the openings of the oceans are the only movements that can be determined from a study of the magnetic anomaly pattern. Consequently, deformations within continents or possible exchange of land between continents (for example, the Riff zone of Morocco in the Mediterranean region between Africa and Europe) are not considered. In particular, in the case of the Alpine-Himalayan belt, the reconstructions will give the relative movements between the continital masses as they are now. They are not meant to be detailed paleogeographical maps. Special attention will be paid to the Atlantic and the South Pacific-South Indian oceans, for which the data are easiest to interpret. The problem of the right-angle bend of the magnetic pattern south of the Aleutian trench will not be treated here (Peter, 1966). Pitman and Hayes (1968) show that this magnetic pattern can be understood in terms of the migration of the ridge crests toward the east and north, away from the central Pacific. This migration of the crests implies that the main trench system was then along the eastern border (that is, along the west coast of America) and not along the western border (coast of eastern Asia) as it is now.

The World at the Time of Anomaly 5 (Later Miocene)

Figure 9–7 has been obtained by assuming that Antarctica has not moved during the last 10 m.y. and that the movements of all the other blocks (as defined in Figure 9–6) were determined by the rotations listed in Table 9–2. The movement of a continent with respect to Antarctica was determined by estimating (1) the effect of the Pacific-Antarctica rotation, (2) the America-Pacific rotation, and (3) the rotation of Eurasia-America or Africa-America, etc. The total time since anomaly 5 is assumed to have been 10 m.y., so that the finite angle of rotation is 10.8° in the South Pacific, 3.7° in the Atlantic, etc. (see Table 9–2). To obtain the total movement of any continent with respect to Antarctica, we used the method described in the preceding section. The only difference is that, as we are summing finite rotations and not instantaneous velocities, it is necessary to rotate the centers of rotation before summing the rotations in such a way that they conserve their spatial relations to the blocks being moved. Most of these angles are small, however, and the directions and relative amplitudes of the vectors of differential movement between blocks will be very similar to those shown in Figure 9–6. For example, the total amount of compression since the end of the Miocene near vector 9 of Figure 9–6 is about 470 km.

There is no compelling reason to assume that Antarctica has not moved since late Miocene time. The constant position was assumed for convenience, as we have no way of knowing the absolute movement. However, it is fairly clear that, whichever block is chosen to be fixed, all the others will be displaced at rates of the order of 5 cm/yr (500 km in 10 m.y.). The general movement can be considered to be one of counterclockwise drift relative to Antarctica.

The principal interest of Figure 9–7 is to complement Figure 9–6 by showing the gen-

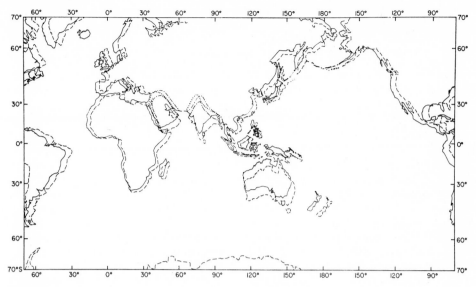

Figure 9-7
The positions of the continents at the time of anomaly 5 ($\approx$ 10 m.y. ago) with respect to
Antarctica are shown by dashed lines, the present positions of the continents by full lines. The
present northern boundaries of Arabia and India are shown by dotted lines, and the dash-dot lines
define the gap which results from the rotations back to 10 m.y. ago

eral pattern and amplitudes of displacements
which have occurred relative to Antarctica in
the last 10 m.y. Note in particular the movement of rotation of Africa and the amount of
crustal shortening (from 200 to 500 km) which
has occurred in the Alpine-Himalayan belt.
Note also that no attempt was made to close
the Gulf of California in Figure 9–7. The
movement assumed for the computation is
somewhat too large ($\approx$500 km) to explain the
formation of the Gulf. Because it was difficult
to account for the spreading pattern in the
North Pacific, which has progressively
changed from 10 to 5 m.y. ago, a detailed analysis is not given in this paper.

The World at the Time of Anomaly 31 (Paleocene)

The maximum movement of opening which
has occurred since late Miocene time is 10.8°.
However, on the basis of the distance between
the crests of the ridges and anomaly 31 in each
ocean, we can determine that the angles of
opening have been as large as 52° in the south

Pacific Ocean, 25° in the Atlantic Ocean, etc.
(see Table 9–6). This poses an interesting geometrical problem. As was mentioned earlier,
the sum of two finite rotations on a sphere is a
rotation, the parameters of which will in general depend on the amplitudes of the finite
angles of rotation. If we consider the Indian
Ocean, for example, the sum of the rotation of
the Indian block away from Africa and of the
rotation of the Africa block away from Antarctica will be a rotation of the Indian block
away from Antarctica, the parameters of
which will progressively change with time as
the two other openings progress. This means
that the trends of the transform faults and the
amplitudes of the spreading rates will change
with time, if the two other rotations keep
constant parameters.

   Of course, there is no way of knowing which
of the three rotations will adjust itself with
time to the two others. It is even conceivable
that all three rotations will progressively adjust
to each other with time. Also, it is possible
that, instead of a gradual modification of the
parameters of rotation, there is a discontinuous
change of the parameters when the stresses
produced by the deformation have reached a

Table 9–6
Relative rotations used for Figures 9–8, 9–9, and 9–10[a].

Blocks	Latitude	Longitude	Angle of rotation
New Zealand–Antarctica	70N	62W	+52° (anomaly 31), +21° (anomaly 18)
America–Africa	58N	37W	+25° (anomaly 31)
Eurasia–America[b]	78N	102E	−17° (to resorb Reikjanes ridge)
Australia–Antarctica	36S	53E	−31° (anomaly 18)[c]
Arabia–Africa	26N	21E	−7° (to close Gulf of Aden)
India–Africa	26N	21E	−17° (anomaly 18)[d]

[a]Antarctica and Africa fixed in their present positions.
[b]America in its present position.
[c]Anomaly 18 is at the foot of the Australian continental rise; there are no older anomalies.
[d]No anomalies older than anomaly 18 were identified.

certain level. This might eventually lead to the 'death' of a ridge and the 'birth' of a new one along completely new lines. The great complexity of the Indian Ocean, and perhaps also of the Norwegian Sea-Arctic Ocean region, is probably caused by these facts. One can consider that the role of these two oceans is to resolve the geometrical problems caused by the progressive openings of the Atlantic and Pacific oceans. This problem, of course, introduces great complexity and considerable uncertainty in the attempt to reconstruct the world in the distant past because the set of parameters given by the present magnetic coverage of the oceans is incomplete and inaccurate.

We do not know the movement of America relative to the Pacific block before the time of anomaly 5, as there was presumably a trench system along the west coast of North America. Consequently, we cannot follow the paths used in the preceding section to obtain the relative movement of any continent with respect to Antarctica. We can instead obtain separately the movements of Australia and New Zealand with respect to Antarctica and of all the other continents with respect to Africa. Africa and Antarctica are then assumed to have remained in their present positions throughout Cenozoic time. This leads us to make two assumptions. The first and most important one will be that the movement of Africa relative to Antarctica has on the average been small since Paleocene time. This is supported by the paleomagnetic data (see, for example, Gough et al., 1964), magnetic data (i.e., the

absence of clear magnetic pattern, Le Pichon and Heirtzler, 1968), and sediment data (i.e., the large sediment cover of the ridge flanks up to the crestal zone, M. Ewing, personal communication). With this approximation, we do not need to know the movement of America with respect to the Pacific block. The second assumption involves the choice of an average center of rotation of Australia with respect to Antarctica. This center of rotation should satisfy the probable trends of the fracture zones south of Australia. It should also satisfy the progressive increase in the distance between anomaly 13 and the ridge crest from west to east in the southwest mid-Indian Ocean ridge (Le Pichon and Heirtzler, 1968). The center of rotation obtained (36°S, 53°E) may be in error by as much as 10°. However, an error of even this magnitude will not seriously alter the results. The angles of rotation chosen (Table 9–6) are such that they bring into coincidence the two anomaly 31 lines on each side of the crest (when only one is known, we assume that the other is symmetric with respect to the crest). In the Indian Ocean no anomalies older than anomaly 18 have been identified. Anomaly 18 is at the base of the Australian continental rise (see Figure 9–4). Before the time of anomaly 18, Australia and Antarctica must have formed a single continent. On the other hand, anomaly 18 is not positively identified over the Carlsberg ridge. Over the ridge southeast of Mauritius no anomalies were identified beyond anomaly 18, and there are indications that a major readjustment in the pattern of spreading occurred at this time. Consequently, in Figures

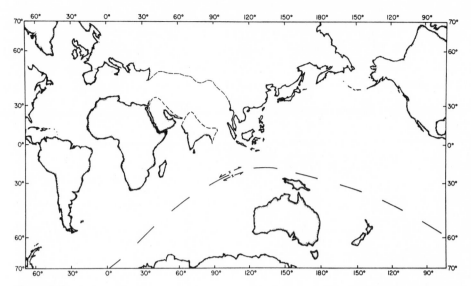

Figure 9-8
The positions of the continents at the time of anomaly 31 (Paleocene). The dashed line
indicates that the relative positions of the two groups of continents on each side of it were
not known; Antarctica and Africa are assumed to have been in their present positions. See
text and Table 9-7 for details

9–8 and 9–10, India was rotated by the amount
necessary to superpose the two anomaly 18
lines south of Mauritius, the movement before
the time of anomaly 18 being unknown. Finally, the Arctic opening was assumed to have
started 60 m.y. ago, on the basis of the
Paleocene-Eocene basalt eruptions presumably indicative of the initial breakup (see, for
example, Vine, 1966). Table 9–6 summarizes
the rotations used to obtain Figure 9–8.

Figure 9–8 is the reconstruction of the world
in Paleocene time obtained by resorbing the
amount of earth's surface created by sea-floor
spreading at the axes of the mid-ocean ridges
after the time of anomaly 31. The exact position of the Australia-New Zealand-Antarctica
block with respect to the rest of the continents
is not known. In regions of Tertiary mountain
belts the surface of the crust was presumably
much larger before folding than after. A large
part of the gaps appearing in these regions
(e.g., between India-Arabia and Eurasia, and
possibly also between Europe and Africa) may
have been occupied by continental crust.
Eurasia was treated as a unit, and the differential movement between eastern Asia and

North America implies that the Pacific Ocean
was larger than it is now. If, instead, eastern
Asia and North America are assumed to have
moved as a unit, a large gap would have existed
between eastern Asia and the remaining part
of Eurasia along what is now the Verkhoyansk
Mountains. It is quite possible that the opening
of the Labrador Sea is an early Cenozoic phenomenon, in which case the whole Eurasia-
Greenland block should be moved toward the
west with respect to Africa in Figure 9–8.

In spite of the many limitations which apply
to this reconstruction, its main features are a
necessary consequence of the magnetic anomaly pattern described by Heirtzler et al. (1968).
By Paleocene time, the present South Pacific
Ocean did not exist and the present North and
equatorial Pacific were larger. Either a large
'Tethys' sea existed or a minimum amount of
crustal shortening, ranging from 500 km in the
west to 1000 km north of India, must have occurred since. This may have been accompanied in early Cenozoic time by very large
shearing between the north and south shores of
the Tethys if the Labrador sea opened as late
as Paleocene time.

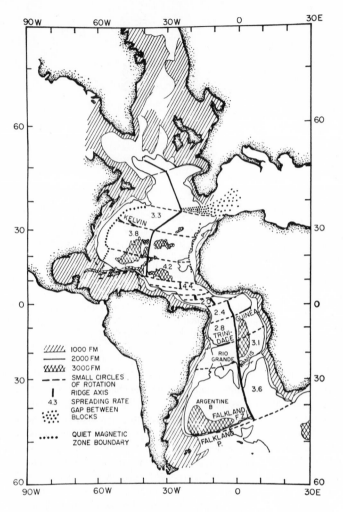

Figure 9-9
The Atlantic at the time of anomaly 31. The possible locations of major Mesozoic fracture zones, shown by dashed lines, were obtained by assuming that the pre-Cenozoic drifts can be described for South and North America by a single rotation each. The average spreading rate in centimeters per year for each block is obtained by assuming a constant rate of spreading between 120 and 70 m.y. ago. The bathymetry is not shown for the Caribbean and Gulf of Mexico. Notice that the trends of the Falkland plateau and fracture zone, Rio Grande and Walvis ridges, Trinidad ridge, Guinea ridge, and Kelvin seamount chain agree well with the predicted trends. See text and Table 9-7 for details

The Atlantic Opening

To illustrate in more detail the implications of this reconstruction, we show in larger scale (Figure 9–9) the reconstruction of the Atlantic Ocean in Paleocene time. The tracings of the present-day isobaths in the basins help us to visualize the changes resulting from this opening during Cenozoic time. A first conclusion obtained by an examination of Figure 9–9 is that the effect of applying the rotation determined by the distance to anomaly 31 in the South Atlantic has been to resorb the present mid-Atlantic ridge area in the North Atlantic as well as in the South Atlantic. In Figure 9–9 the ocean basins, as we know them, still have their present extent. This observation

supports the hypothesis that the present mid-Atlantic ridge, in the North Atlantic as well as in the South Atlantic, was created by spreading during the Cenozoic era, whereas the basins are pre-Cenozoic structures. Consequently, anomaly 31 should be found at the boundary ridge-basins in the North Atlantic. In fact, according to this hypothesis, we can predict the location of any magnetic anomaly over the whole mid-Atlantic ridge and south of the Azores-Gibraltar ridge.

The Mesozoic Fracture Zones. Figure 9–9 was obtained by rotating North and South America as a unit by 25° around 58°N, 37°W. Bullard et al. (1965) have given the rotations necessary to bring North America and South

Table 9–7
Rotations used for figure 9–9[a].

Blocks	Pole of rotation		
	Latitude	Longitude	Angle of rotation
BULLARD ET AL. (1965) ROTATIONS			
North America–Africa	67.6N	14.0W	+74.8°
South America–Africa	44.1N	30.3W	+56.1°
ROTATION FOR ANOMALY 31			
America–Africa	58N	37W	+25°
RESIDUAL ROTATIONS FOR TIME BEFORE ANOMALY 31			
North America–Africa	73.9N	4.8W	+50.8°[b] or 42.0°[c]
South America–Africa	35.4N	21.0W	+32.5°

[a]Africa in its present position.
[b]Fit to 500-fathom line.
[c]Fit to "quiet magnetic zone" boundary

America into best fit with Africa. If we subtract the Cenozoic rotation from the Bullard et al. rotations, the residual rotations obtained describe the movements involved in the pre-Cenozoic drift, provided that they occurred during one single episode of spreading. If, on the other hand, the pre-Cenozoic drifts occurred during two or more episodes of spreading, the movements of North and South America would probably be described by two or more successive rotations each. This idea can be tested by comparing the lines of flow corresponding to a single pre-Cenozoic rotation with the trends of the major fracture zones in the basins.

The residual rotations obtained by subtracting the Cenozoic rotation from the Bullard et al. rotations are listed in Table 9–7. These residual rotations have been used to draw lines of flow (Figure 9–9) which should be compared with the trends of major features in the basins. As noted by Morgan, the lines of flow defined by these rotations do not agree with the lines of flow of the Cenozoic rotations. A major change in the pattern of spreading must have occurred in the Atlantic between the Mesozoic and Cenozoic eras.

The northern boundary of the Falkland plateau and its eastern extension, the Falkland fracture zone, nearly follows the line of flow predicted for the tip of South Africa. No clear break in the trend of this fault zone can be detected. Le Pichon et al. (1971a) have described this topographic feature as a steep scarp 2 to 3 km high which extends for more than 2000 km. They have shown in particular that the Falkland fracture zone ends abruptly to the east near 28°W, the location of anomaly 31. The ending of the Falkland fracture zone at the boundary of the Cenozoic mid-Atlantic ridge is another confirmation that the Cenozoic episode of spreading was not a simple continuation of the Mesozoic episode of spreading.

Wilson (1965b) had noted that the terminations of the Walvis and Rio Grande ridges correspond to two conjugate points in the fit of South America to Africa. Figure 9–9 shows that the northern boundaries of the Walvis and Rio Grande ridges follow the same flow line. Le Pichon et al. (1968) have described the northern boundary of the Rio Grande rise as a large continuous fault scarp trending nearly east-west, with numerous outcrops of upper Mesozoic and Cenozoic sediments. Like the Falkland plateau scarp, the northern scarps of the Walvis and Rio Grande ridges mark the locations of former Mesozoic fracture zones. The subsequent history of these features was controlled by these structural trends.

Figure 9–9 shows that, similarly, the ridge connecting the island of Trinidad to the Brazilian continental rise, the Guinea ridge, and the Kelvin seamount chain closely follow the trends predicated by the flow lines. Accordingly, a fracture zone should extend along the trend followed by the southeast Newfoundland ridge, marking the upper limit of the drift of North America. This evidence suggests that the pre-Cenozoic drift in the Atlantic was produced during a single episode of spreading, presumably during the late Mesozoic time. If there was an early breakup of the continents during Permian time, as suggested by McElhinny (1968), the movements involved must have been very limited or must have occurred along exactly the same flow lines.

The 'Quiet' Magnetic Zones.   A complication is introduced by the existence of the band of 'quiet' magnetic anomaly field paralleling the northeast American shelf (Heirtzler and Hayes, 1967). A similar but much narrower zone exists along the west coast of North Africa within the 2000-fathom isobath. Heirtzler and Hayes attribute the quiet zones to spreading during the long period of the Permian without reversals of the geomagnetic field. They could be portions of continental margins which subsided to their present depths when the active spreading started, in Aptian-Albian time. The history of subsidence of the adjacent continental margin is well known, and the base of the Lower Cretaceous (from drilling data) is now found at depths equivalent to the adjacent oceanic depths in parts of the Florida platform and Bahama banks (Sheridan et al., 1966).

Support for this hypothesis comes from cores raised from 5 km water depth from the Lower Cretaceous (Barremian to Albian) seismic reflector β, which is just above horizon B in the sedimentary column (Ewing et al., 1966). According to T. Saito (personal communication), these cores indicate a shallow-water environment. Houtz et al. (1968) report that this reflector is marked by a large increase in sediment compressional velocity and can be recognized, on this basis, at several locations near the continental margin, Heezen and Sheridan (1966) reported the dredging of shallow-water Lower Cretaceous (Neocomian) rocks at a depth of 5 km on the Blake plateau scarp. They interpreted it as indicating rapid subsidence during earliest Cretaceous

time. This hypothesis would necessitate a conversion from continental crust to oceanic crust by a process similar to the one presumably responsible for the subsidence of the Gulf of Mexico (according to the interpretation given by Ewing et al. (1962) of the diapirs as salt domes). One possible process may be the subcrustal erosion advocated by Van Bemmelen (1966). If this hypothesis is accepted, the fit between Africa and North America should be made along these magnetic boundaries as has been done by Drake and Nafe (1967).

Mesozoic Spreading Rates: The Caribbean. If we assume that the active Mesozoic drifts of North and South America occurred simultaneously and at a steady rate, starting 120 m.y. ago and ending 70 m.y. ago, it is possible to compute the corresponding spreading rate at any location along the axis of the ridge. In Figure 9–9 the average Mesozoic spreading rate within each block between pairs of fracture zones is given in centimeters per year (the total movement of drift is equal to twice this number). Note the change in spreading rate by more than a factor of 2 in the equatorial region. The divergence of the flow lines in this boundary region between North and South American drifts resulted in the creation of the Caribbean, presumably by subsidiary spreading. The general direction of Mesozoic spreading in the Caribbean is perpendicular to the trend of the Beata ridge.

Most of the Mesozoic drift probably occurred in Early and Middle Cretaceous times, and by the Late Cretaceous the movement was coming to a stop, as western North America was progressively overriding the adjacent trench system (Hamilton and Myers, 1966). Thus the actual spreading rates may have been as high as 6 to 8 cm/yr during the Early Cretaceous and much smaller during the Late Cretaceous.

Europe and Greenland Movements: The Azores-Gibraltar Ridge.   At the time Figure 9–9 was drawn, we accepted Vine's (1966) assumption that the breakup between Greenland and Great Britain started 60 m.y. ago at the time of the Paleocene-Eocene basalt eruption. The Labrador mid-ocean ridge being now a 'dead' ridge, we assumed that it corresponded to an earlier breakup. Consequently, in Figure 9–9 the Labrador Sea has its present extent, whereas the Reikjanes ridge area is closed. This resulsts in the creation of a gap southwest of Spain, which indicates the amount of compression that should have occurred in the region during Cenozoic time along what is now the Azores-Gibraltar ridge. Thus, in this reconstruction, Spain and Africa may have been in contact all the time.

Mayhew and Drake (1968) have identified the magnetic pattern over the Labrador Sea and over the basin west of the Reikjanes ridge. Their results confirm the fact that the rift between Greenland and Eurasia started near anomaly 31 (60 m.y. ago). However, they indicate, apparently, that the Labrador Sea was produced by spreading during the interval of time corresponding to anomalies 31–27, that is, during Paleocene time. If this is so, the Labrador Sea should be closed in Figure 9–9, and Spain should be separated from Europe by as much as 1500 km. the movement during the Cenozoic thus should have resulted mostly in strong shear during the Paleocene, then mostly in compression from the Eocene to the recent between Eurasia and Africa.

The general paleogeography of the Atlantic Ocean at the end of the Mesozoic era was such that the deep-water circulation must have been very limited. The northern source of North Atlantic deep water did not exist, as the Labrador Sea was closed. The deep-sea communications between North and South Atlantic were probably closed by the mid-ocean ridge at the equator. Finally, the Falkland fracture zone was closing the southwest Atlantic to circulation of deep water coming from the south. The present deep-sea circulation must have established itself at the beginning of the Cenozoic episode of spreading (Ewing et al., 1971).

Asymmetry of the Ridge in the South Atlantic. The location of anomaly 31 in the South Atlantic is such that the ridge is not median. The western basins are much wider than the eastern basins. Yet all the observations made to this day indicate that spreading occurs at nearly equal rates on each side of the crest of the mid-ocean ridges. If spreading is supposed to have occurred symmetrically with respect

to the crest of the ridge, the western two-thirds of the Argentine basin should represent older crust produced in earlier spreading. As no change in the direction of flow is indicated by the trend of the Falkland plateau scarp, it is probable that an eastward shift in the position of the crest occurred at the beginning of Mesozoic time, during the same episode of spreading. This probable evolution is shown in Figure 9–10, where the main phases of Mesozoic drift have been reconstructed using the rotations listed in Table 9–7. The great thickness of Mesozoic sediments in the Argentine basin (Ewing et al., 1971) may be explained in this way, as the sedimentation rates must have been much larger when the ocean was still very narrow.

History of the Opening. The following still very tentative history of the opening of the Atlantic Ocean, which is in agreement with some of the conclusions of Dickson et al. (1968) for the South Atlantic, is outlined:

1. The active phase of drift began in Aptian-Albian time, about 120 m.y. ago (King, 1962; Wilson, 1965a). The North Atlantic 'quiet' magnetic zones in the continent are included with the original continent of Bullard et al. (1965). The initiation of drift was accompanied by the subsidence of a large part of the northeast American continental margin. possibly related to the earlier (Jurassic) Gulf of Mexico subsidence. McElhinny (1968) has presented paleomagnetic evidence for some opening of the South Atlantic since Permian time. Although there is no evidence for considering parts of the South Atlantic basins to be 100 m.y. older than the others, it is probable that early rifting and subsidence began to appear along the line of parting in both the North and South Atlantic in Jurassic time. This may have resulted in the creation of marine gulfs of limited width, long before the episode of fast spreading began in the Early Cretaceous. For example, the first signs of uplift in the Red Sea−Gulf of Aden region appeared at least 50 m.y. ago. Yet most of their deeps were probably created after late Miocene time (10 m.y. ago). This early rifting could explain the few examples of deposition of marine sediments during the Late Jurassic (e.g., Colom, 1955).

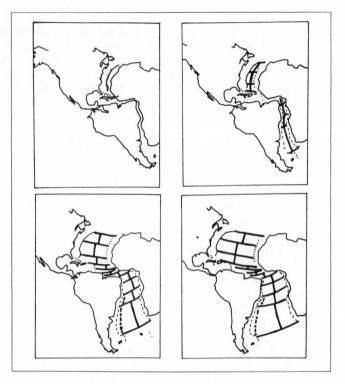

Figure 9-10
The Mesozoic episode of spreading in the Atlantic. These
reconstructions were obtained by assuming that North and South
America parted simultaneously from Africa 120 m.y. ago (upper
left) and reached the position corresponding to anomaly 31 (Figure
9-9) about 90 m.y. ago (lower right) after a single movement of
rotation for each. The asymmetry of the ridge crest in the South
Atlantic at the time of anomaly 31 is explained by a shift in the
ridge crest positions in the Early Cretaceous. The fit is made to the
"quiet" magnetic zone boundaries in the North Atlantic and to
the 500 fathom line in the South Atlantic (Bullard et al., 1965). See
Table 9-7. The active ridge crests and fracture zones are
represented by a thick continuous line, the abandoned ridge crest
by a thick dash-dot line

2. A fast episode of spreading, possibly
30 m.y. long, followed. Figure 9–10 shows the
probable evolution of this episode of drift.
Notice that after the first 10 m.y. of drift the
crest of the ridge in the South Atlantic must
have migrated to a more easterly position to
reach the nonmedian line along which it was
lying at the end of the Mesozoic era. The
original positions of the continents at the end
of Jurassic and beginning of Cretaceous time
were obtained by using the Bullard et al. fit
for South America and making the fit of North
America along the boundary of the 'quiet'

magnetic zone (Drake and Nafe, 1967). Notice that some distortion is implied in the
Caribbean, as Puerto Rico overlaps Africa.
Also the southern tip of Central America
overlaps South America. However, a great
part of this overlap corresponds to Cretaceous
or more recent geological formations (Edgar,
1968).

It is probable that by Middle Cretaceous
time the fast episode of drift was ending. The
Caribbean had then essentially its present
extent. Great fracture zones were occupying
the positions of the northern boundaries of

the Walvis and Rio Grande ridges and of the Falkland plateau. The Guinea ridge, Trinidade ridge, Kelvin seamount group, and southeast Newfoundland ridge were probably fracture zones too. As North America and Eurasia moved as a unit, a large extensional shear zone was created between Africa and Eurasia.

During this period the deep-water circulation was of the closed-basin type, and the sedimentation rates, especially in the early stages of breakup, must have been very large, as the ratio of coast periphery to area of sea was much larger than now. By Late Cretaceous time, the spreading must have already slowed down considerably, coming progressively to a complete stop. This may have been due to resistance encountered in the overriding of the adjacent trench system by the western border of America (Hamilton and Myers, 1966). The deposition of horizon A, (Ewing et al., 1966) seems to be related to the end of this episode of spreading. Perhaps the flow of turbidites was triggered by the regression associated with the subsidence of the Mesozoic mid-ocean ridges (Menard, 1964, p. 238).

3. By Paleocene time, the movement of opening resumed, following a new pattern of flow. North and South America were moving as a unit at this time, except possibly for some small differential movement. The actively spreading ridge extended north to the Labrador Sea and the Norwegian and Arctic seas, according to the identification of magnetic anomalies by Mayhew and Drake. The southeastward movement of Eurasia relative to Africa must have created a zone of shear between Eurasia and Africa. This inversion of tectonic forces, from extensional shear in Mesozoic time to shear and their compression in Cenozoic, is a necessary consequence of the interpretation of the magnetic anomalies in the Labrador Sea and Reikjanes ridge area. It might explain the complexity of the phenomenon named by Van Hilten and Zijderveld (1966) the 'Tethys Twist.' By Eocene time, the Labrador Sea was opened and the mid-Labrador Sea ridge had died.

By Oligocene time, the spreading rate had begun to slow down, and it came to a complete stop in the Miocene. This is the time during which northwest America probably overrode first the trench system and then the ridge crest situated along its border (Vine, 1966). It seems that the terminations of both Mesozoic and early Cenozoic episodes of spreading coincide with the times at which western America overrode the adjacent trench system. The episodes may have been terminated because the lithosphere, which was sinking along the trench (Oliver and Isacks, 1967), had sunk to the maximum depth it could reach. These episodes of spreading also correlate well with the compressive period which culminated in the Oligocene and Miocene in the Mediterranean (Glangeaud, 1961), when North Africa and Europe came into close contact. The block faulting on the northern mid-Atlantic ridge and the renewed uplift of the Walvis and Rio Grande ridges (Ewing et al., 1966) and possibly of the Falkland plateau along old fracture zone lines may be related to this stoppage of spreading. Le Pichon et al, (1968) have reported the recovery of lower Eocene turbidites from the summit of the Rio Grande rise, now at 800 m depth. The beginning of the early Cenozoic episode of spreading opened the Atlantic basin to deep-water circulation and started a change in the pattern of sedimentation (Ewing et al., 1968).

4. In late Miocene time, the third cycle of spreading started apparently along the same pattern of flow as the previous one in the Atlantic, but at a slower rate, and produced the crestal region devoid of sediments (Ewing and Ewing, 1967). In the North Pacific, however, a new pattern of spreading had established itself, in which the Pacific Ocean floor was now rotating away from North America instead of being thrust below it along a trench system.

This reconstruction does not explain why the part of the mid-Atlantic ridge situated north of the line formed by the northern boundaries of the Walvis and Rio Grande ridges is so much more fractured and devoid of sediments than the part immediately south of it (Ewing et al., 1966). The hypothesis of Ewing et al. of a large Miocene tectonic disturbance which disrupted the crust north of this boundary at the end of the early Cenozoic cycle of spreading still seems to be the only one available. The reason for this drastic change on each side of the boundary is not clear, but

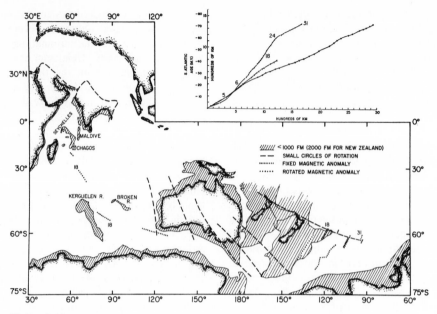

Figure 9-11
The Indian and South Pacific oceans at the times of anomalies 31 (Paleocene) and 18 (late Eocene). See text and Table 9-7 for explanations. The curves of distance of a magnetic anomaly from the crest in a given ocean versus distance of the same anomaly from the crest in the Atlantic Ocean are, respectively, from left to right for the South Pacific, South Indian, and North Pacific oceans (from Heirtzler et al., 1968)

there is little doubt that it has been and still is an important structural division in the South Atlantic.

The South Pacific and Indian Ocean Openings

Figure 9–11 shows in greater detail the evolution of the South Pacific and Indian oceans since Paleocene time. The value of this reconstruction can be judged from how well the anomaly lines, shown previously in Figures 9–2 and 9–4, fit together. The positions of the New Zealand plateau and of the South Pacific ridge crest are shown at two different times: anomaly 31 (Paleocene) and anomaly 18 (late Eocene). As Antarctica is assumed to be fixed, the ridge crest was migrating away from it at half the speed of the New Zealand plateau. The positions of Australia relative to Antarctica and of India relative to Africa are plotted for the time of anomaly 18 (late Eocene). As mentioned previously, the position

of India before the time of anomaly 18 is unknown. Eurasia is shown in the same position as in Figures 9–8 and 9–9 (i.e., at the time of anomaly 31, the Labrador Sea having already opened). The dashed lines represent the lines of flow of Australia and of the New Zealand plateau during the Cenozoic.

The Indian Ocean. The evolution of the Indian Ocean during the Cenozoic is till incompletely understood, and the tentative conclusions of Le Pichon and Heirtzler (1968) will be summarized here. Before early Eocene time, the movement of India had to be due north, corresponding to a rotation around a point in central Africa (see Figure 9–11). This movement probably resulted from spreading along an east-west ridge in the then much greater basin north of Broken ridge. Broken ridge is probably continental in origin. Eurasia, in late Eocene time, would have been a few hundred kilometers south of the position it occupies in Figure 9–11. Consequently, the present northern border of India was then within 1000 km of the present border of Eurasia. As this was the time of the first large orogenic phase, a total compression of the order of 1000 km must be assumed for this area since Eocene time.

In any case, the pattern of spreading in the Indian Ocean readjusted itself at the time of anomaly 18, when a northwest-southeast axis became active for the first time, from the then closed Red Sea and Gulf of Aden to Australia. In the beginning there must have been considerable differential motion along the Ninetyeast ridge, which probably came into evidence as a result of the motion. Some differential motion probably also occurred along the southwest branch of the ridge to reconcile the direction of spreading south of Australia with the direction of spreading on the Carlsberg ridge. By Miocene time the spreading had completely stopped, at the time of the peak phase of the Himalayan orogeny. It resumed at the time of anomaly 5 (latest Miocene), creating the Red Sea and Aden axial troughs in particular.

Australia and New Zealand. The evolution of the northern and central Indian Ocean is difficult to obtain from a study of the magnetic pattern. It is much easier to decipher the history of spreading in the South Pacific and South Indian oceans. We see in Figure 9–11 that the New Zealand plateau separated from Antarctica in Paleocene time at a relatively fast spreading rate (see inset, Figure 9–11). By the time of anomaly 24 (early Eocene), the spreading rate slowed by a factor of 3, as the New Zealand plateau was approaching eastern Australia. By the time of anomaly 18 (late Eocene), Australia began to detach itself from Antarctica, along flow lines parallel to those of the New Zealand plateau. Australia was then moving twice as fast as New Zealand. Consequently, during the period of extension between Australia and New Zealand the present Tasman Sea was created. When the spreading resumed, after the Miocene interruption, the velocity of spreading became greater in the South Pacific than in the South Indian Ocean, resulting in general compression between New Zealand and Australia. Thus, on the basis of the magnetic anomaly pattern, we can infer a very complex history

for New Zealand during the Cenozoic, a period of compression in Paleocene and Eocene times, a period of extension in Oligocene and early Miocene times, and a period of compression again in the Pliocene. Furthermore, according to this reconstruction, Australia and the New Zealand plateau were probably in contact in late Eocene time.

CONCLUSIONS

Morgan (1968b) has suggested that the surface of the earth can be approximated by a small number of rigid blocks in relative motion with respect to each other. This assumption has been confirmed by a geometrical analysis of the data on sea-floor spreading. In accordance with this concept, we have given a simplified but complete and consistent picture of the global pattern of surface motion. We have shown that all movements are interrelated, so that no spreading mid-ocean ridge can be understood independently of the others. Thus any major change in the pattern of spreading must be global.

The results support Elsasser (1968), McKenzie (1967), Morgan (1968b), and Oliver and Isacks' (1967) model of a rigid and mobile lithosphere (tectosphere) several tens of kilometers thick on top of a weak asthenosphere. The mechanism of sea-floor spreading at the mid-ocean ridges, then, corresponds to the breaking apart of a plate, preferably along lines of weakness, in response to a stress pattern.

We have attempted to apply this concept to obtain a reconstruction of the history of spreading during Cenozoic time. Three main episodes of spreading are recognized—late Mesozoic, early Cenozoic, and late Cenozoic. The beginning of each cycle of spreading is marked by the reorganization of the global pattern of motion. A correlation is made between slowing of spreading at the ends of the two previous cycles and paroxysms of orogenic phases. This history of spreading follows closely one advocated by Ewing et al. (1968) to explain the sediment distribution.

The results presented by Oliver and Isacks suggest a possible mechanism to account for the episodicity of sea-floor spreading. The main zones along which the lithosphere is sinking are the deep-sea trenches. Because of thermal inertia, the lithosphere apparently maintains its identity to a depth as great as 700 km. Let us assume that there is a part of the mantle, possibly below 800 km, into which the lithosphere cannot penetrate further. At first, the adjacent continental block will probably be forced to override the island arc and trench system. Then the spreading will have to stop until a new trench system has been formed. As the rate of thrusting of the oceanic crust along the trenches is of the order of 6 cm/yr, the average length of the active part of a cycle of spreading is unlikely to exceed 30 m.y., corresponding to 1800 km of crust. This length roughly corresponds to the length of a Gutenberg fault zone having a dip of 30° and extending to a depth of 800 km.

Finally, in this paper, regions in which earth's surface is shortened or destroyed are called regions of compression. Most geologists would agree that the Alpine-Himalayan belt is indeed a region of compression. It has been argued convincingly, however, that deep-sea trenches are regions of tensile stress (e.g., Worzel, 1965). Elsasser (1968) has suggested that the tectosphere becomes denser as it slides down the Gutenberg fault zone and acquires a sufficiently greater density than surrounding mantle material to sink on its own. 'The motion . . . then leads to a tensile pull in the adjacent part of the tectosphere.' Consider the India-Pacific boundary, for example. From Macquarie Island to the Kermadec trench the differential movement of compression between the two plates results in compressional surface features, but along the Kermadec and Tonga trenches, where the differential movement is larger, the Pacific tectosphere has decoupled itself from the adjacent block and sinks on its own, creating the surface tensional features associated with the trenches.

As the differential movement of compression between two blocks increases, the associated surface compressional features apparently become larger and reach a maximum for a rate of movement of about 5–6 cm/yr (Himalayas). At larger rates, the lithosphere sinks along an active trench, and the associated surface features are tensional instead of compressional. The narrow range of

rates of differential movement associated with the trenches where active thrusting of the tectosphere occurs (6–9 cm/yr) may be one of the significant results of this study.

APPENDIX

Determination of Center of Rotation From Azimuths of Fracture Zones. We assume that the latitude Lat_i, longitude $Long_i$, and azimuth Azi_i of the fracture zones at N points along the crest of the ridge are known.

1. Start with an estimated position of the center of rotation (latitude $Plat$ and longitude $Plong$).

2. Compute the theoretical azimuth $Tazi_i$ at the N points. $Tazi_i$ is the aximuth of the tangent to the small circle having for center $Plat$ and $Plong$ and passing by Lat_i and $Long_i$.

3. Compute the sum of the squares of the deviations

$$Sum = \sum_{i=1}^{N} (Azi_i - Tazi_i)^2$$

4. Modify $Plat$ and $Plong$ until Sum is minimum.

The actual search is made through a grid centered on the original $Plat$ and $Plong$. The process is repeated until the point is located to better than 1°. The convergence of the computation can be estimated from the sizes of the ellipses in Figure 9–1.

Determination of Center of Rotation From Spreading Rates. Let R be the spreading rate measured along a perpendicular to the crest of the ridge having the azimuth Azi. Let Spr be the true spreading rate measured along a tangent to the small circle (centered on the center of rotation) which has the azimuth $Tazi$. Let $Rmax$ be the maximum true spreading rate (along the equator of rotation) and $Sprn = Spr/Rmax$ be the normalized true spreading rate.

Then $Sprn$ is equal to the sine of the distance (in degrees of arc) from the center of rotation and $Sprn = R/[Rmax \cdot \cos(Azi - Tazi)]$

We assume that the latitude Lat_i, longitude $Long_i$, and spreading rate R_i (measured along a perpendicular to the crest having the azimuth Azi_i) at N points along the crest of the ridge are known.

1. Start with an estimated position of the center of rotation (*Plat, Plong*) and an estimated maximum true spreading rate *Rmax*.

2. Compute $Sprn_i$ and $Tazi_i$ at the N points.

3. Compute the sum of the squares of the deviations.

$Sum =$

$$\sum_{i=1}^{N} \left[\frac{R_i}{Rmax \cdot \cos (Azi_i - Tazi_i)} - Sprn_i \right]^2$$

4. Modify *Plat* and *Plong* until *Sum* is minimum.

5. When the location is obtained to better than 1°, the process is resumed with a different value of *Rmax*.

The method is very sensitive to the value of *Rmax*. It is not very sensitive to the difference between true spreading rate and spreading rate measured along a perpendicular to the crest (the latter varies as the inverse of the cosine of the difference in azimuths).

For example, for the South Pacific, for *Rmax* = 6.3 cm/yr, *Plat* = 70.9°S, *Plong* = 127.1°E, and the standard deviation *SD* is 0.059. For *Rmax* = 5.7 cm/yr, *Plat* = 66.3°S, *Plong* = 118.7°E, and *SD* = 0.065. The minimum *SD* = 0.058 is found for *Rmax* = 6.0, *Plat* = 68.3°S, and *Plong* = 123.3°E.

For the Atlantic Ocean, for *Rmax* = 2.25 cm/yr, *Plat* = 69.6°N, *Plong* = 54.7°W, and *SD* = 0.082. For *Rmax* = 1.85, *Plat* = 64.4°N, *Plong* = 10.5°W, and *SD* = 0.096. And the minimum *SD* = 0.065 is found for *Rmax* = 2.05, *Plat* = 68.6°N, and *Plong* = 31.8°W.

The location is poorly determined in longitude, and a cause of error lies in the determination of the trend of the crest and the associated magnetic anomalies. For this reason, in this paper, the adopted positions of the centers of rotation are those obtained from the azimuths of the fracture zones.

10

Evolution of Triple Junctions

DAN P McKENZIE
W. JASON MORGAN

From *Nature*, v. 224, p. 125–133, 1969. Reprinted with permission of the authors and Macmillan Journals, Ltd.

A precise version of the hypothesis of sea floor spreading (Hess, 1962; Dietz, 1961) has recently been suggested. This new formulation (McKenzie and Parker, 1967; Morgan, 1968a; Le Pichon, 1968; Isacks et al., 1968) requires that all aseismic areas of the Earth's surface move as rigid spherical caps and for this reason it is often called "plate tectonics." The instantaneous relative motion of any two plates on the surface of a sphere can be represented by a rotation about an axis, and so problems of present day tectonics reduce to determining the plate boundaries and relative rotation vectors of all plates on the Earth's surface. There are various methods of obtaining such information. If two plates are separating and new oceanic crust is being generated, the rate can be obtained from the magnetic lineations (Vine, 1966; Pitman et al., 1968), which can also be used to map the plate boundaries (Vine, 1966). Where the ridge axis is offset by transform faults the relative motion vector must be parallel to the strike of the faults (Wilson, 1965c). The most general method of mapping plate boundaries, however, is by their seismicity (Isacks et al., 1968; Barazangi and Dorman, 1969), and earthquakes can also be used to measure the direction (McKenzie and Parker, 1967; Isacks et al., 1968; Sykes, 1967) and magnitude (Brune, 1968) of the relative velocity between the two plates involved. The agreement between these methods is striking, especially in oceanic areas (Isacks et al., 1968), and demonstrates that aseismic regions are indeed rigid. It is now clear that the principal features of ridges, trenches and transform faults are a direct consequence of the relative motion of rigid plates.

There are two main reasons why plate tectonics does not yet provide a complete theory of global tectonics. The first is that the mechanism by which the motions are maintained is still unknown, though it now seems that some form of thermal convection can provide

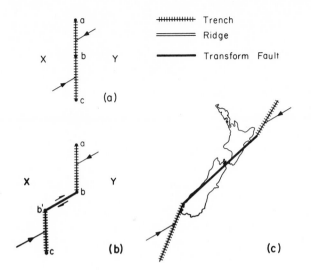

Figure 10-1
The evolution of a trench. The arrows show the relative motion vector and are on the plates being consumed. Thus Y is consumed between a and b, X between b and c. The trench evolves to form two trenches joined by a transform fault. (c) is a sketch of New Zealand showing that the Alpine fault is a trench-trench transform fault of the type in (b)

sufficient energy (McKenzie, 1969b). This problem will not be discussed further. The other is that the original ideas only apply to motions at present taking place, and are not concerned with either the slow evolution of plate boundaries or with changes in their relative motion through geological time. For example, the break-up of Gondwanaland was presumably caused by stresses within the original plate, and cannot be understood using geometry alone. Two causes of plate evolution, however, are the geometric consequences of the motion of rigid plates, and it is with these that this article is concerned.

The simplest example of such evolution is that of the trench shown in plan view in Fig. 10–1,a and occurs because a trench consumes lithosphere on only one side. The upper part of the trench ab consumes the plate Y, whereas the lower part bc in the figure consumes X. The arrows show the relative motion vector between X and Y, and are on the plate which is being consumed. As the motions continue in the directions of the arrows, Y is consumed between a and b, but not between b and c. bc must therefore be steadily offset from ab to form two trenches joined by a transform fault (Fig. 10–1,b), the length of which increases at the consumption rate. The Alpine fault in New Zealand is an example of such a transform fault joining two trenches which consume different plates (Isacks et al., 1968;

Hamilton and Evison, 1967) (Fig. 10–1,c). To the north of North Island the Kermadec trench consumes the Pacific plate, whereas the shallow and intermediate earthquakes beneath South Island and Macquarie ridge demonstrate that the Tasman Sea is being consumed in this region. This example is particularly simple because the slip vector does not change anywhere on the boundary between X and Y. More complicated effects can occur when the point at which three plates meet moves along plate boundaries. Evolution of such triple junctions can produce many of the changes which would otherwise appear to have been caused by a change in the direction or magnitude of the relative motion between plates. In particular, sudden changes in tectonic style are more likely to be caused by the movement of such junctions than by a change in relative velocity. It is especially difficult to alter the motion of large plates in a short time ($\sim$ 1 million years) because the long thermal time constant ($\sim$ 50 million years or more) of any mantle convection driving the plates (McKenzie, 1969b).

The discussion which follows is easier to follow if trenches, ridges and transform faults are defined in terms of destruction and creation of plates, rather than in terms of topographic features. Trenches are therefore defined as structures which consume the lithosphere from only one side, and ridges as

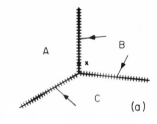

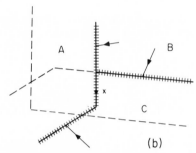

Figure 10-2
The juction of three trenches. (a)
shows the geometry before evolution,
with the arrows as in Figure 10-1. The
arrows on B point in two directions
because the relative motion between
B and C is different from that between
A and B. At some later time (b), the
positions B and C would have reached
if they had not been consumed are
shown as dashed lines. The trench
between B and C must move with C,
and therefore moves away from the
apex of A

structures which both produce lithosphere
symmetrically and lie at right angles to the
relative velocity vector between two separating plates. Transform faults are defined as
active faults parallel to the relative slip vector.
It is easy to modify the arguments which follow to take account of the complications of
the real Earth where these definitions are not
exactly true (Menard and Atwater, 1968;
Laughton, 1966). This is not done in general
because the basic principles would then become obscure.

For the purposes of plate tectonics, the surface of the Earth is completely covered by a
mosaic of interlocking plates in relative motion. There are many points where three plates
meet, but, except instantaneously, none where

four or more boundaries meet. The relations
between the relative velocities of the plates
at triple junctions have been discussed previously (McKenzie and Parker, 1967); they
are a consequence of the rigidity of the plates
and do not impose any restrictions on the
orientation of plate boundaries or on the relative velocity vectors. If, however, the triple
junction is required to look the same at some
later time, there are important restrictions on
the possible orientations of the three plate
boundaries. Unless these conditions are satisfied the junction can exist for an instant only,
and for this reason is defined as unstable. If
evolution is possible without a change in geometry, then the vertex is defined as a stable
junction. The distinction between the two
types is important because movement of stable
junctions alone permits continuous plate
evolution.

An example of a triple junction is the point
at which three trenches meet (Fig. 10–2,*a*).
The arrows are on the plates which are being
consumed, and show the relative velocity vector between plates. These vectors are not in
general perpendicular to the boundaries. Consider the evolution of this junction relative to
A which is taken as fixed. The positions of the
plates *B* and *C* at some later time are shown in
Fig. 10–2,*b*. The dashed boundaries show
where the plates would have extended if they
had not been consumed by the trenches *AC*
and *AB*. The trench *BC*, however, has migrated up the boundary, consuming *B* during
the process, to reach the position shown. A
point such as *x* on the boundary *AB* will show a
sudden change in motion direction as the triple
junction passes. This apparent change in
spreading direction of the plates is easily distinguished from a real change in motion of one
of the plates because it takes place at different
times at different places along the plate boundary. Fig. 10–2 also shows how an unstable
junction may become a stable one. The original
orientation in Fig. 10–2,*a* is unstable unless
the slip vector $_A\mathbf{v}_C$ is parallel to the boundary
BC. This condition is satisfied if *BC* does not
move relative to *A*, and does not require any of
the trenches to turn into transform faults. The
slip vector $_A\mathbf{v}_C$ in Fig. 10–2,*a* does not satisfy
this condition, and therefore the trench *BC*
does not remain on the apex of *A* but moves

upward. Once the geometry in Fig. 10–2,*b* occurs, the junction is stable and further evolution causes no change in geometry as *BC* moves along *AB* with a velocity with respect to *A* given by v_T. If therefore the triple junction is watched by an observer moving with no rotation and with velocity v_T with respect to *A*, the triple junction will be stationary in his frame of reference. If all the boundaries are straight, the angles between them will not change. In the first example of a stable junction of three trenches the geometry remained unchanged in a frame fixed to *A*. In the second example all plates move in the frame fixed to the junction.

It is not easy to discover the general stability conditions for all possible junctions by the method used in constructing Fig. 10–2. Perhaps the simplest example of the general method is the triple junction between three ridges, which we shall call an *RRR* junction (Fig. 10–3). An example of such a junction is the meeting of the east Pacific rise and the Galapagos rift zone in the equatorial east Pacific (Herron and Hiertzler, 1968; Raff, 1968). The Great Magnetic Bight in the north Pacific was probably formed by another such junction which has now ceased to exist (Pitman and Hayes, 1968; Vine and Hess, 1970). In this and all other examples the relative velocity vectors at the junction are required to satisfy (McKenzie and Parker, 1967) (Figure 10–3):

$$_A v_B + _B v_C + _C v_A = 0 \qquad (1)$$

This equation must be satisfied if the plates are rigid.

In Fig. 10–3 the lengths *AB, BC* and *CA* are proportional to and parallel to the velocities $_A v_B$, $_B v_C$ and $_C v_A$ respectively. The triangle is therefore in velocity space, and represents the condition imposed by equation 1. Because ridges spread symmetrically at right angles to their strike, a point on the axis of the ridge *AB* will move with a velocity $_A v_B / 2$ relative to *A*. This velocity corresponds to the mid point of *AB* in Fig. 10–3. Consider a reference frame moving with a velocity corresponding to some point on the perpendicular bisector *ab* of *AB*. *ab* is parallel to the ridge *AB*, so in this frame the ridge will move along itself and will have no velocity at right angles to *AB*. The same is true of the plate boundaries *BC* and *CA* when

observed from reference frames whose velocities lie on *bc* and *ac* respectively. The perpendicular bisectors of the sides of any triangle meet at a point called the centroid, and this point *J* in velocity space gives the velocity with which the triple junction moves. It is therefore always possible to choose a reference frame in which the triple junction does not change with time. From the velocity triangle the relative velocity v of all plates relative to the triple junction is:

$$v = \frac{|_B v_C|}{2 \sin \alpha} = \frac{|_C v_A|}{2 \sin \beta} = \frac{|_A v_B|}{2 \sin \gamma} \qquad (2)$$

Also the angle between *AB* and *AJ* is 90-γ. Such a junction between three ridges is therefore stable for all ridge orientations and spreading rates. If the ridges spread symmetrically, but not at right angles to the relative slip vectors, the lines *ab, bc* and *ac* must still be drawn through the mid points of the sides, but not at right angles to the velocity vectors. Certain simple geometric conditions must then be satisfied if the triple junction is to be stable in these conditions.

A more complicated junction is that of three trenches, *TTT*(a) in Fig. 10–3, which has already been discussed. An example of such a junction occurs in the north-west Pacific (McKenzie and Parker, 1967; Gutenberg and Richter, 1954), where the Japan trench branches to form the Ryukyu and Bonin arcs (Fig. 10–9). The arrows are on the plates being consumed and show the relative vector velocities between plates. The velocity triangle is formed as before, but points in velocity space corresponding to reference frames in which the position of the plate boundaries is fixed no longer lie on the perpendicular bisectors of the sides of the triangle. Consider, for example, the trench between plates *A* and *B*. Because *A* is not consumed, the trench does not move relative to *A*. Clearly this condition is also satisfied by any reference frame with a velocity parallel to the plate boundary *AB*. Such velocities correspond to points on *ab*, a line through *A* parallel to the trench *AB*. The lines *bc* and *ac* are constructed in the same way. Unlike the triple junction of three ridges, *ab, ac* and *bc* do not intersect at a point unless certain conditions are satisfied. The velocity triangle

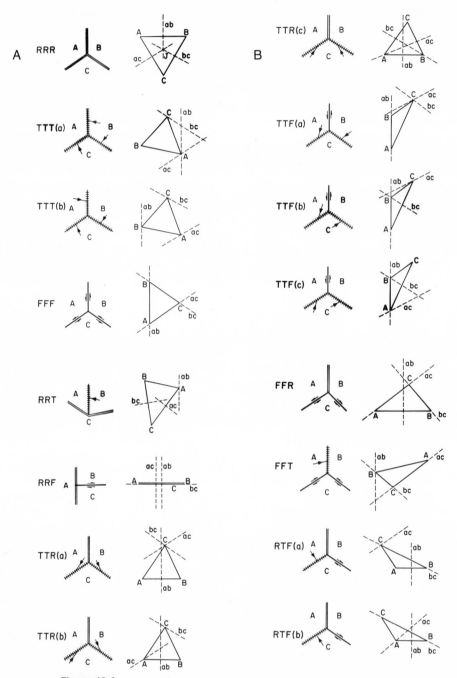

Figure 10-3
The geometry and stability of all possible triple junctions. Representation
of structures is the same as in Figure 10-1. Dashed lines ab, bc and ac in
the velocity triangles join points the vector velocities of which leave the
geometry of AB, BC and AC, respectively, unchanged. The relevant
junctions are stable only if ab, bc and ac meet at a point. This condition is
always satisfied by RRR; in other cases the general velocity triangles are
drawn to demonstrate instability. Several of the examples are speculative

shows that if *bc* goes through *A*, and therefore the plate boundary *BC* is parallel to $_A\mathbf{v}_C$, the junction is stable and fixed to plate *A*. Another possible stable arrangement occurs if *ab* and *ac* are the same line. This requires the boundaries *AB* and *AC* to form one straight line. Thus a triple junction between three trenches can be stable. These stability conditions have already been obtained (Fig. 10–2), but it was difficult to prove that these were the only possible stable junctions by using the previous method. There is another possible junction between three trenches, *TTT*(b), which has a rather complicated general condition for stability.

The junction between three transform faults is unstable in all circumstances (Fig. 10–3) because *ab, bc* and *ac* can never meet at a point. Though an unstable junction can occur, it immediately changes into one or more stable junctions and is therefore not useful in understanding plate evolution.

A collection of all sixteen possible triple junctions (Fig. 10–3) shows that all except two are stable in certain conditions which can easily be obtained from the vector velocity diagrams. The most important of these stable junctions are those with two boundaries in a straight line, bounding a plate which is neither being generated nor consumed. Such junctions are easy to form, and do not depend for their stability on the exact values of the relative velocities. The only important exception to this rule is *RRR*.

There are at least four examples of triple junctions active at present in the north Pacific. Three of these were discussed by McKenzie and Parker (1967) before the importance of stability was understood, and as a result two of the junctions they describe are unstable. The two junctions concerned are probably both stable at present. Three of the active junctions occur along the west coast of North America, and their evolution demonstrates how complicated the interaction of three plates can be even without any changes in their relative motion. Fig 10–4 is taken from Menard and Atwater (1968), and shows diagrammatically the magnetic lineations in the north-east Pacific. They point out that there are two striking changes in the trend of both transform faults and of magnetic lineations. The first is between

anomalies 23 and 21 throughout the north-east Pacific north of the Pioneer fracture zone, and is best explained by a change in the motion direction of one of the plates. Most probably the plate between the Main Pacific plate and the American plate was the one involved. This plate is called the Farallon plate throughout the rest of this discussion, after the Farallon Islands off the coast of central California. The second change in trend occurs at anomaly 10 near San Francisco, but not until after anomaly 5 north of the Mendocino. It is thus not possible to produce this change by a change in the motion of any of the plates at a given time. Such apparent changes in spreading direction are, however, easily explained by the evolution of the triple junctions formed at the time of anomaly 10.

The main features of the north-eastern Pacific before the time anomaly 10 was formed have now largely vanished. Except near the Gorda and Juan da Fuca ridges, and south of Baja California, only one half of the anomaly pattern remains. Thus there must have been a trench between the ridge and the coast of North America which consumed the Farallon plate with its anomalies. This trench must have existed as a continuous feature up to about the time of anomaly 10, or the Middle Oligocene. Fig. 10–5,*a* shows the arrangement of plates at about the time of anomaly 13. If we assume that all relative plate motions remain constant from the time of anomaly 13 onwards we can deduce the motion of all the junctions relative to any plate. In Fig. 10–5 all fracture zones except the Mendocino and the Murray have been omitted. The offsets in the ridge show that it will first meet the trench just south of the Mendocino fracture zone to form two triple junctions, *FFT* in the north and *RTF*(a) in the south (Fig. 10–5,*b*). Fig. 10–5,*c* shows that the first of these is stable if the transform fault between *A* and *C* and the trench between *A* and *D* lie in a straight line. The triple junction is then at rest relative to *C*, and therefore *J* moves north-westward relative to *A*, changing the trench into a transform fault. Similarly the second junction is stable if the trench and the transform fault are in a straight line (Fig. 10–5,*d*). Clearly this junction can move north-west or south-east relative to *A*, depending on the magnitude and direction of the relative

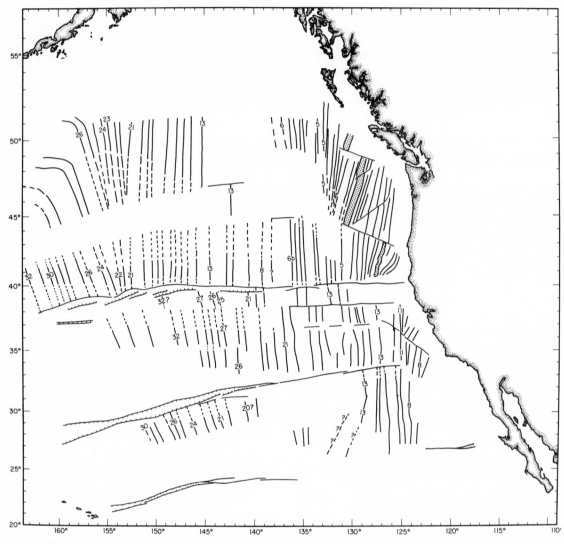

Figure 10-4
Fracture zones and magnetic anomalies in the north-eastern Pacific (From Menard and Atwater, 1968)

velocities. Unless *bc* lies to the east of *A* in Fig. 10–5,*d*, however, the ridge axis will move away from the trench, and they could never have met. Thus *J* lies to the south-east of *A*, and the junction will move down the boundary of *A*. This southward migration stops when the junction *RTF*(a) reaches the Murray fault (Fig. 10–6,*a*) where it must change to *FFT* and move rapidly north-westward relative to the American plate, for it must then be fixed to *C* (Fig. 10–6,*b*). The stability condition is again that the trench and transform fault between *A*

and *C* form a straight line. The north-west motion then regenerates the trench on the western margin. During this period (Fig. 10–6,*b*) the trench along the west coast continues to consume the two remaining pieces of the Farallon plate except between the Mendocino and the Murray faults. Thus whether or not the oceanic transform faults possess continental extensions they can influence the tectonics of the continental margin. The geometry of the plate boundaries in Fig. 10–6,*b* changes back to Fig. 10–5,*b* when the ridge south of

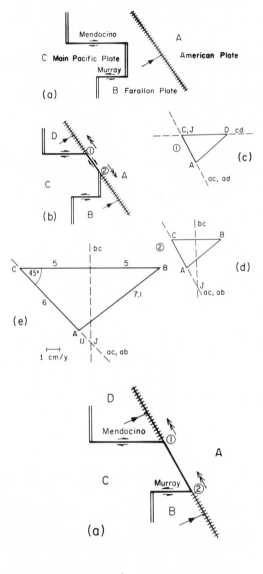

(a)

(b)

(c)

(d)

(e)

(a)

(b)

Figure 10-5
(a) The geometry of the north-east Pacific at about the time of anomaly 13. All fracture zones except the Mendocino and the Murray have been omitted for simplicity. (b) Stable triple junctions at about the time of anomaly 9, formed when the east Pacific met the trench off western North America. The double headed arrows show the motion of the two junctions (1) and (2) relative to the American place A. (c) is a sketch of the vector velocity diagram for junction (1) and shows it will move north-west with the Main Pacific plate. (d) is a similar diagram for (2). If the relative plate motions have not changed since at least the Middle Oligocene, the magnetic lineations and the present motion on the San Andreas may be used to draw the velocity diagrams to scale. (e) is such a drawing of (d), and shows that the triple junction J will slowly move to the south-east relative to A. The numbers are in cm/yr and the vector AB shows the direction and rate of consumption of the Farallon plate by the American

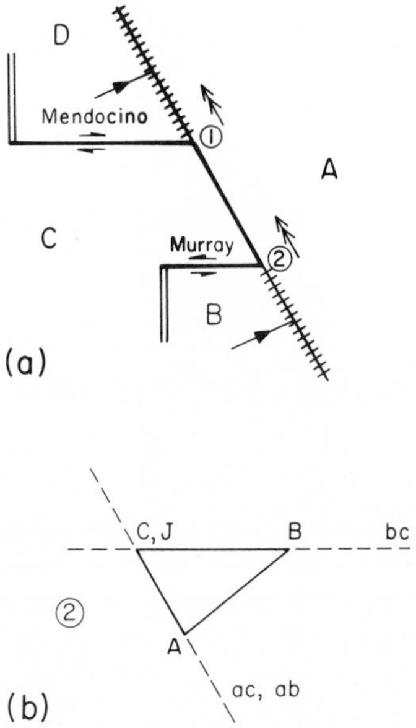

Figure 10-6
When the southern junction (2) in Figure 10-5,b reaches the Murray, the right lateral offset of the ridge must change it from RTF(a) to FFT, and also cause the junction to move north-west, because it is then fixed to plate C. (b) gives the velocity triangle for junction (2) in (a)

the Murray migrates east to meet the trench. The resulting triple junction $RTF(a)$ then continues the earlier slow movement to the southeast relative to the American plate.

The stability of all the junctions on the west coast of North America depends on the trench which was originally on the west coast being parallel to the slip vector between the American and the Main Pacific plates. Fig. 3 of McKenzie and Parker (1967) clearly shows that this condition would have been satisfied by a trench along the continental margin, and is still satisfied by the northern part of the Central American trench. This curious and important coincidence is not easily explained.

The complex series of events described is the inescapable consequence of the motion of rigid plates whose relative velocity remains constant. The geological history of the west coast of America and of the surrounding sea floor since the Cretaceous is in general compatible with the evolution of the plates and triple junctions outlined here. The magnetic lineations off the west coast (Menard and Atwater, 1968; Peter, 1965; Larson et al., 1968; Vacquier et al., 1961; Mason and Raff, 1961; Raff and Mason, 1961) show that the Farallon plate remained intact until the ridge first met the trench at the time of anomaly 10 (Figs. 10–4 and 10–7) or in the Middle Oligocene (Heirtzler et al., 1968; Maxwell, 1969). As expected, the anomalies to the north and south show no change in spreading direction at this time. To the south the anomalies in contact with the continental margin become progressively younger to the present active ridge axis at the mouth of the Gulf of California, with probably a short interruption in the steady progression at the Murray fracture zone. If the motion of the Main Pacific and American plates has remained unchanged since at least the Oligocene, it is possible to determine the relative velocities of all three plates and their associated triple junctions. The relative velocity between the main Pacific plate and the Farallon may be obtained from magnetic lineations older than anomaly 10 (Menard and Atwater, 1968; Heirtzler et al., 1968; Maxwell, 1969), and is 5.0 cm/yr half rate. The fracture zones show that the ridge was at right angles to the east-west relative motion, shown to scale as BC in Fig. 10–5,e. The present motion between the Main Pacific and Ameri-

can plates is close to 6 cm/yr (Morgan, 1968; Vine, 1966; Menard and Atwater, 1968), with the slip vector parallel to the San Andreas. This vector is shown as AC in Fig. 10–5,e. The motion of the Farallon plate towards the American plate is then found to be 7 cm/yr. The motion vector is almost at right angles to the trench ab, with a small left-handed component of 1 cm/yr. This consumption rate is similar to that of the eastern end of the Aleutian arc.

Fig. 10–5,e also determines the relative motion of both triple junctions with respect to all plates. The southern junction (Fig. 10–5,d) moves south-east with a velocity of 1.1 cm/yr relative to the American plate, whereas the northern junction moves north-west with a velocity of 6 cm/yr. The length of the strike slip fault between these junctions therefore increases at 7.1 cm/yr, and if the junctions first formed in the Oligocene 32 million years ago, they should now be 2,270 km apart. This estimate agrees remarkably well with the observed separation of about 2,300 km between the triple junctions at Cape Mendocino and at the mouth of the Gulf of California. This simple calculation is successful because the right handed offset on the Murray fracture zone is almost the same as the left-handed offset on the Molokai (Raff, 1966). Thus the effective velocity of the southern junction has been constant since the time it was formed.

The success of this calculation supports the original assumption that the relative velocities of the major plates have remained unchanged during the Tertiary. It also suggests that the point at which the ridge first met the trench was at the southern end of Baja California, about 350 km north-west of where the southern triple junction is now. It is difficult to understand how the right lateral motion on the San Andreas and related faults can have begun on a large scale before the Oligocene (Hill and Dibblee, 1953), for the small strike slip motion on the trench was left lateral. This evolution of the west coast also suggests that a considerable part of the Franciscan may have been removed from Baja California, where it exists only as isolated outcrops, and added to that of the Coast Ranges.

These observations agree well with the evolutionary outline. In detail, however, the history is much more complicated, principally

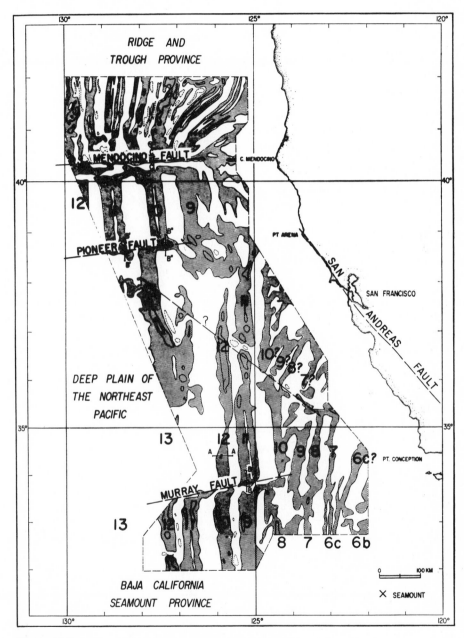

Figure 10-7
Magnetic lineations off central California (Mason and Raff, 1961).
The identification of anomalies followed by a question mark is somewhat in doubt

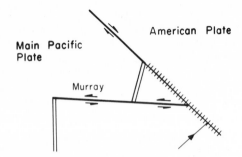

Figure 10-8
The geometry of the plates at about the time of anomaly 8. The magnetic lineations in Figure 10-7 show that the plate to the east of the ridge axis broke into at least two parts, and the simplest explanation is that the eastward continuation of the Murray fault became active

because the Farallon plate did not remain intact during the events described, nor did the resulting pieces continue to move in the same directions, for the Farallon plate was moving before the time of anomaly 10. One such break is apparent in Fig. 10–7, which shows the lineations off central California (Mason and Raff, 1961). If the Murray was active only between ridge crests during the period shown by Fig. 10–5,*b*, then the spreading rate to the north must be the same as that to the south. The fracture zone must then offset each anomaly by the same distance. This is true for anomalies 10 to 12 inclusive, but not for those which are perhaps 8 and 9. Thus some fault to the east became active as a left lateral ridge trench transform fault (Fig. 10–8). This fault may well have been the eastward extension of the Murray, though it is not possible to prove this from the remaining magnetic anomalies. This activity east of the ridge started at the time of anomaly 10, or the same time as the ridge first met the trench south of the Pioneer. This behaviour is very similar to the present activity in the mouth of the Gulf of California, where the extension of the east Pacific rise is in the process of changing strike (Larson et al., 1968), and in so doing has broken a small plate containing Las Tres Marias islands from the Cocos plate to the south (Sykes, 1967). Because the spreading rate to the north of the Murray was less than that to the south, the offset decreased after anomaly 10 was formed, and may have vanished by the time the ridge to the south

reached the trench. Thus the geometry in Fig. 10–6,*a* may well never have existed in this area.

The land geology is at least as complicated as that of the ocean floor, but is less complete because the stratigraphy and tectonics have to be painstakingly reconstructed from careful observations (McKenzie and Parker, 1967; Morgan, 1968a; Le Pichon, 1968; Isacks et al., 1968; Vine, 1966; Pitman et al., 1968; Wilson, 1965c; Barazangi and Dorman, 1969; Sykes, 1967; 1968; Brune, 1968; McKenzie, 1969b; Hamilton and Evison, 1967; Menard and Atwater, 1968; Laughton, 1966; Herron and Heirtzler, 1968; Raff, 1968; Pitman and Hayes, 1968; Vine and Hess, 1970; Gutenberg and Richter, 1954; Peter, 1965; Elvers, 1967a; Matthews, 1966; Morgan et al., 1969; Larson et al., 1968; Vacquier et al., 1961; Mason and Raff, 1961; Raff and Mason, 1961; Raff, 1966; Heirtzler, et al., 1968; Maxwell, 1969), and cannot be obtained simply by towing a magnetometer behind an aeroplane or ship. It is tempting to identify the transform fault between plates *A* and *C* in Figs. 10–5 and 10–6 with the San Andreas. There is, however, an important objection to such a choice. If, as seems likely from palaeomagnetic and other evidence, the motion of the American plate relative to the Main Pacific plate has remained approximately unchanged at 6 cm/yr since anomaly 10 time or for the last 32 million years (Heirtzler et al., 1968; Maxwell, 1969), then the total displacement between the plates since their first contact must be about 2,000 km. The largest postulated displacement on the San Andreas since the Oligocene is about 350 km (Hill and Dibblee, 1953), and therefore the remainder must have been taken up on other faults. Some of these are offshore, and formed a series of transform faults joined by ridges south-west of San Francisco. The youngest of the anomalies produced by these ridges which is visible in Fig. 10–7 is probably between 6 and 7. Such movement can therefore account for 500 km and perhaps more, leaving about 1,200 km still unaccounted. Some of this remaining displacement may be on the Nacimiento, and some on offshore faults. It is not, however, possible to use some of this displacement to create the Central Valley by moving the coast ranges away from the Sierra Nevada. Though such movement after the Oligocene could account for a large part of the missing displacement, and is not in conflict with the seismic (Bateman and Eaton, 1967) and magnetic evidence (Grantz and Zietz, 1960; Griscom, 1966) for an oceanic basement beneath the sediments, it is not consistent with the presence of large thicknesses of Cretaceous sediment found throughout the Valley (Hackel, 1966). Thus the evolution discussed here is no help in understanding how the Central Valley was formed, nor why it extends from the extension of the Murray to that of the Mendocino. It is particularly difficult to understand the relation of the Valley to the oceanic transform faults if the San Andreas in the coast ranges has a displacement of at least 300 km. Thus the evolution of the triple junctions in Figs. 10–5 and 10–6 provides a simplified but probably useful guide to the evolution of the ocean floor off California, but not as yet to that of the land. Perhaps the geology of Baja California will be more revealing.

A different type of triple junction evolution occurs in Japan where the Japan trench divides into the Ryukyu and Bonin arcs (McKenzie and Parker, 1967; Gutenberg and Richter, 1954). Fig. 10–9 is from Gutenberg and Richter (Gutenberg and Richter, 1954) and demonstrates that the deep earthquakes do not follow this division but occur beneath the Bonin arc alone. Along this eastern arc the deep earthquakes occur considerably closer to the trench and island arc than they do farther north beneath the seas of Japan and Okhotsk (Uyeda and Vacquier, 1968). This difference is even more marked if the recent more accurate locations of earthquakes are used (Katsumata and Sykes, 1969). These show that the Benioff zone beneath the Bonin arc is not plane but has a steeply inclined part at intermediate depths, with less steeply inclined parts above and below. Farther south beneath the Marianas islands the earthquake zone becomes vertical. In contrast to this remarkable geometry of the deep earthquake zone beneath the eastern branch, that of the western branch is planar and dips at 45°. This difference between the two zones is probably caused by the evolution of the triple junction in central Japan. If the Ryukyu arc were inactive and the Bonin arc were about 200 km east of its present position, the present position of the deep earthquakes would lie on a plane dipping at 45° into the mantle. The Philippine Sea would have been joined to the plate containing America and Kamchatka. Crustal consumption must then have started along the Ryukyu arc, perhaps about 3 million years ago, to form a $TTT(a)$ junction off northern Japan. Such a junction must migrate southwards, carrying the Bonin arc westwards towards its deep earthquake zone.

Though both examples are from the north Pacific, there seems no reason to believe that other regions are essentially different, though they are less well studied. It is therefore expected that stable triple junctions and their evolution will provide an outline of the evolution of many areas, especially oceanic ones. Indeed, many of the examples suggested in Fig. 10–3 show the effects expected, though few of these have as yet been studied in detail.

Throughout our discussion, velocity, rather than angular velocity triangles, were used. This simplification is justified (McKenzie and Parker, 1967) because the behaviour of a triple junction depends only on the relative motion of the three plates at the point where they meet. A quite different cause of plate evolution does, however, depend on the relative motions being rotations. This type of evolution produces real changes in the relative motion of three plates at a triple junction, and depends on the observation that finite rotations, unlike infinitesimal ones, do not add vectorially. Fig. 10–10,a shows such a junction between three plates A, B and C, and the three axes of relative rotations $_A\omega_B$, $_B\omega_C$ and $_C\omega_A$ which satisfy:

$$_A\omega_B + {_B\omega_C} + {_C\omega_A} = 0 \qquad (3)$$

By definition the points a, b and c where these axes intersect the plates B and C are fixed with respect to A and B, B and C, and C and A respectively. If finite rotations of B and C relative to A take place about the axes $_A\omega_B$ and $_C\omega_A$ and at a rotation rate given by the values of the two vectors, the orientation and magnitude of $_B\omega_C$ will remain constant and fixed relative to A, $_A\omega_B$ and $_C\omega_A$. The original point at which $_B\omega_C$ cuts plate C, b, is not, however, fixed to A but to C, which rotates about $_C\omega_A$ relative to A. Thus the final position of b after finite rotations of B and C relative to A will not be at the intersection of $_B\omega_C$ with C (Fig. 10–10,b). Thus it is not possible for all three plates to rotate through finite angles

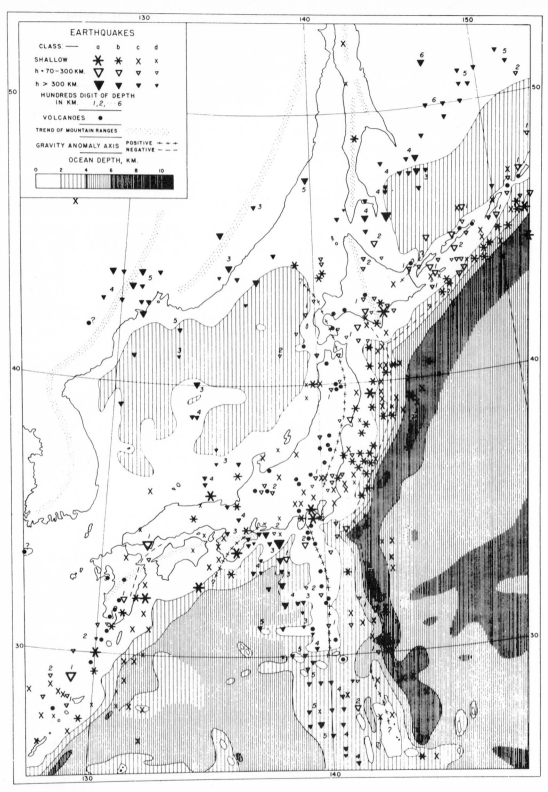

Figure 10-9
The seismicity, vulcanism and water depth in the region of Japan (Gutenberg and Richter, 1954). The branching of all surface features in central Japan does not extend to the deep earthquakes, which follow the Bonin and not the Ryukyu arc. The horizontal separation between the deep earthquakes and the arc is less in the region of the Bonin islands than it is in that of the Kuriles

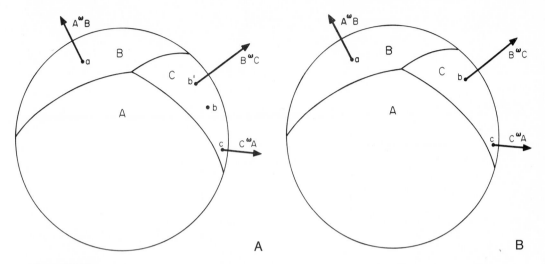

Figure 10-10
(a) shows the angular velocity vectors of plates forming a triple junction. If $A^\omega B$ and $A^\omega C$ are taken to be fixed relative to A throughout the evolution, and their magnitudes also remain constant, then by (3) the direction and magnitude of $B^\omega C$ are constant. The point b is fixed to C and is the point at which $B^\omega C$ initially intersects C. As B and C rotate through finite angles with respect to A the axis of rotation $B^\omega C$ moves with respect to both B and C, and does not continue to pass through the point b on C. (b) shows the geometry after finite rotations

about their instantaneous relative rotation axes. The only special case occurs if all ω vectors lie along the same axes, when finite rotations add as scalars and no changes need take place.

The geometry of plate boundaries suggests that there will be wide variations in their resistance to changes in spreading direction required by Fig. 10–10,*b*. In general, ridges offset by both right and left handed transform faults can change their spreading direction only if one or other of the plates breaks. Such structures will therefore strongly oppose any changes in spreading direction, though not in spreading rate. Ridges with only left handed faults only oppose clockwise changes in the motion direction, because anticlockwise changes can change all transform faults into ridges. The opposite is true of right handed systems. Because the slip vector across trenches is rarely without a strike slip component, and because there seems to be no par-

ticular preferred direction of the slip vector relative to the strike of the trench, they probably offer little resistance to a change in the direction or magnitude of the slip vector. Thus the changes in relative motion caused by finite rotations may often be accommodated by the trenches and not by the ridges. There is no obvious method of separating the consequences of finite rotations from the other causes of velocity changes, so it is not known if it is this mechanism which produced the change in relative motion between the Main Pacific plate and the Farallon plate observed by Menard and Atwater (1968).

These extensions of plate tectonics are geometric results, and are not concerned with the driving forces. The examples illustrate how a complex series of events can be produced by geometry alone, and also how sea floor spreading and plate tectonics may be used to provide a framework for the understanding of the tectonic evolution of continents and oceans.

GEOMAGNETIC REVERSALS:
THE STORY ON LAND

11

Introduction and Reading List

The Founding Fathers

To a paleomagnetist a volcano is a giant magnetic tape recorder. As each lava flow cools, small grains of magnetite become aligned magnetically in the direction of the earth's field. By measuring the remanent magnetism of the lava flows, the paleomagnetist can determine the direction of the earth's field at the time of formation of each lava flow.

The idea that the earth's magnetic field changes its polarity was advanced in the early decades of this century by paleomagnetists who were studying the magnetic properties of volcanic rocks. They noticed that some rocks were naturally magnetized in what is now called a "reversed" direction, 180° from the present direction of the earth's field. They proposed that at the time the rocks were formed, the direction of the earth's field had been opposite to its direction today (Brunhes, 1906; Mercanton, 1926). These early paleomagnetists would have been astonished to learn that half a century later this speculative interpretation of their somewhat shaky data would provide firm observational evidence for the reality of ocean-floor spreading.

The first paleomagnetist to investigate the important question of the timing of magnetic reversals was Motonari Matuyama. During the 1920's he was a professor at the Kyoto Imperial University. In making surveys of magnetic anomalies in Japan, Matuyama found that the magnetic field was less intense over certain volcanic formations than it was elsewhere. This surprised him because volcanic formations contain magnetite, which might be expected to enhance the strength of the local magnetic field. To try to understand his puzzling negative anomalies, Matuyama collected samples from many parts of Japan and measured their magnetism. He found that the magnetism of the rocks fell into two distinct groups. One was directed toward the north and downward, approximately parallel to the present geomagnetic field in Japan. The other group was directed toward the south and upwards in what we would now call a reversed direction.

Then Matuyama took a step that earlier workers had not taken: he looked carefully at the ages of the rocks. Although these were not known very precisely, he was able to determine that the rocks magnetized to the south were *all* of early Pleistocene age or older. From this he deduced that the earth's field had reversed polarity early in the Pleistocene (Chapter 12). Matuyama's estimate of the age of the most recent reversal still looks surprisingly good.

Matuyama's proposal that the magnetic field had once pointed south might seem to be as revolutionary as a proposal that the gravitational field had once pointed up. In 1929, however, when Matuyama published his proposal, geophysicists still did not have a sensible theory for the origin of the geomagnetic field. And so long as geophysicists did not know why the field was pointing north, there was no reason to reject the idea that it might at another time have pointed south. So the response of Matuyama's fellow scientists was neither disbelief nor outrage, but rather silence. (Matuyama eventually left geophysics to become the president of a university and to pursue his other great interest, acting in the No theater.)

The Modern Quest for a Time Scale of Magnetic Reversals

By the late 1950's, experimental research in paleomagnetism had accelerated greatly, as had theoretical research into the origin of the magnetic field. Edward Bullard (1955) and Walter Elsasser (1955) had developed the theory that the magnetic field was produced by the interaction of fluid motions and electrical currents in the earth's core of molten iron. There is no intrinsic reason why a generator of this type cannot reverse polarity. Moreover, Tsuneji Rikitake (1958) had proposed a mechanical analogue of the earth's generator in the form of a two-disc feedback dynamo that had the property of reversing its polarity.

Most of the paleomagnetic research at that time was directed toward the problems of polar wandering and continental drift. Emphasis was placed on rocks of Cretaceous age or older because the amount of continental drift in post-Cretaceous time was too small to be easily detected paleomagnetically.

However several paleomagnetists continued research during the 1950's on extremely young rocks with a view toward determining the age of the most recent reversals, notably Hospers (1953) and Sigurgeirsson (1957) working in Iceland, and Khramov (1957) working in the Soviet Union. I began my own research on reversals in the United States (Cox, 1959) at this time. The main result of this research was to confirm Matuyama's conclusion that the earth's field had reversed at some time in the early Pleistocene. The dating tools then available did not permit more precise determination of the time of this reversal. The main difficulty was that classical paleontologic techniques began to break down when applied to dating within the Pleistocene because the times required for the processes of evolution and dispersal are a sizable fraction of the Pleistocene. An uncertainty of 0.2 million years is trivial in the stratigraphic correlation of Permian rocks 250 million years old, but it is not trivial in correlating early Pleistocene rocks, which are only 1.5 million years old. On the other hand, the radiometric-dating techniques available in the late 1950's were not suitable for dating early Pleistocene rocks because the time constants of the decay processes were either too short (carbon-14 dating) or too long (potassium-argon, strontium-rubidium, and lead-lead dating).

The precise dating of reversals using the potassium-argon method became possible in the late 1950's because of a breakthrough in techniques for cleaning the mass spectrometers used for potassium-argon dating. This is an "hour-glass" type of dating based on the accumulation in the rock of argon produced by the decay of radioactive potassium. Previously rocks younger than 10 million years had not been datable using this method because the mass spectrometers used to measure argon were contaminated with the small amounts of argon in the atmosphere. During an experiment, this atmospheric argon would be released from the inner surfaces of the spectrometer and would overwhelm the small amount of argon released from a young rock. In the mid–1950's John Reynolds at the University of California at Berkeley built an all-glass spectrometer (Reynolds, 1956) that could be heated while under vacuum to drive off the absorbed atmospheric argon. Once decontaminated in this way, Reynolds' spectrometer could measure extremely minute amounts of radiogenic argon very accurately. Garniss Curtis and Jack Evernden at Berkeley were quick to apply this new technology to the dating of young rocks.

Most of the dating used to determine the time scale for geomagnetic reversals was later done by the U.S. Geological Survey and at the Australian National University. However, the paths of most of the scientists involved crossed in Berkeley in the late 1950's and early 1960's, just as the paths of workers in plate tectonics crossed at Cambridge and Princeton in the early 1960's. John Verhoogen on the Berkeley faculty was doing research on paleomagnetism and on the solid-state physics of magnetic self reversal; Richard Doell, Brent Dalrymple, Sherman Gromme, and I were at Berkeley as graduate students; and Ian McDougall was at Berkeley as a post-doctoral fellow. Again, personal contacts among a small group of people seem to have been important in getting a new line of research started.

As a graduate student I began working on reversals in 1956, when James Balsley of the U.S. Geological Survey sent me a dozen samples of basalt from the Snake River Plain, some of which were reversed. At about that time Seiya Uyeda (1958) discovered a volcanic rock in Japan that reproducibly became reversely magnetized when cooled in the present field of the earth. Having two explanations for the single phenomenon of reversed magnetism seemed rather untidy, so for a while many geophysicists thought that all the observed reversed magnetism in rocks was due to mineralogical self reversal. In 1957 I collected

Figure 11-1
Allan Cox collecting samples in the Galapagos Islands using a
portable diamond drill

several hundred samples from different time horizons in the Snake River basalts,
expecting to find, as Uyeda had in Japan, that the magnetic polarity was controlled
by the mineralogy of the rocks. After several years of laboratory work, I found
that the polarity was unrelated to the mineralogy but depended on the age of the
rocks. My volcanic rocks from Idaho were like Matuyama's volcanic rocks from
Japan and not like Uyeda's. Moreover, from the available paleontological evidence, the time of the most recent reversal in Idaho seemed to be within the
Pleistocene, as Matuyama had found in Japan. Similar results were being found
by scientists in other countries. Apparently self reversal was a rare phenomenon.
Most reversely magnetized rocks had formed at a time when the field was reversed.

The obvious next step (Chapter 13) was to try to date the rocks more accurately, using the potassium-argon method. By 1961 Doell and I were working

Figure 11-2
Richard Doell collecting samples on the Hawaiian Islands

together on paleomagnetism at the U.S. Geological Survey at Menlo Park, while Dalrymple, Gromme, and McDougall were doing graduate and post-graduate work at Berkeley, about an hour's drive away. At that time the Berkeley laboratory was the only laboratory in the world capable of dating young volcanic rocks precisely and there were a number of more important problems to work on. Dalrymple, for example, was learning how to date young volcanic rocks for the purpose of dating the uplift of the Sierra Nevada mountain range and McDougall was learning dating for the purpose of dating the migration of volcanism along the Hawaiian Island chain. Not having direct access to a dating laboratory, Doell and I began measuring the polarity of rocks that had been dated for other purposes, which resulted in the first tentative time scale for reversals (Chapter 14).

During late 1960 and early 1961 the Menlo Park and Berkeley groups had discussions about the need to obtain more potassium-argon and paleomagnetic data from young volcanic rocks. On his return trip to the Australian National University at Canberra, McDougall collected samples for dating from the Hawaiian Islands. He then set up a potassium-argon dating laboratory and began

Figure 11-3
G. Brent Dalrymple and brown boobies on Nihoa Island

working on the reversal problem with the paleomagnetists Don Tarling and Francois Chamalaun at the Australian National University. Dalrymple joined Doell and me at the U.S. Geological Survey in 1963 to concentrate more fully on the reversal problem and, like McDougall, constructed a dating laboratory modeled after the laboratory at Berkeley.

For four years beginning in 1963, there was a friendly spirit of competition between the Menlo Park and Canberra groups in trying to unravel the history of reversals, with both groups contributing about equally to the final solution. At times the competition was rather lively, resembling a long distance chess game in which the two sides communicated via letters to *Nature* and *Science*. Something of the spirit of this game may be seen in the succession of reversal time scales shown in Fig. 11–5. The two groups quickly amassed an enormous amount of data from Alaska, California, New Mexico, the Hawaiian Islands, Cocos Island, the Galapagos Islands, Mauritius Island, Reunion Island, Australia, and Iceland. After ironing out some initial discrepancies the two data sets began to agree remarkably well, and by 1966 successive versions of the time scale were clearly converging.

An unexpected result to come from this work was the discovery of short polarity events. At first a fairly regular period between reversals of about one million

Figure 11-4
Ian McDougall (left) and Francois Chamalaun

years seemed likely on the basis of earlier studies of undated rocks. In the first few studies using radiometrically dated rocks, this seemed to be the pattern. Then in 1964 discrepancies began showing up. Gromme and Hay (1963) found a 1.9 million year old normally magnetized lava flow in Olduvai Gorge that "should" have been reversed. At first it seemed that this discrepancy might be due to a dating error or self reversal. However in 1964 our group at Menlo Park discovered a second normally magnetized lava flow with an age of 1.9 million years from the other side of the world on the Pribilof Islands in the Bering Sea (Cox et al., 1964b). This was either an odd coincidence or it meant that a heretofore unsuspected fine-structure was present in the reversal sequence. We decided to trust our data and proposed that short polarity events existed, splitting up the previously recognized long epochs (see the time scale labeled June 26, 1964, in Fig. 11–5). Evidence for these and other short events were soon found by McDougall and his colleagues in Australia. The reality of these short events is now generally accepted.

With the discovery of evidence for more and more reversals, it soon became clear that the length of time between reversals varies continuously over a wide range from very short events to long polarity epochs. There is no natural group-

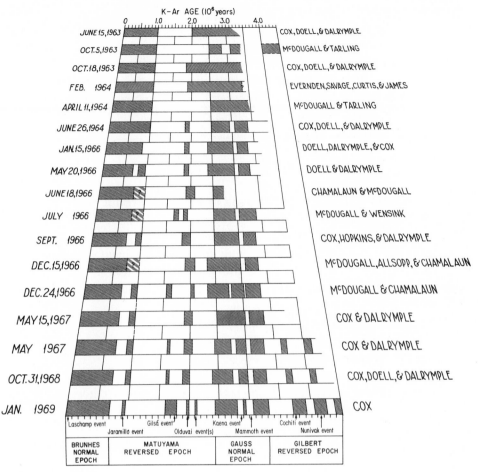

Figure 11-5
Successive versions of the geomagnetic-reversal time scale as determined from the potassium-argon dating of volcanic rocks from the continents. Shaded rectangles, period of normal polarity. Unshaded rectangles, periods of reversed polarity. (From Dalrymple, G. B., Potassium-argon dating of geomagnetic reversals and North American glaciations, in Bishop, W. W., and Miller, J. A., eds., *Calibration of Hominid Evolution.* Reprinted with permission of the author and the publisher, Scottish Academic Press, Edinburgh. Copyright 1972)

ing into a bimodal distribution of events and epochs. For stratigraphic purposes, however, it has proved useful to retain the two-term nomenclature. In most deep-sea sediments, for example, the main polarity epochs are recognizable in the paleomagnetic record, but the short events are not.

By the late 1960's the time scale of reversals had been extended back about 4.5 million years, and research to extend it back further was discontinued. As the time scale of reversals has a number of useful applications, the question naturally arises, why hasn't it been extended further back? The answer is that for rocks older than about 4.5 million years the precision of potassium-argon dating isn't adequate for the requisite age determinations (Cox and Dalrymple, 1967b).

There have been two important applications of the chronology of reversals. The first was to provide the calibration needed for the interpretation of marine magnetic anomalies, as discussed in Section V; the second was to provide the calibration for the stratigraphic correlation of late Pliocene and Pleistocene rocks. The latter has proved extremely useful for correlating and dating deep-sea sediments. The leader in this important research has been Neil Opdyke at Lamont-Doherty Geological Observatory. Articles on this new branch of research have not been included in the present volume, but a list of them is given at the end of this chapter.

The Cause of Reversals and Changes in Reversal Frequency

Why does the earth's field reverse? Before answering this question it is necessary to consider the more basic question, why does the earth have a magnetic field? The answer is that magnetic fields are generated by magnetohydrodynamic (MHD) processes not only in the earth but in other planets and in many stars that have the following properties:
1. They are partially or entirely in the fluid state.
2. The fluid is a good electrical conductor.
3. The fluid is in a turbulent or convective state.
4. The star or planet is rotating.

The basic idea of the MHD dynamo is that an electrical current will be induced in the conducting fluid as it moves across any magnetic field. Under certain conditions the current will, in turn, produce a magnetic field that reinforces the original field. The feedback system is a very complicated one and is, as yet, understood only for the simplest types of fluid motion. Mathematical solutions have not yet been obtained to describe the way a MHD dynamo would operate in the earth's core.

Geophysicists like to obtain some sort of mathematical model to describe their observations. Since the MHD dynamo problem has proved to be rather intractable, they have tried to make a model of geomagnetic reversals indirectly in two ways.

The first is to obtain a complete mathematical solution for a simple mechanical analogue of the earth's dynamo. The particular mechanical dynamo most commonly analyzed consists of two rotating metal discs interconnected with coils of wire (Rikitake, 1958; Allan, 1962). The mathematical descriptions of these mechanical dynamos show that under some conditions the current flowing in the wires changes direction; that is, it reverses. A second approach to creating a model of reversals is to regard them as a random or stochastic process (Cox, 1968; Parker, 1969; Nagata, 1969). In this approach, reversals are linked to large-scale convection cells in the earth's core, which are analogous to cyclones in the earth's atmosphere. The link between these cyclones and reversals is based on two assumptions. (1) At any instant, cyclonic convection cells are randomly distributed through the earth's liquid core. (2) Reversals occur when cyclones in the liquid, through random processes, arrive at certain critical configurations. The probability of occurrence of such a configuration need not be high. In fact, given the short lifetime of individual cyclones (about a thousand years) the probability of a reversal during any particular random configuration *must* be low.

The exact nature of the configuration of cyclones needed to produce a reversal is still subject to speculation. I have suggested (Cox, 1968) that the reversal process may be linked both to the configuration of cyclones and to the well-known large fluctuations in the intensity. Nagata (1969) has suggested that when the cyclones randomly assume a highly symmetrical configuration there may be a collapse of the field and a reversal. Parker (1969) suggests that a reversal will occur if the cyclones all happen to be located near the geographic poles. All of these variations of the basic stochastic model use the assumption that reversals are triggered by the independent, random movement of convection cyclones in the core (Cox, 1970). Research is continuing actively in an attempt to obtain better physical and mathematical models for the reversal process (see reading list).

Although the length of time between successive polarity changes is now known to be highly irregular, it is still meaningful to speak of the "frequency of reversals" in the sense of the average number of polarity changes per million years. Has this frequency remained constant throughout geologic time? The answer is no, it has clearly changed. About 45 million years ago the reversal frequency doubled to its present rate of 5 reversals per million years (Fig. 25–5). There was a similar increase at the end of the Cretaceous Period. Looking at the entire paleomagnetic record, the frequency of reversals has fluctuated widely, both increasing and decreasing many times during the past two billion years. Similarly there have been long intervals when the field was in one direction most of the time and long intervals when it was predominantly in the other. These shifts in the predominant polarity of the field appear to occur at intervals varying from about 50 million years to several hundred million years.

Although individual reversals are probably due entirely to fluid motions in the earth's core, the shifts in *average* reversal frequency are probably due to changes in the boundary conditions at the core-mantle interface. Convection in the lower mantle, for example, could bring cold mantle material in a slowly descending convection column to the core-mantle interface. This cold spot would partially control the fluid motion in the core (Cox and Doell, 1964). Alternatively the denser material in the descending column might produce a bump, which also might partially control the fluid motion in the core (Hide, 1967). The number of such hot spots or bumps and their positions relative to each other and to the rotation axis would determine the average reversal frequency and the proportion of the time the field is normal. If this interpretation is correct, there is the intriguing possibility of a direct link between large-scale tectonic processes such as ocean-floor spreading at the earth's surface and geomagnetic reversals in the earth's core. The link would be provided by slow convective overturn in the earth's lower mantle.

READING LIST

References of Historical Interest

Brunhes, B., 1906, Recherches sur la direction d'aimentation des roches volcaniques (1): J. Physique, 4e sér., v. 5, p. 705–724.

Förstemann, F. C., 1859, Ueber den Magnetismus der Gesteine: Ann. Physik, ser. 4, v. 106, p. 106–136.

Melloni, M., 1853, Sur l'aimentation des roches volcaniques: Acad. Sci. Paris, C. R., v. 37, p. 229–231.

Mercanton, P. L., 1910, Physique du globe: Acad. Sci. Paris, C. R., v. 151, p. 1092–1097.

Mercanton, P. L., 1926a, Inversion de l'inclinaison magnétique terrestre aux âges géologiques: J. Geophys. Res., v. 31, p. 187–190.

Mercanton, P. L., 1926b, Magnétisme terrestre—Aimentation de basaltes groenlandais: Acad. Sci. Paris, C. R., v. 182, p. 859–860.

Radiometric Dating of Reversals

Chamalaun, F. 'I., and McDougall, I., 1966, Dating geomagnetic polarity epochs in Reunion: Nature, v. 210, p. 1212–1214.

Cox, A., and Dalrymple, G. B., 1967a, Geomagnetic polarity epochs—Nunivak Island, Alaska: Earth Planet. Sci. Lett., v. 3, p. 173–177.

Cox, A., Doell, R. R., and Dalrymple, G. B., 1963b, Geomagnetic polarity epochs: Sierra Nevada II: Science, v. 142, p. 382–385.

Cox, A., Doell, R. R., and Dalrymple, G. B., 1964a, Geomagnetic polarity epochs: Science, v. 143, p. 351–352.

Cox, A., Doell, R. R., and Dalrymple, G. B., 1968, Radiometric time scale for geomagnetic reversals: Geol. Soc. London, Quart. J., v. 124, p. 53–66.

Cox, A., Hopkins, D. M., and Dalrymple, G. B., 1966, Geomagnetic polarity epochs: Pribilof Islands, Alaska: Geol. Soc. Amer., Bull., v. 77, p. 883–910.

Dalrymple, G. B., 1963, Potassium-argon dates of some Cenozoic volcanic rocks of the Sierra Nevada, California: Geol. Soc. Amer., Bull., v. 74, p. 379–390.

Dalrymple, G. B., Cox, A., Doell, R. R., and Grommé, C. S., 1967, Pliocene geomagnetic polarity epochs: Earth Planet. Sci. Lett., v. 2, p. 163–173.

Doell, R. R., and Dalrymple, G. B., 1966, Geomagnetic polarity epochs: A new polarity event and the age of the Brunhes-Matuyama boundary: Science, v. 152, p. 1060–1061.

Doell, R. R., Dalrymple, G. B., and Cox, A., 1966, Geomagnetic polarity epochs: Sierra Nevada data, 3: J. Geophys. Res., v. 71, p. 531–541.

Doell, R. R., Dalrymple, G. B., Smith, R. L., and Bailey, R. A., 1968, Paleomagnetism, potassium-argon ages, and geology of rhyolites and associated rocks of the Valles Caldera, New Mexico: Geol. Soc. Amer., Mem. 116, p. 211–248.

Evernden, J. F., Savage, D. E., Curtis, G. H., and James, G. T., 1964, Potassium-argon dates and the Cenozoic mammalian chronology of North America: Amer. J. Sci., v. 262, p. 145–198.

Grommé, C. S., and Hay, R. L., 1963, Magnetizations of basalt of Bed I, Olduvai Gorge, Tanganyika: Nature, v. 200, p. 560–561.

Grommé, C. S., and Hay, R. L., 1967, Geomagnetic polarity epochs – new data from Olduvai Gorge, Tanganyika: Earth Planet. Sci. Lett., v. 2, p. 111–115.

Grommé, C. S., and Hay, R. L., 1971, Geomagnetic polarity epochs: age and duration of the Olduvai normal polarity event: Earth Planet. Sci. Lett., v. 10, p. 179–185.

McDougall, I., Allsopp, H. L., and Chamalaun, F. H., 1966, Isotopic dating of the newer volcanics of Victoria, Australia, and geomagnetic polarity epochs: J. Geophys. Res., v. 71, p. 6107–6118.

McDougall, I., and Chamalaun, F. H., 1966, Geomagnetic polarity scale of time: Nature: v. 212, p. 1415–1418.

McDougall, I., and Chamalaun, F. H., 1969, Isotopic dating and geomagnetic polarity studies on volcanic rocks from Mauritius, Indian Ocean: Geol. Soc. Amer., Bull., v. 80, p. 1419–1442.

McDougall, I., and Tarling, D. H., 1964, Dating geomagnetic polarity zones: Nature, v. 202, p. 171–172.

McDougall, I., and Wensink, H., 1966, Paleomagnetism and geochronology of the Plio-cene-Pleistocene lavas in Iceland: Earth Planet. Sci. Lett., v. 1, p. 232–236.

Watkins, N. D., Gunn, B. M., Baksi, A. K., York, D., and Ade-Hall, J., 1971, Paleo-magnetism, geochemistry, and potassium-argon ages of the Rio Grande de Santiago volcanics, central Mexico: Geol. Soc. Amer., Bull., v. 82, p. 1955–1968.

Reversal Stratigraphy

Berggren, W. A., 1969a, Cenozoic chronostratigraphy, planktonic foraminiferal zonation and the radiometric time scale: Nature, v. 224, p. 1072–1075.

Berggren, W. A., Phillips, J. D., Bertels, A., and Wall, D., 1967, Late Pliocene-Pleisto-cene stratigraphy in deep sea cores from the south-central North Atlantic: Nature, v. 216, p. 253–255.

Bucha, V., 1970, Geomagnetic reversals in Quaternary revealed from a paleomagnetic investigation of sedimentary rocks: J. Geomag. Geoelec., v. 22, p. 253–272.

Burek, P. J., 1970, Magnetic reversals: their application to stratigraphic problems: Amer. Ass. Petrol. Geol., Bull., v. 54, p. 1120–1139.

Clark, D. L., 1970, Magnetic reversals and sedimentation rates in the Arctic Ocean: Geol. Soc. Amer., Bull., v. 81, p. 3129–3134.

Cox, A., and Doell, R. R., 1968, Paleomagnetism and Quaternary correlation, in Morri-son, R. B., and Wright, H. E., Jr., eds., Means of correlation of Quaternary succes-sions: Int. Ass. Quater. Res., 7th Congr. (1965), Proc., v. 8, p. 253–265.

Cox, A., Doell, R. R., and Dalrymple, G. B., 1965, Quaternary paleomagnetic stratig-raphy, in Wright, H. E., Jr., and Frey, D. G., eds., The Quaternary of the United States: Princeton, N. J., Princeton Univ. Press, p. 817–830.

Creer, K. M., 1971, Mesozoic paleomagnetic reversal column: Nature, v. 233, p. 545–546.

Dickson, G. O., and Foster, J. H., 1966, The magnetic stratigraphy of a deep-sea core from the North Pacific Ocean: Earth Planet. Sci. Lett., v. 1, p. 458–462.

Evans, A. L., 1970, Geomagnetic polarity reversals in a late Tertiary lava sequence from the Akaroa volcano, New Zealand: Roy. Astron. Soc., Geophys. J., v. 21, p. 163–183.

Foster, J. H., and Opdyke, N. D., 1970, Upper Miocene to Recent magnetic stratigraphy in deep-sea sediments: J. Geophys. Res., v. 75, p. 4465–4473.

Glass, B., Ericson, D. B., Heezen, B. C., Opdyke, N. D., and Glass, J. A., 1967, Geomagnetic reversals and Pleistocene chronology: Nature, v. 216, p. 437–442.

Goodell, H. G., and Watkins, N. D., 1968, The paleomagnetic stratigraphy of the Southern Ocean: 20° West to 160° East longitude: Deep-Sea Res., v. 15, p. 89–112.

Harrison, C. G. A., 1966, The paleomagnetism of deep sea sediments: J. Geophys. Res., v. 71, p. 3033–3043.

Hays, J. D., and Berggren, W. A., 1971, Quaternary boundaries and correlations, *in* Funnell, B. M., and Riedel, W. R., eds., Micropaleontology of oceans: Cambridge, Cambridge Univ. Press, p. 669–691.

Hays, J. D., Saito, T., Opdyke, N. D., and Burckle, L. H., 1969, Pliocene-Pleistocene sediments of the equatorial Pacific: their paleomagnetic, biostratigraphic, and climatic record: Geol. Soc. Amer., Bull., v. 80, p. 1481–1514.

Helsley, C. E., 1969, Magnetic reversal stratigraphy of the Lower Triassic Moenkopi formation of western Colorado: Geol. Soc. Amer., Bull., v. 80, p. 2431–2450.

Hoare, J. M., Condon, W. H., Cox, A., and Dalrymple, G. B., 1968, Geology, paleomagnetism, and potassium-argon ages of volcanic rocks from Nunivak Island, Alaska: Geol. Soc. Amer., Mem. 116, p. 377–413.

Hospers, J., 1954, Magnetic correlation in volcanic districts: Geol. Mag., v. 91, p. 352–360.

Irving, E., 1971, Nomenclature in magnetic stratigraphy: Roy. Astron. Soc., Geophys. J., v. 24, p. 529–531.

Johnson, A. H., Nairn, A. E. M., and Peterson, D. N., 1972, Mesozoic reversal stratigraphy: Nature, v. 237, p. 9–10.

Kennet, J. P., Watkins, N. D., and Vella, P., 1971, Paleomagnetic chronology of Pliocene-Early Pleistocene climates and the Plio-Pleistocene boundary in New Zealand: Science, v. 171, p. 276–279.

Khramov, A. N., 1957, Paleomagnetism—the basis of a new method of correlation and subdivision of sedimentary strata: Acad. Sci. USSR, Dokl., Earth Sci. Sect., v. 112, p. 129–132.

McElhinny, M. W., and Burek, P. J., 1971, Mesozoic palaeomagnetic stratigraphy: Nature, v. 232, p. 98–102.

Morrison, R. B., and Wright, H. E., Jr., eds., 1968, Means of correlation of Quaternary successions: Int. Ass. Quater. Res., 7th Congr. (1965), Proc., v. 8: Salt Lake City, Utah Univ. Press, 631 p.

Nakagawa, H., Niitsuma, N., and Hayasaka, I., 1969, Late Cenozoic geomagnetic chronology of the Boso Peninsula: Geol. Soc. Japan, J., v. 75, p. 267–280.

Opdyke, N. D., 1972, Paleomagnetism of deep-sea cores: Rev. Geophys. Space Phys., v. 10, p. 213–249.

Opdyke, N. D., and Foster, J. H., 1970, The paleomagnetism of cores from the North Pacific, *in* Hays, J. D., ed., Geological investigations of the North Pacific: Geol. Soc. Amer., Mem. 126, p. 83–120.

Opdyke, N. D., and Glass, B. P., 1969, The paleomagnetism of sediment cores from the Indian Ocean: Deep-Sea Res., v. 16, p. 249–261.

Opdyke, N. D., Glass, B. P., Hays, J. D., and Foster, J. H., 1966, Paleomagnetic study of Antarctic deep-sea cores: Science, v. 154, p. 349–357.

Opdyke, N. D., Ninkovich, D., Lowrie, W., and Hays, J. D., 1972, The paleomagnetism of two Aegean deep-sea cores: Earth Planet. Sci. Lett., v. 14, p. 145–159.

Pecherski, D. M., 1970, Paleomagnetism and paleomagnetic correlation of Mesozoic formations of north-east USSR (in Russian): Akad. Nauk SSSR, Sib. Otd., Sev.-Vost. Kompleks. Nauch.-Issled. Inst., Tr., v. 37, p. 58–114.

Pavzner, M. A., 1970, Paleomagnetic studies of Pliocene-Quaternary deposits of Pridniestrovie: Palaeogeogr. Palaeoclimatol. Palaeoecol., v. 8, p. 215–219.

Phillips, J. D., Berggren, W. A., Bertels A., and Wall, D., 1968, Paleomagnetic stratigraphy and micropaleontology of three deep sea cores from the central North Atlantic Ocean: Earth Planet. Sci. Lett., v. 4, p. 118–130.

Picard, M. D., 1964, Paleomagnetic correlation of units within Chugwater (Triassic) formation, west-central Wyoming: Amer. Ass. Petrol. Geol., Bull., v. 48, p. 269–291.

Rutten, M. G., 1960, Paleomagnetic dating of younger volcanic series: Geol. Rundschau, Sonderdr., v. 49, p. 161–167.

Rutten, M. G., and Wensink, H., 1960, Paleomagnetic dating, glaciation and the chronology of the Plio-Pleistocene in Iceland: Int. Geol. Congr., 21st., Proc., pt. 4, p. 62–70.

Smith, J. D., and Foster, J. H., 1969, Geomagnetic reversal in Brunhes normal polarity epoch: Science, v. 163, p. 365–367.

Steuerwald, B. A., Clark, D. L., and Andrews, J. A., 1968, Magnetic stratigraphy and faunal patterns in Arctic Ocean sediments: Earth Planet. Sci. Lett., v. 5, p. 79–85.

Tarling, D. H., 1962, Tentative correlation of Samoan and Hawaiian Islands using "reversals" of magnetization: Nature, v. 196, p. 882–883.

Van Montfrans, H. M., 1971, Paleomagnetic dating in the North Sea basin: Earth Planet. Sci. Lett., v. 11, p. 226–235.

Van Montfrans, H. M., and Hospers, J., 1969, A preliminary report on the stratigraphical position of the Matuyama-Brunhes geomagnetic field reversal in the Quaternary sediments of the Netherlands: Geol. Mijnbouw, v. 48, p. 565–572.

Watkins, N. D., 1972, Review of the development of the geomagnetic polarity time scale and discussion of prospects for its finer definition: Geol. Soc. Amer., Bull., v. 83, p. 551–574.

Watkins, N. D., and Abdel-Monem, A., 1971, Detection of the Gilsa geomagnetic polarity event on the island of Madeira: Geol. Soc. Amer., Bull., v. 82, p. 191–198.

Watkins, N. D., and Goodell, H. G., 1967a, Confirmation of the reality of the Gilsa geomagnetic polarity event: Earth Planet. Sci. Lett., v. 2, p. 123–129.

Reversals and Faunal Extinctions

Black, D. I., 1967, Cosmic ray effects and faunal extinctions at geomagnetic field reversals: Earth Planet. Sci. Lett., v. 3, p. 225–236.

Harrison, C. G. A., 1968a, Evolutionary processes and reversals of the earth's magnetic field: Nature, v. 217, p. 46–47.

Hays, J. D., 1971, Faunal extinctions and reversals of the earth's magnetic field: Geol. Soc. Amer., Bull., v. 82, p. 2433–2447.

Hays, J. D., and Opdyke, N. D., 1967, Antarctic radiolaria, magnetic reversals, and climatic change: Science, v. 158, p. 1001–1010.

Kennet, J. P., and Watkins, N. D., 1970, Geomagnetic polarity change, volcanic maxima and faunal extinction in the southern ocean: Nature, v. 227, p. 930–934.

Sagan, C., 1965, Is the early evolution of life related to the development of the earth's core?: Nature, v. 206, p. 448.

Simpson, J. F., 1966, Evolutionary pulsations and geomagnetic polarity: Geol. Soc. Amer., Bull., v. 77, p. 197–203.

Uffen, R. J., 1963, Influence of the earth's core on the origin and evolution of life: Nature, v. 198, p. 143–144.

Waddington, C. J., 1967, Paleomagnetic field reversals and cosmic radiation: Science, v. 158, p. 913–915.

Watkins, N. D., and Goodell, H. G., 1967b, Geomagnetic polarity changes and faunal extinction in the Southern Ocean: Science, v. 156, p. 1083–1089.

Changes in Reversal Frequency

Cox, A., 1968, Lengths of geomagnetic polarity intervals: J. Geophys. Res., v. 73, p. 3247–3260.

Cox, A., and Cain, J. C., 1972, International conference on the core-mantle interface: EOS (Amer. Geophys. Union, Trans.), v. 53, p. 591–623.

Harrison, C. G. A., 1969, What is the true rate of reversals of the earth's magnetic field?: Earth Planet. Sci. Lett., v. 6, p. 186–188.

Heirtzler, J. R., Dickson, G. O., Herron, E. M., Pitman, W. C., III, and Le Pichon, X, 1968, Marine magnetic anomalies, geomagnetic field reversals, and motions of the ocean floor and continents: J. Geophys. Res., v. 73, p. 2119–2136.

Helsely, C. E., and Steiner, M. B., 1969, Evidence for long intervals of normal polarity during the Cretaceous period: Earth Planet. Sci. Lett., v. 5, p. 325–332.

McElhinny, M. W., 1971, Geomagnetic reversals during the Phanerozoic: Science, v. 172, p. 157–159.

Stochastic Models for Reversals

Cox, A., 1968, Lengths of geomagnetic polarity intervals: J. Geophys. Res., v. 73, p. 3247–3260.

Cox, A., 1970, Reconciliation of statistical models for reversals: J. Geophys. Res., v. 75, p. 7501–7503.

Cox, A., and Cain, J. C., 1972, International conference on the core-mantle interface: EOS (Amer. Geophys. Union, Trans.), v. 53, p. 591–623.

Crain, I. K., and Crain, P. L., 1970, New stochastic model for geomagnetic reversals: Nature, v. 228, p. 39–41.

Crain, I. K., Crain, P. L., and Plaut, M. G., 1969, Long period Fourier spectrum of geomagnetic reversals: Nature, v. 223, p. 283.

Nagata, T., 1969, Length of geomagnetic polarity intervals (discussion of papers by A. Cox, 1968, 1969): J. Geomag. Geoelec., v. 21, p. 701–704.

Naidu, P. S., 1971, Statistical structure of geomagnetic field reversals: J. Geophys. Res., v. 76, p. 2649–2662.

Parker, E. N., 1969, The occasional reversal of the geomagnetic field: Astrophys. J., v. 158, p. 815–827.

Deterministic Models for Reversals

Allan, D., 1958, Reversals of the earth's magnetic field: Nature, v. 182, p. 469–470.

Allan, D. W., 1962, On the behaviour of systems of coupled dynamos: Cambridge Phil. Soc., Proc., v. 58, p. 671–693.

Bullard, E. C., 1955, The stability of a homopolar dynamo: Cambridge Phil. Soc., Proc., v. 51, p. 744–760.

Bullard, E. C., 1968, Reversals of the earth's magnetic field: The Bakerian Lecture, 1967: Roy. Soc. London, Phil. Trans., ser. A, v. 263, p. 481–524.

Hide, R., 1967, Motions of the earth's core and mantle, and variations of the main geomagnetic field: Science, v. 157, p. 55–56.

Hide, R., and Roberts, P. H., 1961, The origin of the main geomagnetic field: Phys. Chem. Earth, v. 4, p. 27–98.

Mathews, J. H., and Gardner, W. K., 1963, Field reversals of "paleomagnetic" type in coupled disk dynamos: U.S. Naval Res. Lab., Rep. 5886, p. 1–11.

Rikitake, T., 1958, Oscillations of a system of disk dynamos: Cambridge Phil. Soc., Proc., v. 54, p. 89–105.

A more complete set of references on geomagnetic dynamos is given in the following article:

Cox, A., and Cain, J. C., 1972, Internation conference on the core-mantle interface: EOS (Amer. Geophys. Union, Trans.), v. 3, p. 591–623.

12

On the Direction of Magnetisation of Basalt in Japan, Tyôsen and Manchuria

MOTONORI MATUYAMA
1929

From *Japan Academy Proceedings*, v. 5, p. 203–205, 1929. Reprinted with permission of the Japan Academy.

Early in April, 1926, a specimen of basalt from Genbudô, Tazima, a celebrated basalt cave, was collected for the purpose of examining its magnetic properties. Its orientation was carefully measured in its natural position before it was removed. When this block was tested by bringing near to a freely suspended magnetic needle, its magnetic north pole was found to be directed to the south and above the horizontal direction. This is nearly opposite to the present earth's magnetic field at the locality. In May of the same year, four specimens of basalt were collected from Yakuno, Tanba, with the similar care. When tested, their magnetic axes were found to have an easterly declination of some 20° and a downward inclination of some 50°.

Since the time of Melloni (Förstemann, 1859) it is believed that lava gets its magnetism in cooling in the direction of the earth's magnetic field. This was also proved experimentally by Prof. Nakamura (Nakamura and Kikuchi, 1912). The above mentioned places are not much distant from each other and have nearly the same magnetic field. These basalts are described as the lavas of probably Quarternary eruptions.

Since that time 139 specimens of basalt were collected from 36 places in Honsyû, Kyûsyû, Tyôsen and Manchuria, of which 38 specimens were already examined more accurately. The method depended upon Gauss's principle of analysing the earth's magnetism, which was also used by Prof. Nakamura. The specimen was enclosed and fixed in a spherical surface in such a way that its orientation could be read from outside. Distribution of the normal component of magnetic force on the surface of the sphere due to the enclosed basalt was determined by means of a magnetometer and the direction of magnetic axis and the intensity of magnetisation were determined by the method of harmonic analysis.

To see the degree of reliability, 4 specimens from Genbudô were examined. As the result

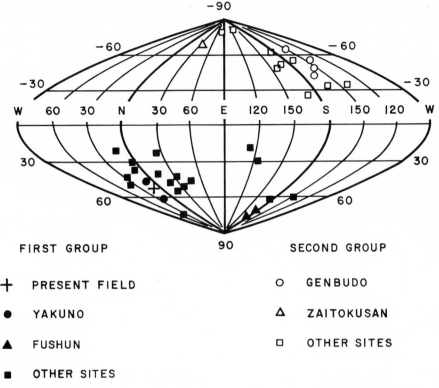

FIRST GROUP

+ PRESENT FIELD

● YAKUNO

▲ FUSHUN

■ OTHER SITES

SECOND GROUP

○ GENBUDO

△ ZAITOKUSAN

□ OTHER SITES

Figure 12-1
Direction of north magnetic pole of basalt

their magnetic axes were found to have a mean westerly declination of 150° and a mean upward inclination of 49°, the largest deviations being 12° in the former and 14° in the latter.

Care was also paid to the tilting of the crust after the basalt sheet was laid out. At Fushun, Manchuria, the specimens were taken from a dolerite sheet which was conformably under and overlaid by a shale strata dipping 25° to the north with a nearly eastwesterly strike. Hence due corrections were made for the measured direction of magnetic axes of these specimens.

The measured directions of magnetic north pole are plotted in the figure. The earth's magnetic field in the related area has a westerly declination varying from 4.5° to 6.5° and a downward inclination from 40.5° to 60°. Since much larger variations of angles are concerned here, the mean direction was considered to prevail over this area.

From the figure, a peculiar arrangement can easily be noticed. There is a group of specimens, including that from Yakuno, whose directions of magnetisation falls around the present earth's field. A number of other specimens, including that from Genbudô, forms another group almost exactly antipodal to the former. The remaining specimens which belong to neither of these groups are not so numerous so far as it was examined.

The age of eruption of the collected basalts are not always clearly known. Among the specimens of the first group, those from the northwestern Kyûsyû are described to be of post-Tertiary and that from Tansen, Tyôsen, is believed to be of Quarternary period. None of them is known to be older than the beginning of the Quarternary; they might be younger. In the second group, the basalt of Genbudô is considered to be also of Quarternary eruption. One of the specimens from Kissyû, Tyôsen, is reported to be of Pleistocene while the other is definitely older, since it is surrounded and a part covered by the first. Thus we may consider that in the earlier part of the Quarternary period the earth's magnetic field in the area under consideration was probably in the state represented by the second group, which gradually changed to the state represented by the first group.

There are three examined specimens of basalt known as of Tertiary eruption. Two from Fushun, Manchuria, had mean westerly declination of 168° and downward inclination of 77°. They are described to be of Miocene period. A specimen from Zaitokusan, Tyôsen, is reported to form the base of the Sitihôzan group unconformably covering the probably Oligocene group, and may be of Miocene.

When examined this specimen had easterly declination of 35° and upward inclination of 70°. Thus there is also a roughly antipodal relation.

According to Mercanton the earth's magnetic field was probably in a greatly different or nearly opposite state in Permo-carboniferous and Tertiary ages compared to the present. From my results it seems as if the earth's magnetic field in the present area has changed even to opposite direction in comparatively shorter duration, in Miocene and also in Quarternary periods.

13

Paleomagnetic Reconnaissance of Mid-Italian Volcanoes

MARTIN G. RUTTEN
1959

From *Geologie en Mijnbouw*, v. 21, p. 373–374, 1959.
Reprinted with permission of the publisher.

In the centre of Italy a series of five large extinct quaternary volcanoes is found. Four of them lie to the north of Rome, and one to the south of this city. Rome is built on the southern-most products of number four and the northern-most fan of number five, occupying an exceptional position in regard to this quaternary volcanism.

Four of these volcanoes have beautiful crater lakes. Their names are easily found on all topographic maps, but some of the volcanoes have received in literature names different from their crater lakes. So they will be enumerated here for the sake of convenience.

In the north, we find volcano Vulsini with the central crater lake Lago di Bolsena. Further south, the volcano Cimini still forms a volcanic cone, without a crater lake. In conjunction to Cimini, volcano Vico with the Lago di Vico is found. Then follows volcano Sabatini with the central crater lake Lago di Bracciano. At last, to the south of Rome, we find the Volcano Laziale with two lateral crater lakes, Lago di Albano and Lago di Nemi.

It is stated that volcanic activity began in the north and was displaced southward step by step. This area, consequently, seemed well suited to study the paleomagnetism of the early Quaternary. In particular, to date the last reversal of the earth's magnetic field from Reversed to Normal.

For the paleoagnetic studies the field method, using a normal geological compass. was used (Rutten and Den Boer, 1954; Rutten, 1959).

In addition, earlier, presumably Pliocene, volcanism is found west of the Vico and Sabatini volcanoes. This occupies two very close but separate districts, north of Cervéteri and around Tolfa.

For an introduction to the geology of this area, and for extremely helpful guidance, I am very much indebted to Mrs. A. C. Blanc and Professor U. Ventriglia of the University of

Rome, and to Dr. P. Marquardt of the Geological Survey.

RESULTS

In regional paleomagnetism, it was found that all products of the quaternary volcanoes Vulsini, Vico, Sabatini and Laziale showed Normal magnetisation. All products studied of the Cimini volcano, on the other hand, had Reverse remanent magnetisation.

This poses a problem in the geologic dating. If, as stated, the northernmost volcano, the Vulsini with the Lago di Bolsena, is indeed the earliest of the series, it is possible that Vulsini belongs to the last-but-one paleomagnetic period (N_2), Cimini to the ultimate Reversed period (R_1) and Vico, Sabatini and Laziale to the ultimate or present period of Normal magnetisation (N_1).

It is, however, geologically impossible to attribute the Vulsini volcano to the N_2 paleomagnetic period. This would place its activity in the Upper Pliocene. Now the massive tuff flows from the Lago di Bolsena have spread out eastwards till the region of Orvieto. Here they have filled deep river valleys which are connected with the present Tiber river system. So a tertiary age for the Vulsini volcano is out of the question. Consequently, it must belong to the Quaternary, and is then identified with the present Normal period N_1. The tuff flows from the Vico volcano also descend into an early Tiber river and are of more or less the same age as those originating from Vulsini.

The paleomagnetic dating of the mid-Italian quaternary volcanoes consequently is as summarised in table 13–1.

In a general way, this gives a minimum age for the present Normal magnetic period N_1. Products of the Sabatini volcano have been dated by Everenden and Curtis (cf. Blanc, 1958) as from 230,000 to 470,000 yrs. old. As all products from the Sabatini volcano belong to the present Normal magnetic period, the minimum duration of this period is 470,000 years.

Also, Tolfa volcanics were dated by Everenden as 2.4 m.yrs. old. These were found to present Normal remanent magnetism, so there was a period of Normal earth magnetism at that date. This possibly is the paleomagnetic period N_2 of the Upper Pliocene. The geology of these Tolfa volcanics is, however, not well known, so this is not to be taken as certain.

Another result of a more general character stems from the fact that acid volcanics of this region show only weak remanent magnetisation, notwithstanding their slight age. With the field method one must, consequently, place the remanent magnetic pole in the sample studied very close indeed to the compass needle, to get a reaction at all. So it is easy to find random orientation from inclusions in a mudflow or a re-sedimented tuff, which have moved after cooling. The inclusions in the tuffs, which evidently were of a temperature above their Curie point when arriving at their present place, all show the same magnetic orientation, either N or R, as the case may be. As some of the re-sedimented tuffs are difficult to distinguish from primary tuffs in the field, this paleomagnetic method proved to be of great help. Near Bracciano and Anguilare, along the steep western side of the Lago di Bracciano, for instance, such re-sedimented tuffs and mudflows are wide spread. In their most typical form they

Table 13–1
Paleomagnetism of mid-Italian volcanoes.

Volcanoes	Paleomagnetism	K-Ar dating
Post-Villafranchian Laziale Sabatini Vico Vulsini	N_1	230,000–470,000 yrs.
Villafranchian Cimini	R_1	

may look like tillites, but there are many intermediate types which still resemble the primary tuffs very much.

This difference in magnetic orientation in primary tuffs and re-sedimented tuffs or mudflows was checked at the historical sites of Herculaneum and Pompei. The 79 A.D. eruption of Vesuvius covered Pompei with a hot ash, whereas Herculaneum was destroyed by a thick mudflow. The consolidated mud from fresh excavations at Herculaneum consequently gives no magnetic reaction. In Pompei the main body of the tuff consists of "marbles" of pumice, one to several cm in diameter, each with a distinct chilling surface. These do not give a reaction with the hand compass, either because they are too light in weight and do not contain enough magnetic material, or because they were chilled below their Curie point before settling. Small lava inclusions, however, some 5 cm across, all give consistent magnetic orientation. This is, of course, Normal, as it ought to be.

14

Geomagnetic Polarity Epochs and Pleistocene Geochronometry

ALLAN COX
RICHARD R. DOELL
G. BRENT DALRYMPLE
1963

From *Nature*, v. 198, no. 4885, p. 1049–1051, 1963. Reprinted with permission of the authors and Macmillan Journals, Ltd.

The magnetic properties of six radiometrically dated volcanic flows and plugs of Pleistocene and late Pliocene age from California have been investigated as a step toward determining the radiometric ages of recent geomagnetic polarity epochs in western North America. Three of these flows have normal remanent magnetizations, which by definition are north-trending and, in the northern hemisphere, dip below the horizontal. Three of the flows are reversed, that is, the direction of their remanent magnetization trends southward and is inclined upward.

The theory that the magnetic field of the Earth reversed its polarity at least once during the Pleistocene and many times during the Tertiary was originally advanced (Brunhes, 1906; Matuyama, 1929; Hospers, 1951) to explain occurrences of reversed remanent magnetization similar to those described in this article. The abundance and distribution of reversely magnetized rocks preclude their being dismissed as rare, unexplained accidents: these rocks exist on all continents; they occur among many petrological rock types, and they constitute about half of all Tertiary and Pleistocene rocks. Moreover, their stratigraphic distribution is not random. Normal and reversed rocks usually occur in stratigraphic groups of like polarity, and in areas of late Pleistocene volcanism the youngest group is invariably normal.

Because the geomagnetic field is a global phenomenon, the field reversal theory requires that transitions from one magnetic polarity to another must occur at exactly the same time over the entire Earth. However, most paleomagnetic studies of Pleistocene rocks have been made on volcanics which cannot be correlated reliably over long distances. Therefore it has proved difficult to put the field reversal theory to a definitive stratigraphic test. As pointed out by Rutten (1959), physical dating techniques may provide the key to this crucial question, and in the present work we have collected palaeomagnetic data from rocks for which potassium – argon ages have also been calculated.

Palaeomagnetic investigations of reversals would be relatively simple were it not for the phenomenon of mineralogically controlled self-reversal, whereby lava flows containing certain minerals may acquire a reversed remanent magnetization when they cool in a normal field. One case of a self-reversing pyrrhotite (Everitt, 1962) and perhaps a dozen examples of self-reversing hemo-ilmenite (Uyeda, 1958; Carmichael, 1961) have been reported. Self-reversals due to these minerals can be recognized in the laboratory by thermomagnetic experiments, although the dependence of these self-reversals on a specific cooling history sometimes makes it difficult to reproduce the self-reversal. Self-reversal is at least theoretically possible for some other minerals and might not always be reproducible because of mineralogical changes which have occurred since the rock originally cooled (Néel, 1955; Verhoogen, 1962).

Several approaches were used in the present work to test the reliability of the polarity determinations. The sampling of each flow was designed to include a variety of ferromagnetic minerals and initial cooling rates. For one flow ($S1$, Table 14–1) we were able to include a study of xenolithic inclusions and in others we found more than one ferromagnetic mineral present. In all cases, concordant polarities were found between all the different ferromagnetic minerals within each flow, indicating that self-reversal has probably not occurred. In addition, the following laboratory investigations were made. (1) The absence of hemo-ilmenite and pyrrhotite was established by microscopic examination of polished sections and confirmed by thermomagnetic experiments. (2) Representative specimens from each flow were partially demagnetized in alternating magnetic fields up to 800 oersteds. In all cases the magnetic polarity remained unchanged, which eliminated the possibility that the natural thermo-remanent magnetization might be masked by soft magnetization acquired subsequent to the initial cooling of the flow. (3) The mode of thermal decay of the natural remanent magnetization was found by

heating at $50°$ C increments and cooling in field-free space. The change of polarity on heating which has been found to be typical of known self-reversed rocks (Nagata et al., 1954) did not occur in any of these flows. (4) In specimens from flows $S3$, $S5$ and $S6$ (Table 14–1) which were observed under the microscope to have both the usual magnetite or titanomagnetite and, in addition, minor amounts of hamatite, remanent magnetization could still be measured after heating above $600°$ C. This component of magnetization, undoubtedly due to the hamatite, invariably had the same direction as the component due to the magnetite, making self-reversal unlikely. (5) Specimens from all flows, on being heated above their Curie temperatures and cooled in a normal field in the laboratory, always acquired normal magnetization. We conclude from these experiments that self-reversals are unlikely in any of these flows.

While the geomagnetic field is changing polarity, its direction is intermediate between normal and reversed, and on the basis of several palaeomagnetic studies in other parts of the world the time required for the transition has been estimated at about 10,000 years (Doell and Cox, 1961). Although the opposing polarities and similar radiometric ages of flows $S1$ and $S2$ suggest that they lie close to a polarity transition, neither flow shows any tendency to have an intermediate direction of magnetization. However, this is not surprising when the 4 per cent standard deviation of the analytical methods used in dating is considered. This corresponds to an age uncertainty of 40,000 years for each flow, so $S2$ may actually be the younger flow, and both may easily lie more than 5,000 years away from the mid-point of the polarity transition.

These data, although obviously insufficient to define a time-scale for magnetic polarity epochs, establish some limits for the possible ranges of past polarity alternations. The first question to be answered is whether the more recent polarity epochs are all the same length. Khramov's studies (1957) suggest that they are, since in several sediment well-cores the stratigraphic thicknesses of the upper three magnetic zones, including the present normal epoch, are equal or nearly so; periodic polarity epochs are implied unless, as seems unlikely,

Table 14–1

Petrographic descriptions, magnetic properties, and radiometric ages.

No.	Locality*	Age‡ (10^6 y.)	Reference for age	Polarity	Paleomagnetic sampling extent	Petrographic description of magnetic samples	Ferromagnetic minerals	Curie temp.
S1	Owens Gorge	0.98 (sanidine)	a	Normal	30 miles laterally, 600 ft. vertically, 90 oriented samples	Ignimbrite, marked increase in degree of welding with depth; abundant lithic fragments and inclusions	Magnetite or titanomagnetite, grain diameter 2–500μ. Minor alteration to hematite	570° C
S2	Big Pine†	0.99 ± 0.04 (obsidian)		Reversed	1 mile laterally, 400 ft. vertically, 17 oriented samples	Pumiceous to dense and banded rhyolite flow with varying degrees of perlitization	Magnetite or titanomagnetite, grain diameter 1–40 μ. Opaque grains rare	530° C
S3	Sutter Buttes	1.57 ± 0.24 (biotite)	b	Reversed	35 ft. laterally, 10 ft. vertically, 4 oriented samples	Vesicular hornblende andesite flow with large poikilitic phenocrysts of plagioclase	Isotropic grains 2–45μ in diameter, pinkish brown to grey in colour, probably titanomagnetite. Minor deuteric hematite in biotite	575° C 675° C
S4	Sutter Buttes	1.69 ± 0.10 (biotite)	b	Reversed	110 ft. laterally, 25 ft. vertically, 7 samples	Slightly vesicular dacite intruded at shallow depth; phenocrysts of biotite, plagioclase, and quartz	Earthy red hematite dust, rare grains of magnetite 3–10μ diameter	560° C
S5	McGee Mtn.	2.6 ± 0.1 (whole rock)	c	Normal	430 ft. laterally, 90 ft. vertically, 20 samples	Fine-grained holocrystalline olivine and biotite-bearing basalt flow	Magnetite grains 1–60μ in diameter with occasional rims and veinlets of hematite; both minerals also appear as rims around olivine	520° C 675° C
S6	Owens Gorge	3.2 ± 0.1 (whole rock)	c	Normal	100 ft. laterally, 80 ft. vertically, 8 samples	Vesicular porphyritic holocrystalline olivine basalt flow	Magnetite or titanomagnetite, grain diameter 1–35μ; extensively replaced by hematite	575° C 675° C

a. Evernden et al., 1964.
b. Evernden et al., 1957.
c. Dalrymple, 1963.
d. Bateman, 1956.

*These localities (all in California) serve to identify the lava flows in the references cited, which give complete geographic and stratigraphic descriptions.

†This age, not previously published, is based on obsidian sample $KA\,1069$ from a rhyolite flow (ref d) in the N.W. ¼ Sec. 30, T. 10 S., R. 34 E. Big Pine 1 : 62,500 quadrangle. Sample weight 11.38 g, % $K = 3.74$, % $Ar_{40}^{atm} = 67$, $Ar_{40}^r/K_{40} = 0.578 \times 10^{-4}$. Absence of perlitization is indicated by low weight loss on heating to 900° C ($<0.25\%$) and by refractive index (1.486 before heating, 1.483 after heating).

‡(±) figures are standard deviations for the precision of the analyses.

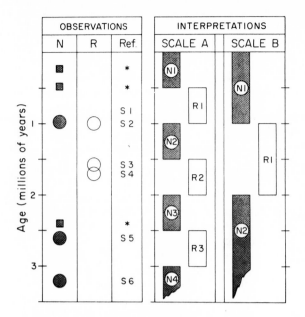

Figure 14-1
Magnetic polarity epochs consistent with
paleomagnetic and radiometric observations.
For Ref. column, see Table 14-1. (asterisk,
data from European rocks; Rutten, 1959)

irregular rates of sedimentation compensate
for irregular polarity epochs. If the more recent polarity epochs are equal or nearly equal
in length, the data of the present study indicate
that these epochs must be either $\frac{1}{2}$ or 1 million
years long. Three previously published determinations from Europe (Rutten, 1959) (asterisk in Fig. 14-1) are consistent with either of
these two periods, whereas the combined data
are inconsistent with periods less than $\frac{1}{2}$ million
years in length, greater than 1 million years in
length, or between $\frac{1}{2}$ and 1 million years in
length. At present, we favour scale B (Fig.

14-1) because we have been unable to discover any reversals in flows younger than $S1$
with the possible exception of $S2$ as previously
noted. (*Note added in proof.* Additional work
on several flows from California also strongly
favours scale B.)

Magnetic epochs of unequal length are also
consistent with these results—for example,
$N2$ (Scale B) may be considerably longer than
$R1$. Additional measurements are being made
to clarify the important problem of periodicity
and also to aid in evaluating the accuracy of
the techniques used for dating young rocks.

15

Dating of Polarity Zones in the Hawaiian Islands

IAN McDOUGALL
DON H. TARLING
1963

Periodicity of reversal of the Earth's magnetic field is of some geophysical and geological interest. Study of the polarity of the Earth's field throughout geological time provides much information relating to the origin of the field, and the stratigraphical zonation of polarities in rock sequences may offer a valuable means of geological correlation.

Brunhes (1906) observed that the natural remanent magnetization of some rocks has a polarity opposed to the present sense of the Earth's field, and Mercanton (1910) showed that rocks of different ages from different parts of the world are magnetized in this sense. These observations led to the supposition that periodic reversals of the Earth's magnetic field may have occurred during geological time. Hospers (1953, 1954) found that in a sequence of Cainozoic lavas in Iceland polarity zones occur, and that some of these zones are separated by a thin zone of rocks with intermediate directions. Since then, several levels of reversal have been recognized, and it appears that reversals are frequent in the Cainozoic in which rocks magnetized in the same sense as the present Earth's field (that is, 'normal') are about as common as rocks with a direction of magnetization opposite to that of the present field (that is, 'reversed').

Self-reversal (Néel, 1955; Verhoogen, 1956a) and reversals of the Earth's magnetic field appear to be the only mechanisms able to account for the common occurrence of rocks with exactly opposed polarities. It is possible to test the hypothesis of reversal of the Earth's field by comparing palaeomagnetic observations from various parts of the world, as one of the consequences is that rocks of the same age, irrespective of composition and geographical location, possess the same polarity. It is therefore of interest to obtain accurate dates on the reversal sequences. This article attempts to give a stratigraphic table of polarity zones in the Upper Cainozoic on palaeomagnetic measurements and isotopic age determinations on

From *Nature,* v. 200, p. 54–56, 1963. Reprinted with permission of the authors and Macmillan Journals, Ltd.

Table 15–1
Paleomagnetic and age data on rocks from the Hawaiian Islands.

Formation	Polarity[a]	No. of paleomagnetic sites	K-Ar age (m.y.)	No. of samples dated, or sample no.
Hana (East Maui)	N	3	<0.4	—
Kula (East Maui)	N	6	0.86, 0.43	2
Honomanu (East Maui)	R	1	>0.86	—
Honolua (West Maui)	R	1	1.15 ± 0.02	4
Wailuku (West Maui)	R	5	1.29 ± 0.03	3
Lanai	R	3	—	—
East Molokai	R	6	1.3–1.5	5
West Molokai	R	4	1.85 ± 0.01	1
Koolau (East Oahu)	R	8	2.2–2.5	5
Waianae (West Oahu)	N	1	2.76 ± 0.02	GA 556
	N	1	2.84 ± 0.02	GA 809
	R	8	2.95 ± 0.06	GA 557
	N	2	3.27 ± 0.04	GA 810
Koloa (Kauai)	N and R	2	—	—
Napali (Kauai)	N	3	4.5 – 5.6	3

[a]N = normal; R = reversed.

samples of lavas from the Hawaiian Islands. A summary of the magnetic observations was given earlier (Tarling, 1962), and a description of the dating methods and results is given elsewhere (McDougall, 1963, 1964). These results are consistent with palaeomagnetic and age measurements made (Cox et. al., 1963a) on rocks from six sites in California, but they suggest the presence of two additional polarity changes and allow more precise dating of each level of polarity change. This work goes some way toward establishing a stratigraphical sequence of polarity zones during the Upper Cainozoic against which other palaeomagnetic results can be compared to test the hypothesis of reversal of the Earth's field.

Sampling for palaeomagnetic and age measurements was undertaken separately in the Hawaiian Islands, although in a number of cases the same sample, or samples from the same outcrop, were used for both measurements. Hence, there is some uncertainty in the correlation between the measured ages and polarities observed in the same formation; but it is considered that the pattern that emerges is essentially correct, particularly as the age measurements show that many of the formations (as defined by Stearns and Macdonald and summarized by Stearns, 1946) were erupted over periods of 200,000 years or less.

The dates were determined on whole-rock samples by the potassium-argon method. Argon was measured by isotope dilution and potassium by flame photometry. Determinations were made in duplicate in most cases, and the plus and minus figures quoted after an age indicate the range. The ages generally are precise to better than ± 3 per cent at the 95 per cent confidence-level.

A total of 164 samples were collected for palaeomagnetic study from 59 sites in five of the Hawaiian Islands. The samples appear to be magnetically stable as the directions of magnetization remain constant in alternating fields (Thellier and Rimbert, 1954) up to 750 oersteds (peak), and only directions observed in samples after treatment in a magnetic field of 150 oersteds (peak) (Irving et al., 1961) are considered here. Stability is also indicated by a high average coercivity of maximum isothermal remanence (445 oersteds) and the consistency of results both within and between different sites in the same formation. The direction observed is thought to reflect the actual direction of the past magnetic field as 16 specimens that were heated and then cooled in the present Earth's field at Canberra (0.59 oersteds) acquired directions parallel to the applied field. The results are listed in Table 15–1 and illustrated in Fig. 15–1.

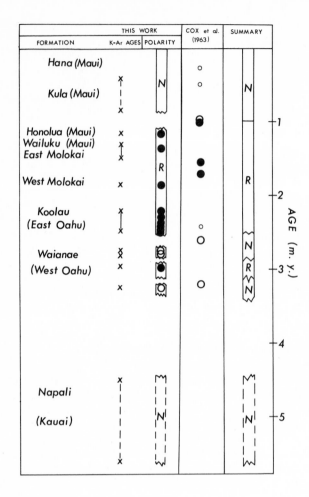

Figure 15-1
Summary of paleomagnetic and age data (filled circles, "reversed" polarity; open circles, "normal" polarity)

The polarity of the Kula and Hana Volcanic Series of East Maui is 'normal'. Potassium-argon age measurements on two samples from the Kula Series yield ages of 0.45 ± 0.02 and 0.84 ± 0.04 m.y. The older age was obtained on a sample from a lava immediately overlying the Honomanu Series; hence, this age probably closely dates the beginning of the eruption of the Kula Series. It is therefore likely that the specimens on which the magnetic measurements were made are not older than 0.84 m.y. The Hana Volcanic Series, which overlies the Kula Series, must be younger than 0.45 m.y. One site in the Honomanu Series is 'reversed', but the age of these lavas only is known to be older than 0.84 m.y.

Samples from a single site in the Honolua Series of West Maui are 'reversed'. A speci-men from the same lava flow yields an age of 1.16 ± 0.02 m.y. Hence, the results suggest that between 0.84 and 1.15 m.y. a change in polarity occurs.

The Wailuku Series of West Maui, underlying the Honolua Series, has 'reversed' polarity, and three samples from this formation yield ages agreeing at 1.29 ± 0.03 m.y. One of the six sites sampled for magnetic measurements has 'normal' polarity, but as the inclination is negative (−49) compared with the average of all other 'normal' sites (+35) from Hawaii this result is clearly anomalous and is not included.

The seven East Molokai sites show 're-versed' polarity, and age measurements of five samples indicate that the formation is 1.3 to 1.5 m.y. old. Two of the samples used for the age determinations are probably from the same lava flows as two of the sites sampled for magnetic measurements. The West Molokai lavas also have 'reversed' polarity, and the single specimen dated from this formation has an age of 1.85 ± 0.01 m.y.

Sites on the Island of Lanai are all 'reversed', but no age measurements were made on these rocks. As Lanai lies between Molokai and Maui it may be inferred that the age of these volcanics is probably about 1.3 m.y.; this is consistent with the 'reversed' polarity of the Lanai rocks.

Samples from five of the eight sites in the Koolau Series of East Oahu were also used for the age determinations. All the sites are 'reversed', and the ages range from 2.2 to 2.5 m.y. Two sites sampled in the Honolulu basalt are 'normal', but these were not dated, and only are known to be less than 2.2 m.y.

Two samples from the Upper Waianae Series of West Oahu yield ages of 2.76 ± 0.02 and 2.84 ± 0.02 m.y., and have 'normal' polarity. A lava of the Lower Waianae Series dated at 2.95 ± 0.06 m.y. has 'reversed' polarity, and a further sample from the same formation at a different locality has 'normal' polarity and is dated at 3.27 ± 0.04 m.y. Other sites in the Waianae Series also have 'reversed' polarity.

Three samples from the Napali Formation of Kauai (MacDonald et al., 1960) yield ages ranging from 4.5 to 5.6 m.y., and the polarity at three different sites is 'normal'. However, as the sites from which the samples for magnetic and age measurements were collected are widely separated, it is not possible to make precise correlations. Samples also were dated from the Makaweli Formation of Kauai, but no rocks were collected for palaeomagnetic measurement. In the Koloa Series of Kauai both 'normal' and 'reversed' polarities were found. The single specimen dated from this formation is not from the same locality as any of the palaeomagnetic sites, hence no correlation is possible.

It may be assumed that the 'reversed' polarities observed in Hawaiian rocks are caused by reversal of the Earth's magnetic field. This is supported by the observation that rocks of each polarity occur in discrete stratigraphic zones, and that in each of these zones the composition of the lavas ranges from picrite basalt to trachyte. In addition, there is no evidence of self-reversal in specimens examined in the

laboratory. However, these observations do not refute the possibility of self-reversal, and it is necessary to compare the results from Hawaii with results from rocks of similar age in other parts of the world if the hypothesis of reversal of the Earth's field is to be substantiated.

Our data are consistent with the change from 'normal' to 'reversed' polarity observed (Cox et al., 1963a) in California (Fig. 15–1) at about 1 m.y., but suggest that a return to 'normal' polarity occurred between 2.5 and 2.75 m.y., rather than at about 2 m.y. as inferred by Cox et al. (1963a). They recorded 'normal' polarity in a rock dated at 2.6 ± 0.1 m.y. and refer to Rutten (1959), who reported 'normal' polarity in a sequence in Italy dated at about 2.4 m.y. These data, combined with our measurements, suggest that the change from 'reversed' back to 'normal' polarity occurred at 2.5 ± 0.1 m.y. The 'reversed' polarity found in a rock from the Waianae Series with a measured age of 2.95 m.y. indicates that the behaviour of the magnetic field is more complex than has been suggested previously. Owing to the lack of data it is not possible to state whether 'normal' polarities found in rocks dated at 3.3 and about 4.5 m.y. belong to the same magnetic epoch or whether a zone of 'reversed' polarity occurs during this period. Our results are consistent with varying frequency of change in the polarity of the Earth's magnetic field in the Upper Cainozoic, rather than a regular change about every 1 m.y. (Cox et al., 1963a).

Roche (1951; 1956) showed that in France lavas of Upper Pleistocene age are normally magnetized, and that lavas of Villafranchian (Lower Pleistocene) age have 'reversed' polarity. Hence, the change in polarity at about 1 m.y. found in the Hawaiian Islands and in California may be correlated with that found in France. The stratigraphic position of the return to 'normal' polarity has not been established in France; but it is either in the lower part of the Villafranchian or in the Astian, although Rutten (1960) suggested that it occurs at the base of the Villafranchian. Clearly it is of importance to establish the position of this boundary in order that correlations may be made with greater certainty throughout the world. Fourteen polarity zones were reported by Khramov (1958) from sequences of Upper Cainozoic age in Russia, but at this stage it would be premature to attempt direct correlations.

The evidence from Hawaii suports, but does not prove, the hypothesis of changes in polarity of the Earth's magnetic field. Further study on rocks of similar age clearly is necessary to confirm the hypothesis, and to place closer limits as to the times when polarity changes occurred. The present work suggests that models of the Earth's magnetic field should account for irregular periodicity of reversal.

16

Reversals of the Earth's Magnetic Field

ALLAN COX
RICHARD R. DOELL
G. BRENT DALRYMPLE
1964

From *Science*, v. 144, p. 1537–1543, 1964. Reprinted with permission of the authors and the American Association for the Advancement of Science. Copyright 1964 by the AAAS.

The idea that the earth's main magnetic field changes its polarity was first advanced during the early decades of this century by geophysicists who were investigating the remanent magnetization of volcanic rocks and baked earth (Brunhes, 1906; Chevallier, 1925; Matuyama, 1929; Mercanton, 1926b). They found that these materials, when heated to their Curie temperatures and cooled in the present weak field of the earth, acquire a weak but extremely stable remanent magnetization parallel to the present direction of the geomagnetic field. They further found that bricks and baked earth from archeological sites, as well as rock samples from young lava flows, possess a similar natural magnetization, undoubtedly acquired when the baked earth and lava flows originally cooled. The direction of this magnetization was found to be close to that of the present geomagnetic field at the sampling site, with differences of the order of 10 degrees attributable to secular variation of the geomagnetic field. However, when the early paleomagnetists studied rocks of early Pleistocene age or older, they discovered that a substantial proportion were magnetized in a direction more nearly 180 degrees from that of the present field. In explanation of these reversed remanent magnetizations they proposed 180-degree reversals of the direction of the geomagnetic field.

During the two decades (1928 to 1948) that followed this early work there was surprisingly little scientific reaction to the hypothesis of geomagnetic field reversal. This silence reflects the embarrassing lack, even at so late a date, of a theory adequate to account for the present geomagnetic field, let alone reversed magnetic fields which may or may not have existed earlier in the earth's history. The problem of field reversals was taken up again around 1950 by scientists who became involved in problems of geomagnetism from theoretical as well as observational starting points. The theory of magnetohydrodynamic motions in an incompressible fluid was interpreted to show how magnetic fields may be generated in rotating spheres of electrically conducting fluid. It was further shown that reversals in polarity are a possible, if not a necessary, property of simplified dynamo models for the magnetohydrodynamic motions. Moreover, conditions necessary for the maintenance of such a magnetic system were shown to be consistent with conditions thought to exist in the earth's core (Hide and Roberts, 1961; Jacobs, 1963).

The observational basis for field reversals has been greatly extended since 1950, partly as a result of improvement in techniques of determining the stability and reliability of rock magnetism for determining past geomagnetic field directions, and partly as a result of the vast increase in paleomagnetic data available. Although many of these paleomagnetic studies have focused on the problems of continental drift and polar wandering, interest in the reversal problem has also continued up to the present, and the observations of the early workers have been amply confirmed. The directions of remanent magnetization in many tens of thousands of rock samples, with ages in the range 0 to 30 million years, show a strikingly bimodal distribution; the directions are grouped about either the present geomagnetic field directions at the sampling sites or about antiparallel (reversed) directions. Remarkably few intermediate directions have been observed.

Solid-state physicists have also made important contributions to the study of reversals. In 1950 Graham (*in* Tuve, 1951) concluded that reversals he had observed in sediments were due not to reversals of the earth's field but, rather, to mineralogically controlled self-reversal along the lines earlier reported by Smith et al. (1924). Informed of this result, Néel (1951) was able to predict, on the basis of contemporary advances in ferromagnetism and ferrimagnetism, that rock-forming minerals might acquire remanent magnetizations at an angle 180 degrees from the direction of the ambient magnetic field in which they cool. This prediction was almost immediately confirmed by Nagata et al. (1951) and Uyeda's (1958) discovery of a volcanic rock from Japan that possessed this predicted property of self-

reversal. As we discuss more fully later on, this discovery of mineralogically controlled self-reversal, in offering a possible alternative explanation for reversed magnetization, considerably clouded the experimental evidence for geomagnetic field reversal.

What, then, is the present state of knowledge about geomagnetic field reversals? Our conclusion, based upon recent paleomagnetic-geochronometric evidence presented in this article, is that after a period of considerable uncertainty the balance has finally tipped in favor of geomagnetic field reversals as the cause of the reversals in most rocks. Recent experimental evidence not only confirms this interpretation but also tells us exactly when some of the most recent switches in polarity occurred.

EARLY DISCOVERIES

Some of the most compelling evidence favoring the field-reversal hypothesis appears in the earliest investigations. Brunhes (1906) made the experimental observations that are summarized in Fig. 16–1. Faced with these results, and the experimentally verifiable fact that bricks and other baked earths acquire remanent magnetization parallel to the ambient magnetic field in which they cool, Brunhes concluded that the reversed natural remanent magnetization had been acquired in a reversed magnetic field which existed during the epoch when the lava flow originally cooled.

Although Brunhes's classic work was done decades prior to the discovery of mineralogically controlled self-reversals, his results from baked sediments have a direct bearing on this problem. When lava flows and other igneous bodies are emplaced, the sediments or other rocks in contact with them are heated and also commonly are altered chemically. The ferromagnetic minerals in the igneous body are generally different from those in the surrounding baked country rock. As the igneous body and the baked rocks cool in the same ambient geomagnetic field, they become magnetized in the same direction unless magnetization of the ferromagnetic minerals in one of them is self-reversing. If the fraction of self-reversing ferromagnetic minerals in all rocks is x and if these self-reversing min-

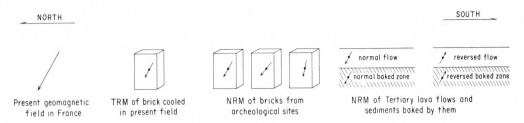

Figure 16-1
Brunhes' observations of experimentally produced thermoremanent magnetization (TRM) and of natural remanent magnetization (NRM), which led him to propose geomagnetic field reversals

erals are randomly distributed between igneous and baked rocks, then the fraction of pairs of igneous rock and baked rock with opposing polarities will be $2(x - x^2)$. In a recent review of paleomagnetic studies of igneous and baked rocks, Wilson (1962) found 50 pairs with the same polarity and three pairs for which the original paleomagnetic data were ambiguous and which may or may not have opposing polarities. On the basis of these data, it appears that, at most, 3 percent of the reversals in igneous rocks and baked sediments are due to mineralogically controlled self-reversals.

The next important step in the study of reversals came with the stratigraphic investigations of Mercanton (1926) and Matuyama (1929), in which an attempt was made to delineate the times when the geomagnetic field was reversed. Matuyama's study of volcanic rocks from Japan and Korea is especially important because it was the first successful attempt to determine the age of the most recent switch from a reversed-polarity epoch to the present epoch of normal polarity. He found that "there is a group of specimens whose directions of magnetization falls around the present earth's field. A number of other specimens forms another group almost antipodal to the former. [Although] the ages of the collected basalts are not always clearly known. . . . we may consider that in the earlier part of the Quaternary Period the earth's magnetic field in the area under consideration was probably in the state represented by the second group, which gradually changed to the state represented by the first group." This result, like that of Brunhes, has been amply confirmed by subsequent investigations.

WORLDWIDE EXTENT OF REVERSALS

If we accept for the moment the hypothesis that most reversely magnetized rocks were formed in reversed magnetic fields, the question arises, Were such reversed fields anomalies restricted to the paleomagnetic sampling sites or were they worldwide phenomena? A theory of field reversals due to local concentrations of ferromagnetic minerals in the earth's crust may be eliminated at the outset on several grounds, one of them magnitude: the magnetizations of rock formations are generally too small by a factor of 10 to produce a reversed field. From the theoretical point of view, if field reversals occur at all, they should be worldwide events. This follows from the very nature of the earth's magnetic field. Since the time of Gauss we have known from spherical harmonic analysis of the field that its source lies within the earth, and from physical reasoning it is equally clear that this source is within the earth's fluid core. The only proposed physical mechanism for producing the field which at present appears tenable is the magnetohydrodynamic mechanism, in which generation of the field is attributed to the interaction between fluid motions and electrical currents in the core.

The shape of the geomagnetic field is approximately that of a dipole at the earth's center, the direction of the field observed at any locality today differing from that of the dipole field by 5 to 25 degrees. The worldwide fit to the dipole-field configuration is vastly improved if average rather than instantaneous field directions at each locality are considered. With paleomagnetic techniques it is possible to obtain the direction of the earth's field at a

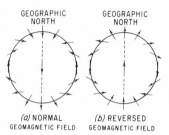

Figure 16-2
Directions of an axial dipole
geomagnetic field over the earth
for (a) normal polarity (present
configuration) and (b) reversed
polarity

given locality averaged over thousands of
years, and when this is done it is found that
young rocks from all over the world fit a dipole-
field configuration with the dipole axis parallel
to the earth's present axis of rotation (Cox
and Doell, 1960). Two such rotationally symmetrical dipole configurations are possible,
one normal (Fig. 16–2,*a*) and one reversed
(Fig. 16–2,*b*). Only two processes appear
capable of explaining why the remanent magnetizations of rocks from all over the world
are tightly grouped about these two field configurations. Either 180-degree self-reversals
controlled mineralogically are responsible,
or else the fluid and current systems in the
earth's core have changed in such a way as
to produce 180-degree reversals in the direction of the main dipolar component of the
earth's magnetic field.

SELF-REVERSAL

Self-reversal is defined as the acquisition of
thermoremanent magnetization in a direction
180 degrees from that of the ambient magnetic
field in which a rock or mineral specimen cools.
Knowing as we do that some rocks, such as
the Haruna dacite (Nagata et al., 1951; Uyeda,
1958), are reproducibly self-reversing in the
laboratory, the crucial question is whether all
reversely magnetized rocks have undergone
self-reversal. If they have, the experimental
basis for seeking evidence of geomagnetic
field reversal vanishes, and the hypothesis

must be discarded. If not, we are faced with
the difficult task of sorting out reversals due
to mineralogy from those due to reversals of
the geomagnetic field.

The obvious experiment might seem to be
to heat and cool rocks in known magnetic
fields and determine whether these rocks are
self-reversing. This experiment has been performed by numerous investigators on thousands of rocks from many localities; in the
laboratory, only a small fraction of 1 percent of
the rocks have been found to be self-reversing.
However these experiments do not tell the
entire story or completely rule out the possibility of self-reversal. The inconclusiveness of
simple heating and cooling experiments is
illustrated in the studies by Uyeda (1958) and
Ishikawa and Syono (1963) on minerals of
the ilmeno-hematite series. Minerals of this
series were found to be self-reversing only
over a limited range of chemical composition
and only for certain rates of cooling. Thus, a
natural self-reversal in a rock containing these
minerals might not be reproduced if the rate
of cooling in the laboratory were different
from that when the rock originally formed.

Self-reversal requires the coexistence and
interaction of two ferromagnetic constituents.
Of the many types of interaction that may produce self-reversal, the simplest involves the
magnetostatic energy of interaction between
two closely intergrown ferromagnetic minerals,
A and B, having different Curie temperatures.
If the geometry of the mineral intergrowths
is appropriate, and if the intensity of the magnetization of the A constituent is sufficiently
large when the rock has cooled to the Curie
temperature of the B constituent (the Curie
temperature of A being greater than that of
B), B will become reversely magnetized. The
rock will undergo reproducible self-reversal
if, after complete cooling, the remanent magnetization of B is greater than that of A. Even
if the remanent magnetization of B is initially
less than that of A, the rock may still become
self-reversed if A is selectively dissolved during the geologic history of the rock, or if the
remanent magnetization of A undergoes spontaneous decay to a value less than that of B.

The two constituents, A and B, need not
be different ferromagnetic minerals. In ferrites
they are the two interpenetrating cubic sub-

lattices which make up the crystal structure. Because the exchange interaction between the cation sites in the two sublattices is negative, the B sublattice acquires a spontaneous magnetization exactly antiparallel to that of the A sublattice when the ferrite cools through its Curie temperature. Given appropriate temperature coefficients for the two spontaneous magnetizations, reproducible self-reversal will occur on further cooling. Self-reversing minerals of this type have been synthesized (Gorter and Schulkes, 1953) but have not been discovered in rocks.

Ferrites may also undergo self-reversal through cation migration between adjacent lattice sites. At temperatures above the Curie temperature the equilibrium distribution of cations on lattice sites is disordered, but at low temperatures it is more highly ordered. On rapid quenching, however, such as may occur when a lava flow cools, a ferrite could temporarily retain a disordered cation distribution and later undergo a self-reversal through slow diffusion of cations toward the more ordered low-temperature distribution. Verhoogen (1956a) has shown that a self-reversal of this type would not take place in magnetites containing impurities or vacancies on certain lattice sites until 10^5 or 10^6 years after cooling. These self-reversals would not be reproducible in the laboratory.

In view of the complexity and nonreproducibility of many types of self-reversal, what approaches are open to the investigator attempting to assess the prevalence of self-reversal in rocks? One is to look for correlation between mineralogic properties and magnetic polarity. If all the reversely polarized members of a suite of rocks possess a unique mineralogy, self-reversal is probable even if it cannot be reproduced in the laboratory.

Balsley and Buddington (1954) have discovered such a correlation in their study of metamorphic rocks in the Adirondack Mountains. Rocks of reversed polarity invariably contain ilmeno-hematite; normal rocks invariably contain magnetite. These mineralogical differences indicate self-reversal and also indicate that the geomagnetic field was not reversing its polarity during the long interval when these rocks were acquiring their remanent magnetizations.

Significant as these results from the Adirondack Mountains are, they appear to be the exception rather than the rule. In our own petrologic and paleomagnetic studies of many hundreds of samples from normal and reversely polarized lava flows, no significant correlation between polarity and petrography has emerged. However, some uncertainty lingers even after the most careful search fails to reveal a physical property correlated with reversals. New theoretical mechanisms which might produce reversals are still being advanced, and it is possible that the significant variables controlling self-reversals have not been isolated and identified.

A much stronger and more direct approach to the problem of self-reversal is available where two different ferromagnetic minerals have become magnetized at the same place and the same time. The magnetization of igneous bodies and sedimentary deposits baked by them was discussed earlier. Recently we have been able to extend this technique to some igneous rocks not associated with baked sedimentary rocks by demonstrating that the natural remanent magnetization of these rocks resides in two distinct ferromagnetic minerals —hematite, with a Curie temperature of 680°C, and magnetite or titanomagnetite, with a Curie temperature below 580°C. The remanent magnetization of the two minerals has thus far always been found to have the same polarity. These results, like those from the baked sedimentary rocks, suggest that self-reversals in nature are rare.

STRATIGRAPHIC RELATIONS OF REVERSALS

If the field-reversal theory is correct and if self-reversals are rare, then the same magnetic polarity should be found in rocks of the same age all over the globe. Conversely, a demonstration that groups of strata of normal and reversed polarity are exactly correlative in age all over the earth would constitute the strongest possible demonstration of the validity of the theory of geomagnetic field reversal. The success of the stratigraphic approach is contingent, however, on one principal condition: the duration of epochs in which magnetic polarity is constant must be sufficiently long to be resolved by means of the available geologic techniques for establishing global contemporaneity of events. For example, if polarity epochs lasted only 50,000 years, the classic techniques of paleontology could hardly be used to establish contemporaneity. The pace of evolution and the rates of dispersal of organisms are not sufficiently rapid to produce significant changes in widely separated fossil assemblages in so short a time.

Matuyama's discovery that the youngest reversely magnetized rocks are early Quaternary in age has been confirmed by paleomagnetic studies in many parts of the world. Thus, in a general way, these stratigraphic relations lend support to the theory of geomagnetic field reversal. If, however, a substantial proportion of the rocks studied contain minerals which undergo self-reversal after an interval about equal to the duration of the Pleistocene, as suggested by Verhoogen (1956a) for impure magnetite, then a similar stratigraphic distribution of rocks of normal and reversed polarity is to be expected without a field reversal's having occurred. If this explanation is correct, one might expect variations, from place to place, in the age of the youngest reversely polarized rocks, reflecting variations in magnetite composition, cooling rate, and other factors which might control the time required for self-reversal. However, if field reversal is the correct explanation, the transition from reversed to normal polarization must occur at exactly the same stratigraphic horizon everywhere in the world.

For the past 5 years we have attempted to determine whether the most recent transitions from reversed to normal polarization in volcanic rocks from the Snake River Plain of Idaho, from Alaska, from California, from Hawaii, and from New Mexico are contemporaneous. In a general way, our results agree with those from other continents. Reversals are lacking in rocks from the upper Pleistocene and appear first in strata designated middle or lower Pleistocene. However, our attempts to determine whether the most recent transitions at all these localities are exactly contemporaneous have proved in-

conclusive. Much of the difficulty lies in the fact that this last transition occurs during the Pleistocene Epoch. Due to both the slow rate at which evolution proceeds and the time required for plant and animal migrations, paleontological techniques are not capable of precisely resolving separate events occurring within such a short time interval as the Pleistocene. Independent indicators of difficulties encountered in attempts to obtain early Pleistocene correlations are the well-known problems of correlating the Pliocene-Pleistocene boundary from place to place, and the equally great difficulty of correlating glaciations over even moderate distances. In working with volcanic rocks, as we do, the difficulties are compounded by the paucity of intercalated sedimentary deposits bearing fossils or indicators of glacial climatic fluctuations.

Our stratigraphic studies of reversals indicate that the adjustments one must make in local stratigraphic assignments to make all data consistent with the hypothesis that a worldwide transition from reversed to normal polarity occurred during the Pleistocene fall well within the uncertainties of the original assignments. However, this is far from proving stratigraphically that the transition occurred everywhere at the same time. Indeed, if the most favored stratigraphic assignments at all localities are taken at face value, then it must be concluded that the most recent transitions were not contemporaneous.

PALEOMAGNETIC–GEOCHRONOMETRIC STUDIES

An obvious way to resolve this ambiguity is to obtain dates by radiometric methods from groups of rocks of normal and reversed polarity. On the basis of general stratigraphic considerations, the duration of geomagnetic polarity epochs is estimated to have been between 0.25 and 0.5 million years (Hospers, 1954; Khramov, 1957; 1958; 1960). Until recently, none of the radiometric methods used for dating carbonaceous materials or for dating minerals could be used in this time range. The development, by J. Reynolds at the University of California, of a gas mass

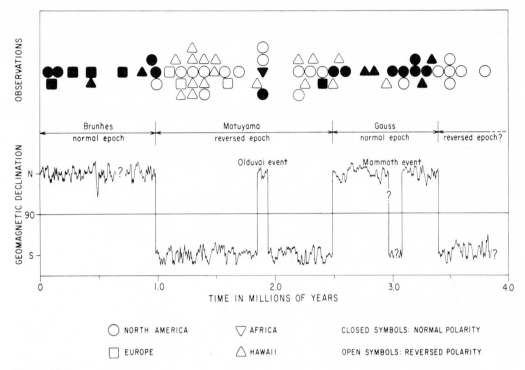

Figure 16-3
Magnetic polarities of 64 volcanic rock and their potassium-argon ages (see footnote 1, page 175).
Geomagnetic declination for moderate latitudes is indicated schematically

spectometer with improved sensitivity and low background has made it possible to date young rocks by means of the decay of potassium-40 to argon-40, despite the low concentration of radiogenic argon (a 10-g sample of a half-million-year-old rock containing 2 percent of potassium contains only 1.78×10^{-11} mole of radiogenic argon-40). The group at the University of California led by G. H. Curtis and J. F. Evernden has been particularly successful in dating young volcanic rocks by using refined extraction techniques to reduce contamination from atmospheric argon and using the Reynolds spectrometer for argon analysis. For several years the time has been ripe for applying the potassium-argon geochronometric method to the study of reversals of the magnetic field.

The polarities of volcanic rocks which have been dated by radiometry are shown in Fig. 16–3. Both the paleomagnetic and the geochronometric data represent the results of independent investigations by several laboratories.[1] We obtained the paleomagnetic data for rocks from North America; Tarling obtained the results for Hawaii; Rutten, the results for Europe; and Grommé and Hay, the results for Africa. The radiometric analyses were made by Curtis and Evernden at the University of California, by McDougall at the Australian National University, and by Dalrymple at the University of California and later at the U.S. Geological Survey.

Except for the samples from Europe, all the samples in these studies were treated in alternating magnetic fields to test stability of

[1]Cox et al. (1963a, 1963b, 1964a); McDougall and Tarling (1963); Rutten (1959); Gromme and Hay (1963); Evernden et al. (1957); Evernden et al. (1964); Dalrymple (1963). Also included are 13 new data recently determined in our laboratory. The dates are preliminary and subject to revision of ±5 percent; paleomagnetic analysis for self-reversal was made on all samples with the procedure described in the text and in our earlier publications, cited above.

the remanent magnetization, and in several instances other reliability tests were made. For example, in our paleomagnetic investigations of radiometrically dated volcanic rocks we have been especially concerned with the problem of detecting possible self-reversals, because even a few undetected self-reversals would greatly obscure the age relations of rocks of normal and reversed polarity. In an attempt to detect reversals dependent on cooling rate, we have collected from each outcrop multiple samples that cooled at different rates. Where available, baked zones were sampled, as were xenolithic inclusions containing ferromagnetic minerals different from those in the parent lava flow. Never in the course of collecting several thousand samples from several hundred lava flows and intrusive bodies have we encountered both normal and reversed polarity in the same igneous cooling unit. [A trivial exception is the occasional sample that has been struck by lightning; the magnetization effects of lightning are easily detected and removed (Cox, 1961; Graham, 1961).] When heated and cooled in the laboratory, none of these samples is self-reversing—a finding which suggests, as does the field evidence, that self-reversals dependent on cooling rate have not occurred.

In addition, a variety of thermomagnetic, crystallographic, and chemical experiments and observations were made in an attempt to find a correlation between magnetic polarity and other physical properties. None was found. It is therefore very unlikely that lavas of self-reversed polarity contributed to the date for North America cited in Fig. 16–3. Moreover, the agreement of all the data from the diverse sources is remarkably good. The ten volcanic rocks with ages between 0 and 1.0 million years all have normal polarization, whereas over 30 reversely magnetized volcanic units have ages in the range 1.0 to 2.5 million years. The only normally magnetized rocks in this interval are two normally magnetized lava flows, both with ages of 1.9 million years; their significance is discussed later. Some of the dates are for mineral separates of sanidine, biotite, and plagioclase; two are for glass; and some are for total rock basalt, this being by far the largest category. The ferromagnetic minerals have a wide range of composition, including magnetite, titanomagnetite

and oxidized titanomagnetite, and hematite, and they also show a wide range of intergrowth textures. In the samples we have studied, ferromagnetic minerals of all types occur in both the normal and the reversely polarized groups. Faced with these data, one must conclude (i) that the geomagnetic field has reversed polarity repeatedly; (ii) that self-reversals in young volcanic rocks are rare; and (iii) that the potassium-argon technique for dating young volcanic rocks is remarkably precise.

DISCUSSION

The succession of young volcanic rocks at any one locality rarely if ever is sufficiently continuous to give a complete record of the history of the geomagnetic field. Nonetheless, because the polarity of the field is a global phenomenon, data from all over the world may be synthesized, as has been done in Fig. 16–3, to indicate the declination that would be observed at any sampling site that possessed a continuous record. The short-period deviations shown schematically about the north and south directions are typical of the amplitude of secular variation in moderate latitudes; little is known as yet about the actual periods of the variation. The general character of the geomagnetic field appears to change abruptly at the boundaries of certain time intervals which we have termed polarity epochs. Epochs of two types may be recognized— those represented wholly or predominantly by rocks of normal polarity and those represented wholly or predominantly by rocks of reversed polarity.

Within the youngest reversed-polarity epoch there occurs a short interval of normal polarity identified in Fig. 16–3 as the Olduvai event. This event is now supported by data from both Africa and North America and may be considered well established. Less securely documented is the Mammoth event (Fig. 16–3), based on two points of reversal near 3.0 million years ago; the time interval between these points is about equal to the precision of the potassium-argon dating method; thus, additional data will be needed to determine whether these two points represent one or two events. At several other places in the reversal time scale there are gaps in the data sufficiently broad to suggest the possibility of additional events, but the pattern of epochs appears well established.

The accuracy with which the boundaries of epochs and events may be drawn depends on the accuracy of individual datum points, on their density and distribution, and on the time required for transitions between polarity epochs. For this time interval, it is estimated that the standard deviation of ages obtained by potassium-argon dating averages about 5 percent, although the precision of individual dates may vary considerably. We estimate that the boundary shown at 1.0 million years is precise to within 0.05 million years; that at 2.5 million years, to within 0.2 million years; and that at 3.4 million years, to within 0.1 million years.

The time required for transitions between polarity epochs is not known from direct measurement. Estimates of the order of 10^4 years have been made on the basis of the proportion of all strata that are intermediate stratigraphically between rock sequences of normal polarity and sequences of reversed polarity and that also have intermediate directions of magnetization (Doell and Cox, 1961; Picard, 1964). Another estimate may be made on the basis of the observation that of the 64 flows for which we have radiometrically determined dates, none has an intermediate direction of magnetization. Under the assumption that each of the seven transitions shown in Fig. 16–3 lasted 50,000 years, the probability that 64 randomly distributed dates would all fall outside all of the transition zones is $3\frac{1}{2}$ percent, suggesting that the transitions are shorter than 50,000 years. The corresponding probability for transitions lasting 10,000 years or less is 28 percent or more; thus such transitions are more compatible with the radiometric and paleomagnetic data. The rapidity of the transitions makes it difficult to study the important problem of the global shape of the field and the intensity of the field during the transitions, both of which factors control the magnetic shielding of the earth from cosmic rays and hence may have significant biological effects (Uffen, 1963). On the other hand, the sharpness of the epoch boundaries makes them especially useful for worldwide stratigraphic correlation and also for assessing the precision of radiometrically determined dates. For example, the overlap, near the 2.5-million-year

transition, of normal and reversed polarity, although conceivably due to rapid oscillation of the geomagnetic field, more probably reflects the precision of the individual dates. Reversals may thus provide a convenient method for directly comparing radiometric measurements on different types of rocks from all parts of the world.

We have given names rather than numerical or sequential designations to the polarity epochs for the following reasons. The previously used numerical systems count back sequentially from the present at each change in polarity (first normal, first reversed, second normal, and so on). However, if even a short-lived polarity interval is missed when a numerical system is set up, all older designations must be changed when the new polarity interval is identified. This would undoubtedly introduce considerable confusion into attempts to use the epochs for purposes of stratigraphic correlation. For example, in the first four articles linking paleomagnetic and radiometric results, most of the polarity data in the period 1.0 to 2.5 million years ago were reported, yet the significance of the short normal-polarity event at 1.9 million years ago was missed. The numerical systems also preclude designation of any given interval between polarity changes until all later intervals have been recognized.

The magnetohydrodynamic theory for the origin of the geomagnetic field, generally considered the only reasonable theory yet proposed, seems capable of accommodating reversals (Hide and Roberts, 1961; Jacobs, 1963). Complete hydrodynamic analyses of the homogeneous-type dynamos that could be operating within the fluid core of the earth have not yet been made. In fact, the very formidable mathematics required has precluded all but the most simple analyses. However, analogies, suggested by Bullard (1955) and Rikitake (1958) to the self-excited single-disc dynamo and the mutually excited two-disc dynamo may be analyzed more completely, and several solutions have been obtained for certain configurations in which the oscillating currents in the models change sign.

An axially symmetric two-disc dynamo recently analyzed by Mathews and Gardner (Mathews and Gardner, 1963) is depicted in Fig. 16–4,*a*. One of their solutions for this model (Fig. 16–4,*b*) suggests the features

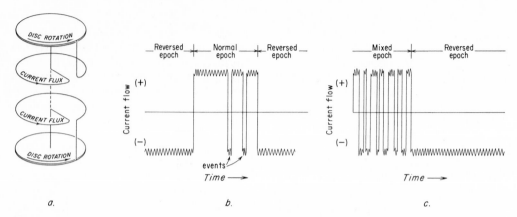

Figure 16-4
(a) Two-disc dynamo model with two solutions (b and c) for current flow (after Mathews and Gardner, 1963)

shown in Fig. 16–3 for the geomagnetic field. Another of their solutions (Fig. 16–4,*c*) infers a third type of epoch in which the field rapidly and repeatedly changes polarity. Epochs of this type, which may be termed mixed, have not yet been recognized.

Striking as these similarities are, the two-disc dynamo model is vastly different from a more realistic homogeneous dynamo model, and care should be used in interpreting the solutions for current flow as indicative of geomagnetic-field behavior. Nonetheless, the solutions do suggest that the magnetohydrodynamic theory is capable of explaining polarity reversals of the types observed, as well as the origin of the geomagnetic field itself. Conversely, any complete model or theory for the geomagnetic field must be capable of producing reversals of irregular duration.

17

Geomagnetic Polarity Epochs: a New Polarity Event and the Age of the Brunhes-Matuyama Boundary

RICHARD R. DOELL
G. BRENT DALRYMPLE
1966

From *Science*, v. 152, p. 1060–1061, 1966. Reprinted with permission of the authors and the American Association for the Advancement of Science. Copyright 1966 by the AAAS.

Earlier quantitative investigations of the geomagnetic polarity epoch time scale have placed the last change in the polarity of the earth's magnetic field (from the Matuyama reversed epoch to the Brunhes normal epoch) at 1.0 million years ago (McDougall and Tarling, 1963; Cox et al., 1963b, 1964b; Evernden et al., 1964). In a recent publication (Dalrymple et al., 1965) we pointed out that the revision of the age of the normally magnetized Bishop Tuff (Pleistocene), of California, from 1.0 to 0.7 million years made the age of the Brunhes-Matuyama boundary less certain. Over 15 reversely magnetized volcanic rocks with potassium-argon ages in the range between 2.5 and 1.0 million years clearly define the boundary as being younger than 1.0 million years. Between 1.0 and 0.7 million years, however, the only datum was that from the Bishop Tuff, and thus the revision of its age raised the possibility that the boundary might be as young as 0.7 million years.

We have now completed paleomagnetic measurements and potassium-argon age determinations of 19 Pleistocene volcanic units from the Valles Caldera, Sandoval County, New Mexico. Six of these units, all rhyolite domes that were emplaced in the caldera after the extrusion of the Bandelier Tuff (Smith et al., 1961), have potassium-argon ages between 0.7 and 1.0 million years (Table 17–1), and are therefore important for defining the age of the Brunhes-Matuyama boundary. The ages are based on replicate potassium and argon determinations, and the paleomagnetic data were obtained from multiple samples. Sampling and measurement techniques are essentially the same as those described previously (Cox et al., 1963b; Doell et al., 1966). Ages were measured on sanidine, except for unit 3X194, for which obsidian was used, and were calculated with $\lambda_\epsilon = 0.585 \times 10^{-10}$ yr^{-1}, $\lambda_\beta = 4.72 \times 10^{-10}$ yr^{-1}, $K^{40}/K = 1.19 \times 10^{-4}$ mole/mole. The atmospheric corrections ranged from 20 to 58 percent for the five

Table 17–1
Potassium-argon ages and polarities of six volcanic units from the Valles Caldera, New Mexico.

Unit no.	K-Ar age (millions of years)	Polarity
4D049	0.71	Reversed
3X122	0.72	Reversed
4D074	0.73	Reversed
4D057	0.88	Intermediate
3X187	0.89	Normal
3X194	1.04	Reversed

younger units and from 76 to 77 percent for 3X194. The standard deviation of the calculated ages is 6 percent for 3X194 and 4 percent for the other units. This calculated standard deviation is based on the results of replication studies (Dalrymple and Hirooka, 1965) and on the effect on precision of the atmospheric argon correction as calculated from the formula given by Lipson (1958).

The discovery of both normal and reverse remanent magnetizations (as well as an intermediate direction) in these rocks is not surprising because they were formed near the time of the last polarity transition. Finding the normal and intermediate directions bracketed between reversed directions was, however, not anticipated and suggests three possibilities: (i) The precision of the potassium-argon age measurements is not sufficiently high to distinguish between the ages of the units, that is 4D057 and 3X187 are really younger than the other domes; (ii) One or more of the domes may have self-reversed remanent magnetization (Uyeda, 1958; Verhoogen, 1962; Ade-Hall, 1964); or (iii) There may be a short polarity event near the Brunhes-Matuyama boundary.

There is no stratigraphic evidence concerning the relative ages of the first five domes listed in the table, although all are known to be younger than 3X194. Four other domes in the Valles Caldera that are normally magnetized and are stratigraphically younger than any of the domes discussed here give ages between 0.43 and 0.54 million years. Thus, the ages in Table 17–1 are not inconsistent

with the known stratigraphy. These relations, and the fact that at least two of the calculated ages would have to be in error by more than four times their standard deviations, lead us to reject the first hypothesis.

To investigate the possibility of self-reversal, several petrographic, thermomagnetic, and other paleomagnetic "reliability" investigations were made on these rocks. We have not found any evidence that might be construed as indicating self-reversal, and, in fact, these rocks all have quite similar intrinsic magnetic properties. Finally, we note that a self-reversal hypothesis for these data cannot explain the intermediate direction of magnetization for 4D057.

Because polarity intervals of short duration are known in other parts of the polarity epoch time scale—the Olduvai normal event at 1.9 million years and the Mammoth reversed event at about 3.0 million years, during the Matuyama reversed and Gauss normal epochs, respectively (Doell et al., 1966)—the third of the three hypotheses appears most likely.

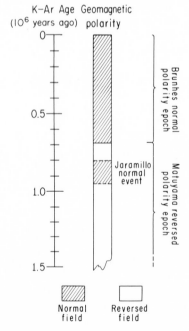

Figure 17-1
Suggested sequence of the most recent changes in polarity of the earth's magnetic field

Thus, these data and those previously published suggest the sequence for the more recent polarity changes shown in Fig. 17–1.

The placement of the Brunhes-Matuyama boundary is more or less arbitrary in view of the present data. It could be placed between 0.9 and 1.0 million years ago, in which case the three reversely magnetized domes with ages between 0.71 and 0.73 million years would represent a reversed polarity event in the Brunhes normal epoch; or the boundary could be placed at 0.7 million years ago with 4D057 and 3X187 representing a normal event in the Matuyama reversed epoch at about 0.9 million years. For purposes of stratigraphic correlation, the last transition of polarity will undoubtedly be the most useful, and we therefore prefer to assign the epoch boundary at 0.7 million years. Accordingly we here name the normal event near 0.9 million years the "Jaramillo normal event," after Jaramillo Creek, which is approximately 3 km south of the locality of unit 3X187. From the present data it is not possible to tell whether the intermediate direction represents the transition to or from the Jaramillo normal event, nor, therefore, whether the event occurred just before or just after 0.9 million years ago.

Details of these studies as well as those on the other 13 units investigated in this region will be published shortly. Meanwhile, stratigraphers and other scientists making use of the geomagnetic polarity epoch time scale for geological correlation or other purposes may find these recent data valuable and timely.

18

Dating Geomagnetic Polarity Epochs in Reunion

FRANCOIS H. CHAMALAUN
IAN McDOUGALL
1966

From *Nature*, v. 120, p. 1212–1214, 1966. Reprinted with permission of the authors and Macmillan Journals, Ltd.

In most palaeomagnetic investigations of Tertiary rocks the directions of natural remanent magnetization (NRM) are found to be approximately parallel or antiparallel to the direction of the present geomagnetic field. The rocks are said to possess normal or reversed polarity respectively. This bimodal distribution of polarities may be caused either by reversals in polarity of the geomagnetic field or by the rocks possessing the property of self-reversal.

Theoretical considerations indicate that rocks containing certain magnetic minerals may spontaneously acquire a direction of magnetization opposite to that of the ambient field. There are no laboratory tests at present available which can establish whether a rock had the property of self-reversal at the time it became magnetized. Other tests have to be used to distinguish between the two possible explanations of the observations.

If it can be shown that rocks of the same age have the same polarity throughout the world then this would constitute strong support for the hypothesis of geomagnetic field reversal. By contrast, if self-reversal were of importance in producing the bimodal distribution of polarities, then no consistent pattern of age versus polarity would be expected. Hence, a combination of isotopic dating and palaeomagnetic studies should provide a means of distinguishing between the two hypotheses.

Recently, by using these two techniques, a surprisingly consistent pattern of age versus polarity has emerged (McDougall and Tarling, 1963; 1964; Cox et al., 1963b; 1964b; 1965); this is strong evidence in favour of the hypothesis of geomagnetic field reversal. The investigations have been directed mainly at establishing the relations during the past 4 m.y., and for this period four polarity epochs have been named (Cox et al., 1964b; 1965).

In addition, at least two events have been recognized (Cox et al., 1964b; 1965), an event being a short-period change in polarity during an otherwise normal or reversed epoch. The duration of an event may be of the order of 0.1 m.y. (Cox et al., 1965).

These investigations are important for a proper understanding of the origin and behaviour of the geomagnetic field; they should also form the basis for using polarity measurements as a means of correlating rocks that cannot be dated by isotopic or palaeontological methods.

Most of the available data have been obtained on rocks from the northern hemisphere; it is of some importance to make similar investigations on rocks from the southern hemisphere. This article presents palaeomagnetic and age results on rocks from the island of Réunion in the Indian Ocean. We are concerned here mainly with geomagnetic polarity epochs; full discussion of other aspects of this work will be given elsewhere.

Réunion is situated at 55° 30′ E., 21° S. at the southwest extremity of the Mauritius–Seychelles Ridge. The island is built of two volcanoes. Piton des Neiges, an extinct volcano, forms the north-western two-thirds of the island, and Piton de la Fournaise, which is still active, occupies the south-eastern part. Various aspects of the geology of Réunion have been described previously (Lacroix, 1936; Rivals, 1950; Upton and Wadsworth, 1965; McDougall and Compston, 1967). Both volcanoes are built predominantly of basaltic lavas which have tholeiitic affinities. The main basaltic shield of Piton des Neiges is covered by an extensive veneer of feldspar-phyric basalt, hawaiite and mugearite, termed the "differentiated series" by Upton and Wadsworth (Upton and Wadsworth, 1965).

More than one hundred samples were collected from unweathered, well-exposed outcrops on both volcanoes. Specimens were oriented by means of a Brunton compass. Isotopic dating was carried out on more than thirty-five of the samples used in the palaeomagnetic investigations.

The whole-rock technique of potassium-argon dating was used as previously described (McDougall, 1964; 1966). Argon was determined by isotope dilution and potassium by flame photometry. The experimental uncertainty in the measured ages generally is less than ± 5 per cent at the 95 per cent confidence level, except in those cases where the air argon correction exceeds 85 per cent, when the error may increase to more than ± 10 per cent. Estimates of the experimental error are given in the list of results, based on the quality of the mass spectrometer runs, the air correction, and the reproducibility of both argon and potassium determinations, as most results were measured in duplicate. Only essentially holo-crystalline, unaltered rocks were used in the dating to minimize the possibility of loss of radiogenic argon. The consistency of dates determined on rocks the relative stratigraphic positions of which are known provides good evidence that the measured ages are correct within the limits of experimental error.

Palaeomagnetic measurements were made with a short-period astatic magnetometer. A series of fifteen test specimens were thermally demagnetized in steps up to 600° C, and by alternating fields up to 700 oersted (peak). These tests showed that the rocks generally are magnetically somewhat unstable. The thermal decay curves are variable, and commonly show low blocking temperatures. In many cases the intensity of NRM was reduced to less than one-half at 150 oersted (peak) during alternating current demagnetization. Furthermore, when left in the laboratory, the direction changed by as much as 15° over a period of 6 weeks. Most samples had a substantial component of temporary viscous magnetization which, however, could be removed after alternating current treatment to 150 oersted (peak). After alternating current demagnetization, the directions fall into two reasonably well-grouped clusters of normal and reversed polarity which approximate closely to the magnetic dipole field direction, when sign is disregarded. There are two exceptions: the magnetic direction of $RU13$ moves towards the reversed group on alternating current demagnetization, but does not quite reach it. Sample $RU80$ has an anomalous direction of magnetization, but would fall within the normal group if it is assumed that there was a misorientation of 180° in the strike line. This appears to be confirmed

Table 18–1

Potassium-argon dates and paleomagnetic polarity
results on rocks from Reunion (N–normal polarity;
R–reversed polarity). (Errors quoted for dates are
estimates only, equivalent to about the 95 percent
confidence level.)

Field no.	Sample no. (GA)	Potassium-argon age (m.y.)	Polarity
PITON DE LA FOURNAISE			
RU 97	1287	0.036 ± 0.03	N
RU 80	1292	0.05 ± 0.01	(N)
RU 86	1283	0.087 ± 0.02	N
RU 87	1262	0.088 ± 0.01	N
RU 88	1263	0.18 ± 0.02	N
RU 70	1290	0.33 ± 0.04	N
RU 76	1291	0.36 ± 0.04	N
DIFFERENTIATED SERIES–PITON DES NEIGES			
RU105	1264	0.104 ± 0.003	N
RU106	1265	0.137 ± 0.007	N
RU 68	1277	0.15 ± 0.02	N
RU 95	1286	0.20 ± 0.015	N
RU 66	1279	0.212 ± 0.015	N
RU 31	1257	0.227 ± 0.007	N
RU 90	1284	0.25 ± 0.01	N
RU 34	1258	0.29 ± 0.03	N
RU 49	1294	0.29 ± 0.03	N
RU 45	1260	0.35 ± 0.02	N
OCEANITE SERIES–PITON DES NEIGES			
RU 41	1259	0.43 ± 0.07	N
RU 53	1295	0.43 ± 0.03	N
RU 85	1282	0.45 ± 0.04	N
RU 36	1293	0.50 ± 0.15	N
RU 82	1281	0.54 ± 0.03	N
RU 46	1261	0.58 ± 0.02	N
RU 92	1299	1.05 ± 0.10	N ⎫
RU 93	1285	1.01 ± 0.03	N ⎬
RU182	2000	0.97 ± 0.05	N ⎭
RU 13	1255	1.02 ± 0.03	(R)
RU 9	1271	1.07 ± 0.04	R
RU 65	1276	1.10 ± 0.10	R
RU 64	1275	1.15 ± 0.03	R
RU 63	1280	1.18 ± 0.07	R
RU 62	1274	1.18 ± 0.15	R
RU 10	1278	1.96 ± 0.07	R
RU 7	1270	1.97 ± 0.07	R
RU 5	1269	1.97 ± 0.06	N
RU 27	1256	2.00 ± 0.10	R
RU 17	1272	2.03 ± 0.05	R
RU 23	1273	2.03 ± 0.05	R
RU 1	1268	2.04 ± 0.05	N ⎫
RU 2	1289	1.98 ± 0.06	N ⎬

by the fact that another sample from the same lava flow has normal polarity.

The results are listed in Table 18–1, and plotted in Fig. 18–1 together with previously published data. The nomenclature of the polarity epochs follows that of Cox et al. (1964b; 1965). Only those results are given for which the magnetic and age measurements were made on the same sample. Those samples measured palaeomagnetically, but not dated, can readily be correlated with the dated samples from the stratigraphy and polarity.

All the samples from Piton de la Fournaise have normal polarity. The indicated ages on seven samples from the volcano range from 0.04 to 0.35 m.y. Owing to the low proportion of radiogenic argon in the runs the dates have uncertainties of between ± 10 and ± 80 percent (Table 18–1) the larger error relating to the younger samples. Similarly, all specimens from the differentiated series of Piton des Neiges possess normal polarity, and the measured dates, which are generally of good precision, lie in the range 0.10–0.35 m.y. The age data confirm that Piton de la Fournaise was erupting before Piton des Neiges became extinct.

The lavas of the main dome building phase of Piton des Neiges, termed the "oceanite series" (Upton and Wadsworth, 1965), can be divided into three distinct groups on the basis of the age measurements. It is not known whether this indicates three major periods of eruption or whether it is the result of biased sampling because of the restricted accessibility of the outcrop of the oceanite series.

The youngest group of lavas of the oceanite series are dated at between 0.43 and 0.58 m.y. (six samples), and indicate that the hiatus between this series and the differentiated series was of short duration. Lavas of this group consistently have normal polarity.

The results from the lavas of Piton de la Fournaise, the differentiated series and the younger lavas of the oceanite series of Piton des Neiges provide a large body of data that show that in the age range 0.04–0.58 m.y. the polarity was normal. Fig. 18–1 emphasizes that, prior to the present work, very few results were available for this part of the time scale. These rocks clearly belong to the Brunhes normal polarity epoch.

A second group of lavas from the oceanite series have measured ages lying between 0.97 and 1.18 m.y., based on samples from seven flows. Five of the specimens yield dates in the range 1.07–1.18 m.y., and have reversed polarity. These results are consistent with previously published data; the rocks can confidently be assigned to the Matuyama reversed epoch. Sample RU13 (GA 1255), dated at 1.02 ± 0.03 m.y., was magnetically somewhat unstable, but also appears to belong to the reversed polarity group. The remaining lava of this age group is a distinctive columnar olivine basalt that crops out on the road to Takamaka. Field observations suggest that the lava occurs adjacent to the edge of a valley eroded in the oceanite series; the valley was afterwards filled by lavas of the differentiated series. The directions of magnetization measured on three different samples from this particular olivine basalt agree to within a few degrees and indicate normal polarity. Two of the specimens were collected by one of us, and the third was obtained at a later time by Dr. M. W. McElhinny. All three samples were dated (Table 18–1), and the results agree closely at 1.01 ± 0.05 m.y. The rocks are holocrystalline; RU93 and RU182 are quite free of alteration, but RU92 showed incipient alteration which may account for the poor reproducibility of the age measurements on this sample. Because of the consistency of the results we are very confident of both the polarity and age measurements.

These results suggest that a change from normal to reversed polarity occurred at about 1.0 m.y. The data available are too few to decide whether this is the boundary between the Brunhes normal and Matuyama reversed epochs. The results from Réunion and elsewhere indicate that normal polarity prevailed from the present back to about 0.7 m.y. ago, and that the polarity was reversed from about 1.0 m.y. to at least 1.9 m.y. ago (Fig. 18–1). The boundary between the two epochs must lie between 0.7 and 1.0 m.y., but because of somewhat conflicting evidence the relationships are not clear in this time range. The relevant data may be briefly reviewed. The normally magnetized Bishop Tuff, California, is dated at about 0.71 m.y. (Dalrymple et al., 1965), and basalts with reversed polarity from

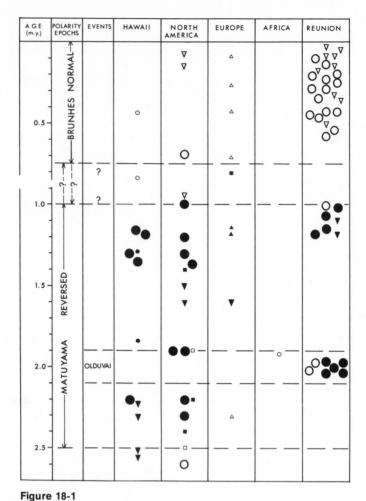

Figure 18-1
Plot of potassium-argon dates versus magnetic polarity for rocks from
Reunion, together with previously published data from elsewhere
within the age range 0 to 2.6 m.y. See refs: McDougall and Tarling
(1963), Cox et al. (1963a, 1963b, 1964b, 1965), for sources of data.
Open symbols, normal polarity; closed symbols, reversed polarity.
Large circle, data appear to be satisfactory. Small circle, date determined on different sample from that on which paleomagnetic
measurement was made, although generally from the same formation.
Triangle, polarity determined using a hand compass. Inverted triangle,
uncertainty in age determination greater than 10 per cent. Small
square, detailed information not published

Victoria and France are dated at 0.81 ± 0.03
m.y. (McDougall et al., 1966), suggesting that
a change from normal to reversed polarity occurred at 0.75 ± 0.07 m.y. A date of $0.84 \pm$
0.04 m.y. was recorded on a rock from East
Maui, Hawaii (McDougall, 1964); it was inferred that this rock had normal polarity because other samples from the same volcano
had this polarity (Tarling, 1965). Basalt of the
Devils Postpile, California, has normal polarity and was dated at 0.94 ± 0.16 m.y. (Cox et
al., 1963b), and the Big Pine obsidian is reversely magnetized and has a measured age of
0.99 ± 0.04 m.y. (Cox et al., 1963a).

These data suggest that either the boundary between the Brunhes normal and Matuyama reversed epochs occurred at 0.75 ± 0.07 m.y., with a short interval of normal polarity (that is, an event) at close to 1.0 m.y., or that the boundary lies at 1.01 ± 0.03 m.y. and the reversed polarity of rocks dated at 0.81 m.y. records an event in an otherwise normal epoch. Another possibility is that self-reversal is partly responsible for the somewhat confused picture. Clearly many additional data are needed in the age range 0.7–1.0 m.y. to distinguish between the alternative explanations.

One of the most important findings from our work in Réunion concerns the third and oldest group of lavas from the oceanite series of Piton des Neiges. The eight samples which were used for the age determinations all give dates agreeing closely at 2.0 ± 0.1 m.y. Of these rocks three have normal polarity and five show reversed polarity. Two of the rocks of normal polarity were from the same lava flow about 30 m apart. However, several other samples from the same sequence also have normal polarity, and although they were not dated it is obvious that they must have essentially the same age. The age data are not precise enough to distinguish between the two groups, but at one locality it appears that rocks with reversed polarity conformably overlie rocks with normal polarity. According to the dates these samples belong to the Matuyama reversed epoch, which extends from 1.0 m.y. or younger to 2.5 ± 0.1 m.y. (McDougall and Tarling, 1963; 1964; Cox, Doell and Dalrymple, 1963b; 1964b). Hence, from Réunion we have evidence for normal polarity within the Matuyama reversed epoch, apparently indicating a normal event of short, but uncertain, duration. This can be correlated with the Olduvai event (Cox et al., 1964b; 1965), which was first suggested from palaeomagnetic studies (Grommé and Hay, 1963) on a basalt from Olduvai Gorge, Tanganyika, dated at 1.9 m.y. (Evernden and Curtis, 1965). Our data provide strong confirmatory evidence for the reality of the Olduvai event, which we confidently date at 2.0 ± 0.1 m.y. Its existence also seems to be confirmed by work in the Pribiloff Islands, Alaska (Cox et al., 1965), but the data have not been published, so that no accurate comparison with the present investigation can be made.

The combined palaeomagnetic and age data from Réunion generally are in good agreement with results from other parts of the world. The results lend additional strong support for the hypothesis that geomagnetic field reversals are the main cause of the bimodal distribution of polarity. Our investigations emphasize the need for detailed work on rocks the ages of which lie between 0.7 and 1.0 m.y.

Although the main purpose of this article is to discuss polarity epochs a brief summary of some of the conclusions from the dating seems desirable. Piton des Neiges was active at least 2.0 m.y. ago and eruptions continued until as recently as 0.1 m.y. ago. The marked change in composition of the lavas of Piton des Neiges from basalts of tholeiitic affinities to alkali lavas occurred over a short interval of time of the order of 0.1 m.y. or less. This suggests a close genetic association between the two lava series, in agreement with conclusions reached from strontium isotope studies (McDougall and Compston, 1967; Hamilton, 1965). The data also indicate that Piton de la Fournaise was active at least during the period when the lavas of the differentiated series of Piton des Neiges were being erupted.

19

Reversals of the Earth's Magnetic Field

ALLAN COX
G. BRENT DALRYMPLE
RICHARD R. DOELL
1967

When molten volcanic rocks cool and solidify, the magnetic minerals in them are magnetized in the direction of the earth's magnetic field. They retain that magnetism, thus serving as permanent magnetic memories (much like the magnetic memory elements of a computer) of the direction of the earth's field in the place and at the time they solidified. In 1906 the French physicist Bernard Brunhes found some volcanic rocks that were magnetized not in the direction of the earth's present field but in exactly the opposite direction. Brunhes concluded that the field must have reversed. Although his observations and conclusion were accepted by some later workers, the concept of reversals in the earth's magnetic field attracted little attention. In the past few years, however, it has has been definitely established that the earth's magnetic field has two stable states: it can point either toward the North Pole as it does today or toward the South Pole, and it has repeatedly alternated between the two orientations.

There was no basis in theory for anticipating this characteristic of the earth's magnetic field. Moreover, theory on the whole subject of the earth's magnetism is so rudimentary that the mechanism of reversal is still far from being understood. Nevertheless, the magnetic memory of volcanic rocks, together with the presence in the same rocks of atomic clocks that begin to run just when their magnetism is acquired, has made it possible to draw up a time scale that shows no fewer than nine reversals of the earth's field in the past 3.6 million years. This time scale is a valuable tool for dating events in the earth's history and may help earth scientists to deal with such large questions as how much the continents have drifted.

The earth's magnetic field is the field of an axial magnetic dipole, which is to say that it is equivalent to the external field of a huge bar magnet in the core of the earth aligned approximately along the planet's axis of rotation (or to the external field of a uniformly magnetized sphere or of a loop of electric current in the plane of the Equator). The lines of force in such a field are directed not toward the geographic poles but toward the magnetic poles, and the angle at any point between true north and the direction of the field is called the declination. The lines of force are also directed, except at the Equator, toward or away from the center of the earth, and the angle above or below the horizontal is called the inclination. It is along these lines of force that the memory elements in volcanic rocks have been oriented.

The memory elements themselves are magnetic "domains": tiny bodies in which magnetism is uniform. These bodies consist of various iron and titanium oxides that can be recognized quite easily under the microscope because, unlike most rock-forming minerals, they are opaque to transmitted light and are excellent reflectors of incident light (Fig. 19–1).

At high temperatures the iron and titanium oxides are nonmagnetic. They become magnetic only after they cool to a critical point called the Curie temperature, which for the common minerals in volcanic rocks may be as high as 680 degrees centigrade or as low as about 200 degrees, depending on chemical composition. These temperatures are well below those at which rocks crystallize (about 1,000 degrees C.), so that it is clear that rocks are not magnetized by the physical rotation and orientation in the earth's field of previously magnetized grains in the molten lava, as was once thought. As the minerals begin to cool through the Curie temperature, even the earth's weak field of less than one gauss is adequate to partly magnetize them. That is because this initial magnetization is "soft," like that of iron or ordinary steel, both of which are easily magnetized by weak magnetic fields. As the rocks continue to cool, the minerals undergo a second abrupt change: the initially soft magnetism acquired in the earth's field is frozen in and becomes "hard," like the magnetism of a man-made permanent magnet.

The pertinent question for the geophysicist is how well these magnetic memory elements function as recorders of the earth's field. Do they record its direction accurately? The most direct way to assess the accuracy of volcanic rocks as recorders is to measure the magnetism of such rocks that flowed out and cooled recently in places where the magnetic field that existed at the time of flow is known. We have made such measurements on three lava flows that formed on the island of Hawaii in the years 1907, 1935 and 1955.

To obtain samples of undisturbed rock from the solid parts of a lava, a hollow cylindrical diamond drill is generally used. From five to eight cylindrical "cores" are taken from each lava flow to obtain a representative magnetic direction for the entire flow rather than for one isolated sample (Fig. 19–2). Each core's orientation with respect to the horizontal and to true north is accurately recorded before it is removed. Back in the laboratory the sample's magnetic vector is determined with a magnetometer (Fig. 19–3). The results of the measurements on the three formations indicate that lava flows record the direction of the earth's magnetic field with an accuracy of several degrees (Fig. 19–4). This is ample for most geophysical applications.

If rock magnetism is to provide a record of the ancient earth's field, the magnetic record must also be stable. Is the magnetism of rocks soft like that of iron and ordinary steel or is it hard like that of permanent magnets? This question of stability is so critical that laboratory tests to deal with it have become an integral part of paleomagnetic research. The usual technique is to place a sample from a rock formation in a kind of magnetic "washing machine," subject it to a rapidly alternating magnetic field and determine the amount of magnetism that survives. The natural magnetism of most volcanic rocks turns out to be comparable in stability to the magnetism of the hardest permanent magnets (Fig. 19–5). Once the magnetic hardness of the rocks from a given flow is established, the magnetic cleaning process can be used to strip away from each sample whatever soft magnetism has been acquired (from such sources as lightning strokes) since the rock solidified, leaving only the hard magnetism that reflects the direction of the original ambient field.

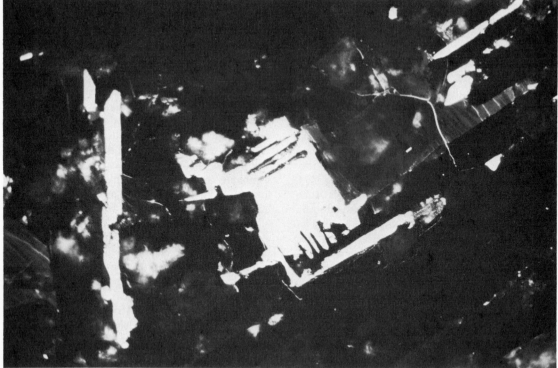

Figure 19-1

Basalt from the Pribilof Islands, seen in two photomicrographs made by Norman Prime of the U.S. Geological Survey, is one of the samples studied by the authors. The magnetic minerals, complex inter-growths of iron and titanium oxides, are opaque and therefore appear dark in transmitted light (top) and very bright in reflected light (bottom). The large clear minerals that are pale green in the top photograph and black in the bottom one are feldspars and contain the radioactive potassium isotope used for dating. The age of this rock is 1.95 million years; its magnetism approximately parallels that of the present field. Magnifications are about 600 diameters

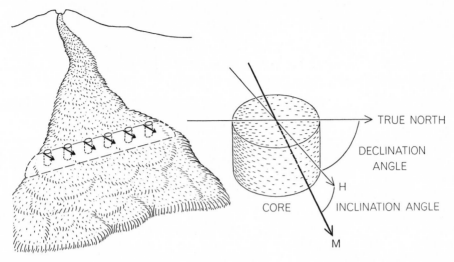

Figure 19-2
Samples for paleomagnetic studies are cores drilled from volcanic formations. The direction
of magnetization (M) is expressed as the declination angle between true north and the
horizontal projection (H) of M and the inclination angle of M above or below horizontal

It is clear, then, that paleomagnetism is
accurate and stable enough to provide information about past states of the earth's magnetic field. In assessing such information one
must of course take into consideration the
movement of rock masses that takes place
over a period of geologic time; the deviation
of a sample's magnetism from the direction of
the present field could reflect mountain-
building, warping along faults or continental
drift. Our studies of magnetic reversal have
been restricted, however, to relatively young

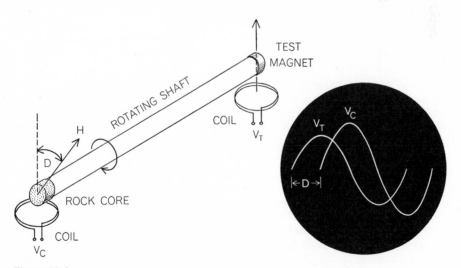

Figure 19-3
Core is mounted on a magnetometer. As the shaft rotates, electrical signals (V_C and V_T)
are induced in the coils by the core and a test magnet and can be displayed on an ocillo-
scope. The intensity and direction of the core's magnetism are determined by comparing
the magnitudes of the signals and their phase shift, which is equal to the declination (D)

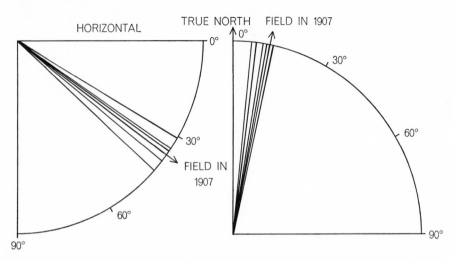

Figure 19-4
Angles of inclination (left) and declination (right) of six cores from a 1907 Hawaiian
lava flow cluster about the known angles of the 1907 field. Although angles obtained from
individual cores vary, the average values are accurate measures of the historic field

rocks and to volcanic formations we can be fairly sure are still oriented as they were when they solidified.

We began our paleomagnetic research on the island of Hawaii, where we had tested the technique and where the superb lava flows exposed on the flanks of volcanoes provide magnetic records going back about half a million years. We collected samples from 107 of these flows and found that their declination angles clustered at around 10 degrees east of true north and their inclination angles at around 30 degrees below the horizontal (Fig. 19-6). This was just about what we expected on the basis of the dipole nature of the earth's field.

Studies of other young volcanic rocks along the eastern edge of the Pacific ocean basin have yielded similar results (Fig. 19-7). At the high latitude of the Pribilof Islands in the Bering Sea the magnetic vectors of the lava flows are inclined steeply downward, as one would expect for a dipole field in high latitudes, whereas in the Galápagos Islands on the Equator the magnetic vectors are almost horizontal. Measurements in many parts of the world indicate that during the time spanned by these young lava flows (roughly half a million years) the earth's field was essentially dipolar and was aligned as it is today.

Quite different results are obtained when paleomagnetic techniques are applied to somewhat older lava flows. Only about half of these flows are magnetized in the same direction as the younger ones; the remainder are magnetized in the opposite direction. For example, some volcanic rocks at middle latitudes in the Northern Hemisphere are magnetized toward the south and upward, rather than toward the north and downward (Fig. 19-7). In recent years this "antiparallel" magnetism has been found in thousands of samples of volcanic rock from all over the world by scores of investigators working independently. Sampling has been particularly intensive in the range of ages between 3.5 million years ago and the present, and the paleomagnetic results obtained are always remarkably similar. The magnetic vectors fall into two groups: "normal" vectors nearly parallel to the present field of the earth and "reversed" ones that are nearly opposite. Most of the data are clustered within 30 degrees of these two directions, with very few vectors oriented in intermediate directions (Fig. 19-8).

The immediate implication is that the earth's magnetic field has indeed reversed its direction in the past. Brunhes so interpreted his results in France in 1906, although he cautiously restricted field-reversal to the area from which

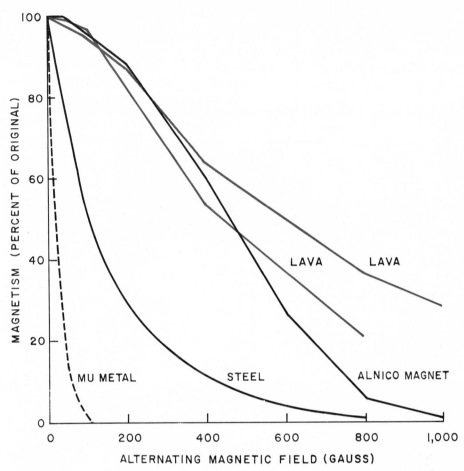

Figure 19-5
"Hardness" of the magnetism of basalt samples is expressed as the amount of magnetism remaining after "washing" in alternating magnetic fields of various strengths. The stability of the rocks' magnetism is far greater than that of the alloy "mu metal" or of steel; it is comparable to that of the alloy Alnico, of which permanent magnets are often made

he collected his samples. In 1929 Motonori Matuyama also found evidence that the field has reversed, but he too restricted his conclusions to the area in Japan from which his samples had come. The accumulating evidence that reversed magnetic directions are invariably opposite to the present field direction at the sampling site led in time to the hypothesis that the sample reversals are not local but global; in other words, that the entire field reverses.

An important alternative explanation must be considered before the field-reversal hypothesis can be accepted. The alternative is that rocks magnetized in reverse may possess some special mineralogical property that causes them to become so magnetized in a normal field. The existence of such "self-reversal" in rocks was suggested in 1950 by John Graham, then at the Carnegie Institution of Washington's Department of Terrestrial Magnetism, as an explanation for the occurrence of both normal and reversed magnetism in rock samples that had formed simultaneously. Graham's suggestion stimulated the French physicist Louis Néel to examine the problem from the viewpoint of solid-state physics, and Néel soon discovered several

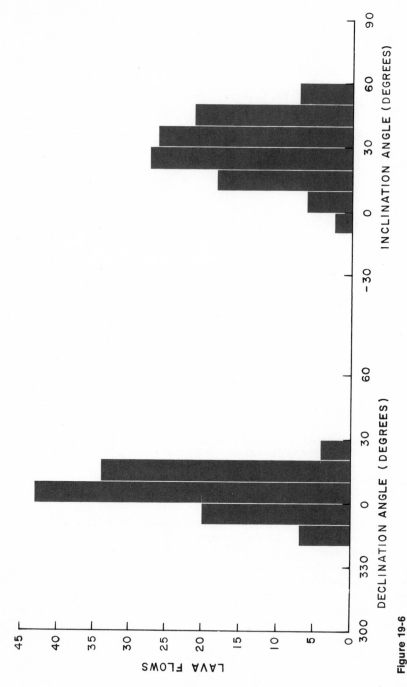

Figure 19-6

Paleomagnetic results are shown for 107 volcanic formations on the island of Hawaii. Each angle represented in these histograms is an average derived from five or more cores in one lava flow. The data are grouped in a mean northerly direction and at a mean inclination of about 30 degrees below the horizontal. That is, they roughly parallel the present direction of the earth's field

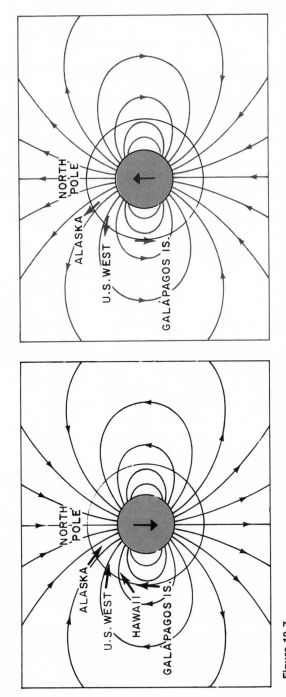

Figure 19-7
Inclination angles from flows in Alaska, the U. S. West (California, Idaho and New Mexico), and island of Hawaii and the Galápagos Islands are shown by the heavy inclined arrows. The flows range up to three million years in age. The angles fall into two distinct groups: a "normal" group aligned with the earth's present field, that of a bar magnet pointed toward the South Pole (left), and a "reversed" group appropriate to an oppositely oriented field (right). All the flows on Hawaii had normal magnetism

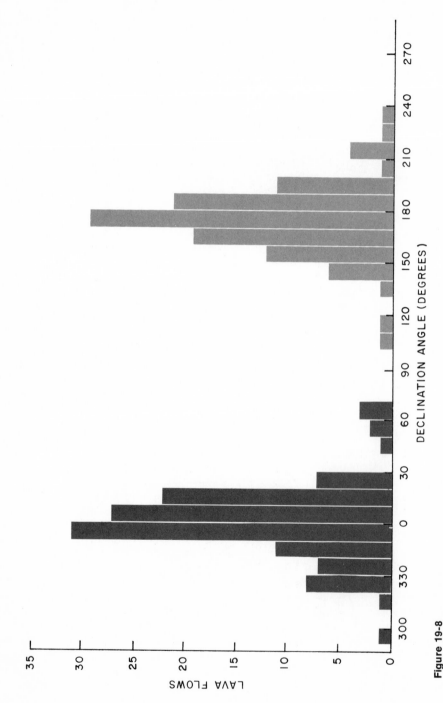

Figure 19-8
Declination angles from 229 flows up to three million years old in Alaska, the western U.S., Hawaii and the Galapagos Islands display a similar twofold grouping: northerly (normal) and southerly (reversed). Intermediate directions are seldom observed

ways in which self-reversal might occur. Experimental confirmation came almost immediately. At the Philips Research Laboratories in the Netherlands, E. W. Görter synthesized an iron-chromium-manganese compound that underwent self-reversal, and S. Uyeda and T. Nagata of the University of Tokyo found a self-reversing volcanic rock.

It is thus apparent that at least some volcanic rocks are not infallible magnetic recorders. Like laboratory recorders that are hooked up backward, they sometimes record a signal that is not only wrong but is exactly wrong by 180 degrees. If all reversed magnetism could be explained in this way, the experimental evidence for reversals in the earth's magnetic field would vanish. An obvious experiment is to heat and then cool rock samples in a known field and measure their acquired magnetization. This operation has been performed on many hundreds of rock samples with reversed magnetism, and fewer than 1 percent have turned out to be self-reversing.

Therefore the laboratory evidence favors the field-reversal hypothesis. Like many rock-forming processes, however, the acquisition of natural magnetism cannot be reproduced with complete fidelity in the laboratory. The missing ingredient is time, and for certain of the theoretical self-reversing processes this ingredient is crucial. For example, John Verhoogen of the University of California at Berkeley has shown theoretically that whereas certain iron oxides containing impurities of aluminum, magnesium or titanium would be magnetized normally when cooled rapidly in a normal magnetic field, the magnetism could be reversed as the atoms in the cooled oxide reordered themselves toward an equilibrium distribution. The calculated time required for this self-reversal is on the order of 100,000 to a million years, so that it could hardly be reproduced in the laboratory. The theoretical studies by Néel and Verhoogen showed that the fact that self-reversal is rare in the laboratory does not make it safe to conclude that it is equally rare in nature. How, then, could one determine the geophysical significance of reversed magnetism? Two main lines of experimental attack have been pursued during the past decade, each closely related to one of the two proposed reversal-producing processes.

One approach was to search for a correlation between the magnetism of rocks and their mineralogy. Even though self-reversals may not always be reproducible in the laboratory, if all reversed magnetism is due to a process occurring on the mineralogical level, rocks with reversed magnetism should be somehow different from those with normal magnetism; chemical processes being the same the world over, the unique mineralogical properties associated with reversed magnetism should appear in rocks from all over the world.

This approach has been pursued most actively by P. M. S. Blackett at the Imperial College of Science and Technology in London and Rodney Wilson at the University of Liverpool. In some sequences of rocks Wilson has found a correlation between reversed magnetism and mineralogical properties, but in other rocks he finds no such correlation. Like Wilson, we have occasionally noted a correlation between mineralogy and magnetism within a sequence from one locality, but such a local correlation may well stem from the tendency of volcanic flows to occur in pulses. Between two successive pulses separated by a long time interval the mineralogical character of the lavas commonly changes; if the polarity of the earth's field also happens to change in this interval, there will be an apparent correlation between mineralogy and polarity. In short, mineralogical investigations have not yielded evidence that all or even most reversed magnetism is produced by self-reversal.

The second experimental approach followed from an implication of the field-reversal theory: If the earth's magnetic field alternates between intervals when it is normal and intervals when it is reversed, the geologic ages of normal and reversed rocks should fall into corresponding intervals. Data bearing on the age and magnetism of rocks should provide a yes-or-no answer to the validity of the field-reversal theory and, if the theory is valid, should yield a time scale for reversals. Matuyama had noted in 1929 that the geologic age of all the rocks with reversed magnetism in Japan was early Pleistocene (about a million years ago), whereas younger rocks invariably had normal magnetism. The strongest possible evidence in support of the field-reversal theory would be to extend Matuyama's study to show that rocks from all parts of the world,

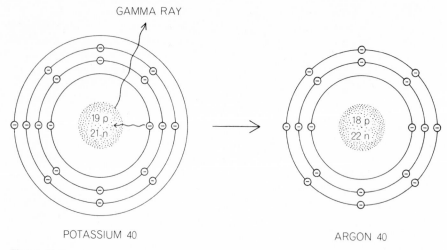

Figure 19-9
Potassium-argon clock by which reversals of the earth's field are dated is based on the decay of potassium 40 to yield argon 40. In the decay an extranuclear electron captured by the potassium nucleus converts a proton to a neutron and a gamma ray is emitted

regardless of mineralogy, occur in similar normal and reversed sequences that are time-dependent.

The difficulty lay in finding a sufficiently precise method for establishing the age relations of normal and reversed rocks. Many techniques that yield fairly precise age relations when applied to older rocks are based on plant and animal fossils; these techniques begin to break down when applied to the past million years or so because of the slow rate at which evolution proceeds and the time required for plant and animal migrations. A solution that suggested itself was some kind of radioactive clock, and our search quickly narrowed to the potassium-argon clock first suggested in 1940 by Robley D. Evans of the Massachusetts Institute of Technology and now widely applied in geological investigations (Fig. 19–9). Potassium 40, a radioactive isotope that constitutes .012 percent of all potassium, can be found as a chemical constituent in most rock-forming minerals. It decays at a known and constant rate to argon 40, a gas that forms no known compounds.

The argon is trapped within the crystal structure of the minerals, and if the minerals are not heated or changed in some way, it accumulates there. Its amount is a function of the amount of potassium present and the length of time since the decay and entrapment processes began. Therefore by measuring the amount of potassium 40 and argon 40 in a rock one can calculate its age. Argon will not accumulate as long as the rock is in a molten state, so for volcanic rocks the potassium-argon clock is started only when the rock solidifies.

The amount of potassium 40 in a sample is usually determined by measuring all the potassium in the sample by standard chemical methods and then calculating the potassium 40 from its known relative abundance. The argon determination is more difficult because the amounts are extremely small. In a typical 10-gram sample of basalt a million years old the amount of argon 40 from potassium 40 is 10^{-9} (.000000001) gram, and the accuracy of the dating depends on the accuracy with which this argon can be measured. A sample of the rock or mineral is placed in a gas-extraction apparatus and melted to release the accumulated argon 40. Reactive gases such as oxygen, nitrogen and water are removed. During the extraction a known amount of isotopically enriched argon, called the tracer, is mixed with the gas from the sample, so that the final argon gas consist of three components: the argon 40 whose amount is to be determined; the tracer, which is mostly argon 38 but which also contains some argon 36 and argon 40, and contaminating argon from the atmosphere, for which a correction must be made. This argon mixture is analyzed with a mass spectrometer that gives the relative amounts of the three isotopes of argon. Knowing the amount of the enriched tracer and its isotopic composition, and the relative composition of atmospheric argon and of the total gas mixture, one can calculate the amount of argon derived from potassium 40. This information is used with the results of the potassium analysis to determine the age of the rock.

For the reversal problem the potassium-argon method has several distinct advantages over other dating methods. It can be applied to a wide variety of volcanic rocks. It is also the only dating method that can be applied in the range from a few thousand to several million years ago. And, as we have noted, the potassium-argon clock starts to run at exactly the same time the magnetic record is frozen into a volcanic rock.

The potassium-argon dating method has now been successfully applied to rocks from nearly 100 magnetized volcanic formations with ages ranging from the present back to 3.6 million years (Fig. 19–10). This work has been done primarily by ourselves at the U.S. Geological Survey laboratory in Menlo Park, Calif., and by Ian McDougall, D. H. Tarling and F. H. Chamalaun at the Australian National University. Relevant data have also been contributed by M. Rutten of the University of Utrecht and by C. S. Grommé, R. L. Hay, J. F. Evernden and G. H. Curtis of the University of California at Berkeley. The rocks that were investigated came from different parts of the world and are of different types, so that the available data come from heterogeneous sources.

As Fig. 19–10 shows, the ages of these magnetically normal and magnetically reversed rocks are well grouped in distinct sequences, leaving little room for doubting the reality of geomagnetic field reversals. To explain them by self-reversal would require an unreasonable kind of coincidence involving synchronous worldwide changes in the nature of the processes by which minerals are formed and magnetized.

Four major normal and reversed sequences are defined by the paleomagnetic and radioactive-clock data for the past 3.6 million years. We call these major groupings geomagnetic polarity epochs and have named them for people who made significant contributions to our knowledge of the earth's magnetic field. Superimposed on the polarity epochs are brief fluctuations in magnetic polarity with a duration that is an order of magnitude shorter. We call these occasions polarity events and have named them for the localities where they were first recognized.

The polarity events are important for theories of the earth's magnetism because they emphasize the irregular nature of reversals of the earth's field. The first polarity event to be discovered was the "Olduvai" normal event, which is recorded in a flow in Olduvai Gorge in Tanzania that was investigated in 1963 by Grommé and Hay. At first the Olduvai flow was thought to lie within the "Gauss" normal polarity epoch and hence was not recognized as an anomaly. When better dating of the epochs placed the date of the Olduvai flow within the "Matuyama" reversed epoch, it appeared to be an unexplained anomaly in an otherwise coherent picture.

The explanation that the Olduvai result represents a brief, worldwide fluctuation in polarity was first advanced by us after we discovered in the Pribilof Islands three lava flows that are normally magnetized, like the Olduvai flow, and that have similar ages of about 1.9 million years. These flows were sandwiched between reversed flows that gave slightly older and slightly younger ages, providing the evidence that confirmed the existence of polarity events. Since then we have recognized and named two additional events: a reversed one that was recorded 3,050,000 years ago at Mammoth, Calif., and a normal one recorded about 900,000 years ago in some rocks near Jaramillo Creek in New Mexico. The Jaramillo event was recently confirmed by Chamalaun and McDougall in their study of lava flows on Réunion Island in the Indian

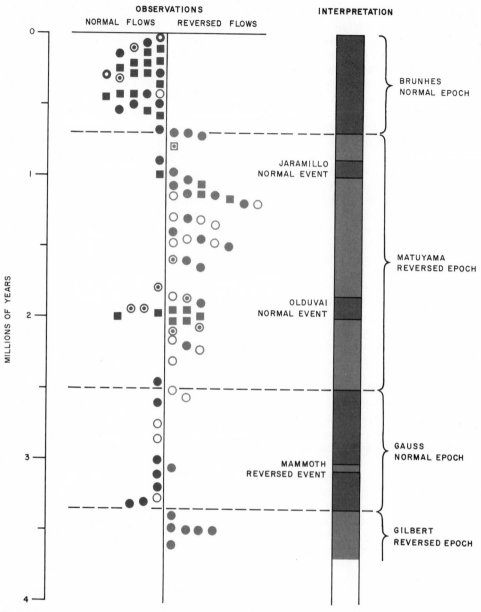

Figure 19-10

Time scale for reversals of the earth's magnetic field was established on the basis of paleomagnetic data and radiometric age obtained for nearly 100 volcanic formations in both hemispheres. Here the flows with "normal" (blue) and "reversed" (red) magnetism are arranged by their age (left). It is clear that the data fall into four principal time groupings, or geomagnetic polarity "epochs," during which the field was entirely or predominantly of one polarity. Superimposed on the epochs are shorter polarity "events."

Ocean, where they also found two additional flows that represent the Olduvai event.

Only rarely does a sequence of lava flows succeed in capturing a record of a polarity transition. This indicates that the time required for a complete change of the earth's magnetic field from one polarity to another is amazingly short; our best estimate of the transition time is 5,000 years. This is based on the ratio between the number of lava flows that happen to have recorded the earth's field during a transition and the number of flows with clearly defined normal or reversed directions. An indirect estimate of this kind is necessary because the potassium-argon dating method is unable to resolve age differences as small as 5,000 years. On the scale of geologic time, polarity transitions appear to be almost instantaneous, and they therefore provide sharp time markers indeed.

The idea that the earth's magnetic field reverses at first seems so preposterous that one immediately suspects a violation of some basic law of physics, and most investigators working on reversals have sometimes wondered if the reversals are really compatible with the physical theory of magnetism. The question is meaningful only within the context of a broader question: Why does the earth have a magnetic field? Geophysicists are simply not sure. After centuries of research the earth's magnetic field remains one of the best-described and least-understood of all planetary phenomena. The only physical mechanism that has been proposed as the basis of a tenable theory is the mechanism of a magneto-hydrodynamic dynamo. According to this theory, which has been developed primarily by Walter M. Elsasser, now at Princeton University, and Sir Edward Bullard of the University of Cambridge, the molten iron-and-nickel core of the earth is analogous to the electrical conductors of a dynamo. Convection currents in the core supply the necessary motion, and the resulting electric currents create a magnetic field. The entire regenerative process presumably began with either a stray magnetic field in the earth's formative period or with small electric currents produced by some kind of battery-like action [see "The Earth as a Dynamo," by Walter M. Elsasser; SCIENTIFIC AMERICAN, May, 1958].

The mathematical difficulties of this theory are immense. It is impossible to predict what the intensity of the earth's field should be or whether it fluctuates or remains stationary. Certainly the theory is in too rudimentary a state for one to predict whether reversals should or should not occur or should occur only under certain conditions. On the other hand, complete mathematical solutions have been obtained for simple theoretical models of dynamos, and these models do show spontaneous reversals of magnetic field; some of the models show sequences of reversals that are strikingly similar to the geomagnetic polarity time scale. These results at least demonstrate that magnetic reversals are possible in self-regenerating dynamos. The fact remains that observations are leading theory in this area of investigation, and any complete theory of geomagnetism will eventually have to accommodate the observed reversals of the field.

Meanwhile geologists are applying the reversal time scale to establish age relations among rocks they would be hard put to date any other way. An especially important application is in determining the ages of deep-sea sediments, which are very difficult to date beyond the short range of 200,000 years. It has long been recognized that fine-grained sediments may become magnetized in the earth's field as they drift slowly downward in quiet water. Recently C. G. A. Harrison and B. M. Funnell of the Scripps Institution of Oceanography and N. D. Opdyke and J. D. Hays and their colleagues at the Lamont Geological Observatory of Columbia University have observed magnetic reversals in the sediments of deep-sea cores (Fig. 19–11). In one core in particular (from the Bellingshausen Sea near Antarctica) Opdyke and Hays found a polarity record going back to the "Gilbert" epoch, or 3.6 million years, in which the pattern of reversals is remarkably similar to the pattern of our polarity time scale. Even the brief polarity events are clearly discernible. These findings confirm the reversal time scale determined from volcanic rocks and suggest that polarity studies can provide a method for determining rates of sedimentation and for establishing worldwide correlations among various deep-sea sediments, two

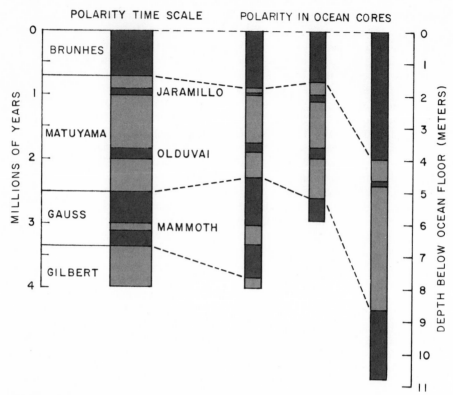

Figure 19-11

Deep-sea sediments confirm the field-reversal time scale. Magnetic particles become oriented in the direction of the earth's field as they settle through the water; a core that samples many layers of sediments may record a series of normal (gray) and reversed (light) epochs and events. Here cores from antarctic waters are correlated with the time scale

problems that have long perplexed oceanographers. Magnetic studies are also helping to establish stratigraphic links between marine and continental rocks. Magnetic-reversal stratigraphy has shown, for example, that sediments of glacial origin on Iceland and at the bottom of the Bellingshausen Sea were both deposited at about the end of the "Gauss" normal polarity epoch, or about 2.5 million years ago—a fact of considerable importance for Pleistocene geology.

Reversals may explain certain puzzling magnetic anomalies characteristic of many oceanic areas, particularly those adjacent to the mid-ocean rises, or ridges [see "The Magnetism of the Ocean Floor," by Arthur D. Raff; SCIENTIFIC AMERICAN, October, 1961]. These anomalies are parallel bands, extending

for hundreds and even thousands of miles, in which the intensity of the earth's magnetic field is higher or lower than the average for the region. It is easy to see how the presence of normal and reversed magnetized rock formations in the crust of the earth, which would add to and subtract from the earth's main dipole field, could account for such findings. Many of the magnetic-anomaly patterns, however, display a striking symmetry around the crests of certain mid-ocean ridges (Figs. 19–12 and 19–13) this is difficult to explain on the basis of familiar volcanic processes.

Recently F. J. Vine, now at Princeton University, and J. H. Matthews of the University of Cambridge have pointed out that ideas advanced by Harry H. Hess of Princeton and by the Canadian geophysicist J. Tuzo Wilson

Figure 19-12
Magnetic anomalies have been discovered in ocean floors,
particularly along mid-ocean rises. One pattern was mapped
in an area (dark gray) on the mid-Atlantic ridge

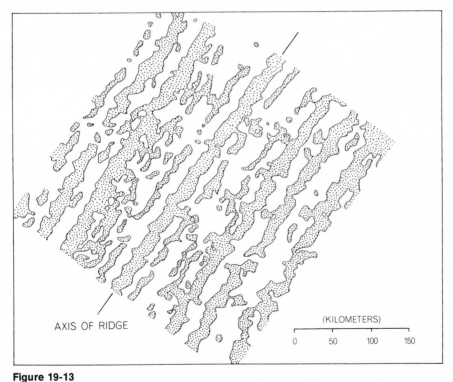

AXIS OF RIDGE

(KILOMETERS)

0   50   100   150

Figure 19-13
Anomaly pattern in the area delineated on the map at the top of the page is strikingly
symmetrical. The parallel bands in which the earth's field is stronger (stippled) or weaker
(white) than the regional average are oriented along the ridge's axis. The magnetic bands
are presumably produced by bands of rock with normal and reversed magnetism

Figure 19-15
"Magnetic washing machine" devised by the authors subjects a sample placed in the chamber (center) to an alternating magnetic field while rotating it about three perpendicular axes. The sample's hard magnetism is unaffected, but the soft component, forced to change direction repeatedly, is destroyed as the alternating current is reduced to zero

to account for certain characteristics of ocean basins and their margins and also for the drifting of continents may shed light on the symmetrical anomalies. Hess and Wilson had suggested that convection currents in the earth's mantle, the layer below the crust, may bring material up to form a mid-ocean ridge and then move the material outward, away from the ridge [see "Continental Drift," by J. Tuzo Wilson; SCIENTIFIC AMERICAN, April, 1963]. If successive bands solidified and were magnetized during successive polarity epochs, Vine and Matthews reported, the symmetry of the patterns could be explained on this basis. So could the particular spacing of the bands along the mid-Atlantic ridge, for exam-

ple, provided that the sea floor is spreading at the rate of about one centimeter per year (Fig. 19–14). This rate is consistent with earlier estimates by Wilson. Although the hypothesis of sea-floor spreading seems to be inconsistent with some other lines of evidence and has been resisted by many oceanographers and geologists, the magnetic evidence seems to reinforce it.

Reversals of the earth's magnetic field may even have implications for the history of life on our planet. R. J. Uffen of the University of Western Ontario pointed out in 1963 that if the magnetic field of the earth disappears or is greatly attenuated during a reversal in polarity, the earth would lose some of its

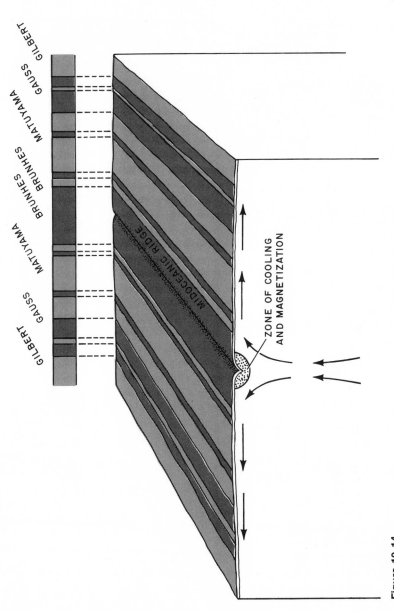

Figure 19-14
Spreading of the ocean floor could explain the magnetic-anomaly patterns. According to one theory (see text) convection currents bring molten material up under the mid-ocean ridge, where it cools, becomes magnetized and then spreads laterally away from the ridge. Symmetrical bands of normal and reversed rocks would be produced by the combined effect of field reversal and spreading

magnetic shielding against cosmic rays; with the resulting increase in radiation dosages, mutation rates should increase. Paleomagnetic evidence for the behavior of the earth's field during polarity transitions is fragmentary, but there are indications that the field may be only about a fifth as intense as in normal times. Uffen argues on paleontological grounds that rates of evolution were exceptionally high at times when the earth's magnetic field was undergoing many changes in polarity, although the support for this conclusion in the paleomagnetic record is rather weak. Cores examined by Opdyke and Hays do provide some support for Uffen's theory in that major changes in the assemblages of microfossils appear near two of the magnetic-polarity changes. Much additional information is needed, however, before it will be possible to judge the extent to which field reversals may have affected life on the earth.

20

Geomagnetic Reversals

ALLAN COX
1969

The periods of the earth's magnetic field that are most crucial to an understanding of the geomagnetic dynamo occur at the ultralow-frequency end of the spectrum. Yet, until recently, this part of the spectrum received relatively little attention, mainly because the spectrum extends to periods longer than the productive lifetimes of individual scientists. The longest periods are, indeed, longer than the entire history of scientific observation. The first hint of the existence of periods greater than 100 years came from observations of changes in field intensity. In 1835—the first year for which C. F. Gauss was able to assemble enough worldwide data to analyze the field through the use of spherical harmonic functions—the earth's dipole moment was 8.5×10^{25} gauss cm^3. By 1965, the moment had decreased to 8.0×10^{25}.[1] The dipole field decreased during this time at a remarkably uniform rate, and it will, if the rate remains constant, pass through a zero point about 2000 years from now and then reverse its polarity (Leaton and Malin, 1967; McDonald and Gunst, 1967). However, other interpretations are equally consistent with the intensity data— for example, the field might oscillate without changing polarity. Therefore, although there is little question that the geomagnetic spectrum extends to periods considerably greater than the few centuries spanned by the records of magnetic observatories, the nature of the spectrum has long remained uncertain.

The length of the available magnetic record has now been increased by more than six orders of magnitude through the study of the natural magnetism of rocks and baked clay which retain a magnetic memory of the earth's

[1]Recent analyses of the field include those of McDonald and Gunst (1967), Hurwitz et al. (1966), and Cain et al (1965). Cain and Hendricks (1968) report that the rate of decrease has dropped since 1900. A summary of the older analyses is given by Vestine (1953).

field in the past. This paleomagnetic research indicates that the earth's field has undergone numerous fluctuations of intensity and that it has also undergone many (although less numerous) reversals in polarity. Recent work suggests that these two phenomena may occupy adjacent parts of the geomagnetic spectrum, the division between them being at periods of about 10^4 years. This part of the spectrum, which is very difficult to resolve experimentally, may hold the key to understanding how polarity reversals are related to intensity fluctuations.

CHANGES IN THE EARTH'S DIPOLE MOMENT

The technique for measuring ancient geomagnetic field intensity is based on the observation that, when volcanic rocks and pieces of pottery are cooled in weak magnetic fields, they acquire thermal remanent magnetization which is parallel in direction and proportional in intensity to the applied field. If the natural thermal remanence of rocks is magnetically stable, the ancient field acting on a sample when it originally cooled may be found by reheating the sample and cooling it in a known field. The ancient field intensity F_0 is then given by

$$F_0 = F_a \frac{J_0}{J_a} \qquad (1)$$

where J_0 is the natural remanence and J_a is the remanence acquired in the known applied field F_a (Thellier and Thellier, 1959; Coe, 1967a; Smith, 1967a).

This technique, although simple in concept, is difficult to carry out experimentally because, on being reheated in the laboratory, the ferromagnetic minerals contained in rocks and baked clay commonly undergo chemical changes to form new ferromagnetic minerals. Clearly, the magnetization acquired when the altered samples are cooled in a known field does not provide a measure of the ancient field. However it has proved possible experimentally to identify samples which are chemically stable on heating. These yield values for ancient field intensity which are internally very consistent and appear to be reliable.

In all, 127 paleomagnetic intensity determinations have now been made on samples with ages $0 < t \le 10^4$ years. This work, which was carried out in different laboratories on samples from many parts of the world, was summarized recently by P. J. Smith (1967b; 1967c), who reduced each ancient field intensity to a virtual dipole moment, defined as the moment of the dipole needed to produce the observed ancient intensity if the earth's field had been entirely dipolar. Virtual dipole moments for samples of the same age from different parts of the world are scattered, with a standard deviation of about 20 percent, because the earth's field consists in part of an irregular nondipole component. This scatter has been reduced in Fig. 20–1 by averaging virtual dipole moments for different parts of the world. The decrease in dipole moment since 1885, as shown by the slanting bar at left in Fig. 20–1, is seen to be but the most recent part of a well-defined half cycle with a maximum of 12×10^{25} gauss cm^3, which began about 4000 years ago. An earlier half cycle is weakly suggested by the few older intensity measurements. The important question of whether changes in field intensity are periodic cannot be answered from the present data, but there can be no question that dipole fluctuations occur, with durations of the order of 10^4 years.

POLARITY REVERSALS

Extension of the known geomagnetic spectrum to even longer periods has come about with the discovery that the earth's field undergoes reversals in polarity. Through paleomagnetic research it has been found that the earth's dipole moment alternates between two antiparallel polarity states, a *normal* state, in which the field at the earth's surface is directed northward, and a *reversed* state in which the field has the opposite direction. In either state the dipole undergoes an irregular wobble about the earth's axis of rotation, with an angular standard deviation of 12 degrees (Cox and Doell, 1964). The time required to complete a transition between polarity states is estimated to be from 10^3 to 10^4 years (Harrison and Somayajulu, 1966; Ninkovich et al., 1966; Cox and Dalrymple, 1967b); during this time

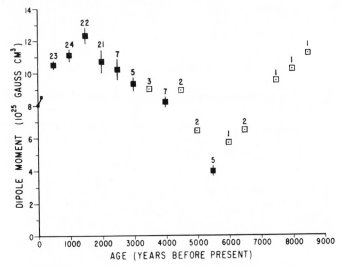

Figure 20-1
Variations in geomagnetic dipole moment (Cox, 1968). Changes
during the past 130 years, as determined from observatory measurements, are shown by the slanting bar at left. Other values were
determined paleomagnetically (Smith, 1967b, 1967c). The number of
data that were averaged is shown above each point, and the standard
error of the mean is indicated by the vertical lines (Cox, 1968)
except for points (open squares) with too few data to provide
meaningful statistics

the field does not go to zero but undergoes an
intensity decrease of from 60 to 80 percent
(Smith, 1967b; 1967c; Cox and Doell, 1964;
Cox and Dalrymple, 1967b; Coe, 1967b;
Harrison and Somayajulu, 1966; Ninkovich
et al., 1966). Averaged over a long sequence
of reversals, the total amount of time spent by
the dipole in the reversed state is equal to that
spent in the normal state (Doell and Cox,
1961; Cox, 1968; Heirtzler et al., 1968).
Moreover, the average intensity of the field
in the two states is the same (Smith, 1967b;
1967c), suggesting that the two states have
equal energy levels. In all of these characteristics the earth's field operates like a remarkably
symmetrical oscillator or, more precisely, like
a bistable flip-flop circuit.

The greatest element of irregularity in reversals is the length of time between successive changes in polarity. The longest known
polarity interval lasted for 5×10^7 years (Irving, 1966b; McMahon and Strangway, 1967);
intervals with lengths of the order of 10^6 years
are common, and short polarity intervals oc-

cur, with durations of less than 10^5 years. The
latter have proved to be the most difficult to
measure. This article deals mainly with recent
experimental work on short polarity intervals.

DISTRIBUTION FUNCTION
FOR POLARITY INTERVALS

The results from paleomagnetic studies of
reversals may be presented either in the form
of a time scale for reversals or, more com-
pactly, in the form of a histogram showing the
frequency of occurrence of polarity intervals
as a function of their length. The question of
what distribution function should be used
to describe the observed frequency distribution depends on the more fundamental questions of why reversals occur and what controls
the immense variation in the length of time
between them.

It is now a generally accepted theory that
the earth's field is generated by hydromagnetic
processes in the earth's fluid core and that the

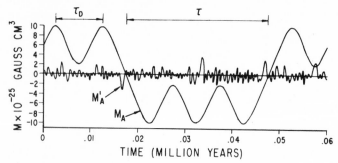

Figure 20-2

Model for reversals used to derive the distribution function for reversals. τ_D is the period of the dipole field and τ is the length of a polarity interval. A reversal occurs whenever the quantity M'_A, which is a measure of the nondipole field, becomes sufficiently large relative to the dipole moment M_A (Cox, 1968)

energy which maintains the field against ohmic dissipation is provided either by thermal convection or by turbulence generated by the earth's precession (Malkus, 1968). Unfortunately, however, the theory of the homogeneous fluid dynamo is not sufficiently developed to account quantitatively either for fluctuations in dipole intensity or for polarity reversals. Particularly difficult is the problem of reconciling the long times between polarity reversals, which may exceed 10^6 years, with the much shorter time constants of dipole fluctuation (10^4 years) and fluctuations of the nondipole field (3 years to 10^3 years).

In the absence of a complete theory of geomagnetism, some interesting and possibly relevant analogies are provided by the behavior of nonfluid self-excited dynamos. The simplest one that provides a model for geomagnetic intensity fluctuations and polarity reversals consists of two mutually coupled Faraday disk dynamos (Rikitake, 1958; Allan, 1962; Matthews and Gardner, 1963). In some solutions for this dynamo the current oscillates about a mean current flowing first in one and then in the opposite direction. The distribution function for the lengths of time between successive reversals in current direction depends on (i) the period of the oscillations and (ii) the number of cycles between successive reversals. Both are sensitive to changes in the physical parameters of the model, some solutions for the dynamo having sharply peaked distributions and others having monotonic decreasing functions, the shortest polarity intervals being the most frequent.

The earth's field is produced by a much more complex fluid dynamo, so that detailed comparisons with particular solutions for the disk dynamos are of little value. Unlike the rigid disks of the dynamo, the earth's core resembles the atmosphere in possessing a large random component of motion due to turbulence. A direct measure of this component is provided by the rapidly varying nondipole field, the average intensity of which is 20 percent of the dipole field. The sensitivity of the timing of reversals in a solid disk dynamo to small changes in the physical conditions of the model suggests that in fluid dynamos the timing of reversals would also be sensitive to a large random element in the pattern of fluid motions and magnetic fields.

The foregoing considerations suggest the following model, from which a distribution function for polarity intervals can be derived (Cox, 1968) (Fig. 20–2). The main geomagnetic dynamo is assumed to be a steady dipole oscillator which undergoes a polarity reversal only when this is triggered by random fluctuations of the much more rapidly varying nondipole field. The probability P that a reversal will occur during one cycle of dipole oscillation is assumed to be the same for all cycles and is determined by the spectrum of nondipole fluctuations and the amplitude of dipole oscillations. The resulting distribution function for variations in the length T of polar-

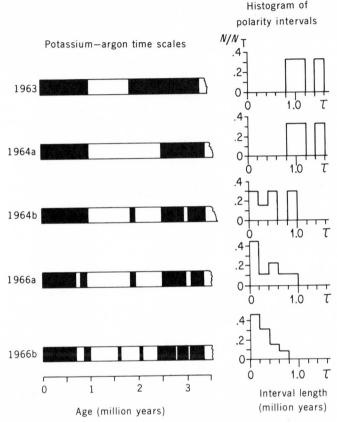

Figure 20-3
Successive versions of the radiometric time scale for reversals,
showing how the discovery of polarity events changed the apparent
distribution of polarity intervals (Cox et al., 1963b, 1964b; McDougall
and Tarling, 1964; Doell and Dalrymple, 1966; McDougall and
Chamalaun, 1966). In the corresponding histograms, N_T is the total
number of polarity intervals and N is the number in each class
interval of the histogram

ity intervals depends only on the probability
P and the period T_D of dipole oscillation:

$$f[T_C < T \leqslant T_C + T_D] = P(1 - P)^{T_C/T_D} \quad (2)$$

where T_C is an integral number of dipole
periods. For small values of P, this may be
approximated by a simple exponential distribution

$$f'(T) = \lambda \exp(-\lambda T) \, dT \quad (3)$$

where $\lambda = P/T_D$ and $f'(T)$ is the probability
for the interval dT.

Before comparing this function with experimental data, one should note the extreme
sensitivity of the distribution function to the
proportion of short events. This sensitivity
is much greater than that of the spectrum of a
continuously varying signal to a high-frequency
component. Because the earth's field has only
two polarity states, inserting a short polarity
interval near the middle of a long interval not
only adds the short interval to the distribution
but also removes a long interval and adds two
intervals of intermediate length. This effect
may be seen (Fig. 20–3) in the marked change
that occurred in the apparent distribution of
polarity intervals as several short polarity
intervals were discovered during the course
of research between the years 1963 and 1966.

RADIOMETRIC TIME SCALE
FOR REVERSALS

The first quantitative time scale for reversals was achieved in 1963 by measuring the ages (using the potassium-argon technique) and the magnetic polarities of young volcanic rocks (Cox et al., 1963a; 1963b). This work appeared to confirm earlier paleomagnetic results from undated sedimentary rocks which indicated that all polarity intervals were of nearly equal length (Fig. 20–3, 1963 scale). These polarity intervals were termed "polarity epochs" and given the names of early workers in the field of geomagnetism.[2]

As more data were obtained (McDougall and Tarling, 1964; Doell and Dalrymple, 1966; McDougall and Chamalaun, 1966), some of the ages and polarities were found to be inconsistent with the simple pattern of epochs. This inconsistency led in 1964 (Cox et al., 1964b) to the discovery of intraepoch "polarity events" with unexpectedly short durations of about 10^5 years (Fig. 20–3). The first polarity events to be recognized were the Olduvai event ($t = 1.9 \times 10^6$ years ago) and the Mammoth event ($t = 3.1 \times 10^6$ years ago), named after the sites of their discovery in Tanzania and California. A paleomagnetic record of polarity events has been found in rocks with similar ages from many parts of the world, demonstrating that the events, like the epochs, are the result of rapid switching of the main dipole field.

With the discovery of polarity events it became apparent that complete resolution of the fine structure of reversals would require an immense number of paleomagnetic and radiometric data. Ideally, such data should be obtained from formations whose ages are uniformly spaced at intervals no greater than 10^4 years; this would require at least 450 precisely determined radiometric

dates for the interval $0 < t \leq 4.5$ million years. The task of making this many age determinations is formidable, and that of locating volcanic formations with the required ages is even more formidable. The age of a volcanic formation is rarely known to within a factor of 2 prior to determination of the radiometric age. Thus the best sampling scheme that can be hoped for is a relatively inefficient one in which the ages are randomly rather than uniformly distributed. Even this is difficult to achieve because of the episodic character of volcanic activity.

For the interval $0 < t \leq 4.5$ million years there are now 150 radiometric ages and polarity determinations which meet reasonable standards of reliability and precision.[3] The main contributions from the data acquired since 1964 have been more accurate determination of the ages of polarity changes (Cox and Dalrymple, 1967b) and the identification of several hitherto unknown events, summarized in Fig. 20–4. Attempts to extend the radiometric time scale for reversals back beyond 4.5 million years have been unsuccessful because the errors in the radiometric ages of the older rocks are too large. A dating error of 5 percent in a 5-million-year-old sample is 2.5×10^5 years, which is larger than many polarity intervals (Cox and Dalrymple, 1967b; Heirtzler et al., 1968; Dalrymple et al., 1967).

Have all the polarity events more recent than 4.5 million years ago been discovered? All of the longer ones appear to have been, to judge both from the density of the present age determinations and from the convergence of the more recent versions of the reversal time scale. However, there are gaps of 10^5 years or more in the present data, and the

---

[2]The alternatives of using the radiometric ages of the boundaries to identify the epochs or of numbering the epochs in sequence were considered. However it was felt that each modification of the time scale would require a confusing change in the nomenclature if numbers were used, especially if (as has happened) additional reversals were discovered.

[3]The criteria used are as follows. (i) The paleomagnetic study includes laboratory measurements of magnetic stability. (ii) The precision of the potassium-argon age determination is <0.1 million years for ages $0 < t < 2.0$ million years and < 5 percent for ages > 2.0 million years. (iii) The magnetic and age measurements were made on rocks and minerals of types known to yield reliable results, and the samples for both measurements were collected from the same volcanic cooling unit (see Cox and Dalrymple, 1967b).

recent discovery by Bonhommet and Babkine[4] of a previously unsuspected reversed event near the end of the Brunhes normal-polarity epoch demonstrates that additional events may exist in gaps even shorter than 10^5 years.

MIDOCEANIC MAGNETIC ANOMALIES

Additional information about reversals is provided by the magnetic anomalies over the midoceanic ridges (Heirtzler et al., 1968; Pitman and Heirtzler, 1966; Vine, 1966). These anomalies are produced by igneous rocks which become magnetized as they solidify and cool in a narrow zone along the ridge axis. As new material forms, the previously magnetized material spreads to either side. If the rate of spreading is the same on both sides of the ridge, the result is a bilaterally symmetrical pattern of normally and reversely magnetized strips with widths proportional to the lengths of the corresponding polarity intervals. The magnetic anomalies do not in themselves determine an independent reversal time scale because the ages of the igneous rocks are usually not known. However, after being calibrated against known points on the radiometric time scale for reversals, the profiles provide a nearly continuous record of polarity intervals. Of special interest are events that may exist within the gaps in the radiometric data.

The value for the minimum duration of a detectable event depends on the width of the strip of crust that was formed during the time of the event, the distance of the strip from the magnetometers at the sea surface (usually about 3 kilometers), and the level of background noise due to irregularities in the formation and magnetization of the crust. The strong dependence of the minimum value for duration on the rate of crustal spreading, v, may be seen from variations in the size of the anomaly due to the Jaramillo event ($T = 5 \times 10^4$ years). This anomaly is quite large on the profiles across the East Pacific Rise ($v = 4$ to 5 centimeters per year) and the Juan de Fuca Ridge ($v = 3$ centimeters per year) but is not visible on profiles across the Reykjanes Ridge ($v = 1$ centimeter per year) and is at about the limit of resolution on the profile across the Indian Ocean ($v = 2$ centimeters per year) (Heirtzler et al., 1968; Pitman and Heirtzler, 1966; Vine, 1966). The minimum strip width ($v \times T$) detectable at the sea surface is therefore about 1 kilometer, and the shortest detectable event is about 2×10^4, 5×10^4, or 10^5 years long, depending on whether the rate of lateral spreading is 5, 2, or 1 centimeter per year[5]

On the profiles one can distinguish small peaks due to polarity events from those due to magnetic noise only by determining which peaks are consistent from profile to profile. A difficulty in doing this arises from the fact that variations in the rate of lateral spreading along one profile may, over long distances, produce cumulative displacements equal to half the wavelength of anomaly peaks, producing a large loss of signal on cross-correlation. In looking for short events it is desirable to tie the magnetic profiles as closely as possible to well-determined points on the radiometric time scale. Displacements due to variable rates of spreading may then be minimized by interpolating between closely spaced points.

Figure 20–5 shows the total magnetic field anomaly along the Eltanin 19 profile, which crosses the East Pacific Rise at latitude 52°S

---

[4]The results of Bonhommet and Babkine (1967) were based on reversed magnetization of the Laschamp volcanic cone in France. The available radiometric data do not precisely determine the age and duration of this event but suggest that it was late Quaternary or Recent and that it was very short. This event is less well documented than the Jaramillo, Gilsá, and Olduvai events, each of which is supported by independent results from two or more volcanic formations from different parts of the world.

[5]A boundary zone with low average intensity of magnetization and containing rocks of mixed polarity probably exists between the magnetized strips as the result of irregularities in the process by which the crust is magnetized (Vine and Wilson, 1965; Harrison, 1968). The width of the strip of magnetized crust which produces the anomaly is therefore smaller than the quantity ($v \times T$), but the estimate for the minimum length of an event is not changed if the thickness of the boundary zone is independent of spreading rate.

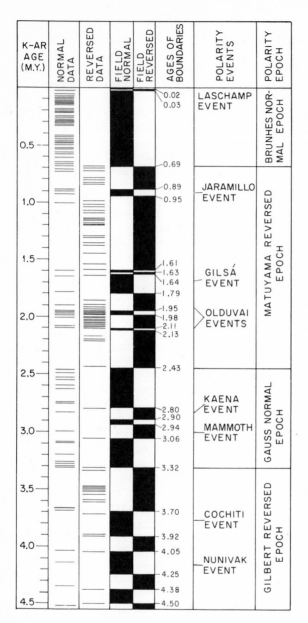

Figure 20-4
Time scale for geomagnetic reversals. Each short horizontal line shows the age as determined by potassium-argon dating and the magnetic polarity (normal or reversed) of one volcanic cooling unit. Included are all published data (Cox et al., 1968) which meet reasonable standards of reliability and precision (see footnote 3, page 212). Normal-polarity intervals are shown by the solid portions of the "field normal" column, and reversed-polarity intervals, by the solid portions of the "field reversed" column. The duration of events is based in part on paleomagnetic data from sediments (Ninkovich et al., 1966) and magnetic profiles (Heirtzler et al., 1968; Pitman and Heirtzler, 1966; Vine, 1966, 1968)

and longitude 118°W. As is typical of mid-oceanic ridges, the positive anomaly over the central zone (M-B to M-B in Fig. 20–5) is complex, for reasons not yet understood. Elsewhere, many of the polarity transitions can be correlated unambiguously with those of the radiometrically determined time scale. For example, the transition from the Matuyama reversed epoch to the Brunhes normal epoch can

be easily recognized (M-B in Fig. 20–5), as can the transition from the Gauss to the Matuyama epochs (G-M) and the transition from the Gilbert reversed to the Gauss normal epoch (G-G). At high latitudes the steepest gradients in the magnetic profiles occur almost exactly above the boundaries between the normally and the reversely magnetized strips, so that distances between polarity transitions and

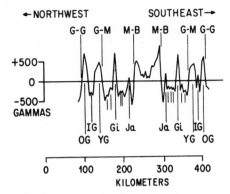

←NORTHWEST **SOUTHEAST→**

G-G G-M M-B M-B G-M G-G

+500
0
-500
GAMMAS

IG Gi Ja Ja Gi IG
OG YG YG OG

0 100 200 300 400
KILOMETERS

Figure 20-5
Magnetic profiles across the East Pacific
Rise at latitude 52°S, longitude 118°W
(Heirtzler et al., 1968; Vine, 1966). (G-G,
G-M, M-B) Boundaries between the
Gilbert, Gauss, Matuyama, and Brunhes
polarity epochs. (OG, IG, YG, Gi, Ja) The
larger and more consistent positive anomaly
peaks. Other small positive peaks are shown
with unlabeled lines

events may be read directly from the magnetic
profiles. It may be seen that the anomalies to
the northwest of the rise are more widely
spaced; this indicates that the rate of spreading
was greater in the direction of the Pacific than
toward Antarctica. Therefore the interpola-
tions must be made separately for the two
sides of the rise.

The age of the boundary between the Gauss
and Matuyama epochs was first found, as a
check on the method, by interpolating between
the Gilbert-to-Gauss boundary (3.32 million
years ago) and the Matuyama-to-Brunhes
boundary (0.69 million years ago), both of
which have well-determined radiometric ages.
Fifteen half profiles were used, all from high
latitudes in the South Pacific, North Pacific,
and Indian oceans.[6] The mean interpolated age
is 2.41 ± 0.03 million years (standard error).
This age is consistent with the two radiometrically determined ages of 2.43 million years
(reversed magnetization) and 2.45 million
years (normal magnetization) which bracket
the boundary. When both radiometric and in-

terpolated results are taken into account, the
best estimate of the age of the boundary is 2.43
million years.

Three positive anomaly peaks corresponding to the three normal polarity intervals (YG,
IG, and OG in Fig. 20–5) of the Gauss normal
epoch appear on those profiles for which
spreading rates are rapid. The two reversed
events, the Mammoth and the Kaena, which
separate the normal intervals have been recognized by radiometric dating (McDougall and
Tarling, 1964; Doell and Dalrymple, 1966;
McDougall and Chamalaun, 1966), and from
magnetic profiles (Heirtzler et al., 1968; Pitman and Heirtzler, 1966; Vine, 1966). The
ages of the midpoints of the two larger anomalies, found by interpolating 13 profiles (Table
20–1) between the boundaries of the Gauss
epoch, are in good agreement with the radiometric time scale. The short central anomaly
peak has not been dated radiometrically but is
bracketed by ages of 3.0 and 2.8 million years
obtained for samples from the adjacent reversed events, in agreement with the interpolated age of the normal event of 2.94 ± 0.04
million years.

Two positive anomalies (Ja and Gi in Fig.
20–5) appear in the Matuyama epoch on almost all magnetic profiles. The smaller and
younger anomaly (Ja) has usually been identified with the Jaramillo normal event; the older
and larger anomaly with the Olduvai event
(Heirtzler et al., 1968; Pitman and Heirtzler,
1966; Vine, 1966). The age of 0.93 million
years (Table 20–2) found for the younger
anomaly by interpolating between the boundaries of the Matuyama reversed epoch agrees
well with the radiometric age of 0.92 million
years. However the interpolated age of 1.70
million years for the larger anomaly does
not agree with the ages obtained for the Olduvai normal event by potassium-argon dating,
most of which are from 1.95 to 2.00 million
years. It agrees more nearly with the two ages
of 1.60 and 1.65 million years for normally
magnetized lava flows from Iceland and
Alaska (McDougall and Wensink, 1966; Cox
and Dalrymple, 1967), which constitute the
evidence for the existence of a normal event
termed the Gilsá (McDougall and Wensink,
1966; Cox and Dalrymple, 1967). Vine (1968)
concluded independently, from detailed study

[6]The profiles used are from the South Pacific
(EL20, EL19, S16, S18, MN5, S15) (Hamilton et al.,
1968), the Indian Ocean (V16) (Hamilton et al.,
1968) and the Juan de Fuca Ridge.

Table 20–1
Ages of the three normal intervals (YG, IG, and OG) in the Gauss normal epoch, as found by interpolation between the Gilbert-to-Gauss boundary (3.32 million years ago) and the Gauss-to-Matuyama boundary (2.43 million years ago) (See footnote 6).

Interval	Number of profiles	Age found by interpolation (million years)	Standard deviation	Standard error	Age of interval midpoint, from K-Ar dating (million years)
YG	13	2.64	0.03	0.01	2.62
IG	8	2.94	.10	.04	2.9
OG	13	3.19	.03	.01	3.19

of the Eltanin 19 profile, that the main anomaly in the Matuyama reversed epoch lasted from 1.80 to 1.64 million years ago. Previously Ninkovich et al. (1966) had concluded from their paleomagnetic study of marine sediments that the main normal event in the Matuyama reversed epoch began 1.79 million years ago and ended 1.65 million years ago, but they correlated this event with the Olduvai event, as it had been defined previously from radiometric dating (Cox et al., 1964b). It now appears that the main normal-polarity event in the Matuyama epoch is not the Olduvai but rather the Gilsá event. Its younger boundary is determined from radiometric dating to be at 1.61 million years ago. Its older boundary, although poorly determined radiometrically, appears from the magnetic anomalies and from the paleomagnetism of sediment cores (Ninkovich et al., 1966) to be at about 1.79 million years ago.

To determine whether the profiles contain evidence for shorter events, ages were found by interpolation for all the small positive peaks in the Matuyama reversed epoch (Fig. 20–5) on 17 half profiles from high latitudes. The most consistent anomaly appears on 11 half profiles at positions having interpolated ages between 1.94 and 2.00 million years (Fig. 20–6). The radiometric ages associated with the Olduvai event are almost all in this same range, a fact which indicates that the two are correlative. From the size of the anomaly the duration of the event is estimated to be 3×10^4 years, an estimate which agrees with Vine's (1968) interpretation of the Eltanin 19 profile. On the other hand, in paleomagnetic studies of deep-sea sediments no consistent evidence has been found for the existence of

both the Gilsá and the Olduvai events.[7] Apparently the processes by which sediments become magnetized are sufficiently irregular and integrative in nature as to make it difficult to consistently resolve events as short as 3×10^4 years.

Why have so many normally magnetized rocks been found from the short Olduvai event and so few from the much longer Gilsá event? The answer appears to lie in the uneven distribution of the ages of both the normally and the reversely magnetized samples. Few samples of either polarity have ages in the range between 1.6 and 1.8 million years, whereas many samples of both polarities have ages between 1.9 and 2.1 million years. The overlap in the ages of normally and reversely magnetized samples with ages between 1.9 and 2.1 million years is due to the fact that the duration of the Olduvai event is shorter than many of the discrepancies due to dating errors. The high ratio of reversely to normally magnetized samples is in accord with the conclusion that the Olduvai normal event was short.

Rocks with ages of from 2.2 to 1.9 million years used for these studies are from volcanic formations from Africa, Alaska, the western United States, Cocos Island, Australia, Re-

---

[7]Watkins and Goodell (1967) report that both the Gilsá and Olduvai events are recorded in marine sediments from the Pacific-Antarctic Rise, whereas Hays and Opdyke (1967) dispute this interpretation. However, in the three cores studied by Hays and Opdyke, the two reversed events in the Gauss normal epoch are resolved on only one of the cores (Eltanin 13, core 3), and this same core also shows a clearly defined normal event with an interpolated age between the Gilsá event and the Gauss-Matuyama boundary of 1.99 million years ago. in agreement with the radiometric age of the Olduvai event.

Table 20–2
Age of the Ja and Gi events, as found by interpolation between the Gauss-to-Matuyama boundary (2.43 million years ago) and the Matuyama-to-Brunhes boundary (0.69 million years ago) (See footnote 6).

Interval	Number of profiles	Age found by interpolation (million years)	Standard deviation	Standard error	Age from K-Ar dating[a]
Ja	17	0.93	0.05	0.01	0.92
Gi	17	1.70	.10	.02	1.60–1.86

[a]See Fig. 20–4 for range of uncertainty.

union Island in the Indian Ocean, and Iceland —a range which suggests that an interval of unusually intense volcanic activity occurred 2 million years ago. It was also at about this time that a marked evolutionary change occurred in marine microorganisms, including Radiolaria (Hays and Opdyke, 1967) and Foraminifera (Berggren et al., 1967), marking the beginning of the Pleistocene. Such faunal changes have been noted near several polarity transitions and, in particular, at the time of or slightly before the Olduvai event (Hays and Opdyke, 1967; Berggren et al., 1967), lending support to earlier suggestions (Uffen, 1963; Simpson, 1966) that, when the earth's field decayed during a reversal, the increase in radiation would have been large enough to produce a sudden increase in the rate of evolution. However, recent quantitative studies indicate that the shielding provided by the earth's atmosphere is so great that the increase in radiation on complete collapse of the field

would be no more than 12 percent (Sagan, 1965; Waddington, 1967; Harrison, 1968; Black, 1968). Such an increase is comparable with the variations that occur during a normal sunspot cycle, or with the normal variation between equatorial and polar regions. The accompanying increase in mortality rate and in the rate of spontaneous mutation is too small to account for the extinction of a species. An alternative explanation is that the faunal extinctions at the beginning of the Pleistocene are related to volcanic activity which occurred at that time, a possible causal link being increased absorption and reflection of sunlight by volcanic dust in the atmosphere.

An anomaly slightly older than the Olduvai (the X anomaly of Heirtzler et al., 1968) appears on some profiles. The interpolated ages, although less well grouped than those of the Olduvai anomaly, provide weak evidence that a short event may exist. Support for this is provided by two radiometric ages slightly

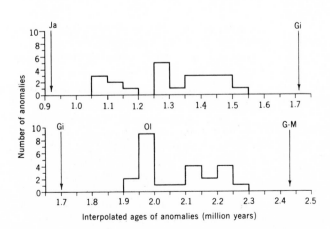

Figure 20-6
Histogram of ages of small peaks on magnetic profiles found by interpolating between the Jaramillo event (0.92 million years ago), and the Gilsa event (1.71 million years ago), and the Gauss-Matuyama boundary (2.43 million years ago). The most consistent anomaly peak corresponds to the Olduvai event (01)

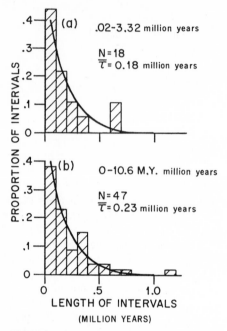

Figure 20-7
Histograms of lengths of polarity intervals,
(a) from the reversal time scale of Figure
20-4 based on ages obtained by potassium-
argon dating, and (b) from the reversal time
scale of Vine (1968) based on the Eltanin 19
magnetic profile. (Solid curves) Distribution
function of Eq. 2 for $P = 0.05$. N is the
number of intervals

greater than those of the main group of normally magnetized Olduvai samples (Fig. 20–4),
but again the evidence is rather weak. There
are weak suggestions of possible events at
1.26 and 2.24 million years ago, but whether
they are real cannot be determined radiometrically because both occur where there
are gaps in the data.

In summary, all of the events with dura-
tions $T \geqslant 3 \times 10^4$ years that have been found
from radiometric dating have also been identified on the magnetic profiles. The radiometric
ages of the events and the ages derived through
interpolation from the profiles are in excellent
agreement. The profiles further establish that
there are no additional, undetected events
longer than 3×10^4 years, even where there
are gaps longer than this between known
radiometric ages. Events shorter than 3×10^4
years are below the level of experimental
noise and have not been resolved convincingly
from the magnetic profiles.

SEDIMENTARY RECORD OF SHORT EVENTS

The paleomagnetism of marine sediments
provides a third possible source of information
about short events (Hays and Opdyke, 1967;
Berggren et al., 1967). In general, the record of
polarity epochs in sediments agrees with the
radiometric time scale. However, the record
of short events is obscured by noise due to
stratigraphic gaps, variations in the rate of
deposition, and delays in the time between
deposition and magnetization of the sediments.
The amount of delay varies, depending on
variations in the amount of reworking by organisms, on the rate of compaction, and on
authigenic chemical changes in the ferromagnetic minerals. In addition, there is a loss of
information about short polarity intervals
because the magnetizing processes in sediments are integrative over time intervals comparable with the durations of the shorter
events. As a result, the research on sediments
has not produced consistent evidence for the
existence of events shorter than those detectable by other methods ($T = 3 \times 10^4$ years) —
with one notable exception: Ninkovich et al.
(1966) report rather convincing evidence for
two short polarity intervals ($T \sim 10^4$ years) at
the end of the Gilsá event. Thus, while the
paleomagnetic research on sediments has provided valuable information about the duration
of the longer events and has helped extend
the time scale back into the Gilbert reversed
epoch, it has not resolved the question of the
frequency of very short events.

DISCUSSION

Over the interval $0.02 < t \leqslant 3.32$ million years
ago (the part of the reversal time scale for
which the number of short events is most completely known), a good fit to the observed
frequency of polarity intervals (Fig. 20–7) is
obtained from Eq. 2 when P is set equal to
0.055. The observed number of very small
events is consistent with the number predicted
by the model. Reversal times scales going
back several tens of millions of years have
been obtained from magnetic profiles on the
assumption that the rate of sea-floor spreading
was constant (Heirtzler et al., 1968; Pitman

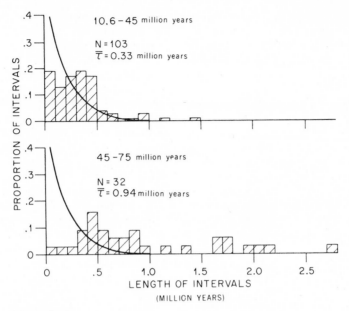

Figure 20-8
Histograms of lengths of polarity intervals. Both histograms are from
the time scale for intervals of Heirtzler et al. (1968) based on
magnetic profiles. Solid curves are as in Figure 20-7

and Heirtzler, 1966; Vine, 1966). The best
fit of Eq. 2 to the data of Vine (1966) for the
past 10.6 million years is obtained with $P = 0.043$. The observed proportion of very short
events ($T < 5 \times 10^4$ years) is slightly smaller
than the proportion found for the interval
$0.02 < t \leq 3.32$ million years; this suggests
that, as had been suspected, some of the shortest events were not resolved on the magnetic
profiles.

It is concluded from the analysis discussed
here that, on the average, 20 fluctuations in
intensity occur between successive polarity
reversals. This conclusion, although based
on a model in which dipole intensity fluctuations are periodic with a period of 10^4 years,
would also be valid if the intensity changes
were nonperiodic, with an average time between minima of 10^4 years. The corresponding
probability that a geomagnetic reversal will
result from the decrease in dipole moment
currently in progress is 5 percent.

If sea-floor spreading has occurred at a
constant rate, the marine magnetic profiles
may be interpreted to yield a reversal time
scale going back 75 million years (Heirtzler
et al., 1968). The apparent average duration
of the polarity intervals was greater during

the time $10.6 < t \leq 45$ million years than during the past 10.6 million years, and during the
time $45 < t \leq 75$ million years the average
length was still greater (Fig. 20–8). Part of
the difference may be due to variations in
spreading rate (Heirtzler et al., 1968; Ewing
and Ewing, 1967; Schneider and Vogt, 1968).
However, this would account only for an apparent change in the probability P and not for
the observed change in the shape of the distribution. The latter change may be more
apparent than real, however, if, as is quite
possible, a small number of additional short
events occurred which have not yet been
detected. Numerical experiments indicate
that about 30 randomly distributed short
events would change the observed distribution
to one fitted by Eq. 2 when $P = .02$.

As one goes farther back in time, an average
length of 10^7 years has been reported for
polarity intervals in the early Paleozoic
(Khramov et al., 1965), and the Kiaman
reversed-polarity interval during the late
Paleozoic was 5×10^7 years long (Irving,
1966b; McMahon and Strangway, 1967).
Clearly, the average length of polarity intervals and hence the value of P have changed
during the earth's history, reflecting changes

in the physical conditions which control the geomagnetic dynamo. The possible importance to magnetohydrodynamic processes of changes in the properties of the core-mantle interface has been pointed out by Hide (1967) and by Irving (1966b).[8] These changes may result from large-scale tectonic processes, such as polar wandering accompanied by mass transport in the lower mantle. Hence it is conceivable that individual changes in polarity are produced by individual tectonic events. The present analysis suggests the alternative possibility that the timing of reversals is controlled on two levels. The *average* length of polarity intervals is controlled by conditions at the core-mantle interface and hence may

be related to tectonic processes in the mantle, whereas the length of individual polarity intervals is determined by random processes in the fluid core.

Observationally, the main difference between the two interpretations is in the cutoff anticipated for short periods. In the present model, the cutoff is 10^4 years, the assumed period of fluctuations in the intensity of the dipole field. In Hide's model the cutoff is estimated to be from 10^5 to 10^6 years, depending on the strain rate of the lower mantle and the minimum displacement of the core-mantle interface needed to produce a change in polarity. With the discovery of polarity events at least as short as 3×10^4 years, it appears increasingly likely that the timing of individual reversals is controlled by processes occurring in the fluid core.

---

[8] Irving (1966b) has also suggested that reversals may be due to shifts in the earth's axis of rotation.

REVERSALS AT SEA: THE MAGNETIC STRIPES

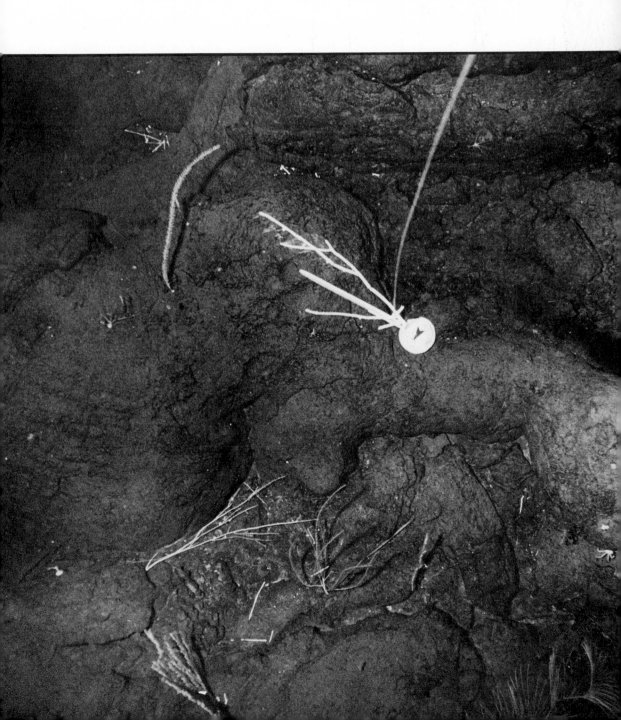

21

Introduction and Reading List

Discovery of Magnetic Stripes

During the 1950's oceanographic ships exploring the ocean basins began routinely to measure the strength of the earth's magnetic field. This had become much easier to do with the invention of the proton precession magnetometer, an instrument which does not need to be leveled by hand and which can be operated remotely. It is possible to tow this instrument behind the ship away from shipboard magnetic noise and to record the intensity of the field automatically. Fortunately for science, oceanographers from Scripps Institution of Oceanography at La Jolla, California, during their exploration of the eastern Pacific basin began routinely to tow magnetometers. When they plotted their data on a map and drew contours showing the intensity of the earth's field, the results looked like ridges and valleys on a topographic map (Mason, 1958; Vacquier et al., 1961; Mason and Raff, 1961; Raff and Mason, 1961). The "ridges" of high magnetic intensity alternated with "valleys" of low intensity. These came to be known as "magnetic stripes." The stripes varied in width from a few kilometers to about eighty kilometers and were several thousand kilometers long. Where they intersect fracture zones, they were found to be truncated and offset by varying distances up to more than a thousand kilometers (Vacquier, 1962). These remarkably regular and continuous magnetic stripes were eventually found over almost all ocean floors. Some of the stripes were traced half way around the world, offset only by the fracture zones.

Figure 21-1
Drummond Matthews (left) and Fred Vine

The Birth of a Theory

When scientists look at data that have a pattern as regular and simple as that of the magnetic stripes, they usually get a gut feeling that there must be a simple explanation. In 1963 Frederick Vine and Drummond Matthews provided the simple explanation. They did this (Chapter 22) by combining Hess's theory about sea-floor spreading with the research then being done on the time scale for geomagnetic reversals.

In 1961 Drummond Matthews had completed his Ph.D. research on dredged basalts from the northeast Atlantic ocean. This had convinced him that the mid-Atlantic ridge must be volcanic. Fred Vine, during his undergraduate studies at Cambridge, became convinced that continental drift had occurred and felt that some of the crucial evidence for drift would somehow be found in the study of marine geology and geophysics. In 1962 Vine became a graduate student with the oceanography group of Edward Bullard and Maurice Hill at Cambridge. Together with other members of the group, including Matthews, who had now joined the staff, Vine sailed to the Indian Ocean in 1962 to help make magnetic surveys of the Carlsberg Ridge as part of the International Indian Ocean Expedition. On returning, Matthews became Vine's research supervisor and the two young scientists shared a room in the old stables at Cambridge. They spent many hours reading Hess's paper (Chapter 4) and discussing the volcanic nature of the sea floor. They also began interpreting the magnetic results from the Indian Ocean. In doing computer modeling to fit the anomalies, they realized that the central part of the ridge was normally magnetized, whereas seamounts on the flank of the ridge were reversely magnetized. They knew from earlier work by Ewing et al. (1959) that a positive magnetic stripe was generally observed over mid-oceanic ridges. They were also aware of current research on the timing of reversals. Their final synthesis was remarkably simple: new sea floor, as it forms continuously at the mid-ocean ridges, becomes magnetized in the earth's field. The positive magnetic stripes formed when the field was normal, as it is today; the negative stripes formed when the field was reversed. Matthews writes (personal communication) "I believe the fitting together of those three ideas came to Fred during a discussion over our tea table, but I am not too sure about that."

When the time is ripe for the formation of a new hypothesis, it is not uncommon for different scientists working completely independently to think of it at about the same time. As Vine and Matthews were developing their theory to account for the magnetic stripes, a Canadian scientist, L. W. Morley, was independently arriving at the same conclusions. During the early 1960's Canadian geophysicists had been very active in making aeromagnetic surveys of Canada and of the North Atlantic Ocean. Morley was trying to interpret the magnetic highs and lows on the maps and saw that there were strong similarities between the magnetic anomalies at sea and those over Canada. Since he had done his own thesis research in the field of paleomagnetism, he was aware of reversals and realized that many of the negative anomalies required the presence of reversely magnetized rocks in the crust. Then, shortly after reading of Raff and Mason's (1961) discovery of the magnetic stripes in the eastern Pacific, Morley read an article by Robert Dietz (1961) on sea-floor spreading. The light immediately dawned, and Morley proposed the following explanation for the magnetic stripes, which is essentially the same as that of Vine and Matthews:

> Several investigators and authors writing on the subject of continental drift and convection currents in the earth's mantle have referred to the puzzling linear magnetic anomalies in the Eastern Pacific Ocean basin reported by scientists of the Scripps Institution of Oceanography.
>
> If one accepts in principle the concept of mantle convection currents rising under ocean ridges, traveling horizontally under the ocean floor and sinking at ocean troughs, one cannot escape the argument that the upwelling rock under the ocean ridges, as it rises above the Curie Point geotherm, must become magnetized in the direction of the earth's field prevailing at the time. If this portion of rock moves upward and then horizontally to make room for new upwelling material, and if, in the meantime, the earth's field has reversed, and the same process continues, it stands to reason that a linear magnetic anomaly pattern of the type observed would result. This explanation has the advantage over many of the others put forward that it does not require a petrologically, structurally, thermally, or strain-banded oceanic crust. It requires a convection cell whose axis of rotation is at least as long as the linear magnetic anomalies and whose horizontal distance of travel stretches from ocean rise to ocean trough. In addition to this, it requires a large number of reversals of the earth's magnetic field from at least the Cretaceous period (which ended 68,000,000

to 72,000,000 years ago) to the present (since no rocks older than Cretaceous have been found in the ocean basins; that is, no rocks older than 140 million years).

R. L. Wilson reported that Mrs. J. Cox, in a recent search of the paleomagnetic literature, was able to find 136 normally polarized cases and 141 reversely polarized from the Carboniferous to the present. Since there is no evidence to suggest that the earth's field should have been "normally" polarized for any more periods or for longer periods than it has been reversely polarized, it is entirely possible that there may have been as many as 180 reversals since the Lower Cretaceous. This would be one reversal about every half million years on the average (a figure which T. Einarsson gives from his investigations of Icelandic lavas. He also suggests that the time taken for a reversal of the field is geologically very short—a few centuries to 10,000 years).

From an examination of the Scripps magnetic maps, the width of a complete positive and negative cycle, averaged, over the widest part of the available surveyed section is about 35 kilometers. To travel this distance in 1,000,000 years (time of two reversals) the convection current must have a rate of about 3.5 centimeters per year. This figure is only good to an order of magnitude because no accurate data are available on the length of the periods of reversals. A better way to arrive at the rate of convection travel and the reversal period would be to measure the ages of rocks at widely-spaced locations in the Pacific and to count the number of reversals occurring between these points.

R. G. Mason and A. D. Raff (who made the magnetic survey for Scripps) report that some of the many guyots (flat-topped undersea islands) which were detected on the echo sounder produced magnetic anomalies, while others apparently had little or no effect. It seems unlikely that these guyots would be divided into two classes — those containing magnetite and those containing little or none. A more likely explanation would be that the ones which give little or no effect are negatively polarized to an intensity which nearly equalized their magnetization induced by the present earth's field. If the "non-magnetic" guyots always occur in the negative anomaly bands and the magnetic ones in the positive bands, this would be evidence that they cooled below the Curie Point at approximately the same time as the rock surrounding them because they were magnetized in the same direction. Indeed, since at that time they would have been in the shallow water of the oceanic ridge, they would have protruded above the surface and have had their tops flattened by erosion. As they proceeded along with the mantle convection current they would pass into deeper water. This is an alternative explanation of origin to that suggested by Darwin for the flat-topped guyots in the deep Pacific.

The purpose of this letter is to point up the possibility of calibrating the frequency and duration of reversals of the earth's magnetic field in geologic history from a study of magnetic surveys of the ocean basins, and the idea presented is considered to add support to the theory of convection in the earth's mantle.

This passage is quoted from an article by Morley that, by a curious set of circumstances, was never published in a scientific journal. It was submitted to the British journal *Nature* in January, 1963, and rejected. It was then submitted to the *Journal of Geophysical Research* in the United States and rejected as being too speculative for publication and more suitable for discussion at a cocktail party. (It was eventually published in a literary magazine[1]) At the time Morley wrote the paper, he was the manager of a group of sixty geophysicists engaged in a major geophysical survey of Canada. Like many good scientific administrators, he was frustrated in not being able to find very much time for science and he did not follow up on his original idea.

Development of a Theory

Initially few geophysicists felt that the magnetic stripes provided very strong support for the ocean-floor-spreading hypothesis. In part this was simply scientific conservatism, as possibly indicated by the repeated rejection of Morley's article. However there were also several good, objective reasons for being skeptical. First, the magnetic profiles shown in the original paper (Chapter 22) are rather irregular whereas the theory predicts they should be regular and symmetrical on both sides of the ridge. Secondly, for the magnetic recorder to work it would be necessary that new sea floor be formed in a very narrow zone by a steady, continuous process. This seemed unlikely because on land volcanism usually occurs irregularly both in space and time. Thirdly, the time scale of reversals available to Vine and Matthews in 1963 (Fig. 11–5, June 15, 1963, scale) was still not accurate enough to provide a good model with which to try to compare magnetic profiles.

By 1965 when Vine and Wilson wrote the article that is included in this book as Chapter 23, better data were available and the case for the Vine and Matthews hypothesis was much stronger. Recognizing that the Juan de Fuca ridge was a spreading center, they were able to show that the magnetic profile was rather symmetrical and that models based on the revised reversal time scale of Cox et al. (1964b) fit the profile remarkably well. Another advance over their original model was made with the realization that the magnetic anomalies were mainly produced not by a thick layer of serpentine and dike rock but rather by a thin layer of strongly magnetized basalt that had been extruded onto the sea floor. This interpretation is now generally accepted (Marshall and Cox, 1971; Talwani et al., 1971).

By 1966 magnetic profiles of even better quality were available and the reversal time scale was well determined except for its ultrafine structure. The fit between the profiles and the time scale was amazingly good. The final piece of data which convinced many geophysicists of the reality of ocean-floor spreading was a magnetic profile measured across the East Pacific rise south of Easter Island by the National Science Foundation ship *Eltanin* (Fig. 24–7). All of the polarity epochs and events that had painstakingly been determined from paleomagnetic and radiometric dating studies on hundreds of volcanic rock samples may be seen in the form of the peaks and troughs on this one magnetic profile. Moreover the profile

---

[1] It was included as a letter from Morley in an article by John Lear, "Canada's Unappreciated Role as Scientific Innovator," in the September 2, 1967, *Saturday Review* (Copyright 1967 Saturday Review, Inc.).

Figure 21-2
James R. Heirtzler

is almost perfectly symmetrical out to a distance of 2000 km on both sides of the ridge (Fig. 25–1), as predicted by the theory. The Eltanin-19 profile is surely one of the most important pieces of data in the history of geophysics. With the appearance of Vine's masterful paper in 1966 (Chapter 24), it was clear that the Vine and Matthews hypothesis was correct. Here was a powerful new tool for determining how and when the ocean basins of the world had formed.

Magnetic Dating of the World Ocean Basins

Within two years the age of a large fraction of the world's ocean basins had been determined using magnetic stripes. Most of this research was done at Lamont-Doherty Geological Observatory. It was no accident that the group at this facility was in a unique position to apply the concept of plate tectonics on a global scale. Under the leadership of Maurice Ewing, Lamont had become one of the world's most productive oceanographic institutions. With the development of the proton magnetometer, Lamont began in the early 1960's to collect magnetic data routinely on almost all of its many cruises. James R. Heirtzler had premonitions in the early 1960's that the magnetic data would eventually prove important. One early clue was the large positive anomaly (Ewing et al., 1959) consistently found over the mid-Atlantic ridge. By the late 1960's the bank of magnetic data available was so enormous as to be almost overwhelming. However these data were not in a readily usable form. On a typical oceanographic cruise the ship does not record a profile directly across a ridge, but changes course and speed many times. In order to plot a profile it is necessary to merge the navigational and magnetic data, which are recorded separately. James Heirtzler, taking on the task of organizing these data, was aided by a timely technological development: the medium-sized computer that could be operated by the researcher himself. Using such a computer Heirtzler was able to store great quantities of oceanographic data and then plot profiles at any desired scale. This permitted him to compare data from different areas. He writes (personal communication) "I can distinctly remember the night when, for the first time, I could lay 13 trans-Atlantic magnetic profiles on the top of my desk for cross comparison. I had never even had one complete profile on my desk before."

Scientists at Lamont, as elsewhere, initially had been skeptical of Vine and Matthews' theory. The conversion at Lamont came in 1966 when Walter Pitman suddenly realized that a magnetic profile across the Pacific-Antarctic ridge was symmetrical and could be accounted for by the same bodies needed to explain the anomalies over the Reykjanes ridges south of Iceland, on the opposite side of the world. All that was necessary was to expand the Reykjanes ridge bodies laterally by a factor of four, in accord with the hypothesis that the sea floor had been spreading faster in the South Pacific.

Searching their magnetic-data archives, the Lamont scientists found that the same magnetic anomalies could be identified over most of the world's oceans with the notable exception of the northwest Pacific. For reasons discussed in Chapter 25, they adopted profiles across the South Atlantic Ocean as their standard. They determined the rate of spreading using the younger radiometrically dated reversals. Then, making the daring assumption that the opening of the Atlantic had occurred at a constant rate, they proposed a reversal time scale covering the past 76 million years (Table 25–1). Subsequent research has indicated that this time scale is essentially correct (Chapter 44). Marine geophysicists working in many parts of the world have been able to match new magnetic profiles with parts of this time scale and thereby determine the age of formation of the ocean floor in their survey areas. Using this technique, marine geophysicists were able to map the age of the major basins with an accuracy that, on the continents, had taken geologists a century to achieve. There were two reasons for this. The structure of the sea floor is as simple as a set of tree rings, and like a modern bank check it carries an easily decipherable magnetic signature.

READING LIST

Marine Magnetic Surveys and Their Interpretation

Adams, R. D., and Christoffel, D. A., 1962, Total magnetic field surveys between New Zealand and the Ross Sea: J. Geophys. Res., v. 67, p. 805–813.

Allan, T. D., Charnock, H., and Morelli, C., 1964, Magnetic gravity and depth surveys in the Mediterranean and Red Sea: Nature, v. 204, p. 1245–1248.

Atwater, T., and Menard, H. W., 1970, Magnetic lineations in the north-east Pacific: Earth Planet. Sci. Lett., v. 7, p. 445–450.

Avery, O. E., Burton, G. D., and Heirtzler, J. R., 1968, An aeromagnetic survey of the Norwegian Sea: J. Geophys. Res., v. 73, p. 4583–4600.

Backus, G. E., 1964, Magnetic anomalies over oceanic ridges: Nature, v. 201, p. 591–592.

Bassinger, B. G., DeWald, D. E., and Peter, G., 1969, Interpretations of magnetic anomalies off central California: J. Geophys. Res., v. 74, p. 1484–1487.

Bullard, E. C., and Mason, R. G., 1963, The magnetic field over the oceans, *in* Hill, M. N., ed., The sea, v. 3: New York, Wiley-Interscience, p. 175–217.

Christoffel, D. A., and Ross, D. I., 1965, Magnetic anomalies south of the New Zealand plateau: J. Geophys. Res., v. 70, p. 2857–2861.

Dickson, G. O., Pitman, W. C., III, and Heirtzler, J. R., 1968, Magnetic anomalies in the South Atlantic and ocean floor spreading: J. Geophys. Res., v. 73, p. 2087–2100.

Erickson, B. H., and Grim, P. J., 1969, Profiles of magnetic anomalies south of the Aleutian island arc: Geol. Soc. Amer., Bull., v. 80, p. 1387–1390.

Godby, E. A., Baker, R. C., Bower, M. E., and Hood, P. J., 1966, Aeromagnetic reconnaissance of the Labrador Sea: J. Geophys. Res., v. 71, p. 511–517.

Grim, P. J., and Erickson, B. H., 1969, Fracture zones and magnetic anomalies south of the Aleutian trench: J. Geophys. Res., v. 74, p. 1488–1494.

Harrison, C. G. A., 1968b, Formation of magnetic anomaly patterns by Dyke injection: J. Geophys. Res., v. 73, p. 2137–2142.

Hayes, D. E., and Heirtzler, J. R., 1968, Magnetic anomalies and their relation to the Aleutian island arc: J. Geophys. Res., v. 73, p. 4637–4646.

Hayes, D. E., and Pitman, W. C., III, 1970, Magnetic lineations in the North Pacific, *in* Hays, J. D., ed., Geological investigations of the North Pacific: Geol. Soc. Amer., Mem. 126, p. 291–314.

Heirtzler, J. R., Dickson, G. O., Herron, E. M., Pitman, W. C., III, and Le Pichon X., 1968, Marine magnetic anomalies, geomagnetic field reversals, and motions of the ocean floor and continents: J. Geophys. Res., v. 73, p. 2119–2136.

Heirtzler, J. R., and Hayes, D. E., 1967, Magnetic boundaries in the North Atlantic Ocean: Science, v. 157, p. 185–187.

Heirtzler, J. R., and Le Pichon, X., 1965, Crustal structure of the mid-ocean ridges, 3. Magnetic anomalies over the mid-Atlantic ridge: J. Geophys. Res., v. 70, p. 4013–4033.

Heirtzler, J. R., Le Pichon, X., and Baron, J. G., 1966, Magnetic anomalies over the Reykjanes ridge: Deep-Sea Res., v. 13, p. 427–443.

Herron, E. M., and Heirtzler, J. R., 1967, Sea-floor spreading near the Galapagos: Science, v. 158, p. 775–780.

Keen, M. J., 1963, Magnetic anomalies over the mid-Atlantic ridge: Nature, v. 197, p. 888–890.

King, E. R., Zietz, I., and Alldredge, L. R., 1966, Magnetic data on the structure of the central Arctic region: Geol. Soc. Amer., Bull., v. 77, p. 619–646.

Larson, R. L., 1970, Near-bottom studies of the east Pacific rise crest and tectonics of the mouth of the Gulf of California, Ph.D thesis, University of California at San Diego, 164 p.

Larson, R. L., and Spiess, F. N., 1969, East Pacific rise crest: a near-bottom geophysical profile: Science, v. 163, p. 68–71.

Le Pichon, X., and Heirtzler, J. R., 1968, Magnetic anomalies in the Indian Ocean and sea floor spreading: J. Geophys. Res., v. 73, p. 2101–2117.

Luyendyk, B. P., 1969, Origin of short-wavelength magnetic lineations observed near the ocean bottom: J. Geophys. Res., v. 74, p. 4869–4881.

Luyendyk, B. P., and Fisher, D. E., 1969, Fission track age of magnetic anomaly 10: A new point on the sea-floor spreading curve: Science, v. 164, p. 1516–1517.

Marshall, M., and Cox, A., 1971, Magnetism of pillow basalts and their petrology: Geol. Soc. Amer., Bull., v. 82, p. 537–552.

Mason, R. G., 1958, A magnetic survey off the west coast of the United States between latitudes 32° and 36° N, longitudes 121° and 128° W: Roy. Astron. Soc., Geophys. J., v. 1, p. 320–329.

Mason, R. G., and Raff, A. D., 1961, A magnetic survey off the west coast of North America 32° N to 42° N: Geol. Soc. Amer., Bull., v. 72, p. 1259–1265.

Matthews, D. H., and Bath, J., 1967, Formation of magnetic anomaly pattern of mid-Atlantic ridge: Roy. Astron. Soc., Geophys. J., v. 13, p. 349–357.

Morgan, W. J., Vogt, P. R., and Falls, D. F., 1969, Magnetic anomalies and sea-floor spreading on the Chile rise: Nature, v. 222, p. 137–142.

Peter, G., 1966, Magnetic anomalies and fracture pattern in the northest Pacific Ocean: J. Geophys. Res., v. 71, p. 5365–5374.

Peter, G., Erickson, B. H., and Grim, P. J., 1970, Magnetic structure of the Aleutian and northeast Pacific basin, in Maxwell, A. E., ed., The sea, v. 4, pt. 2: New York, Wiley-Interscience, p. 191–222.

Phillips, J. D., 1967, Magnetic anomalies over the mid-Atlantic ridge near 27° N: Science, v. 157, p. 920–922.

Pitman, W. C., III, and Heirtzler, J. R., 1966, Magnetic anomalies over the Pacific-Antarctic ridge: Science, v. 154, p. 1164–1171.

Pitman, W. C., III, Herron, E. M., and Heirtzler, J. R., 1968, Magnetic anomalies in the Pacific and sea floor spreading: J. Geophys. Res., v. 73, p. 2069–2085.

Pitman, W. C., III, Talwani, M., and Heirtzler, J. R., 1971, Age of the North Atlantic from magnetic anomalies: Earth Planet. Sci. Lett., v. 11, p. 195–200.

Raff, A. D., 1966, Boundaries of an area of very long magnetic anomalies in the northeast Pacific: J. Geophys. Res., v. 71, p. 2631–2636.

Raff, A. D., 1968, Sea-floor spreading—Another rift: J. Geophys. Res., v. 73, p. 3699–3705.

Raff, A. D., and Mason, R. G., 1961, Magnetic survey off the west coast of North America, 40° N latitude to 50° N latitude: Geol. Soc. Amer., Bull., v. 72, p. 1267–1270.

Talwani, M., Windisch, C. C., and Langseth, M. G., 1971, Reykjanes ridge crest: a detailed geophysical survey: J. Geophys. Res., v. 76, p. 473–517.

Uyeda, S., Vacquier, V., Yasui, M., Sclater, J., Sato, T., Lawson, J., Watanabe, T., Dixon, F., Silver, E., Fukao, Y., Sudo, K., Nishikawa, M., and Tanaka, T., 1967, Results of geomagnetic survey during the cruise of R/V Argo in western Pacific 1966 and the compilation of magnetic charts of the same area: Tokyo, Univ., Earthquake Res. Inst., Bull., v. 45, p. 799–814.

Vine, F. J., 1968, Magnetic anomalies associated with mid-ocean ridges, in Phinney, R. A., ed., The history of the earth's crust: Princeton, N. J., Princeton Univ. Press, p. 73–89.

Vine, F. J., and Hess, H. H., 1970, Sea floor spreading, in Maxwell, A. E., ed., The sea, v. 4, pt. 2: New York, Wiley-Interscience, p. 587–622.

Vogt, P. R., Anderson, C. N., and Bracey, D. R., 1971, Mesozoic magnetic anomalies, sea-floor spreading, and geomagnetic reversals in the southwestern North Atlantic: J. Geophys. Res., v. 76, p. 4796–4823.

Vogt, P. R., Anderson, C. N., Bracey, D. R., and Schneider, E. D., 1970a, North Atlantic magnetic smooth zones: J. Geophys. Res., v. 75, p. 3955–3967.

Vogt, P. R., and Higgs, R. H., 1969, An aeromagnetic survey of the eastern Mediterranean and its interpretation: Earth Planet. Sci. Lett., v. 5, p. 439–448.

Vogt, P. R., and Johnson, G. L., 1971, Cretaceous sea floor spreading in the western North Atlantic: Nature, v. 234, p. 22–25.

Vogt, P. R., Johnson, G. L., Holcombe, T. L., Gilg, J. G., and Avery, O. E., 1972, Episodes of sea-floor spreading recorded by the North Atlantic basement: Tectonophysics, v. 12, p. 211–234.

Vogt, P. R., and Ostenso, N. A., 1970, Magnetic and gravity profiles across the Alpha Cordillera and their relation to Arctic sea floor spreading: J. Geophys. Res., v. 75, p. 4925–4937.

Vogt, P. R., Ostenso, N. A., and Johnson, G. L., 1970b, Bathymetric and magnetic data bearing on sea-floor spreading north of Iceland: J. Geophys. Res., v. 75, p. 903–920.

Watkins, N. D., 1968a, Comments on the interpretation of linear magnetic anomalies: Pure Appl. Geophys., v. 69, p. 170–192.

Watkins, N. D., and Richardson, A., 1971, Intrusives, extrusives, and linear magnetic anomalies: Roy, Astron. Soc., Geophys. J., v. 23, p. 1–13.

22

Magnetic Anomalies over Oceanic Ridges

FRED J. VINE
DRUMMOND H. MATTHEWS
1963

Typical profiles showing bathymetry and the associated total magnetic field anomaly observed on crossing the North Atlantic and North-West Indian Oceans are shown in Fig. 22–1. They illustrate the essential features of magnetic anomalies over the oceanic ridges: (1) long-period anomalies over the exposed or buried foothills of the ridge; (2) shorter-period anomalies over the rugged flanks of the ridge; (3) a pronounced central anomaly associated with the median valley. This pattern has now been observed in the North Atlantic (Heezen et al., 1953; Keen, 1963), the Antarctic (Adams and Cristoffel, 1962), and the Indian Oceans (Heirtzler, 1961; Matthews et al., 1963). In this article we describe an attempt to account for it.

The general increase in wave-length of the anomalies away from the crest of the ridge is almost certainly associated with the increase in depth to the magnetic crustal material (Heezen et al., 1953). Local anomalies of short-period may often be correlated with bathymetry, and explained in terms of reasonable susceptibility contrasts and crustal configurations; but the long-period anomalies of category (1) are not so readily explained. The central anomaly can be reproduced if it is assumed that a block of material very strongly magnetized in the present direction of the Earth's field underlies the median valley and produces a positive susceptibility contrast with the adjacent crust. It is not clear, however, why this considerable susceptibility contrast should exist beneath the median valley but not elsewhere under the ridge. Recent work in this Department has suggest a new mechanism.

In November 1962, H.M.S. *Owen* made a detailed magnetic survey over a central part of the Carlsberg Ridge as part of the International Indian Ocean Expedition. The area (50 × 40 nautical miles; centred on 5° 25′ N., 61° 45′ E.) is predominantly mountainous, depths ranging from 900 to 2,200 fathoms, and the

From *Nature*, v. 199, p. 947–949, 1963. Reprinted with permission of the authors and Macmillan Journals, Ltd.

233

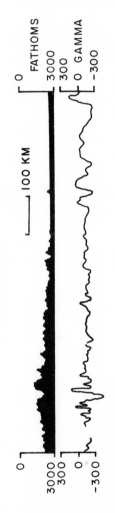

MID–ATLANTIC RIDGE 1960

100 KM

0
3000
FATHOMS

300
0 GAMMA
–300

CARLSBERG RIDGE 1962

100 KM

0
3000
FATHOMS

300
0 GAMMA
–300

0
3000
300
0
–300

Figure 22-1
Profiles showing bathymetry and the associated total magnetic field anomaly observed on crossing the North Atlantic and the northwest Indian Oceans. Upper profile from 45° 17′ N. 28° 17′ W. to 45° 19′ N. 11° 29′ W. Lower profile from 30° 5′ N. 61° 57′ E. to 10° 10′ N. 66° 27′ E.

TRUE NORTH

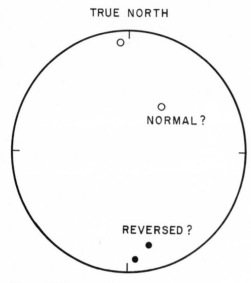

Figure 22-2
Directions of the magnetic vectors obtained by
the computer programme plotted on a stereographic
projection, together with the present field vector
and its reverse. Bearings and inclinations: present
field vector 356°; −6° (up); computed vectors
038°; −40° (up); 166°30′; +13° (down)

topographic features are generally elongated
parallel to the trend of the Ridge. This elongation is more marked on the total magnetic field
anomaly map where a trough of negative
anomalies, flanked by steep gradients, separates two areas of positive anomalies. The
trough of negative anomalies corresponds to a
general depression in the bottom topography
which represents the median valley of the
Ridge. The positive anomalies correspond to
mountains on either side of the valley.

In this low magnetic latitude (inclination
−6°) the effect of a body magnetized in the present direction of the Earth's field is to reduce the
strength of the field above it, producing a negative anomaly over the body and a slight positive anomaly to the north. Here, over the
centre of the Ridge, the bottom topography indicates the relief of basic extrusives such as
volcanoes and fissure eruptives, and there is
little sediment fill. The bathymetry, therefore,
defines the upper surface of magnetic material
having a considerable intensity of magnetization, potentially as high as any known igneous
rock type (Hill, 1963), and probably higher,
because it is extrusive, than the main crustal

layer beneath. That the topographic features
are capable of producing anomalies is immediately apparent on comparing the bathymetric
and the anomaly charts; several have well-
defined anomalies associated with them.

Two comparatively isolated volcano-like
features were singled out and considered in detail. One has an associated negative anomaly
as one would expect for normal magnetization,
the other, completely the reverse anomaly pattern, that is, a pronounced positive anomaly
suggesting reversed magnetization. Data on
the topography of each feature and its associated anomaly were fed into a computer and
an intensity and direction of magnetization
for each obtained. Fig 22–2 shows the directions of the resulting vectors plotted on a stereographic projection. Having computed the
magnetic vector by a 'best fit' process, the
computer recalculated the anomaly over the
body, assuming this vector, thus giving an indication of the accuracy of fit. The fit was good
for the case of reversed magnetization but poor
for that of approximately normal magnetization. The discrepancy is scarcely surprising
since we have ignored the effects of adjacent
topography, and the interference of other
anomalies in the vicinity. In addition, the example of normal magnetization is near a corner
of the area where the control of contouring is
less precise. The other example is central
where the control is good. In both cases the intensity of magnetization deduced was about
0.005 e.m.u.; this is equivalent to an effective
susceptibility of ± 0.0133; (effective susceptibility = total intensity of magnetization (remanent + induced)/present total magnetic field
intensity : mean value for basalts of the order
of 0.01).

In addition, three profiles, perpendicular to
the trend of the Ridge, have been considered.
Computed profiles along these, assuming infinite lateral extent of the bathymetric profile,
and uniform normal magnetization, bear little
resemblance to the observed profiles (Fig.
22–3). These results suggested that whole
blocks of the survey area might be reversely
magnetized. The dotted curve in Fig. 22–3, *B*
was computed for a model in which the main
crustal layer and overlying volcanic terrain
were divided into blocks about 20 km wide, alternately normally and reversely magnetized.

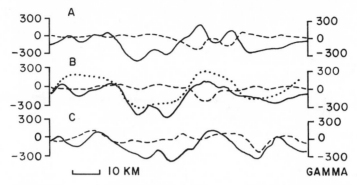

Figure 22-3
Observed and computed profiles across the crest of the Carlsberg
Ridge. Solid lines, observed anomaly; broken lines, computed profile
assuming uniform normal magnetization and an effective susceptibility
of 0.0133; dotted line, assuming reversals—see text. The computed
profiles were obtained assuming infinite lateral extent of the bathymetric
profiles

The blocks were given the effective susceptibility values shown in the caption to Fig.
22–4(3).

Work on this survey led us to suggest that
some 50 per cent of the oceanic crust might be
reversely magnetized and this in turn has suggested a new model to account for the pattern
of magnetic anomalies over the ridges.

The theory is consistent with, in fact virtually a corollary of, current ideas on ocean floor
spreading (Dietz, 1961) and periodic reversals
in the Earth's magnetic field (Cox et al.,
1963a). If the main crustal layer (seismic layer
3) of the oceanic crust is formed over a convective up-current in the mantle at the centre
of an oceanic ridge, it will be magnetized in the
current direction of the Earth's field. Assuming
impermanence of the ocean floor, the whole of
the oceanic crust is comparatively young,
probably not older than 150 million years, and
the thermo-remanent component of its magnetization is therefore either essentially normal, or reversed with respect to the present
field of the Earth. Thus, if spreading of the
ocean floor occurs, blocks of alternately normal and reversely magnetized material would
drift away from the centre of the ridge and
parallel to the crest of it.

This configuration of magnetic material
could explain the lineation or 'grain' of magnetic anomalies observed over the Eastern Pacific to the west of North America (Hill, 1963)

(probably equivalent to the long-period anomalies of category (1)). Here north-south highs
and lows of varying width, usually of the order
of 20 km, are bounded by steep gradients. The
amplitude and form of these anomalies have
been reproduced by Mason (Mason, 1958;
Mason and Raff, 1961), but the most plausible
of the models used involved very severe restrictions on the distribution of lava flows in
crustal layer 2. They are readily explained in
terms of reversals assuming the model shown
in Fig. 22–4 (1). It can be shown that this type
of anomaly pattern will be produced for virtually all orientations and magnetic latitudes,
the amplitude decreasing as the trend of the
ridge approaches north-south or the profile
approaches the magnetic equator. The pronounced central anomaly over the ridges is
also readily explained in terms of reversals.
The central block, being most recent, is the
only one which has a uniformly directed magnetic vector. This is comparable to the area of
normally magnetized late Quaternary basics
in Central Iceland (Hospers, 1954; Thorarinsson et al., 1959–1960) on the line of the Mid-
Atlantic Ridge. Adjacent and all other blocks
have doubtless been subjected to subsequent
vulcanism in the form of volcanoes, fissure
eruptions, and lava flows, often oppositely
magnetized and hence reducing the effective
susceptibility of the block, whether initially
normal or reversed. The effect of assuming a

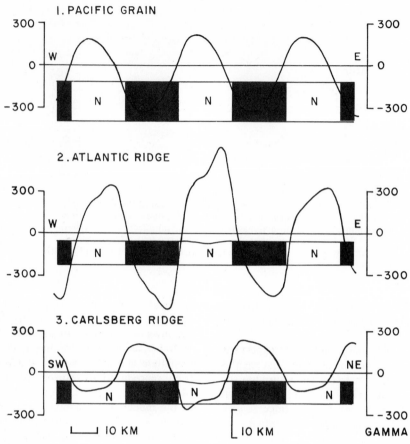

Figure 22-4
Magnetic profiles computed for various crustal models. Crustal blocks marked N, normally magnetized; diagonally shaded blocks, reversely magnetized. Effective susceptibility of blocks, 0.0027, except for the block under the median valley in profiles 2 and 3, 0.0053. (1) Pacific Grain. Total field strength, $T = 0.5$ oersted; inclination, $I = 60°$; magnetic bearing of profile, $\theta = 073°$. (2) Mid-Atlantic Ridge, $T = 0.48$ oersted; $I = 65°$; $\theta = 120°$. (3) Carlsberg Ridge, $T = 0.376$ oersted; $I = -6°$; $\theta = 044°$

reduced effective susceptibility for the adjacent blocks is illustrated for the North Atlantic and Carlsberg Ridges in Fig. 22–4(2,3).

In Fig. 22–4, no attempt has been made to reproduce observed profiles in detail, the computations simply show that the essential form of the anomalies is readily achieved. The whole of the magnetic material of the oceanic crust is probably of basic igneous composition; however, variations in its intensity of magnetization and in the topography and direction of magnetization of surface extrusives could account for the complexity of the observed profiles. The results from the preliminary Mohole drilling (Cox and Doell, 1962; Raff, 1963) are considered to substantiate this conception. The drill penetrated 40 ft. into a basalt lava flow at the bottom of the hole, and this proved to be reversely magnetized (Cox and Doell, 1962). Since the only reasonable explantion of the magnetic anomalies mapped near the site of the drilling is that the area is underlain by a block of normally magnetized crustal material (Raff, 1963), it appears that the drill penetrated a layer of reversely magnetized lava overlying a normally magnetized block.

In Fig. 22–4 it will also be noticed that the effective susceptibilities assumed are two to five times less than that derived for the isolated features in the survey area described. Although no great significance can be attached to this derived intensity it is suggested that the fine-grained extrusives (basalts) of surface features are more highly magnetized than the intrusive material of the main crustal layer which, in the absence of evidence to the contrary, we assume to be of analogous chemical composition (that is, gabbros). This would appear to be consistent with recent investigations of the magnetic properties of basic rocks (Hill, 1963).

The vertical extent of the magnetic crust is defined by the depth to the curie-point isotherm. In the models this has been assumed to be at 20 km below sea-level over the deep ocean but at a depth of 11 km beneath the centre of the ridges where the heat flow and presumably the thermal gradient are higher. These assumptions are questionable but not critical because the amplitude of the simulated anomaly depends on both the thickness of the block and its effective susceptibility, and, although the thickness is in doubt by a factor of two, the susceptibility is in doubt by a factor of ten.

Present magnetic declination has been assumed throughout the calculations: it would probably have been better to have ignored this, as in palaeomagnetism, assuming that true north approximates to the mean of secular variations; but this is unimportant and in no way affects the essential features of the computations.

In order to explain the steep gradients and large amplitudes of magnetic anomalies observed over oceanic ridges all authors have been compelled to assume vertical boundaries and high-susceptibility contrasts between adjacent crustal blocks. It is appreciated that mgnetic contrasts within the oceanic crust can be explained without postulating reversals of the Earth's magnetic field; for example, the crust might contain blocks of very strongly magnetized material adjacent to blocks of material weakly magnetized in the same direction. However, the model suggested in this article seems to be more plausible because high susceptibility contrasts between adjacent blocks can be explained without recourse to major inhomogeneities of rock type within the main crustal layer or to unusually strongly magnetized rocks.

23

Magnetic Anomalies over A Young Oceanic Ridge off Vancouver Island

FRED J. VINE
J. TUZO WILSON
1965

Surveys of the earth's total magnetic field have been made along closely spaced lines over large areas in the northeastern Pacific Ocean. (Mason, 1958; Vacquier et al., 1961; Mason and Raff, 1961; Raff and Mason, 1961; Peter and Stewart, 1965). These show a surprisingly regular, linear pattern of anomalies, often hundreds of kilometers long and tens of kilometers wide, and usually aligned approximately north-south. Vine and Matthews (1963) have suggested that these anomalies, together with the central magnetic anomaly observed over certain oceanic ridges, might be explained in terms of ocean-floor spreading (Holmes, 1929; Hess, 1962) and periodic reversals of the earth's mgnetic field. The idea proposes that as new oceanic crust is formed over a convective upcurrent in the mantle, at the center of an oceanic ridge, it will be magnetized in the ambient direction of the earth's magnetic field. If the earth's field reverses periodically as ocean-floor spreading occurs, then successive strips of crust paralleling the crest of the ridge will be alternately normally and reversely magnetized, thus producing the linear anomalies of the northeastern Pacific. These anomalies are not obviously parallel to any active oceanic ridge, but it seems possible that they are related in this way either to the East Pacific Rise, as suggested by Menard (1964), or to the extinct Darwin Rise, as suggested recently by Hess (1965).

At the time it was put forward, the Vine and Matthews hypothesis was particularly speculative in that no large-scale magnetic survey was thought to be available for an oceanic ridge, and results regarding the periodicity, or even confirmation, of possible reversals of the earth's magnetic field were very preliminary (Cox et al., 1963a). Recently the evidence suggesting possible reversals of the earth's field has been examined more critically and a periodicity suggested for the past 4 million years (Cox et al., 1964b). Furthermore, it has been suggested in the preceding report that the area of detailed magnetic survey in the northeastern Pacific might include one or more short lengths of a young and active oceanic ridge (Wilson, 1965b). This suggestion was originally based on the concept of transform faults and the distribution of earthquake epicenters along the western coast of North America (Wilson, 1965c). Only subsequently was reference made to the magnetic survey to find that it lends convincing additional support to the proposal.

Clearly, if these interpretations are correct we now have information with which to reexamine the original suggestion of Vine and Matthews (1963). If one assumes that there have been major reversals of the earth's field at 1, 2.5, and 3.4 million years and short-lived reversals at about 1.9 and, possibly, 3 million years, as suggested by Cox, Doell, and Dalrymple (1964b) and if one assumes two commonly suggested rates of spreading, 1 and 2 cm per year per limb of the convecting system, one obtains the models and calculated anomalies shown in Fig. 23–1, a and b. The models are directly comparable with those originally suggested by Vine and Matthews, that is, the normal and reversed blocks extend from a depth of 3 to 11 km below sea level and have effective susceptibilities of ±0.0025, except for the central block, for which the value is assumed to be +0.005.[1]

The model for the 1 cm per year rate (Fig. 23–1,a) suggests a possible explanation for the central high-amplitude anomaly observed over certain oceanic ridges, notably the Mid-Atlantic Ridge and the northwestern Indian Ocean (Carlsberg) Ridge, as discussed previously (Vine and Matthews, 1963). For this rate of spreading, the anomalies resulting from the normal-reverse contacts on either side of the central block reinforce each other to produce

the central anomaly. For the faster rate of spreading, giving rise to a wide central block, the reinforcement is much less, and a rather broad central anomaly is produced which is not so easily distinguished from its neighbors (Fig. 23–1,b). This might possibly be the case over the East Pacific Rise and the new, Juan de Fuca Ridge (Wilson, 1965b) (see Fig. 23–2). One would hardly expect the rate of spreading to be constant throughout the lifespan of an oceanic ridge, and therefore it is unlikely to be the same for all ridges at the present time.

The essential feature of the Vine and Matthews hypothesis is that the normal-reverse contacts produce the steep, often isolated, magnetic gradients over ridges. On the basis of this criterion the steepest gradients over the Juan de Fuca Ridge have been assumed to delineate normal-reverse boundaries, and a crude model has been drawn up, again along the lines originally proposed by Vine and Matthews. The models and calculated anomalies are presented in Figs. 23–1,c and 23–2,c. Fig. 23–1,c shows the central part of Fig. 23–2,c. Despite the simple nature of this model it agrees well with the observed anomalies (Fig. 23–2,b). Assuming that the times of the field reversals are the same as those suggested by Cox, Doell, and Dalrymple (1964b), dates have been added to the section and an average rate of spreading deduced (Fig. 23–1,c). In Fig. 23–1, comparison of c with a and b suggests that the rate of spreading of the ridge has been rather irregular, as one might expect, but averages about 3 cm per year (1.5 cm per year per limb of the cell). This implies that the central 120 km or so of the crustal material over the crest of this new ridge has been formed within the past 4 million years. Thus if it is assumed that the rifting and associated faulting has continued without interruption and at this average rate, then the whole ridge (a total width of about 350 km) would be no more than 11 or 12 million years old. These deductions agree well with the rate of movement observed at present along the San Andreas fault (Whitten, 1948) and with the total displacement across it, as discussed by Wilson (1965b).

Clearly, the models shown in Figs. 23–1 and 23–2,c are oversimplified. However, they express the basic tenet of the Vine and Matthews

---

[1]For all three models in Fig. 23–1 the intensity of the earth's field has been taken as 54,000 gama, its dip as +66°, and the magnetic bearing of the profile as 087°, that is, as for the Juan de Fuca Ridge (46°30'N, 129°30'W). Normal or reverse magnetization is with respect to an axial dipole vector; axial dipole dip taken as +65°. Effective susceptibility of the blocks, ± 0.0025, except for the central block, + 0.005. (1 gamma = 10⁻⁵ oersted).

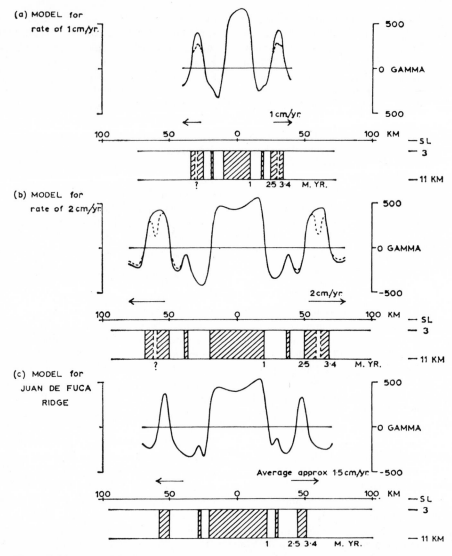

Figure 23-1
Models and calculated total field magnetic anomalies resulting from a combination of suggested recent polarities for the earth's magnetic field (Cox et al., 1964) and ocean floor spreading. Normally magnetized blocks are shaded; reversely magnetized blocks unshaded. Portions (a) and (b) assume uniform rates of spreading. Portion (c) was deduced from the gradients on the map of observed anomalies. The dashed parts of the computed profiles show the effect of including the possible reversal at 3 million years (Cox et al., 1964) (see footnote 1, page 239)

idea that the steep magnetic gradients so obvious from any detailed magnetic survey over the oceans might delineate the boundaries between essentially normally and essentially reversely magnetized crust, thus reproducing the observed gradients without recourse to improbable structures or lateral changes in petrology. If this basic principle is accepted, there is no difficulty in explaining the anomalies but only in deciding on the distribution of magnetization within the various layers of the oceanic crust (Cann and Vine, 1966). As ever in the interpretation of magnetic anomalies, there is no unique solution, and the various

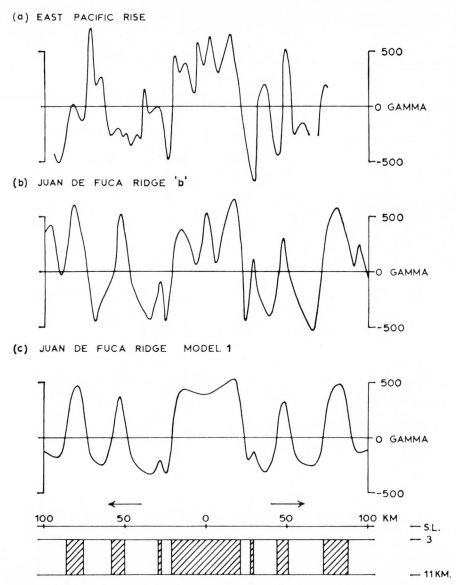

Figure 23-2
(a) Observed profile across the East Pacific Rise at 59°S, 149°W (Heirtzler, 1961). (b) Observed profile "b" across the Juan de Fuca Ridge (see Figure 23-4). (c) Model and calculated anomaly for Juan de Fuca Ridge, assuming generalized crustal blocks (compare Figure 23-1,c)

parameters are so "flexible" that, having assumed normal and reverse strips, the model can be fitted to any existing concept of the structure of oceanic ridges.

Ocean-floor spreading implies that the oceanic crust is a surface expression of the mantle; it must therefore be generated from the mantle and be capable of being resorbed by it, as emphasized by Hess (1965). Basalt is the most common outcropping hard rock on the ocean floor; if this is regarded as the lowest melting fraction of the material of the upper mantle then it probably represents only a small percentage of this material by volume. Hess

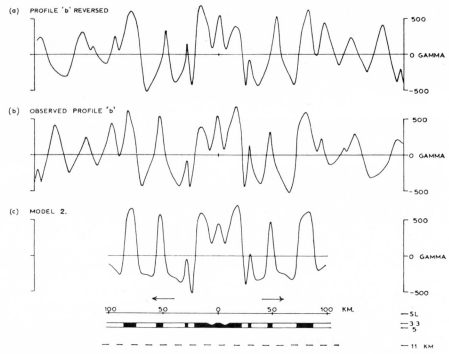

Figure 23-3
(a) and (b) Observed profile "b" across the Juan de Fuca Ridge together with its mirror image about its midpoint, to demonstrate its symmetry. (c) Model and calculated anomaly for Juan de Fuca Ridge assuming a strongly magnetized basalt layer only. Black, normally magnetized material; unshaded material of this layer, reversely magnetized. Normal or reverse magnetization is with respect to an axial dipole vector; axial dipole dip taken at +65°. Effective susceptibility taken as ±0.01, except for the central block, +0.02

considers, therefore, that basalt accounts for only a thin veneer 1 or 2 km thick on top of a main crustal layer of serpentinite, that is, hydrated mantle. The great thickness of basalt lavas in central Iceland (Bodvarsson and Walker, 1964) is clearly anomalous in that the whole crustal section is thicker, and away from the center the volcanics have been subjected to erosion, unlike those of the submarine ridges. Assuming the validity of the model for oceanic ridges proposed by Hess (1965), the "magnetic" material of the crust would be largely confined to the basalt layer (layer 2). Hess envisages that the serpentinite layer (layer 3) is emplaced in the solid state and it would therefore acquire its remanent magnetization at depth on passing through the Curie point isotherm. By the time it is emplaced beneath the central rift it might well be highly sheared, fractured, and randomly orientated. Serpen-

tinite would appear to be weakly magnetized and to have a Konigsberger ratio of approximately 1 (Cox et al., 1964c). All in all it would probably be capable of contributing little to the observed magnetic anomalies. The basalt, however, cools through the Curie point in place in the form of lava flows or intrusives. It is strongly magnetized, and its remanent magnetization probably predominates, since its susceptibility would appear to be comparatively low (Ade-Hall, 1964). In Fig. 23–3,c the magnetic anomalies have been computed over a model in which the magnetic material is confined entirely to layer 2. As previously, the central block is assumed to be more strongly magnetized because it is the only block composed exclusively of young material which is magnetized normally, except for the minor possibility of self-reversals. Volcanism probably occurs over a wider zone than the central

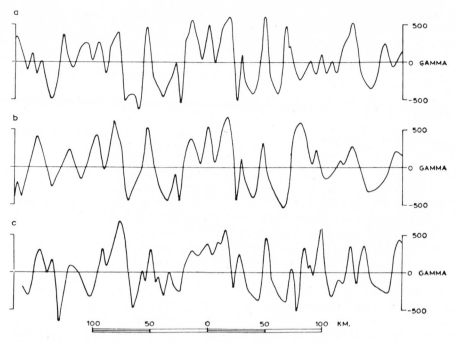

Figure 23-4
Observed profiles (a), (b), and (c) at intervals of 45 km along Juan de Fuca ridge, north to south. Midpoint of profile "b" is 46°39′N. 129°24′W. True bearing of profiles is 110

block, and all other blocks will therefore be contaminated with younger material, often of reverse polarity to that of the initial block, and hence lowering or modifying its resultant magnetic effect. The serpentinite of layer 3 is almost certainly riddled with basaltic feeders for the flows and intrusives of layer 2. If these feeders are taken into account they will have the effect of slightly lowering the effective susceptibility assumed for layer 2 in Fig. 23–3,c. This susceptibility is, as it stands, comparatively high but not unreasonable.

Comparison of Figs. 23–2,c and 23–3,c confirms that the essential feature of the Vine and Matthews hypothesis is the normal-reverse contacts; the actual distribution of magnetization within layers 2 and 3 of the oceanic crust is a matter of speculation at the present time (Vine and Matthews, 1963; Cann and Vine, 1966). However, the comparison also suggests that the second model is a considerable improvement on the first, despite the fact that it is still very simple. The original, generalized

model of Vine and Matthews (Model 1) and the specific model after Hess (Model 2) have been chosen to illustrate what are possibly the two extremes, but it seems increasingly probable that the observed anomalies can best be reproduced by strongly magnetized, basalt material in layer 2 and less strongly magnetized material at depth, whether this decrease in magnetization be due to a general increase in grain size, a change in rock type, or metamorphic effects.

A literal interpretation of the Vine and Matthews hypothesis implies that the magnetic anomalies observed over ridges at certain latitudes and orientations should be roughly symmetrical (for example, as in Fig. 23–1), but the simplicity of this model when compared with the probable complexity of the real situation makes a high degree of symmetry improbable. Iceland, although atypical in some ways, must give certain pointers as to the nature of the crestal province of the ridge system. Work on Iceland (Bodvarsson and Walker,

1964) suggests that the structure is symmetrical about the central rift or tension crack and has been formed by crustal drifting away from it. Although the rift is persistent, it seems likely that its location might change occasionally, and multiple cracks may occur, as is the case in the south of Iceland at present. Different conditions might prevail over different parts of the ridge system. Quiet, steady conditions of crustal emplacement and associated volcanism might produce a considerable degree of symmetry.

A surprisingly high degree of symmetry is exhibited by the Juan de Fuca Ridge (Wilson, 1965b). Three profiles across the ridge are shown in Fig. 23–4; they are at 45-km intervals along the ridge. In Fig. 23–3 the central profile is shown together with its mirror image to demonstrate the symmetry. This suggests a quiet growth for the ridge, and it is possibly significant that no active submarine volcanism has been reported from this area despite the fact that the ridge is presumed to be active because of the occurrence of recent earthquakes along the bounding transform faults (Wilson, 1965b). Symmetry of ridge profiles might be sought elsewhere. As with correlation of magnetic anomalies on adjacent profiles, the symmetry of the anomalies about the axis of a ridge is probably much less obvious from profiles than from a detailed survey.

It is becoming increasingly apparent that both the linearity of oceanic, magnetic anomalies and the steep gradients which bound them are obvious from a detailed magnetic survey but much more difficult to see from comparatively random profiles. Even the marked linearity of the anomalies over the Juan de Fuca Ridge is not necessarily obvious from adjacent profiles across it. Although profiles *a* and *b* in Fig. 23–4 correlate well, their correlation with profile *c* is not so good, and one might easily not anticipate the pronounced "grain" of the survey map (Mason, 1958; Vacquier et al., 1961; Mason and Raff, 1961; Raff and Mason, 1961; Peter and Stewart, 1965). We should therefore like to reiterate the recent plea by Peter and Stewart (1965) that magnetic surveys are of so much greater value than random profiles. Aeromagnetic surveys would appear to be perfectly adequate. It has been known for several years, but more convincingly established recently (Mason, 1958; Vacquier et al., 1961; Mason and Raff, 1961; Raff and Mason, 1961; Peter and Steward, 1965; Cann and Vine, 1966), that the magnetic anomalies are, in general, quite unrelated to the bathymetry except over isolated seamounts or apparent transcurrent faults [all of which are, possibly, transform faults (Wilson, 1965b)].

24

Spreading of the Ocean Floor: New Evidence

FRED J. VINE

1966

From *Science*, v. 154, p. 1405–1415, 1966. Reprinted with permission of the author and the American Association for the Advancement of Science. Copyright 1966 by the AAAS.

Controversy regarding continental drift has raged within the earth sciences for more than 40 years. Within the last decade it has been enlivened by the results of paleomagnetic research and exploration of the ocean basins (Bullard et al., 1965). Throughout, one of the main stumbling blocks has been the lack of a plausible mechanism to initiate and maintain drift. Recently, however, the concept of spreading of the ocean floor, as proposed by Hess (1962), has renewed for many the feasibility of drift and provided an excellent working hypothesis for the interpretation and investigation of the ocean floors. The hypothesis invokes slow convection within the upper mantle by creep processes, drift being initiated above an upwelling, and continental fragments riding passively away from such a rift on a conveyor belt of upper-mantle material; movements of the order of a few centimeters per annum are required. Thus the oceanic crust is a surface expression of the upper mantle and is considered to be derived from it, in part by partial fusion, and in part by low-temperature modification. This model, as developed by Hess (1965) and Dietz (1966), can be shown to account for many features of the ocean basins and continental margins.

It seems reasonable to assume that, if drift has occurred, some record of it should exist within the ocean basins. Heezen and Tharp (1965; 1966) have delineated north-south topographic scars on the floor of the Indian Ocean that may well be caused by the northward drift of India since Jurassic time. Wilson (1965d) has suggested that drift and ocean-floor spreading in the South Atlantic and East Pacific may be recorded in the form of fracture zones and aseismic volcanic ridges. It has also been postulated that the history of a spreading ocean floor may be recorded in terms of the permanent (remanent) magnetization of the oceanic crust.

Vine and Matthews (1963) have suggested that variations in the intensity and polarity of Earth's magnetic field may be "fossilized" in the oceanic crust, and that this condition in turn should be manifest in the resulting short-wavelength disturbances or "anomalies" in Earth's magnetic field, observed at or above Earth's surface. Thus the conveyor belt can also be thought of as a tape recorder. As new oceanic crust forms and cools through the Curie temperature at the center of an oceanic ridge, the permanent component of its magnetization, which predominates, will assume the ambient direction of Earth's magnetic field. A rate of spreading of a few centimeters per annum and a duration of 700,000 years for the present polarity (Cox et al., 1964b; Doell and Dalrymple, 1966) imply a central "block" of crust, a few tens of kilometers in width, in which the magnetization is uniformly and "normally" directed. The adjacent blocks will be of essentially reversed polarity, and the width and polarity of blocks successively more distant from the central block will depend on the reversal time scale for Earth's field in the past.

Vine and Wilson considered that the bulk of the magnetization resides in a comparatively thin layer, 1 or 2 kilometers of basaltic extrusives and intrusives, coating a main crustal layer of serpentinite (Vine and Wilson, 1965). If the frequency of extrusion and intrusion of this material is approximately normally distributed about the axis of the ridge (Loncarevic et al., 1966), all blocks other than the central block will be contaminated with younger material, possibly of opposite polarity, in which case their bulk resultant or effective magnetization will be reduced. In this way a model has been developed in which the magnetization of the central block is assumed to be twice that of the others. This model derives from work on a very small but detailed survey of an area on the crest of Carlsberg Ridge in the northwest Indian Ocean (Cann and Vine, 1966).

The idea was an attempt to explain two interesting and enigmatic features of oceanic magnetic anomalies: the well-known central anomaly associated with the axes of ridges, first observed by Ewing, Hirshman, and Heezen (1959); and the remarkable striped pattern of anomalies revealed by surveys of Earth's magnetic field in the northeastern Pacific (for example, Fig. 24–1). These anomalies are known to retain their characteristic shape and spacing for thousands of kilometers along their length (Vacquier, 1965) and are quite unlike anomalies observed over the continents. They are very difficult to simulate if one assumes any reasonable lithologic contrasts within the oceanic crust, or plausible geologic structures. Intuitively it was felt that in some way the anomalies must be a surface expression of convection within the mantle. The thesis assumes therefore that the linear anomalies of the northeastern Pacific are quite ubiquitous over the deep ocean basins; that they are interrupted only by anomalies associated with isolated seamounts or volcanic ridges, and by fracture zones which offset the anomaly pattern — as was shown by Vacquier in the northeast Pacific (Vacquier, 1965).

Of the three basic assumptions of the Vine and Matthews hypothesis, field reversals (Cox et al., 1964b; Doell and Dalrymple, 1966) and the important of remanence (Ade-Hall, 1964) have recently become more firmly established and widely held; thus in demonstrating the efficacy of the idea one might provide virtual proof of the third assumption: ocean-floor spreading, and its various implications.

DIFFICULTIES

At the time this concept was proposed there was very little concrete evidence to support it, and in some ways it posed more problems than it solved. There were, for example, at least three rather awkward points that it did not explain:

1) Many workers felt and feel that the northeast Pacific anomalies do not parallel any existing or preexisting oceanic ridge (Peter and Stewart, 1965).

2) Whereas one can visually correlate anomalies on widely spaced profiles in the northeastern Pacific, one cannot do this over ridge crests, except for the central anomaly. Vacquier (1965) maintained, therefore, that there are no linear anomalies paralleling the central anomaly over the crests of ridges.

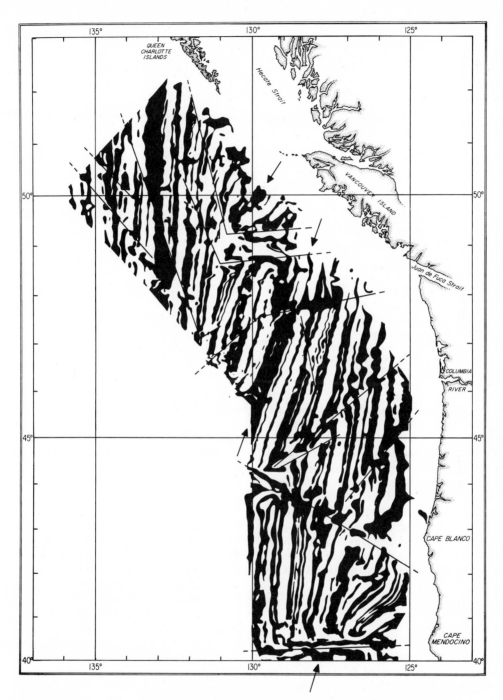

Figure 24-1
Summary diagram of total magnetic-field anomalies southwest of Vancouver Island. Areas of
positive anomaly are shown in black. Straight lines indicate faults offsetting the anomaly pattern:
arrows, the axes of the three short ridge lengths within this area—from north to south. Explorer,
Juan de Fuca, and Gorda ridges. See also Figure 24-15. (Based on fig. 1 of Raff and Mason (1961);
courtesy Geol. Soc. Amer.)

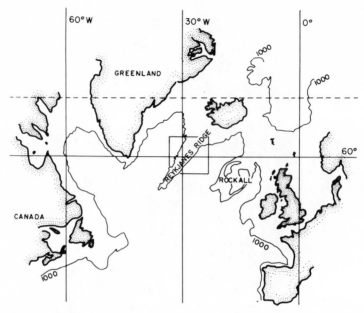

Figure 24-2
The location of Reykjanes Ridge, southwest of Iceland, and the area of
Figure 24-3. The 1000-fathom submarine contour is shown, together
with the 500-fathom contours for Rockall Bank

3) The idea did not, very obviously, explain the fact that the low-amplitude, short-wavelength anomalies observed on either side of the axis of a ridge give way to higher-amplitude, long-wavelength anomalies over the more distant flanks—an observation originally made by Vine and Matthews (1963) and emphasized by Heirtzler and Le Pichon (1965). With the increase in depth of the magnetic material as one moves from the ridge crest to the flanks, one would expect disappearance of shorter wavelengths but not an increase in amplitude.

COROLLARIES

The second difficultly is clearly rather fundamental, but has persisted because until recently no large, detailed survey of the crest of a midocean ridge was thought to be available. However, in 1963 the U.S. Naval Oceanographic Office[1] made a detailed aeromagnetic

---

[1]At the suggestion of Lamont Geological Observatory.

survey of Reykjanes Ridge, southwest of Iceland (Fig. 24-2) (Heirtzler et al., 1966). The ridge was chosen because it clearly forms part of the northerly extension of the Mid-Atlantic Ridge through Iceland, and because earlier traverses had indicated a typical central anomaly over its crest (Avery, 1963). A diagram summarizing the anomalies revealed by this survey appears in Fig. 24–3. The area summarized, approximating a 400-kilometer square, shows a pattern of linear anomalies paralleling the central anomaly and symmetrically disposed about it. This finding, together with the symmetry and linearity of the magnetic anomalies about the Juan de Fuca and Gorda ridges (Fig. 24–1), recently described by Wilson (1965b), provides convincing confirmation of the two most obvious corollaries of a literal interpretation of the Vine-Matthews hypothesis: (i) linear magnetic anomalies should parallel or subparallel ridge crests, and (ii) for many latitudes and orientations the anomalies should be symmetric about the axis of the ridge.

If one pursues a literal application of the idea, a further possibility is simulation of

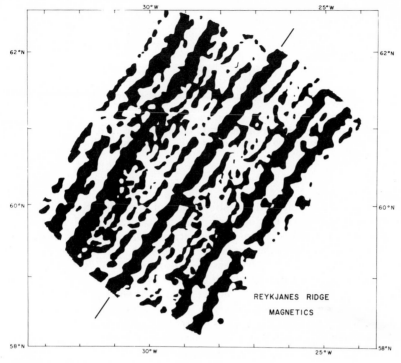

Figure 24-3
Summary diagram of the magnetic anomalies observed over Reykjanes Ridge
(see Figure 24-2). Straight lines indicate the axis of the ridge and the central
positive anomaly (Heirtzler et al., 1966)

anomalies at ridge crests by assuming the
reversal time scale for the last 4 million years
proposed by Cox, Doell, and Dalrymple
(1964b), the only additional parameter being
the rate of spreading; the scale (Fig. 24–4)
has recently received striking independent
confirmation from the work of Opdyke et al.
(1966) on deep-sea sedimentary cores.

REYKJANES RIDGE

Observed anomaly profiles obtained during
four crossings of the crest of Reykjanes Ridge
are compared (Fig. 24–5) with simulations
obtained by assumption of reversal time scales
for the last 4 million years and a rate of spreading of 1 centimeter per annum for each limb
of the ridge. The model assumed is analogous
to the one I have described (Vine and Wilson,
1965 – model 2), but the depths have been
made compatible with the depth to the ridge

crest in this area and with the altitude at which
the survey was flown.[2] In performance of the
survey, 58 parallel courses were flown normal to the ridge axis, but the crest was not
traversed by the first four and last five courses:
thus crossings 15, 25, 35, and 45 are shown as
being representative. The correlation between
the observed and computed anomalies is very
encouraging and suggests a rate of spreading
of rather less than 1 centimeter per annum.

When one applies the concept of continental
drift to this region, it seems reasonable to
assume that Rockall Bank, southeast of the
ridge (Fig. 24–2), is a continental fragment,
as was assumed by Bullard, Everett, and Smith

---

[2]For the two models in Fig. 24–5 the intensity of
Earth's field was taken as 51,600 gamma; its dip,
$\pm74.3°$; the magnetic bearing of the profile, 153°.
Normal or reverse magnetization is with respect to
an axial dipole vector. Effective susceptibility assumed, ±0.01 – except for the central block ($+0.02$).

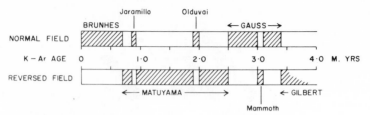

Figure 24-4
Geomagnetic-polarity epochs deduced from paleomagnetic results and potassium-argon dating (based on Cox et al. 1964b; and Doell and Dalrymple, 1966)

(1965) in reconstructing the fit of the continents around the Atlantic. In this instance the deep to the southeast of Rockall may represent an initial abortive split; the oceanic area to the northwest, centered on Reykjanes Ridge, a subsequent and more persistent site of spreading of the ocean floor. There is every indication from the existing bathymetry (Heirtzler et al., 1966) that the ridge crest is linear and not interrupted or offset by transverse fractures.

This area therefore, 1200 kilometers in width, may well record a comparatively simple and straightforward example of drifting and spreading. The oldest rocks in the Thulean or Brito-Arctic Tertiary Igneous province occur in northwestern Scotland and eastern Greenland. Preliminary potassium-argon dates from Arran, Mull, and other centers in the British Isles suggest an age of approximately 60 million years (perhaps slightly greater) (Miller and Brown, 1965). If it is assumed that this igneous activity indicates the initiation of drift in this area, then the implied average rate of spreading from Reykjanes Ridge (half width, 600 km) is approximately 1 centimeter per annum—that is, the rate of "drifting" is approximately 2 centimeters per annum.

OTHER RIDGES

The model proposed by Vine and Matthews (1963) and developed by Vine and Wilson (1965c) has been applied to four widely separated areas on the midoceanic ridge system (Figs. 24–6–24–9) by assumption of the reversal time scale shown in Fig. 24–4 and a rate of spreading compatible with the width of the central anomaly.[3] An observed profile across Juan de Fuca Ridge, southwest of Vancouver Island (Fig. 24–1) (Vine and Wilson, 1965; Wilson, 1965b), is compared (Fig. 24–6) with a simulated profile based on a rate of spreading of 2.9 centimeters per annum per limb of the spreading system. A profile across the East Pacific Rise, just north of the Eltanin Fracture Zone (Pitman and Heirtzler, 1966), is compared (Fig. 24–7) with a computed profile based on a rate of spreading of 4.4 centimeters per annum. Clearly this rate, implying a rate of separation of nearly 9 centimeters per annum, is an order of magnitude greater than the commonly quoted rates of 1 or 2 centimeters per annum; the significance and possible implications of this difference are discussed later.

Observed profiles across Carlsberg Ridge in the northwest Indian Ocean and across the Mid-Atlantic Ridge in the South Atlantic are compared (Figs. 24–8 and 24–9) with simulations based on a rate of spreading of 1.5 centimeters per annum. As Backus has pointed out (Backus, 1964), one may expect the width of the anomalies in the South Atlantic to increase southward, reflecting a progressive increase

[3]For the models shown in Figs. 24–6–24–9 the intensity and dip of Earth's field and the magnetic bearing of the profile assumed in each case, are, respectively: Juan de Fuca, 54,000 gamma, +66°, 087°; East Pacific Rise, 48,700 gamma, −62.6°, 102°; N.W. Indian Ocean, 37,620 gamma, −6°, 044°; South Atlantic, 28,500 gamma, −53.5°, 114°. Intensity and direction of magnetization as for Fig. 24–5 (see Vine and Wilson, 1965, Fig. 3).

REYKJANES RIDGE   60° N

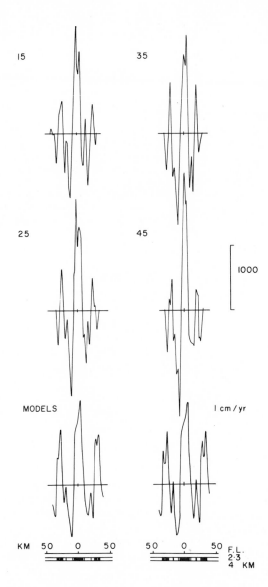

Figure 24-5
Profiles observed across Reykjanes Ridge,
together with computed profiles. The model
to the left assumes the reversal time scale of
Figure 24-4; that to the right, the "revised"
time scale of Figures 24-12 and 24-13 (see
footnote 2, page 249). All observed and
computed profiles have been drawn to the
same proportion: 10 kilometers horizontally is
equivalent to 100 gamma vertically (1 gamma
$= 10^{-5}$ oersted). F.L., flight level

in the rate of spreading, because of the rotation of South America relative to Africa
(Bullard et al., 1965) and the resultant southward increase in separation. The increase
southward, in the width of the envelope of
the central magnetic anomalies indicated by
Heirtzler and Le Pichon (1965, fig. 3), may
well be an expression of this phenomenon.

THE RED SEA

In Fig. 24–10 a modification of the model has
been applied at two points on the Red Sea
rift. If the axial depression and zone of magnetic anomalies in the Red Sea are considered
to indicate the initiation of continental drift
by spreading of the ocean floor (Girdler, 1962),

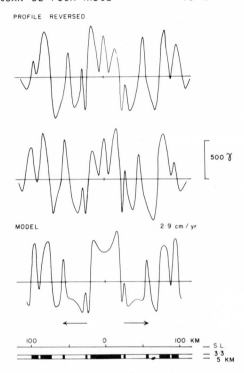

PROFILE REVERSED

500 γ

MODEL   2·9 cm/yr

Figure 24-6
Observed magnetic profile over the Juan de Fuca Ridge compared with a simulated profile based on the reversal time scale of Figure 24-4 assuming a constant rate of spreading and the model described in footnote 3, page 250. The observed profile is taken from Raff and Mason (1961). See also Vine and Wilson (1965), Figure 3. S.L., sea level

100   0   100 KM   S.L.
                                        3·3
                                        5 KM

EAST PACIFIC RISE   51° S

PROFILE REVERSED

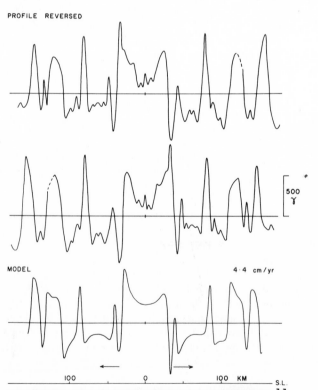

500
γ

MODEL   4·4 cm/yr

Figure 24-7
Observed magnetic profile over the East Pacific Rise compared with a simulated profile based on the reversal time scale of Figure 24-4 assuming a constant rate of spreading and the model described in footnote 3, page 250. The observed profile is the Eltanin-19 profile of Pitman and Heirtzler (1966). S.L., sea level

100   0   100 KM   S.L.
                                        3·3
                                        5 KM

N.W. INDIAN OCEAN   5° N

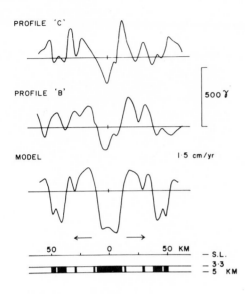

Figure 24-8
Observed magnetic profile over the northwest
Indian Ocean compared with a simulated profile
based on the reversal time scale of Figure 24-4
assuming a constant rate of spreading and the
model described in footnote 3, page 250. The
observed profiles are the Owen profiles of
Matthews, Vine, and Cann (1965), Figure 2.
S.L., sea level

SOUTH ATLANTIC   38° S

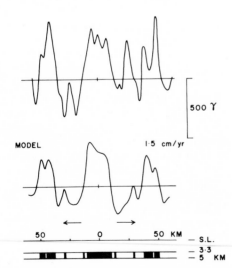

Figure 24-9
Observed magnetic profile over the South
Atlantic Ocean compared with a simulated
profile based on the reversal time scale of
Figure 24-4 assuming a constant rate of
spreading and the model described in footnote
3, page 250. The observed profile is taken
from Heirtzler and Le Pichon (1965), Figure 1.
S.L., sea level

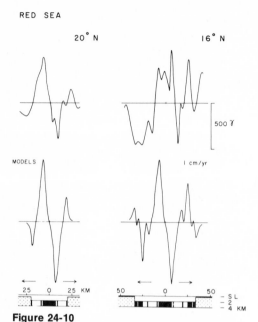

RED SEA

20° N   16° N

MODELS   1 cm/yr

25   0   25 KM   50   0   50

Figure 24-10
Observed profiles across the Red Sea (from Allan
et al., 1964; Drake and Girdler, 1961) compared
with computed profiles based on a constant rate
of spreading and a truncated model and time scale
according to the width of the central depression
and zone of magnetic anomalies. (See footnote 4,
page 254). Stippled area, "nonmagnetic" continental material

then clearly such a simulation should be attempted. However, the depth to this embryonic
ocean floor and its crustal section (Drake and
Girdler, 1961) are not typical of an oceanic
ridge, and this floor almost certainly includes
some assimilated or foundered continental
material.

In Fig. 24–10 a slightly thickened "volcanic" layer has been truncated against non-
magnetic continental material according to
the width of the central depression and anomalies at each point. The same rate of spreading, 1 centimeter per annum, has been assumed
at both points, hence the different lengths of
the reversal time scale involved.

One would not expect the anomaly pattern
in the Red Sea rift to be as clear-cut as that
over the more mature Juan de Fuca or Reykjanes ridges; nevertheless the approximation
of the simulated to the observed anomalies is

encouraging.[4] The dates of rifting implied from
these models should not be taken to indicate
the initiation here of crustal extension. In
initiating drift in a typical shield area (that is,
beneath possibly 35 kilometers of continental
crust, in this area), an upwelling in the mantle
may well start by producing "necking" (thinning) of the crust, normal faulting, and intrusion and extrusion of basic igneous material
—all effects producing extension and thinning
of the crust and the possibility of marine transgression prior to the initiation of drift and the
emplacement of quasi-oceanic crust.

THE REVERSAL TIME SCALE

In simulating the anomalies observed centrally
over oceanic ridges, in terms of normal-
reverse boundaries within the oceanic crust,
one can deduce these boundaries independently without reference to a reversal time
scale (Vine and Wilson, 1965). In Fig. 24–11
boundaries inferred from the observed profiles
across the East Pacific Rise and Juan de Fuca
and Reykjanes ridges (Figs. 24–5–24–7) are
plotted against the reversal time scale of Cox,
Doell, and Dalrymple (1964b), according to
their distances from the axis of the ridge. The
dashed line in this graph indicates a similar
plot for the boundaries at the Juan de Fuca
Ridge and the time scale assumed by Vine
and Wilson (1965).

In this earlier time scale the Jaramillo event
(Fig. 24–4) had not been differentiated, and
the most recent reversal was placed at 1 million years ago. Consequently the narrow peaks
on either side of the central positive anomaly
in Fig. 24–6 were correlated with the Olduvai
event. This correlation implied a very erratic
rate of spreading and a much slower average
rate of 1.5 centimeters per annum (Vine and
Wilson, 1965).

It will be seen that, had the authors had
more faith in the idea and the probability of a

---

[4]Parameters assumed for the Red Sea models in
Fig. 24–10; intensity and dip of Earth's field, 38,500
gamma and +24° respectively; magnetic bearing of
profiles, 054°. Intensity and direction of magnetization as for Fig. 24–5.

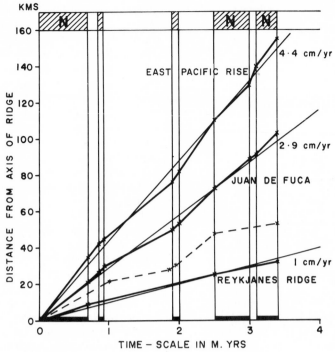

Figure 24-11
Inferred normal-reverse boundaries within the crust plotted against
the reversal time scale of Figure 24-4. The dashed line represents a
similar plot for Juan de Fuca Ridge, if one assumes the earlier time
scale—as did Vine and Wilson (1965). Note the similar deviations
from linearity for the East Pacific Rise and Juan de Fuca Ridge

more constant rate of spreading, for inertial
reasons, they could have predicted the Jaramillo event. The recent detailing of this event
by Doell and Dalrymple (1966) and its independent discovery by Opdyke et al. (1966)
are therefore of considerable interest and importance in interpretation of the magnetic
anomalies.

On correlation of the crustal boundaries
with the new time scale (Fig. 24–11), the implied rates of spreading are more constant
and much faster than before. Moreover, rather
remarkably, the deviations from linearity for
the East Pacific Rise and Juan de Fuca Ridge
—11,000 kilometers apart—are exactly analogous. The only discrepancy occurs in the
region of the Mammoth event, concerning
which there is a suggestion, from the profile
resulting from the very fast rate of spreading
in the South Pacific (Fig. 24–7), that this event

is multiple; that is, it may include a short period of normal polarity (see Fig. 24–13).

If, therefore, one assumes constant rates of
spreading for these two areas (the averages
obtained from Fig. 24–11), one can replot the
inferred boundaries on lines of constant
spreading rates and suggest slight revisions
of the reversal time scale, as are shown in Fig.
24–12. Clearly, much more data should be
analyzed in this way to confirm or invalidate
this type of revision, but it presents an interesting possibility.

In discussing the magnetic anomalies associated with Juan de Fuca Ridge, Vine and Wilson
compared a profile across this ridge with the
only available profile across the East Pacific
Rise (Vine and Wilson, 1965—fig. 2). This
rather rash comparison has now been vindi-
cated by the publication of four new profiles
across the East Pacific Rise by Pitman and

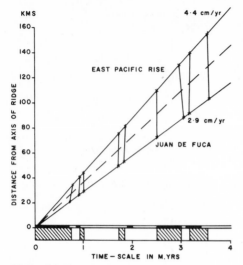

Figure 24-12
Crustal boundaries replotted along lines of
constant rates of spreading (that is, best-fitting
straight lines in Figure 24-11): slight revision
of the reversal time scale is suggested. The
"revised" time scale is the lower, ruled scale

Heirtzler (1966) (see also Figs. 24–6 and
24–7). Of these new profiles, the *Eltanin*-19
profile (Fig. 24–14) shows, as Pitman and
Heirtzler emphasized, remarkable symmetry
about its midpoint; it is presumably most suitable for deduction of the frequency and occurrence of reversals in the Pliocene. Thus, if
one assumes the rate of spreading obtained
centrally from this profile (Fig. 24–11) and
deduces normal-reverse boundaries in the
crust out to a distance of 500 kilometers, one
can suggest a reversal time scale for the last
11.5 million years (Fig. 24–13).

This extrapolation clearly depends on the
continued applicability of the model and a
constant rate of spreading, but it presents at
least two very interesting possibilities: (i) it
can be compared with the time scales obtained
by other techniques as these scales are extended back into the Pliocene (Cox et al.,
1964b; Doell and Dalrymple, 1966; Opdyke
et al., 1966); and (ii) if the Vine-Matthews
hypothesis is applicable to all active oceanic
ridges as has been suggested, one can use this
time scale to simulate and predict central
anomalies at other latitudes and orientations
of the ridge system, as is illustrated for the

time scale out to 4 million years in Figs. 24–5–
24–10. Pitman and Heirtzler's (1966) simulation, of the type suggested, for Reykjanes
Ridge agrees very well with the observed
profile.

THE EAST PACIFIC RISE

In the Pacific, where the spreading rate appears to be much faster, many details of the
reversal time scale are apparent: For example,
the four positive peaks associated with the
periods of normal polarity between 4 and 5
million years ago, and the broad positive resulting from the period between 9 and 10
million years ago, are clearly identifiable in
Fig. 24–1, together with other details of the
time scale, centered on Juan de Fuca and
Gorda ridges.

The area covered by this survey appears
to be transected by many transcurrent or
transform faults, or both, separating apparently
rotated blocks, as Raff and Mason (1961) and
Wilson (1965b) suggested. However, if one
takes into consideration the offset of the anomaly pattern across these faults, one can reconstruct a profile across and to the northwest
of Juan de Fuca Ridge (Fig. 24–14), and this
profile remarkably resembles the one obtained
in the South Pacific, 11,000 kilometers away.
Furthermore, if one deduces boundaries within
the crust from the Juan de Fuca profile and
plots these on the vertical time lines obtained
from the South Pacific profile in Fig. 24–13,
there is a suggestion that the rate of spreading
in the Juan de Fuca area may have decreased
within the Pliocene from a rate of 4 or 5 to
2.9 centimeters per annum for the last 5.5
million years.

Clearly one cannot distinguish between an
acceleration of the East Pacific Rise in the
South Pacific and a deceleration of Juan de
Fuca Ridge in the north, but the latter is considered more likely and leads to an interesting
speculation.

The anomalous width and unique features
of the American Cordillera in the western
United States were emphasized by Wise
(1963), who considered that these features
may be related to, or at least reflected in, the
apparently rotated oceanic crustal blocks

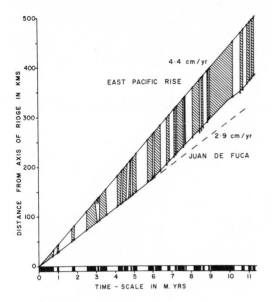

Figure 24-13
Magnetic boundaries across the East Pacific Rise, deduced out to 500 kilometers from the crest and plotted on a line representing a constant spreading rate of 4.4 centimeters per annum. Similar boundaries from Juan de Fuca Ridge are plotted out to 150 kilometers on the assumption of a constant spreading rate of 2.9 centimeters per annum. The time scale out to 5.5 million years is based on both plots. Beyond that time the time scale is based on the East Pacific Rise boundaries, and those deduced from Juan de Fuca Ridge are simply plotted on these time lines

revealed by the magnetic survey (Fig. 24–1). Wise felt that his hypothesis was outrageous, but I should like to develop it further in the light of this reinterpretation of the magnetic anomalies.

I suggest that, to a first-order approximation, the more recent geologic history and structures of the western United States can be ascribed to the progressive westward drift of the North American continent away from the spreading Atlantic Ridge, and to the fact that the continent has overridden and partially resorbed first the trench system and more recently the crest of the East Pacific Rise.

As Hess has noted on the basis of ocean-floor spreading (1965), only ridges that have been initiated beneath continents, and that are therefore actively causing continents to drift apart, should be approximately median within the ocean basins. Pacific ridges such as the extinct Darwin Rise and the active East Pacific Rise are not constrained in this way, having been initiated presumably in oceanic areas. Therefore I follow Menard (1964) and Wilson (1965d) in maintaining that the northeast Pacific basin represents a flank of the East Pacific Rise, the crest of which has been modified, and arrested by the encroachment of North America.

The former east-west direction of spreading from the East Pacific Rise, reflected in the north-south magnetic anomalies of the northeast Pacific, has apparently been replaced within the Pliocene so that the present direction of motion is northwest-southeast, paralleling the San Andreas fault, as Wilson (1965c) suggested.

In the area off Washington and Oregon and to the north of the Mendocino Fracture Zone, this change in direction has been accommodated by faulting and a gradual stifling and reorientation of the ridge crest to form Juan de Fuca and Gorda ridges. This stifling is illustrated by profiles in Fig. 24–14, but it is most graphically shown by coloring the anomaly bands of Fig. 24–1 with a spectrum of colors according to their ages. Such a diagram reveals another short and less obvious ridge to the north of Juan de Fuca Ridge, as is indicated in Figs. 24–1 and 24–15; the short ridge, like the Gorda, has a pronounced topographic expression (McManus, 1965), particularly near the continental margin, where the central magnetic pattern is also clearest. The proximity of this new ridge to the Explorer seamount and the Explorer trench (McManus, 1965) has led to its name: Explorer Ridge (see Fig. 24–15).

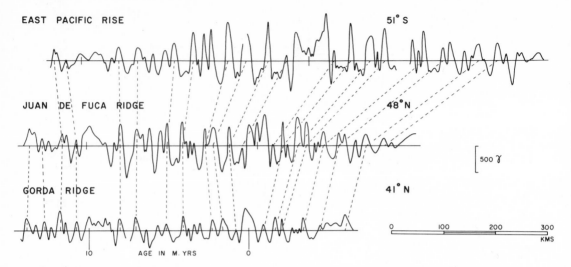

Figure 24-14
The East Pacific Rise profile Eltanin-19 (Pitman and Heirtzler, 1966) compared with a composite profile across and to the northwest of Juan de Fuca Ridge, and with a profile normal to the strike of the anomalies across and to the west of Gorda Ridge. (The last two profiles from Raff and Mason, 1961, and Vacquier et al., 1961)

South of Cape Mendocino, current crustal spreading appears to be accommodated along the San Andreas fault, as was proposed by Wilson (1965c). In this area, between the Mendocino and Murray fracture zones, the former ridge crest has presumably been overridden and damped out, perhaps not without attempts at modification as suggested by the northeasterly trending anomalies, near the continental margin, in the magnetic survey of this area (Mason and Raff, 1961) (see Fig. 24–15). However, it is interesting to reconstruct the ridge crest as it would be had it not been overridden and modified. If one calculates the position of the ridge crest north of the Mendocino (had it not been stifled) and assumes the offsets measured further west on the Mendocino and Pioneer fractures (Vacquier, 1965), the reconstructed ridge crest lies beneath Utah and Arizona—the area of the Colorado Plateau uplift (see Fig. 24–15).

South of the Murray fracture zone the picture is less clear; possible clues from the oceanic magnetic anomalies are still confused because of the lack of an extensive survey. However, from the very nature of the Gulf of California, from its close analogy with the Gulf of Aden (Rusnak et al., 1964; Laughton, 1966), and in the light of the important observation by Menard that the Clipperton frac-

ture zone does not offset the present crest of the East Pacific Rise (Menard, 1966), it seems probable that this length of the present crest, at least as far south as the Clipperton fracture, is new and not a modification of the former crest, as are Juan de Fuca and Gorda ridges.

This interpretation of the present crustal motion in the northeast Pacific, involving transform faults (Wilson, 1965c) and northwest-southeast movement, seems to accord with the anomalous nature of the circum-Pacific belt between Mexico and Alaska. This region lacks trench systems and their associated planes of deeper-focus earthquakes—an observation underlined by Girdler (1964), and a fact perhaps precluding east-west or northeast-southwest compression. I must emphasize that my evidence (essentially contained in Fig. 24–14) suggests that this change in direction occurred within the last 10 million years and that earlier, for example, quasi-transform faults may have existed along the continental extension of the Mendocino and Pioneer fracture zones, producing the right lateral offset of the various tectonic belts indicated by Wise (1963) (see Fig. 24–15). Thus the north-south magnetic anomalies of the northeast Pacific are considered to be related to a former crest of the East Pacific Rise. Further support for this hypothesis

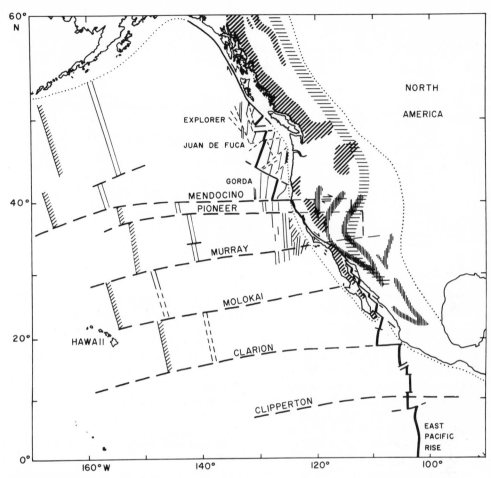

Figure 24-15
The East Pacific Rise in the North Pacific. Solid black lines indicate the present crest and active transform faults (the crest south of the Gulf of California, from Menard, 1966). Thin lines represent key magnetic anomalies; in particular, the pair of anomalies traced by Peter (1966) between 160° and 140°W. The half-herringbone pattern to the west suggests a possible boundary of the rise and its associated north-south magnetic anomalies. Broken lines indicate inactive faults or fractures. Dotted lines enclose the circum-Pacific cordillera within which the tectonic belts of Wise (1963) are shown by ruled shading

comes from the flank anomalies. New data[5] have enabled Peter (1966) to trace a particular pattern of flank anomalies approximately from north to south from the Aleutian trench to just south of the Murray fracture zone, a distance of 2800 kilometers; the top two profiles in Fig. 24–16 illustrate the pattern along latitudes 35°40′ and 36°30′N (Raff, 1966; Peter, 1966). Christoffel and Ross (1965),

working south of New Zealand, have similarly correlated flank anomalies on adjacent north-south profiles at approximately 173°E (Fig. 24–16). It is suggested that the two patterns are the same except for difference in the rate of spreading that formed them. The two areas are 11,000 kilometers apart.

In addition, the pattern to the south of New Zealand bears the same relation to the New Zealand Plateau, the presumed northern boundary of the East Pacific Rise in this area, as the northeast Pacific pattern bears to the

---
[5]Recorded by the U.S. Coast and Geodetic Survey.

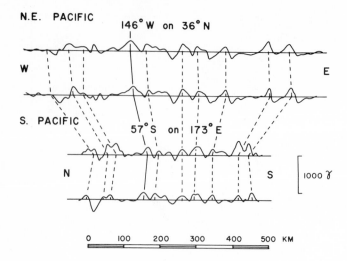

N.E. PACIFIC   146° W on 36° N

W   E

S. PACIFIC   57° S on 173° E

N   S   ⎡ 1000 γ

0   100   200   300   400   500 KM

Figure 24-16
The anomaly pattern correlated between profiles by Peter and Raff (1966) in the North Pacific and Christoffel and Ross (1965) in the South Pacific. It is suggested that the only difference between the two is the rate of spreading that formed them

western boundary of the north-south anomalies suggested by Raff (1966) and Peter (1966). Judged from the lengths of the patterns in the two areas (Fig. 24–16), the ratio of the rates of spreading was 3 : 2 at the time of formation. In the South Pacific and to the south of New Zealand the width of the East Pacific Rise decreases southwestward, the decrease implying slower rates of spreading. This implication was supported by crossings of the ridge in this area by U.S.S. *Staten Island* in 1961 (U.S. Naval Oceanographic Office, 1962), showing that the width of the central anomalies also decreases southwestward.

The fast rate of spreading that I suggest for the East Pacific Rise has two important implications regarding its heat budget: (i) any systematic variation in the heat flow, caused by convection beneath, should be much clearer here than in the Atlantic or Indian Oceans; and (ii) the elevation of the Rise, which is presumably related to thermal expansion or partial fusion within the upper mantle, should persist to greater distances from its crest — as it appears to do.

EARTH'S PALEOFIELD

I have demonstrated that the fast rate of spreading in the Pacific shows up incredible details in the reversal time scale out to at least 4 and perhaps 11 million years ago. If one assumes that the rate has always been high (4 to 5 cm/year) and that the hypothesis continues to apply, changes in the intensity and polarity of Earth's magnetic field during the remainder of the Tertiary should be recorded in the oceanic crust and associated magnetic anomalies, out to the boundary of the East Pacific Rise.

Immediately north of the Mendocino fracture zone one can reconstruct such a profile from the crest of Gorda Ridge to the boundary of the north-south anomalies at 168°W. This profile (Fig. 24–17) has been calibrated with a suggested time scale beyond 10 million years, on the basis of a constant rate of spreading of 4.5 centimeters per annum. Thus it is implied that the East Pacific Rise, at least in the north and south, and perhaps throughout the length of the Pacific, was initiated in the late Cretaceous, possibly at the time of the extinction and beginning of subsidence of the Darwin Rise (Menard, 1964).

Furthermore, the profile of Fig. 24–17, since it implies a particular sequence of normal-reverse boundaries and changes in bulk magnetization within the crust, may enable one to predict and correlate anomalies in other oceanic areas, just as one could, from the South Pacfic profile, for the North Pacific (Fig. 24–14) and North Atlantic (Pitman and Heirtzler, 1966); in fact, until this is done,

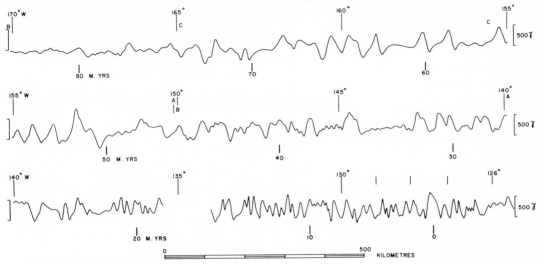

Figure 24-17
East-west profile immediately north of the Mendocino Fracture Zone (approximately 41°N). Lengths AA and BB are from Raff (1966); the length CC is equivalent to those shown in Figure 24-16. The lower section is taken from the contour maps of Vacquier et al. (1961) and Raff and Mason (1961). The profile has been calibrated with a possible time scale from the crest of Gorda ridge out to the boundary of the East Pacific rise at 168°W

the speculative nature of this time scale cannot be overemphasized.

Ridges in other oceanic areas have been initiated beneath continents, and this fact may well be recorded in the form of marine transgressions or basic igneous intrusions on the present-day continental margins (Du Toit, 1937). By use of these criteria the main part of the Mid-Atlantic Ridge is perhaps 150 to 200 million years old; the northwest Indian Ocean Ridge, 80 to 100 million years old. These ages are compatible with the rates of spreading that I have deduced at the centers of these ridges—approximately 1.5 centimeters per annum.

Of the three points originally unexplained by the Vine-Matthews hypothesis, two have been answered; the remaining difficulty concerns the change in character of the anomalies as one moves from the axial zone of a ridge to the flanks. This change to higher-amplitude and longer-wavelength anomalies is often rather abrupt (Heirtzler and Le Pichon, 1965). If the Vine and Matthews hypothesis is applicable beyond the central, axial zones of ridges, this change in character may reflect a change in the intensity or frequency, or both, of reversals

of Earth's magnetic field. If the frequency of reversals is high, the resulting "blocks" of material of a particular polarity will be narrow; their width will depend on the rate of spreading, but if they are a few kilometers in width they will give rise to a considerably reduced anomaly (Fig. 24–18). Narrower blocks may well have no obvious individual expression in the magnetic anomaly but will tend to lower the bulk resultant magnetization of the surrounding block. Thus this boundary between the flank and axial-zone anomalies may reflect an increase in the frequency of reversals of Earth's field, together possibly with a decrease in its intensity.

Clearly, if this is the case, the boundary should occur at different distances from ridge axes according to the average rate of spreading in that region. A preliminary investigation of many ridge profiles suggests that this change may have occurred approximately 25 million years ago. Changes in the frequency of reversal seem quite probable when one bears in mind that for the whole of the Permian and part of the Upper Carboniferous the field appears to have been of a single polarity (Irving, 1966).

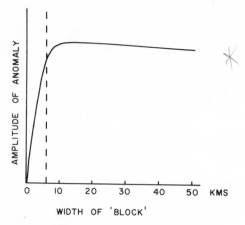

Figure 24-18
A schematic diagram illustrating the way in which the amplitude of the anomaly, associated with a "block" of oceanic crust of a particular polarity, is considerably reduced if the width of the block is less than 5 or 6 kilometers

The fact that Vacquier (1965) was unable to correlate axial-zone anomalies on profiles in the equatorial Atlantic is considered to result partly from this difference in character. Moreover, the model for this latitude and orientation would predict zero or near-zero anomalies, so that one is recording only background noise caused by topography and second-order magnetic contrasts within the crust.

The concept of transform faults (Wilson, 1965c) very neatly explains many oceanic fracture zones, especially in the Atlantic and Indian oceans. There the faults appear to accommodate changes in direction of the ridge crest in splitting the continents, and parts of these oceans are probably riddled with minor transform fractures that obscure the magnetic symmetry on any random profile. In the northeast Pacific, however, the fractures seem to be rather different in character and are often not very obviously accommodating to a general change in direction of the ridge crest (Fig. 24–15). It seems essential to assume ocean-floor spreading to explain the large offsets on these fractures at all, but within this framework there seem to be two possibilities: transform faults, or a different velocity of spreading on either side of the fracture zone (Hess, 1965; Dietz, 1966). By matching anomalies across

the fractures one will be able to distinguish between these two possibilities, according to whether the offset remains constant or changes along the length of the fault. As yet no evidence indicates pronounced differential spreading.

Finally, the Vine and Matthews hypothesis may provide the best criterion for distinguishing between active and inactive ridges. Ridges that have actively spread during the last 1 million years should be characterized by a central magnetic anomaly of appropriate sign and shape. Investigation of approximately 100 available crossings of the worldwide ridge system indicates good agreement between observed and predicted anomalies, two sectors excepted.[6] Magnetic profiles across the Labrador Sea (Godby et al., 1966) show a certain possible symmetry about the center but do not reveal a central anomaly (Fig. 24–19). If actively spreading, this ridge should be characterized by a very pronounced central anomaly because of the high latitude (compare Fig. 24–5); this lack of central anomaly fits well with the concept of an extinct and buried ridge, as was revealed by the reflection seismic data (Drake et al., 1963), although occasional shallow-focus earthquakes indicate some residual activity.

The second area is the ridge system to the southeast of South Africa (Fig. 24-19); it appears that the central anomaly is consistently absent, although more data are required for confirmation. It is further suggested that, with the exception of Prince Edward Island Rise, the depths in this area are not typical of a mature oceanic ridge. Such mountainous bathymetry and seismic activity as there is can probably be attributed to transverse fractures and residual shearing and igneous activity.

If this observation is significant and the ridge systems in these areas are not actively spreading, the active part of the ridge system at present appears to form two isolated lengths, each traversing half the circumference of Earth: one extends from the Red Sea and the Gulf

---

[6]M. Talwani has drawn my attention to the fact that the central anomaly may also be absent over the extension of the mid-Atlantic Ridge across the Eurasian Basin of the Arctic; N. A. Ostenso et al., 1965; E. R. King et al., 1966.

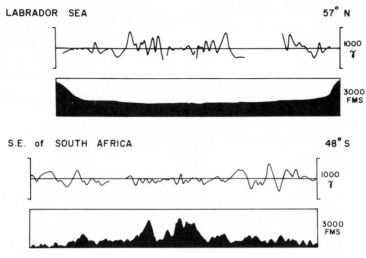

Figure 24-19
Observed magnetic profiles and bathymetry recorded by Vema across the
Labrador Sea and to the southeast of South Africa (Heirtzler, 1961). Note
the absence of a central magnetic anomaly in each case

of Aden, south of Australia, and across the
east Pacific to the Gulf of California; the other,
down the whole length of the Atlantic. This
concept contrasts with the continuous nature
of the worldwide ridge system stressed by
most authors, but fits well with the "tennis
ball" pattern of convection within the upper
mantle (Hess, 1965; Dietz, 1966), in which the
seam of the ball should approximate down
currents and compressional features on Earth's
surface.

SUMMARY

It is suggested that the entire history of the
ocean basins, in terms of oceanfloor spreading, is contained frozen in the oceanic crust.
Variations in the intensity and polarity of
Earth's magnetic field are considered to be
recorded in the remanent magnetism of the
igneous rocks as they solidified and cooled
through the Curie temperature at the crest of
an oceanic ridge, and subsequently spread
away from it at a steady rate. The hypothesis
is supported by the extreme linearity and continuity of oceanic magnetic anomalies and
their symmetry about the axes of ridges.

If the proposed reversal time scale for the
last 4 million years is combined with the model,
computed anomaly profiles show remarkably
good agreement with those observed, and one
can deduce rates of spreading for all active
parts of the midoceanic ridge system for which
magnetic profiles or surveys are available. The
rates obtained are in exact agreement with
those needed to account for continental drift.

An exceptionally high rate of spreading
(approximately 4.5 cm/year) in the South
Pacific enables one to deduce by extrapolation
considerable details of the reversal time scale
back to 11.5 million years ago. Again, this
scale can be applied to other parts of the ridge
system. Thus one is led to the suggestion that
the crest of the East Pacific Rise in the northeast Pacific has been overridden and modified
by the westward drift of North America, with
the production of the anomalous width and
unique features of the American cordillera in
the western United States. The oceanic magnetic anomalies also indicate that there was a
change in direction of crustal spreading in
this region during Pliocene time from eastwest
to southeast—northwest.

A profile from the crest to the boundary of
the East Pacific Rise, and the difference between axial-zone and flank anomalies over

ridges, suggest increase in the frequency of reversal of Earth's magnetic field, together, possibly, with decrease in its intensity, approximately 25 million years ago.

Within the framework of ocean-floor spreading, it is suggested that magnetic anomalies may indicate the nature of oceanic fracture zones and distinguish the parts of the ridge system that are actively spreading. Thus data derived during the past year lend remarkable support to the hypothesis that magnetic anomalies may reveal the history of the ocean basins.

25

Marine Magnetic Anomalies, Geomagnetic Field Reversals, and Motions of the Ocean Floor and Continents

JAMES R. HEIRTZLER
G. O. DICKSON
ELLEN M. HERRON
WALTER C. PITMAN, III
XAVIER LE PICHON
1968

From *Journal of Geophysical Research*, v. 73, p. 2119–2136, 1968. Reprinted with permission of the authors and the American Geophysical Union. Copyright 1968.

The three previous papers of this series (Pitman et al., 1968; Dickson et al., 1968; Le Pichon and Heirtzler, 1968) have shown that a magnetic anomaly pattern, parallel to and bilaterally symmetric about the mid-oceanic ridge system, exists over extensive regions of the North Pacific, South Pacific, South Atlantic, and Indian oceans. It has been further demonstrated that the pattern is the same in each of these oceanic areas and that the pattern may be simulated in each region by the same sequence of source blocks. These blocks comprise a series of alternate strips of normally and reversely magnetized material, presumably basalt (Figure 25–1).

The symmetric and parallel nature of the anomaly pattern and the general configuration of the crustal model conform to that predicted by Vine and Matthews (1963) and may be regarded as strong support for the concept of ocean floor spreading (Dietz, 1961; Hess, 1962).

Figure 25–1 shows sample anomaly profiles from the various regions under discussion. Also shown are the model blocks, derived for the North Pacific profile and adjusted to fit the other oceans. The North Pacific profile is a composite using sections at different latitudes because of complex structure.

It is our purpose in this paper to derive a time scale for the sequence of magnetic polarity events predicted by the basic model and to discuss the concepts of continental drift and ocean floor spreading in terms of the anomaly pattern and the time scale.

Following Vine and Matthews (1963), we will assume that all the linear magnetic anomalies, sub-parallel to the ridge axis, are due to geomagnetic field reversals. If that theory is basically in error, the conclusions in this paper do not apply. Yet, if that theory is applied carefully, many geophysical and geological observations, previously thought to be unrelated or casually related, are seen to have a simple unified explanation.

CHOICE OF TIME SCALE

The most obvious method for deriving a detailed time scale for the magnetic anomaly is
by extrapolation from well-dated paleomagnetic events. Pitman and Heirtzler (1966),
Vine (1966), Pitman et al. (1968), Dickson et
al. (1968), and Le Pichon and Heirtzler (1968)
have demonstrated that, for the various sections of mid-oceanic ridge under discussion,
the magnetic anomaly pattern associated
with the axial zone may always be related to
the known sequence of magnetic polarity
events of the past 3.35 m.y. by assuming a
constant spreading rate. The fault plane solutions of Sykes (1967) show that the movement
between offset ridge axes is in agreement with
that predicted by Wilson (1965c) for a transform fault, thus indicating that the spreading
process is active in these regions today.

In Figure 25-1 are displayed observed magnetic profiles from the North and South Pacific,
the South Indian, and the South Atlantic
Oceans and model bodies that could account
for the observations. The latitude and longitude of the end points of each profile are noted
on the figure. Study of the theoretical profiles
for the three models shows that these models
can account for the observed profiles immediately above them. With each model an age
scale is given. Each scale is based on the date
of 3.35 m.y. for the beginning of the Gauss
normal polarity epoch (Doell et al., 1966). The
time scales obviously differ and are not linearly
related, since the models for the South Pacific
and the South Atlantic were derived from the
basic model for the North Pacific by adjusting
the set of model blocks according to the curve
for the relative spreading rate. The relative
spreading rate curves (Figure 8, Pitman et al.,
(1968) and Figure 25-2 of this paper) though
nonlinear, are continuous. The problem is to
choose the most reasonable time scale.

T. Saito (personal communication) has paleontologically determined the age at the bottom
of a core (V 20–80) taken at 46°30′N, 135°W
as Lower to Middle Miocene (13 – 26 m.y.).
The core location is on anomaly 6. By the
North Pacific and South Atlantic standards
the magnetic basement at anomaly 6 should be
20 – 22 m.y. old. The South Pacific time scale
suggests an age of 14 m.y. for this basement,
much younger than the age of the core.

The core was taken on an outcrop where
no sediment was detected by seismic reflection
methods. The seismic reflection technique
employed, however, was unable to resolve
reflectors less than 50 meters in thickness.
Thus, allowing for some sediment beneath the
core, we can accept the older date for anomaly
6 as more reasonable.

Ewing et al., (1968) have found the eastern
boundary of layer A′ (as detected by reflection
techniques) in the North Pacific to be just west
of anomaly 32. They have proposed an Upper
Cretaceous age (70 m.y.) for this layer. The
South Pacific time scale would suggest an age
of 47 m.y. for anomaly 32. Unless there has
been a major discontinuity in spreading rates,
the South Pacific time scale is in error at this
point by a factor of 2.

A Cretaceous core (Le Pichon et al., 1966)
taken at the base of the Rio Grande rise in the
South Atlantic just west of anomaly 31 supports the South Atlantic or the North Pacific
time scale.

As stated previously, the curves shown
plotted in Figure 8 of Pitman et al. (1968) and in
Figure 25–2 of this paper may be regarded as
indicative of the relative spreading rates between several major sections of the mid-oceanic ridge system. In Figure 8 of Pitman et al.
(1968) the North Pacific has been used as a
standard and in Figure 25–2 of this paper the
South Atlantic (V-20 SA) has been used. In
both cases the South Pacific (EL-19S) follows

Figure 25-1 *(facing page)*
Sample magnetic profiles from various oceans. The South Atlantic (S.A.) profile is from the paper by
Dickson et al. (1968), the South Indian Ocean (S.I.O.) profile is from Le Pichon and Heirtzler (1968) and
the remaining profiles (north Pacific, SI-6, EL-19S) are from Pitman et al. (1968). Beneath each of the
observed profiles is a theoretical profile calculated from the normally magnetized (black) and reversely
magnetized (white) bodies shown. Each body is 2 km thick. With each model is a time scale constructed
by assuming an age of 3.35 m.y. for the end of the Gilbert reversed epoch. Dashed vertical lines connect
similarly shaped anomalies identified by the numbers at the top of the dashed lines

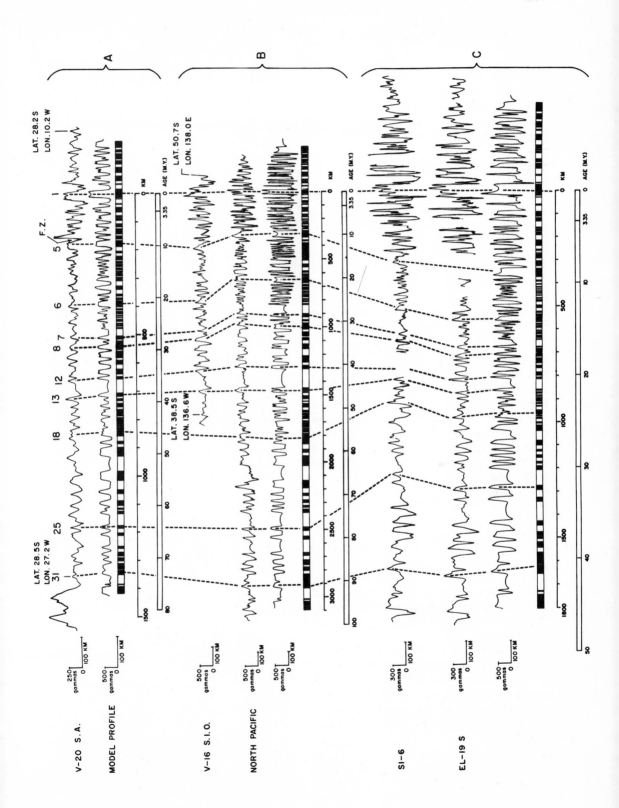

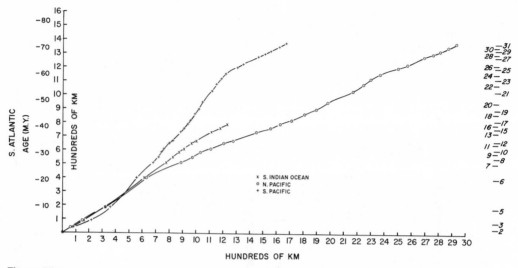

Figure 25-2
The distance to a given anomaly in the South Atlantic (V-20 S.A.) versus distance to the same anomaly in the South Idian, North Pacific, and South Pacific oceans. Numbers on right refer to anomaly numbers

the same general curvature, the points of inflection occuring in the vicinity of anomalies 5 and 24. The spreading rates for both the North Pacific and the South Atlantic relative to the spreading rate for the South Pacific may have varied in the same way with respect to time, the spreading rates in both the North Pacific and the South Atlantic may have remained relatively constant and the South Pacific spreading rate may have varied with time, or both processes may have occurred simultaneously. To us it seems most likely that the South Pacific spreading rate has varied with time. For these reasons the South Pacific can be eliminated as a standard.

The Indian Ocean (V-16 SIO) has been eliminated because the anomalies extend only to number 16, at which point the profile ends at the base of the south Australian continental slope.

It has been demonstrated (Pitman and Heirtzler, 1966) that the ELT-19 N axial model may be applied to the Reykjanes ridge if a constant spreading rate of 1 cm/yr is assumed for this region. Dickson et al. (1968) have applied this model to two regions of the mid-Atlantic ridge in the South Atlantic, again assuming a constant spreading rate. One case, V-20 S.A., is shown here (Figure 25–1). The other, V-18 (latitude 30.5°S, longitude 13.7°W), is shown

in Figure 7 of Dickson et al. (1968). In both cases, the fit between the measured and the computed profiles is good. The applicability of this model to the North Atlantic (Reykjanes ridge) and the South Atlantic (mid-Atlantic ridge) indicates that the spreading rate in each of these regions has been constant for the past 10 m.y.

Vine (1966), in comparing the details of the axial zone pattern of ELT-19 N and Juan de Fuca ridge, suggested that the spreading rate at the Juan de Fuca ridge has not been constant over the past 10 m.y. but has decreased from 4.4 to 2.9 cm/yr.

The time scale extrapolated from the V-20 S.A. profile has been selected as a standard because:

1. The relative spreading in the South Pacific appears to have varied considerably with time and the paleontological evidence suggests that the South Pacific time scale is too young by a factor of 2.

2. The North Pacific profile is distorted at the ridge axis.

3. The anomaly pattern for the South Indian Ocean is not sufficiently long.

The bottom of Figure 25–1, *A* shows the extrapolated time scale based on the 3.35-m.y. date for the end of the Gilbert reversed epoch. The date of 80 m.y. just beyond anomaly 32 is

Table 25–1
Intervals of normal polarity (m.y.).

0.00– 0.69	11.93–12.43	25.25–25.43	40.03–40.25	67.77–68.51
0.89– 1.93	12.72–13.09	26.86–26.98	40.71–40.97	68.84–69.44
1.78– 1.93	13.29–13.71	27.05–27.37	41.15–41.46	69.93–71.12
2.48– 2.93	13.96–14.28	27.83–28.03	41.52–41.96	71.22–72.11
3.06– 3.37	14.51–14.82	28.35–28.44	42.28–43.26	74.17–74.30
4.04– 4.22	14.98–15.45	28.52–29.33	43.34–43.56	74.64–76.33
4.35– 4.53	15.71–16.00	29.78–30.42	43.64–44.01	
4.66– 4.77	16.03–16.41	30.48–30.93	44.21–44.69	
4.81– 5.01	17.33–17.80	31.50–31.84	44.77–45.24	
5.61– 5.88	17.83–18.02	31.90–32.17	45.32–45.79	
5.96– 6.24	18.91–19.26	33.16–33.55	46.76–47.26	
6.57– 6.70	19.62–19.96	33.61–34.07	47.91–49.58	
6.91– 7.00	20.19–21.31	34.52–35.00	52.41–54.16	
7.07– 7.46	21.65–21.91	37.61–37.82	55.92–56.66	
7.51– 7.55	22.17–22.64	37.89–38.26	58.04–58.94	
7.91– 8.28	22.90–23.08	38.68–38.77	59.43–59.69	
8.37– 8.51	23.29–23.40	38.83–38.92	60.01–60.53	
8.79– 9.94	23.63–24.07	39.03–39.11	62.75–63.28	
10.77–11.14	24.41–24.59	39.42–39.47	64.14–64.62	
11.72–11.85	24.82–24.97	39.77–40.00	66.65–67.10	

close to the 73-m.y. date given by Vine (1966) for the same point.

The possible error inherent in such an extrapolation cannot be over emphasized. By adopting the South Atlantic time scale we are assuming that the ocean floor has been spreading at a constant rate in the locality of the V-20 profile for the last 80 m.y.

Dividing the distance to the South Atlantic model magnetized bodies by the calculated axial zone spreading rate of 1.9 cm/yr (see Dickson. et al., 1968) gives the times of the magnetic reversals and the corresponding ages for the normal and reversed magnetic intervals. The ages of the normally polarized intervals are given in Table 25-1. The intervening times are reversely polarized.

In Figure 25-3 anomaly numbers are shown with their corresponding geologic and absolute ages. If the ages of these anomalies are placed on a map for the areas where anomalies had previously been identified only by number, the age map of Figure 25-4 results.

There are three principal observations that conflict with the proposed time scale. The first of these is the age of 27 m.y. determined by the potassium-argon method for a basalt boulder from Cobb seamount on the East Pacific rise (Budinger and Enbysk, 1967). According to

the magnetic anomaly pattern, the seamount should be no older than 3.5 m.y. (Vine, 1966). The boulder may not have developed as part of the seamount but may be a glacial erratic. Further, Dymond et al. (1968) have dated two basalt samples drilled from the top of Cobb seamount as less than 2 m.y. Ewing et al. (1966) and Saito et al. (1966) have pointed out that a Lower Miocene boulder was dredged at the intersection of the Atlantis fracture zone and the rift valley of the mid-Atlantic ridge. Although the magnetic anomalies near the Atlantis fracture zone in the North Atlantic have not been studied, one would assume that spreading has occurred there just as it has immediately to the south (Phillips, 1967; Van Andel et al., 1967). L. R. Sykes (personal communication) has pointed out that although the boulder was located within 10 km of the axis on the north side of the fracture zone, it is about 60 km from the axis on the south side. Hence, it is impossible to predict the age of such a boulder if we assume the existence of sea floor spreading.

Recently, Ewing and Ewing (1967) have documented the existence of a significant increase in sediment thickness occurring near anomaly 5, approximately 10 m.y. ago, and have interpreted this as probably marking an

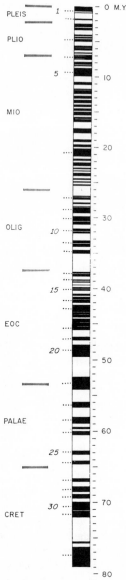

Figure 25-3
The geomagnetic time scale.
From left to right: Phanerozoic
time scale for geologic eras,
numbers assigned to bodies and
magnetic anomalies, geomagnetic
field polarity with normal
polarity periods colored black

interruption in spreading. Our data cannot exclude the possibility of such an interruption, provided that it occurred simultaneously everywhere. Thus, this fact introduced a further restriction in the interpretation of our time scale.

CHARACTERISTICS OF SEQUENCE OF GEOMAGNETIC FIELD REVERSALS

Although it is important to substantiate this history of geomagnetic reversals by other, independent means, it will be difficult to do so in any detailed fashion. The numerous and relatively rapid reversals make it difficult to distinguish any particular magnetic event from its neighbors. Even if samples of the magnetic basement can be recovered by deep drilling operations, age determinations by the present techniques probably cannot be made with sufficient precision.

For example, a dating technique that can ascertain age to within 10% could identify several events that are less than 10 m.y. old, but this technique would have insufficient accuracy to identify a particular event older than 10 m.y. If the dating technique could identify an event whose length is only 5% of its age, we might be able to identify an event as old as the long reversed period at 49.6 to 52.4 m.y. but nothing older. With drilling, too, one can never be sure of sampling the magnetic basement instead of a local magnetic inhomogeneity.

Figure 25–3 shows the geomagnetic time scale with geologic ages according to the Phanerozoic time scale (Harland et al., 1964). Among the 171 reversals there are many short events throughout the geomagnetic time scale, some as short as 30,000 years. As shown in Figure 25–5, however, in which the frequency of reversals has been plotted as a function of time, the older events are generally longer. The graphs were constructed by determining the average number of reversals that occurred during periods of 2, 5, and 10 m.y., respectively. The frequency of reversals was appreciably lower prior to 40 m.y. ago.

The distribution of the lengths of normal and reversed intervals is illustrated by the

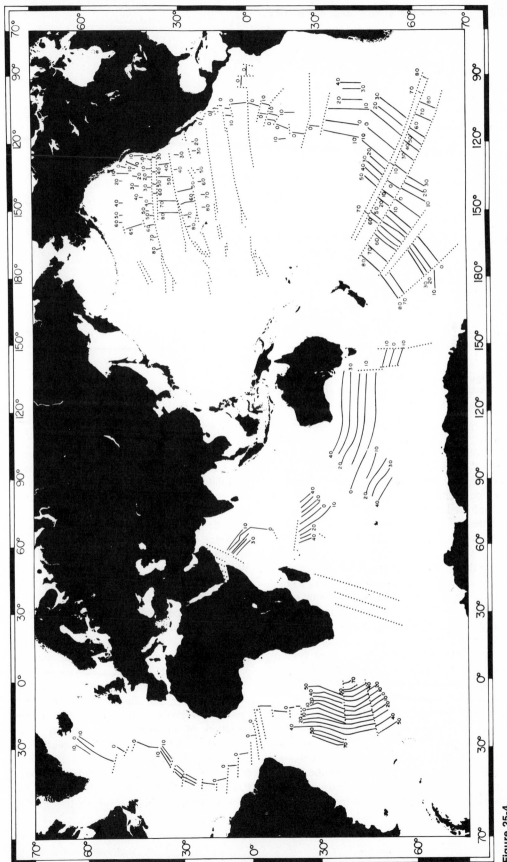

Figure 25-4
Isochron map of the ocean floor according to the magnetic anomaly pattern. Numbers on isochron lines represent age in millions of years. Dotted lines represent fracture zones

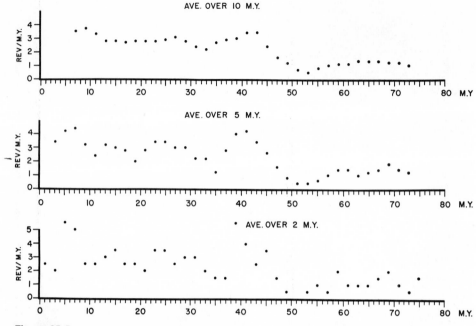

Figure 25-5
Frequency of geomagnetic field reversals as a function of time averaged over intervals of 2, 5, and 10 m.y.

histograms of Figure 25-6. The average normal interval is 0.42 m.y. and the average reversed interval is 0.48 m.y. These nearly identical average lengths and the similar shapes of the normal and reversed distributions suggest that there is no fundamental difference in the stability of the normal and reversed magnetic states. Only 15% of the normal intervals were longer than the present normal 700,000-year interval, although some existed for nearly 3 m.y.

The completeness with which the time scale accounts for details in the marine anomaly profile near the axis is illustrated in Figure 25-7. Ten observed profiles are compared with the theoretical profile for the axial region of the South Pacific and Indian oceans, where the spreading is relatively fast and small anomalies are well displayed. Anomalies that occur with some consistency are labeled 'X' and 'Y.' These anomalies would presumably be respectively due to splits in the Olduvai and Mammoth events within the Matuyama and Gauss epochs. Since they are identified with more difficulty and less consistency than other anomalies, we hesitate to interpret them

as due to magnetic polarity events. McDougall and Chamalaun (1966) and Vine (1966) have postulated a normal magnetic polarity event corresponding to Y named by McDougall and Chamalaun the 'Kaena' event. N. D. Opdyke (personal communication) has found evidence for the Kaena from paleomagnetic studies of deep-sea cores. McDougall and Wensink (1966) have proposed a Gilsa event, a period of normal magnetic polarity, between the Jaramillo and Olduvai. We find no consistent evidence for such an event in the magnetic profiles studied. To date no paleomagnetic evidence has been presented for an event of positive polarity corresponding to the anomaly X.

Comparison of the geomagnetic time scales presented here and the time scale as determined by Hays and Opdyke (1967) from the direction of magnetization of ocean cores emphasizes that the magnetic anomalies are due to magnetic reversals. Figure 25-8 illustrates the magnetic stratigraphy of three Pacific-Antarctic cores. It also shows the history of the magnetic reversals that Pitman

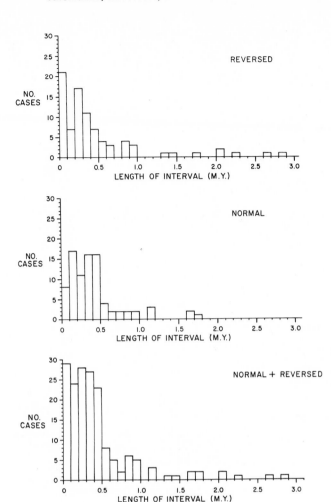

Figure 25-6
Distributions of the lengths of magnetic states for normal polarity and reversed polarity and for the length of intervals regardless of polarity

and Heirtzler (1966) found from marine anomalies and the polarity history according to this paper. The scale used by Pitman and Heirtzler (1966) for these axial and near-axial anomalies differs slightly from the scale reported here in the first part of Table 25–1 because more profiles have since been examined.

Dalrymple et al. (1967) have studied Pliocene geomagnetic epochs from forty-five rock samples of the western United States. The polarities and ages that they found are illustrated in Figure 25–9 along with the geomagnetic time scale of this paper. When the probable error of the age determinations, shown by the bar, is considered, the correspondence is generally good. Contradictions

in the two scales are evident with the specimens W 10, W 23, and W 24. These discrepancies might be explained by the existence of magnetic polarity events of very short duration. It is also possible that the spreading has been discontinuous (Ewing and Ewing, 1967). As pointed out previously, however, this section of the model (from 10 m.y. B. P. to the present) has been shown to be applicable to the South Pacific and Reykjanes ridge (Pitman and Heirtzler, 1966), to the Juan de Fuca ridge (Vine, 1966), and to the South Atlantic (Dickson et al., 1968), suggesting that the spreading has been continuous and at a constant rate in each region or that, if discontinuities have occurred, they have been simultaneous in all these regions.

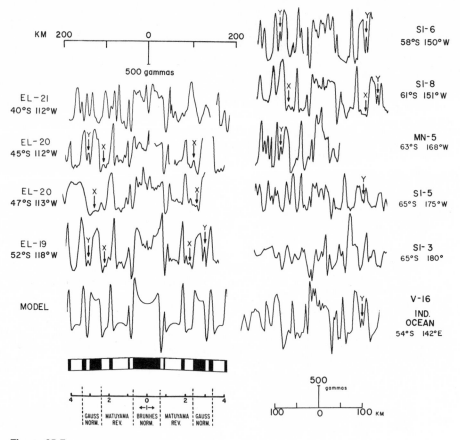

Figure 25-7
Ten observed axial magnetic profiles with theoretical profile for known reversals during the last 4 m.y. Anomalies X and Y are not accounted for with the model

COUPLING BETWEEN MOVING OCEAN FLOOR AND CONTINENTAL MASSES: RELATION OF SEA FLOOR SPREADING TO CONTINENTAL DRIFT

It is most important to determine if coupling is present between the moving ocean floor and the continental blocks. If such coupling exists, the magnetic pattern in the ocean floor can be used to reconstruct the detailed motion of the continents. Not only would the magnetic pattern give the detailed paths followed by the continents but also the extent of marine magnetic data available today would permit a study of the synoptic motions of continents.

It was assumed by Hess (1962) and Dietz (1961) that sea floor spreading was responsible for the drift of some of the continents. They

assumed that the moving ocean floor in all oceans, except the Pacific, rafted the bordering continents ahead of it. In the Pacific, however, the moving floor did not push the continents; on the contrary, the continents encroached upon the ocean floor. The cirum-Pacific trenches were interpreted as the places where the moving ocean floor turned down under the advancing blocks of crust. Although this study does not answer the question of why most cirum-Pacific continental masses are encroaching on the Pacific instead of drifting away from it, the distribution of relative ocean floor spreading rates suggests that the oceanic crust that is rafting or pushing a continental block has the slower spreading rate.

If the southern continents have been drifting away from a fixed Antarctica while main-

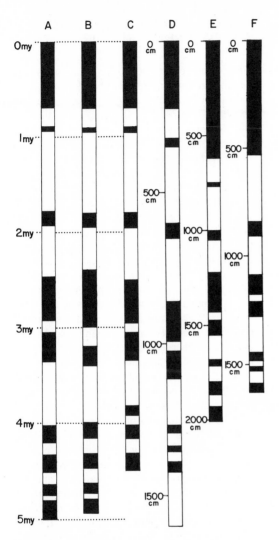

Figure 25-8
Geomagnetic field reversals for the last 5 m.y. from ocean sediment cores compared with field reversals according to the magnetic anomaly pattern. Column A represents the geomagnetic time scale according to this paper; column B, according to Pitman and Heirtzler (1966); column C, according to sediment cores (Hays and Opdyke, 1967). Columns D, E, and F show the magnetic states of the three cores used by Hays and Opdyke

taining a median ridge, the ridge must also have been drifting (Wilson, 1965c) and the apparent ocean floor spreading rates are only relative to the moving axis of this ridge. Various data support the thesis that the New Zealand plateau was once against Antarctica: the direction and length of fracture zones east of New Zealand, the continental nature of the plateau (Brodie, 1964), the match of the 80-m.y. isochron lines (Figure 25-4) against the plateau, and the correlation of structural units and petrographic provinces (Wright, 1966).

If only the continents have moved, the amount of offset along any fracture zone should be constant. Study of Figure 25-4 shows, however, that the offset along the fracture zones at the base of the New Zealand plateau gradually increases to the southeast. If ocean floor spreading does not involve large-scale distortion of the oceanic crust (Morgan, 1968b) and if the direction of motion of the continents is parallel to the fracture zones, the increase in offset of the anomalies to the southeast from the New Zealand plateau suggests that the Pacific-Antarctic ridge has migrated to the southeast away from New Zealand.

It is also possible that the variation in offset results from different spreading rates across

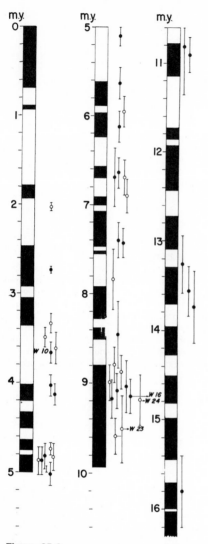

Figure 25-9
Geomagnetic field polarity of Pliocene
rock specimens (Dalrymple et al., 1967)
compared with field polarity according
to the magnetic anomaly pattern

PATTERN OF RECENT OCEAN FLOOR AND CONTINENTAL MOTION

The pattern of recent motion in the oceans is summarized in Figure 25–10 and Table 25–2. Considered on a worldwide scale, one can recognize four active lines of spreading in the Atlantic Ocean, the Arctic Ocean, the Indian Ocean, and the Pacific Ocean. With the exception of the Arctic Ocean, both spreading rates parallel to the fracture zones for the last 10 m.y. and the strikes of fracture zones have now been determined along their whole length. Although in many places the knowledge of the pattern of spreading is still rather poor, no major gap in geographic coverage exists. It is now possible to test critically the spreading floor hypothesis by examining whether the geometry of this pattern of spreading is compatible with the observation that, in general, no major large-scale distortions of continental blocks or of oceanic blocks are currently observed. For example, if the Atlantic Ocean is opening at the mid-Atlantic ridge, the movement should occur in such a way that it does not deform or distort the large bodies of horizontally stratified sediments lying in its basins and on the continental margins. Advocates of continental drift have long recognized that the continents should be displaced as rigid crustal blocks. Recent studies of the physiography of the ocean floor and the distribution of sediment have demonstrated that most oceanic crustal blocks do not show evidence of major compression or distortion. Thus, as pointed out by Morgan (1968b), the motion can be described as the rotation of rigid blocks on the spherical surface of the earth, with the only modification of these blocks occurring at the crests of the ridges, along the trenches, and within regions of major compressive folding. Within a rigid block, motion everywhere should parallel the fracture zones, which should form small circles centered on the pole of rotation of this movement. Also, the angular velocity of rotation about some pole should be the same everywhere in the absence of relative movement along fracture zones outside the region between the offset ridge axes. In the region between the offset ridge axes, the motion is that of a transform fault (Wilson, 1965c). This implies that the spreading velocity increases as the sine of

the fracture zones. If the rate of sea floor spreading in an area between two fracture zones has varied independently of that in adjacent areas, the observed isochron displacement pattern shown in Figure 25–4 could result without eastward migration of the ridge axis. Such a pattern of spreading would result, however, in seismically active fracture zones along their whole length and in distortion of the adjacent continents. There is no evidence that this is so.

Table 25–2
Spreading rates.

Pacific Ocean			Atlantic Ocean			Indian Ocean		
Lat.	Long.	Rate	Lat.	Long.	Rate	Lat.	Long.	Rate
48°N	127°W	2.9	28°N	44°W	1.25	19°N	40°W	1.0
17°S	113°W	6.0	22°N	45°W	1.4	13°N	50°W	1.0
40°S	112°W	5.1	25°S	13°W	2.25	7°N	60°W	1.5
45°S	112°W	5.1	28°S	13°W	1.95	5°N	62°W	1.5
48°S	113°W	4.7	30°S	14°W	2.0	22°S	69°W	2.2
51°S	117°W	4.9	38°S	17°W	2.0	30°S	76°W	2.4
58°S	149°W	3.9	41°S	18°W	1.65			
58°S	149°W	3.7	47°S	14°W	1.60			
60°S	150°W	4.0	50°S	8°W	1.53			
63°S	167°W	2.3						
65°S	170°W	2.0						
65°S	174°W	2.8						

the distance from the pole of rotation and reaches a maximum at the equator of rotation. Morgan (1968b) has shown that the fracture zones in the Atlantic Ocean between 30°N and 10°S approximate small circles centered at a pole near the southern tip of Greenland and that the previously determined spreading velocities roughly agree with the velocities predicted by opening of the Atlantic Ocean about this pole. Similarly, Morgan showed that the sets of faults along the west coast of North America (e.g., the San Andreas and Denali fault systems) are centered about a pole of rotation situated near the southern tip of Greenland.

The more evenly distributed information now available on the spreading rates and strikes of fracture zones along three of the four active lines of spreading enable us to carry this analysis further. For the Atlantic, Indian, and Pacific oceans one can determine whether one pole of rotation can account for the variation of spreading rates and strikes of fracture zones along the active line of spreading.

Using least squares, we have independently obtained the pole best fitting the spreading rates and the pole best fitting the strikes of the fracture zones (see Le Pichon, 1968, for details). Ideally, these two poles should coincide. Table 25–3 lists the positions of the poles of rotation so obtained and Figure 25–10 shows their location. In general, the distribution of spreading rates and fracture zones within each ocean agrees well with a simple rotation about a pole. The disagreement is largest in the Indian Ocean, where the fracture zones east of the Owen fracture zone do

Table 25–3
Poles of rotation.

Location	Method of determination[a]	Position		Number of observations	Standard deviation (deg)
		North	South		
North Pacific	F.Z.	53°N, 47°W	53°S, 133°E	32	5.7
South Pacific	F.Z.	70°N, 62°W	70°S, 118°E	6	4.5
South Pacific	S.R.	69°N, 57°W	69°S, 123°E	11	0.06
North Atlantic	F.Z.[b]	58°N, 37°W	58°S, 143°E	18	2.9
South Atlantic	S.R.	70°N, 65°W	70°S, 115°E	9	0.07
Indian Ocean	F.Z.	26°N, 21°E	26°S, 159°W	5	5.1
Arctic Ocean	F.Z.	78°N, 102°E	78°S, 78°W	5	10.

[a]S.R. indicates spreading rate; F.Z., fracture zone.
[b]From Morgan (1968).

not obey a simple pattern and where the spreading rates continue to increase along the southern part of the ridge south of Australia at a distance greater than 90° from the pole of rotation at 30°N, 10°E. By contrast, in the South Pacific the whole pattern is simply explained by a single rotation. Similarly, in the equatorial and northern Atlantic Ocean the strike of the fracture zones is satisfied by a single north pole of rotation located near Greenland. Although the distribution of spreading rates in the South Atlantic agrees essentially with this rotation, it implies a north pole situated about 10° farther north. Similarly, the south pole for the South Pacific seems to lie about 10°S of the south pole for the North Pacific (see Figure 25–10). Within the accuracy of the determination the poles for the North Pacific and the equatorial and northern Atlantic (south of 30°N) coincide, and the poles for the South Pacific and South Atlantic coincide. All poles for both Atlantic and Pacific oceans lie within a circle centered in the Labrador Sea having a radius less than 10°. This result indicates that the two major movements of spreading in the Atlantic and Pacific oceans open about the same axis of rotation. Consequently, the general picture emerging from this analysis is one of great simplicity with the Atlantic and Pacific oceans opening about the same axis, which is slightly inclined to the rotational axis of the earth, and the Indian Ocean opening about an axis between the Pacific and Atlantic oceans.

Figure 25–11 illustrates this opening by using a Mercator projection centered on the axis passing through the South Pacific pole of rotation (69°N, 57°W). On a Mercator projection, latitudes are horizontal lines, longitudes are vertical lines, and the scale is everywhere proportional to the inverse of the sine of the distance from the pole. Thus,

if the projection is centered on the pole of rotation, fracture zones should coincide with latitudes, ridge crests should coincide with longitudes, and the distance between the ridge crest and a nearby magnetic anomaly (on the map, anomaly 5, about 10 m.y. ago) should be everywhere constant. Figures 25–10 and 25–11 should be compared to see how all these requirements are met. Because the pole of rotation chosen was the one for the South Pacific, the agreement is best in the South Pacific. The present ridge crest and associated set of fracture zones (e.g., the San Andreas fault system) in the equatorial and northern Pacific Ocean also fit this pole of rotation well. The Juan de Fuca and Gorda ridges agree particularly well, and the San Andreas also essentially agrees with this pattern of movement. By contrast, the set of fracture zones in the northern and equatorial Pacific (e.g., the Mendocino and Clipperton fracture zones), which, according to Morgan (1968b), fits a pole near 79°N, 111°E, completely disagree with the present pattern of movement (see Figure 25–11). Because the position of anomaly 5 in the northern and equatorial Pacific agrees with the old pattern of movement shown by these fracture zones, the present pattern must have established itself since 10 m.y. ago. Prior to 10 m.y. the South Pacific was moving about the same pole as now, but the North Pacific was moving about a different pole. Perhaps as a result of the shift in the North Pacific pole in the Pliocene the area near the Galapagos Islands has been subjected to extension, and the observed symmetric magnetic pattern records crustal material upwelling along an east-west line on the east side of the ridge axis for at least 5 m.y. (Herron and Heirtzler, 1967).

In the Atlantic Ocean, the agreement is not as good because the pole chosen was the

Figure 25-10 (facing page)
Location of mid-ocean ridge axis, fracture zones, and trenches. The double line defines the axis; the adjacent dashed lines define the 10 m.y. isochron. Numbers near the axis give the half spreading rate in centimeters per year. Small circles indicate poles of spreading for the North Atlantic (N.A.), South Atlantic (S.A.), North Pacific (N.P.), South Pacific (S.P.), Arctic (A), and Indian Ocean (I.O.). A plus within these circles indicates poles according to fracture zones; a cross within the circles, poles according to spreading rates

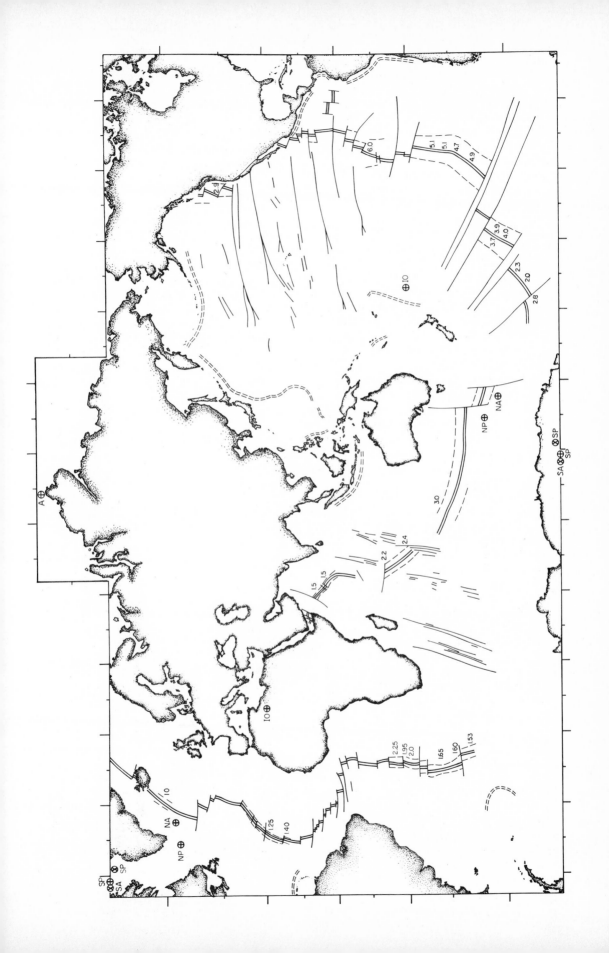

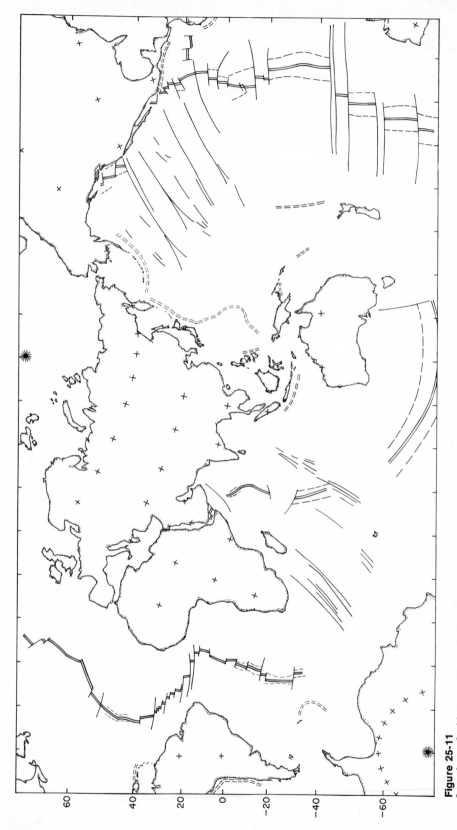

Figure 25-11

Locations of mid-ocean ridge axis, fracture zones, and trenches. The poles of this mercator projection are the South Pacific poles of rotation (north pole located at 69°N, 57°W). The stars show the present location of the geographic poles, and the plus signs indicate the intersections of present latitudes and longitudes at 20° intervals. Note how, on this projection, the fracture zones in the South Pacific are along latitude lines, the ridge crest is along longitude lines, and the distance to anomaly 5 does not change with distance from the pole of rotation. This confirms that the movement of opening of the South Pacific Ocean corresponds to a rotation around a point situated near 69°N, 57°W.

one for the South Pacific. South of 30°N, however, the agreement is fair. North of 30°N, there is disagreement as noted by Morgan (1968b), and this should be expected as the relative movement of Eurasia away from America cannot be the same as the movement of Africa away from America. Africa and Eurasia are now involved in the opening of the Indian Ocean around a pole situated near 26°N and 21°E. The small but systematic difference between the poles determined for South Pacific and South Atlantic and the poles determined from North Pacific and North Atlantic might suggest a small differential movement of North America and South America.

The fact that the spreading in the Indian Ocean does not obey a single pattern of rotation accounts for the complexity of the physiography of this ocean and can be taken into account when one recognizes the presence of the Java trench north of the only fast spreading ridge segment running essentially east-west (Le Pichon and Heirtzler, 1968). In spite of these complications, however, the main movement in the Indian Ocean is one of opening around a pole situated somewhere near the present location of Libya. The Alpine fault and Macquarie ridge must result from the conflicting movement of the Indian and Pacific Ocean crustal plates. The absence of a fast spreading ridge around the southern tip of Africa removes one of the major geometrical difficulties posed by the distribution of the mid-ocean ridges.

To summarize, Figure 25–10 illustrates how the present pattern of relative movement over the world can be described by a few rotations. The main rotations are in the Atlantic and the Pacific (about slightly different poles), with a third major rotation in the Indian Ocean. The rates of angular opening are, however, quite different, 3.6×10^{-7} deg/yr in the Atlantic Ocean and 4.0×10^{-7} deg/yr in the Indian Ocean compared with 10.8×10^{-7} deg/yr in the South Pacific Ocean. Figure 25–11 also shows that the active trenches run either north-south or east-west in this new system of coordinates, essentially absorbing the movements of the Pacific crustal blocks and of the south Indian Ocean crustal blocks.

RÉSUMÉ OF SPREADING HISTORY AND CONTINENTAL MOVEMENTS IN SOUTHERN HEMISPHERE SINCE LOWER MESOZOIC

The magnetic evidence for ocean floor spreading presented in the companion papers places several restrictions on the allowed paths of continental movements since the break up of Gondwanaland. It is thus possible to synthesize a picture of continental movement in the southern hemisphere since the early Mesozoic.

Paleomagnetic reconstructions by Irving (1964) and Creer (1965) have the continents of Africa, Antarctica, Australia, India, and South America united and close to the south geographic pole in the Early Permian. Gough et al. (1964) show that there has been little change in the paleomagnetic latitude of Africa since the middle of the Mesozoic. They also show evidence of rapid northward movement of Africa during the early Mesozoic and Permian. Studies by Creer (1965) suggest that there has not been appreciable change in the paleolatitude of South America since Triassic-Jurassic times but that large northward movements did take place in the Lower and Middle Permian. Investigations on the Triassic-Jurassic lavas that appear in all the Gondwanic continents indicate that the breakup of Gondwanaland had already occurred by this time.

From the study of the magnetic pattern in the Indian Ocean the initial northward movement of the Africa-South America block in Mesozoic-Permian times occurred as a result of spreading about the southwest branch of the mid-Indian ridge. The break from the initial mass started at the Horn of Africa and then opened down the east coast of Africa. At present there is little or no spreading about this branch of the mid-Indian ridge, and, on the basis of paleomagnetic evidence, it appears that it was active during the lower Mesozoic and Permian but that by Jurassic the spreading had mainly ceased. The split in the South American and African continents may have begun in early Mesozoic almost simultaneously with the break away of the African-South American block. Thus, the Argentine and Cape basins were born. The cessation of spreading about the southwest branch of the mid-Indian ridge in Cretaceous and the development of a major part of the Argentine and Cape basins ended the first major phase of spreading in the southern Indian and South Atlantic oceans.

The second phase of spreading involved the further separation of South America from Africa and the northward movement of India in the Indian Ocean. The direction of spreading in the South Atlantic underwent a major change in the Upper Cretaceous (80 m.y.), becoming nearly east-west, and there was little change in the paleolatitudes of either South America or Africa. The Upper Cretaceous also saw the separation of the New Zealand block from Antarctica. The spreading on the west side of the mid-Atlantic ridge was smooth and continuous, and South America was able to drift westward over what was then a large Pacific Ocean. Spreading on the east side of the ridge was not as smooth, and the east-west spreading on that side is not evident until the Early Eocene. Perhaps the restrictions placed on the African block by activity in the Indian Ocean caused this hiatus in the spreading history and resulted in the initial formation of the Walvis ridge. The northward trek of India was accommodated by a major change in the spreading pattern of the Indian Ocean. Spreading on the southwest branch of the mid-Indian Ocean ridge had almost ceased, and it now began to act as a large series of fracture zones that essentially marked the locus of the west coast of India in the northward movement. Le Pichon and Heirtzler (1968) suggest that the movement of India was accomplished by spreading about a now subsided east-west striking ridge situated in the Java basin.

The third major readjustment to the spreading in the Indian Ocean occurred at the end of the Eocene (40 m.y.) with the commencement of spreading about the southeast and northwest branches of the mid-Indian Ocean ridge. The readjustment was occasioned by the abutting of the Indian subcontinent against the Asian block. This new stage of spreading resulted in the rapid separation of Australia from Antarctica and Broken ridge from the Kerguelen plateau. Much of the differential movement resulting from different directions and unequal rates of spreading on these two limbs has been absorbed by the Amsterdam fracture zone. Ninetyeast ridge was formed by the differential movement of the blocks on either side of it. A result of the various stages of spreading in the Indian Ocean is that India moved much farther north than Antarctica moved south. Spreading about the southeast branch of the mid-Indian ridge alone does not satisfy the paleomagnetic evidence of high latitudes for Australia during the Permian; it seems likely that the ridge itself has migrated northward. Interaction between the northward movement of Australia and the northwest movement of the New Zealand plateau has probably resulted in the Alpine fracture zone in western New Zealand. Finally, during the last 10 m.y., the Atlantic, south of 30°N, and the Pacific oceans have been rotating about the same pole, while the Indian Ocean has been rotating about a different pole.

EARTHQUAKES AT THE EDGES OF PLATES

26

Introduction and Reading List

Once the idea of plate tectonics had been clearly formulated in the years 1963–1967, it was amazing how quickly and how neatly a large body of seismological data fell into place. In part this was because seismology was one of the lines of research that had contributed to the original development of plate tectonics. Moreover, seismology was important in confirming some specific predictions made by plate-tectonic theory. Plate-tectonic theory, in turn, has provided a new rational framework for several lines of research in seismology.

Trenches and Benioff Zones

Long before plate tectonics had been formulated, seismologists recognized that a large proportion of the world's earthquakes, including almost all of the deep ones, occur near the deep oceanic trenches. As early as 1939, David Griggs wrote: "A tenuous argument in favor of oceanic rising currents is the distribution of deep-focus earthquakes. Visser, Leith, Sharpe, Gutenberg and Richter all agree that foci of deep earthquakes in the circum-Pacific region seem to be on planes inclined about 45° toward the continents. It might be possible that these quakes were caused by slipping along the convection-current surfaces." It is clear from the literature of the time that seismologists were becoming increasingly aware of the planar distribution of earthquake foci beneath oceanic trenches.

These ideas were brought together by Hugo Benioff (1949) in a paper on deep-focus earthquakes in the Tonga-Kermadec region and western South America. He concluded that earthquakes originated on great faults that dipped under the continents, the oceanic deeps being the result of downwarping of the oceanic blocks. Benioff continued to develop this idea until, in 1954, he presented his classic work (included in this book as Chapter 27) on what are now called Benioff Zones. In this paper he describes the Benioff Zones of all the circum-Pacific oceanic deeps. He not only recognized that earthquakes tend to originate on dipping planes, but he presented arguments to show that, where there was an earthquake, the direction of slip was such that the oceanic side was being thrust under the overlying continental (or shallower oceanic) block, and cited the first motion studies of Honda and Masatsuka (1952) in support of this interpretation. Benioff's illustration that is reproduced here as Fig. 27–11 would seem to lead inevitably to the conclusion that the sea floor was spreading and that the mantle was overturning convectively. Although Benioff refrained from speculating about this, he was aware as he was writing the paper under discussion that continued underthrusting was required to maintain a trench, and he speculated that a trench may have existed off the west coast of North America in earlier times and had then disappeared as the motion changed from underthrusting to horizontal slip. This interpretation anticipates some of the latest ideas about the plate-tectonic history of the west coast of North America (Chapter 45).

Focal Mechanism of Earthquakes

During the 1920's the seismologists H. Nakano (1923) and P. Byerly (1926, 1928) developed a new way of analyzing earthquakes that, decades later, was to make a very important contribution to plate tectonics. Their idea was a simple one. Consider first an explosion taking place in a cavity. The first waves caused by it arriving at a distant station will be compressional and the first motion of the ground will be in a direction away from the source. Consider next elastic waves resulting from a collapsing cavity. These waves will be dilatational and the first motion observed would be toward the source. Nakano and Byerly found that the elastic waves generated by an earthquake are more complicated than the two sorts of wave just described, with both compressions and dilatations being generated by a single earthquake. Imagine a sphere drawn around the point on a fault where the earthquake motion starts. Pass two perpendicular planes (called nodal planes) through the center of the sphere, dividing the sphere into quadrants. Then the motion in two diagonally opposed quadrants will be compressional and the motion in the other two will be dilatational. Nakano and Byerly demonstrated that one of the nodal planes is parallel to the plane of the fault. The other is perpendicular to the "slip direction," which can be visualized as the direction of a scratch made in one of the fault blocks by a sharp rock imbedded in the other fault block. From first-motion studies, the seismologist is able to determine the orientation of faults and the slip directions of earthquakes, even where the actual faults are inaccessible to direct observation. However

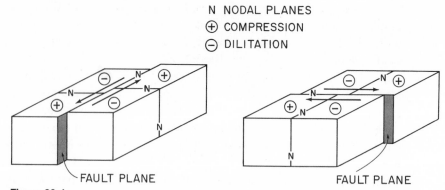

Figure 26-1
Focal-mechanism solutions. Regions of compression and dilation are separated by two perpendicular planes termed nodal planes. Either of the two nodal planes may be the fault plane as indicated above

the method is inherently ambiguous in the sense that there is nothing in the seismic data to determine which of the two nodal planes is parallel to the fault plane and which is perpendicular to the direction of slip. For this, the seismologist must use geologic information and some measure of common sense.

The technique of Nakano and Byerly came to be known as "focal-mechanism," "fault-plane," or "first-motion" research. After a few decades its value became apparent and it began to be used on a substantial scale. By the mid-1950's enough first-motion solutions were available to permit the seismologists to begin to look for regional patterns. Oddly enough, one of the first conclusions drawn from this work led geophysicists away from plate tectonics and into a blind alley. For reasons that are mainly of historical interest now, seismologists wrongly concluded from these regional studies that the slip direction on the major circum-Pacific earthquakes was "strike-slip"—that is, nearly horizontal (Hodgson, 1957). Geologists, on the other hand, preferred Benioff's earlier interpretation that many of the major faults were thrusts and that among these thrust faults were included the great faults associated with the deep trenches.

The key event that finally convinced many earth scientists that the geologists were correct was the great earthquake in Alaska on March 27, 1964. Initially, the seismologists favored the interpretation that the earthquake had occurred on a nearly vertical fault. It was the careful geologic mapping of George Plafker (Chapter 28) that convinced them that the earthquake occurred on a thrust fault that dipped northward at a shallow angle. Along this fault, the crust of the Pacific Ocean was being thrust under Alaska.

To measure the amount of uplift or depression caused by the Alaska earthquake, Plafker used what he called the "barnacle line," which was the upper limit of growth of barnacles that are present ubiquitously along the Alaska coast. The barnacle line is at or near the mean high tide level of the sea. Using this as his reference datum line, Plafker found uplifts as great as 8 meters produced by the earthquake. The regional pattern of uplift and subsidence, together with

Figure 26-2
Peter Molnar

the location of the foci of earthquake aftershocks, indicated that of the two nodal planes found for the main shock, the one dipping shallowly beneath Alaska was probably the fault plane. The slip direction corresponding to the other nodal plane indicated that the Pacific Ocean sea floor was being thrust downward along this fault. Plafker's interpretation of thrusting was supported by analysis of the aftershock sequence (Stauder and Bollinger, 1966a) and is now generally accepted. It is recognized that thrusting is the dominant type of fault movement along circum-Pacific trenches (Plafker, 1965).

First-motion research was also used by Brian Isacks and Peter Molnar (Chapter 31) to determine from earthquakes whether plates descending under the trenches are under compression or tension. They find that the upper parts of the plates are under tension whereas the lower parts are under compression. Their interpretation in terms of plate tectonics is that the downgoing slabs exert a pull on the upper parts of the plates. However as the downgoing slabs descend through the lithosphere and asthenosphere, they eventually "hit bottom" and the stresses become compressive.

Rises, Fracture Zones, and Earthquakes

Since the 1920's earthquakes have been known to occur in a broad band along the mid-Atlantic ridge (Tams, 1927). However, in the mid-1950's Bruce Heezen and Maurice Ewing at Lamont Geological Observatory began to recognize that the earthquake epicenters were concentrated *exactly* along the median valley of the ridge (Ewing and Heezen, 1956). We know now that they were correct and that earthquake epicenters provide some of the most precise information available about the location of plate boundaries. However it took courage to make their suggestion at the time they did because the epicenters appeared to .

Figure 26-3
Maurice Ewing

be rather scattered as a result of experimental errors due to the small number of seismographic stations supplying data. The early work of Ewing and Heezen was especially important in focusing attention on the ridges as zones of tectonic activity.

During the next decade the observational situation greatly improved with the creation of more than 125 seismographic stations constituting the World Wide Standardized Seismograph Network (WWSSN). This network was originally sponsored by the Defense Department of the United States to provide a data base for research into how to discriminate between nuclear tests and natural earthquakes in order to make feasible a nuclear test ban treaty. Not only was this research successful, but the WWSSN also provided quality seismograms from well-distributed stations for use in basic research. By the early 1960's a large new data base had become available in seismology both from the WWSSN and from the many new independent seismological observatories that had been established.

Seismology has long been our main source of information about processes taking place in the earth. It is not surprising, therefore, that in the mid-1960's there was an almost explosive growth of research in seismology related to the new paradigm of plate tectonics. The key research in this field, including the

Figure 26-4
Lynn Sykes

already classic work described in Chapters 29, 30, and 31, was done at Lamont Geological Observatory by faculty and graduate students including Bryan Isacks, Peter Molnar, Jack Oliver, and Lynn Sykes.

Lynn Sykes became interested in seismic activity along fracture zones in 1962 while doing his Ph.D. thesis research on the propagation of seismic surface waves across oceanic areas. He analyzed the newly available worldwide seismological data using computer programs that had become available for the precise and rapid determination of epicenters, and observed that nearly all earthquakes along the East Pacific rise followed the ridge crest, just as they did in the Atlantic (Sykes, 1963). Near major fracture zones, however, the earthquakes were offset in a Z-shaped pattern, the amount of offset being as much as 1000 km. Sykes' most unexpected discovery was that seismicity along the fracture zones was restricted almost exclusively to parts located between the offset crests of the oceanic ridge.

In his classic paper introducing transform faults, Wilson (1965c) pointed out that his concept of transform faults neatly explained Sykes' (1963) observation that earthquakes occurred on fracture zones only between ridges. After reading Wilson's paper, Sykes realized that another test of Wilson's hypothesis was possible because Wilson predicted exactly the opposite direction of motion on the two sides of a fracture zone as that predicted from traditional geologic reasoning (Fig. 29–1). Sykes realized that focal-mechanism studies might be able to settle the question, but he didn't pursue the idea until May, 1966, when Walter Pitman and James Heirtzler showed him the amazingly symmetrical Eltanin-19 magnetic profile across the East Pacific rise. Sykes was not the first skeptic to be converted by that "magic profile." The next morning he stopped everything else he was doing and started to work on focal mechanisms of earthquakes along the mid-ocean ridge system. Fortunately, a vast amount of new data from stations of the WWSSN had just become available on microfilm. Within a few weeks he was able to convince himself that earthquakes along the fracture zones had horizontal slip directions, in agreement with Wilson's concept of transform faults. He found focal mechanisms for earthquakes along the ridges that were consistent with sea-floor spreading from the ridges. These results (Sykes, 1967), which were presented at the meeting of the Geological Society of America in November, 1966, suddenly made it very difficult for anyone not to take the idea of transform faults seriously.

A second important line of research at Lamont began in 1964 when Jack Oliver and Bryan Isacks, then a new Ph.D. at Lamont, decided to study the deep earthquakes of the Tonga-Fiji region. They undertook this program as a contribution to the U.S. Upper Mantle Program and chose the region because it included the world's greatest concentration of earthquakes occurring deep in the mantle. Initially they were prepared to do the conventional kinds of seismological analysis, but they also took the novel approach of searching for differences between the deep seismic zone and the surrounding mantle. They anticipated that seismic waves would be propagated more slowly and less efficiently within the seismic zone. A network of seismometers was installed in the Tonga-Fiji area by Isacks, Oliver, and Sykes in 1965 and much to their surprise and ultimate delight, Oliver and Isacks discovered that indeed there was a gross difference in propagation. However the observed effect was the opposite of the one expected: high frequency seismic waves traveled *faster* and *more efficiently* along the seismic zone. Similar results had been obtained in Japan by Utsu (1967) and Katsumata (1967), so apparently this effect was not confined to the Tonga-Fiji trench. In 1966 Oliver and Isacks arrived at their now familiar interpretation that the earthquakes were being generated in a slab of rigid lithosphere thrust down into a softer aesthenosphere. Stimulated by conversations with their Lamont colleagues, including James Heirtzler, Walter Pitman, and Lynn Sykes, they pointed out the beautiful consistency of this model with plate tectonics (Oliver and Isacks, 1967).

The year 1967 was a banner year for plate tectonics, just as 1966 had been for geomagnetic reversals. By the time of the exciting April, 1967, meeting of the American Geophysical Union in Washington, news had spread of the then unpublished work on plate tectonics by Morgan (Chapter 8) and McKenzie and Parker (Chapter 7). Stimulated by this and by a seminar presented by Walter Elsasser at Lamont, Jack Oliver, Lynn Sykes, and Bryan Isacks decided to attempt a comprehensive test of the hypotheses of sea-floor spreading and plate tectonics based on the data of seismology. The result was the article included in this book as Chapter 30, which became an instant classic. Jack Oliver writes (personal communication):

> Our paper did indeed support the hypothesis and it emphasized, almost to the exclusion of the other versions, the moving plate model. It was very exciting to write that paper for we sensed that we were involved in a major upheaval in the earth sciences in general, and in seismology in particular. Even if time should prove the entire concept in error, I am sure I would continue to think of those years as highlights of my scientific career, for surely the events transpiring then marked the start of a new era in the study of the earth.

READING LIST

Earthquakes and Plates

Barazangi, M., and Dorman, J., 1969, World seismicity map compiled from ESSA Coast and Geodetic Survey epicenter data, 1961–1967: Seismol. Soc. Amer., Bull., v. 59, p. 369–380.

Benioff, H., 1962, Movements on major transcurrent faults, *in* Runcorn, S. K., ed., Continental drift: New York, Academic Press, p. 103–134.

Bolt, B. A., Lomnitz, C., and McKevilly, T. V., 1968, Seismological evidence on the tectonics of central and northern California and the Mendocino Escarpment: Seismol. Soc. Amer., Bull., v. 58, p. 1725–1767.

Brune, J. N., 1968, Seismic moment, seismicity, and rate of slip along major fault zones: J. Geophys. Res., v. 73, p. 777–784.

Brune, J. N., and Allen, C. R., 1967, A low-stress-drop, low-magnitude earthquake with surface faulting: the Imperial California earthquake of March 4, 1966: Seismol. Soc. Amer., Bull., v. 57, p. 501–514.

Davies, D., and McKenzie, D. P., 1969, Seismic traveltime residuals and plates: Roy. Astron. Soc., Geophys. J., v. 18, p. 51–63.

Molnar, P., and Sykes, L. R., 1969, Tectonics of the Caribbean and middle America regions from focal mechanisms and seismicity: Geol. Soc. Amer., Bull., v. 80, p. 1639–1684.

Sykes, L. R., 1970a, Earthquake swarms and sea-floor spreading: J. Geophys. Res., v. 75, p. 6598–6611.

Sykes, L. R., 1970c, Seismicity of the Indian Ocean and a possible nascent island arc between Ceylon and Australia: J. Geophys. Res., v. 75, p. 5041–5055.

Tobin, D. G., and Sykes, L. R., 1968, Seismicity and tectonics of the northeast Pacific Ocean: J. Geophys. Res., v. 73, p. 3821–3845.

Benioff Zones

Benioff, H., 1949, Seismic evidence for the fault origin of oceanic deeps: Geol. Soc. Amer., Bull., v. 60, p. 1837–1856.

Benioff, H., 1951, Earthquakes and rock creep, part 1: Creep characteristics of rocks and the origin of aftershocks: Seismol. Soc. Amer., Bull., v. 41, p. 31–62.

Benioff, H., 1955, Seismic evidence for crustal structure and tectonic activity, *in* Poldervaart, A., ed., Crust of the earth—a symposium: Geol. Soc. Amer., Spec. Pap. 62, p. 61–75.

Fedotov, S. M., Bagdasarova, A. M., Kuzin, I. P., Tarakanov, R. Z., and Shmidt, O. Yu., 1963, Seismicity and deep structure of the southern part of the Kurile Island arc: Acad. Sci. USSR, Dokl., Earth Sci. Sect., no. 153, p. 71–73.

Hamilton, R. M., and Evison, F. F., 1967, Earthquakes at intermediate depths in southwest New Zealand: N. Z. J. Geol. Geophys., v. 10, p. 1319–1329.

Hamilton, R. M., and Gale, A. W., 1968, Seismicity and structure of the North Island, New Zealand: J. Geophys. Res., v. 73, p. 3859–3876.

Holmes, M. L., Von Huene, R., and McManus, D. A., 1972, Seismic reflection evidence supporting underthrusting beneath the Aleutian arc near Amchitka Island: J. Geophys. Res., v. 77, p. 959–964.

Kanamori, H., 1971a, Great earthquakes at island arcs and the lithosphere: Tectonophysics, v. 12, p. 187–198.

Kanamori, H., 1971b, Seismological evidence for a lithospheric normal faulting: the Sanriku earthquake of 1933: Phys. Earth Planet. Interiors, v. 4, p. 289–300.

Kanamori, H., and Abe. K., 1968, Deep structure of island arcs as revealed by surface waves: Tokyo, Univ., Earthquake Res. Inst., Bull., v. 46, p. 1001–1025.

Katsumata, M., and Sykes, L. R., 1969, Seismicity and tectonics of the western Pacific; Izu-Mariana-Caroline and Ryukyu-Taiwan regions: J. Geophys. Res., v. 74, p. 5923–5948.

Mitronovas, W., Seeber, L., and Isacks, B., 1971, Earthquake distribution and seismic wave propagation in the upper 200 km of the Tonga island arc: J. Geophys. Res., v. 76, p. 7154–7180.

Oliver, J., and Isacks, B., 1968, Structure and mobility of the crust and mantle in the vicinity of island arcs: Can. J. Earth Sci., v. 5, p. 985–991.

Plafker, G., 1972, Alaskan earthquake of 1964 and Chilean earthquake of 1960: implications for arc tectonics: J. Geophys. Res., v. 77, p. 901–925.

Scheidegger, A. E., 1966, Tectonics of the Arctic seismic belt in the light of fault-plane solutions of earthquakes: Seismol. Soc. Amer., Bull., v. 56, p. 241–245.

Stauder, W., 1968a, Mechanism of the Rat Island earthquake sequence of February 4, 1965, with relation to island arcs and sea-floor spreading: J. Geophys. Res., v. 73, p. 3847–3858.

Stauder, W., 1968b, Tensional character of earthquake foci beneath the Aleutian trench with relation to sea-floor spreading: J. Geophy. Res., v. 73, p. 7693–7701.

Sykes, L. R., 1966, The seismicity and deep structure of island arcs: J. Geophys. Res., v. 71, p. 2981–3006.

Sykes, L. R., 1971, Aftershock zones of great earthquakes, seismicity gaps, and earthquake prediction for Alaska and the Aleutians: J. Geophys. Res., v. 76, p. 8021–8041.

Sykes, L. R., Isacks, B., and Oliver, J., 1969, Spatial distribution of deep and shallow earthquakes of small magnitudes in the Fiji-Tonga region: Seismol. Soc. Amer., Bull., v. 59, p. 1093–1113.

Focal Mechanisms

Banghar, A., and Sykes, L. R., 1969, Focal mechanism of earthquakes in the Indian Ocean and adjacent areas: J. Geophys. Res., v. 74 p. 632–649.

Benioff, H., 1964, Earthquake source mechanisms: Science, v. 143, p. 1399–1406.

Fitch, T. J., and Molnar, P., 1970, Focal mechanisms along inclined earthquake zones in the Indonesia-Philippines region: J. Geophys. Res., v. 75, p. 1431–1444.

Hodgson, J. H., 1957a, Current status of fault-plane studies—a summing up: Canada, Dominion Observ., Publ., v. 20, p. 413–418.

Hodgson, J. H., 1957b, Nature of faulting in large earthquakes: Geol. Soc. Amer., Bull., v. 68, p. 611–644.

Honda, H., 1962, Earthquake mechanism and seismic waves: J. Phys. Earth, v. 10, p. 1–97.

Ichikawa, M., 1966, Mechanism of earthquakes in and near Japan, 1950–1962: Pap. Meteorol. Geophys., v. 16, p. 201–229.

Isacks, B., and Molnar, P., 1971, Distribution of stresses in the descending lithosphere from a global survey of focal-mechanism solutions of mantle earthquakes: Rev. Geophys. Space Phys., v. 9, p. 103–174.

Isacks, B., Sykes, L. R., and Oliver, J., 1969, Focal mechanisms of deep and shallow earthquakes in the Tonga-Kermadec region and the tectonics of island arcs: Geol. Soc. Amer., Bull., v. 80, p. 1443–1470.

Stauder, W., and Bollinger, G. A., 1964a, The S-wave project for focal mechanism studies: earthquakes of 1962: Seismol. Soc. Amer., Bull., v. 54, p. 2198–2208.

Stauder, W., and Bollinger, G. A., 1964b, The S-wave project for focal mechanism studies, earthquakes of 1962: Air Force Off. Sci. Res., Grant AF-AFOSR 62–458, Rep.

Stauder, W., and Bollinger, G. A., 1965, The S-wave project for focal mechanism studies —earthquakes of 1963: Air Force Off. Sci. Res., Grant AF-AFOSR 62–458, Rep.

Stauder, W., and Bollinger, G. A., 1966a, The focal mechanism of the Alaska earthquake of March 28, 1964, and of its aftershock sequence: J. Geophys. Res., v. 71, p. 5283–5296.

Stauder, W., and Bollinger, G. A., 1966b, The S-wave project for focal mechanism studies, earthquakes of 1963: Seismol. Soc. Amer., Bull., v. 56, p. 1363–1371.

Stauder, W., and Udias, A., 1963, S-wave studies of earthquakes of the North Pacific, part II: Aleutian Islands: Seismol. Soc. Amer., Bull., v. 53, p. 59–77.

Attentuation of Seismic Waves and Mantle Structure

Barazangi, M., and Isacks, B., 1971, Lateral variations of seismic wave attenuation in the upper mantle above the inclined earthquake zone of the Tonga island arc: deep anomaly in the upper mantle: J. Geophys. Res., v. 76, p. 8493–8516.

Barazangi, M., Isacks, B., and Oliver, J., 1972, Propagation of seismic waves through and beneath the lithosphere that descends under the Tonga island arc: J. Geophys. Res., v. 77, p. 952–958.

Molnar, P., and Oliver, J., 1969, Lateral variation of attenuation in the upper mantle and discontinuities in the lithosphere: J. Geophys. Res., v. 74, p. 2648–2682.

Oliver, J., and Isacks, B., 1967, Deep earthquake zones, anomalous structures in the upper mantle, and the lithosphere: J. Geophys. Res., v. 72, p. 4259–4275.

Utsu, T., 1967, Anomalies in seismic wave velocity and attenuation associated with a deep earthquake zone 1: Hokkaido Univ., Fac. Sci., J., ser. 7, Geophys., v. 3, p. 1–25.

Wadati, K., Hirono, T., and Yumura, T., 1969, On the attenuation of S-waves and the structure of the upper mantle in the region of Japanese islands: Pap. Meteorol. Geophys., v. 20, p. 49–78.

27

Orogenesis and Deep Crustal Structure: Additional Evidence from Seismology

HUGO BENIOFF
1954

From *Bulletin of the Geological Society of America*, v. 65, p. 385–400, 1954. Reprinted with permission of the Geological Society of America.

The physical science of seismology is based almost entirely on observations made with seismographs and clocks. The principal observed data refer to the origin times of earthquakes, the depths and geographic distribution of their foci, and the amplitudes, frequencies, and propagation characteristics of seismic waves. In the past, seismic contributions to the problem of orogenesis included chiefly information as to locations of regions of tectonic activity and evidence for crustal structure exhibited by the speeds and transmission discontinuities of seismic waves. Publication of the monumental catalog of earthquakes by Gutenberg and Richter (1950) has made it possible to augment these early contributions with knowledge derived from magnitudes, time sequences, and accurate hyperfocal locations. Thus an investigation (Benioff, 1949) of the elastic strain-rebound characteristics and related spatial distributions of foci of the seismic sequences of South America and the Tonga-Kermadec region indicated that the orogenic features associated with these two structures represent surface expressions of great faults extending 650 km in depth and up to 4500 km in length. These preliminary findings were thus at variance with those older concepts in which orogenesis is considered a result of sinking and subsequent rising of weakened portions of a thin (35 km ±) crust floating on a viscous or plastic substratum. In the present discussion, studies of an additional number of orogenically active regions are presented. These include the Sunda arc, the Kurile-Kamchatka segment, Mexico and Central America, the New Hebrides, the Philippines, the Bonin-Honshu segment, the Aleutian arc, as well as revisions of the South American and Tonga-Kermadec presentations. Magnitudes, origin times, depths, and geographic locations of the earthquakes and the locations of active volcanoes are taken from Gutenberg and Richter (1950).

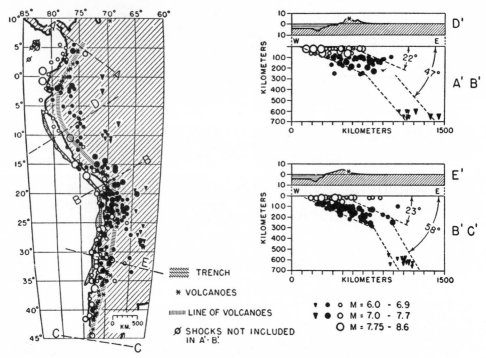

Figure 27-1
Map and composite profiles, South American earthquake sequences

REGIONAL SEISMIC SEQUENCES— OBSERVATIONS

South America

A revised map with composite profiles of the South American earthquake sequences is shown in Figure 27–1. Epicenters and foci of shallow, intermediate, and deep earthquakes are represented by circles, circular dots, and triangular dots respectively. Active volcanoes are represented by stars. The linear distribution of volcanoes exhibited here is characteristic of all the sequences represented in this study except the Philippines. The oceanic trench lying approximately parallel to the line of volcanoes is also characteristic. In constructing the composite profiles of these sequences and of the following ones, the plotted horizontal positions of foci represent their perpendicular distances from the volcano lines taken as reference co-ordinates. The region has been divided as shown into northern and southern sections AB and

BC, for geometric simplicity of the projections A′B′ and B′C′. The complex South American orogenic fault is a good example of the type which occurs along the margins of continents. Moreover, as the author has pointed out (Benioff, 1949), the strain-rebound characteristics and the distribution of foci indicate that this fault structure is made up of three components having three distinctly different types of tectonic movements. The first component involves the shallow layer of the crust above what may possibly be the Mohorovičić discontinuity. Near the fault this layer is assumed to extend to a depth of approximately 60 km. The second component defines an intermediate layer extending from the bottom of the shallow layer to a depth of 250–300 km. The third component extends from the lower surface of the intermediate layer to a depth of approximately 650 km. In the figure, D′ and E′ are vertically exaggerated profiles taken along the lines D and E of the map to show the spatial relations of the oceanic trench, the mountain range, and the line of volcanoes (indicated by the star).

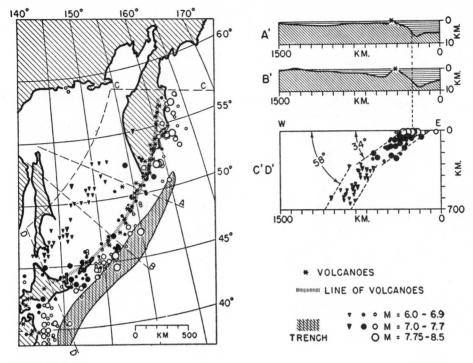

Figure 27-2
Map and composite profile, Kurile-Kamchatka earthquake sequences

In the composite sections A′B′ and B′C′, the vertical and horizontal scales are equal. In the northern (Peru-Ecuador) segment, A′B′, the intermediate component extends to a depth of approximately 250 km with a dip of 22°. In the southern segment (Chile), B′C′, the dip is 23°. Unlike other sequences of this type, South America has no shocks with depths between 300 and 550 km so that the dip of the lower component is not accurately defined. Assuming that there is no horizontal displacement between the intermediate component and the deep component, the fault-zone lines can be extended upward to intersect the intermediate-zone lines as shown, in which case the angles of dip of the deep components of the two segments are 47° and 58° respectively.

Kurile-Kamchatka Segment

Figure 27–2 shows the orogenic elements of the Kurile-Kamchatka segment. The composite profile C′D′ represents the portion of

the map bounded by CC and DD. A′ and B′ are vertically exaggerated profiles taken along the lines A and B. The Kurile-Kamchatka segment is thus a marginal orogenic fault complex with three components similar to the South American structure. The intermediate component extends to a depth of 300 km and dips under the continent at an angle of 34°. The deep component, with a dip of 58°, appears to have undergone a horizontal displacement of approximately 100 km westward (toward the continent) relative to the intermediate component as indicated by failure of the parallel dashed fault-zone lines to intersect at the 300-km level.

Bonin-Honshu Segment

The region of the Bonin-Honshu segment is shown in Figure 27–3. The composite profile B′C′ represents the portion of the map bounded by BBCC. A′ is a vertically exaggerated profile taken along the line A of the map. This orogenic complex is actually a

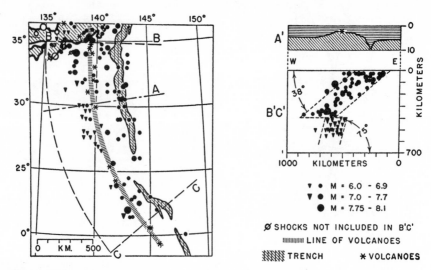

Figure 27-3
Map and composite profile, Bonin-Honshu earthquake sequences

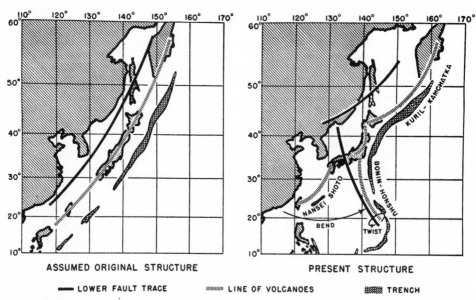

Figure 27-4
Maps showing present and assumed original configurations of Bonin-Japan-Kamchatka arc

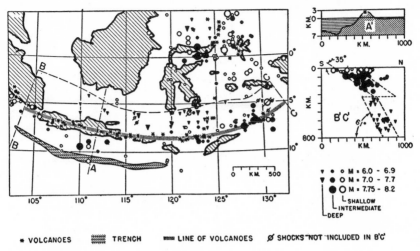

Figure 27-5
Map and composite profile, Sunda arc earthquake sequences

part of the Kurile-Kamchatka-Japan structure, but has been separated here to avoid interpretation difficulties arising from distortion of the shallow layer in the vicinity of the Japanese islands. The high shallow layer activity and absence of foci deeper than 150 km along the Nansei Shoto branch to the west, the paucity of shallow earthquake activity in this (Bonin-Honshu) segment, the large horizontal displacements between its two lower components, and its reverse curvature all strongly suggest that the whole fault complex from northern Kamchatka to the southern portion of the Bonin-Honshu segment was once continuous and convex toward the Pacific over its whole extent, as indicated on the left in Figure 27–4. The present form appears to be a result of an approximately 90-degree counterclockwise bending about a vertical axis in Manchuria, of the two lower components of the original arc relative to the original shallow component now forming the Nansei Shoto branch, as shown on the map on the right in Figure 27–4. Consequently, in the present Bonin-Honshu segment the shallow component is missing, and the intermediate component extends from the surface to a depth of 400 km. It is thus the deepest continental structure observed in this study. The fault dips westward under the continent at 38°. The deep compo-

nent is displaced horizontally eastward, away from the continent, some 200 km relative to the intermediate component. This displacement, and the 75-degree dip, the largest of any studied to date, suggest that the forces which originally reversed the curvature of the Bonin-Honshu segment also twisted the deep component fault surface about a horizontal axis, as indicated on the right in Figure 27–4. It appears therefore that the common component of the couples, responsible for the bend and the twist, acted horizontally eastward along the lower boundary of the lower layer.

Sunda Arc

Figure 27–5 shows the region of the Sunda arc. The portion of the map represented in the composite profile B′C′ is that bounded by the dashed lines BBCC. This structure is also a triple marginal fault. The intermediate component dips northward under the continent at an angle of 35°. It has a maximum depth of approximately 300 km. The dip of the deep component is 61°. It has been displaced horizontally approximately 200 km southward (away from the continent) relative to the intermediate component, as indicated by the dashed lines in B′C′.

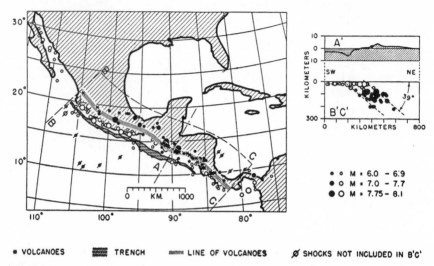

Figure 27-6
Map and composite profile, earthquake sequences of Mexico and Central America

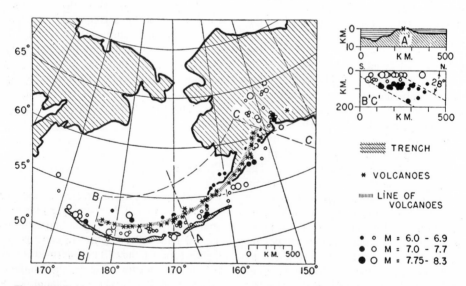

Figure 27-7
Map and composite profile, Aleutian earthquake sequences

Mexico and Central America

Figure 27–6 shows the dual marginal Acapulco-Guatemala fault complex. The intermediate component extends (under the continent) to a depth of only 220 km with a dip of 39°. The shocks northwest of the boundary BB are shallow and presumably represent horizontal movements on the shallow clockwise transcurrent San Andreas fault system. There is no evidence for the existence of a deep component in this region.

Aleutian Arc

Figure 27–7 refers to the Aleutian arc, also a marginal dual fault. The composite profile B′C′ shows only those features bounded by the dashed lines BBCC of the map. The intermediate component dips northward 28° and extends to a depth of 175 km, thus forming the shallowest continental structure treated in this study, with the possible exception of the Philippines.

New Hebrides Region

Figure 27–8 shows the region of the New Hebrides fault. This orogenic structure is also a marginal fault principally of the dual type, although the occurrence of two shocks at the 350-km level suggests the possibility of a deeper component of small activity. The intermediate component extends to a depth of 300 km and dips northeast 42°.

Tonga-Kermadec Region

A revised map and composite profile of the Tonga-Kermadec fault complex is shown in Figure 27–9. In the earlier paper (Benioff, 1949) the author assumed that the structure consisted of a single continuous fault. However, the existence of two lines of volcanoes and the asymmetrical distribution of epicenters are better explained with two faults, even though the strain-rebound characteristics indicate that the two are actuated by a single

stress system. The composite profile A′B′ represents the map area AABB associated with the Tonga segment. The vertically exaggerated profile C′ is taken along the line C of the map. The composite profile D′E′ represents the area DDEE of the map, associated with the Kermadec segment. The latter extends well into the region of North Island of New Zealand. The Tonga-Kermadec structure is the principal representative of the oceanic form of orogenic fault complex. The elastic strain-rebound characteristics of these faults (Benioff, 1949) and their composite profiles show that they are of dual type with shallow components extending to a depth of approximately 70 km. The deep component of the Tonga fault dips 59° NW. and extends from 70 km to 650 km in depth. The deep component of the Kermadec fault dips approximately 64°NW. and extends to a depth of 550 km.

Philippine Islands

The region of the Philippine Islands (Fig. 27–10) is the most disturbed area in this series. The volcanoes exhibit no well-defined line. The projection line B′C′ was constructed with the reference line OX drawn approximately parallel to the strike of the trench. The structure is considered here an oceanic fault complex, although the concentration of foci in the depth range 70–150 km may represent an intermediate component of corresponding thickness and having a dip westward of approximately 36°. If this is the correct interpretation, the fault is a marginal one of triple type. The deep component extends to a depth of 700 km with a dip of roughly 60°.

SUMMARY OF FAULT MEASUREMENTS

Depth ranges and dips of the orogenic fault components for each of the regions studied are assembled in Tables 27–1 and 27–2. In the marginal faults the average values of the dips for the intermediate and lowest components are 33° and 60° respectively. In the oceanic faults the average dip of the lower

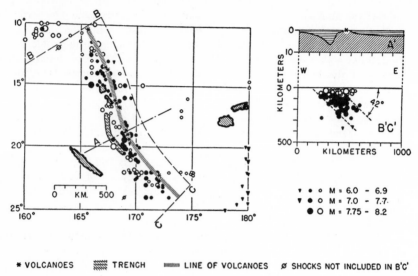

Figure 27-8
Map and composite profile, New Hebrides earthquake sequences

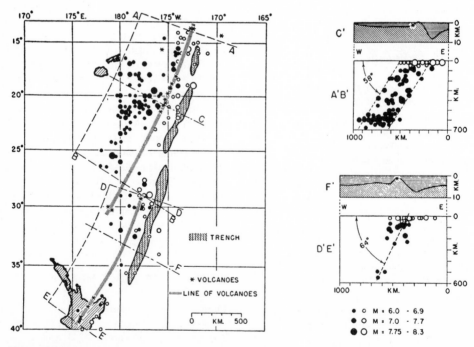

Figure 27-9
Map and composite profiles, Tonga-Kermadec earthquake sequences

Table 27-1
Depth ranges and average dip of orogenic marginal faults.

Fault	Intermediate comp.		Deep component	
	Depth range, km.	Dip	Depth range, km.	Dip
Peru–Ecuador segment	70–250	22°	600–650	47°
Chile segment	70–290	23°	550–650	58°
Bonin–Honshu segment	0–400	38°	400–550	75°
Sunda	70–300	35°	300–700	61°
Kuril–Kamchatka segment	70–300	34°	300–650	58°
Acapulco–Guatemala	70–220	39°		
Aleutian	70–175	28°		
New Hebrides	70–300	42°		
Average dips		33°		60°

Table 27-2
Depth ranges and average dip of orogenic oceanic faults.

Fault	Depth range, km.	Dip
Mindanao	70–700	60°
Tonga	70–650	58°
Kermadec	70–550	64°
Average dip		61°

component is 61°. The observational precision (approximately ±50 km in any direction) is not high enough to permit any conclusions regarding the dip of the shallow components nor to differentiate between dip slip and strike slip faulting in their movements.

IMPLICATIONS CONCERNING CRUSTAL STRUCTURE AND OROGENY

Principal Orogenic Mechanism, Reverse Fault

This extended study of tectonically active regions supplements the conclusions regarding the mechanism of orogenesis found in the earlier investigation of the Tonga and South American sequences. Thus it appears that in those great linear and curvilinear tectonic structures, now seismically active, the principal orogenic mechanism involves a complex reverse fault extending to a depth which varies from 175 km to 720 km depending upon the region.

Shallow and Intermediate Crustal Structure Indicated by Faults

The principal features of the faults and the crustal structure which they define are shown in the generalized section of the crust in Figure 27-11. The continental marginal structure is indicated on the right, and the oceanic type is shown on the left. The shallow continental layer A, above the Mohorovičić discontinuity M, is approximately 35 km thick as determined by many investigators from wave-propagation studies. It is assumed here to be the same as the layer defined by the shallow earthquake sequences. The strain-rebound characteristics studied so far indicate that, in the vicinity of the faults, shocks belonging to the shallow sequences occur down to a depth of approximately 60 or 70 km in accordance with the classification of shallow, Gutenberg and Richter (1950). Accordingly, in the figure the shallow layer is shown thickened to this extent in the vicinity of the fault. Presumably this thickening is a result of drag and plastic flow of the strained fault lip. Thus, in part at least, the *roots* of mountains may be expressions of this fault-generated distortion. In the triple marginal faults the discontinuity in angle of dip at 300 km ±, and the marked dissimilarity between the strain-rebound characteristics of the shocks occurring in the fault components above and below this level, are interpreted as evidence for an additional tectonic discontinuity between B and C, as shown in the figure. In the dual

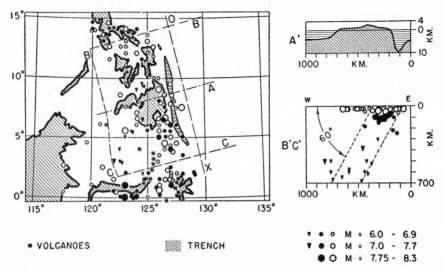

Figure 27-10
Map and composite profile, Philippine earthquake sequences

structures the lower fault component C is either absent or inactive. The intermediate component with the 32-degree dip occurs in marginal structures, but not in oceanic structures; consequently, the layer B defined by this component is here assumed to be a part of the continental structure. If this assumption is correct, the continents extend downward to an average depth of 280 km, with 175 km and 400 km extreme values.

Deep Crustal Structure Indicated by Faults

In the triple marginal structures the fault dip increases suddenly at the lower boundary of the intermediate component to an average value of approximately 60° and continues down with this value to the 700± km level, below which earthquakes do not occur. In the oceanic structures the fault extends all the way from this level to the 60-km discontinuity, and presumably to the surface, with a constant average dip of approximately 61°. Since the dips of the oceanic faults and the deep components of the continental faults are essentially equal, it is assumed that both types are fractures in a single continuous medium (Fig. 27–11, C) subject to a single

stress system. The measured seismic wave speeds near the upper boundaries of the continental layer B and the oceanic layer C are very nearly equal (8.2± km/sec for compressional waves). This suggests that they may be identical in composition. Bullard (1952), on the other hand, cites evidence to the contrary. However, whether or not they are of identical composition, the layers are separated by a tectonic discontinuity and are subject to different stress systems.

RIGIDITY IN DEEP LAYER

Since earthquakes occur down to depths of 550–720 km in both marginal and oceanic faults, it must be assumed that at these depths the crustal rocks can maintain elastic shearing strains of sufficient duration to accumulate the necessary strain energy to generate earthquakes. Morover, the strain-rebound characteristics of the deep sequence of South America provide evidence for accumulation of elastic strain without creep or flow over a period of 5½ years at a depth of 550–650 km (Benioff, 1949). The Tonga deep sequence exhibited elastic (recoverable) creep without flow over a period of 15 years down to a depth of 650 km (Benioff, 1949). Thus, for

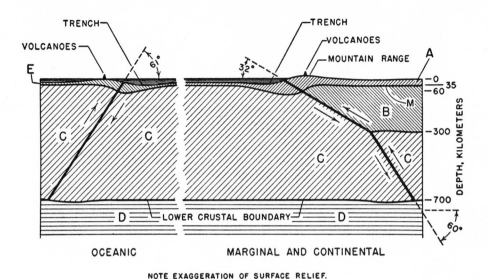

Figure 27-11
Generalized oceanic and continental deep crustal sections with orogenic fault types

time intervals at least of the order of a decade, the earth's crust in the vicinity of the orogenic faults is a rigid solid down to the 700± km level. Accordingly, this depth limit is here taken as the lower boundary of the solid crust. For geologically long time intervals the crust may, of course, behave as a viscous or plastic substance at these (and shallower) depths. The question as to whether or not the crustal structure, defined in this manner by orogenic faults, extends over the whole earth to regions not now seismically active, is one which cannot be answered by studies of this kind alone. In this connection, Birch's (1951) discovery, derived from calculations of the rate of change of the bulk modulus with pressure, of a substantial departure from physical or chemical homogeneity in the depth range 200–800 km may be additional evidence for the discontinuities and structures derived from these studies.

OROGENIC FAULT-GENERATING STRESSES

The intermediate components are reverse faults which dip under the continents at an average angle of 33°. This value falls within

the range calculated and observed by Hubbert (1951) for faults produced by stress patterns in which the greatest principal stress is a horizontal compression. However, there is reason to believe that the Coulomb-Mohr theory of fracture on which Hubbert bases his reasoning may not be applicable to triaxial stress-conditions of extreme magnitude which must exist at the depths of the intermediate and deep fault components. David Griggs has suggested (Personal communication) that under these extreme conditions the relative value of the intermediate principal stress affects the angle which the plane of fracture makes with the direction of the greatest or least principal stress. Thus, in Figure 27–12 the block represents portions of a fault structure subjected to the three principal stresses σ_1 (vertical), σ_2 (horizontal parallel to the fracture) and σ_3 (horizontal perpendicular to σ_2). In order that the direction of slip be as indicated by the arrows, the relations of the principal stresses must be such that

$$\sigma_3 > \sigma_2 > \sigma_1$$

If σ_2 is less than the average of σ_1 and σ_3, the angle of dip is less than 45°. If σ_2 is greater than the average of the other two, the angle of dip is greater than 45°. On this basis it would

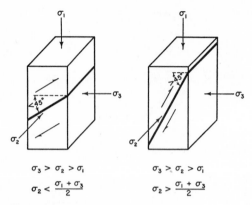

$$\sigma_3 > \sigma_2 > \sigma_1$$

$$\sigma_2 < \frac{\sigma_1 + \sigma_3}{2}$$

$$\sigma_3 > \sigma_2 > \sigma_1$$

$$\sigma_2 > \frac{\sigma_1 + \sigma_3}{2}$$

Figure 27-12
Stress relations indicated by fault dips

appear that both the 33-degree and 61-degree faults are produced by stress patterns in which the greatest principal stress is a horizontal compression oriented effectively at right angles to the orogenic axis. The continental margins are thus moving relatively toward the adjacent oceanic domains. In the 33-degree intermediate continental faults the horizontal intermediate stress σ_2 is nearer in value to the vertical stress σ_1 than to the horizontal stress σ_3. On the other hand, in the 60-degree oceanic faults and the deep components of the continental faults, σ_2 is closer in value to σ_3. In other words, the horizontal stress pattern is more nearly symmetrical in the deep layer than in the intermediate layer.

DIRECTION OF SLIP

The direction of slip assumed for all the faults in which the underside moves down is dictated by the geometry of the trenches and adjacent uplifts. It refers to the fault-generating conditions only. There is no reason to believe that these original conditions necessarily remain unaltered in time, and consequently, once the fault structure has been formed, subsequent stress patterns may change. Later movements may thus involve horizontal slip or even reversals of sip direction. Such alterations may be temporary or permanent. In the latter case, the trench eventually disappears. Perhaps this has been the history of the Pacific coast of North America

between Alaska and Lower California and of the Himalayan-Indian arc. A reverse slip may thus be the evidence for decline in the life of an orogenic fault. Instances of reversed slip have been observed by Ritsema (1952, thesis, Univ. Utrecht) for two deep shocks in the Sunda and Philippine regions respectively, and by Koning (1942) for one shock of the Sunda arc. However, the method employed by these authors for measuring the strike and direction of slip is of questionable reliability. It depends upon correct determination from seismograms of the displacement direction of the initial ground motion for each of a rather large number of seismograph stations. Particularly in deep earthquakes, the initial wave motion is nearly always made up of relatively high-frequency components for which most seismographs have inadequate magnification. Moreover, for those critically situated stations to which the initial waves travel in or near the plane of faulting, the direction of initial displacement is either indeterminate or else too small in relation to the ground unrest to be reliably recorded by an instrument situated outside the epicentral region. Evidence that the results of this method should be accepted with caution is provided by calculations reported by J. H. Hodgson (1952) using a modified form which, nevertheless, depends solely on correct observations of the initial displacements. In an application to the Ancash, Peru, earthquake of November 10, 1946, the method indicated that the faulting was of the transcurrent type, whereas observations on the ground reported by Silgado (1951) showed clearly that the faulting was of the dip-slip type in accordance with the known tectonic character of the region.

In a recent paper, H. Honda and A. Masatuka (1952) have studied the direction of motion of some 145 intermediate and deep earthquakes of Japan using observations of a large number of Japanese stations. Since these stations are all in or near the epicentral region of the shocks, the reliability of first-motion determinations is relatively high. Their results indicate clearly that for these shocks the dip of the faults and the direction of slip are in substantial agreement with the assumptions and findings of the present investigation.

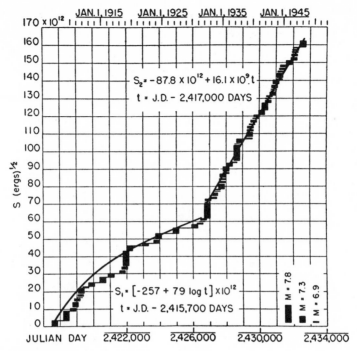

Figure 27-13
Elastic strain-rebound characteristic, Tonga-Kermadec deep earthquake
sequence, for focal depths 70–680 km

RELATIONSHIP BETWEEN SHALLOW OCEANIC AND INTERMEDIATE MARGINAL LAYERS

The continental surface layer of 35 km thickness in which the shallow marginal components occur is generally reduced to approximately 5 km under the oceans insofar as the evidence from wave-propagation studies indicates. On the other hand, the strain-rebound characteristics of the Tonga-Kermadec sequences show that shallow earthquakes occur to a depth of 60 or 70 km in the oceanic structures. Hence, in the vicinity of the oceanic faults we must assume some sort of shallow discontinuity having approximately the same depth as the shallow discontinuity of the continents. Such a discontinuity is therefore shown on the left in Figure 27–11.

As indicated in Figure 27–11, it is assumed that in general the intermediate layer B terminates at the fault. There is some evidence from the strain-rebound characteristics, however, for the conclusion that in the New Hebrides-Tonga region the intermediate continental layer extends across the intermediate fault to become the shallow layer of the Tonga-Kermadec oceanic fault. The elastic strain-rebound characteristic of the Tonga-Kermadec deep components is shown in Figure 27–13. That for the shallow component of the New Hebrides fault is shown in Figure 27–14. It is obvious that these two curves are dissimilar and that they must be derived from movements of two distinct mechanical structures. On the other hand, the characteristics of the Tonga-Kermadec shallow sequence and the New Hebrides intermediate sequence (Fig. 27–15) are strikingly similar, except for a time delay of approximately 2000 days of the former relative to the latter, and both are unlike either the New Hebrides shallow characteristic or the Tonga deep characteristic. It appears, therefore, that the Tonga shallow-fault component and the New Hebrides intermediate component are fractures in a single mechanical structure subject to a single stress system. The situation can best be understood by reference

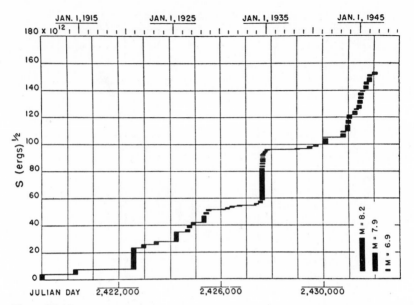

Figure 27-14
Elastic strain-rebound characteristic, New Hebrides shallow earthquake sequence,
h < 70 km

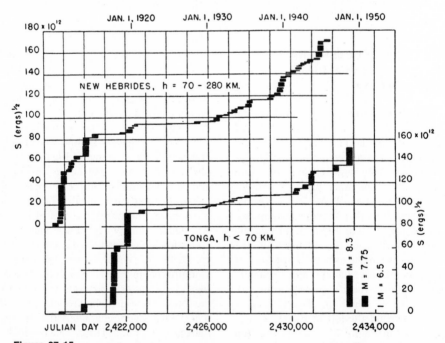

Figure 27-15
Elastic strain-rebound characteristics, Tonga-Kermadec shallow earthquake sequence and
New Hebrides intermediate sequence, Tonga-Kermadec (h < 70 km); New Hebrides
(h = 70–280 km)

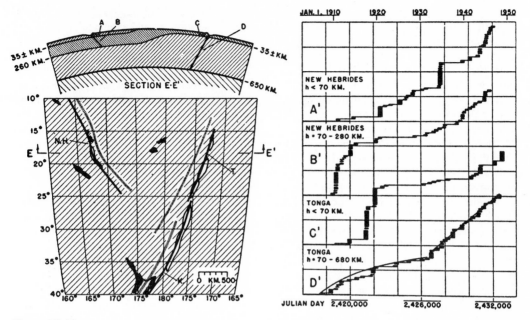

Figure 27-16
Map and elastic strain-rebound characteristics. New Hebrides-Tonga earthquake sequences and local crustal structure

to Figure 27–16 in which the left portion is a map and section of the New Hebrides-Tonga region. On the right, the four rebound characteristics are roughly sketched with coinciding time coordinates. Counting from the bottom, the first curve – the Tonga deep characteristic – refers to the component D in the section E-E'. The second curve, which is of the Tonga shallow component, refers to the component C of the section. Since these two curves are so very dissimilar, movements in the two involved components are unrelated, and consequently a discontinuity must be drawn between C and D. The third curve is the New Hebrides intermediate characteristic. It refers to the component B in the figures. Since the curves for B and C are so nearly alike we must assume that these two components are members of a single stress system, and consequently the intermediate layer at B is shown continuous with the shallow layer at C. Since the rebound characteristic of the shallow New Hebrides component A (the fourth curve on the right) is unlike that of the deeper component below it, a discontinuity must be drawn between the two. The structure deduced here for the Tonga-New Hebrides

region may represent a special configuration of this part of the earth only and may not be valid for other regions in which the structure outlined in Figure 27–11 must be considered representative.

OROGENESIS AND VOLCANOES

One of the most remarkable features of the orogenic faults discussed in this paper is the linear array of volcanoes lying parallel to the fault strike on the uplifted block. Indeed the line of volcanoes has served as the principal reference co-ordinate for drafting the composite profiles. Volcanoes are clearly one of the manifestations of orogeny. In an attempt to explain this close relationship the writer is offering a hypothesis of the origin of the volcanoes which at least will direct thinking toward this interesting problem. By referring to the composite profiles shown in Figures 27-1-27-9 it will be noted that in general the volcano line coincides with the highest elevation of the overhanging fault block and thus appears to be situated along the line of maximum bending of the block. In recent studies the writer (Benioff, 1951) found evidence that the aftershock sequences which follow significant earthquakes are generated by elastic after working or creep recovery of the fault rock. During the interval between principal earthquakes, strains accumulate in the rocks. A portion of this strain is purely elastic, and a portion is time-dependent. When the fault slips to produce the principal earthquake, the purely elastic strain only is involved. The time-dependent or creep strain can be released only relatively slowly in accordance with the creep-recovery characteristics of the rocks. It is this time-dependent elastic recovery which produces the aftershocks. However, only a small portion of the energy stored in the creep-elastic element can be converted into seismic waves in aftershocks. The main portion is liberated as heat. A rough calculation of the energy liberated as heat in a number of aftershock sequences shows that it averages approximately half the amount of the energy liberated as seismic waves in the principal shock. In the orogenic faults on which earthquakes occur repeatedly, the fault rock undergoes a cycle of alternate strain and relief for each earthquake. This repeated to-and-fro bending generates heat in much the same way that quick bending of a wire warms it. The thermal time constant of the great rock masses below the surface is so long that heat generated by these cycles of strain escapes very slowly and consequently accumulates over long intervals of time. If the rock being heated has components with different melting points, there comes a time when one or

more will melt. The region of most intense heat generation should coincide with the regions of most intense bending—the highest elevations of the overhanging block and the lowest depths of the neighboring trenches. The low incidence of volcanoes in the trenches may be ascribable to a higher melting point induced by the increased pressure and to the fact that the trench (in the form of a syncline) is not so effective a structure for concentration of the molten rock as the complementary anticline of the uplifted block on the upper side of the fault.

A rough idea of the magnitude of energy released, say per year, by the aftershock sequences in a region on one side of the fault can be obtained by taking one fourth of the energy released in the same time by seismic waves in the principal earthquakes. Thus, in the case of South America, the shallow and intermediate earthquake sequences each liberate approximately 4×10^{23} ergs per year. Thus roughly 10^{24} ergs per year is being released in the fault blocks. The writer has no knowledge of the amount of energy per year required to maintain the South American system of volcanoes, and consequently it is not possible to say whether or not the energy requirements are met on this hypothesis. Moreover there must be a large time lag between the liberation of heat in the depths and its appearance in the form of volcano output. Thus the present rate of volcanic energy release should be equated to a phase of seismic-heat generation which occurred long ago, rather than to the present rate.

28

Tectonic Deformation Associated with the 1964 Alaska Earthquake

GEORGE PLAFKER

1965

From *Science*, v. 148, p. 1675–1687, 1965. Reprinted with permission of the author and the American Association for the Advancement of Science. Copyright 1965 by the AAAS.

The epicenter of the earthquake of 27 March 1964 (28 March 1964 Universal Time) and its zone of aftershocks lie within a well-defined belt of shallow and intermediate-depth earthquakes which follows the Aleutian Trench and Volcanic Arc from Kamchatka to south-central Alaska, where the arc enters the North American continent (Fig. 28–1). The active and dormant volcanoes of the Wrangell Mountains, shown on Fig. 28–1, may be an eastward extension of the Volcanic Arc. The Aleutian Arc is one of the most seismically active areas of the festoons of volcanic arcs and associated deep ocean trenches which bound the Pacific Ocean from Alaska to New Zealand and from Antarctica to Central America. The seismic and volcanic activity of the arcs of the Pacific, as well as arc structures elsewhere, are manifestations of one type of active diastrophism, or mountain building. The geologic record indicates that comparable arcuate structures in the remote past played a significant role in the formation of mountains and the growth of continents. Consequently, the structure of modern arcs and the nature of the deformations that occur along them are of interest to geologists and geophysicists concerned with the origin of mountains and continents. Most of our knowledge of the deformations that produce large earthquakes along active arcs comes from analysis of the elastic waves they generate. Direct observation of the surface displacements that sometimes accompany these deformations is commonly limited, because all or much of the displacement is submarine.

Perhaps the most notable aspect of the Alaskan earthquake was the great extent and amount of the changes in land level that accompanied it. From the epicenter in northern Prince William Sound, the zone of surface deformation extends for 800 kilometers roughly parallel to the trends of the Aleutian Volcanic Arc and Trench and the coast of

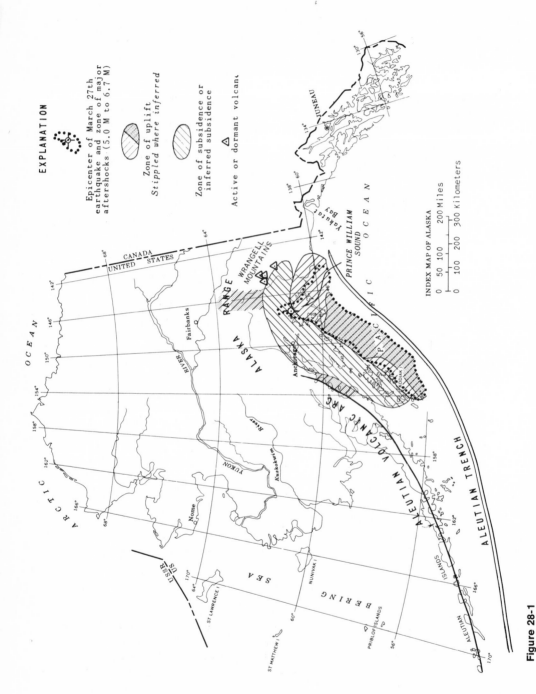

Figure 28-1
Index map, showing the relationship of the Aleutian Arc to the earthquake of 27 March 1964, its belt of aftershocks, and zones of change in land level

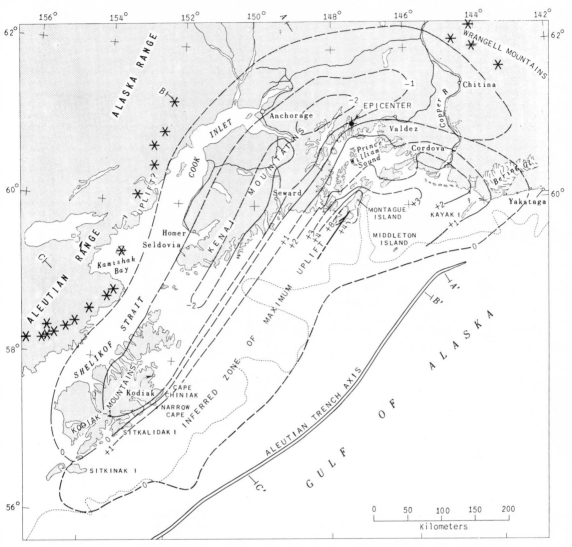

Figure 28-2
Tectonic uplift and subsidence in south-central Alaska. The land-level change, in meters, is shown by the contours, which are dashed where approximate or inferred. The outer edge of the continental shelf, at −2000 meters, is indicated by a dotted line, and active or dormant volcanoes are shown by stars. A-A′, B-B′, and C-C′ are lines of sections shown in Figure 28-6

the Gulf of Alaska. Where the northeastern end of the arc intersects the continent at an oblique angle, the deformation can be observed in an almost complete section across the arc from Middleton Island, near the seaward edge of the continental shelf, to the west shore of Cook Inlet (Fig. 28–2).

This article makes available a summary of the basic data acquired during the 1964 field season on the tectonic deformations that accompanied the earthquake. It substantially enlarges upon the information on land-level changes published in a preliminary U.S. Geological Survey report based on a 2-week reconnaissance made immediately after the earthquake (Grantz et al., 1964). Some tentative interpretations of the mechanism of the earthquake are advanced here on the basis of

Figure 28-3
Measuring the height of the pre-earthquake upper limit of barnacle growth, or barnacle line, above the present water level at Glacier Island in Prince William Sound. The sharply defined upper limit of barnacle growth is typical of much of the Prince William Sound area

the data now available. They may require modification or revision as additional results of continuing and planned investigations into the manifold aspects of this major tectonic event are made known.

METHODS

The vertical tectonic movements in coastal areas were determined mainly by making more than 800 measurements of displacement of intertidal sessile marine organisms along the long, intricately embayed coast (Fig. 28-3). These measurements were supplemented at 16 tidal bench marks by coupled pre- and post-earthquake tide-gauge readings made by the U.S. Coast and Geodetic Survey, and by numerous estimates of relative changes in tide levels by local residents. The amount and distribution of the vertical tectonic movements inland from the coast were defined along

the highways connecting the cities of Seward, Anchorage, Valdez, and Fairbanks by the U.S. Coast and Geodetic Survey's prompt releveling of previously surveyed first-order level lines tied to tidal bench marks at Seward, Anchorage, and Valdez.

In measuring the displacement of sessile marine organisms use is made of the zonation of plants and animals between tide marks that has long been recognized in different parts of the world by marine ecologists. The intertidal zone along the predominately steep and rocky coastline of south-central Alaska is inhabited by certain species of sessile organisms, notably barnacles, mussels, and algae, whose vertical growth limits are commonly well defined. In particular, the common acorn barnacle, *Balanus balanoides* (Linnaeus), is widely distributed on the rocky coast and forms a prominent band with a sharply defined upper limit which is readily recognizable on most shores. This upper limit corresponds roughly to the top of the Balanoid or Midlittoral Zone of Stephenson and Stephenson; to Zone 2, the High Tide Region, or the Upper Horizon of Ricketts and Calvin; and to the Upper Intertidal Zone of Rigg and Miller (Stephenson and Stephenson, 1947; Doty, 1957; Ricketts and Calvin, 1962; Rigg and Miller 1949). No data on the normal height of barnacle growth relative to tide levels could be found for south-central Alaska. However, experience at places on the Pacific Coast of North America in areas where both the prevailing type of tide and the tidal range are similar to those in parts of Prince William Sound and on the coast of the Gulf of Alaska indicates that this height is close to annual mean high water (Ricketts and Calvin, 1962; Rigg and Miller, 1949; Rice, 1930).

At localities where barnacles do not occur, both the common olive-green rockweed (*Fucus furcatus* Agardh), whose upper growth limit is near that of the barnacles, and the dark gray encrusting alga [*Ralfsia verricosus* (Areschoug) J. Agardh], which commonly occupies the splash zone immediately above the barnacles, served as useful datums for measuring landlevel changes. The growth limit of the algae, however, appears to be strongly influenced by exposure to sunlight as well as by the tides; thus their vertical range is generally more variable than that of the barnacles.

The field procedure was to measure the height of the upper limit of barnacle growth (the "barnacle line") above or below water level at any stage of tide (Fig. 28–3). On steep rocky slopes that are sheltered from heavy surf this line is sharply defined and can be readily determined to within 15 centimeters or less; on sloping shores or shores exposed to heavy surf it tends to be less regular. In the usual case, where the barnacle line was above water, it was measured with a hand level or surveyor's level and stadia rod. Where the barnacle line was visible under water, its depth below the surface was measured directly with the stadia rod. The stage of tide at the time of measurement was then determined from the U.S. Coast and Geodetic Survey table of predicted tides for the closest reference station, and the position of the barnacle line relative to lower low water was calculated. For those stations close to the

16 U.S. Coast and Geodetic Survey tide gauges that were installed in the area following the earthquake, we later made corrections to the actual, rather than the predicted, tides. During the period of field work, tides rarely deviated by as much as ½ meter from predictions. Abrupt changes in the height of the barnacle line in local areas of similar tide range provided a fairly precise method of detecting vertical fault displacements of more than 1 to 2 meters in any given locality.

Absolute uplift or subsidence at any given locality was taken as the vertical difference between the measured elevation of the pre-earthquake barnacle line and the "normal" upper growth limit for the barnacles, as determined empirically at 12 tidal bench marks where the amount of vertical displacement was known from tide-guage readings (U.S. Coast Geodetic Survey Office, 1964). The upper limit of barnacle growth depends mainly on the ability of yearling barnacles to survive prolonged exposure to air and on the tidal characteristics at any given locality (Kaye, 1964). To a lesser extent it depends upon a number of other factors which may locally cause the barnacle line to deviate as much as ⅓ meter from its "normal" height. Wave action during the lowest annual neap tides and protection from desiccation in shady locations tend to elevate the upper growth limit; exposure to fresh water, near large streams or tidewater glaciers, tends to depress it. Annual variations in sea level may cause further slight upward or downward shifts of the upper growth limit of the barnacles. It was found that the pre-earthquake barnacle line is at mean high water, or no more than 15 centimeters below it, for the lower tidal ranges of 1.9 to 3.3 meters which prevail along most of the coast of the Gulf of Alaska and in Prince William Sound. For the higher mean ranges of up to 5.1 meters in Cook Inlet and Shelikof Strait, the barnacle line lowers progressively to as much as 45 centimeters below mean high water.

The validity of using the empirically determined upper limit of barnacle growth was confirmed by the agreement found between the observed heights to which new post-earthquake barnacles were growing after the spring neap tides and the predicted heights for those

Figure 28-4
Road along the shore of Middle Bay on Kodiak Island. The road is inundated daily, due to tectonic subsidence of about 1 2/3 meters and to an unknown, but probably, substantial, amount of local settling of unconsolidated deposits. The photograph was taken 20 July 1964 at 1.2 meter tide

localities. Land-level changes determined by the barnacle-line method are generally within $1/3$ meter of changes estimated by local residents or found by means of other techniques; even under the least favorable combination of circumstances the error of the barnacle-line method is probably less than $2/3$ meter.

Throughout the area of land-level change a new post-earthquake line of yearling barnacles was well established above or below the normal growth limit by the end of the field season. The height of this new barnacle line below the pre-earthquake limit of barnacle growth furnished a direct measure of the maximum amount of uplift; its height above the older barnacles indicated the minimum amount of subsidence.

This same technique of measuring elevation differences between the upper limits of dead and living barnacles was used by Tarr and Martin (1912), 8 years after the Yakutat Bay earthquake of 1899, as a criterion of the uplift that occurred during that quake.

There are many other indications of subsidence and uplift along the affected shorelines. In subsided areas vegetation has been killed by immersion in salt water, and beach berms have been shifted landward and are built up to the new, relatively higher sea levels. Stream deltas are also building up to the higher sea levels, and some former beach-barred lakes have become tidal lagoons. Inundation of coastal roads and shoreline installations and property on the Kenai Peninsula and the Kodiak Islands is a direct result of the regional subsidence (Fig. 28-4).

The amount of uplift could be determined directly in some areas by measuring the height of the zone of dead algae and barnacles which were raised above the reach of the tides. Qualitative indications of uplift include new reefs and islands, raised sea cliffs and sea caves, and drained lagoons. Figure 28-5 is a photograph, taken 3 days after the earthquake, of the surf-cut platform that was exposed at the southern end of Montague

Figure 28-5
The southwest tip of Montague Island at half tide, 3 days after the earthquake. The surf-cut platform shown here has been exposed by about 8 meters of uplift. Prior to the earthquake the approximate half-tide line was close to the base of the prominent white zone of barnacles below the sea cliff. The entire reef later turned white because of desiccation of calcareous organisms in the former intertidal zone

Island by about 8 meters of uplift. Uplift resulted in mass extermination of sessile intertidal fauna and flora. The effects of changes in land level on intertidal organisms, as well as other biological effects of the earthquake, have been summarized by Hanna (1964). Navigability of waterways and harbors throughout the eastern part of Prince William Sound and the coastal area to the east of Cordova was impaired by the uplift. At many localities, canneries, docks, waterfront homes, and piers are now inaccessible by boat except at extreme high tides.

REGIONAL UPLIFT AND SUBSIDENCE

The areal distribution and approximate amounts of tectonic changes in land level that accompanied the earthquake are shown in Fig. 28–2 and the profile lines of Fig. 28–6. The coastal area south of a line extending along the southeast coast of Kodiak Island through the western part of Prince William Sound to the vicinity of Valdez has been elevated, and the area north and west of this line has been lowered. To the east the zone of deformation appears to die out between the Bering Glacier and Yakataga. The seaward limits are not known, although parts of the continental shelf as far south as Middleton Island and southwest to Sitkinak Island have been elevated. The northwestern limit of deformation extends at least to the west side of Shelikof Strait and Cook Inlet. Its inland limit is known only along the highway between Valdez and Fairbanks, where it extends northward at least to the latitude of the Wrangell Mountains, and possibly into the Alaska Range.

The major area of uplift is about 800 kilometers long and trends northeast from southern Kodiak Island to Prince William Sound, and east-west to the east of Prince

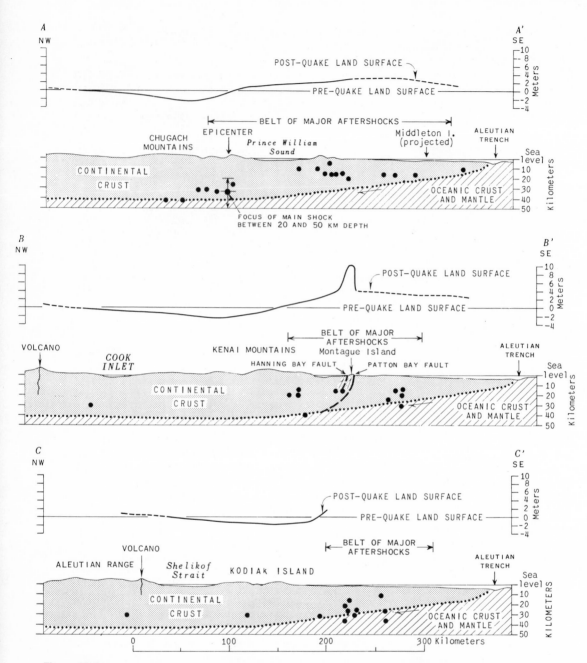

Figure 28-6
Schematic sections across the Aleutian Arc, showing approximate (solid) and inferred (dashed) land-level
changes, aftershock distribution, and geologic interpretations along the section lines. Heavy dots are earthquake hypocenters (U.S. Coast and Geodetic Survey) within 30 kilometers of the section lines. Section
locations are shown in Figure 28-2

William Sound. It includes the southern and eastern parts of Prince William Sound, the coastal area as far east as Bering Glacier, and part of the contiguous continental shelf, as shown by uplift on Kayak and Middleton islands. The average uplift within the zone in Prince William Sound, where the most detailed study was made, is about 2 meters. The maximum measured uplift on land is 10 meters at the southwest end of Montague Island and more than 15 meters offshore from Montague Island (Malloy, 1964). The uplift of Middleton Island demonstrates that the tectonic deformation extended at least 195 kilometers south-southeast of the epicenter almost to the margin of the continental shelf. The exact amount of uplift on the island, however, is uncertain: available estimates range from 1 to 3 meters. Uplift also occurred along the extreme southeastern coasts of Kodiak Island. Sitkalidak Island, and part or all of Sitkinak Island (Fig. 28–2). The maximum measured uplift on Sitkalidak Island, as determined from displacement of barnacles, was $^2/_5$ meter. The estimated uplift on Sitkinak Island is $^1/_3$ to $^2/_3$ meters. The maximum known uplift on Kodiak Island probably occurs at Narrow Cape, 50 kilometers due south of Kodiak; it is estimated to be at least $^2/_3$ meter, and possibly as much as $1^1/_2$ meters.

The areal distribution and initial direction of water motion of the seismic sea waves strongly suggest that the belt of uplift also embraces a large segment of the continental shelf and slope.

Changes in land level were measured in detail between the zones of uplift and subsidence in Prince William Sound and, where possible, along the southeast coast of Kodiak Island, to determine whether the uplifted and lowered areas are separated by a fault. These measurements showed no abrupt changes of level indicative of vertical fault displacement but, rather, a north and northwestward tilting about the axis of zero change in elevation (Fig. 28–6). The slope on the tilted surface in the vicinity of the zero line in Prince William Sound is northward to northwestward at a maximum rate of 11 centimeters per kilometer. Data on the configuration of the warping in the vicinity of the zero line on Kodiak Island although less conclusive, suggest that here, too, warping results from pronounced northwestward tilting and possible local flexure without detectable surface faulting.

There are two areas north and northwest of the zone of subsidence where minor amounts of uplift may have occurred during the earthquake (Fig. 28–1). One is in the Alaska Range, where post-earthquake re-leveling along the Richardson Highway indicates as much as $^1/_3$ meter of uplift relative to the determinations of earlier surveys made in 1944 and 1952.[1] It is not possible to determine how much, if any, of the change occurred during the earthquake of 27 March. The second area of possible uplift is along the northwest shore of Cook Inlet, where residents of Iliamna and Tuxedni bays report anomalously low tide levels following the earthquake which suggest uplift on the order of $^1/_3$ to $^2/_3$ meter.

The main area of known uplift includes at least 60,000 square kilometers. However, the trend of the contours of uplift in the northeastern part of the area, the presence of a fringe of uplift along the southeast coast of Kodiak Island, and the distribution of seismic sea waves and aftershocks suggest that the zone of uplift also includes the continental shelf and part of the continental slope within the belt of major aftershocks, or a total area of about 90,000 square kilometers. Furthermore, it is possible that part, or all, of the continental slope between the edge of the continental shelf and the axis of the Aleutian Trench may also have been included in the uplift.

The zone of subsidence includes the northern and western parts of Prince William Sound, the western segment of the Chugach Mountains and portions of the lowlands north of the mountains, most of the Kenai Peninsula, and almost all of the Kodiak Islands group. It forms an asymmetrical downwarp, 800 kilometers long and approximately 150 kilometers wide, whose axis is roughly along the crest of the coastal mountains. The axis of subsidence plunges gently, northeastward

[1]C. A. Whitten, address presented 19 Nov. 1964 at the annual meeting of the Geological Society of America. I am indebted to Mr. Whitten for making available the preliminary results of re-triangulation carried out during 1964.

from the Kodiak Mountains and southwestward from the Chugach Mountains, to a low of 2⅓ meters on the south coast of the Kenai Peninsula. The total area of probable subsidence is about 110,000 square kilometers, and the average amount of subsidence is roughly 1 meter. The volume of crust that has been depressed below its pre-earthquake level is about 115 cubic kilometers.

From profiles A–A' and B–B' of Fig. 28–6 it may be seen that the vertical displacements about the zero axis of tilting are strongly asymmetrical. Along these lines, the volume of crust elevated above its pre-earthquake position is at least double the volume of the subsidence in the northern part of the deformed area.

The area of observable crustal deformation, or probable deformation, that accompanied the Good Friday earthquake — between 170,000 and 200,000 square kilometers — is larger than any such area known to be associated with a single earthquake in historic times. Comparable tectonic deformations probably have occurred during other great earthquakes, but where they occurred beneath the sea, along linear coastlines, or inland, it generally has not been possible to determine their areal extent with any degree of confidence. The 10 meters of absolute vertical displacement measured on Montague Island is known to have been exceeded only by the 14.3 meters (47 feet, 4 inches) of uplift that occurred during the earthquake of 1899 at Yakutat Bay, 320 kilometers to the east (Fig. 28–1). Substantially larger subsidences are known to have accompanied earthquakes in the past, although in many instances determination of the absolute amount of tectonic subsidence is complicated by surficial effects.

SUBMARINE UPLIFT INDICATED BY SEISMIC SEA WAVES

That the belt of uplift embraces a large segment of the continental shelf and slope, as shown in Fig. 28–2, is inferred from the areal distribution and initial direction of water motion of the seismic sea waves that accompanied the earthquake. The term "seismic sea waves" (also known as tsunamis or "tidal" waves)

refers to the train of long-period waves generated in the Gulf of Alaska which caused extensive damage to the outer coast of the Kodiak Islands group, the Kenai Peninsula, and the coasts of British Columbia and the Pacific Northwest.[2]

Seismic sea waves are gravity waves set up in the ocean by vertical disturbances of the sea bottom. The relative displacement of the sea bottom in the generative area can be determined from the initial water motion at suitably situated tide stations. Tide gauge records of the seismic sea waves outside the immediate area affected by the earthquake show an initial rise, indicating a positive wave resulting from upward motion of the sea bottom[3] (Van Dorn, 1965). The initial direction of water movement along the coast of the Gulf of Alaska within the area affected by the earthquake is less clear, because there were no operative tide gauges and in many localities the water movements were complicated by (i) uplift and subsidence of the coast, (ii) local waves generated by numerous submarine and subaerial landslides, and (iii) seiches. However, most observers along the coast of the Kenai Peninsula and the Kodiak Islands group report that the first strong motion of the waves that came in from the Gulf of Alaska was upward. An approximation to the amount of submarine uplift that generated the waves is suggested by half-wave amplitudes for the highest waves of about 5 meters at Kodiak, 7 to 8 meters at Seward, and 6 meters at Cordova (Grantz et al., 1964). Along segments of the coast exposed to the open sea, run-up of the seismic sea waves was substantially higher.

The shape of the source area within which the train of seismic sea waves was generated can be approximated from an envelope of imaginary, wave fronts projected back toward the wave source from observation stations

---

[2]The seismic sea waves are differentiated from locally destructive waves, most of which were generated along the rugged coast of Prince William Sound and the fiords of the Kenai Peninsula as a result of massive submarine landslides and possibly seiches. The local waves struck during the earthquake at Chenega, Valdez, Whittier, Seward, and other smaller communities in Prince William Sound, causing most of the fatalities that resulted from the earthquake in Alaska.

along the shore at which arrival times are known. The distance traveled by the wave to any shore station is calculated from the velocity of propagation, which conforms closely to Lagrange's equation $V = (gh)^{1/2}$ (where g is the gravitational constant and h is the depth of water as determined from nautical charts), and from the elapsed time between the main shock and the reported arrival of the wave. As computed independently by Van Dorn of the Scripps Institution of Oceanography and Spaeth of the U.S. Coast and Geodetic Survey, from data of observation stations outside the area of deformation, the wave source lies in a broad area, between the Aleutian Trench axis and the coast, that extends from the northeastern limits of the zone of uplift on land southwestward to about the latitude of Kodiak[3] (Van Dorn, 1965). The direction of travel and reported arrival times of the initial wave crest, which struck the shores of the Kenai Peninsula within 19 minutes and Kodiak Island within 34 minutes after the start of the earthquake, indicate that the wave crest was generated along one or more line sources within an elongate belt that extends southwestward from the axis of maximum uplift on Montague Island (Fig. 28–2).

It is assumed in the computations of travel distance that the waves were generated at the time the earthquake began or shortly thereafter. This assumption appears to be justified by the numerous reports of immediate withdrawals of water from uplifted coastal areas, which suggest that much, if not all, of the deformation occurred during the most violent tremors, estimated to have lasted 1½ to 4 minutes. Two apparently reliable accounts further suggest that under conditions favorable for close observation, the vertical displacements were perceptible as distinct upward accelerations in the area of uplift or downward accelerations in the subsided area. The vertical displacements occurred at least fast enough to generate atmospheric waves that were recorded shortly after the earthquake on microbarographs at both the Uni-

[3]M. G. Spaeth, address presented 7 May 1964 at a special meeting of the American Society of Photogrammetry, Washington, D.C.

versity of California at Berkeley and the Scripps Institution of Oceanography at La Jolla (Van Dorn, 1965; Bolt, 1964).

From long experience with seismic sea waves, Japanese seismologists have found that the generative area of seismic sea waves for a given earthquake broadly corresponds to the distribution of major aftershocks (Lida, 1963). In the absence of direct information, the seaward extent of uplift indicated in Figs. 28–1 and 28–2 is arbitrarily inferred to coincide with the belt of major aftershocks. The actual area probably is no smaller, but could be somewhat larger, than the area outlined.

SURFACE FAULTS

Faulting associated with the earthquake of 27 March 1964 was found at the two localities shown in Fig. 28–7, through combined air reconnaissance and measurement of barnacle-line displacements. The fault along the southeast side of the island has been informally named the Patton Bay Fault; that on the northwest side of the island, the Hanning Bay Fault.

The longer of the two faults, the Patton Bay Fault, can be traced on the ground for 16 kilometers from near Patton Bay westward to the place where it strikes out to sea at Neck Point. Inland it is marked by a discontinuous line of landslides along the base of the rugged main ridge that forms the axis of the island (Fig. 28–8). Where the fault crosses level or gently sloping ground overlain by unconsolidated surficial deposits, there is commonly a line of open fissures and stepped topography along the fault trace. Vertical offset could be measured only where the fault intersects the coastline west of Neck Point. Here the displacement of barnacles indicated a net vertical movement of 5.2 meters, with the northwest side relatively upthrown and both sides of the fault uplifted relative to sea level. About 2½ meters of the displacement occurred along a fault which cuts beach gravels and the reef at the shore (Fig. 28–9). The remainder of the offset results from a pronounced seaward bending of the upthrown block within 300 meters of the fault, as illustrated in section A–A' of Fig. 28–7. The fault

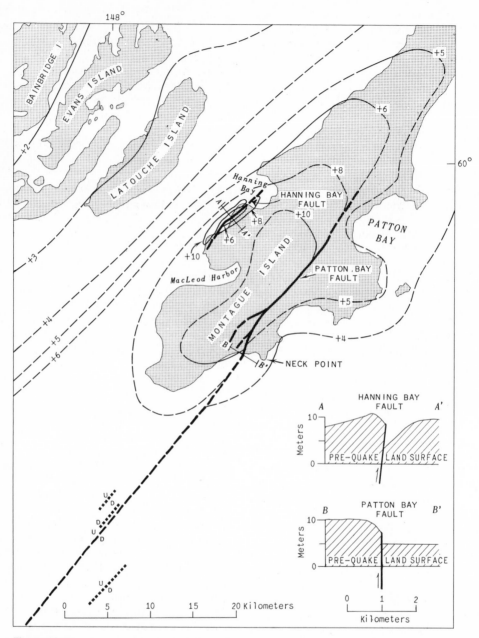

Figure 28-7
Uplift and faulting on Montague Island and in the vicinity. The contours indicate approximate uplift, in meters: they are dashed where inferred

trace is vertical where it can be seen in the sea cliff (Fig. 28–9). It is not known whether there is any component of strike-slip displacement. No evidence was found, during our brief reconnaissance investigation, of lateral displacement of roots or fallen trees across numerous surface fissures in unconsolidated deposits that mark the fault trace. Displacement along the fault dies out to the northeast in the vicinity of Patton Bay. Fathometer profiles taken southwest of Montague Island are interpreted as indicating that the fault

Figure 28-8
Aerial photograph, looking northeastward, showing the line of landslides along the trace of the Patton Bay
Fault. Patton Bay is in the background

may extend southwestward as a zone of prominent scarps for at least 28 kilometers. Comparisons made by Malloy of bathymetric data
obtained before and after the earthquake in
the area 15 to 28 kilometers southwest of
Neck Point show that displacement along the
fault increases seaward to a maximum of
about 10 meters (Malloy, 1964).

The shorter of the two faults found on
Montague Island is exceptionally well exposed for 4.8 kilometers, extending from the
south shore of Hanning Bay almost to Mac-
Leod Harbor. As with the Patton Bay Fault,
the northwest side is upthrown relative to
the southeast side, but both blocks are uplifted relative to sea level (Fig. 28-7, section
A-A'). The fault plane dips at an angle of
70 to 80 degrees toward the northwest, and
the movement is essentially dip-slip, with a
maximum-measured left-lateral strike-slip
component of 15 centimeters in surficial deposits near the southern limit of exposure.
Vertical displacement varies rapidly along

the fault strike; it reaches a maximum of 4
meters in bedrock and 5 meters in a beach
ridge at the cove between Hanning Bay and
MacLeod Harbor (Fig. 28-10 and 28-11).
The fault trace inland from the coast is marked
by a continuous line of toppled trees, ponded
streams, landslides, and fissures. The fault
dies out to the south before reaching Mac-
Leod Harbor, and it could not be traced northward beyond the south shore of Hanning Bay.

The pattern of near-vertical and reverse
faulting on southern Montague Island suggests
that the displacements resulted from compressive stress oriented approximately normal to the trend of the Arc. Preliminary results
of the U.S. Coast and Geodetic Survey's
re-triangulation of part of the network of
primary horizontal control stations (triangulation net) in the vicinity of Montague Island
indicate that the shortening resulting from this
compression may be substantially greater
than that indicated by the surface ruptures.
Survey points less than 15 kilometers apart

Figure 28-9
Vertical shear zone in sea cliff, and offset along the beach where the
Patton Bay Fault intersects the shoreline near Jeanie Point. The
approximate trace of the principal fault break is shown by the dashed line,
and the vertical offset in talus at the base of the cliff is indicated by
arrows. The figure of a man standing at the base of a 3 meter high scarp
is circled. Dip of bedding is 45° to the southwest (right)

on southern Montague Island and the islands
immediately to the north and northwest have
apparently been displaced horizontally toward one another by as much as 5 meters
(Footnote 1, p. 319).

Both the Patton Bay and the Hanning Bay
faults on Montague Island occurred along

prominent linear breaks in slope or linear
stream valleys clearly visible on aerial photographs taken before the earthquake. One of
these, the Patton Bay Fault, had been delineated on a photogeological map of part of
Prince William Sound made prior to the earthquake (Condon and Cass, 1958). Although

Figure 28-10
View to the northeast along the Hanning Bay Fault. The northwest block (left) has been relatively upthrown 4 to 5 meters along a high-angle reverse fault. The white coating on the reef rock of the upthrown block is the bleached remains of calcareous algae and bryozoans that lived below mean tide level. The X marks the location of Figure 28-11

the vertical displacements that occurred along these two faults during this earthquake are large, our reconnaissance observations did not reveal any significant lithologic differences in the rock sequences on the two sides of the faults, or wide zones of brecciated rock along them, such as commonly occur along faults that form major tectonic boundaries.

There are a number of abrupt submarine scarps in the vicinity of Montague Island and Hinchinbrook Island, which lies immediately northeast of Montague Island, and on the continental shelf between Montague and Kodiak islands (Barnes and Von Huene, personal communication). Some of these may be the surface expression of faults along which dip-slip displacement occurred during the earthquake.

Figure 28-11
The Hanning Bay Fault scarp, looking northeast. Vertical displacement in the foreground, in rock, is about 4 meters; the maximum measured displacement of 5 meters is at the beach ridge in the trees

SEISMOLOGICAL DATA

Large earthquakes at shallow and intermediate depths are thought by most North American geologists and seismologists to result from the sudden rupture of strained rocks (Hodgson, 1957b). According to the elastic rebound theory, the energy released in seismic waves and in other ways is derived from accumulated elastic-strain energy in deformed blocks of rock as they snap back toward equilibrium on either side of a fracture or fault (Reid, 1910). The instrumental epicenter of the main shock marks the surface projection of the point at which the rupture begins.

The epicenter of the Alaskan earthquake, which has a Richter magnitude variously estimated as 8.4 (Pasadena seismograph station) to 8.5 (U.S. Coast and Geodetic Survey), is located (with an uncertainty of 0.2 degree, or a radius of error of 12 km) on the east shore of Unakwik Inlet in northern Prince William Sound at latitude 66.1°N, longitude 147.7°W.[4] The hypocenter (focal depth) could not be determined more closely than in the range between 20 and 50 kilometers.

Aftershocks which follow large earthquakes are thought to be generated by continuous adjustments of the strained volume of rock, or focal region, within which the rupture occurred (Wilson, 1936; Benioff, 1951). Thus, the spatial distribution of the aftershocks approximately delimit the focal region of the main shock in all three dimensions, even where there is no surface breakage.

The horizontal extent of the focal region is roughly delineated by the areal distribution

[4]S. T. Algermissen, address presented 19 November 1964 at the annual meeting of the Geological Society of America.

of 83 percent of the first 132 large aftershocks (shocks of magnitude greater than 5.0, as determined by the Coast and Geodetic Survey), which account for most of the release of elastic-strain energy during the aftershock sequence. The aftershocks lie in a well-defined belt, 100 to 200 kilometers wide and about 800 kilometers long, that roughly parallels the trend of the Aleutian Trench and includes the known area of uplift and the adjacent continental shelf and parts of the continental slope. The most intense aftershock activity is concentrated toward the northeastern and southeastern ends. The northwestern limits of this belt are close to the boundary between the major areas of uplift and subsidence. The only part of the belt of major aftershocks that lies within the zone of subsidence is a small area immediately north of the epicenter: seven aftershocks occurred in this area. Major aftershocks that lie outside the belt shown in Fig. 28–1 were widely distributed beneath the Kenai-Kodiak Mountains, along the west shore of Cook Inlet, and seaward from the Aleutian Trench. The more numerous smaller aftershocks follow the same general pattern of distribution, although they are spread over a somewhat larger area.

The vertical extent of the inferred focal region is less perfectly defined because of inherent errors in the determination of focal depths in areas of uncertain crustal structure and seismic velocity. Hypocenters of the major aftershocks occur at depths between 5 and 40 kilometers, averaging about 20 kilometers.[4] They show a general tendency toward deepening of their lower limit beneath the continent: the deeper large aftershocks occur at depths of 30 to 40 kilometers and are situated, in general, approximately along the axis of the Chugach-Kenai Mountains (Fig. 28–2). There is no clear indication of a regular increase in depth of the upper limit of aftershocks within the zone of uplift; this presumably means that fractures were occurring throughout the thickness of the continental crust.

According to a theory advanced by Benioff, based on both experimental and observational data, the characteristics of the strain-release pattern of the aftershock sequence provide a means of distinguishing shear from compressional strain (Benioff, 1951; 1955; 1964). The strain-release curve derived from the aftershock sequence of the earthquake of 27 March is of the form that indicates dominantly compressional deformation within the focal region,[4] in accord with the observed pattern of surface warping and faulting.

The initial direction of rupture at an earthquake focus may be derived from the worldwide distribution of initial compressions and dilatations of seismic waves recorded at seismograph stations, an elastic-rebound fault source being assumed (Byerly and Stauder, 1957; Hodgson, 1957b). Solutions based upon compressional waves, or P-waves, define a pair of orthogonal planes at the focus, one of which presumably contains the active fault surface. Theoretically, an unambiguous solution for the fault plane may be obtained by analysis of shear waves and surface waves in addition to the P-waves, although the records of these phases may be considerably more difficult to interpret than those of the P-waves. Inherent in the fault-plane solutions is the basic assumption made by earthquake seismologists that the initial displacement, at least in the larger earthquakes, reflects the regional stress field.

A fault-plane solution based on P-waves, made by the Seismological Division of the U.S. Coast and Geodetic Survey, yields two planes that strike N64°E and dip about 82° SE and 8°NW through the focus.[4] Another solution, by Berg, of the seismological laboratory of the University of Alaska, yields one well-defined plane that strikes N72°E, with an almost vertical dip, and a poorly controlled second plane with a low-angle dip (Berg, 1964). If the steep plane is taken as the fault plane, the solutions indicate predominantly dip-slip movement, with the southeast side relatively upthrown and a left-lateral strike-slip component. The alternative low-angle plane would yield a thrust fault with the northwest block upthrown. The near-vertical fault is indicated by a study that was made of the surface-wave spectra (Press and Jackson, 1965). As Berg points out, however (Berg, 1964), "the other possibility of a low angle thrust can not be disregarded, and in fact, was strongly suggested in one of the aftershock solutions, which in other respects was very similar to the main shock." It is significant that, regardless of the orientation of the primary faulting, the preliminary solutions indicate that the major stress axis is oriented normal to the structural trend of the Aleutian Arc, in agreement with the general trend suggested by mechanism studies of previous large earthquakes elsewhere along the Arc (Hodgson, 1964).

GEOLOGIC RECORD OF PRE-EARTHQUAKE DEFORMATION

The tectonic movements that occurred on 27 March were the most recent pulse in an episode of deformation that began in south-central Alaska during late Cenozoic time and has continued intermittently ever since. The available geologic evidence reveals a history of complex postglacial vertical displacements relative to sea level, in which areas of net uplift or subsidence appear to correspond in general with areas in which uplift and subsidence occurred during the 27 March earthquake.

The structure of strata of Tertiary age along the coast of the Gulf of Alaska within the zone of uplift and the adjacent area to the east is dominated by asymmetric folds and north-dipping thrust faults that strike roughly parallel to the coastline (Miller et al., 1959). Late Cenozoic diastrophism along this same trend is evidenced by 30-degree dips in marine strata of late Pliocene and early Pleistocene age on Middleton Island, well out on the continental shelf. Postglacial deformation has left multiple raised beaches and terraces along the southeast coast of Kodiak Island, in Prince William Sound, along the mainland to the east of the sound, on Kayak Island and adjacent islands, and on Middleton Island. Recent spasmodic vertical displacements with net uplift of the continental-shelf margin relative to sea level are recorded in five steplike marine terraces to an elevation of 30 meters on Middleton Island (Miller, 1953). Radiocarbon dating of driftwood from the highest terrace[5] demonstrates that 30 meters of net uplift relative to sea level has occurred in the past 4470 ± 250

---

[5]The dated sample was No. W-4405 (M. Rubin, written communication.)

years. The actual amount of uplift is slightly greater than 30 meters because sea level was also rising to its present level until about 3500 years ago (Coleman and Smith, 1964). Within most of Prince William Sound and the Controller Bay area to the east, deformation immediately prior to the earthquake was a general subsidence relative to sea level, as shown by a fringe of drowned forests and intertidal peat bogs along the shores of the mainland and the islands of eastern and southern Prince William Sound (Twenhofel, 1952).

In contrast to the complex structural deformation and uplift of the Cenozoic strata along the coast, Tertiary sediments in the Cook Inlet region within the zone of subsidence and along the Aleutian Volcanic Arc to the southwest are flat-lying or only mildly deformed (Miller et al., 1959; Coats, 1962). During postglacial time that part of the zone of subsidence which is within the Cook Inlet area has been a region of relative crustal stability (Karlstrom, 1964). Postglacial high-angle reverse-fault movement is reported at one point along the Castle Mountain–Lake Clark fault system, which extends for nearly 320 kilometers along the northwestern and northern boundary of the Cook Inlet basin (Kelly, 1963). Local subsidence of the land relative to sea level along the axis of the Kenai and Kodiak mountains is indicated by Wisconsin-age cirques that are now well below sea level along the south shore of the Kenai Peninsula.

Surface deformation has been recorded in Alaska after three previous major earthquakes that occurred along the coast of the Gulf of Alaska. In 1889 and 1958 displacement occurred along segments of the Chugach-St. Elias-Fairweather fault system, which roughly parallels the south coast of Alaska (Dutro and Payne, 1957; Gates and Gryc, 1963). The maximum known vertical displacement associated with the 1899 earthquake at Yakutat Bay (Fig. 28–1) was 14.3 meters of uplift on bedrock and as much as 2 meters of subsidence in unconsolidated deposits. There was also un unknown amount of oblique-slip movement during this earthquake at one locality in Nunatak Fiord (Tarr and Martin, 1912). Measured displacement along the Fairweather Fault at a point 200 kilometers southeast of Yakutat Bay after the 1958 earthquake showed a predominant right-lateral strike-slip offset of as much as 6½ meters, with a slight upward movement of the southern block (Tocher, 1960). A large earthquake in 1880 on Chirikof Island, roughly 160 kilometers southwest of Kodiak Island (Fig. 28–1), was accompanied by vertical fault displacement of 1.8 meters (Moore, 1962). In my opinion the limited historic record of faulting on Chirikof Island, on Montague Island, and possibly at Yakutat Bay is consistent with the geologic record, which suggests late Cenozoic movement that is predominantly dip-slip along the coast of south-central Alaska. Significant strike-slip displacement along faults parallel to the coast of the Gulf of Alaska during late Cenozoic time has not yet been recognized west of Yakutat Bay, although it is possible that such evidence may have been overlooked or underestimated in the geological mapping of these areas.

SPECULATION ON ORIGIN OF THE EARTHQUAKE

A characteristic of the earthquake belt associated with the Aleutian Arc, and of others associated with island arcs throughout the world, is that the earthquake epicenters lie on the concave, generally continental side of the associated oceanic trench and the hypocenters deepen toward the adjacent continent (Gutenberg and Richter, 1954). As stated elsewhere (Grantz et al., 1964) "this distribution of earthquakes in the Aleutian Arc and Trench is believed by many geologists and geophysicists to indicate that the earthquakes originate in a fault or perhaps a zone of movement which extends with a moderate northward dip from the Aleutian Trench northward beneath the Aleutian Arc." That the earthquake of 27 March occurred along the postulated zone of faulting is strongly suggested by the occurrence of its belt of seismic activity and surface deformation within, and parallel to, the Aleutian Arc and Trench. The orientation and sense of displacement on this postulated fault or zone of faulting, however, can only be deduced indirectly from the seismological data, the residual surface dis

placements, and the geologic record of past deformation.

If it is assumed that the earthquake originated by rupture along one or more faults, and that the uplift and subsidence resulted from elastic rebound, it is important to consider the possible orientation and sense of movement that might best explain the observed pattern of surface displacement. The two most plausible alternatives consistent with the available fault-plane solutions are (i) dip-slip movement on a near-vertical fault between the major zones of uplift and subsidence, and (ii) thrusting along a fault or zone of faulting that dips northwestward from the Aleutian Trench beneath the Aleutian Arc.

According to the first hypothesis, a near-vertical fault strikes along the line of zero change in land level, with the southeast block up and the northwest block down relative to sea level. The greatest attraction of this hypothesis is that it provides a simple model to account for the distribution of surface uplift and subsidence in the two major zones. It is supported by the occurrence of the earthquake epicenter close to the zero line between the two zones, and by a preliminary unambiguous fault-plane solution based on surface waves (Press and Jackson, 1965).

Upon detailed examination, however, the hypothesis appears to pose more problems than it answers. Most serious among these are the absence of surface fault displacement at or in the vicinity of the zero line and the lack of evidence that the line corresponds to major geologic boundaries, as might be expected if it marked the trace of a major fault along which vertical movement has occurred in the past. An alternative possibility is that the displacement represents flexure above a fault at depth. In an elegant analysis of the displacements by application of dislocation theory, Press and Jackson (1965) have shown that the observed vertical surface displacements of less than 6 meters could be accounted for by a steeply dipping subsurface fault extending from 15 kilometers below the surface to the considerable depth of 100 to 200 kilometers. There are four serious objections to this conclusion, however. (i) It seems improbable that a near-vertical fault 800 kilometers long and from 85 to 185 kilometers deep, which generated one of the greatest earthquakes in history, should fail to reach the surface anywhere along its length or show detectable surface offset, especially since 5 meters of vertical offset occurred at the surface along faults on Montague Island which are presumed to be subsidiary ruptures. (ii) The postulated fault is more than twice the depth of the hypocenters of the initial shock or of the deepest large aftershocks of this earthquake or of previously recorded major earthquakes in the same general region (Gutenberg and Richter, 1954). (iii) The focal region, as inferred from aftershock distribution, is a belt 160 to 320 kilometers wide that lies mainly to the south of the postulated fault rather than being more symmetrically disposed with respect to its surface trace, as might be expected for a steeply dipping fault (Fig. 28–1). (iv) The sense of displacement on the postulated fault is opposite to the up-to-the-north movement of known faults that strike parallel to the regional structural trend and the surface faulting that occurred on Montague Island during the earthquake.

The second hypothesis, which proposes that the earthquake originated along a fault or zone of movement that extends northward from the Aleutian Trench, seems to be more promising, perhaps because it postulates a primary fault that is safely concealed from view beneath the sea. According to this model, the zone of fault slippage is within the belt of major aftershocks (Figs. 28–1 and 28–6). Dominantly compressive stress, oriented approximately normal to the Arc, is indicated by the pattern of folds and faults in rocks of late Cenozoic age as well as by the pattern of surface deformation that accompanied the earthquake and the strain-release characteristics of the aftershock sequence. A downward vertical component of pre-earthquake strain is also suggested by the fringe of drowned forest along much of the coast in the northwestern part of the uplifted zone. The postulated stress pattern could result from progressive underthrusting of the oceanic crust and mantle beneath the continental margin, as illustrated diagrammatically in Fig. 28–6. The crustal configuration shown in Fig. 28–6 is largely speculative, for crustal thicknesses in this part of Alaska have been deter-

mined only along short seismic-refraction lines in the Kodiak Island area and in northern Prince William Sound (Shor, 1962; Tatel and Tuve, 1956). Elastic rebound during the earthquake resulted in relative seaward thrusting of the continental margin along one or more primary northward-dipping faults (not shown in Fig. 28–6) with accompanying uplift and warping of the upper continental block. The surface faults on Montague Island and their seaward extensions are situated in the zone of maximum uplift, where renewed movement occurred along preexisting vertical or high-angle reverse faults. Comparable subsidiary faults could well occur elsewhere on the continental shelf.

The most serious limitation of the thrust-fault hypothesis lies in its attempts to account for the observed subsidence to the north of the zone of uplift. The relative scarcity of large aftershocks within the zone of subsidence, except in the immediate vicinity of the epicenter of the initial shock, is interpreted as indicating that this zone was largely outside the area of the primary fault rupture along which the earthquake occurred. In the absence of surface-fault displacement, it is tentatively suggested that the subsidence may be a secondary effect resulting from elastic deformation immediately adjacent to the postulated zone of thrusting. Preliminary results of resurveys of small portions of the triangulation net within the zone of subsidence near Anchorage show as much as $2\frac{2}{3}$ meters of horizontal elongation within a north-south distance of 48 kilometers,[1] indicating that crustal extension could have been a significant factor in producing the observed subsidence.

The hypothesis outlined above is generally consistent with most modern theories which relate arc structures in the circum-Pacific region and elsewhere to master thrust faults along the unstable interface between the oceanic and continental crusts. It can also account for the following observed features: (i) the areal distribution of the major zones of uplift and subsidence; (ii) the marked asymmetry in the volumes of uplift and subsidence in the two major zones; (iii) vertical or reverse surface faults in the zone of uplift with up-to-the-north displacement; (iv) occurrence of the belt of major aftershocks mainly within the zone of uplift and its inferred offshore extension; (v) the shallow depths of the initial shock and aftershocks and the tendency toward deepening of aftershock hypocenters beneath the continent: (vi) the geologic record of late Cenozoic folding, reverse faulting, and net uplift of the continental margin and shelf, in contrast to the history of relative stability or slight subsidence in the adjacent area to the north.

Neither of the two hypotheses outlined above attempts to account for the possibility that there may be a second zone of slight uplift adjacent to the zone of subsidence. Additional field investigations planned for the 1965 season by the U.S. Geological Survey and the U.S. Coast and Geodetic Survey are aimed at resolving this problem, and at filling gaps in the existing picture of horizontal and vertical surface deformations that accompanied the earthquake.

SUMMARY

Alaska's Good Friday earthquake of 27 March 1964 was accompanied by vertical tectonic deformation over an area of 170,000 to 200,000 square kilometers in south-central Alaska. The deformation included two major northeast-trending zones of uplift and subsidence situated between the Aleutian Trench and the Aleutian Volcanic Arc; together they are 700 to 800 kilometers long and from 150 to 250 kilometers wide. The seaward zone is one in which uplift of as much as 10 meters on land and 15 meters on the sea floor has occurred as a result of both crustal warping and local faulting. Submarine uplift within this zone generated a train of seismic sea waves with half-wave amplitudes of more than 7 meters along the coast near the source. The adjacent zone to the northwest is one of subsidence that averages about 1 meter and attains a measured maximum of 2.3 meters. A second zone of slight uplift may exist along all or part of the Aleutian and Alaska ranges northwest of the zone of subsidence.

The studies made to date demonstrate that great earthquakes such as the earthquake of 27 March may be accompanied by regional deformation on a larger scale than has been generally recognized. Perhaps better than any previous seismic event, this earthquake indicates that vertical displacement of the sea bottom can generate destructive seismic sea waves, even where the epicenter of the main shock is as much as 100 kilometers inland from the coast.

The focal region of the earthquake, as inferred from the spatial distribution of the major aftershocks, lies almost entirely within the seaward zone of uplift and extends from close to the surface to a depth of about 50 kilometers. The primary fault or zone of faulting along which the earthquake presumably occurred is not exposed at the surface on land. The only known surface breakage is along two preexisting faults on Montague Island, within the area of maximum uplift, that trend northeast and are near-vertical or dip steeply northwest. The displacement, which was subsidiary to the regional uplift, was essentially dip-slip, with the northwest blocks relatively upthrown. The maximum measured vertical displacement on land is 5 meters, and along the submarine extension of one of these faults the vertical displacement may exceed 10 meters.

It is postulated that the earthquake is genetically related to the Aleutian Arc and probably resulted from regional compressive stress oriented roughly normal to the Arc. Neither the orientation nor the sense of movement on the primary fault along which the earthquake occurred is known. Available fault-plane solutions based on P-waves indicate that the primary fault could be either a northwest-dipping thrust or a northeast-striking near-vertical fault with the southeast side upthrown. Which, if either, of these alternatives represents the primary fault cannot be determined without additional field data on the horizontal and vertical displacements that accompanied the earthquake and detailed analyses of the innumerable seismographic records written by the earthquake and its aftershocks.

29

Mechanism of Earthquakes and Nature of Faulting on The Mid-ocean Ridges

LYNN R. SYKES
1967

From *Journal of Geophysical Research*, v. 72, p. 2131–2153, 1967. Reprinted with permission of the author and the American Geophysical Union. Copyright 1967.

The discovery of a large number of fracture zones on the mid-oceanic ridges (Menard, 1955, 1965, 1966; Heezen and Tharp, 1961, 1964) has led to a renewed interest in the existence of large horizontal displacements on the ocean floor and to reconsiderations of various hypotheses of sea-floor growth, continental drift, and convection currents. Evidence that magnetic anomalies on the ocean floor can be identified with past reversals in the earth's magnetic field (Vine and Matthews, 1963; Vine and Wilson, 1965; Cann and Vine, 1966; Pitman and Heirtzler, 1966; Vine 1966) has added strong support to the hypothesis of ocean-floor spreading as postulated by Hess (1962) and Dietz (1961).

The offset of magnetic anomalies and bathymetric contours at prominent fracture zones have been used as arguments for transcurrent fault displacements as great as 1000 km (Vacquier, 1962; Menard, 1965). Nevertheless, several problems are raised about interpretations of this type. The inferred magnitude and sense of displacement are not always constant along a given fracture zone (Menard, 1965; Talwani et al., 1965b), and some fault zones appear to end quite abruptly (Wilson, 1965c). In addition, seismic activity along fracture zones is concentrated almost exclusively between the two crests of the mid-oceanic ridge (Sykes, 1963, 1965); very few earthquakes are found on the other portions of fracture zones.

Wilson (1965b, c) has presented arguments for a new class of faults—the transform fault. As an alternative proposal to transcurrent faulting, Talwani et al. (1965b) have assumed that major faulting on fracture zones is normal faulting. Wilson and Talwani et al. assume that the various segments of the mid-oceanic ridge were never displaced at all but developed at their present locations. Hence, these theories and the transcurrent fault hypothesis predict different types of relative motion on the seismically active portions of fracture zones.

Since considerable interest now centers on the problem of large-scale deformation of the sea floor, it seemed imperative to inquire if the nature of relative displacements could be ascertained from an analysis of first motions of earthquakes that occur on the mid-oceanic ridges. Fortunately this interest in large horizontal displacements coincided with the advent of a significantly new source of seismological data. Long-period seismograph records from more than 125 stations of the World-Wide Standardized Seismograph Network (WWSSN) of the U. S. Coast and Geodetic Survey, from about 25 stations of the Canadian network, and from about 20 stations that cooperate with the Lamont Geological Observatory now furnish data of greater sensitivity, greater reliability, and broader geographical coverage than were available in most previous investigations of the mechanisms of earthquakes.

This paper presents data from 17 earthquakes on the mid-oceanic ridges and their extensions into East Africa and the Arctic. Nine of these events were located on the mid-Atlantic ridge, and three were situated on the east Pacific rise. This study of the mechanisms of earthquakes is in agreement in every case with the sense of motion predicted by Wilson (1965b, c) for transform faults. The results support the hypothesis of ocean-floor growth at the crest of the mid-oceanic ridge. The sense of displacement indicated from these studies of earthquakes is opposite to that expected for a simple offset of the ridge crest.

These investigations revealed two principal types of mechanisms for earthquakes on the mid-oceanic ridges. Earthquakes on fracture zones are characterized by a predominance of strike-slip motion; earthquakes located on the crest of the ridge but apparently not situated on fracture zones are characterized by a predominance of normal faulting. The inferred axes of maximum tension for these latter events are approximately perpendicular to the local strike of the ridge.

PREVIOUS MECHANISM SOLUTIONS FOR EARTHQUAKES ON THE MID-OCEANIC RIDGES

Although more than a thousand mechanism solutions have been presented in the seismological literature, only about thirty are for earthquakes that occurred on the mid-oceanic ridges or on the extensions of this system into East Africa and the Arctic. Many of these solutions were based on a poor distribution of recording stations; many of the events were poorly recorded by the less sensitive network of instruments that existed before the advent of the WWSSN. Hodgson and Stevens (1964) concluded that almost none of these earthquakes on the ridge have yielded an unambiguous solution.

From an analysis of data from 13 shocks on the mid-Atlantic ridge, Misharina (1964) concluded that the mechanisms were characterized by a predominance of strike-slip faulting. Unlike the results presented in this paper, however, the strikes of many of the nodal planes for P waves do not agree very closely with the strikes of the various fracture zones. Most of the data were obtained from seismological bulletins; the original records were not examined by the investigator in most cases. The percentage of inconsistent readings is rather high.

Lazareva and Misharina (1965) performed a similar analysis for 15 earthquakes along the continuation of the mid-oceanic seismic belt into the arctic basin and northern Siberia. Most of the solutions were predominantly of a strike-slip nature. Although the inferred axes of maximum tension were approximately perpendicular to the ridge in the Greenland Sea, the inferred axis of maximum compression occupies a similar position in the arctic basin and in northern Siberia. Scheidegger (1966) reached similar conclusions from a statistical analysis of the results of Lazareva and Misharina.

Stauder and Bollinger (1964a, b; 1965) used data from the WWSSN and other stations in an investigation of the larger earthquakes of 1962 and 1963. A combination of data from P and S waves was employed in their solutions. Seven of the earthquakes examined in the present investigation were also studied by Stauder and Bollinger. The percentage of inconsistent P-wave readings in their results ranged from 5 to 30% for these seven earthquakes. Their results indicate a predominance of strike-slip motion in five cases and a predominance of normal faulting in another. A seventh event for which only a tentative solution was presented was characterized by a predominance of thrust faulting. Stefánsson

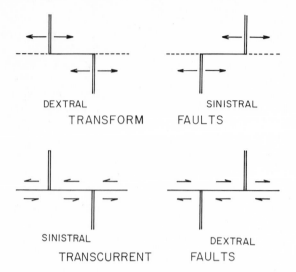

DEXTRAL   SINISTRAL
TRANSFORM   FAULTS

SINISTRAL   DEXTRAL
TRANSCURRENT   FAULTS

Figure 29-1
Sense of motion associated with transform faults and transcurrent faults (after
Wilson, 1965c). Double line represents
crest of mid-oceanic ridge; single line,
fracture zone. The terms "dextral" and
"sinistral" describe the sense of motion of
the fracture zones; for the transform faults
they do not denote the relative configuration
of the two segments of ridge on either side
of the fracture zone

(1966) found, however, that a foreshock preceded the latter event by a few seconds. Stefánsson's analysis and the results reported in
this paper indicate a predominance of the
strike-slip motion in the main shock.

Stauder and Bollinger (1966b) concluded
that the inferred axes of maximum tension
were approximately normal to the trend of the
ridge. They apparently failed to realize, however, that in each of the solutions dominated
by strike-slip motion the epicenter was located
on a fracture zone and one of the two nodal
planes for *P* waves nearly coincides with the
strike of the fracture zone. Stauder and Bollinger's analysis indicated that all their solutions were of the double-couple type; hence,
from *P*- and *S*-wave data alone it would not be
possible to choose which of the two nodal
planes was the fault plane.

Several of the strike-slip solutions of
Stauder and Bollinger (1966b) are in agreement with the sense of motion predicted by
Wilson (1965b, c) for transform faults. Unfortunately, a full test of the hypothesis is not
possible with these data, since all the solutions
for the mid-oceanic ridges were of the dextral
transform type (Figure 29–1).

Stefánsson (1966) conducted an extensive
investigation of the focal mechanisms of two
shocks on the mid-Atlantic ridge. Amplitude
measurements and *P*- and *S*-wave data were in
agreement with a double-couple mechanism.
For his best solution only 2 of 85 stations were

inconsistent with a quadrant distribution of
first motions. Sutton and Berg (1958a) examined the distribution of first motions for
earthquakes in the western rift valley of East
Africa. Although the data were not extensive
enough to give a unique determination of the
predominant type of faulting, the first motions
were consistent either with dip-slip faulting on
steeply dipping planes parallel to known faults
or with strike-slip faulting along near-vertical
faults with the eastern sides moving north.

ANALYSIS OF DATA

Data Used. In this investigation first motions
of the phases *P* and *PKP* were examined for 17
earthquakes that occurred between March
1962 and March 1966. Epicentral and other
pertinent data are listed in Table 29–1. Reliable solutions probably could be obtained for
several additional earthquakes on the mid-
oceanic ridges, the west Chile and Macquarie
ridges, and the seismically active zone that extends from the Azores to Gibraltar. However,
with the exception of one earthquake on the
Macquarie ridge emphasis in this paper was
placed on an analysis of earthquakes on features that are more generally accepted as mid-
oceanic ridges.

Seismograms used in this study were supplied by the World-Wide Standardized Seismograph Network, the Canadian network, and

Table 29–1
Summary of earthquake locations and other pertinent data.

Event number	Figure number	Region	Latitude	Longitude	Date	Origin time h	m	s	Magnitude[a]
1	5	Romanche fracture zone, equatorial Atlantic	00.17°S	18.70°W	Nov. 15, 1965	11	18	46.8	5.6
2	6	Equatorial Atlantic, fracture zone 'Z' of Heezen, Gerard, and Tharp (1964)	07.45°N	35.82°W	Aug. 3, 1963	10	21	30.4	6.1
3	7	Equatorial Atlantic, fracture zone 'Z' of Heezen, Gerard, and Tharp (1964)	07.80°N	37.35°W	Nov. 17, 1963	00	47	58.8	5.9
4	8	Vema fracture zone, equatorial Atlantic	10.77°N	43.30°W	March 17, 1962	20	47	33.4	
5	10	North Atlantic unnamed fracture zone	23.87°N	45.96°W	May 19, 1963	21	35	45.4	6.0
6	11	North Atlantic	31.03°N	41.49°W	Nov. 16, 1965	15	24	40.8	6.0
7	12	North Atlantic	32.26°N	41.03°W	Aug. 6, 1962	01	35	27.7	
8	13	North Atlantic unnamed fracture zone	35.29°N	36.07°W	May 17, 1964	19	26	16.4	5.6
9	14	Off north coast of Iceland unnamed fracture zone	66.29°N	19.78°W	March 28, 1963	00	15	46.2	
10	16	Laptev Sea, near arctic shelf of Siberia	78.12°N	126.64°E	Aug. 25, 1964	13	47	13.8	6.1
11	18	Lake Kariba, Southern Rhodesia	16.64°S	28.55°E	Sept. 23, 1963	09	01	51.1	5.8
12		Lake Kariba, Southern Rhodesia	16.70°S	28.57°E	Sept. 25, 1963	07	03	48.8	5.8
13	19	Western rift, East Africa	00.81°N	29.93°E	March 20, 1966	01	42	46.8	6.1
14	20	Rivera fracture zone, east Pacific rise	18.87°N	107.18°W	Dec. 6, 1965	11	34	48.9	5.9
15	22	Easter fracture zone, east Pacific rise	26.87°S	113.58°W	March 7, 1963	05	21	56.6	
16	23	Eltanin fracture zone, east Pacific rise	55.35°S	128.24°W	April 3, 1963	14	47	50.4	5.8
17	24	Macquarie ridge, south of New Zealand	49.05°S	164.26°E	Sept. 12, 1964	22	06	58.5	6.9

[a]Body-wave magnitude, U.S. Coast and Geodetic Survey. All computations for focal depth of 0 km, except event 4 for which depth taken as 31 km.

about twenty stations that operate long-period seismographs in cooperation with the Lamont Geological Observatory. The geographical distribution of stations is generally more complete for earthquakes that occurred after 1962.

Reliability of Data. An important factor in the use of these new data is the availability of seismograph records. All the readings used in this study were made by the author. A better distribution of stations both in distance and in azimuth can now be obtained with the long-period seismographs in the three networks. Because the networks have a greater sensitivity to small earthquakes, reliable solutions can be determined for a large number of earthquakes

with magnitudes as small as about 6. This is an important factor in investigations of earthquakes on the mid-oceanic ridges, since an event of magnitude 7 occurs in these regions only about once during a period of a few years.

First motions of the *P* phase are usually more reliable on the long-period records. The example in Figure 29–2, although not typical of all long- and short-period readings, illustrates that in a number of instances the quality of first-motion data is considerably better on the long-period seismograms. Most of the first motions used in this paper were read from long-period records; short-period data were employed only when other records were not available and when the sense of first motion

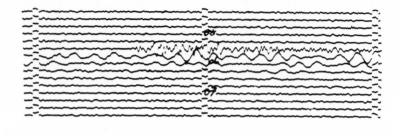

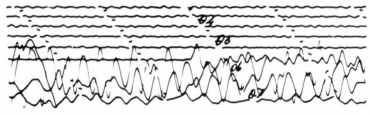

Figure 29-2
Comparison of first motions on short-period vertical component (above) and long-period vertical component (below) at Arequipa, Peru, for earthquake of March 7, 1963 (event 15 in Table 29-1). *P* wave arrived at 05 h 59m. First motion is uncertain on the short-period record but is clearly compressional (upward motion on record) on the long-period seismogram. Deflections of trace occur once per minute

was clearly evident. For earthquakes of magnitude near 6 the response of the long-period instruments is such that the recorded signals approximate those from a point source. A point source is probably not a good approximation for waves of periods between 0.5 and 2 sec.

In addition to the factors already mentioned, the calibration and polarity of the WWSSN stations are better known. In this study the number of inconsistent readings of first motion is less than 1%. In many previous investigations 15 to 20% of the data were often inconsistent with the inferred quadrant distribution of first motion (Hodgson and Adams, 1958).

The reliability of the long-period networks is such that readings from stations with polarity reversals are clearly indicated. In this study and in an investigation of earthquake mechanisms for the southwest Pacific (Isacks et al., 1968) the four stations GIE, PEL, TAU, and PHC consistently exhibited first motions that were inconsistent with determinations from other nearby stations. These reversals were confirmed by the Seismology Data Center of the WWSSN for the first three stations; the polarity reversals were corrected during mainte-

nance visits to these stations (T. A. Modgling, personal communication). These reversals were taken into account in the presentation of the data in this paper.

Method of Projection. An equal-area projection (Friedman, 1964) of the lower hemisphere of the focal sphere was used throughout this investigation. The two coordinates used to describe a point are the azimuth of the station at the epicenter and the radial distance R, where R is proportional to $\sqrt{2} \cdot \sin (i/2)$ and i is the angle of incidence at the source as measured from the downward vertical.

Values of i for a given distance and depth were computed from the tables of Hodgson and Storey (1953) and Hodgson and Allen (1954). The assumed values of i may be in error because the tables do not make allowance for velocities at the focus that are less than about 7.8 km/sec (Romney, 1957; Sutton and Berg, 1958b). Nevertheless, since the computed depths of the earthquakes in this study may be uncertain by as much as 50 km, a more complex model for the velocity distribution did not seem to be justified at the present time. Mechanism solutions that represent a predominance of strike-slip motion on steeply dipping faults are not influenced significantly by these uncertainties. Solutions representing a predominance of dip-slip motion on faults of dip less than about 60°, however, may be affected to a much greater extent. Nonetheless, for the latter class of solutions the principal point made in this paper is that a large dip-slip component did, in fact, exist.

TRANSFORM AND TRANSCURRENT FAULTS

Wilson (1965b, c) has recently proposed a separate class of horizontal shear faults—the transform fault. Dextral and sinistral transform faults of the ridge-ridge type and their transcurrent counterparts are illustrated in Figure 29-1. In the transcurrent models it is tacitly assumed that the faulted medium is continuous and conserved, whereas in the transform hypothesis the ridges expand to produce new crust (Wilson, 1965c). Thus, a transform fault may terminate abruptly at both ends even though great displacements may have occurred either on the central portions of the fracture zone or, at some time in the past, on portions of the fault that are no longer located between the two ridge crests (Figure 29-1, dashed segments).

Another significant difference noted by Wilson (1965c) was that the motion on transform faults (upper half Figure 29-1) is the reverse of the motion required to offset the ridge (lower half Figure 29-1). Also present seismicity should be largely confined to the region between the ridge crests for transform faulting but should not exhibit such an abrupt decrease for transcurrent faulting.

In this paper the term 'fault plane' is used to describe a zone of shear displacement or shear dislocation. No attempt is made to define the physical mechanism of deformation more specifically.

PRESENTATION OF DATA FOR THE MID-OCEANIC RIDGES

Figure 29-3 illustrates the locations of the 17 earthquakes for which mechanism solutions were obtained. The numbers beside the

epicenters refer to the data in Table 29-1. Events 1 through 9 are located on the mid-Atlantic ridge, earthquake 10 is on the extension of the mid-oceanic ridge system in the Arctic, events 11 through 13 are in East Africa, 14 through 16 are on the east Pacfic rise, and event 17 is on the Macquarie ridge.

Equatorial Atlantic

Fracture Zones. Between 15°N and 5°S the crest of the mid-Atlantic ridge is displaced to the east a total of nearly 35° (Figure 29-4). The apparent displacement is such that the ridge crest maintains its median character throughout the North and South Atlantic oceans. Hess (1955) identified the Romanche trench (near event 1 in Figure 29-4) as a part of a major fracture zone and noted that St. Paul's Rocks appear to be located at the western end of a great fault scarp. Heezen and Tharp (1961), Heezen et al. (1964a), and Heezen et al. (1964b) more recently identified a whole series of fracture zones in the equatorial Atlantic (Figure 29-4). Heezen and Tharp (1965) concluded that in this area the ridge crest had been offset by sinistral transcurrent faults. They noted that the fracture zones appeared to be parallel to inferred flow lines for the continental drift of Africa relative to South America.

Seismicity of the Equatorial Atlantic. Earthquake epicenters in the equatorial Atlantic were relocated for the period 1955 to 1965; the results are presented in Figure 29-4. The methods and computer programs used in these relocations were similar to those described in previous seismicity studies (Sykes, 1963, 1965, 1966; Sykes and Ewing, 1965; Sykes and Landisman, 1964; Tobin and Sykes, 1966).

The distribution of earthquakes is similar to that found in other portions of the mid-oceanic ridge (Sykes, 1963; 1964; 1965)—nearly all the activity on each fracture zone is confined to the region between the ridge crests. This is particularly well illustrated for the Chain fracture zone near 1°S, 15°W. The distribution of epicenters also shows that Heezen and Tharp's (1961, 1965) interpretations of the pattern of ridges and fracture zones are largely correct. Although a few

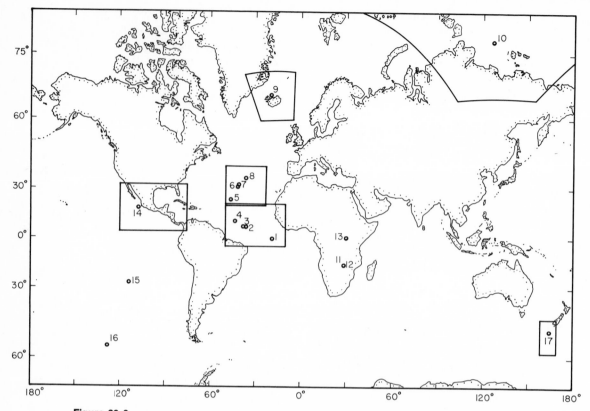

Figure 29-3
Earthquakes on mid-oceanic ridge system for which mechanism solutions are presented in this paper. Numbers beside epicenters refer to designations and data in Table 25-1. Inserts denote areas that are shown in greater detail in Figures 29-4, 29-9, 29-15, 29-17, 29-21, and 29-25

earthquakes are found on other portions of fracture zones, the number is relatively small. The abrupt cessation of activity at the ridge crests appears to be a strong argument for the transform-fault hypothesis.

Menard (1965) stated that several fracture zones are active far out on the flanks of oceanic rises. As examples he cited the west Chile ridge, an active zone that extends from the Azores to Gibraltar, and a few additional epicenters on fracture zones. Nonetheless, the important generalization from this and other seismicity studies seems to be that on *most* fracture zones *most* of the activity is confined to the region between the two ridge crests. On the contrary, seismic activity between the Azores and Gibraltar and on the Macquarie and west Chile ridges suggests that at the present time these features are quite different from most of the fracture zones of the mid-

oceanic ridge. Investigations of the mechanism of earthquakes and analyses of magnetic anomalies should provide more definitive information about these more unusual features. A small amount of differential spreading on the two sides of a fracture zone could account for the isolated earthquakes that occur off the ridge crest on fracture zones. In any case the seismicity of the basins on either side of the ridge appears to be extremely low.

Most of the seismic activity that is associated with fracture zones and with the crest of the ridge is confined to remarkably narrow zones. In some cases these active zones are less than 20 km. wide. In contrast, activity along continental extensions of the mid-oceanic ridges usually exhibits a greater areal scatter (Sykes and Landisman, 1964; Sykes, 1965). Likewise, in California the epicenters of earthquakes smaller than magnitude 6 are

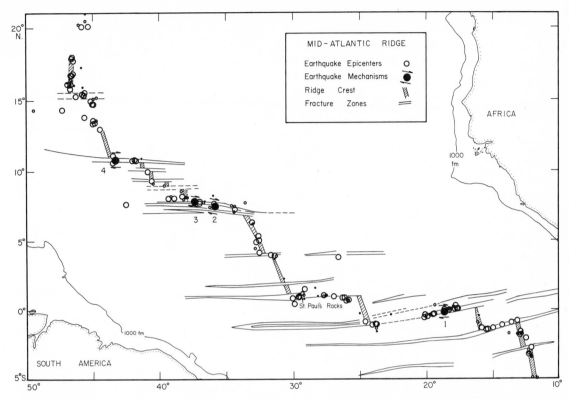

Figure 29-4
Relocated epicenters of earthquakes (1955–1965) and mechanism solutions for four earthquakes along the equatorial portion of the mid-Atlantic ridge. Ridge crests and fracture zones from Heezen et al. (1964a), and Heezen et al., (1964b). Sense of shear displacement and strike of inferred fault plane are indicated by the orientation of the set of arrows beside each mechanism. Numbers beside mechanism solutions refer to data in Table 29-1. Large circles denote more precise epicentral determinations: smaller circles, poorer determinations

also scattered throughout a zone more than 200 km wide (Richter, 1958; Allen et al., 1965). Hence, the pattern of seismic dislocations appears to be extremely simple in most of the oceanic portions of the ridge system. Analyses of earthquake periodicities in time and in space and investigations of triggering mechanisms might yield more interesting results if a relatively simple system such as that associated with a particular fracture zone were analyzed. Many of the larger earthquakes listed by Gutenberg and Richter (1954) for the mid-Atlantic ridge are located on or near major fracture zones.

Strike-slip Mechanisms. Mechanism solutions for four earthquakes that were located on equatorial fracture zones are presented in

Figures 29–5 through 29–8. Solid circles denote compressions; open circles, dilatations. When long-period records were available, it was often possible to tell from the size of the P wave relative to that of other phases whether the station was in the vicinity of one of the two nodal planes. Arrivals of this type are denoted by a cross on the figures. As an examination of the figures reveals, these arrivals are often a good qualitative indication that the station was near a nodal plane. Even when the first motion of such a signal cannot be ascertained, this information is an additional constraint on the orientation of the nodal planes.

A quadrant distribution of first motions seems to be an excellent approximation to the data in Figures 29–5 through 29–8. Only

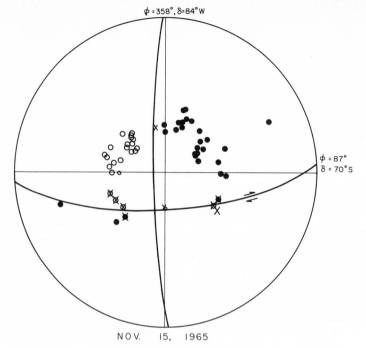

Figure 29-5
Mechanism determination for the shock of November 15, 1965, on the mid-Atlantic ridge (event 1 in Table 29-1 and in Figures 29-3 and 29-4). Diagram is an equal area projection of the lower hemisphere of the radiation field. Solid circles represent compressions; open circles, dilatation; crosses, wave character on seismograms, indicating station is near nodal plane. Smaller symbols represent poorer data. ϕ and δ are strike and dip of the nodal planes. Arrows indicate sense of shear displacement on the plane that was chosen as the fault plane

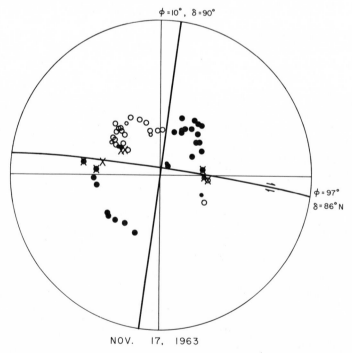

Figure 29-6
Mechanism solution for event 2. Table 29-1 and Figures 29-3 and 29-4 give epicentral and other pertinent data. Symbols same as Figure 29-5

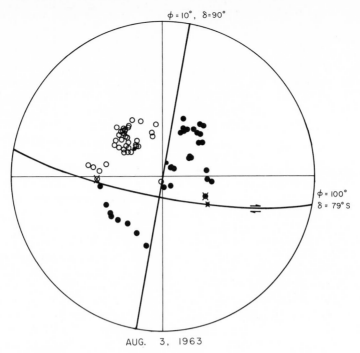

$\phi = 10°, \ \delta = 90°$

$\phi = 100°$
$\delta = 79° S$

AUG. 3, 1963

Figure 29-7
Mechanism solution for event 3. Table 29-1 and Figures 29-3 and 29-4
give position and other pertinent data. Symbols same as Figure 29-5.
Note added in proof: strikes of two planes should read 8° and 98°
rather than 10° and 97°

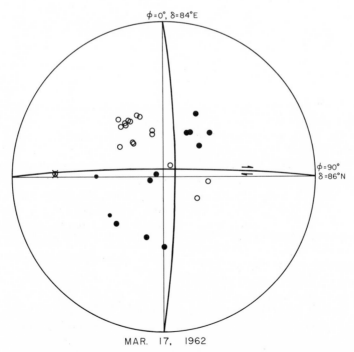

$\phi = 0°, \ \delta = 84° E$

$\phi = 90°$
$\delta = 86° N$

MAR. 17, 1962

Figure 29-8
Mechanism solution for event 4, Table 29-1 and Figures 29-3 and
29-4 contain position and other pertinent data. Symbols same as
Figure 29-5

four readings are inconsistent with this interpretation. Seismograms from these four stations are of marginal quality and should not be taken as evidence against a quadrant distribution.

In each of the four cases illustrated in these figures one nodal plane strikes approximately east and the other strikes nearly north. From analyses of the polarization of the *S* waves for several shocks on the mid-oceanic ridge Stauder and Bollinger (1966) showed that the mechanisms are of the double-couple type. Thus, *S* waves do not furnish criteria for choosing which of the two nodal planes is the fault plane. In each of these four cases the choice is between a predominance of dextral strike-slip motion on a steeply dipping plane that strikes approximately east or a predominance of sinistral strike-slip motion on a steeply dipping plane that strikes nearly north.

Choice of the Fault Plane. Although a unique choice of the fault plane cannot be made from the first motion data or from an analysis of *S* waves, the east-striking plane seems to be overwhelmingly favored for the following reasons:

1. For each earthquake the epicenter is located on a prominent fracture zone; the strike of one of the two nodal planes nearly coincides with the strike of the fracture zone (Figure 29–4).

2. Earthquake epicenters are aligned along the strike of the fracture zones.

3. The linearity of fracture zones suggests a strike-slip origin (Menard, 1965). The bathymetry and other morphological aspects are similar to the morphology of the great strike-slip fault zones on continents.

4. Many of the rocks from St. Paul's Rocks (Figure 29–4) are described as dunite-mylonites (Washington, 1930a, b; Tilley, 1947; Wiseman, 1966). Rocks dredged from other fracture zones (Shand, 1949; Tolstoy, 1951; Quon and Ehlers, 1963), as well as samples from cores taken in fracture zones (Heezen et al., 1964a), exhibit a similar petrology and provide evidence for intense shearing stresses in the vicinity of fracture zones.

5. The choice of the north-striking plane would indicate strike-slip motion nearly parallel to the ridge axis. On the contrary, earthquakes on the ridge axis but not on fracture zones are characterized by a predominance of normal faulting.

6. If the east-striking plane is chosen, the sense of relative motion on the fracture zone reverses when the apparent offset of the crests of the ridge is interchanged (Figure 29–1).

Agreement with Strike of Fracture Zones. Figure 29–4 demonstrates that the observed strike of one of the two nodal planes agrees very closely with the trend of the fracture zones on which the earthquakes occurred. Heezen et al. (1964a) report that the over-all trend of the Romanche trench (a fault trough in the Romanche fracture zone) is about 80°. The strike determined for event 1, an earthquake that occurred in the trench, is about 87°. Fortunately, in this solution and in the other solutions for the mid-Atlantic ridge the east-striking nodal plane can be estimated with greater precision than the north-striking plane. The discrepancy between the two strikes could be explained by variations in the strikes of the individual fractures in the trench or by uncertainties in the determination of the mechanism.

Events 2 and 3 are both located on a large fracture zone near 8°N; their mechanisms appear to be very similar. The Guinea fracture zone near 9°N, 16°W (Figure 29–4) may represent the eastward continuation of this fracture zone rather than the continuation of the Vema fracture zone (near 11°N, 43°W) as Krause (1964) suggested. Wilson (1965c) suggested that the offsets of the crest of the ridge are a reflection of the shape of the initial break between the continental blocks of Africa and South America. The pattern of ridges and fracture zones between 8°N and 1°S appears to match the configuration of the African coast (or the shape of the continental slope) between 9°N and 2°N.

North Atlantic

Figure 29–9 indicates the locations of four mechanism determinations for the mid-

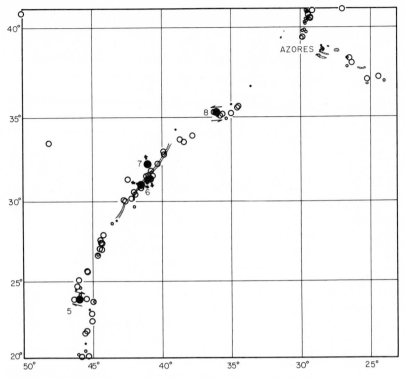

Figure 29-9
Relocated epicenters of earthquakes (1955-1965) and mechanism solutions for
four earthquakes along a portion of the mid-Atlantic ridge. Rift valley in mid-
Atlantic ridge is denoted by diagonal hatching and is from Heezen and Tharp
(1965b). Other symbols same as Figure 29-4. For events 6 and 7 thick arrows
denote inferred axes of maximum tension

Atlantic ridge in the North Atlantic. The
mechanisms are indicated in Figures 29–10
through 29–13. Relocated epicenters for the
period 1955 to 1965 are also shown in Figure
29–9. Two of the solutions are characterized
by a predominance of strike-slip motion on
steeply dipping planes; the others are characterized by a large component of normal faulting.

Strike-slip Mechanisms. The observed radiation field for event 5 (Figures 29–9 and 29–
10) is similar to that for events 1 through 4.
The pattern of epicenters near 24°N and the
observed mechanism for event 5 are indicative
of a dextral transform fault. Nearly all the
first motions observed for event 8 (Figure
29–13), however, are opposite in sense to
those observed for events 1 through 5. The

over-all strike of the ridge does not appear to
change appreciably between events 5 and 8.
The east-west alignment of epicenters near
35°N, the bathymetry in this region (Tolstoy
and Ewing, 1949; Tolstoy, 1951; Heezen et
al., 1959), and the mechanism of event 8 are
indicative of a sinistral transform fault that
strikes approximately east. Heezen and Ewing
(1963) indicated a major fracture zone on the
ridge near 33°N. The configuration of the two
ridge crests and the distribution of epicenters
suggest that this fracture zone is also a sinistral
transform fault.

Normal Faults. Solutions for events 6 and
7 (Figures 29–11 and 29–12) display an entirely different pattern of first motions; all the
more distant stations recorded dilatations.
The inferred solutions are characterized by a

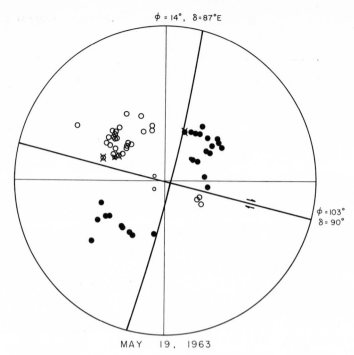

Figure 29-10
Mechanism solution for event 5. Table 29-1 and Figures 29-3 and
29-9 contain position and other pertinent data. Symbols same as
Figure 29-5

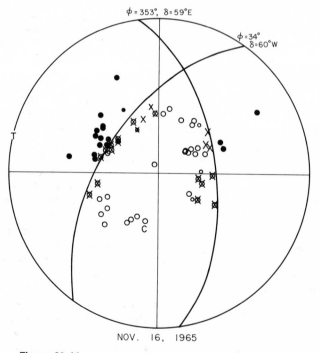

Figure 29-11
Mechanism solution for event 6. Table 29-1 and Figures 29-3 and
29-9 give epicentral and other data. T and C are inferred axes of
maximum tension and compression, respectively. Other symbols
same as Figure 29-5

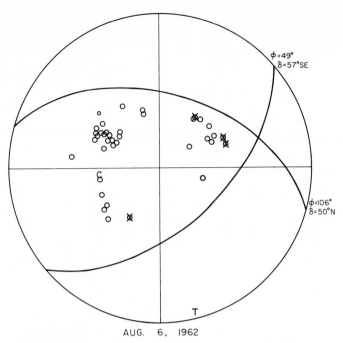

Figure 29-12
Mechanism solution for event 7. Table 29-1 and Figures 29-3 and
29-9 give epicentral and other data. Other symbols same as Figures
29-5 and 29-11

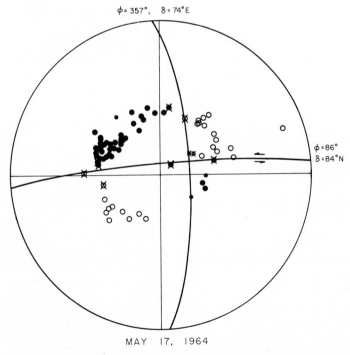

Figure 29-13
Mechanism solution for event 8. Table 29-1 and Figures 29-3 and
29-9 contain epicentral and other pertinent data. Symbols same as
Figure 29-5

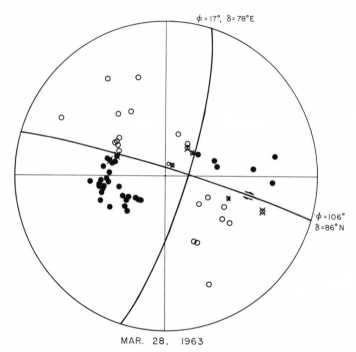

$\phi = 17°, \ \delta = 78°E$

$\phi = 106°$
$\delta = 86° N$

MAR. 28, 1963

Figure 29-14
Mechanism solution for Icelandic earthquake of March 28, 1963 (event
9 in Table 29-1). Location of earthquake is illustrated in Figures 29-3
and 29-15

large component of normal-fault motion on
planes that strike approximately parallel to
the axis of the ridge. Although a fracture zone
has been mapped near 30.0°N (Tolstoy and
Ewing, 1949; Tolstoy, 1951; Heezen and
Tharp, 1965), no major fracture zones seem
to intersect the ridge near events 6 and 7. In
this region the position of the rift valley was
mapped with enough precision so that offsets
greater than a few tens of kilometers should
have been resolvable.

Talwani (1964) suggested that all earthquakes on the mid-oceanic ridge may be related to fracture zones that are not prominent
enough to have been detected yet on the basis
of bathymetry. The existence of two types of
earthquake mechanisms suggests, however,
that this is not the case. Heirtzler et al. (1966)
concluded that, although earthquakes have
been detected south of Iceland on the Reykjanes ridge (Sykes, 1965), there is no bathymetric or magnetic evidence for offsets of the
ridge between 60°N and 63°N.

Inferred Axis of Maximum Tension. The
directions of principal stress difference were
estimated by assuming that these directions
bisect the angle between the nodal planes and
are located in a plane perpendicular to the
two nodal planes. This approximation is probably sufficiently accurate for the purposes of
the discussion in this paper. The term 'maximum tension' as used in this discussion refers
to the state of stress relative to a constant
mean normal stress (hydrostatic pressure).
The inferred axes of maximum tension for
events 6 and 7 are nearly horizontal and are
approximately perpendicular to the local strike
of the ridge. The orientation of these planes
for event 7 (Figure 29–12) may be in error by
several tens of degrees. For each of the earthquakes examined in this study the pattern of
first motions is similar to that predicted for a
shear displacement. Radiation patterns that
resemble those calculated for extension fractures (Honda, 1962; Savage, 1965) were not
observed.

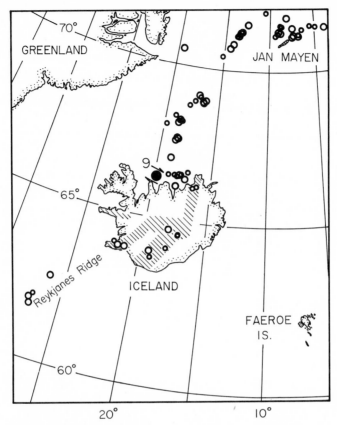

Figure 29-15
Relocated epicenters of earthquakes along a portion of the mid-
Atlantic ridge near Iceland (after Sykes, 1965). Diagonal hatching
denotes regions of postglacial volcanic activity in Iceland (after
Bodvarsson and Walker, 1964). Other symbols same as Figure 29-4

Iceland and Jan Mayen

First-motion data for the Icelandic earthquake
of March 28, 1963, are shown in Figure 29–
14. Epicenters near this shock (Figure 29–15)
and the configuration of the zone of postglacial
volcanic activity suggest that this earthquake
occurred near the western end of a dextral
transform fault. The mechanism for this earthquake is in agreement with this interpretation.

The bathymetry and the pattern of epicenters (Sykes, 1965) indicate that a major
fracture zone striking nearly east is present
near the island of Jan Mayen (Figure 29–15).
Since no large earthquakes occurred on this
fracture zone during the period 1962 to 1965,

data from the WWSSN could not be used to
investigate this feature. Solutions presented
by Lazareva and Misharina (1965) suggest
that the fracture zone is of the sinistral transform type.

EXTENSIONS OF THE
MID-OCEANIC RIDGE

Arctic Extension

Event 10, which is located near the edge of
the continental shelf of northern Eurasia, is
characterized by a large component of normal faulting (Figure 29–16). The epicenter of

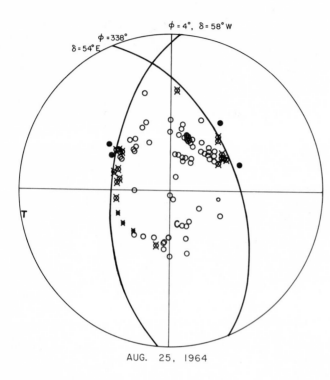

AUG. 25, 1964

Figure 29-16
Mechanism solution for earthquake (event 10) near the continental shelf of northern Siberia. Symbols same as Figures 29-5 and 29-11. Location of event shown in Figures 29-3 and 29-17

this event is located on the extension of the mid-oceanic ridge system into the Arctic basin and into northern Siberia (Figure 29–17). The seismic zone (Sykes, 1965), which is believed to define the crest of the ridge in this region, is nearly perpendicular to the inferred axis of maximum tension.

Balakina et al. (1960) reported that earthquakes in northeastern and central Prebaikalye (the region south of the area shown in Figure 29–17) are also characterized by a large component of normal faulting. Nevertheless, Lazareva and Misharina (1965) found that earthquakes in the area shown in Figure 29–17 are characterized by a predominance of strike-slip faulting and that the inferred axes of maximum compression are approximately perpendicular to the trend of the ridge. Many of these solutions are based on a rather poor distribution of stations, and a relatively large number of the observations are inconsistent with the inferred mechanisms. Solutions based on a poor distribution of stations and on poor data may be biased so that the computed displacements appear to be largely strike-slip (Adams, 1963; Ritsema, 1964; Isacks et al.,

1967). In this paper the distribution of stations in distance and azimuth is more extensive. Relatively little difficulty was encountered in distinguishing dip-slip from strike-slip motion. Until other high-quality solutions are available it does not seem possible to ascertain whether faulting in the Arctic is predominantly strike-slip faulting, normal faulting, or, as Wilson (1965c) suggests, a combination of extensional tectonics in the arctic basin and compressional tectonics in the Verkhoyansk mountains.

Orthogonality Criterion. In most experimental studies of earthquake mechanisms it is assumed that the two nodal planes are orthogonal. Even with some of the best published data it is not possible, however, to state that the two nodal planes are orthogonal with a precision better than about 10° or 20°. Data for events 6 and 10 (Figures 29–11 and 29–16) are extensive enough so that the validity of the orthogonality criterion may be tested. The fewest inconsistencies were encountered when the dilatational quadrants subtended an angle of about 70°. The use of a velocity model

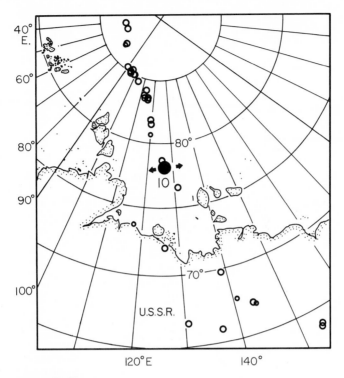

Figure 29-17
Epicenters along a portion of the mid-oceanic ridge in the Arctic
(after Sykes, 1965). Thick arrows denote inferred axis of maximum
tension for event 10. Other symbols same as Figure 29-4

more appropriate for crustal earthquakes would result in a somewhat smaller angle. Nevertheless, orthogonal nodal planes would not violate very many of the data for these two examples, and the case for nonorthogonal solutions does not seem to be established with certainty. The other solutions in this paper were fitted with orthogonal nodal planes. An angle of 70° between the dilatational quadrants would also satisfy the observations equally well. This uncertainty, however, mainly affects the orientation of the north-striking plane for the solutions characterized by a predominance of strike-slip motion; the orientation of the east-striking nodal plane, the plane chosen as the fault plane, is not noticeably affected by this uncertainty.

East Africa

The mechanisms of three earthquakes in East Africa all indicate a large component of nor-

mal faulting (Figures 29–18 and 29–19). Since the number of stations in Africa is limited, the position of the two nodal planes cannot be fixed with great precision; a strike-slip component nearly as large as the dip-slip component would still be compatible with the observed data. Studies of shear waves may help to define the mechanisms more precisely.

The inferred axis of maximum tension for event 13, an earthquake in the western rift of East Africa on March 20, 1966, is nearly perpendicular to the strike of the rift in the vicinity of the epicenter. Wohlenberg (1966) described fault displacements associated with this earthquake that start at about 0.7°N, 29.8°E and continue for about 40 km in a northerly direction. He reports a vertical displacement of 30 to 40 cm with the western side moving up relative to the eastern side. No horizontal displacement was observed. Loupekine (1966) reported a fault break of

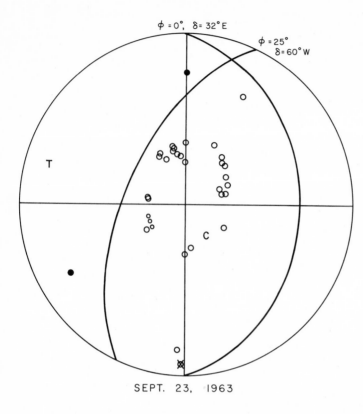

SEPT. 23, 1963

Figure 29-18
Mechanism solution for east African earthquake of September 23, 1963 (event 11). Symbols same as Figures 29-5 and 29-11. Table 29-1 and Figure 29-3 indicate location and other pertinent data. Solution for another earthquake in this swarm (event 12 on September 25, 1963) is nearly identical to that for event 11

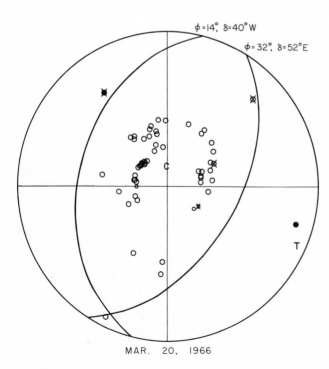

MAR. 20, 1966

Figure 29-19
Mechanism solution for east African earthquake of March 20, 1966 (event 14). Symbols same as Figures 29-5 and 29-11. Table 29-1 and Figure 29-3 give location and other data

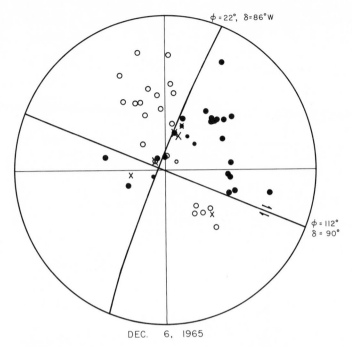

Figure 29-20
Mechanism solution for earthquake on the Rivera fracture zone
(event 14). Symbols same as Figure 29-5. Location and other
pertinent data shown in Figures 29-3 and 29-21 and in Table 29-1

15 to 20 km striking NNE. He indicated a throw of 2 meters with the downthrow on the east side. These observations are in close agreement with the inferred mechanism (Figure 29–19) if the nodal plane that strikes about 14° is chosen as the fault plane. Geological work and drilling in this portion of the western rift (Davies, 1951) are strongly indicative of a system of extensional tectonics; no evidence of compression was obtained.

Events 11 and 12 were part of a swarm of earthquakes that occurred beneath Lake Kariba near the border of Zambia and Rhodesia. The first motions for the two events are identical except that data were not available from two of the close stations for event 12. The inferred state of stress was such that a large water load would have increased the stress difference and hence may have triggered the swarm of earthquakes. Carder (1945) described a similar phenomenon in the Hoover Dam area.

Gough and Gough (1962, personal communication) recorded several thousand small earthquakes near Lake Kariba between 1961

and 1963. The large events in September 1963 occurred a few months after the lake reached its maximum level.

EAST PACIFIC RISE

General. First-motion data and inferred mechanisms for three earthquakes on the east Pacific rise are shown in Figures 29–20, 29–22, and 29–23. All three shocks were located on known fracture zones; the mechanisms are each characterized by a predominance of strike-slip motion. Solutions with a large component of dip slip were not detected on the east Pacific rise. Event 14 was located off the west coast of Mexico on the Rivera fracture zone (Figure 29–21); event 15 occurred on the Easter fracture zone near 27°S, 114°W. In both cases the inferred mechanisms, the bathymetry (Menard, 1966), and the seismicity (Sykes, 1963; Sykes, 1967) are indicative of dextral transform faults.

Because of its small size and remote location, it was not possible to obtain a unique solution for event 16, an earthquake on the

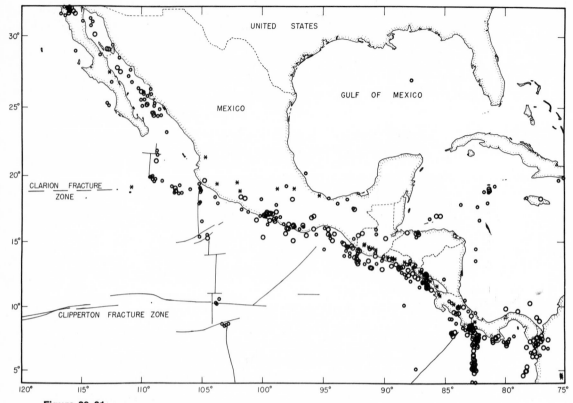

Figure 29-21
Epicenters of shallow earthquakes (depth less than about 70 km) in Mexico, Central America, and adjacent oceanic areas for the period 1954-1962 (after Sykes, 1967). Bathymetry of middle America trench after Fisher (1961); water depths in kilometers. Fracture zones and crests of East Pacific rise after Menard (1966). Mechanism solution indicated for event 14 on the Rivera fracture zone. Asterisks indicate historically active volcanoes (after Gutenberg and Richter, 1951); circles, epicenters

Eltanin fracture zone. If, however, strike-slip faulting on a steeply dipping plane with the same strike as the fracture zone is assumed, the sense of motion can be inferred from the first motions (Figure 29–23). The bathymetry (Menard, 1964), seismicity (Sykes, 1963), and mechanism are oriented in the correct sense for a sinistral transform fault.

It is interesting that the pattern of epicenters and the inferred mechanism of an earthquake on the Rivera fracture zone are rotated about 20° to 30° with respect to the strike of the Clarion zone further west (Figure 29–21). Menard (1966) and Chase and Menard (personal communication) have deduced a similar strike for the Rivera fracture zone. The configuration of the active seismic zone and the mechanism for event 14 support Menard's

(1966) interpretation that the Rivera zone and the Clarion fracture zone may be distinct tectonic features. Likewise, the Clipperton fracture zone (Figure 29–21) does not appear to be seismically active for more than a few tens of kilometers near the crest of the east Pacific rise (if at all). The data for the Rivera fracture zone support Vine's (1966) contention that the direction of spreading from the east Pacific rise has changed during the Pliocene from east-west to approximately northwest-southeast.

Recently a system of mid-oceanic ridges and fracture zones has been recognized off the coasts of British Columbia, Washington, Oregon, and northern California (Menard, 1964; Talwani et al., 1965b; Wilson, 1965b). Although no solutions of earthquake mecha-

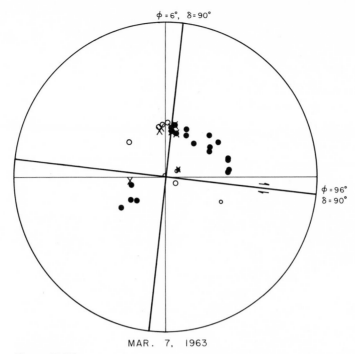

MAR. 7, 1963

Figure 29-22
Mechanism solution for earthquake on Easter fracture zone (event 15).
Location and other pertinent data in Figure 29-3 and Table 29-1.
Symbols same as Figure 29-5

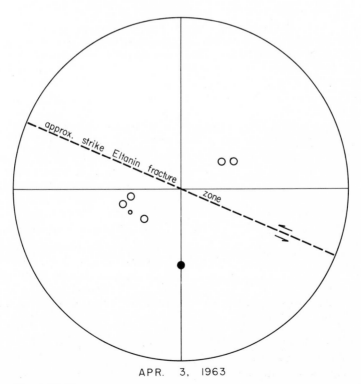

APR. 3, 1963

Figure 29-23
Mechanism solution for event 16 on the Eltanin fracture zone in the
southeast Pacific. Epicenter indicated in Figure 29-3. Symbols same as
Figure 29-5

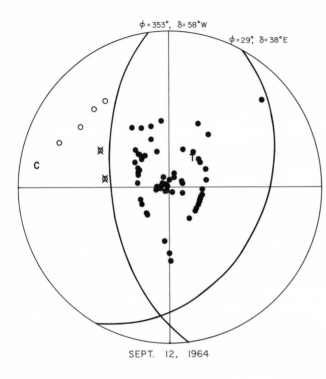

SEPT. 12, 1964

Figure 29-24
Mechanism solution for event 17
on the Macquarie ridge (Figures
29-3 and 29-25). Symbols same as
Figures 29-5 and 29-11

nisms were made for shocks in this region, an
analysis of future events should be of great
value in deciphering the tectonic interaction
of these features both with one another and
with the San Andreas system. A reconsideration of the Mendocino, Murray, and other
fracture zones in this region as transform
faults could lead to an interpretation wherein
the sense of movements inferred for the ocean
floor would no longer be the reverse of those
observed on land.

Macquarie Ridge. In the previous solutions
emphasis was placed on examining data for
features that are generally recognized as mid-
oceanic ridges or the continental extensions
of these ridges. Seismically active features,
such as the Macquarie, west Chile, and
Azores-Gibraltar ridges, have not been universally interpreted either as mid-oceanic ridges
or as fracture zones (Heezen et al., 1959;
Menard, 1965, 1966).

The inferred mechanism for event 17, an
earthquake located on the Macquarie ridge

(Figures 29–24 and 29–25), is unlike any of
the other solutions described in this paper.
Although interpretations involving greater
or lesser amounts of strike-slip motion than
shown in Figure 29–24 are possible, still a
large component of thrust faulting seems to
be demanded by the compressional arrivals
at the more distant stations.

The inferred compressional axis for this
solution is nearly horizontal and is approximately perpendicular to the trend of the Macquarie ridge (Brodie and Dawson, 1965) and
the strike of the seismic belt (Sykes, 1963;
Cooke, 1966). The orientation of the compressional axis is nearly the same as the direction of principal horizontal stress deduced by
Lensen (1960) for southern New Zealand.
This suggests that the tectonics of this portion of the Macquarie ridge may be more similar to the tectonics of New Zealand than to
the tectonics of the mid-oceanic ridge. Because
Macquarie Island is the only island in the deep
oceans to be well folded, prophyllitized, and
to have veins of sulphides, it seems to be
partly continental in character (Wilson, 1963).

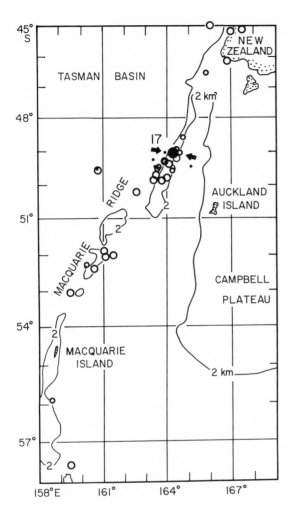

Figure 29-25

Epicenters of earthquakes for period 1955-1965 on the Macquarie ridge. Inferred axis of maximum compression shown for event 17. Bathymetry after Brodie and Dawson (1965). Other symbols same as Figure 29-4

COMPARISON OF INFERRED MECHANISMS WITH THOSE DEDUCED BY OTHER INVESTIGATORS

General Comparison. Stauder and Bollinger (1964a, b; 1965) have presented solutions for seven of the earthquakes considered in this paper. Stefánsson (1966) has also made an extensive study of the mechanisms of two of these seven shocks. The various solutions for these earthquakes are compared in Table 29–2. With the exception of event 4, an earthquake preceded by a small forerunner, the strikes and dips of the remaining solutions agree within about 15°. For some of the best solutions these differences are less than 5°. The strikes and dips reported in this paper

were read from an equal-area net and may be in error by 1° or 2°.

Stauder (1964) and Hodgson and Stevens (1964) have mentioned the general lack of agreement among the solutions presented by different workers for the same earthquake. It now seems evident that many of these disagreements arose from the poor quality of the data that were generally available before the installation of the WWSSN. Further studies with data of high quality should lead to a much better understanding of world tectonics and stress release.

Correct Sense of Motion. The determination of the sense of strike-slip motion on fracture zones has been one of the most important top-

Table 29-2
Comparison of mechanism solutions for earthquakes on the
mid-ocean ridges.

Event number	Author	Plane striking easterly		Plane striking northerly	
		Strike (deg)	Dip	Strike (deg)	Dip
2	Sykes	100	79°S	10	90°
	Stauder and Bollinger	101	80°S	12	86°W
3	Sykes	98	86°N	8	90°
	Stauder and Bollinger	87	78°N	−1	81°E
4	Sykes	90	86°N	0	84°E
	Stauder and Bollinger	(81)	(63°S)	(−61)	(34°N)
	Stefánsson	89	84°N	0	84°E
5	Sykes	103	90°	14	87°E
	Stauder and Bollinger	101	76°S	9	80°E
7	Sykes	106	50°N	49	57°E
	Stauder and Bollinger	91	45°N	45	55°E
9	Sykes	106	86°N	17	78°E
	Stauder and Bollinger	103	77°N	17	70°E
	Stefánsson, long-period P	107	88°N	18	79°E
	Stefánsson, short-period P	109	84°N	18	76°E
	Stefánsson, short-period S	104	81°N	12	77°E
15	Sykes	96	90°	6	90°
	Stauder and Bollinger	109	90°	20	90°

ics of this paper. Thus, it is important to inquire if the sense of motion inferred from studies of first motions could be in error by large amounts or even possibly reversed. Although geologic and geodetic evidence for displacements are not available for many earthquakes, evidence of this kind is in close agreement with the solutions inferred from first motions in the case of several prominent earthquakes. Very good agreement with geologic evidence was found in investigations of the mechanisms of the Fairview Peak earthquake of December 1954 (Romney, 1957), the southeast Alaska earthquake of 1958 (Stauder, 1960), and the Parkfield earthquake of 1966 (McEvilly, 1966). Hodgson (1957) has cited other examples of good or fair agreement between first-motion studies and geologic and geodetic evidence. Hence, it seems highly unlikely that the inferred sense of motion on fracture zones is incorrect.

CONCLUSIONS AND DISCUSSION

Because of several remarkable improvements in seismic instrumentation and in the availability of these data, mechanism solutions of high precision can now be obtained for many areas of the world. Earthquakes on the mid-oceanic ridge and the extensions of this ridge system into East Africa and the Arctic seem to be characterized by two principal mechanisms: (1) a predominance of strike-slip motion on steeply dipping planes or (2) a large component of normal faulting with the inferred axis of tension approximately perpendicular to the ridge.

Of the ten events of the first group seven were located on the mid-Atlantic ridge and three were found on the east Pacific rise; all ten occurred on major fracture zones that intersect the crest of the ridge. In each case the inferred sense of displacement is in agreement with that predicted for transform faults; the motion is opposite to that expected for a simple offset of the ridge. Thus, displacements inferred by the use of the assumption of simple offset appear to be incorrect, and worldwide shear nets deduced by such a procedure (e.g., Van Bemmelen, 1964) also indicate the wrong sense of motion on oceanic fracture zones.

Seismic activity on fracture zones is confined almost exclusively to the region between the two crests of the ridge. This distribution of activity is a strong argument in favor of the hypothesis of transform faults. Information obtained from earthquake mechanisms and seismicity is, of course, limited to a time scale of a few or a few tens of years for the mid-oceanic ridge. Since a very similar behavior was found for all the fracture zones investigated in this paper, the results are very likely to approximate closely the tectonics of the ridge system for much longer periods of time.

Of the six events that are characterized by a large component of normal faulting three occurred in East Africa, one was located near the continental shelf of Siberia, and two were situated on the mid-Atlantic ridge in a region where fracture zones have not been found. Although evidence for transform faults in East Africa was not found in this study, Bloomfield (1966) has suggested that a transform fault may connect two portions of the Malawi rift. This zone apparently follows an old line of weakness. Although a zone of weakness might account for the development of separate segments of ridge in cases where the distance between segments is small, it is more difficult to understand how zones of weakness could alone explain the development of separate ridge crests, when this distance is hundreds of kilometers as it is in the equatorial Atlantic and in the southeast Pacific near the Eltanin fracture zone.

As Wilson (1965) has pointed out, transform faults can only exist if there is crustal displacement. The deduced mechanisms and the distribution of earthquakes both seem to demand a process of sea-floor growth at or near the crest of the mid-oceanic ridge system. Transform faults also provide a relatively simple mechanism for continental drift. Although the East Africa rift valleys have been cited as examples of compressional, extensional, or strike-slip tectonics, more recent discussions have largely centered on extensional mechanisms. This interpretation became more prevalent with the realization that the mid-oceanic ridge system existed on a worldwide scale (Ewing and Heezen, 1956). Unfortunately, neither the seismological evidence nor the magnetic evidence for sea-floor growth gives information directly pertinent to the tectonics of the ridges in the third dimension, depth.

Although there seems to be a considerable variation in the orientation of earthquake mechanisms in island arcs, one factor that seems to be common among many of the solutions for these regions is that the horizontal component of compression is approximately perpendicular to the strike of the arc (Balakina et al., 1960., Lensen, 1960; Ichikawa, 1961; Honda, 1962; Balakina, 1962; Ritsema, 1964; Isacks et al., 1967). Thus, results from investigations of the mechanisms of earthquakes seem to be indicative of a system of compressional tectonics in island arcs and of extensional tectonics along the mid-oceanic ridges.

30

Seismology and the New Global Tectonics

BRYAN ISACKS
JACK OLIVER
LYNN R. SYKES
1968

From *Journal of Geophysical Research*, v. 73, p. 5855–5899, 1968. Reprinted with permission of the authors and the American Geophysical Union. Copyright 1968.

This paper relates observations from the field of seismology and allied disciplines to what is here termed the 'new global tectonics.' This term is used to refer in a general way to current concepts of large-scale tectonic movements and processes within the earth, concepts that are based on the hypotheses of continental drift (Wegener, 1966), sea-floor spreading (Hess, 1962; Dietz, 1961), and transform faults (Wilson, 1965c) and that include various refinements and developments of these ideas. A comprehensive view of the relationship between seismology and the new global tectonics is attempted, but there is emphasis on data from earthquake seismology, as opposed to explosion seismology, and on a particular version of the sea-floor spreading hypothesis in which a mobile, near-surface layer of strength, the lithosphere, plays a key role. Two basic questions are considered. First, do the observations of seismology support the new global tectonics in some form? To summarize briefly, they do, in general, give remarkable support to the new tectonics. Second, what new approaches to the problems of seismology are suggested by the new global tectonics? There are many; at the very least the new global tectonics is a highly stimulating influence on the field of seismology; very likely the effect will be one of revolutionary proportions.

The mobile lithosphere concept is based partly on an earlier study (Oliver and Isacks, 1967), but, as presented here, it incorporates ideas from Elsasser (1967), who independently developed a model with many similar features based on entirely different considerations, and ideas from Morgan (1968b) and Le Pichon (1968), who pursued this concept further by investigating the relative motion in plan of large blocks of lithosphere.

Figure 30–1 is a block diagram illustrating some of the principal points of the mobile lithosphere hypothesis. In a relatively undis-

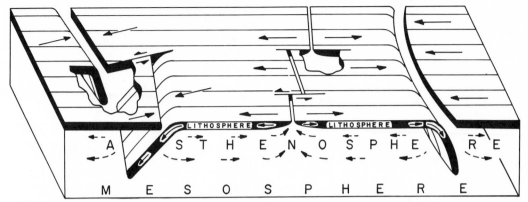

Figure 30-1
Block diagram illustrating schematically the configurations and roles of the lithosphere, asthenosphere, and mesosphere in a version of the new global tectonics in which the lithosphere, a layer of strength, plays a key role. Arrows on lithosphere indicate relative movements of adjoining blocks. Arrows in asthenosphere represent possible compensating flow in response to downward movement of segments of lithosphere. One arc-to-arc transform fault appears at left between oppositely facing zones of convergence (island arcs), two ridge-to-ridge transform faults along ocean ridge at center, simple arc structure at right

turbed section, three flat-lying layers are distinguished: (1) the *lithosphere*, which generally includes the crust and uppermost mantle, has significant strength, and is of the order of 100 km in thickness; (2) the *asthenosphere*, which is a layer of effectively no strength on the appropriate time scale and which extends from the base of the lithosphere to a depth of several hundred kilometers; and (3) the *mesosphere*, which may have strength and which makes up the lower remaining portion of the mantle and is relatively passive, perhaps inert, at present, in tectonic processes. (Elsasser refers to the lithosphere as the tectosphere and defines some other terms somewhat differently, but the terminology of Daly (1940) is retained here. The term 'strength,' which has many definitions and connotations, is used here, following Daly, in a general sense to denote enduring resistance to a shearing stress with a limiting value.) The boundaries between the layers may be gradational within the earth. The asthenosphere corresponds more or less to the low-velocity layer of seismology; it strongly attenuates seismic waves, particularly high-frequency shear waves. The lithosphere and the mesosphere have relatively high seismic velocities and propagate seismic waves without great attenuation.

At the principal zones of tectonic activity within the earth (the ocean ridges, the island arc or island-arc-like structures, and the major strike-slip faults) the lithosphere is discontinuous; elsewhere it is continuous. Thus, the lithosphere is composed of relatively thin blocks, some of enormous size, which in the first approximation may be considered infinitely rigid laterally. The major tectonic features are the result of relative movement and interaction of these blocks, which spread apart at the rifts, slide past one another at large strike-slip faults, and are underthrust at island arcs and similar structures. Morgan (1968b) and Le Pichon (1968) have demonstrated in a general way and with remarkable success that such movement is self-consistent on a worldwide scale and that the movements agree with the pattern of sea-floor spreading rates determined from magnetic anomalies at sea and with the orientation of oceanic fracture zones. McKenzie and Parker (1967) used the mobile lithosphere concept to explain focal mechanisms of earthquakes, volcanism, and other tectonic features in the northern Pacific.

Figure 30–1 also demonstrates these concepts in block diagram form. Near the center of the figure the lithosphere has been pulled apart, leaving a pattern of ocean ridges and transform faults on the surface and a thin lithosphere thickening toward the flanks beneath the ridge as the new surface material cools and gains strength. To the right of the diagram the lithosphere has been thrust, or has settled, beneath an island arc or a continental margin that is currently active. At an inactive margin the lithosphere would be unbroken or healed. The left side of the diagram shows two island-arc structures, back to back, with the lithosphere plunging in a different direction in each case and with a transform fault between the structures. Whereas the real earth must be more complicated, particularly at this back-to-back structure, this figure represents, in a general way, a part of the Pacific basin including the New Hebrides, Fiji, Tonga, the East Pacific rise, and western South America.

The counterflow corresponding to movement of the lithosphere into the deeper mantle takes place in the asthenosphere, as indicated schematically by the appropriate arrows in the figure. To what extent, if any, there is flow of the adjoining upper part of the asthenosphere in the same direction as the overlying lithosphere is an important but open question, partially dependent on the definition of the boundary. A key point of this model is that the pattern of flow in the asthenosphere may largely be controlled by the configurations and motions of the surface plates of lithosphere and not by a geometrical fit of convection cells of simple shape into an idealized model of the earth. It is tempting to think that the basic driving mechanisms for this process is gravitational instability resulting from surface cooling and hence a relatively high density of near-surface mantle materials. Thus, convective circulation in the upper mantle might occur as thin blocks of lithosphere of large horizontal dimensions slide laterally over large distances as they descend; a compensating return flow takes place in the asthenosphere. The process in the real earth must be more complex than this simple model, however. The reader is referred to Elsasser (1967) for a discussion of many points relating to this problem.

Alternatively, the surface configuration might be taken as the complicated response of the strong lithosphere to relatively simple convection patterns within the asthenosphere. Thus, the basic question whether the lithosphere or the asthenosphere may be thought of as the active element, with the other being passive, is not yet resolved. Probably, however, there has been a progressive thinning of the convective zone with time, deeper parts of the mantle having also been involved during early geologic time.

Figure 30–2, adapted from Le Pichon (1968) with additions, shows the plan of blocks of lithosphere as chosen by Le Pichon for the spherical earth and indicates how their movements are being accommodated on a worldwide scale. The remarkably detailed fit between this scheme, based on a very small number of rigid blocks of lithosphere (six) and the data of a number of fields, is very impressive. The number and configuration of the blocks of lithosphere is surely larger than six at present and almost certainly the pattern has changed within geologic time, but the present pattern must, in general, be representative of at least the Quaternary and late Tertiary. The duration of the current episode of sea-floor spreading is not known. Some evidence suggests that it began in the Mesozoic and has continued rather steadily to the present. Other evidence (Ewing and Ewing, 1967) indicates that the most recent episode of spreading began about 10 m.y. ago. This suggestion is considered here because it opens new possibilities for explaining certain seismological observations, particularly the configuration of the deep earthquake zones. Other explanations for such evidence are also considered, however.

With this one very simple version of the new global tectonics as background it is possible to begin considering the data, but in this process it soon becomes evident that much more detailed information on the earth is available and that the hypothesis and the earth model can be developed much further. These developments are presented later in the text as the relevant data are discussed.

This paragraph gives a brief review of some of the developments leading to the new global tectonics. A number of contributions vital to the development of the current position on this topic are cited, but the review is not intended to be comprehensive. The literature bearing on this topic is voluminous, is widespread in space and time, and differs in degree of relevance, so that a thorough documentation of its development is a job for a historian, not a scientist. The hypothesis of continental

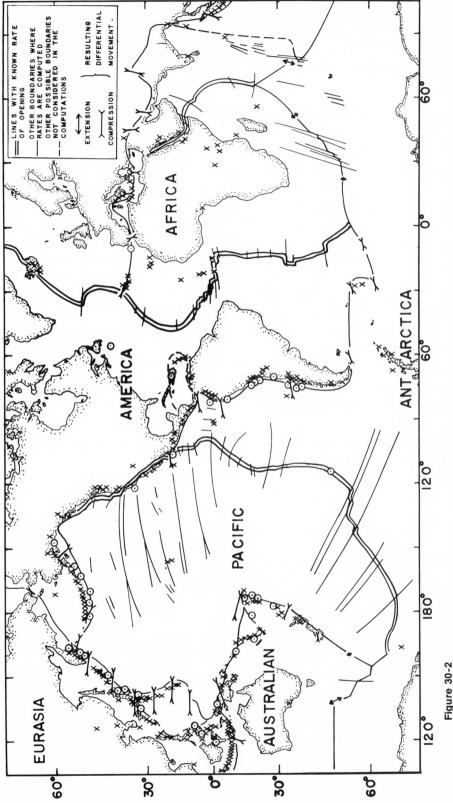

Figure 30-2
Computed rates of compression and extension along boundaries of six lithospheric blocks (after Le Pichon, 1968). Computed movements were derived from rates of spreading determined from magnetic data and from orientations of fracture zones along features indicated by double lines. The extensional and compressional symbols in the legend represent rates of 10 cm/yr; other similar symbols are scaled proportionally. Symbols appearing as diamonds represent small computed rates of extension for which the arrowheads coalesced. Historically active volcanos (Gutenberg and Richter, 1954) are denoted by crosses. Open circles represent earthquakes that generated tsunamis (seismic sea waves) detected at distances of 1000 km or more from the source

drift had a substantial impact on the field of geology when it was proposed in 1929 by Wegener (1966), but until recently it had not received general acceptance, largely because no satisfactory mechanism had been proposed to explain the movement, without substantial change of form, of the continents through the oceanic crust and upper mantle. When many new data became available, particularly in the fields of marine geology and geophysics, Hess (1962) and Dietz (1961) proposed that the sea floor was spreading apart at the ocean ridges so that new 'crust' was being generated there while older 'crust' was disappearing into the mantle at the sites of the ocean trenches. The driving mechanism for this spreading was thought to be convection within the mantle. The remarkable success with which the hypothesis of sea-floor spreading accommodated such diverse geologic observations as the linear magnetic anomalies of the ocean (Vine and Matthews, 1963; Pitman and Heirtzler, 1966), the topography of the ocean floor (Menard, 1965), the distribution and configuration of continental margins and various other land patterns (Wilson, 1965c; Bullard, 1964; Bullard et al., 1965), and certain aspects of deep-sea sediments (Ewing and Ewing, 1967) raised this hypothesis to a level of great importance and still greater promise. The contributions of seismology to this development have been substantial, not only in the form of general information on earth structure but also in the form of certain studies that bear especially on this hypothesis. Two specific examples are Sykes's (1967) evidence on seismicity patterns and focal mechanisms to support the transform fault hypothesis of Wilson (1965c) and Oliver and Isack's (1967) discovery of anomalous zones that appear to correspond to underthrust lithosphere in the mantle beneath island arcs.

There are many important seismological facts that are so apparent that they are commonly accepted without much concern as to their origin; they fall into place remarkably well under the new global tectonics. For example, the general pattern of seismicity, which consists of a number of continuous narrow active belts dividing the earth's surface into a number of stable blocks, is in accord with this concept. In part this agreement is by design, for the blocks were chosen to some extent on this basis, but data from other fields were used as well. That the end result is internally consistent is significant. Zones predicted by the theory to be tensional, such as ocean rifts, are sites of only shallow earthquakes (the thin shallow lithosphere is being pulled apart; earthquakes cannot occur in the asthenosphere), and the general level of seismic activity and the size of the largest earthquakes are lower there than in the more active compressional features. In the compressional features (the arcs) large, deep earthquakes occur and activity is high as the lithosphere plunges into the deeper mantle eventually to be absorbed. Deep earthquakes can occur only where former crustal and uppermost mantle materials are now found in the mantle. Where one block of the lithosphere is moving past another along the surface at the zones of large strike-slip faulting, seismic activity is shallow, but occasional rather large shallow earthquakes are observed. Some zones combine thrusting and strike-slip motion. The general pattern of earthquake focal mechanisms is in remarkable agreement with the pattern predicted by the movements of the lithosphere determined in other ways and provides much additional information on this process. The depth of the deepest earthquakes (about 700 km) has been reasonably well known, but unexplained, for many years. The mobile lithosphere hypothesis offers, at this writing, several possible alternatives to explain this observation. Many similar points are raised in the remainder of this paper. Other hypotheses on global tectonics, for example, the expanding earth and the contracting earth hypotheses, have been far less satisfactory in explaining seismological phenomena.

Certainly the most important factor is that the new global tectonics seem capable of drawing together the observations of seismology and observations of a host of other fields, such as geomagnetism, marine geology, geochemistry, gravity, and various branches of land geology, under a single unifying concept. Such a step is of utmost importance to the earth sciences and will surely mark the beginning of a new era.

In the remainder of this paper, the relationship between the new global tectonics and the field of seismology is discussed for a variety of topics ranging from seismicity to tsunamis, from earth structure to earthquake prediction. In each case what the authors judge to be representative, reliable evidence from the field of seismology is presented. This judgment is based on the quality of raw data and their analysis, not on the relation of the results to the new global tectonics. Reasonable speculation is presented where it seems proper. The organization of the paper is based not on the classical divisions of seismology but on the principal effects predicted by the new global tectonics and relevant to seismology. As a result of the remarkable capacity of the new global tectonics for unification, an obvious division of material among the sections was not completely achieved, however.

The first two sections present seismological evidence that the worldwide rift system and island arcs are the sources and sinks, respectively, for surficial material. The third section on compatibility of movements on a worldwide scale is closely related to the first two sections. Evidence from seismology on the structure of the mantle in terms of a lithosphere, an asthenosphere, and a mesosphere is so voluminous and well known that the section on this topic, the fourth section, presents primarily additional evidence of particular relevance to the new global tectonics. The fifth section, on the impact of the new global tectonics on seismology, is less documented by data than the previous sections partly because the impact of the new global tectonics is quite recent. The natural lag in pursuing this aspect is such that there has been to date relatively little emphasis in this particular field. This section is, then, somewhat speculative and, hopefully, provocative.

Few scientific papers are completely objective and impartial; this one is not. It clearly favors the new global tectonics with a strong preference for the mobile lithosphere version of this subject. In the final section, however, we report an earnest effort to uncover reliable information from the field of seismology that might provide a case against the new global tectonics. There appears to be no such evidence. This does not mean, however, that many of the data could not be explained equally well by other hypotheses (although probably not so well by any other single hypothesis) or that further development or modification of the new global tectonics will not be required to explain some of the observations of seismology. It merely means that, at present, in the field of seismology, there cannot readily be found a major obstacle to the new global tectonics.

MID-OCEAN RIDGES – THE SOURCES

Displacements along Fracture Zones. Recent studies of earthquakes have revealed several important facts about the nature of displacements on the ocean floor (Sykes, 1967, 1968). The recognition of the worldwide extent of the mid-ocean ridge system (Figures 30–2 and 30–3) (Ewing and Heezen, 1956) led to a great interest in the significance of this major feature to global tectonics. Although the ridge system appears to be a continuous feature on a large scale, the crest of the ridge is actually discontinuous in a number of places (Figures 30–1, 30–2, and 30–3). These discontinuities correlate with the intersections of the ridge and the major fracture zones – long linear zones of rough topography that resemble major fault zones on the continents. The apparent displacements along these fracture zones have been explained in at least three different ways, including simple offset of the ridge by strike-slip faulting (Vacquier, 1962), in situ development of the ridge crests at separate locations accompanied by normal faulting along fracture zones (Talwani et al., 1965b), and transform faulting (Wilson, 1965c).

Transform Faults

Although the concept of simple offset tacitly assumes the conservation of surface area, the growth or the destruction of surface area is basic to the definition of the transform fault. In this hypothesis the active portion (BC in Figure 30–4) of a strike-slip fault along which large horizontal displacement has occurred ends abruptly at the crest of a growing ocean ridge. The horizontal displacement along the fault is transformed (or absorbed) by sea-floor growth on the ridge; the growing ridge is, in

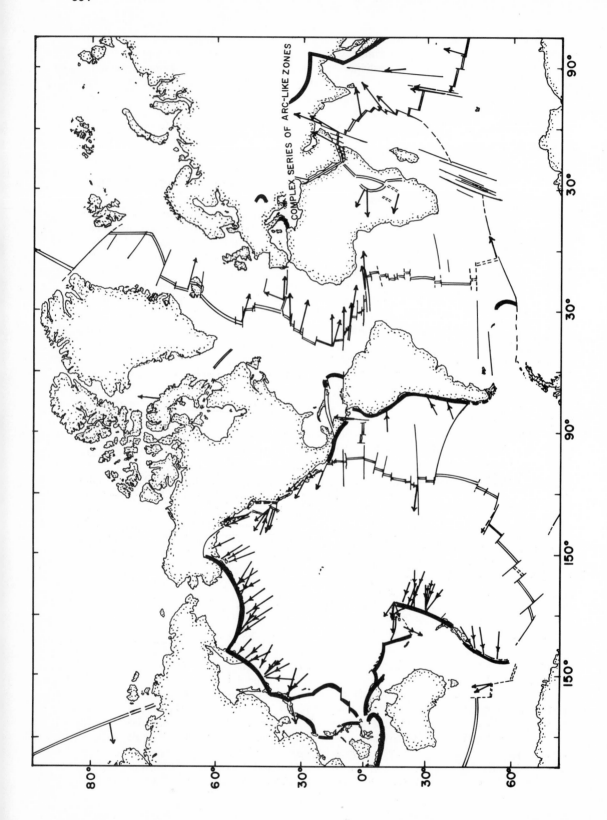

COMPLEX SERIES OF ARC-LIKE ZONES

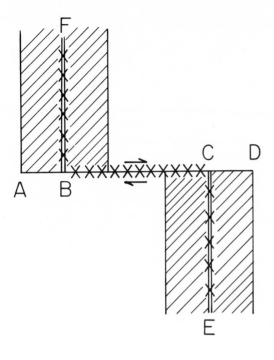

Figure 30-4
An idealized model of sea-floor spreading and transform faulting of the ridge-ridge type. Hatching indicates new surface area created during a given period of sea-floor spreading along the active ridge crests BF and CE. Present seismicity (indicated by crosses) is confined to ridge crests and to segment BC of the fracture zone AD. Arrows denote sense of shear motion along active segment BC

turn, terminated by the fault. Two separate segments of ridge crest can be joined (Figure 30–4) by a strike-slip fault of this type; these faults are called transform faults of the ridge-ridge type.

Wilson (1966c) recognized that the sense of shear displacement along transform faults of the ridge-ridge type would be exactly opposite that required for a simple offset of the two segments of ridge crest. He also pointed out that seismic activity along transform faults should be confined to the region between the two ridge crests (segment BC in Figure 30–4). If the crestal zones are being displaced by simple offset, however, seismic activity should be present along the entire length of the fracture zone.

Earthquake Mechanisms

The first motions of seismic waves from earthquakes offer a means for ascertaining the sense and type of displacements on fracture zones. First-motion studies are often called 'fault-plane solutions' or 'focal-mechanism solutions.' Although many earth scientists have been disappointed by the large uncertainties involved in many first-motion studies, investigations of focal mechanisms were vastly upgraded by the installation of the World-Wide Standardized Seismograph Network (Murphy, 1966). Reliable calibration, availability of data, high sensitivity, use of seismographs of both long and short periods, and greater geographical coverage are some of

Figure 30-3 *(facing page)*
Summary map of slip vectors derived from earthquake mechanism studies. Arrows indicate horizontal component of direction of relative motion of block on which arrow is drawn to adjoining block. Crests of world rift system are denoted by double lines; island arcs, and arc-like features, by bold single lines; major transform faults, by thin single lines. Both slip vectors are shown for an earthquake near the western end of the Azores-Gibraltar ridge since a rational choice between the two could not be made. Compare with directions computed by Le Pichon (Figure 30-2)

the more important characteristics of this network, which commenced operation in 1962. Various studies using data from these stations have confirmed that a double couple (or a shear dislocation) is an appropriate model for the radiation field of earthquakes (Stauder, 1967; Isacks and Sykes, 1968). Hence, the first motions observed at seismograph stations around the world may be used to determine the orientation and the sense of the shear motion at the sources of earthquakes in various tectonic regions. Additional background information on earthquake mechanisms will be introduced in later sections as further clarification is required.

Mechanisms along World Rift System

Sykes (1967) examined the focal mechanisms of seventeen earthquakes along various parts of the world rift system. In his study all the earthquakes located on fracture zones were characterized by a predominance of strike-slip motion. In each case the shear motion was in the correct sense for transform faulting (Figure 30–4), but it was consistently opposite in sense to that expected for simple offset. This is an instance in the earth sciences in which a yes-or-no answer could be supplied by data analysis. The sense of motion (left lateral) along one of the major fracture zones of the East Pacific rise (a branch of the mid-ocean ridge system) is illustrated in Figure 30–1.

Sykes also showed that earthquakes located on the ridge crests (segments BF and CE in Figure 30–4) but not located on fracture zones are characterized by a predominance of normal faulting. Normal faulting on ocean ridges had long been suspected because of the existence of a rift valley near the crest of large portions of the ridge system (Ewing and Heezen, 1956). More than fifty mechanism solutions (Figure 30–3) have now been obtained for the world rift system (Sykes, 1968; Tobin and Sykes, 1968; Banghar and Sykes, 1968); they continue to confirm the pattern of transform faulting and normal faulting described by Sykes. Nearly the same tectonic phenomenon is observed for each of the major oceans.

Seismicity

The distribution of earthquakes is another key piece of seismic evidence for the hypothesis of transform faulting. Nearly all the earthquakes on the mid-ocean ridges are confined either to the ridge crests or to the parts of fracture zones that lie between ridge crests (Sykes, 1967). Seismic activity along a fracture zone ends abruptly (Figure 30–4) when the fracture zone encounters a ridge; only a few earthquakes have been detected from the outer parts (segments AB and CD) of most fracture zones. If the transform fault theory is correct, the areas of sea floor that are now bounded by the outer inactive parts of fracture zones were once located between two ridge crests; these blocks of sea floor moved beyond either crest as spreading progressed. Thus, the age of deformation becomes older as the distance from an active crest increases.

Earthquake Swarms

The occurrence of earthquake swarms along the world rift system suggests that the crestal zone probably is characterized by submarine volcanic eruptions (Sykes et al., 1968). Earthquake swarms are a distinctive sequence of shocks highly grouped in space and time with no one outstanding principal event. Although these sequences sometimes occur in nonvolcanic regions, most of the world's earthquake swarms are concentrated in areas of present volcanism or geologically recent volcanism (Richter, 1958; Minikami, 1960). Large swarms often occur before volcanic eruptions or accompanying them; smaller swarms may be indicative of magmatic activity that failed to reach the surface as an eruption.

From the seismograph records at Palisades, New York, Sykes et al. (1968) recognized more than twenty swarms of earthquakes occurring during the past 10 years. These swarms commonly lasted a few hours or a few days. Although many of the larger earthquakes along the world rift system occur on fracture zones and are characterized by strike-slip faulting, nearly all the swarms are restricted to the ridge crests (segments BF and CE in Figure 30–4) and seem to be characterized by normal faulting. Swarms are commonly (but not always) associated with volcanic eruptions on islands or on or near the crest of the world rift system.

From a simulation of magnetic anomalies Matthews and Bath (1967) and Vine and Morgan (1967) estimate that most of the new surface material along the world rift system is injected within a few kilometers of the axis of the ridge. In Iceland, where the rift may be seen and studied in detail, postglacial volcanism is confined largely to the median rift that crosses the island (Bodvarsson and Walker, 1964). The rift apparently marks the landward continuation of the crest of the mid-Atlantic ridge (Figures 30–2 and 30–3).

The lack of weathering in rock samples, the young ages measured by radioactive and paleontologic dating of rocks and core materials, and the general absence of sediment as revealed by bottom photographs and by reflection profiling all attest to the youthful character of the crestal zones of the mid-ocean ridge (Ewing et al., 1964; Burckle et al., 1967; van Andel and Bowin, 1968; Dymond and Deffeyes, 1968). Thus, the occurrence of earthquake swarms is compatible with the hypothesis that new surface materials are being emplaced magmatically near the axes of the ocean ridges. The large earthquake swarms (and perhaps some of the smaller swarms) may be indicative of eruptions or magmatic processes in progress near the ridge crests. Nonetheless, more work is needed to ascertain if a causal relationship exists between the two phenomena.

Synthesis of Data for Ridges

Seismological evidence of various types seems to provide a definitive argument for the hypotheses of transform faulting and sea-floor spreading on the mid-ocean ridge system. These data are in excellent agreement with evidence of spreading from magnetic anomalies, ages of rocks, and the distribution of sediments (Vine, 1966; Heirtzler et al., 1968; Wilson, 1963a; Burckle et al., 1967). The world rift system must be recognized as one of the major tectonic features of the world. It is characterized nearly everywhere by extensional tectonics, sea-floor growth at its crest, and transform faulting on its fracture zones.

The focal depths and the maximum magnitudes of earthquakes, the narrowness of seismic zones, and the propagation of S_n waves along ocean ridges and transform faults will be described in the sections on worldwide compatability of movements and on additional evidence for the existence of the lithosphere.

Implications for Continental Drift

The similarity of the earthquake mechanisms along nearly the entire length of the ridge system suggests that transform faulting and spreading have been occurring in these regions for extended, but as yet unspecified, periods of time. The distribution of magnetic anomalies, paleomagnetic investigations, and the shapes of continental blocks that supposedly were split apart by spreading furnish a more complete history of the processes of sea-floor spreading and transform faulting. A question of particular interest is: Have the various segments of ridge grown in place; i.e., has the en echelon pattern of ridges and fracture zones prevailed throughout an episode of sea-floor spreading?

Both the Gulf of Aden and the Gulf of California are thought to have opened by continental drift during the last 25 m.y. (Hamilton, 1961; Laughton, 1966). If drift occurred in these areas, the displacements are at most a few hundred kilometers. If continental drift can be confirmed for these features, inferences about drift on an ocean-wide scale are placed on a much firmer basis.

Figure 30–5 shows the distribution of structural features, earthquake epicenters, and earthquake mechanisms for the Gulf of Aden (Sykes, 1968). Nearly all the epicenters are confined either to northeast-striking fracture zones or to the ridge that extends from a branch of the mid-ocean ridge (the Carlsberg ridge) near 9°N, 57°E to the western part of the Gulf of Aden near 12°N, 43°E. This ridge coincides with the rough central zone in Figure 30–5. As in other parts of the world rift system the earthquakes occurring on fracture

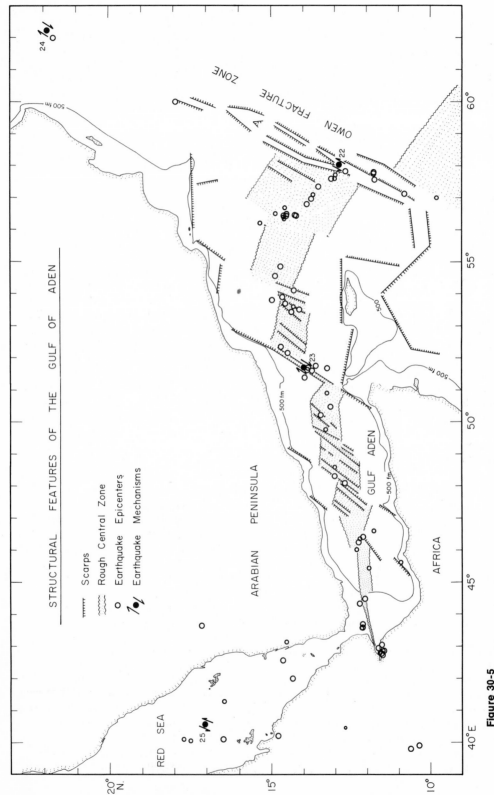

Figure 30-5
Structural features of the Gulf of Aden (after Sykes, 1968). Relocated epicenters of earthquakes for the period 1955 to 1966. Scarps and rough central zone after Laughton (1966). Seismicity and focal mechanisms support hypothesis of spreading by ocean-ridge–transform-fault mechanism

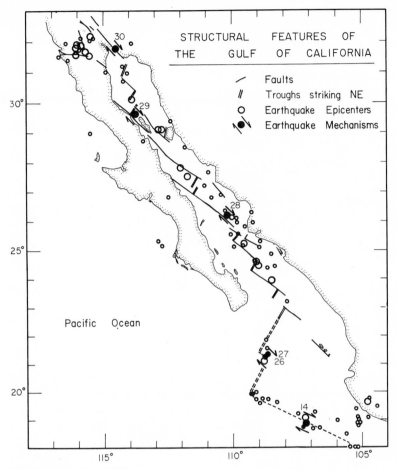

Figure 30-6
Structural features of the Gulf of California (after Sykes, 1968). Relocated epicenters of earthquakes for the period 1954 to 1962. Seismicity and focal mechanisms support the hypothesis of spreading by ocean-ridge–transform-fault mechanism

zones are mostly restricted to the regions between ridge crests. Mechanism solutions for events 22 and 23 (numbers after Sykes, 1968) indicate transform faulting of the ridge-ridge type.

If the opening of the Gulf of Aden was accomplished through a simple process of sea-floor spreading and transform faulting, the fracture zones should join points in Arabia and in Africa that were together before the drifting commenced. Also, the fracture zones should not continue into the two continental plates. Laughton (1966) has shown, in fact, that these faults do not continue inland. In

addition, his pre-Miocene reconstruction, in which the two sides of the Gulf of Aden are moved together parallel to the fracture zones, juxtaposes a large number of older structural features on the two sides of the gulf. The en echelon arrangement of segments of ridge is also mirrored in the stepped shape of the continental margins of Arabia and Africa. Hence, the present en echelon pattern seems to have prevailed since the initial breakup of these two blocks about 5 to 25 m.y. ago.

A similar pattern of en echelon ridges is present in the Gulf of California (Figure 30–6). Earthquake mechanisms from this region are indicative of a series of northwesterly striking transform faults with right-lateral displacement (Sykes, 1968). These transform faults, which are arranged en echelon to the San Andreas fault, connect individual segments of growing ridges in the Gulf of California. Hence, sea-floor spreading and transform faulting also were responsible for the displacement of Baja California relative to the mainland of Mexico. If these two blocks are reconstructed by horizontal displacements parallel to the northwesterly striking fracture zones, the peninsula of Baja California is placed in the indentation or 'nitch' of the mainland of Mexico near 21°N, 106°W. Thus, the two pieces appear to fit together in this reconstruction. Wilson (1965c) has pointed out that the stepped shape of the fracture-zone–ridge pattern in the equatorial Atlantic is mirrored in the stepped shape of the coastlines and the continental margins of Africa and Brazil.

ISLAND ARCS – THE SINKS

Almost anyone who glances casually at a map of the world is intrigued by the organized patterns of the island arcs. The close association of the major ocean deeps with these arcs is obvious and suggests exceptional subsidence in these zones, but other facts are equally striking. Nearly all the world's earthquakes in the deep and intermediate range, most of the world's shallow earthquakes, and the largest departures from isostatic equilibrium are associated with island arcs or arc-like structures, as shown by Gutenberg and Richter (1954). Volcanoes, sea-level changes, folding, faulting, and other forms of geologic evidence also demonstrate the high level of tectonic activity of these features. A concept of global tectonics in which the arcs do not play an important role is unthinkable. If crustal material is to descend into the mantle, the island arcs are suspect as sites of the sinks.

The asymmetrical structure of the arcs and the associated pattern of earthquake occurrence in the mantle led many investigators (e.g., Vening Meinesz, 1954; Benioff, 1954; Hess, 1962; Dietz, 1961) to postulate that the structures are the result of compressive stresses normal to the arc and are the sites of

vertical movements in various conventive schemes. Although such ideas were supported by the investigations of focal mechanisms of earthquakes made by Honda et al. (1956) and by the gravity studies of Vening Meinesz (1930) and Hess (1938), later analyses by Hodgson (1957b) for focal mechanisms and by Talwani et al. (1959) and Worzel (1965) for gravity led to different conclusions. This section reviews the data and shows that there is strong support for the compressive nature of island arcs and for their role as sites where surface material moves downward into the mantle. In particular, a variety of evidence supports the model of the arc shown in Figure 30–1. In this model the leading edge of the lithosphere underthrusts the arc and moves downward into the mantle as a coherent body. The proposed predominance of strike-slip faulting in island arcs (Hodgson, 1957b) is not in agreement with this model but appears, in view of recent and vastly improved seismic data, to be based on unreliable determinations of focal mechanisms (Hodgson and Stevens, 1964). The extensional features of structures based on gravity and seismic data appear to be surficial and can be reconciled with, and in fact are predicted by, the new hypothesis.

High-Q and High-Velocity Zones in the Mantle beneath Island Arcs

The gross structure of an idealized island arc as shown in Figure 30–1 is based on the results of Oliver and Isacks (1967). Their study was primarily concerned with the Fiji-Tonga area. Comparison of seismic waves generated by deep earthquakes in the seismic zone and propagated along two different kinds of paths, one along the seismic zone and one through an aseismic part of the mantle, demonstrated the existence of an anomalous zone in the upper mantle. The anomalous zone was estimated to be about 100 km thick and to be bounded on the upper surface by the seismic zone. Thus, the zone dips beneath the Tonga arc at about 45° and extends to depths of almost 700 km. The zone is anomalous in that attenuation of seismic waves is low and seismic velocities are high relative to those of the mantle at comparable depths elsewhere. Recent studies of the Japanese arc (Wadati et al.,

1967; Utsu, 1967) have confirmed the existence of such a structure for that region. Similar zones appear to be associated with other island arcs (Oliver and Isacks, 1967; Cleary, 1967; Molnar and Oliver, 1968).

The presence of a high-velocity slab beneath an island arc introduces a significant azimuthal variation in the travel times of seismic waves. Such variations with respect to source anomalies are shown by Herrin and Taggart (1966), Sykes (1966), Cleary (1967) and are indicated by the data of Carder et al. (1967) all for the case of the Longshot nuclear explosion. With respect to station anomalies such variations are shown by Oliver and Isacks (1967), Utsu (1967), Cleary and Hales (1966), and Herrin (1966) from data from earthquakes. These effects must therefore be taken into account as sources of systematic errors in the locations of earthquakes and the construction of travel-time curves. The large anomaly of Q associated with the slab must play a very important role in the Q structure of the mantle, especially for studies based on body waves from deep earthquakes. Studies in which this effect is ignored (e.g., Teng, 1968) must be reassessed on this basis.

Oliver and Isacks associated the anomalous zone with the layer of low attenuation near the surface to the east of Tonga. In their interpretation of the data, they correlated low attenuation with strength to arrive at the structure of Figure 30–1 in which the lithosphere, a layer of strength, descends into the mantle. This configuration suggests the mobility of the lithosphere implied in Figure 30–1 and described in the introduction. Based on current estimates of lithosphere velocities and other parameters, the down-going slab would be much cooler than its surroundings for a long time interval. Although there is little evidence supporting a direct relation between low attenuation and strength, an indirect relation based on the dependence of each parameter on temperature is reasonable. This point is discussed further in another section.

Bending of Lithosphere Beneath an Island Arc

The evidence supporting the model in which the lithosphere plunges beneath the island arc

is varied. To explore this point further, consider first the configuration of the upper part of the lithosphere in the vicinity of an island arc (Figure 30–7, *top*). Seismic refraction studies of a number of island arcs have been made. Invariably they show the surface of the mantle, which is shallow beneath the deep ocean, deepening beneath the trench, as suggested by Figure 30–7, *top*). Although some authors suggest that the mantle merely deepens slightly beneath the islands of the arc and shoals again behind the arc, evidence for such a structure is incomplete. Mantle velocities beneath the islands, where determined, are low, and there is no case for which the data could not be interpreted as suggested in Figure 30–7, *top* (see, e.g., Badgley, 1965, and Officer et al., 1959). In fact, the difficulty experienced in documenting the model in which the mantle is merely warped beneath the islands is evidence against this model. The main crustal layer as determined from seismic refraction studies seems to parallel the surface of the dipping mantle beneath the seaward slope of the trench. In some interpretations the crustal layer thins beneath the trench; in others it thickens or remains constant. Perhaps these are real variations from trench to trench, but the data are not always definitive.

Thinning of the crust has been interpreted by Worzel (1965) and others as an indication of extension, and there is considerable evidence in the structure of the sediments on the seaward slopes of many trenches supporting the hypothesis of extension (see, e.g., Ludwig et al., 1966). Figure 30–8, one of Ludwig's sections across the Japan trench, demonstrates this point dramatically. Several graben-like structures are seen on the seaward slope of the trench. Although such evidence for extension has been cited as an argument against sea-floor spreading and convection on the basis that down-going currents at the sites of the ocean deeps would cause compression normal to the arcs, the argument loses its force when the role of the lithosphere is recognized. All the evidence for extension relates only to the sediments and crust, i.e., the upper few kilometers of the lithosphere. For the models pictured in Figure 30–7 in which a thick strong layer bends sharply as it passes beneath the trench, extensional stresses are predicted near the surface on the

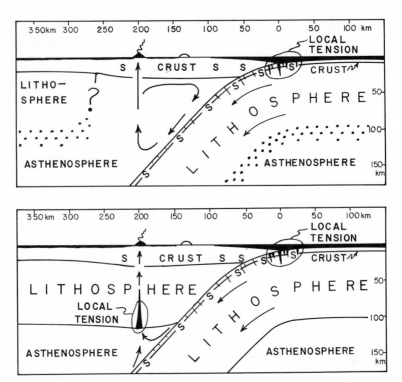

Figure 30-7

This figure shows vertical sections through an island arc indicating hypothetical structures and other features. Both sections show down-going slab of lithosphere, seismic zone near surface of slab and in adjacent crust, tensional features beneath ocean deep where slab bends abruptly and surface is free. (In both sections, S indicates seismic activity.) Above: A gap in mantle portion of lithosphere beneath island arc and circulation in mantle associated with crustal material of the slab and with adjoining mantle (Holmes, 1965). Below: The overriding lithosphere in contact with the down-going slab and bent upward as a result of overthrusting. The relation of the bending to the volcanoes follows Gunn (1947). No vertical exaggeration

convex side of the bend even though the principal stress deeper in the lithosphere may be compressional. Earthquake activity beneath the seaward slope of the trench is, in general, infrequent and apparently of shallow depth. The focal mechanisms that have been determined for such shocks indeed indicate extension as predicted, i.e. normal to the trench, the axis of bending (Stauder, 1968a; T. Fitch and P. Davis, personal communications). Stauder demonstrates this point very well in a paper on focal mechanisms of shocks of the Aleutian arc.

The extensional features also suggest a mechanism for including and transporting some sediments within the down-going rock layers. As implied by Figure 30–7, *top,* sedi-

ments in the graben-like features may be carried down to some depth in quantities that may be very significant petrologically, as suggested by Coats (1962). Probably not all the sediments carried into the trench by motion of the sea floor or by normal processes of sedimentation are absorbed in the mantle, however. There are large volumes of low-density material beneath the inner slope on the island side of most trenches (Talwani and Hayes, 1967) that may correspond to sediment scraped from the crust and deformed in the thrusting. Unfortunately, the structure of these low-density bodies in not well explored, and, in fact, the very difficulty of exploring them may be an indication of their contorted nature, which results from great deformation.

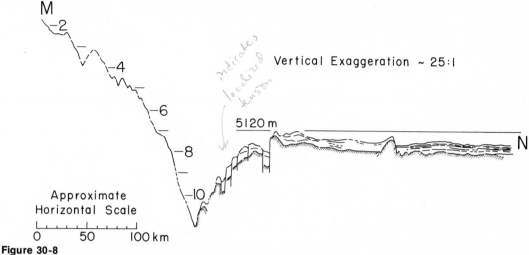

Figure 30-8
Seismic reflection profile across the Japan trench extending easterly along 35°N from point M near Japan
to point N (after Ludwig et al., 1966). Vertical scale represents two-way reflection time in seconds (i.e.,
1 sec = 1 km of penetration for a velocity of 2 km/sec). Note block faulting along seaward slope of trench
demonstrating extension in crust and inclusion of sediments in basement rocks. Also note shoaling of
oceanic basement on approaching trench as suggested by work of Gunn (1937). Vertical exaggeration
~ 25:1

The above arguments apply to trenches that
are relatively free of flat-lying sediments, such
as the Japan or Tonga trenches. The occurrence of substantial quantities of flat-lying
undeformed sediments in some other trenches
has been cited as evidence against underthrusting in island arcs (Scholl et al., 1968b).
Accumulation of underformed sediments
depends on the ratio of rate of sediment accumulation to rate and continuity of underthrusting, and such data must be evaluated
for each area with these factors in mind.
The South Chile trench, for example, has a
large sedimentation rate but no associated
deep earthquakes, suggesting little or no recent thrusting. The results of Scholl et al. must
be considered in this light. The Hikurangi
trench (east of northern New Zealand), another example of a partially filled trench, is
also thought to be in a zone of low convergence rate (see Le Pichon, 1968; and Figure
30–2).

Underthrusting Beneath Island Arcs

The shallow earthquakes mentioned above
that indicate extension normal to the arc occur
relatively infrequently and appear always to
be located beneath or seaward of the trench
axis. The earthquakes that account for most
of the seismic activity at shallow depths in
island arcs are located beneath the landward
slope of the trench and form a slab-like zone
that dips beneath the island arc (Fedotov et
al., 1963, 1964; Sykes, 1966; Hamilton, 1968;
Mitronovas et al., 1968). This point is illustrated in Figure 30–9, which shows a vertical
section through the Tonga arc. Note that, for
a wide range of depths, foci are confined to
a zone 20 km or less in thickness.

In the focal mechanisms of the shallow
shocks along the slab-like zone of the Tonga-
Kermadec arc, Isacks and Sykes (1968) find
consistent evidence for underthrusting of the
seaward block beneath the landward block.
Abundant evidence for a similar process for
various island arcs of the North Pacific is
found by Stauder (1962, 1968a), Udias and
Stauder (1964), Stauder and Bollinger (1964a,
1966a, b), Aki (1966), and Ichikawa (1966).
Critical evaluation of focal mechanism data
by Adams (1963), Hodgson and Stevens
(1964), Stauder (1964), and Ritsema (1964)
shows that the generalization that strike-slip
faulting is predominant in island arcs is based

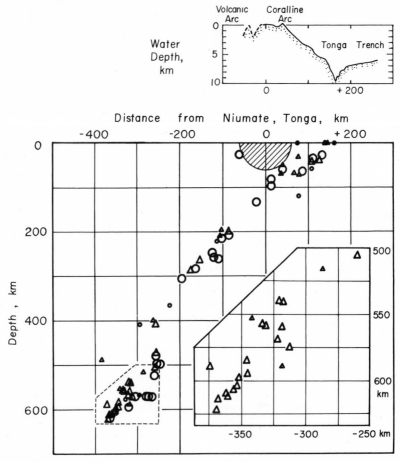

Figure 30-9
Vertical section oriented perpendicular to the Tonga arc. Circles represent
earthquakes projected from within 0 to 150 km north of the section; triangles
correspond to events projected from within 0 to 150 km south of the section.
All shocks occurred during 1965 while the Lamont network of stations in Tonga
and Fiji was in operation. Locations are based on data from these stations and
from more distant stations. No microearthquakes from a sample of 750 events
originated from within the hatched region near the station at Niumate, Tonga
(i.e., for S-P times less than 6.5 sec). A vertical exaggeration of about 13:1 was
used for the insert showing the topography (after Raitt et al., 1955); the horizontal
and vertical scales are equal in the cross section depicting earthquake locations.
Lower insert shows enlargement of southern half of section for depths between
500 and 625 km. Note small thickness (less than ~20 km) of seismic zone for
wide range of depths

on unreliable data and possible systematic
errors in the analyses. The recent data, greatly
improved in quality and quantity, indicate
that, in fact, dip-slip mechanisms are predominant in island arcs. The thrust fault
mechanisms characteristic of shallow earthquakes in island arcs thus appear to reflect
directly the relative movements of the converg-

ing plates of lithosphere and the downward
motion of the oceanic plate. The compatibility
of these motions as determined by focal mechanism data with the worldwide pattern of plate
movements is discussed later and is shown
to be excellent.

Considerable evidence for underthrusting
in the main shallow seismic zone exists in

other kinds of observations. Geodetic and geologic studies of the Alaskan earthquake of 1964 (Parkin, 1966; Plafker, 1965) strongly support the concept of underthrusting. Geologic evidence also indicates the repeated occurrence of such thrusting in this arc during recent time (Plafker and Rubin, 1967). Data from other arcs on crustal movements are voluminous and have not all been examined in light of the hypotheses of the new global tectonics. In fact, in many arcs the principal zone of underthrusting would outcrop beneath the sea and important data would be largely obscured. One important point can be made. It is well known (Richter, 1958) that vertical movements in island arcs are of primary importance. This contrasts with the predominantly horizontal movement in such zones as California, where strike-slip faulting predominates.

Other Shallow Activity in Island Arcs

In some island arcs there is appreciable shallow seismic activity landward of the principal seismic zone. This activity, which is distinct from that of the deep seismic zone below it, appears to be confined mainly to the crust and to be secondary to the activity along the main seismic zone. The Niigata earthquake of June 16, 1964, appears to be located in such a secondary zone of the North Honshu arc. The mechanism of this earthquake (Hirasawa, 1965) indicates that the axis of maximum compressive stress is more nearly horizontal than vertical and trends perpendicular to the strike of the North Honshu arc. It is interesting that this stress is also perpendicular to the trend of Neogene folding in North Honshu (Matsuda et al., 1967). These results might indicate some compressive deformation of the overriding plates in the models of Figure 30–7.

Deep Earthquakes: the Down-going Slab

The shallow seismic zone indicated by the major seismic activity is continuous with the deep zone, which normally dips beneath the island arc at about 45°. The thickness of the seismic zone is not well known in most cases, but it appears to be less than about 100 km and some evidence suggests that it may, at least in some areas, be less than 20 km. Figure 30–9 illustrates this point for a section through the Tonga arc. Although the surface approximating the distribution of hypocenters may be described roughly as above, it is clear that significant variations from this simple picture exist and are important. For example, the over-all dips may vary from at least 30° to 70°, and locally the variation may be greater, as suggested by the data in Figure 30–9. For the Tonga-Kermadec arc, the number of deep events is large and the zone can be defined in some detail (Sykes, 1966). Sykes was able to show, as a result of a marked curvature of the northern part of the Tonga arc, a clear correlation between the configuration of the deep seismic zone and surface features of the arc, thereby demonstrating the intimate relationship between the deep and the shallow processes.

For most arcs, however, the number of deep events, particularly since the World-Wide Standardized Seismograph Network has been in operation, is relatively small; therefore, the deep zones cannot be defined as precisely as one might desire. Nevertheless, sufficient information is available on the pattern of seismic activity so that the concept of the mobile lithosphere can be tested in general, and it must be assumed that subsequent detailed studies of other island arcs may reveal contortions in the seismic zone comparable with the contortions already found in Tonga-Fiji.

Focal Mechanisms

The simple underthrusting typical of the shallow earthquakes of the principal zones does not, in general, persist at great depths. For shocks deeper than about 100 km the orientation of the focal mechanisms varies considerably but exhibits certain clear-cut regularities. To understand these regularities, it is important to recall what is determined in a focal mechanism solution. The double-couple solution, which appears to be the best representation of most earthquakes, comprises two orthogonal nodal planes, either of which may

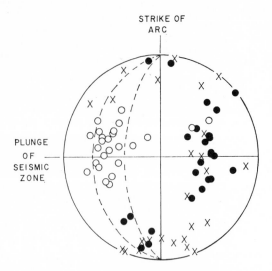

STRIKE OF
ARC

PLUNGE
OF
SEISMIC
ZONE

Figure 30-10
Orientations of the axes of stress as given by the double-couple focal mechanism solutions of deep and intermediate earthquakes in the Tonga arc, the Izu-Bonin arc, and the North Honshu arc. Open circles are axes of compression, P; solid circles are axes of tension, T; and crosses are null axes, B, all plotted on the lower hemisphere of an equal-area projection. The data, selected from available literature as the most reliable solutions, are taken from Isacks and Sykes (1968). Honda et al. (1957), Ritsema (1965), and Hirasawa (1966). The data for each of the three arcs are plotted relative to the strike of the arc (Tonga arc, N 20°E; Izu-Bonin, N 15°W; North Honshu arc, N 20°E). The dips of the zones vary between about 30° and 60° as indicated by the dashed lines in the figure. Note the tendency of the P axes to parallel the dip of the seismic zone and the weaker tendency for the T axes to be perpendicular to the zone

be taken as the slip plane of the equivalent shear dislocation. Bisecting these nodal planes are the axis of compression, P, in the quadrants of dilatational first motions and the axis of tension, T, in the quadrants of compressional first motions. The axis formed by the intersection of the nodal planes is the null, or B, axis, parallel to which no relative motion takes place. If one nodal plane is chosen as the slip plane, the pole of the other nodal plane is the direction of relative motion of the *slip vector*. It is important to realize that the primary information given by a double-couple solution is the orientation of the two possible slip planes and slip vectors. The interpretation of the double-couple mechanisms in terms of stress in the source region requires an assumption about the failure process. The P, T, and B axes correspond to the maximum, minimum, and intermediate axes of compressive stress in the medium *only* if the shear dislocation is assumed to form parallel to a plane of maximum shear stress in the medium, i.e., a plane that is parallel to the axis of intermediate stress and that forms a 45° angle to the axes of maximum and minimum stress.

Patterns of Focal Mechanisms for Deep Earthquakes

The most striking regularity in the orientation of the double-couple focal mechanisms of deep and intermediate earthquakes is the tendency of the P axes to parallel the local dip of the seismic zone. Figure 30–10 illustrates this point for the three zones (Tonga, Izu-Bonin, and North Honshu) for which reliable data are most numerous. This figure also shows that, although the orientation of the axes of tension and the null axes tend to be less stable than the compressional axes, these axes are not randomly oriented. The axis of tension tends to be perpendicular to the seismic zone; the null axis, parallel to the strike of the zone. These generalizations are shown schematically in Figure 30–11. The slip planes and slip directions are thus systematically *nonparallel* to the seismic zones; the orientations are therefore difficult to reconcile with a simple shearing parallel to the seismic zone as suggested by the common concept of the zone as a large thrust fault. Sugimura and Uyeda (1967) sought to reconcile the observations with that concept by postulating a reorientation of crystalline slip planes perpendicular to the axis of maximum compressive stress, such that in the case of horizontal compression the slip planes would tend to be vertical. Alternatively, Isacks and Sykes (1968) show that, if it is assumed that the slip planes form at angles with respect to the axis of maximum compressive stress that are not significantly different from 45°, then a very simple interpretation can be made on the basis of the model of Figure 30–1. In this interpretation the axis of

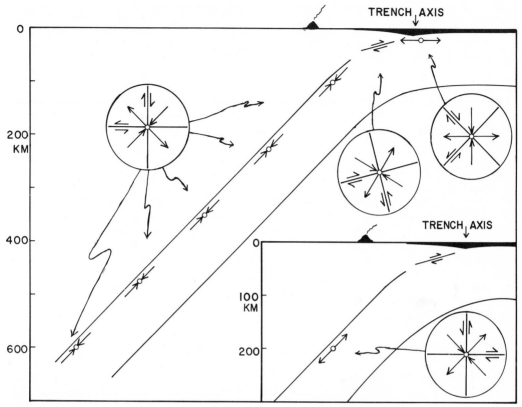

Figure 30-11
Vertical sections perpendicular to the strike of an island arc showing schematically typical orientations of double-couple focal mechanisms. The horizontal scale is the same as the vertical scale. The axis of compression is represented by a converging pair of arrows; the axis of tension is represented by a diverging pair; the null axis is perpendicular to the section. In the circular blowups, the sense of motion is shown for both of the two possible slip planes. The features shown in the main part of the figure are based on results from the Tonga arc and the arcs of the North Pacific. The insert shows the orientation of a focal mechanism that could indicate extension instead of compression parallel to the dip of the zone.

maximum compressive stress is parallel to the dip of the seismic zone (i.e., parallel to the presumed motion of the slab in the mantle), and the axis of least compressive stress is perpendicular to the zone or parallel to the thin dimension of the slab.

The tendency for the compressive axes to be more stable than the other two axes can be interpreted to indicate that the difference between the intermediate and the least principal stresses is less than the difference between the greatest and the intermediate principal stresses. In general, the stress state may be quite variable owing to contortions of the slab, as suggested by Figure 30–9. Possibly large variability in the orientations of the deep

mechanisms would, therefore, be expected, especially near parts of the zone with complex structure.

The important feature of the interpretation presented here is that the deep earthquake mechanisms reflect stresses in the relatively strong slab of lithosphere and do not directly accommodate the shearing motions parallel to the motion of the slab as is implied by the simple fault-zone model. The shearing deformations parallel to the motion of the slab are presumably accommodated by flow or creep in the adjoining ductile parts of the mantle.

Do the stresses in the slab vary with depth? In particular, the axis of least compressive stress, the T axis, may be parallel to the dip of the zone if the material at greater depths were sinking and *pulling* shallower parts of the slab (Elsasser, 1967). In the Tonga, Aleutian, and Japanese arcs, the focal mechanisms indicate that the slab is under compression parallel to its dip at all depths greater than about 75 to 100 km. In these arcs, therefore, any extension in the slab must be shallower than 75 to 100 km.

Very limited evidence from the Kermadec (Isacks and Sykes, 1968), New Zealand (North Island) (Adams, 1963), South American (A. R. Ritsema, personal communication), and Sunda (T. Fitch, personal communication) arcs suggests, however, that mechanisms indicating extension of the slab, as shown in the insert of Figure 30–11, may exist at intermediate depths in some arcs. Further work is required to distinguish such mechanisms from the underthrusting type of mechanism characteristic of earthquakes at shallow depths or from complex mechanisms related to changes in structure or contortions of the slab.

Process of Deep Earthquakes

The idea that deep earthquakes occur in downgoing slabs of lithosphere has important implications for the problem of identifying the physical process responsible for sudden shear failure in the environment of the upper mantle. That deep earthquakes are essentially sudden shearing movements and not explosive or implosive changes in volume is now extensively documented (see Isacks and Sykes, 1968, for references). Anomalous temperatures and composition might be expected to be associated with the down-going slab, either or both of which may account for the existence of earthquakes at great depths.

Several investigators (Raleigh and Paterson, 1965; Raleigh, 1967) concluded that dehydration of hydrous minerals can release enough water to permit shear fracture at temperatures between about 300° and 1000°C. Although Griggs (1966) and Griggs and Baker (1968), assuming normal thermal gradients, suggested that these reactions would not take place for depths greater than about 100 km, rates of underthrusting as high as 5 to 15 cm/yr suggest that temperatures low enough to per-

mit these reactions to occur may exist even to depths of 700 km. Certainly a re-evaluation of these processes is in order.

The lowest temperatures, the largest temperature gradients, and the largest compositional anomalies would probably be most marked near the upper part of the slab, i.e. the part corresponding to the crust and uppermost mantle in the surficial lithosphere. Thus, the seismic activity associated with these anomalies might be expected to concentrate near the upper part of the slab, as is suggested in Figures 30–7, *top* and 30–11 and supported by the data shown in Figure 30–9. Although catastrophic phase changes may be ruled out as direct sources of seismic waves on the basis of the radiation pattern of the waves, the possibility remains that the stresses responsible or partly responsible for shear failure may result from somewhat slower phase changes.

Seismic Activity versus Depth

Frequencies of earthquakes versus depth for several island arcs are shown in Figure 30–12. There are two main results emerging from these analyses. (1) In all island arcs studied the activity decreases in the upper 100 to 200 km approximately exponentially as a function of depth with a decay constant of about 100 km (Sykes, 1966). (2) At greater depths the seismic activity in many (but not all) island arcs increases relative to the exponential decay extrapolated from shallower depths, and the seismic activity shows a fairly well-defined maximum in some depth range in the upper mantle. The variation of seismic activity with depth is thus grossly correlated with the variation of seismic focal mechanisms with depth and supports the generalization that deep earthquake mechanisms have a different relationship to the zone than shallow mechanisms do. In this correlation the earthquakes that define the shallow exponential decay in seismic activity are characterized by the underthrusting type mechanisms, whereas the deeper earthquakes appear to be related to the stresses in the down-going slab.

There is an approximate correlation of the decrease in seismic activity versus depth with

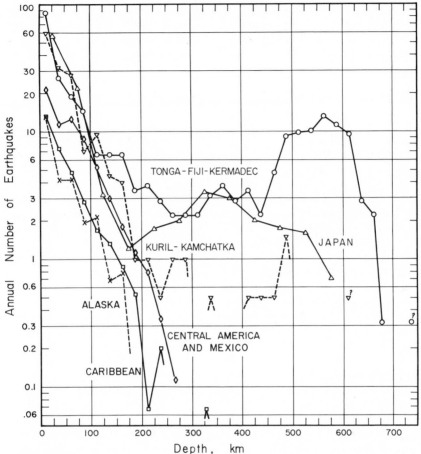

Figure 30-12
Number of earthquakes per 25 km depth intervals as function of depth for several
island arcs. Except for Japan, data are from Sykes (1966). Data for Japan expressed
as percentage of events per 50 km depth intervals (Katsumata, 1967). Since the various
curves were not normalized for the sample lengths and for the lower limit of detectability
in each area, only the relative shapes and not the absolute levels of the various curves
should be compared with one another The number of earthquakes per unit depth within
the upper 200 km of all these island arcs is approximately proportional to exp $(-Z/100)$,
where Z is the depth in kilometers. Peaks in activity below 200 km appear to fluctuate
both in amplitude and in depth among the various arcs

a similar general decrease in seismic velocities,
Q, and viscosity in the upper 150 km. These
effects may be related to a decrease in the
difference between the temperature and local
melting temperature. Thus, the decrease in
activity with depth may correspond to an increase in the ratio of the amount of deformation by ductile flow to that by sudden shear
failure. An implication of this interpretation
is that the exponential decay constant of 100
km may roughly indicate the thickness of the

overthrust plate of lithosphere. This interpretation is illustrated in Figure 30–7, *bottom,*
in which the overriding plate of lithosphere
is in 'contact' with the down-going plate along
the seismic zone. As shown in a later section,
the assumption that the depth distribution of
shallow earthquakes of the mid-ocean rift
system yields a measure of the thickness of
the lithosphere is not an unreasonable one.

Several lines of evidence do not, however,
support the existence of thick lithosphere

directly beneath and behind the arc as shown in Figure 30–7, *bottom*. Oliver and Isacks (1967) and Molnar and Oliver (1968) show that high-frequency S_n does not propagate across the concave side of island arcs, which probably indicates that the uppermost mantle there has low Q values. This result is in agreement with the low P_n velocities generally found beneath islands of many arcs. Also, the active volcanism and high heat flow characteristic of the concave side of island arcs (Uyeda and Horai, 1964; Sclater et al., 1968) suggest that the lithosphere may be thin there. These data are qualitatively fitted by the model shown in Figure 30–7, *top*. One implication of this figure is that at least part of the shallow earthquake zone might not result from the contact between two pieces of lithosphere but might instead indicate an embrittled and weakened zone formed by the downward moving crustal materials (Raleigh and Paterson, 1965; Griggs, 1967). In this case the exponential decay in activity might reflect changes in the properties of the earthquake zone as a function of depth. Thus, both models in Figure 30–7 must be retained for the present.

Although in some arcs such as the Aleutians or Middle America the exponential decay in activity appears to be the only feature present in curves of activity versus depth, most arcs exhibit a more or less well-defined maximum in activity in the mantle, as illustrated in Figure 30–12. The approximate ranges of depth of these maxima are shown in Figure 30–13 for several island arcs. The main point of this figure is to show that the depths of these maxima vary considerably among the various arcs and do not appear to be associated with any particular level of depth in the mantle, contrary to general opinion. As shown in Figure 30–13, the depths of the deep maxima are approximately correlated with the rates of convergence in the arcs as calculated by Le Pichon. As will be shown later (see Figure 30–16), the correlation is considerably better between the rate of convergence and *length* of the zone measured along the dip of the zone. Thus, the simplest explanation, one direct consequence of the model of Figure 30–1, is that the deep maxima are near the leading parts of the down-going slabs.

Two features of the distributions shown in Figures 30–12 and 30–13 may be related to certain levels of depth in the mantle. Although the length of the seismic zone measured along the dip of the zone exceeds 1000 km for several cases, no earthquakes with depths greater than 720 km have ever been documented. The U.S. Coast and Geodetic Survey (USCGS) has located no earthquakes with a depth greater than 690 km during the period 1961–1967. These depths are near the region of the mantle in which gradients in the variation of seismic velocities may be high (Johnson, 1967). Anderson (1967a) argues that this region corresponds to a phase change in the material. These depths may therefore be in some way related to the boundary of the mesosphere as shown in Figures 30–1 and 30–14 and as discussed in the next section. The second feature is the absence of maxima around 300 km. Thus, in a worldwide composite plot of activity versus depth, a minimum in activity near this depth generally appears.

Downward Movement of Lithosphere in the Mantle: Some Hypotheses

Although the concept is so new that it is difficult to make definitive statements, a brief speculative discussion is in order to emphasize the importance of these results in global tectonics. Figure 30–14, four hypothetical and very schematic cross sections of an island arc, illustrates some points that should be considered.

Figure 30–14,*a* shows a case in which the lithosphere has descended into the mantle beneath an island arc. In this model the lithosphere has not been appreciably modified with regard to its potential for earthquakes and the length of the submerged portion, *l*, and hence the depths of the deepest earthquakes are dependent on the rate of movement down dip and the duration of the current cycle of sea-floor spreading and underthrusting.

In Figure 30–14,*b* the leading edge of the descending lithosphere has encountered significant resistance to further descent and has become distorted. In this situation, the depth

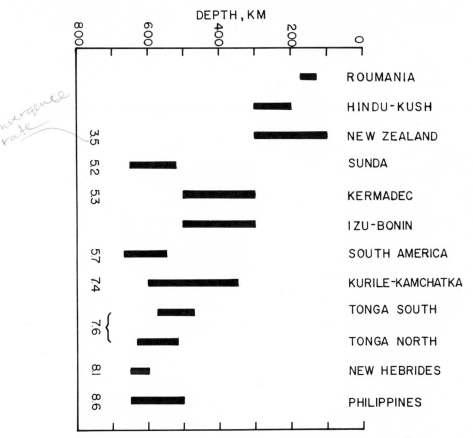

Figure 30-13

Depth range of maxima in the seismic activity (numbers of earthquakes) as a function of depth in island arcs and arc-like structures for which data are sufficiently numerous. The data are from Gutenberg and Richter (1954), Katsumata (1967), Sykes (1966), and listings of earthquakes located by the USCGS in the preliminary determination of epicenters (PDE). The numbers at the bottom of the figure give the rate (in centimeters per year) of convergence for the arc as plotted in Figure 30-2. Note that the maxima occur over a wide range of depths and that the depths appear to correlate, in general, with the calculated slip rate

of the deepest earthquakes in such a zone depends on the depth to the mesosphere. Sykes' (1966) analysis of the relocations of earthquakes in the Tonga arc (see also Figure 30-9) reveals the presence of contortions of the lower part of the seismic zone which might indicate a phenomenon similar to that pictured in Figure 30-14,*b*. This model has the interesting consequence that a cycle of seafloor spreading might be terminated or sharply modified by the bottoming of the lithosphere at certain points.

Figure 30-14,*c* indicates schematically that the depths of the deepest earthquakes might depend on modification of the lithosphere by its environment. In this model the depth of the deepest shocks depends on the rate of descent and the rate of modifications or absorption of the lithosphere.

In Figure 30-14,*d* the lower portion of the descending part of the lithosphere is not connected with the upper portion, possibly because it has pulled away as a result of a large density contrast between the sinking part of

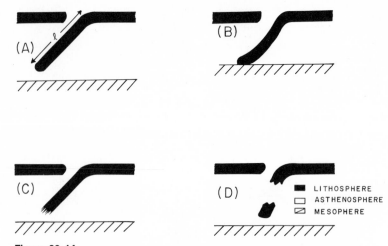

Figure 30-14
Four possible configurations of an underthrust plate of lithosphere in island
arcs. Solid areas indicate lithosphere; white area, asthenosphere; hatched
area, mesosphere. (a) Length 1 is a measure of the amount of underthrusting
during the most recent period of sea-floor spreading. (b) Lithosphere is
deformed along its lower edge as it encounters a more resistant layer (the
mesosphere). (c) Length of seismic zone is the product of rate of underthrusting
and time constant for assimilation of slab by upper mantle. (d) A piece (or
pieces) of the lithosphere becomes detached either by gravitational sinking or
by forces in the asthenosphere

lithosphere and the surrounding mantle. Another possibility is that the lower piece represents a previous episode of movement, so
that the break between the pieces then represents a period of quiescence in the surface
movements. For example, the Spanish deep
earthquake of 1954 (Hodgson and Cock,
1956) and the very deep earthquakes beneath
the North Island of New Zealand (Adams,
1963) might indicate isolated pieces of lithosphere. A variant of Figure 30–14,*d* is the
case in which movements of the ductile material of the asthenosphere, movements that
could be quite different from the movements
of the surficial lithospheric plates, could deform the slabs and possibly break off pieces.
For example, the marked contortions of the
deep seismic zone of Tonga may be explained
by such deformation. Thus, although the evidence is at present only suggestive, such evidence is important because of the implications
of the hypotheses with regard to the dynamics
of the system *within the asthenosphere*. Various combinations of the effects illustrated
in Figure 30–14 may also be considered.

Lateral Terminations of Island Arcs

The discussions above are based on, and apply largely to, the structure of an island arc
taken in a vertical section normal to the strike
of an arc. The three-dimensional configuration
of the arc must also be considered. The plate
model of tectonics provides, in a simple way,
for the termination of an island arc by the
abrupt or gradual transition to a transform
fault, by a decrease in the rate of convergence
to zero, or by some combination of these. In
the first case the relative movement that is
predominantly normal to the zone of deformation changes to relative movement that is
predominantly parallel to the zone. In the
second case the pole governing the relative
motion between the plates may be located
along the strike of the feature. Isacks and
Sykes (1968) describe what may be a particularly simple case of the first possibility. The
northern end of the Tonga arc appears to end
in a transform fault that strikes approximately
normal to the arc. In this case evidence is also
found for a scissors type of faulting in which

the downgoing Pacific plate tears away from the part of the plate remaining at the surface.

Summary of Data on Island Arcs

The lithosphere model of an island arc thus gives a remarkably simple account of diverse and important observed features of island arcs. The existence and distribution of earthquakes in the mantle beneath island arcs, the anomalous transmission properties of deep seismic zones, and the correspondences in the variations of seismic activity and the orientations of the focal mechanisms as functions of depth are all in agreement with the concept of a cooler, relatively strong slab moving through a relatively ductile asthenosphere. The bend in the slab required by this movement provides a simple means of reconciling the conflicting evidence for extension and compressional features of island arcs. The results suggest that there are two basic types of focal mechanisms. The first type is apparently confined to shallow depths and directly accommodates, and therefore indicates the direction of, the movements between the plates of lithosphere. The second type indicates stress and deformation *within* a plate of lithosphere and includes, besides the deep and intermediate earthquake mechanisms, the normal-faulting mechanisms at shallow depths beneath the axis of the trench and, possibly, the shallow earthquake mechanisms located landward of the underthrust zone. The deep earthquake zones may provide the most direct source of information on the movement of material in the asthenosphere and on the basic question of the relative importance of the lithospheric and asthenospheric motions in driving the convective system. The global pattern of motions between the plates, derived in part from the shallow focal mechanism in island arcs, provides a severe test of the hypothesis of plate movements in general and provides in particular key evidence for the conclusion that island arcs are the major zones of convergence and downward movements of the lithospheric plates. This evidence, including observations of directions as well as rates of movement, is discussed in the next section.

COMPATIBILITY OF MOVEMENTS ON A WORLDWIDE SCALE

In this section deformations along the world rift system and along island arcs and major mountain belts are examined for their internal consistency and for their global compatibility. The major finding is that these displacements can be approximated rather precisely by the interactions and the relative movements of large plates of lithosphere, much of the deformation being concentrated along the edges of the plates and relatively little deformation being within the individual plates themselves. It has long been recognized that recent deformations of the earth's surface are concentrated in narrow belts. These belts, which largely coincide with the major seismic zones of the world, include the world rift system, island arcs, and such island-arc-like features as active mountain belts and active continental margins. These major tectonic features do not end abruptly; they appear to be linked together into a global tectonic scheme.

Continuity of Seismic Belts and Distribution of Seismic Activity

Figure 30–15, a compilation of about 29,000 earthquake epicenters for the world as reported by the U.S. Coast and Geodetic Survey for the period 1961 to 1967 (Barazangi and Dorman, 1968), shows that most of the world's seismic activity is concentrated in rather narrow belts and that these belts may be regarded as continuous. Thus, if global tectonics can be modeled by the interaction of a few large plates of lithosphere, this model can account for most of the world's seismic activity as effects at or near the edges of the plates. Figure 30–15 also shows that the earthquakes occur much more frequently, in general, in the zones of convergence, the arcs and arc-like features, than in the zones of divergence, the ocean ridges. Along the ocean ridges, where the less complicated processes of tectonics are apparently occurring, the zones are narrow; on the continents, where the processes are apparently more complex, the zones are broad, and distinctive features are not easily resolved. Deep earthquake zones,

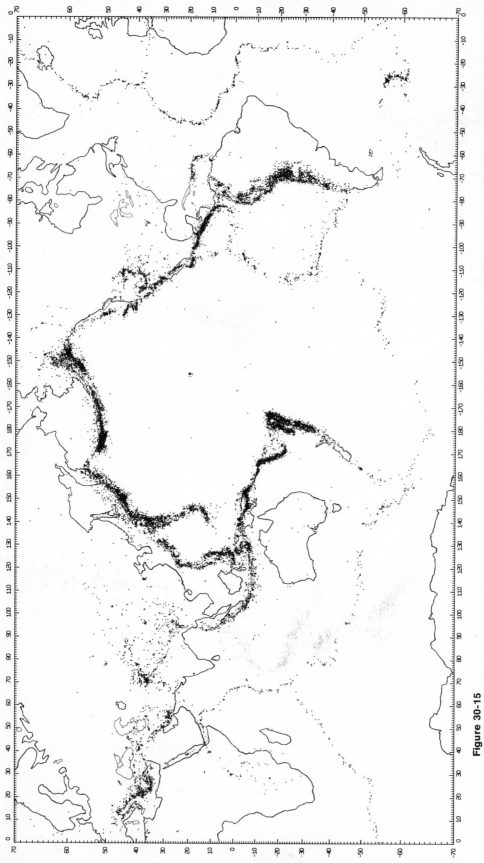

Figure 30-15
Worldwide distribution of all earthquake epicenters for the period 1961 through 1967 as reported by U.S. Coast and Geodetic Survey (after Barazangi and Dorman, 1968). Note continuous narrow major seismic belts that outline aseismic blocks; very narrow, sometimes steplike pattern of belts of only moderate activity along zones of spreading; broader very active belts along zones of convergence; diffuse pattern of moderate activity in certain continental zones

indicated in Figure 30–15 only by the width of epicentral regions behind arcs, correspond to the zones of underthrusting. Thus, all the major features of the map of seismic epicenters are in general accord with the new global tectonics. No other hypothesis has ever begun to account so well for the distribution of seismic activity, which must rank as one of the primary observations of seismology. The details of the configuration of the seismic belts of Figure 30–15 are discussed further in other sections of this paper.

Slip Vectors

Figure 30–3 illustrates the distribution of these major tectonic features and summarizes azimuths of motion as indicated by the slip vectors determined from various studies of the focal mechanisms of shallow-focus earthquakes. Deep and intermediate earthquakes as well as shallow earthquakes with normal faulting mechanisms near trenches were not represented in this figure, since these mechanisms are not thought to involve the relative displacements of two large blocks of lithosphere. Earthquake mechanisms were included in Figure 30–3 only when, by careful examination of the first-motion plots, we could verify that the slip vectors were reasonably well determined. These data were taken from Stauder (1962, 1968a), Stauder and Udias (1963), Stauder and Bollinger (1964a, 1966a, b), Harding and Rinehart (1966), Ichikawa (1966), Sykes (1967, 1968), Banghar and Sykes (1968), Isacks and Sykes (1968), and Tobin and Sykes (1968). Although no attempt was made to ensure that the collection of mechanism solutions represented all the reliable previous work, nonetheless, the data are thought to be representative; no attempt was made to select the data by criteria other than their reliability. In some cases such as the aftershocks of the great Alaska earthquake of 1964 and the aftershocks of the large Rat Islands earthquake of 1965 (Stauder and Bollinger, 1966a; Stauder, 1968a), only representative solutions were included, since the number of solutions for these regions was too large to depict clearly in Figure 30–3. Solutions cited as reliable by Ritsema (1964) as well as those of Honda et al. (1957), for ex-

ample, were not used because they pertain to subcrustal shocks.

From each mechanism solution used one of two possible slip vectors was chosen as indicative of the relative motions of the two interacting blocks of lithosphere. For each slip vector the arrow depicts the relative motion of the block on which it is drawn with respect to the block on the other side of the tectonic feature. Since the double-couple model (or shear dislocation) appears to be an excellent approximation to the radiation field of earthquakes, it is not possible to choose from seismic data alone one of the two possible slip vectors as the actual motion vector (or alternatively to choose one of two possible nodal planes as the fault plane).

Nevertheless, the choice of one vector is not arbitrary but is justified either by the orientation of the vectors with respect to known tectonic features such as fracture zones or by the consistency of a set of vectors in a given region. For earthquakes located on such major transform faults as the oceanic fracture zones, the San Andreas fault, and the Queen Charlotte Islands fault (Figure 30–3), one of the slip vectors is very nearly parallel to the transform fault on which the earthquake was located. Observed surface breakage and geodetic measurements in some earthquakes, the alignment of epicenters along the strike of major transform faults, the linearity of fracture zones, and petrological evidence for intense shearing stresses in the vicinity of fracture zones constitute strong evidence for making a rational choice between the two possible slip vectors (Sykes, 1967). The choice of the other possible slip vector (or nodal plane) for many of the oceanic fracture zones would indicate strike-slip motion nearly parallel to the ridge axis. On the contrary, earthquakes along the ridge crest but not on fracture zones do not contain a large strike-slip component but are characterized by a predominance of normal faulting.

Evidence for Motions of Lithospheric Plates

One of the most obvious features in Figure 30–3 is that the slip vectors are consistent with the hypothesis that surface area is being created along the world rift system and is being destroyed in island arcs. Along the mid-Atlantic ridge, for example, slip vectors for more than ten events are nearly parallel to one another and are parallel to their neighboring fracture zones within the limits of uncertainty in either the mechanism solutions (about 20°) or the strikes of the fracture zones.

Morgan (1968) and Le Pichon (1968) showed that the distribution of fracture zones and the observed directions and rates of spreading on ocean ridges as determined from geomagnetic data could be explained by the relative motions of a few large plates of lithosphere. They determined the poles of rotation that describe the relative motion of adjacent plates on the globe. Our evidence from earthquake mechanisms and from the worldwide distribution of seismic activity is in remarkable agreement with their hypothesis. Although their data are mostly from ridges and transform faults, earthquake mechanisms give the relative motions along island arcs as well as along ridges and transform faults.

Le Pichon used data from ocean ridges to infer the direction of motion in island arcs. His predicted movements (Figure 30–2), which are based on the assumption of conservation of surface area and no deformation within the plates of lithosphere, compare very closely with mechanism solutions in a number of arcs. This agreement is a strong argument for the hypothesis that the amount of surface area that is destroyed in island arcs is approximately equal to the amount of new area that is created along the world rift system. Thus, although modest expansion or contraction of the earth is not ruled out in the new global tectonics, rapid expansion of the earth is not required to explain the large amounts of new materials added at the crests of the world rift system. This approximate equality of surface area is, however, probably maintained for periods longer than thousands to millions of years, but minor imbalances very likely could be maintained for shorter periods as strains within the plates of lithosphere. More exact knowledge of these imbalances could be of direct interest to the problem of earthquake prediction.

Figure 30–3 suggests that nearly all the east-west spreading along the East Pacific rise and the mid-Atlantic ridge is taken up either by the island arcs of the western Pacific or by the arc-like features bordering the west coasts of Central and South America. Much of the north-south spreading in the Indian Ocean is absorbed in the Alpide zone, which stretches from the Azores-Gibraltar ridge across the Mediterranean to southern Asia and then to Indonesia.

Relative Motions in the Southwest Pacific

Le Pichon's computed directions of motion in the Tonga and Kermadec arcs of the southwest Pacific agree very closely with mechanisms we obtained from a special study of that region (Isacks and Sykes, 1968). Mechanisms south of New Zealand along the Macquarie ridge (Sykes, 1967; Banghar and Sykes, 1968) indicate a combination of thrust faulting and right-lateral strike-slip motion. These data suggest that the pole of rotation for these two large blocks is located about 10° farther south than estimated by Le Pichon (1968). Although the Pacific plate is being *underthrust* in the Tonga and Kermadec arcs and in northern New Zealand, this plate is apparently being *overthrust* along the Macquarie ridge (see also Summerhayes, 1967). In this interpretation the Alpine fault is a right-lateral transform fault of the arc-arc type that connects two zones of thrusting with opposing dips. Computed slip vectors for this region also indicate a component of thrust faulting either along the Alpine fault itself or in other parts of the South Island of New Zealand. Wellman's (1955) studies of Quaternary deformation, which indicate a thrusting component as well as a right-lateral strike-slip component of motion along the Alpine and associated faults, seem to be in general accord with this concept.

Likewise, the Philippine fault appears to connect a zone of underthrusting of the Pacific floor near the Philippine trench with a region of overthrusting west of the island of Luzon near the Manila trench. Also, the existence of a deep seismic zone in the New Hebrides arc that dips toward the Pacific rather than away from the Pacific as in the Tonga arc (Sykes, 1966) is understandable if the Pacific plate is being overthrust in the New Hebrides and underthrust in Tonga (Figure 30–1). The ends of these two arcs appear to be joined together by one or more transform faults that pass close to Fiji, but additional complications appear to exist in this area.

North Pacific

The uniformity in the slip directions and the distribution of major faults along the margins of the North Pacific (with perhaps some systematic departure in the slip directions off the coast of Washington and Oregon) indicate that only two blocks are involved in the major tectonics (Tobin and Sykes, 1968; Morgan 1968b; McKenzie and Parker, 1967). In this scheme, the San Andreas fault, the Queen Charlotte Islands fault, and a series of northwesterly striking faults in the Gulf of California are interpreted as major transform faults, (Wilson, 1965 b, c). The observed rates of displacement along the San Andreas as determined geodetically (Whitten, 1955, 1956) are very similar to the rates determined from the seismicity by means of a dislocation model (Brune, 1968). These rates are in close agreement with the rates inferred from magnetic anomalies for the region of growing ridges at the northwestern end of the San Andreas fault (Vine and Wilson, 1965; Vine, 1966). Estimates of the total amount of offset along the San Andreas (Hamilton and Myers, 1966) are comparable to the amount of offset needed to close the Gulf of California and to the width of the zones of northeasterly striking magnetic anomalies off the coast of Oregon and Washington (Vine and Wilson, 1965).

Thus, the present tectonics of much of the North Pacific can be related to the motion of the Pacific plate relative to North America and northeastern Asia. The slip vectors strike northwesterly along the west coast of the United States and Canada, represent nearly pure dip-slip motion in southern Alaska and the eastern Aleutians, and have an increasingly larger strike-slip component along the Aleutian arc as the longitude becomes more westerly. Displacements in the Kurile, Kamchatka, and Japanese arcs are nearly pure dip-slip

and represent underthrusting of the Pacific plate beneath the arcs.

The system of great east-west fracture zones in the northeastern Pacific (Figure 30–2) apparently was formed more than 10 m.y. ago. Since that time the directions of spreading changed from east-west to their present northwest-southeast pattern (Vine, 1966). The more westerly strike of the slip vectors along the Juan de Fuca and Gorda ridges is consistent with the hypothesis that the area to the east of these ridges represents a small separate plate (Morgan, 1968b; McKenzie and Parker, 1967) that was underthrust beneath the coast of Washington and Oregon to form the volcanoes of the Cascade range. A few earthquakes that have been detected in western Oregon and Washington are apparently of a subcrustal origin (Neumann, 1959; Tobin and Sykes, 1968). The tectonics of this block appears to be quite complex. Internal deformation within this block, as indicated by the presence of earthquakes (Tobin and Sykes, 1968), and the small number of subcrustal events previously mentioned suggest that the tectonic regime in this region is being readjusted.

Depths of Earthquakes, Volcanism, and Their Correlation with High Rates of Underthrusting

The depths of the deepest earthquakes, the presence of volcanism, and the occurrence of tsunamis (seismic sea waves) seem to be related closely to the present rates of underthrusting in island arcs. In the series of arcs that stretches from Tonga to the Macquarie ridge the depths of the deepest earthquakes generally decrease from about 690 km in the north to less than 100 km in the south (Hamilton and Evison, 1968). The rates of underthrusting computed by Le Pichon also decrease from north to south. Volcanic activity, which is prominent north of the South Island of New Zealand, dies out when the deepest part of the seismic zone shoals to depths less than about 100 to 200 km. Similarly, the depth of deepest activity in the Aleutian arc and the number of volcanoes (Figure 30–2) appear to decrease from east to west as the rate of underthrusting decreases. In this arc the rate of underthrusting decreases because the slip vectors become more nearly parallel to the arc. Volcanism and the depth of seismic activity increase spectacularly as the slip vectors change from a predominance of strike-slip motion in the western Aleutians to largely dip-slip motion in Kamchatka, the Kurile Islands, and Japan. Most of the world's active volcanoes are located either along the world rift system or in regions that contain intermediate-depth earthquakes. Hence, the latter volcanoes are presumably located in the regions where the lithosphere has been underthrust to depths of at least 100 km.

Tsunamis

Although some prominent seismologists (e.g., Gutenberg, 1939) have argued that some of the largest tsunamis (seismic sea waves) are generated by submarine slides, many other geophysicists maintained that sudden dip-slip motion along faults during large earthquakes is the principal generating mechanism for most of the world's more widespread tsunamis. L. Bailey (personal communication), following a lecture by Sykes, suggested that tsunami generation might correlate with areas of large dip-slip motion. This suggestion may have considerable merit.

Figure 30–2 shows the epicenters of earthquakes for which tsunamis were detected at distances of 1000 km or greater (Heck, 1947; Gutenberg and Richter, 1954; U.S. Coast and Geodetic Survey, 1935–1965). The distance criterion was used to eliminate waves of more local origin, some of which actually may be related to seiches or to submarine slides. Although this compilation undoubtedly is not complete even for the present century, the conclusions drawn here should not be seriously affected by the choice of data.

Since most of the earthquakes generating tsunamis are located in regions that are characterized by a high rate of dip-slip motion (Figure 30–2), the two phenomena appear to be causally related in many cases. Although most of the world's largest earthquakes occur in these areas and are characterized by a large component of thrust faulting, large earth-

quakes along major strike-slip faults near the coasts of southeast Alaska and British Columbia have not generated sea waves that were observable at distances greater than about 100 km. An earthquake located off the coast of California in 1927, however, generated a wave detected in Hawaii (Gutenberg and Richter, 1954). Other generating mechanisms, such as volcanic eruptions and submarine slumping, cannot be excluded as the causative agents for at least some tsunamis.

Gutenberg (1939) argued that the epicenters of several earthquakes generating tsunamis were located on land. Although the epicenter, which represents the point of initial rupture, may be located inland, the actual zone of rupture in great earthquakes is now known to extend for several hundred kilometers. In many of the great earthquakes associated with island arcs and with active continental margins at least a portion of the zone of rupture is located in water-covered areas. Hence, the locating of the epicenter itself is not an argument for the absence of significant vertical displacements in nearby submarine areas. Utsu (1967) has shown that, as a result of the anomalous zones in the mantle beneath island arcs, locations of shallow shocks based on teleseismic data are usually landward of the actual location. Both improved locations based on a better knowledge of seismic wave travel times in anomalous regions such as island arcs and rapid determinations of focal mechanisms may be of substantial value in tsunami warning systems.

Length of Seismic Zones in Island Arcs

The systematic changes in the seismic zones in the Aleutians and in the southwest Pacific suggest that the lengths of zones of deep earthquakes might be a measure of the amount of underthrusting during the last several million years. Using the maps of deep and intermediate-depth earthquakes prepared by Gutenberg and Richter (1954), Oliver and Isacks (1968) estimated the area of these zones and divided the total area by the length of the world rift system (about two great circles) and by 10 m.y., which is the duration of the latest cycle of spreading based on data from ocean-

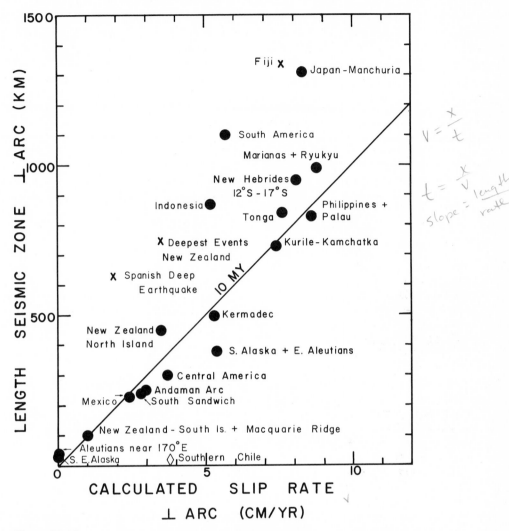

Figure 30-16
Calculated rates of underthrusting (Le Pichon, 1968), and length of seismic zone for
various island arcs and arc-like features (solid circles), for several unusual deep events
(crosses), and for Southern Chile (diamond). The solid line indicates the theoretical
locus of points for uniform spreading over a 10 m.y. interval

floor sediments and from magnetics (Ewing
and Ewing, 1967; Vine, 1966). They obtained
an average rate of spreading for the entire rift
system of 1.3 cm/yr for the half-velocity. This
value is reasonable for the average velocity
of spreading along the world rift system.

The hypothesis that the lengths of deep
seismic zones are a measure of the amount of
underthrusting during the past 10 m.y. is examined in greater detail in Figure 30–16 and
in Table 30–6. Figure 30–16 illustrates the

lengths of the seismic zones in various arcs
and the corresponding rates of underthrusting as calculated by Le Pichon (1968) from
observed velocities of sea-floor spreading
and the orientation of fracture zones along
the world rift system. In nearly all cases the
regions with the deepest earthquakes (and
hence the longest seismic zones as measured
along the zones and perpendicular to the arcs)
correspond to the areas with the greatest
rates of underthrusting; regions with only

Table 30–1
Comparison of estimated slip rates for three island arcs.
Values are in centimeters per year.

Island arc	Slip rate from shallow-focus seismicity, after Brune (1968)	Calculated slip rate, data from ocean ridges, after Le Pichon (1968)	Slip rate[a] assuming length seismic zone measures amount underthrust during last 10 m.y.
Tonga	5.2	7.6	8.4
Japan	15.7	8.8	16.3
Aleutians	3.8	6.1	4.0

[a]Corrected for azimuth of slip vector using data of Le Pichon (1968). Rates given for the three methods are the magnitudes of the total slip vectors and not the magnitudes of the dip-slip components used in Figure 30–16.

shallow- and intermediate-depth events are typified by lower rates of underthrusting. Since the calculated slip rates and some of the measured lengths may be uncertain by 20% or more, the correlation between the two variables is, in fact, surprisingly good.

Although six points fall well above a line of unit slope, which represents an age of 10 m.y., all but one of the lengths are within a factor of 2 of the lengths predicted from the hypothesis that these zones represent materials underthrust during the last 10 m.y. Three of the more discrepant points, which are denoted by crosses on the figure, represent a small number of deep earthquakes that are located in unusual locations with respect to the more active, planar zones of deep earthquakes. They include the unusual deep Spanish earthquake of 1954, three deep earthquakes in New Zealand, and a few deep events under Fiji that appear to fall between the deep zones in the Tonga and New Hebrides arcs.

Other Estimates of Slip Rates in Island Arcs

Slip rates for Tonga, Japan, and the Aleutians (Table 30–1) were calculated (1) from the dislocation theory using data on the past occurrence of earthquakes (Brune, 1968), (2) from the observed spreading velocities along ocean ridges assuming that all of the spreading is absorbed by underthrusting in

island arcs (Le Pichon, 1968), and (3) from the lengths of the seismic zones assuming these lengths are a measure of the amount of underthrusting during the last 10 m.y. For each arc the three rates are within a factor of 2 of one another.

The 10-m.y. Isochron

Two hypotheses may explain the correlation between the length of seismic zones and the computed rate of underthrusting (Figure 30–16). One theory, which was mentioned earlier, assumes that the present seismic zones in island arcs were created during a recent episode of spreading which began about 10 m.y. ago (Figure 30–14,*a*). In the other hypothesis, 10 m.y. is regarded as the approximate time constant for assimilation of the lithosphere by the upper mantle (Figure 30–14,*c*). Since the zone of anomalously high heat flow on ocean ridges appears to be confined to regions less than about 10 m.y. old (McKenzie, 1967; Le Pichon and Langseth, 1969), 10 m.y. is a reasonable time constant for the creation of the normal oceanic lithosphere. Hence, this value is not an unreasonable first approximation to the time constant for assimilation of the lithosphere in island arcs. It should be recognized, however, that the time constant for ocean ridges may be uncertain by more than a factor of 2 because of scatter in heat flow data. At present, it does not seem possible to ascertain which of the two alternative proposals (Figure 30–14,*a* or 30–14,*c*) governs the lengths of seismic zones in island arcs.

With the exception of the data for southern Chile (the diamond in Figure 30–16) there are no points lying significantly below the 10-m.y. isochron. A reduction in slope at the higher spreading rates might occur if the lithosphere were suddenly modified at a given depth through warming or a phase change that would prevent deeper earthquakes from occurring. These processes apparently have not led to significant decreases in the lengths of the seismic zones compared with those predicted by the 10-m.y. isochron in Figure 30–16.

Southern Chile

In southern Chile between 46° and 54°S, the absence of observable deep activity, the near-absence of shallow seismicity, and the presence of a sediment-filled trench (Ewing, 1963; Hayes, 1966) are in obvious conflict with the predicted length of the seismic zone. Le Pichon's omission of two very active features, the West Chile ridge and the Galapagos rift zone (Figure 30–15) in his analysis may, however, explain this discrepancy. Wilson (1965c) argued that since Antarctica is almost surrounded by spreading ridges, the coast of South America south of its juncture with the West Chile ridge at 46°S should not be seismically active. An alternative explanation for the near-absence of shallow seismicity is that activity during the past 70 years of instrumental seismology is not representative of the long-term activity. This explanation is, however, difficult to apply to the hypothetical deep seismic zone that is predicted from Le Pichon's calculations. Hence, we feel that either Wilson's proposal or some other explanation is needed to account for the tectonics of southern Chile.

Departures from the 10-m.y. Isochron

Several factors could explain the six points that fall well above the 10-m.y. isochron in

Figure 30–16: (1) Some pieces of lithosphere became detached from the main dipping zones of deep activity by active processes, such as gravitational settling of slabs of lithosphere or convection currents in the asthenosphere (Figure 30–14,*d*). (2) The initiation of underthrusting and spreading was not simultaneous in all regions. (3) Some or all anomalous deep events may be related to previous episodes of underthrusting. (4) The computed slip rates are not correct. (5) If the thermal time constant for assimilation of the lithosphere is generally about 10 m.y. (Figure 30–14,*c*), the anomalous points may represent pieces of lithosphere with anomalously high time constants so that they are not completely assimilated.

Although some of the computed spreading rates may be in error (particularly the values for Indonesia and South America may be in error because the magnetic data in the Indian Ocean are poor and because the West Chile ridge and the Galapagos rift zone (Figure 30–14) were not included in Le Pichon's analysis), it is difficult to argue that these rates are greatly in error for each of the six anomalous points. The possible ramifications of the various alternative proposals should be exciting enough to encourage further study of these new approaches to the distribution of deep earthquakes.

Interaction of Continental Blocks of Lithosphere

The Alpide belt, which comprises much of the Mediterranean region, the Middle East, and large parts of central and southern Asia, was not included in Figure 30–16 because it is very difficult to define the total amount of underthrusting. Seismic activity in the Alpide zone, unlike that along either the ocean ridges or the typical island arcs, occupies a very broad region (Figure 30–15). Spreading in the Indian Ocean is apparently being absorbed in several subzones within the Alpide belt. This conclusion is supported by the widespread distribution of large shallow earthquakes and by the relatively small number of intermediate- and deep-focus earthquakes (Gutenberg and Richter, 1954).

Although, in principle, it seems reasonable to describe the tectonics of Eurasia by the interaction of blocks of lithosphere, it is not yet clear how successful this idea will be in practice because of the large number of blocks involved. The interaction of blocks of lithosphere appears to be much more complex when all the blocks are continents or pieces of continents than when at least one is an oceanic block. In addition to activity in the Alpide belt, earthquakes in East Africa, northern Siberia, and western North America (including Alaska) are more diffused in areal extent (Figure 30–15). In contrast, seismic zones associated with ocean ridges, island arcs, and many active continental margins appear to be extremely narrow and well defined. Several factors may explain these differences: (1) The lithosphere may be more heterogeneous in some or all continental areas and, hence, may break in a more complex fashion. (2) Old zones of weakness in continental areas may be reactivated. (3) Because of its relatively low density, it may not be possible to underthrust a block of *continental* lithosphere into the mantle to depths of several hundred kilometers. This third point is supported by the relatively large areas of older continental rocks.

As much of the sea floor appears to be relatively young, plates of oceanic lithosphere probably do not contain a large number of zones of weakness, whereas the continental plates, which are older, seem to contain many zones of weakness. Except for a few earthquakes along the extensions of fracture zones, the ocean basins are extremely quiet seismically. Continents, however, exhibit a low level of activity in many areas that do not appear to be undergoing strong deformation. This contrast in activity does not appear to be an artifact of the detection system but rather appears to be related to the different character of the oceanic and continental lithosphere. One exception to this rule is the near-absence of observable activity in Antarctica. Unlike the other continents, Antarctica is almost surrounded by ocean ridges that are probably migrating outwardly (Wilson, 1965c). Hence, the antarctic plate may not be subjected to stresses as large as the stresses occurring in other continents.

Summary of Evidence for Global Movements

The concentration of seismic activity and the consistency of individual mechanism solutions for mid-ocean ridges, for island arcs, and for most of the world's active continental margins indicate that tectonic models involving a few plates are highly applicable for these areas. Since individual mechanism solutions in a given region are highly coherent, each solution may be regarded as an extremely pertinent datum. It is not necessary to resort to complex statistical methods for analyzing the relative motions of these large blocks. For these regions much of the scatter in many previous mechanism studies appears to be largely a result of poor seismic data and errors in analysis. Although the concept of interacting blocks of lithosphere may not be as easily applied to the complex interactions of continental blocks in such areas as the Alpide belt, a consistent (but complex) tectonic pattern may yet emerge when the pattern of seismicity is well defined and when a sufficiently large number of high-quality mechanism solutions is available. Thus, the new global tectonics among other things appears to explain most of the major seismic zones of the world, the distribution and configuration of the zones of intermediate and deep earthquakes, the focal mechanisms of earthquakes in many areas, and the distribution of a large number of the world's active volcanoes.

ADDITIONAL EVIDENCE FOR EXISTENCE OF LITHOSPHERE, ASTHENOSPHERE, AND MESOSPHERE

Complex Tectonic Patterns near the Earth's Surface

A number of objections have been raised to models of sea-floor spreading that involve simple symmetrical convection cells extending from great depths to the surface of the earth. They include such statements as: Since the Atlantic and Indian oceans are both spreading, the tectonics of Africa should be of a compressional nature rather than of an extensional nature, as reported for the East

African rift valleys; if spreading from the eastern flank of the East Pacific rise near Mexico is absorbed only a few hundred kilometers from the rise in the Middle America trench, it is difficult to imagine how materials on the western flank of the rise travel more than 10,000 km before they are absorbed in the trenches of the western Pacific. Although some of the fracture zones on ocean ridges may be as long as 1000 km, some of these zones are less than 50 to 100 km apart. It is almost impossible to believe that these long narrow strips reflect the shape of independent convection cells. These and several similar dilemmas may be resolved if a layer of greater strength (the lithosphere) overlies a region of much lesser strength (the asthenosphere).

The configuration of the asthenosphere and of the various pieces of lithosphere may in some ways be analogous to blocks of ice floating on water. Although the surface pattern of the ice may be very complex, the pattern of convection in the water below or in the air above may be very simple, or it may be complex and of a completely different character than that of the motions of the ice. This analogy may be relevant to the more complex nonsymmetrical tectonic pattern exhibited by the earth's surface. A lithospheric model readily accounts for the asymmetrical structure of island arcs and for the symmetrical configuration of ocean ridges. It also explains the rather smooth variations in rates of spreading in a given ocean.

In this model the ridges and island arcs are dynamic features, and hence they appear to move with respect to one another. Thus, a region of extensional tectonism may be located between two spreading ridges. What is demanded in this model, however, is that surface area be conserved on a worldwide basis. To what extent the motions of large plates of lithosphere are coupled to motions in the asthenosphere remains an unsolved problem.

Depths of Earthquakes along the World Rift System

The existence of a layer of strength near the earth's surface is compatible with the observation that all earthquakes on the ocean ridges are apparently of shallow focus (i.e., less than a few tens of kilometers deep). Using a dislocation model, Brune (1968) estimated that the zone of earthquake generation for some oceanic fracture zones appears to be less than 10 km in vertical extent.

The San Andreas fault of California appears to be a major transform fault connecting growing ocean ridges off the west coast of Oregon and Washington with spreading ridges in the Gulf of California (Figure 30-3). Observed seismic activity along the San Andreas system is confined to the upper 20 km of the earth, and much of this activity is confined to the upper 5 or 10 km (Press and Brace, 1966). Most estimates of the depths of faulting in the great San Francisco earthquake of 1906, the Imperial Valley earthquake of 1940, and the Parkfield earthquake of 1966 are within the upper 10 to 20 km of the earth (Kasahara, 1957; Byerly and DeNoyer, 1958; Chinnery, 1961; McEvilly et al., 1967; Eaton, 1967). Since displacements of at least 100 km (and probably 300 to 500 km) probably have occurred along the San Andreas fault (Hamilton and Myers, 1966), it is almost certain that deformation by creep without observable earthquake activity must be occurring at depths greater than 20 km. Thus, an upper layer of strength in which earthquakes occur associated with a zone of low strength below seems to be demanded for California and for other parts of the world rift system.

Variations in Thickness of the Lithosphere along the World Rift System

Although earthquakes on the mid-Atlantic, mid-Indian, and Arctic ridges occur along the ridge crests themselves as well as along fracture zones, observable seismic activity along much of the East Pacific rise is concentrated almost exclusively along fracture zones (Menard, 1966; Tobin and Sykes, 1968). With the exception of the Gorda ridge off northern California, which appears to be spreading relatively slowly, earthquakes are only rarely recorded from the crest of most of the East Pacific rise. Since evidence from magnetic anomalies indicates a relatively fast rate of spreading (greater than 3 cm/yr for the past

1 m.y.) along much of the East Pacific rise (Heirtzler et al., 1968) and since earthquakes are present on the bounding transform faults, much of the crest of this rise appears to be spreading by ductile flow with little or no observable seismic activity.

Menard (1967a) and van Andel and Bowin (1968) observed that ridges with spreading rates higher than about 3 cm/yr are typified by fairly smooth relief, a thin crest, and the absence of a median rift valley. Ridges with lower spreading rates are, however, characterized by high relief, a thick crust, and a well-developed median valley. Van Andel and Bowin (1968) suggest that materials undergoing brittle fracture may be thin at sites of faster spreading because higher temperatures might occur at shallower depths. The crest of the East Pacific rise is characterized by heat flow anomalies that are broader and apparently of greater amplitude than the anomalies associated with ridge crests in the Atlantic and Indian oceans (Le Pichon and Langseth, 1969). Thus, higher temperatures beneath the East Pacific rise may suppress the buildup of stresses large enough to generate observable seismic activity.

If the occurrence of earthquakes and the presence of a median rift valley correlate with high strength and their absences (in spite of large deformations) correlate with low strength, the lithosphere may be very thin or almost nonexistent near the crest of the East Pacific rise. The lithosphere must be present, however, within a few tens of kilometers or less of the crest to account for the much higher seismic activity along fracture zones. It is possible that seismic activity along the crest of the East Pacific rise may occur mainly as small earthquakes that either are not detected or are rarely detected by the present network of seismograph stations. A microearthquake study using either hydrophones or ocean-bottom seismographs could furnish important information about the mode of deformation at the crest of these submarine ridges.

High-frequency S_n Waves Crossing Ocean Ridges

Molnar and Oliver (1968) studied S_n propagation in the upper mantle for about fifteen

hundred paths; they did not observe any high-frequency S_n waves for ocean paths that either originate at or cross an active ridge crest. Their observations also suggest that the uppermost mantle directly beneath ridge crests is not included in the lithosphere but that the uppermost mantle must be included in the lithosphere beyond about 200 km of the crest in order to explain the propagation of high-frequency S_n from several events on some of the larger fracture zones. The distance for any given ridge may depend on the spreading rate, but the data were inadequate to confirm this assumption. The thinning or near-absence of lithosphere in a zone near the ridge crest would be compatible with the magmatic emplacement of the lithosphere near the crests of ocean ridges and with a gradual thickening of this layer as it cools and moves away from the crest. This thinning may also explain the maintenance of near-isostatic equilibrium over ocean ridges (Talwani et al., 1965a).

Thickness of the Lithosphere from Heat Flow Anomalies

If the spreading rates inferred from magnetic anomalies are accepted, the calculated heat flow anomaly over ridges for a model of a simple mantle convection current that extends to the surface is not compatible with the observed heat flow anomaly (Langseth et al., 1966; Bott, 1967; McKenzie, 1967). These observations are, however, compatible with the computed heat flow for a simple model of cooling lithospheric plate approximately 50 km thick (McKenzie, 1967). In this model the heat flow anomaly results from the cooling of the lithosphere after it is emplaced magmatically near the axis of the ridge. Because of the scatter in the heat flow data and because of the simplifying assumptions used to obtain the estimate of 50 km, the value for the thickness either may be uncertain or may vary by a factor of 2 or more. The occurrence of earthquakes as deep as 60 km beneath Hawaii (Eaton, 1962) may also be used as an estimate of the thickness of the lithosphere. Orowan (1966) tried to fit the observed heat flow pattern for ridges with models in which the crust is stretched on the ridge flanks and the rate of spreading is a function of distance from the ridge. His model does not appear to be compatible either with the symmetry of magnetic anomalies close to the ridge or with the narrow width of the seismic zones on most ocean ridges.

Maximum Sizes of Earthquakes

The sizes of the largest earthquakes along ocean ridges and along island arcs may be related to the thickness of the lithosphere in each of these tectonic provinces. The world rift system (including California, southeast Alaska, and east Africa) accounts for less than 9% of the world's earthquakes and for less than 6% of the seismic energy in earthquakes; island arcs and similar arcuate structures contribute more than 90% of the world's energy for shallow earthquakes and nearly all the energy for deep earthquakes (Gutenberg and Richter, 1954). The relatively small amount of seismic activity on the ridge system and its largely submarine environment probably explain why the significance and the worldwide nature of this tectonic feature were not clearly recognized until about 12 years ago.

Although Richter (1958) lists more than 175 very large earthquakes (magnitude M greater than or equal to 7.9) as originating in arcs and arc-like features, he reports only 5 events of this size for the world rift system. Whereas the largest event on the ridge system was of magnitude 8.4, the largest known earthquakes from island arcs were of magnitude 8.9. This difference in magnitude corresponds to about 8 times as much energy in the larger shocks. Most (and perhaps all) of the earthquakes on the ridge system with magnitudes greater than about 7 appear to have occurred along major transform faults.

Thus, the maximum magnitudes of earthquakes during the last 70 years for various tectonic features may be summarized as follows: island arcs, 8.9; major fracture zones, 8.4; ridge crests in the Atlantic, Indian, and Arctic oceans, about 7; most of the crest of the East Pacific rise (with the exception of the Gorda ridge), few (and possibly no) events larger than 5. The relative frequencies of very large earthquakes and possibly the upper limits to the sizes of earthquakes appear to be related to the area of contact between pieces of lithosphere that move with respect to one another. On the crest of the East Pacific rise the zone of contact between brittle materials is either absent or is confined to a thin superficial layer; at the crests of other ridges the lithosphere appears to be somewhat thicker. For fracture zones the thickness of the lithosphere may be as great as a few tens of kilometers; thus, the maximum magnitudes may be limited by the length of the fracture zone and by the thickness of the lithosphere. In island arcs the thickness of the lithosphere may be greater than that on the ridges because the temperatures beneath ridges are higher and perhaps because some material is added to the bottom of the lithosphere as it moves away from a ridge crest. Since the dip of seismic zones in island arcs usually is not vertical, the area of contact between plates of lithosphere may be increased by this factor alone. For example, the zone of slippage in the great Alaska earthquake of 1964 appears to dip northwesterly at a shallow angle (about 10°) and to extend for about 200 km perpendicular to the Aleutian island arc (Plafker, 1965; Savage and Hastie, 1966). Since the dip is shallow, this zone does not extend more than about 40 km below the surface at its deepest point. Thus, it is not unreasonable that much greater amounts of elastic energy can be stored and released along island arcs than along the mid-ocean ridges.

Since single great earthquakes apparently have not involved a rupture that extended along the entire length of the larger island arcs (Richter, 1958), some other factor appears to limit the length of rupture and hence the maximum sizes of earthquakes. Tear faults (i.e., transform faults of the arc-arc type) within the upper thrust plate may divide island arcs into smaller subprovinces that are not completely coupled mechanically.

Some Properties of the Lithosphere and Asthenosphere

Thus far we have defined the lithosphere as a layer of high strength and the asthenosphere as a layer of low strength. Since the rheological properties of the mantle are not at all well understood, we have purposely used the term strength in a general sense without being more specific about the actual mechanisms of deformation. Thus, it is not at all certain that various techniques for measuring the thickness and the presence of the lithosphere necessarily yield comparable values since these methods probably sample different physical parameters. It is encouraging, however, that estimates of the thickness of the lithosphere from analyses of heat flow anomalies, the absence of earthquakes (in spite of large deformation), the requirements of isostasy and the maintenance of mountains (Daly, 1940), the amplitudes and wavelengths of gravity anomalies (McKenzie, 1967), the propagation of high-frequency S_n waves, and a transition in the types of earthquake mechanisms in island arcs are about 100 km or less. It is not clear, however, to what extent the variations in these estimates represent real differences in thickness or merely differences in the relations between thickness and the various physical parameters measured.

The asthenosphere in the three-layered model shown in Figure 30–1 also roughly coincides with the low-velocity zone for S waves (Gutenberg, 1959; Dorman et al., 1960; Anderson, 1966; Ibrahim and Nuttli, 1967), regions of either low velocity or nearly constant velocity for P waves (Lehmann, 1964a, b) and of low density or nearly constant density (Pekeris, 1966), a region of high attenuation for seismic waves, particularly S waves (Anderson and Archambeau, 1964; Anderson, 1967b; Oliver and Isacks, 1967), and a low-viscosity zone in the upper mantle (McConnell, 1965). Although these observations yield very little information about the actual mechanisms of dissipation, they are consistent with the hypothesis that the asthenosphere is a region of low strength bounded below by the mesosphere, a region of greater strength. The simplest explanation for these phenomena is that the closest approach to melting (or partial melting) occurs in the asthenosphere. That relative displacements of the earth's surface may be modeled by a series of moving plates is a strong argument for a region of low strength (the asthenosphere) in the upper mantle. Nevertheless, the physical properties and configuration of the asthenosphere and the lithosphere may vary from place to place.

In fact, variations of these types seem to be required to account for many of the complexities of the outer few hundred kilometers of the earth. The evidence for such lateral variations is now so strong as to demand re-evaluation of studies of velocity and Q structure based on models that consist only of concentric spherical shells.

CONFLICTING SEISMOLOGICAL EVIDENCE AND SOME PROBLEMS

The new global tectonics, in one form or another, has been remarkably successful in explaining many gross features and observations of geology, but its development must be continued in an effort to establish a theory that is effective throughout the earth sciences at levels of increasingly greater detail. Many difficult problems remain. A quantitative understanding of the processes in the mantle that result in the observed surface features is urgently needed. The roles of the earth's initial heat, gravitational energy, radioactive heat, and phase changes must be understood. The mechanics of flow in the mantle is little known. The history of sea-floor spreading through geologic time and the relation between the vast quantities of geologic observations and the hypotheses must be worked out. A vital unanswered question is why island arcs and ocean ridges have their particular characteristic patterns. These grand questions are of concern to seismologists, not only because there is general interest in the hypotheses but also because the solutions may very well be dependent on evidence from seismology. In the absence of all but the most preliminary attempts to solve these problems, however, it is difficult to speculate on the direction and force of the evidence. At the present there appears to be no evidence from seismology that cannot eventually be reconciled with the new global tectonics in some form. Some traditional views that once would have appeared contradictory (such as, for example, the assignment of substantial rigidity to the entire mantle over a long time interval because of efficient propagation of shear waves with periods of about 10 sec) have long been discounted on the basis of information on glacial rebound, gravity, and, more recently, creep along long strike-slip faults, such as the San Andreas fault. That deep earthquakes generate shear waves and radiate waves in a pattern characteristic of a shear dislocation can hardly be taken as evidence for strength throughout the mantle in that depth range in light of a hypothesis that suggests that the mantle in the deep earthquake zones is much different from the mantle at comparable depths elsewhere.

The arc-like patterns of the active zone is but one of the problems of a subject that might be termed 'lithosphere mechanics.' Others are the shape of the deep zones, which sometimes appear to be near-planar after turning a rather sharp corner near the surface; the distribution of stress and strain in the lithosphere, not merely in active seismic areas but throughout the world; the interaction of lithosphere and asthenosphere; and flow in the asthenosphere.

Some specific evidence from seismology that does not readily fit into the new global tectonics at present does exist, however, For example, the locations of the Spanish deep earthquake and certain deep shocks in New Zealand and Fiji are not easily explained by current thinking, as has been pointed out above. The patterns of minor seismicity, including an occasional large earthquake in certain continental areas such as the St. Lawrence Valley and the Rocky Mountains and broad regional scattering of epicenters as in east Africa, are not simply explained as yet, nor is the almost complete lack of earthquakes in the oceanic lithosphere, which presumably can and does transmit stress over large distances. The occurrence of large active strike-slip faults that trend along the arc structure, such as the Alpine fault in New Zealand, the Philippine fault, the North Anatolian fault in Turkey, and the Atacama fault in Chile, offers some difficulties. Tentative explanations have been proposed for the Alpine and Philippine faults; therefore, this matter may be resolved. Perhaps related to this problem is the occurrence in island arcs of occasional earthquakes with large strike-slip components along faults subnormal to the arc. The tectonics of the arc can hardly be as simple as that implied by Figure 30-1.

A general problem involving seismology concerns contrasts between continental and oceanic areas (MacDonald, 1964, 1966). Some studies (e.g., Brune and Dorman, 1963; Toksöz and Anderson, 1966) show corresponding structural differences to depths of several hundred kilometers. Are such differences contrary to the new global tectonics? Might the observations be equally well satisfied by other models that are compatible with requirements of the new hypotheses? A serious difficulty may arise here if they cannot. Comparable and related problems arise in other disciplines. Observation suggests that heat flow per unit area is about the same under the oceans as under the continents. If it is assumed that the heat flux is largely due to radioactive decay and that radioactivity is heavily concentrated in certain continental rocks, lateral heterogeneity of the upper mantle is required and the mixing predicted by the new tectonics may be so great as to destroy such heterogeneity. Perhaps the amount of radioactivity in the earth has been highly overestimated, as has occasionally been suggested (Verhoogen, 1956), and much of the heat lost at the surface is transported from the deep interior by convection. More accurate determination of deep structure by seismological techniques is thus relevant to the problem of the amount of radioactivity in the earth and its distribution. Seismology is also vitally linked with heat flow in the island arcs, where low heat flow values appear to correlate with zones of descending lithosphere but anomalous high values appear over the deep seismic zones, and at the ocean ridges, where high heat flow may correspond with low strength and hence low seismic activity.

Other disciplines are also involved. The new structures based mainly on seismological information must be tested against gravity data. The record in the ocean sediments surely provides the most complete history of the ocean basins and hence must provide insight into seismological processes. Perhaps the crucial evidence on the validity of new global tectonics will come from cores of the entire sedimentary column of the ocean floor.

In petrology an interesting question concerns the origin of the belts of andesitic volcanoes. Coats (1962) proposed that ocean crust and sediments were thrust into the mantle under the Aleutian arc and subsequently erupted with mantle rock to account for the petrology of those islands. Can all the andesites of the active tectonic belts be explained in this manner? Are there any young (less than 10 m.y.) andesites of the same type that are *not* associated with active deep earthquake zones and arcs? Can this sort of information be used to identify ancient arcs?

The important problems arising from the new global tectonics are many; some are crucial. Evidence from seismology against the new global tectonics appears, however, to lack force.

EFFECTS OF NEW GLOBAL TECTONICS ON SEISMOLOGY

That seismology is providing abundant and important information for testing the new global tectonics is demonstrated in other sections of this paper and elsewhere. To date, most of the seismological work related to this topic has been so directed. The countering impact of the new global tectonics on seismology must also be carefully and thoroughly considered for indications of new directions for, and new attitudes toward, seismological research. This section is largely speculative and some of the points may seem farfetched. If, however, a basic understanding of global tectonics is imminent, even the most imaginative and wisest forecast of its effect on seismology is likely to be too conservative. Nor is seismology alone in this regard, for all branches of geology and geophysics related to the earth's interior will be comparably affected. The assumption that a major advance in our understanding of global tectonics has been achieved is tacit in the following.

Seismicity is an important branch of seismology and one that will be strongly affected by the impact of the new global tectonics. Such basic questions as why earthquakes occur largely in narrow belts separated by large stable blocks, why these belts are continuous on a worldwide scale, why they branch, why certain details of their configuration are as they are, why intermediate and deep earthquakes occur in some areas and not others

are in the process of being answered today. There is every indication that the relation between seismic activity and geology will soon be understood to a much greater extent than contemplated heretofore. Improved accuracy in hypocenter location resulting from a better knowledge of velocity structure will facilitate this development. It will also assist in solving seismology's chief problem of a political nature, distinguishing between earthquakes and underground nuclear explosions. The self-consistent worldwide pattern of focal mechanisms will also be valuable here. Perhaps location of new seismograph stations in the proper relation to high-Q zones within the earth will result in improved detection capability. Important questions still remain. For example, why have the earthquake belts and the associated tectonic belts assumed their present configuration? What can be learned about paleoseismicity and its relation to modern seismicity? How can the pattern of minor seismicity, including the occasional major earthquake outside the established seismic belts, be fit into the new global tectonics?

Closely related to the distribution of seismic events in time and space and to tectonics is the focal mechanism of the earthquake. Data of quality and quantity from modern observing stations can provide information for determination of focal mechanisms that severely test the new hypotheses and are crucial to their development. It appears that with only modest advances in technique properly documented earthquakes will provide reliable detailed information on tectonic activity. Focal mechanism, stress drop, slip, and orientation of principal stresses should be available from individual earthquakes, and singly or cumulatively these data will be integrated with the tectonic pattern.

Implicit in the new global tectonics is a new attitude toward mobility of the earth's strata. Measurements of fault creep and other forms of earth strain over recent short intervals of time are based on a variety of measuring techniques that include geodetic surveying, tide and strain gaging, and measurements of earthquake slip. Various methods of field geology give data extending over varying but greater intervals of time. When all these data are reduced to velocities that describe the motion of one point in the earth relative to another point located in an adjacent relatively underformed block, values are obtained that are in the same range as the velocity values associated with sea-floor spreading determined through analysis of geomagnetic data. This range is from about 1 to perhaps 10 cm/yr or more. These values are considerably higher than the values usually assumed in the past by most earth scientists considering deformation of this type; hence, new attitudes toward old problems must be anticipated. A prime example in seismology, cited above, is the association of the length along the dip of a deep seismic zone with the amount of relatively recent underthrusting in a region.

With the new global tectonics, interrelationships on a large, perhaps worldwide, scale can be predicted and perhaps observed. Thus, major seismic activity in one area could be related to that in an adjoining, or perhaps distant, area associated with the same lithospheric unit of units, for, although the propagation time for the effect may be long, stress may be transmitted over large distances through the lithosphere. Of special interest is the new insight into the subject of earthquake prediction, even prevention. Although an empirical method of prediction could by chance be found in the absence of an understanding of the process, an effective method is much more likely to be achieved if a basic understanding of the earthquake phenomenon is established. The new global tectonics offers great promise for such an achievement. It has already provided a theory that predicts over-all strain rates in tectonic areas throughout the world. It suggests a means for predicting the maximum size of an earthquake in a given region. It provides a framework in which to relate measurements of distortion, such as strains, tilts, and sea-level changes, with observations on the mechanism of the earthquake. It has established the continuity of active zones and has shown that apparently inactive segments of otherwise seismic belts must, indeed, be active, either by subsequent earthquakes or by creep. Refinements and further developments must be anticipated.

Seismology has long been the principal source of information on the structure of the earth's interior and is likely to continue in that role, with or without the new global tectonics. The new hypotheses will, however, certainly stimulate radically new approaches to the exploration of the earth's interior. A common and powerful technique of seismology involves the use of simplified earth models for prediction of certain observed effects. The new global tectonics calls for an entirely new kind of model. Layered models in which the shells are spherically symmetric are now outdated for many areas of the earth. Models based on the effect of spreading and growth of the lithosphere at the rifts and underthrusting at the island arcs must be tested against observation. The conventional division of the earth's surface into oceanic and continental areas, or oceanic, shield, and tectonic areas, requires a new look when the lithosphere, with its lateral variations, is involved. New models in which the age and the thickness of the lithosphere, as well as other properties, are taken into account are required.

In the new tectonics, information on properties and parameters such as attenuation, strength, creep, viscosity, and temperature is critically needed. Further efforts must be made to understand the relation between such properties and the seismic properties that are measured in a more straightforward manner. At the island arcs, where cold exotic materials are descending into the mantle, nature is performing the experiment of subjecting what are normally near-surface materials to the higher temperatures and pressures of the asthenosphere, an experiment in many respects much like the ones being performed in various laboratories. As our techniques for measuring the composition of the material and the dynamic and environmental parameters of this situation improve, this experiment should yield important information on many topics. Of great interest to seismologists is the long-standing problem of the mechanism of deep earthquakes. The radiation patterns of seismic waves from many events present a reasonably consistent pattern but one that cannot readily be explained in detail by existing hypotheses. The variation of seismic activity with depth is another source of information. Are the focal mechanisms of earthquakes at depth really as much like those in near-surface brittle materials as they seem? What is the role of water, of other interstitial fluid, of partial melting? Are phase changes in the exotic mantle materials important, perhaps not as sources of seismic waves but as concentrators of stress that leads ultimately to rupture? How are the seismic observations of focal mechanism, spatial and temporal distribution, energy, etc., related to the material of the mantle and what can we learn about that material? How are these data related to the general configuration of the seismic zone and of the geology of the island arc? These are some of the topics on which this experiment may provide information.

Surely, the most striking and perhaps the most significant effect of the new global tectonics on seismology will be an accentuated interplay between seismology and the many other disciplines of geology. The various disciplines which have tended to go their separate ways will find the attraction of the unifying concepts irresistible, and large numbers of refreshing and revealing interdisciplinary studies may be anticipated. For example, the geomorphology of an area of raised beaches takes on new light for those interested in paleoseismicity; the tectonic significance of a feature of the ocean floor is determined by its seismicity and by the mechanism of the earthquakes; the petrology of a volcano of an island arc is related in a meaningful way to the seismic activity below; the worldwide phenomena of seismology provide crucial evidence on the basic processes of the earth's interior that have shaped and are shaping the surficial features of interest to classical geology. Even if it is destined for discard at some time in the future, the new global tectonics is certain to have a healthy, stimulating, and unifying effect on all the earth sciences.

31

Mantle Earthquake Mechanisms and the Sinking of the Lithosphere

BRYAN ISACKS
PETER MOLNAR
1969

From *Nature*, v. 223, p. 1121–1124, 1969. Reprinted with permission of the authors and Macmillan Journals Ltd.

The concept of the lithosphere as a stress guide (Elsasser, 1969; Isacks et al., 1968; 1969) suggests that information about the major driving forces in global tectonics can be obtained from knowledge of the state of stress within the lithosphere. Most shallow earthquakes seem to occur between plates of lithosphere and therefore yield information about the direction of motion of one plate with respect to the other rather than the stress within a plate. But analysis of focal mechanism solutions from the Tonga and Japanese regions (Isacks et al., 1968) suggested that deep and intermediate depth earthquakes occur within downgoing slabs of lithosphere in response to stresses within the plates. A chief purpose of this article is to show that a comprehensive and worldwide survey (our unpublished results) of focal mechanism solutions supports this interpretation of deep and intermediate depth earthquakes. We can therefore determine the orientations of stress within the various downgoing slabs on a worldwide basis.

GRAVITATIONAL FORCES ON DOWNGOING SLABS

The main result of this survey is that the stress in those portions of the mantle seismic zones that have a relatively simple inclined planar configuration, that is, in those portions removed from remarkable contortions or changes in trend, is often oriented such that either the axis of maximum compressive stress or the axis of minimum compressive stress is approximately parallel to the dip of the inclined seismic zone. We find clear evidence for down-dip extensional stress within slabs at intermediate depths in at least six regions. These regions seem to be also characterized by prominent gaps in the seismicity at depths between about 300 and 500 km or by an absence of earthquakes deeper than about 300 km. On the other hand, down-dip compressional stress

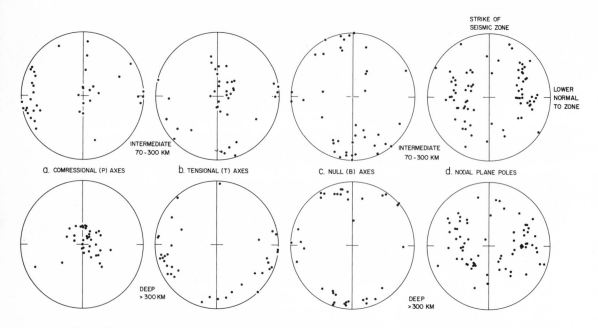

Figure 31-1
Equal area projections of the axes of compression (P), axes of tension (T), null axes (B) and the poles of the nodal planes of the most reliable determined double-couple mechanism solutions of deep and intermediate depth earthquakes. The centre of each plot is in a direction parallel to the dip of the inclined seismic zone: the strike of the zone and the direction normal to the zone (on the lower side) are as shown in Figure 31-1,d. The regions and number of earthquake mechanisms included in this plot are as follows: Tonga (Isacks et al., 1969; Sykes, 1966; Sykes et al., 1969) (15°S–26°S), 14; Kermadec (Isacks et al., 1969; Sykes, 1966) (26°S–36°S), 1; North Island, New Zealand (Adams, 1963; Hamilton and Gale, 1968) (36°S–41°S), 1; New Hebrides (Barazangi and Dorman, 1969) (10°S–22°S), 9; Sunida (unpublished results of B. I. and P. M. and of T. Fitch and P. M.) (100°E–127°E), 9; Marianas (Katsumata and Sykes, 1969) (12°N–24°N), 2; Izu-Bonin (Katsumata and Sykes, 1969; Honda et al., 1956; Hirasawa, 1966; Ritsema, 1965; Katsumata, 1967) (24°N–35°N), 8; Philippines (Barazangi and Dorman, 1969) (3°N–12°N), 3; Ryuku (Katsumata and Sykes, 1969) (24°N–33°N), 2; Northern Honshu (Honda et al., 1956; Hirasawa, 1966; Ritsema, 1965; Katsumata, 1967) (35°N–43°N), 4; Kuriles (Sykes, 1966) (43°N–52°N), 8; Aleutians (Stauder, 1968a) (145°W–170°E), 1; Middle America (Molnar and Sykes, 1969) (14°W–105°W), 5; Chile (Barazangi and Dorman, 1969) (15°S–30°S), 8. The data on focal mechanisms and the structures of the seismic zones are described in more detail in the references indicated here.

is predominant at depths greater than 300 km in all regions studied.

These results are summarized in Figs. 31–1 and 31–2, and a simple model that attempts to account for them is given in Fig. 31–3. In this model a heavy slab sinks into the asthenosphere, exerts a pull on the portion of the plate remaining on the surface, and eventually hits bottom in such a way that the support for the load of excess mass within the slab is transferred from above the load to beneath the load. Although the two-dimensional model is certainly too simple to explain all the data, the general predominance of down-dip stress orientations and, especially, the widespread occurrence of down-dip extensional stress,

lend encouragement to the ideas that gravitational body forces on the downgoing slabs are important forces in determining the stress within the lithosphere and may be important forces in driving the global system of plate movements.

AXES OF STRESS

The data include, primarily, focal mechanism solutions of the double-couple type for deep and intermediate depth earthquakes that are large enough (M ≥ 5.5) to obtain reliable first-motion data from the World-Wide Standardized Seismograph Network (WWSSN).

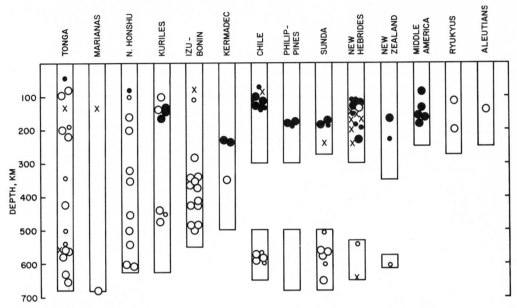

Figure 31-2

Down-dip stress type plotted as a function of depth for fourteen regions (see caption for Figure 31-1 for locations of regions and sources of data). A filled circle represents an orientation such that the axis of tension T is within 20–30 degrees of the local dip of the zone, that is down-dip extension; an unfilled circle represents an orientation such that the axis of compression P is within 20–30 degrees of the local dip of the zone, that is, down-dip compression; and the Xs represent orientations that satisfy neither of the preceding. Smaller symbols represent less reliable determinations. The enclosed rectangular areas approximately indicate the distribution of earthquakes as a function of depth by showing the maximum depths and the presence of gaps for the various zones. In addition to the references listed in the caption to Figure 31-1, data from Barazangi and Dorman (1969) and Gutenberg and Richter (1954) were used for each of the fourteen regions. The zones are grouped (from left to right) according to whether the zone is continuous to depths of 500–700 km, discontinuous with a gap between intermediate depth and deep earthquakes, or continuous but reaching depths less than 300–400 km

Although these data are restricted to earthquakes that occurred since 1962, reliable solutions for previous earthquakes are also used for certain regions. The data are discussed in more detail elsewhere (our unpublished results). In this article we focus attention primarily on the results for fourteen island arcs and arc-like structures for which the orientations of the inclined seismic zones are relatively simple and well defined. This selection is required because the basic analysis consists of comparing the orientation of the double-couple mechanism solutions with the local orientation of the inclined seismic zone.

Some of the excluded earthquakes occur in zones or portions of zones that are remarkably contorted or complex in structure, such as the northern end of the Tonga arc. The mechanism solutions in these and similarly complex regions reflect the complexities of structure and are reported on in detail elsewhere (Isacks et al., 1969 and our unpublished results). Others that are excluded occur in regions where the data are insufficient clearly to define a planar zone, such as the Mediterranean and Himalayan regions. Nevertheless, about 60 per cent of the earthquake mechanism solutions are included, and the results apply to most of the major island-arc structures of the world.

The overall results are shown in Fig. 31–1 in which the stress axes and the poles of nodal planes of the most reliably determined double-couple solutions are plotted relative to the local orientations of the seismic zones. One of the nodal planes of a mechanism solution is the fault plane of the equivalent shear dislocation, and the pole of the other nodal plane gives the direction of slip (Burridge and Knopoff,

1964; Savage, 1965); the seismic data do not, however, distinguish which is which. The compressional (P), tensional (T) and null (B) axes coincide with the axes of maximum, minimum and intermediate compressive stress in the medium only if the nodal planes define planes of maximum shear stress in the medium. This must be taken as an assumption of the analysis. If a Coulomb-Navier type of fracture process is operative for deep and intermediate depth earthquakes and the effective coefficient of internal friction is less than one (Isacks et al., 1968; Raleigh and Paterson, 1965), the P, T and B axes will still be good approximations (within the uncertainites of the analyses) to the principal stress axes.

Fig. 31–1,*d* shows clearly that neither of the possible fault planes is parallel to the seismic zone. This result supports the contention (Isacks et al., 1968; Isacks et al., 1969) that intermediate and deep earthquakes do not indicate slip along a major thrust fault defined by the inclined seismic zone. Figs. 31–1,*a*, 31–1,*b*, and 31–1,*c* show that, instead, the inferred stress axes of the double-couple solutions tend to be parallel or perpendicular to the planar geometry of the zones. This result suggests that the down-going slabs act as stress guides. If the lithosphere can support stresses considerably larger than those in the asthenosphere, so that the shear stresses are relatively small along the upper or lower boundaries of the plate, then the principal axes of stress within a thin plate of lithosphere will be approximately parallel and perpendicular to the plate.

DEVIATORIC STRESSES

This interpretation is strongly supported by recent estimates (Berckhemer and Jacob, 1968; Wyss and Brune, 1969) of the magnitude of the deviatoric stresses involved in deep and intermediate depth earthquakes. These stresses appear to be one to two orders of magnitude greater than the stresses estimated (Wyss and Brune, 1969; Brune and Allen, 1967) for zones of shallow earthquakes where movements between plates of lithosphere are apparently accommodated along zones of weakness. Also, Stacey (1967) has pointed out

that thermodynamic problems arise if stresses much greater than 10 bars are associated with the large strain rates in the mantle that are required by sea-floor spreading and continental drift. These problems may be circumvented if the earthquakes in the mantle indicate large stresses within lithospheric plates rather than the smaller stresses within the weaker asthenospheric material where most of the mechanical work is being done.

The predominance of simple down-dip stress orientations shown in Figs. 31–1,*a* and 31–1,*b* suggests that in those portions of the zones characterized by relatively simple planar geometry the stress may be a result chiefly of the forces directly involved in the downward motion of the plate. The presence of down-dip extensional stress suggests that the slabs may be pulled into the mantle as a result of a negative buoyancy of the slab as suggested by Elsasser (1969). At rates of descent of the order of 10 cm/yr, temperatures in the slab may be of the order of 1,000° C cooler than the adjacent mantle, and positive density anomalies of the order of 0.1 g/ml. would result.

Hatherton (1969) and Morgan and Smith (1969) find support for such excess mass in analyses of gravity data. If this load were supported chiefly by tractions along the large areas beneath the surficial plates of lithosphere, then simple calculations show that the deviatoric stress within the downgoing plate would be extensional, of the order of several kilobars, and would decrease with depth (Fig. 31–3,*a*). As the slab penetrates into stronger material beneath the asthenosphere the support for the load of excess mass is transferred from above to below the load, and the stress in the lithosphere would depend on the variation of excess mass within the slab and the variation of strength with depth. In general, one might imagine a stage in which part of the load is supported from below and part from above (Fig. 31–3,*b*) such that the stress changes from extension at intermediate depths to compression at great depths, with zero deviatoric stress somewhere between. When the load is fully supported from below, the stress becomes compressional throughout. Fig. 31–3,*d* shows an alternative development where, as a result of the extensional stress and a possible stoppage in the movements, a piece of lithosphere

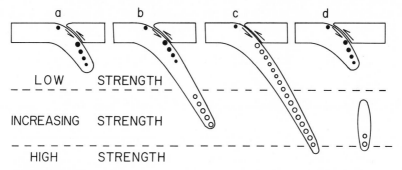

Figure 31-3

A cartoon showing possible distributions of stresses in slabs of lithosphere that sink into the asthenosphere (a) and hit bottom (b and c). (d) represents the case where a piece of lithosphere has broken off. The symbols are the same as in Figure 31-2; the filled circles represent down-dip extension and the unfilled circles represent down-dip compression. The size of the circles qualitatively indicates the amount of activity at the respective depths. In (b) and (d) gaps in the seismicity would be expected. Also shown is the under-thrusting and the extensional stresses near the upper surface of the slabs due to the bending of the slab beneath the trenches. These features are inferred from the mechanisms of shallow earthquakes (Isacks et al., 1968). The lower boundary where the slabs hit bottom might correspond to the transition region or discontinuity near 650–700 km (Anderson, 1967; Engeldahl and Flinn, 1969)

breaks off, sinks independently, hits bottom and thereby comes under compression. In Figs. 31–3,*b*, and 31–3,*d* gaps in the distribution of earthquakes as a function of depth might occur between extensional type mechanisms at intermediate depths and compressional type mechanisms at great depths. Where the seismic zone is continuous, that is, where earthquakes occur throughout the range of depths to 600–700 km, the slab would be under compression throughout.

SEISMICITY AND DEPTH

The model shown in Fig. 31–3 thus predicts a correlation between the respective depth variations of focal mechanism and seismicity. Fig. 31–2 shows that this correlation is partially supported by the data. The regions where down-dip extensional stress predominates, such as Middle America, the New Hebrides and Chile, are characterized by notable gaps in seismicity as a function of depth or, in the case of Middle America, an absence of deep earthquakes. On the other hand, in regions such as Tonga, Izu-Bonin and northern Honshu the

zones are apparently continuous and exhibit down-dip compressional stress at intermediate depths as well as great depths. Thus Tonga, Izu-Bonin and northern Honshu would be represented by Fig. 31–3*c*; Middle America would be represented by Fig. 31–3,*a*; and Kermadec, Chile, North Island of New Zealand, New Hebrides, Sunda and Philippines would be represented by Fig. 31–3,*b* or 31–3,*d*.

Anderson (1967) and others have proposed a phase change at depths near 350–400 km where the density increases by 9–10 per cent. If dT/dP for the phase boundary is positive (as proposed by Anderson), and the change occurs with sufficient rapidity, the transition to greater density in the lithospheric plate might occur above the level of the transition in the adjacent mantle. This added load applied near 300–400 km depth might help to explain the prevalence of compression everywhere below 300 km as well as the prevalence of minima or terminations (Fig. 31–3) in the various distributions of seismicity as a function of depth.

Certain other areas not included in the present analysis because of uncertainties in the configuration of the subcrustal zone may offer further support for the models of Fig. 31–3.

Intermediate depth earthquakes in the Hindu Kush, for example, are characterized by T axes that dip gently northward (Ritsema, 1966; Chander and Brune, 1965). If this zone of intermediate depth earthquakes is a remnant of a larger slab underthrust beneath the Himalayas, then the focal mechanism orientations might be indicative of down-dip extensional stress. The deep earthquakes beneath western Brazil are located nearly vertically beneath intermediate depth earthquakes and are apparently associated with the island arc-like features of the Peru trench. The deep mechanisms are characterized by nearly vertical P axes (Ben-Menahem et al., 1968; Khattri, 1969 and our unpublished results). Although the structure of the seismic zone is not well defined, the locations of earthquakes (Gutenberg and Richter, 1954; Barazangi; and Dorman, 1969) suggest that the deep zone also dips very steeply beneath western Brazil.

LIMITATIONS OF THE TWO-DIMENSIONAL MODEL

Further comparisons of Figs. 31–2 and 31–3 show that the correlation breaks down in certain regions. In particular, the down-dip compressional stresses in the Aleutians and Ryukyus, the presence of both compressional and extensional stresses at intermediate depths in the Kuriles and New Hebrides, and the solutions denoted by "X" are not explained by the two-dimensional models of Fig. 31–3. These discrepancies, as well as the complex mechanisms and structures excluded in the analyses of Figs. 31–1 and 31–2, indicate the unsurprising result that the two-dimensional model of Fig. 31–3 does not explain all the data. Certainly, contortions and disruptions of the lithospheric plates would be expected to affect the stresses, although a search for such effects does not yield any simple relationship between the mechanism orientations and the contortions of the zones.

Nevertheless, one interesting association can be pointed out. The down-dip compressional stresses present in the Aleutians, the New Hebrides and the Ryukyu arc (Fig. 31–2) are located where the curvature of the arc is appreciable. In each case the T axis, instead of the B axis, is nearly parallel to the strike of the seismic zone; this orientation of the extensional stress might result from the type of deformation of the slab suggested by Stauder (1968a). In the less arcuate portion of the New Hebrides the T axes are predominantly parallel to the dip of the seismic zone and the B axes are parallel to the strike. Many of the mechanisms represented by Xs in Fig. 31–2 may reflect other unresolved contortions of the downgoing slab or other sources of stress such as thermal gradients within the plates or local density variations.

Nevertheless, the consistent pattern shown in Fig. 31–2 for zones that are relatively uncontorted offers support for the interpretation that gravitational body forces are a major source of stress in the lithosphere. If we make this interpretation, then the results require that in many regions the slabs are sinking and are exerting a downwards pull. The results also require the complications of Fig. 31–3 in which the sinking slabs are supported from above or below, or both. Thus if we use the results to extract information on the forces that drive the global plate movements, we can infer two important effects: (1) the downgoing slabs can exert a pull on the surface portions of the plates, and (2) as the downgoing slabs "hit bottom" beneath the lithosphere the downward pull on a surface plate is significantly decreased. These interpretations suggest that the pull of the descending slabs may be an important contribution to the driving forces of global tectonics. Moreover, hiatuses or changes in the rates and direction of sea-floor spreading and continental drift might result when the descending slabs reach depths of 500–700 km.

SECOND-GENERATION PLATE TECTONICS

32

Introduction

Like other scientific paradigms, plate tectonics is open-ended in the sense that after the main postulates had been announced, a considerable amount of mopping-up and filling-in remained to be done. Three papers are given in this section that either apply the paradigm, extend it, or modify it.

A Basic Contradiction of Plate Tectonics

Plate tectonics owes much of its simplicity and rigor to Postulate 2 of Chapter 5, which states that poles of motion for two adjacent plates remain fixed relative to the two plates over long periods of time. It is the fixed character of the poles that causes the transform faults and their associated fracture zones to lie along small circles concentric about the pole. In fact, it was the tendency of fracture zones to lie along small circles that first suggested the idea of plate tectonics to Morgan (Chapter 8).

If the earth's lithosphere were divided into only two plates, their pole could theoretically remain fixed and the transforms could continue to propagate along concentric small circles indefinitely. If, however, the earth's lithosphere is divided into three or more plates, this is no longer true. For example, let's assume that the lithosphere is divided into three plates A, B, and C with different poles $\mathbf{P}_{AB}$, $\mathbf{P}_{BC}$, and $\mathbf{P}_{AC}$ (Fig. 32–1). If the transforms between A and B are small circles about $\mathbf{P}_{AB}$ and the transforms between B and C are small circles about $\mathbf{P}_{BC}$, then it is not possible for the transforms between A and C to be small circles, which is equivalent to saying that there is no pole $\mathbf{P}_{AC}$ that remains fixed with respect to both plates A and C. The proof is straightforward. Consider a spherical coordinate

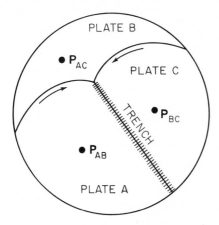

Figure 32-1
Example of three plates A, B, and C
with three poles of relative motion which
cannot maintain fixed angles relative to
each other as the plates move. If pole
$\mathbf{P}_{AB}$ remains fixed relative to plates A
and B as shown and if pole $\mathbf{P}_{BC}$ remains
fixed relative to plates B and C, then
pole $\mathbf{P}_{AC}$ cannot remain fixed relative to
plates A and C

system that remains fixed with respect to plate A and a second that remains fixed relative to plate B. Then $\mathbf{P}_{AB}$ is the only point in the B coordinate system that remains fixed with respect to the A coordinate system. In particular, $\mathbf{P}_{BC}$ is not fixed relative to plate A (except for the trivial case $\mathbf{P}_{AB} = \mathbf{P}_{BC}$). If ω_{AB}, ω_{BC}, and ω_{CA} are the scalar angular velocities of the plates, then by vector addition (Chapters 7 and 8) the instantaneous vector angular velocity of plate C relative to plate A is

$$\omega_{AC}\,\mathbf{P}_{AC} = \omega_{AB}\,\mathbf{P}_{AB} + \omega_{BC}\,\mathbf{P}_{BC}.$$

Since $\mathbf{P}_{AB}$ is fixed relative to plate A whereas $\mathbf{P}_{BC}$ is not, the instantaneous rotation vector $\omega_{AC}\,\mathbf{P}_{AC}$ is moving relative to plate A. In other words, if a boundary between plates A and C is to remain a transform, its locus cannot be a small circle concentric about a point that is fixed relative to plate A. To remain a transform, the boundary must so change direction that the relative displacement on the two sides of the boundary "follows" the moving pole $\mathbf{P}_{AC}$ in the sense of always being tangential to a sequence of circles drawn around $\mathbf{P}_{AC}$ as it moves relative to plate A. The change in the orientation of the transform and the movement of $\mathbf{P}_{AC}$ are both indications that the direction of spreading is changing.

Since the earth's lithosphere comprises about a dozen major and minor plates, at first glance it might seem that the conditions required to maintain a large number of contemporaneous transform faults would be so complex that none would survive. Yet transform faults are very abundant. Something about the dynamics of plate-tectonic processes appears to favor the formation and perpetuation of transforms. Many of them *do*, however, appear to depart from the small circle configuration expected on the basis of first-order plate-tectonic theory.

The way in which this takes place was discovered by H. W. Menard and Tanya Atwater at Scripps Institution of Oceanography soon after the articles reproduced in this book as Chapters 7 and 8 by McKenzie, Parker, and Morgan had appeared. Tanya Atwater was a graduate student at Scripps at the time, and her recollection of working on this problem captures the excitement among the young scientists who were drawn to plate tectonics. She writes (personal communication):

Figure 32-2
Tanya Atwater

Sea floor spreading was a wonderful concept because it could explain so much of what we knew, but plate tectonics really set us free and flying. It gave us some firm rules so that we could predict what we should find in unknown places. At Scripps, Bill Menard had been brewing on the origins of fracture zone offsets for a long time and he immediately began trying to make the rules tell him something about them. The distinct bend in the Mendocino set us off on the direction change. At first Bill and I were catching each other at odd moments, scribbling sketches on envelopes and scraps of paper, but we got more and more excited until we began hunting each other up in the morning to compare the previous night's thoughts. It was wonderful working with Bill because he knows the oceans so incredibly well. Whenever we found a new geometrical relationship, he could think for a moment and draw out of his mind some appropriate examples from the real world. The creation of brand new fracture zones by changes in direction of spreading was a prediction that fell straight out of the sketching games; there was no well-documented case. We went ahead and published it—his enthusiastic optimism overriding my trepidation. I was utterly amazed when we got some new lines near the great magnetic bight and the pattern was there, just as predicted. That day I was converted from a person playing a game to a believer.

The result of their study (Chapter 33) was a demonstration that in the eastern Pacific the older transforms not only changed direction but did so discontinuously. The transforms would move along a set of small circles for a while, then abruptly change direction and move along a slightly different set of small circles.

Bent Stripes

The most striking characteristic of the magnetic stripes is their almost perfect linearity over great distances. In the mid-1960's oceanographers of the Coast and Geodetic Survey began tracing the well-known magnetic lineations off the coasts of Oregon and Washington northward (Elvers et al., 1967b). As they approached the Aleutian trench, they made the surprising discovery that the magnetic stripes bend sharply through an angle of 110 degrees, from a nearly north-south direction to a nearly east-west direction. Moreover, both sets of stripes are offset along fracture zones orthogonal to the stripes. There are two obvious possible explanations: the stripes were formed straight and subsequently bent; or they were formed bent (Peter et al., 1970). In Chapter 34 Walter Pitman and Dennis Hayes show that the latter is true. The stripes were formed by a triple junction (Chapter 10), where three ridges radiated from a common point.

Asymmetrical Spreading

Sea-floor spreading is usually observed to be a symmetrical process in the sense that the rate of crystallization of new lithospheric plate is the same on either side of an oceanic ridge. The observational evidence for this is the symmetry that has been observed in the spacing of magnetic anomalies on either side of most of the mid-oceanic ridges. However, it does not follow from plate-tectonic theory that the formation of new lithosphere is necessarily symmetrical, and on examining most magnetic profiles in detail one finds evidence of short hiatuses, slight departures from symmetry, and other minor irregularities. These simply indicate that the process of sea-floor spreading is not perfectly regular and continuous. In Chapter 35, J. K. Weissel and D. E. Hayes present evidence from the ridge between Australia and Antarctica for a much larger departure from symmetry than has been observed in other profiles.

READING LIST

Additional articles describing the application of plate-tectonic principles to the solution of geologic problems are listed in Section IX.

33

Changes in Direction of Sea Floor Spreading

H. WILLIAM MENARD
TANYA M. ATWATER
1968

The first motion of earthquakes (McKenzie and Parker, 1967) and the orientation of fracture zones (Morgan, 1968b) indicate that very large blocks of the crust move as units away from sea floor spreading centres and towards oceanic trenches. Active transform faults bordering these units follow small circles which are lines of latitude relative to a pole of rotation on the globe. They parallel the direction of motion of one block relative to the adjacent one and thus indicate the spreading direction between the two. Fracture zones preserve the directions of ancient transform faults. At least five trends can be identified in the fracture zones of the north-eastern Pacific and these may indicate that spreading in as many different directions has occurred at different times in the area. Frequent changes in block motion would also commonly change transform faults into spreading centres with profound tectonic effects. Thus consideration of such changes may increase general understanding of the mechanism of sea floor spreading and provide the framework for interpreting the origin of specific tectonic features produced by spreading. We propose here criteria for identifying changes in the direction of spreading and discuss the mode, rate and prevalence of such changes and some of their probable effects and implications.

CRITERIA FOR IDENTIFICATION

Sea floor spreading accounts for various important features of ocean basins, but the ones most indicative of direction of spreading are fracture zones and magnetic anomalies. Active transform faults are required by geometry of motion to be parallel to the direction of spreading (McKenzie and Parker, 1967; Morgan, 1968b). When crust moves away from a spreading system, the offsets and trends of the transform faults are preserved as topography along fracture zones. The fracture zones, then,

From *Nature,* v. 219, p. 463–467, 1968. Reprinted with permission of the authors and Macmillan Journals, Ltd.

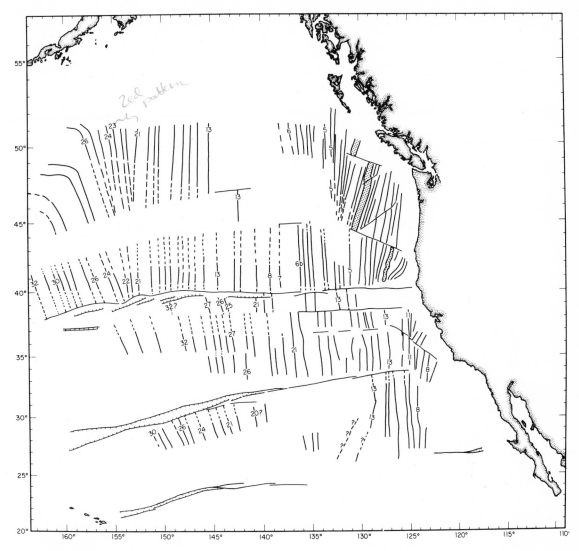

Figure 33-1
Fracture zones and magnetic anomalies of the north-eastern Pacific. Numbers follow system of Pitman et al. (1968)

are parallel to the spreading direction active at the time that they were frozen into the crustal block. Furthermore, midocean ridges and their associated magnetic anomalies usually seem to be perpendicular to the direction of spreading and to the transform faults and fracture zones (Mason, 1958; Vacquier, 1959; Vacquier et al., 1961; Peter, 1966; Raff, 1966). This perpendicular configuration is not required by the geometry of moving blocks but seems to be an intrinsic characteristic of the spreading phe-

nomenon wherever a steady direction of spreading has been established for some time.

Nevertheless, detailed surveys are few and in many places little known fracture zones and magnetic anomalies have been interpreted as trending at various angles to the midocean ridge crest. These interpretations may be correct but, considering the particular importance of fracture zones in the interpretation of crustal block movement, it seems worthwhile to examine the topography and magnetic anomalies of the relatively well known north-eastern Pacific. In this region the data are not imcompatible with a generalization that all large fracture zones and their associated magnetic anomalies are essentially perpendicular (Fig. 33–1). Furthermore, individual anomalies are readily identified and classified (Pitman et al., 1968). The identifiable anomalies are offset along the fracture zones for hundreds to a thousand kilometres (Mason, 1958; Vacquier, 1959; Vacquier et al., 1961; Peter, 1966; Raff, 1966). Each of these anomalies, No. 26, for example, was formed at a given time, the offsets resulting directly from the offsets of the ancient ridge crest by transform faults. At that time, according to block movement tectonics, sea floor spreading in the central north-eastern Pacific was in a uniform direction, forming fracture zones corresponding to small circles about a pole (Fig. 33–2). It is therefore striking that any given anomaly in Fig. 33–1 appears to trend towards the same pole almost everywhere in the region, regardless of offset. Thus the anomaly trends show that all the various segments of the ancient ridge crest were very close to perpendicular to the spreading direction indicated by the appropriate fracture zones.

The most detailed magnetic surveys in the north-east Pacific are those near the coast. They show a number of relatively minor faults which offset magnetic anomalies and are far from perpendicular to them (Mason, 1958; Vacquier, 1959; Vacquier et al., 1961; Peter, 1966; Raff, 1966). This may seem to invalidate our generalization; however, the deformation of this region has been exceptionally complex both because of a continent nearby (Wilson, 1965b) and because of a relatively recent change in direction of spreading. This change has influenced the orientation of both fracture zones and magnetic anomalies and perhaps has been accompanied by unusual minor adjustments between small blocks. Similar non-perpendicular offsets of magnetic anomalies have also been identified in other relatively well surveyed regions such as the Reykjanes ridge (Heirtzler, 1966) and the area north of the Mendocino fracture zone. In any case, our generalization is only that large fracture zones and adjacent magnetic anomalies are essentially perpendicular and does not

apply to relatively minor faults. Furthermore, contiguity is important. The magnetic anomalies produced in one episode of spreading are not necessarily perpendicular to the fracture zones formed during another episode. We have noticed that the orientation of a fracture zone indicates the spreading direction at the time it was frozen into a crustal block. This corresponds to the time that the crust on one side of a transform fault moved past the spreading centre on the other side, so that relative motion across the fault ceased. Thus wherever a fracture zone separates anomalies from two different episodes its orientation is parallel to the spreading direction associated with the younger episode. Matching changes in the orientation of anomalies and fracture zones can be used to identify such episodes.

Changes in direction of spreading can be recognized from the trend of fracture zones alone (Morgan, 1968b), but the topography of the zones in some places is complex and it is difficult to be accurate about relatively minor changes in trend along a single fracture zone. This is particularly true if the zone has been mapped by widely spaced random crossings rather than deliberately surveyed. Consideration of a large system of more or less surveyed fracture zones is more revealing. Block tectonics indicates that the fracture zones produced at a certain time in a certain region have the same trend. Thus subtle changes surveyed in detail on one zone can be extrapolated to adjacent and contemporaneous parallel zones. On the basis of fracture zone trends alone (Menard, 1967a), at least five directions of spreading can be identified in the north-eastern and central Pacific (Fig. 33–2).

Changes in directions of spreading can also be investigated using trends of magnetic anomalies; however, a criterion for identifying a change in direction by means of anomalies is required first. The chief clue is that the sequence of anomalies is continuous in all the ocean basins (Heirtzler et al., 1968) despite prolonged time lapses between anomalies (Ewing and Ewing, 1967). Thus the oceanic crust in any region breaks along any pre-existing spreading centre. Consequently, the older spreading centres are also the spreading centres for a new direction of spreading. Imme-

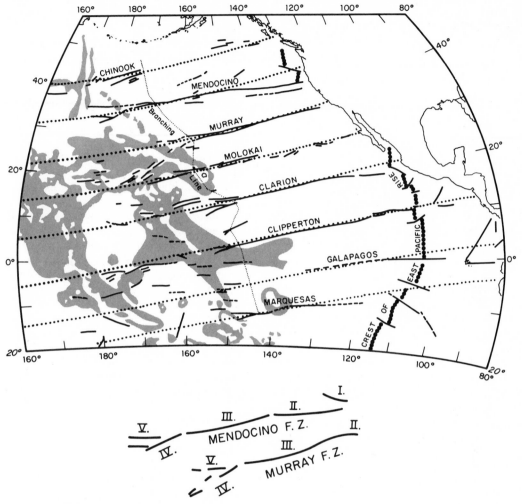

Figure 33-2
Fracture zones in the north-eastern Pacific showing trends corresponding to five possible spreading episodes. Dotted lines are small circles about the pole at 79°N., 111°E. suggested by Morgan (1968b). It is the pole of rotation for episode III

diately after a change in the direction of relative motion of two separating blocks, the transform faults between them become parallel to the new spreading direction. If the spreading centres are intrinsically perpendicular to the transform faults as we propose, then the centres will gradually become realigned. The pattern that would emerge is a "Z" or an "N" or a mirror image, all of which may be called "Zed patterns" (Fig. 33–3). The magnetic anomalies around the Juan de Fuca ridge form just such a Zed pattern slightly split by parallel anomalies and fractured by minor adjust-

ments. The western side of the Gorda ridge also shows the fan of anomaly trends characteristic of the Zed pattern although the eastern side is grossly distorted by interaction with North America and with the Mendocino fracture zone which has resisted the latest change of spreading direction.

It is evident that one side of the Zed pattern may be almost as diagnostic of a change in direction of spreading as the two sides together. Thus it is possible to identify an ancient change in spreading direction by means of an anomaly pattern far out on the flank of a rise

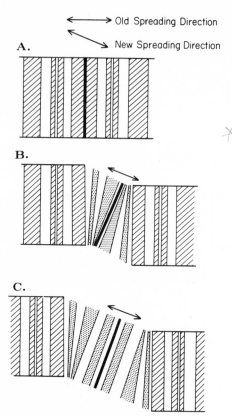

Figure 33-3
Proposed mode of adjustment of spreading
rises to a change in spreading direction.
(A) A segment of rise has been spreading,
forming symmetrical anomalies. The
spreading direction changes. (B) Spreading
occurs in the new direction. The ridge centre
gradually migrates around, forming
anomalies of a Zed pattern as it migrates.
(C) Spreading continues in the new direction. The ridge and its resulting anomalies
are perpendicular to the direction. A split
Zed pattern emerges

or ridge. This is particularly useful in the
north-eastern Pacific because most of the
eastern half of the East Pacific Rise is largely
missing. North of the Mendocino fracture
zone the western group of anomalies have a
different trend from those farther east, and
between No. 21 and No. 23 they form half of
a Zed pattern (Fig. 33–1). South of the Mendocino zone, anomalies 21–23 also appear to
form a transition in trend between the more
regular patterns of higher and lower number.
Data are, however, sparser for this southern

region and the relationships are not everywhere so clear. The existence of minor offsetting fracture zones would cause a significant
misinterpretation of the orientation of anomalies.

Changes in directions of spreading can be
demonstrated more convincingly by consideration of magnetic anomalies and fracture zones
together than by either alone. The essentially
perpendicular orientation of fracture zones
and anomalies helps to resolve some of the
difficulties of identifying changes. The trend
of the Mendocino fracture zone, for example,
changes significantly at the intersection of
anomalies 21 to 23 from north-east to almost
due east. Likewise both anomalies and the
adjacent fracture zones change trend between
anomalies 2 and 5. Thus three of the five directions of spreading suggested by fracture zone
trends are also indicated by patterns of magnetic anomalies.

Heirtzler et al. (1968) have noticed marked
changes in rate of spreading in the South
Pacific at anomalies 5 and 24 (Fig. 33–2).
These would suggest that spreading episodes
I, II and III identified in Fig. 33–2 were distinct motions of the entire central Pacific
block.

An exception to the idea of regional consistency exists in the magnetic anomaly pattern and fracture zones immediately off central
California. Shortly after No. 11 time they both
swing toward the south-east and the magnetic
anomalies form the western half of a Zed
pattern. No similar change in direction can
be seen associated with anomaly No. 11 elsewhere in the north-eastern Pacific, as would
be required if a general change in spreading
direction had occurred. This exceptional pattern may be the result of an edge effect which
occurred as the East Pacific Rise crest between the Pioneer and Murray fracture zones
approached North America in No. 11 time.

It might seem that the change in trend of
the Gorda and Juan de Fuca ridges are also
edge effects (Wilson, 1965b), and we cannot
eliminate this possibility or that of complex
interactions. Several magnetic anomalies
and fracture zones associated with the East
Pacific Rise, however, are known to have
changed trend at about this time both near the
continent and over a large region from it

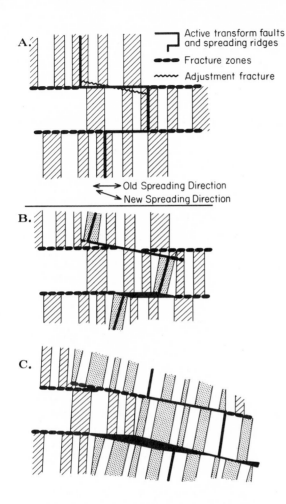

Figure 33-4
Proposed modes of adjustment of transform faults to a change in spreading direction. (A) A rise offset by two transform faults has been spreading in a direction perpendicular to itself. The spreading direction changes as shown. The block corners near the upper transform fault must fracture to adjust. The lower transform fault becomes a centre of slow spreading. (B) Spreading occurs in the new direction. The ridge centre migrates around. At the upper offset the adjustment fractures take on the new transform fault motion. The lower transform fault opens, creating a zone of mixed spreading and shearing. (C) Spreading continues in the new direction. The lower offset has become readjusted to pure transform fault motion. Note that the trends of the fracture zones and magnetic anomalies of the first episode are preserved

(unpublished work of R. L. Larson, H. W. Menard and S. M. Smith). This again suggests a change in the motion of the main Pacific block.

RATE AND MODE OF SPREADING

The Zed pattern of the Juan de Fuca ridge indicates a change of trend of 15°–20° starting at the time of anomaly No. 4 and ending in the time midway between No. 2 and No. 3. The latter time was about 3×10^6 yr ago and by extrapolation No. 4 time was about 7.5×10^6 yr ago. Thus the rate of change of the trend was about 4°–5° per million yr. The amount of change during the times of anomalies No. 21 and 23 was also about 20°, but the rate cannot be determined. The anomaly pattern in 21–23 time, however, is a revealing example of the mode of crustal adjustment to a change in spreading direction. Adjustment occurs along previous lines of weakness, namely, the transform faults and the various segments of ridge. We are preparing an analysis of the origin of fracture zone topography for separate publication. In brief, the transform faults take on the new direction almost as soon as the change begins. This necessarily involves fracturing of the corners of the old blocks if the transform fault is offset in one sense or addition of new crust along them if it is offset in the other (Fig. 33–4). This fracturing of the blocks presumably occurred along the Mendocino zone at approximately the time of anomaly No. 23 breaking off the upper corner of the old block and thus establishing the new direction of the zone. By comparison

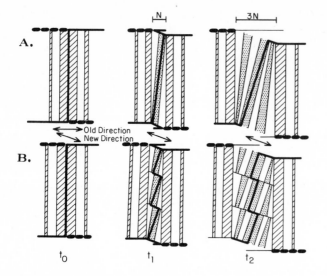

Figure 33-5
Two possible modes for adjustment of a ridge to a change in spreading direction. (A) The ridge readjusts as a unit taking on a new direction by time t_2 and after 3 N km of spreading has occurred. (B) The ridge breaks into segments, each piece becoming re-aligned by time t_1 after N km of spreading

to the complicated adjustments of the transform faults to new spreading directions, the reorientation of the ridges appears to occur relatively smoothly. Each segment of the ridge simply migrates around until it is perpendicular to the new direction. If the ridge segment is very long or if the change in spreading direction is large, the segment may break into shorter sections connected by transform faults as described later. The origin of the Pioneer fracture zone is connected to the change in direction of spreading at the time of anomaly No. 23. It may be the result of a sectioning of the ridge or of a complication in the fracturing of the Mendocino zone.

EFFECTS OF SPREADING

The principal variables influencing the tectonic effects of a change in direction of sea floor spreading are the geometry of the mid-ocean ridge crest, the spacing of transform faults, and the amount and rate of change. We shall consider some simple models to illustrate the effect of these variables. The simplest possible configuration of a ridge crest is a straight line perpendicular to the spreading direction. If the spreading direction then changes, the ridge realigns itself. It might do this as a unit, producing one large Zed pattern (Fig. 33–5, A). The adjustment, however, occurs in a much shorter time and in a much

smaller area if the ridge breaks into a number of sections connected by transform faults (Fig. 33–5, B). Any minor perturbations in the original straight line are areas of weakness where transform faults can begin. Once these transform faults are established, all subsequent changes in spreading direction produce a regularly offset crest similar to that found in parts of the eastern equatorial Pacific (unpublished results of Larson, Menard and Smith) the south Pacific (Pitman et al., 1968) and the equatorial Atlantic (Heezen and Tharp, 1965). This offset crest is reflected in the magnetic anomaly pattern as long as it exists. But, however many changes in direction occur, the apparent offset of magnetic anomalies on all of the fracture zones will have the same sense and this is not what is found in some regions. To explain some observed offsets of magnetic anomalies, such as those in the north-eastern Pacific (Menard, 1964), an initially offset crest is required.

Changes in direction of sea floor spreading have clearly occurred repeatedly in the north-eastern Pacific. They can account for some details of the configuration of fracture zones and magnetic anomalies in the region. In most of this area the magnetic anomalies and fracture zones are very close to perpendicular and change trend about the same time in response to changes in spreading direction. Distinctive anomaly patterns and sometimes the creation of new fracture zones accompany changes in spreading direction. These interacting features presumably are widespread in all the ocean basins and the regularity of the geometrical relationships may provide very useful clues for mapping and interpreting ocean basin tectonics. For example, repeated changes in direction of spreading inevitably produce a jungle of split Zed pattern magnetic anomalies separated by possible intermittent fracture zones. This would be difficult to resolve without some understanding of the origin and characteristics of such patterns.

Identification of changes in direction of spreading in all the ocean basins should be particularly revealing with regard to the interaction between large crustal blocks. Either simultaneous or sequential changes in spreading direction can be recognized by identifying the particular magnetic anomalies in a Zed pattern.

In Hess' original model of sea floor spreading a crust of little strength is created at the mid-ocean ridges and floats away on the convecting mantle. The frequent small changes in block motion which have occurred in the north-east Pacific seem difficult to reconcile with this model, and appear rather to support the hypothesis of strong crustal blocks moving because of relatively variable stresses.

34

Sea-floor Spreading in the Gulf of Alaska

WALTER C. PITMAN, III
DENNIS E. HAYES
1968

From *Journal of Geophysical Research*, v. 73, p. 6571–6580, 1968. Reprinted with permission of the authors and the American Geophysical Union. Copyright 1968.

Recent oceanic geophysical studies strongly support the hypothesis of sea-floor spreading. A magnetic anomaly pattern that is parallel to the axis of the mid-oceanic ridge system has been found in the North Pacific (Vine, 1966), South Pacific (Pitman and Heirtzler, 1966; Pitman et al., 1968), and Indian oceans (Le Pichon and Heirtzler, 1968). The pattern extends from the ridge axis across the flanks to the basins and is bilaterally symmetric with respect to the ridge axis. The pattern in all these regions is similar; the magnetic anomaly profile (L–M) shown at the bottom of Figure 34–2 is considered to be typical for the North Pacific and has been correlated in detail with numerous profiles from the other oceanic areas mentioned. The axial part of the pattern from all these regions plus that of the North Atlantic correlates with the magnetic polarity history for the past 3.4 m.y. as determined by Dalrymple et al. (1967), suggesting that the spreading in these areas has been active in recent times. Fault plane solutions (Sykes, 1967) indicate that the sense of first motion at the fracture zones that offset the ridge axes corresponds to that predicted by the transform fault hypothesis (Wilson, 1965c). This not only supports the concept of spreading but also suggests that the process is active today.

Although there appears to be abundant evidence to substantiate the spreading hypothesis, certain observations remain problematic, such as the sharp, nearly right angle bend in the strike of the magnetic anomaly lineations in the northeast Pacific. The bend, which has been called the Great Magnetic Bight of the northeast Pacific (Elvers et al., 1967a, b), is shown schematically by the broad dashed lines in the lower left-hand corner of Figure 34–3. This paper presents new data that further delineate the magnetic pattern in the northeast Pacific. An explanation of the bight that is consistent with sea-floor spreading is suggested.

NEW MAGNETIC DATA AND INTERPRETATION

Recent cruises of the *Robert D. Conrad* and the *Vema* have enabled us to extend the known pattern of linear anomalies into the Gulf of Alaska. Figure 34–1 shows the tracks of these recent cruises in the Gulf of Alaska. The magnetic anomalies are plotted along the tracks. The shaded areas represent positive anomalies; the unshaded areas, negative anomalies. Figure 34–2 shows east-west projected profiles of the magnetic anomalies. All the profiles except L–M are identified in Figure 34–1. Profile L–M is a composite profile taken from the data of Raff and Mason (1961); the portion from anomalies 5–32 is from just north of the Mendocino fracture zone and the portion from anomaly 5 on the east to anomaly 5 on the west is from a crossing of the axial zone of the Juan de Fuca ridge (Pitman et al., 1968). The profiles have been aligned with respect to anomaly 16. Dashed lines have been drawn between profiles to illustrate some of the correlations. Profile B–D has been broken into the three pieces defined by the inferred fracture zones.

The correlations of anomalies 15 through 24 and anomalies 5 through 6 in profiles A–B, B–E, F–E, H–I, and J–I are considered to be particularly good. Note also the comparison with profile L–M. The correlation in detail of anomalies 6–12 is more difficult; however, the general pattern is discernible in most cases. From anomaly 6 eastward along profiles J–G, F–G, and K–G, the pattern is difficult to correlate in detail, but the wavelengths and amplitudes of the anomaly pattern are similar. This area appears to be complex with major and minor offsets in the anomaly pattern (Vine, 1966).

Our interpretation of the magnetic anomalies is shown in Figure 34–3. The heavy solid lines show the trends of the correlated anomalies. Where these lines are dashed, the trends are regarded as poorly controlled. The strike of the pattern appears to approximately parallel the eastern limb of the Great Magnetic Bight. Three proposed fracture zones are shown. Although their locations are based primarily on magnetic evidence, there is some support in the seismic-profiler and bathymetric data. Important assumptions in our interpretation are that fracture zones are generally continuous through a region of spreading and that the offset of the magnetic anomalies across these fracture zones is constant over short distances. We have chosen the interpretation that requires a minimum number of fracture zones to explain the anomaly correlations. Alternative interpretations are certainly possible in view of the limited control (Peter et al., 1968). It is impossible to distinguish small offsets that might occur between track lines from slight bends or distortions in the lineation pattern.

In the south-central portion of Figure 34–3 the trends appear to have sinuous shapes. This is most noticeable in the region between 138° and 155°W south of 52°N; the small circles in this region indicate data from U.S.S. *Rehoboth* (Obrochta, 1966). This apparent sinuosity is not thought to be particularly significant in relation to the Great Magnetic Bight. Possibly, it can be explained by additional fracture zones.

The trends of the anomaly pattern also appear to be disturbed in the southeast region of Figure 34–3. Here the currently active axis of the Juan de Fuca ridge strikes NNE. This distortion could be due to small differences in spreading rates along the Juan de Fuca ridge or to a change in the direction of spreading as previously suggested (Vine, 1966). This latter possibility will be discussed in greater detail.

By extrapolating a time scale from known magnetic polarity events and observed lineation patterns, Heirtzler et al. (1968) have tentatively dated the entire sequence of anomalies. Their time scale is shown below profile L–M in Figure 34–2. In this context, the lineations in Figure 34–3 can be regarded as isochrons. The age (in millions of years) for selected lineations is shown in parentheses.

GENERATION OF PATTERN BY SPREADING

As a starting point in this study we accept sea-floor spreading as a working hypothesis. We also recognize that the Great Magnetic Bight causes serious problems in terms of the simple patterns of spreading previously proposed. Our purpose is to explain the magnetic and fracture pattern shown in Figure 34–3

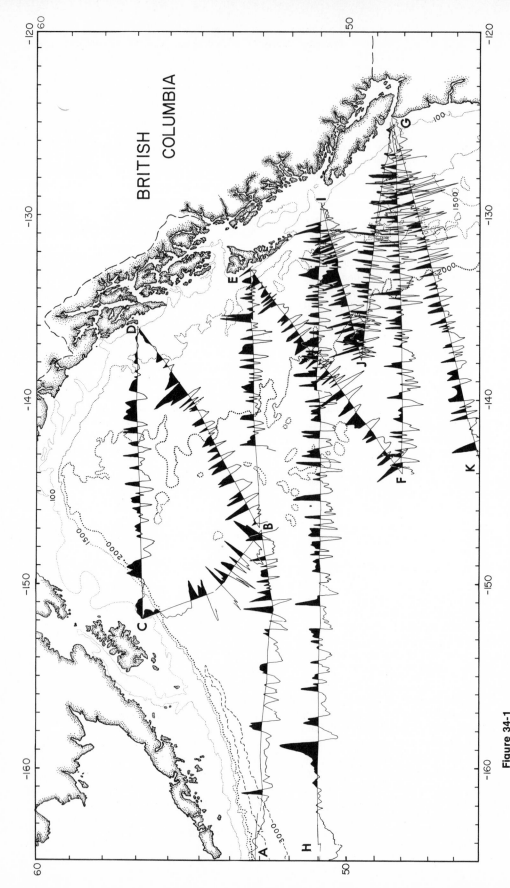

Figure 34-1
Map of the northeast Pacific showing generalized bathymetry and ship tracks. The magnetic anomalies are plotted along the tracks, the solid portions representing positive anomalies

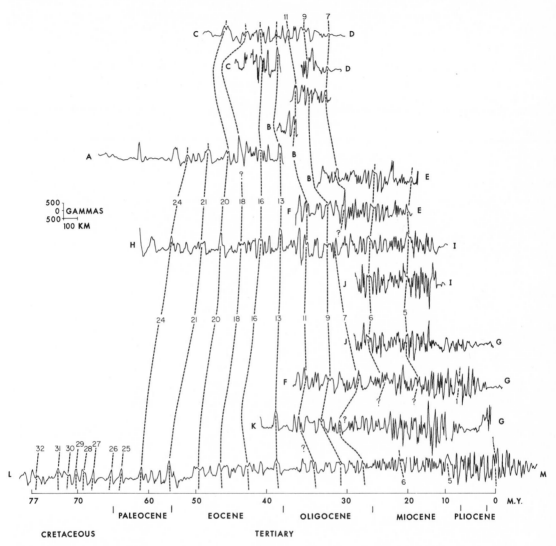

Figure 34-2

Magnetic anomaly profiles from Figure 34-1 projected along east-west azimuths and aligned to illustrate the best correlations. The end points of the profiles are labeled to show their location in Figure 34-1. The bottom-most profile (L-M) is discussed in the text

in terms of spreading and to examine any necessary complications or consequences. Although the details of the lineation pattern as shown in Figure 34–3 are likely to change as more data are acquired, we believe that the basic pattern will remain as one of northerly trending lineations, offset by east-west fracture zones. It is this *generalized* pattern and the pattern described by Hayes and Heirtzler (1968) and Grim and Erickson (1967) to the

east that provide the constraints for arriving at our explanation. We suggest an arrangement of crustal plates, trenches, fracture zones, and ridges through which the observed pattern might have been generated. Convection is still presumed to be the underlying causal mechanism; however, the geometry implied here imposes obvious difficulties regarding any simple relationship of convective cells to the motion of crustal plates.

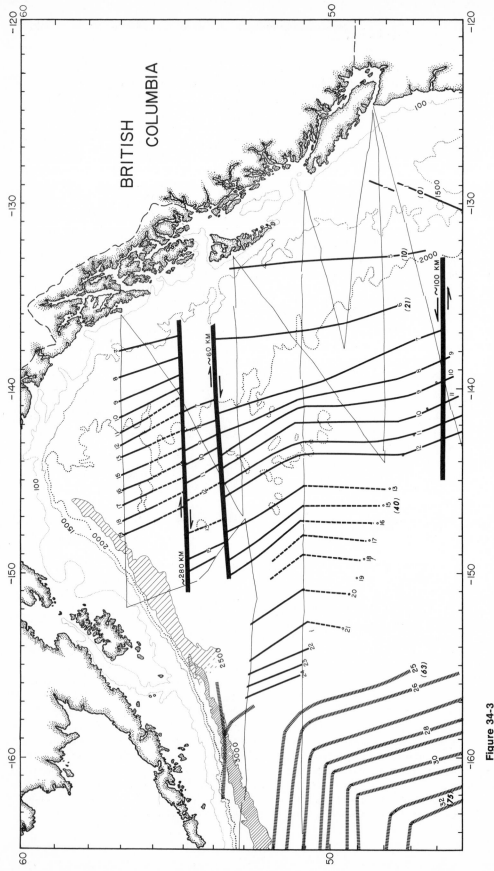

Figure 34-3

Magnetic lineations in the North Pacific. The broad dashed lines are the lineations from Elvers et al. (1967a); the solid lines show lineations inferred from the data of Figure 34-2, where dashed, the lineations are regarded as poorly controlled. The three bold east-west lines show the location of proposed fracture zones

In the following discussion it is tacitly assumed that the time scale shown in Figure 34–3 is correct. The possible errors inherent in the extrapolation of such a time scale are obvious. Other authors have, for example, suggested discontinuities in the spreading process (Ewing and Ewing, 1967; Briden, 1967).

Figure 34–4 *schematically* illustrates four stages in the evolution of the area. Figure 34–4, *A* shows the northeast Pacific in the Late Cretaceous at the onset of spreading. Anomaly 32 is being formed. The heavy lines represent active ridge axes; the fine dashed lines are fracture zones; the crossed regions are trenches or zones of weakness that are being formed into trenches. The shaded regions represent the older portions of oceanic crust. The continental land masses are shown for reference only and are not intended to represent paleogeography.

The region is divided into four rigid plates separated by a combination of incipient fracture zones, ridge axes, and trenches. Plate I is the central Pacific and will be regarded as immobile. It is separated from plate II by an offset east-west-trending active ridge axis plus one or more north-south fracture zones (Hayes and Heirtzler, 1968; Grim and Erickson, 1967). The north-south orientation of the fracture zones indicates that the relative motion between plates I and II must be north-south (Morgan, 1968b). Therefore, plate II is constrained to be moving northward into the trench bordering its northern edge. Plate III consists of continental North America bordered on the west by a now quiescent trench and of the Bering Sea bordered on the south by a trench (not necessarily in the same location as the present Aleutian trench). Plate III is also presumed to be immobile. The validity of this assumption will be discussed later. Plate IV, constrained by the orientation of the fracture zones between it and plates I and III, is presumed to be moving to the east into the trench bordering the western edge of the North American continent.

The system consists of migratory ridge axes defined by the geometry shown in Figure 34–4, *A*. The anomaly pattern would be formed in the fashion described by Vine and Matthews (1963). The strike of the northeast segment

of ridge axis and thus the hinge of the observed magnetic bight is intuitively presumed to be controlled by the relative velocities of plates II and IV.

The configuration of crustal plates shown in Figure 34–4, *A* and 34–4, *B* is similar to that described for the Galapagos region by Herron and Heirtzler (1967) and independently by Raff (1968) for both the Galapagos and the northwest and northeast Pacific.

Figure 34–4, *B* shows the same area after an interval of spreading and the occurrence of several magnetic polarity reversals. The slashed areas represent regions of normally magnetized crustal material; the white areas, regions of reversely magnetized crustal material. Plates II and IV are smaller than in Figure 34–4, *A*. It can be seen that the segments of active ridge axes that intersect must change length with time in order to generate the magnetic bight. As depicted in Figure 34–4, *C*, plate II and the east-west sections of ridge axes have migrated into the east-west trench. Consequently, the north-south spreading has been stifled and this east-west trench has become quiescent. Note on the left-hand edge of Figure 34–3 that anomaly 25 appears to run into the trench, suggesting a maximum age of early Paleocene for the beginning of a period of quiescence. The remaining sections of active ridge shown in Figure 34–4, *C* continue to migrate easterly toward the trench bordering the western margin of North America. Figure 34–4, *D* represents the generalized pattern in early Pliocene. The entire active ridge system except for a small section just north of the southernmost fracture zone in Figure 34–4 has migrated into the trench system along the western margin of Canada, thus stifling the spreading.

In the hypothesis of sea-floor spreading the trenches are generally regarded as regions of convective downwelling. Viewed in this light, the sequence of events described in Figure 34–4 suggests the implausible situation in which the convective source (ridge axis) drifts into the convective sink (trench), the process being stifled by the vanishing lateral dimension of the convection cell. It is possible, however, that the trenches do not mark sites of major convective downwelling. Alternatively, the trenches and perhaps other tectonic

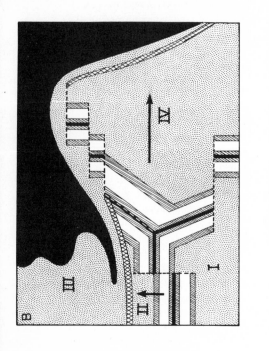

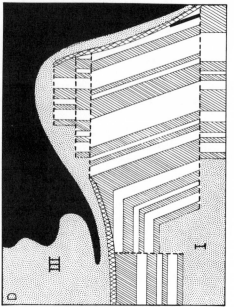

LATE CRETACEOUS

EARLY PLIOCENE

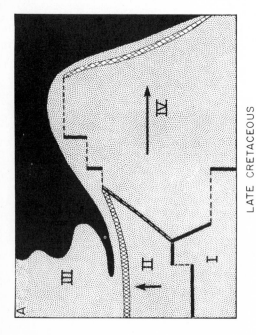

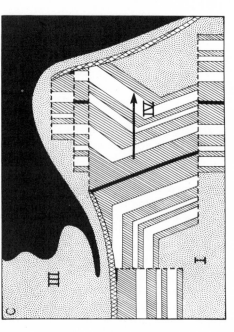

EARLY PALEOCENE

Figure 34-4
Schematic diagrams of four stages in the development of the lineations of the north-east Pacific (see text). The arrows indicate relative motions

features may be the complicated surface expressions of deep-seated convective processes. McKenzie and Parker (1967) have suggested that the convection cells which probably cause spreading are not closely related to the surface boundaries of rigid moving plates. That an active ridge and trench can reside in very close proximity is evidenced by the East Pacific rise and the Middle America trench.

Figure 34–4 does not satisfactorily explain the present day Aleutian trench and its seismic activity or the NW strike of the Blanco fracture zone or the NNE strike of the Juan de Fuca ridge (Pitman et al., 1968). Various authors (Vine, 1966; Morgan, 1968b; Le Pichon, 1968) have suggested, however, that in the late Miocene or early Pliocene the direction of spreading in the northeast Pacific changed from east-west to northwest-southeast. Wilson (1965c), Vine (1966), Morgan (1968b), and Le Pichon (1968) have attributed the opening of the Gulf of California to this northwest-southeast spreading. Tobin and Sykes (1968) have shown that the seismically active fracture zones in the northeast Pacific are transform faults and that the present direction of spreading in this region is northwest-southeast. It has also been demonstrated from earthquake mechanism studies and from other evidence (Stauder and Bollinger, 1966a; Stauder, 1968a; Plafker, 1967; Tobin and Sykes, 1968; McKenzie and Parker, 1967) that, at present, the North Pacific crustal plate is being thrust under the Aleutian trench. Burk (1965) has suggested that the present Aleutian trench is probably as young as late Tertiary and maybe early Pliocene. A change in direction of the spreading process within the Pliocene could have created the present-day Aleutian trench and could explain the limited extent and the northwest strike of the Blanco fracture zone. It also might account for the NNE strike of the axis of the Juan de Fuca ridge.

GEOLOGICAL IMPLICATIONS

Ewing et al. (1968) and Hamilton (1967) have noted the presence of turbidites in the North Pacific east of 165°W. These turbidites are thickest to the north, near the south wall of the Aleutian trench, thus suggesting that the sediment source was to the north and that in this area the Aleutian trench postdates the flow of these turbidites. It is proposed here that the easternmost section of the Aleutian trench began forming in the early Pliocene in response to the more recent northwesterly phase of spreading. To the west of 165°W and immediately south of the Aleutian trench there is no evidence of turbidites (Ewing et al., 1965; Shor, 1964). This would imply that a trench or island arc system persisted in this area throughout the quiescent period, thus preventing the flow of terrigenous material from the north onto the Aleutian abyssal plain.

It was suggested previously that the maximum age of this period of quiescence would be early Paleocene. If, however, it is assumed that the opening of the Gulf of California was entirely the result of about 300-km northwesterly movement of the North Pacific plate, it may be implied that approximately 200 km of oceanic crust has been thrust in a northerly direction into the Aleutian trench. From the time scale shown in Figure 34–2 it can then be inferred that anomaly 24 (200 km from anomaly 25) was the last east-west anomaly to be formed and that the period of quiescence for the east-west trench may have begun as late as middle Paleocene.

Vine (1966), Le Pichon (1968), and others have attempted to relate the process of sea-floor spreading to continental tectonism and orogenesis. The geologic record in the Gulf of Alaska region must provide tests against the geologic implications of this hypothesis. We selectively cite a few geologic facts that appear to be consistent with our hypothesis. Shaw (1963) has suggested that the Eocene-Oligocene crustal shortening that resulted in the formation of the Canadian Rockies might have been caused by the 'rigid simatic Pacific plate' thrusting against the North American continent. Burk (1965) believes that the major deformation of the Alaskan peninsula took place during the Pliocene by vertical uplift. Stoneley (1967) has concluded that the Mesozoic and Cenozoic deposits along the Gulf of Alaska have been deformed by repeated thrusting and folding caused by the 'continuous movement of the Pacific Ocean floor towards and perhaps under the continental margin.' Stoneley thinks that there were two periods during which this movement was most intense, one during the Late Cretaceous-early Tertiary, the second in the Plio-Pleistocene.

Certain other geologic observations imply complications to our simplified explanation for this area. Cenozoic and Mesozoic fold-fault belts parallel the margin of the Gulf of Alaska. Stoneley (1967) and Plafker and Mac-Neil (1966) indicate an episode of orogenic deformation along the margin of the gulf that culminated in late Eocene or early Oligocene. These observations might be explained by considering the possibility that the entire North American continent was moving during the Cenozoic.

CONTINENTAL DRIFT

According to our previous assumption, plate III has been fixed with respect to plate I throughout the sequence of events shown in Figure 34–4. The directions of motion suggested for the plates in Figure 34–4 are relative. Any velocity vector may be superimposed on the entire system. It is possible that plate III drifted to its present position as a single unit or as various separate pieces during the time interval discussed here. In other words, we cannot entirely discount the possibility that the North American continent drifted generally southwestward over a pre-existing ridge and lineation pattern located somewhere in the vicinity of the present Bering Sea. If this assumption is true, this would obviously lead to a more complicated system of trenches and tectonic features along the boundaries of plate III. Seismic reflection and refraction results of Ewing et al. (1965) and Shor (1964) have shown that the Bering Sea contains over 4 km of flat-lying sediments. Ewing et al. (1965) concluded that the large thickness of sediment with its near-horizontal reflectors are suggestive of a long period of tectonic quiescence. Evidence from Scholl et al. (1968a) also suggests that the region of the Bering Sea-Aleutian basin has been relatively stable throughout the Cenozoic. Hayes and Heirtzler (1968) also showed that the Aleutian basin is characterized by relatively low-amplitude

magnetic anomalies, perhaps indicative of long-term stability.

Another possibility is that plate III, consisting of the North American continent and the Bering Sea region, has been in its present position since the onset of the earliest phase of spreading discussed here. Hallam (1967) concluded from paleozoogeographic evidence that a Bering land bridge between Asia and North America must have been established by Middle or Late Cretaceous. This conclusion implies little or no relative motion between Asia and North America since that time. There is evidence that spreading in the North Atlantic, as well as in the South Atlantic, North Pacific, South Pacific, and Indian oceans, has been continuous for the past 10 m.y. (Pitman and Heirtzler, 1966; Vine, 1966; Pitman et al., 1968; Dickson et al., 1968; Le Pichon and Heirtzler, 1968). Ewing and Ewing (1967) suggest, however, from studies of sediment distributions that prior to 10 m.y. B.P. there was a worldwide cessation of spreading lasting many millions of years. The evidence they present is most impressive for the North Atlantic, where there is a large discontinuity in sediment thickness at distances from the ridge axis corresponding to 10 m.y. B.P., as determined by spreading rates. Because this discontinuity appears to be most pronounced for the North Atlantic, it is suggested as a possibility that the interruption in spreading in the North Atlantic began much sooner and therefore lasted longer than in other areas. The presence of Miocene cores within a few tens of kilometers of the ridge axis in the North Atlantic (L. H. Burckle and T. Saito, personal communication, 1967) lends support to this interpretation.

Several problems arise if North America has been essentially in place since the Late Cretaceous. It has been shown by various authors that the anomaly pattern found in the North Pacific is the same as that found in the South Pacific, the Indian, and the South Atlantic oceans, indicating that spreading has been simultaneous in these areas. The time scale derived from this spreading suggests that it began in the Late Cretaceous and has been nearly continuous to the present. The proposition that the North American continent has been in place from the Late Cretaceous until the Pliocene, whereas South America has been drifting during this interval, raises the question of differential motion between the two continents.

Another possibility is that the spreading in the northeast Pacific and North Atlantic has been simultaneous throughout the Cenozoic. If this were the case and if Pacific plates I and III are regarded as geographically fixed, the spreading in the Atlantic might have resulted in the eastward drift of Eurasia over the Pacific plate.

SUMMARY AND CONCLUSIONS

The magnetic data presented here show that the well-known magnetic lineations of the eastern North Pacific extend well into the Gulf of Alaska. The magnetic pattern is offset by one or more east-west fracture zones near 55°N with apparent right-lateral displacements as large as 280 km.

A migratory spreading ridge axis can explain the pattern of observed lineations and fracture zones in a manner consistent with ocean-floor spreading.

The observed magnetic pattern can be identified and related to similar magnetic lineation patterns in the South Pacific, South Atlantic, and Indian oceans. If the chronology as determined by these studies is correct, spreading and migrating of the ridge axes in the northeast Pacific commenced in the Late Cretaceous. The northward motion of a crustal plate into the Aleutian trench was interrupted in the Paleocene. The eastward motion of another crustal plate continued up to the Pliocene. At that time, a second phase of spreading began directed in a NNW–SSE direction, which explains the observed distortion of the present lineation pattern.

We do not contend that the explanation offered here agrees with all aspects of the 'known' geologic history of the adjacent continent. However, the inferred motion of the crustal plates of the northeast Pacific required to generate the observed magnetic pattern may be responsible for the general Cenozoic tectonic structures of northwestern North America.

35

Asymmetric Seafloor Spreading South of Australia

JEFFREY K. WEISSEL
DENNIS E. HAYES
1971

From *Nature*, v. 231, p. 518–521, 1971. Reprinted with permission of the authors and Macmillan Journals, Ltd.

Since mid-1968, the National Science Foundation research vessel Eltanin has operated south of Australia on a systematic geological, geophysical, and oceanographic reconnaissance. North-south tracks at a spacing of 5° longitude or closer transect the area of interest from 140°E to 105°E. The Eltanin data are supplemented by geophysical data from Lamont-Doherty ships, from Australian, French and Japanese sources, and from aeromagnetic tracks.

An earlier account of seafloor spreading between Australia and Antarctica (Le Pichon and Heirtzler, 1968) was based on limited data, confined almost entirely to the north flank of the Indian-Antarctic ridge. A detailed study of the magnetic anomaly lineation pattern for this entire region is now in progress (J. K. W. et al., to be published). The pre-drift reconstruction of Australia and Antarctica has been re-examined (Sproll and Dietz, 1969) by fitting the 2,000 m isobaths of the respective margins in the fashion of Bullard et al. (1965). There is no morphological or geological evidence for crustal subduction zones along the continental margin of Australia or along that of Antarctica; the very thick accumulations of sediment that exist beneath the Wilkes Abyssal Plain and seem to thicken toward the south may, however, be indicative of an ancient buried trench (R. Houtz, private communication). The seismic activity of this region has been reviewed (Barazangi and Dorman, 1969) and relocated epicentres (Sykes, 1970c) are shown in Fig. 35–1.

In this article we will consider briefly the large amount of new data for this area. We recognize the presence of three longitudinal zones, distinctly different in their geophysical and morphological properties, which constitute integral parts of two of the principal crustal plates (Indian Plate and Antarctic Plate). Their contrasting properties suggest that the basic concepts of plate tectonics (that is, large, rigid lithospheric plates, undeformed

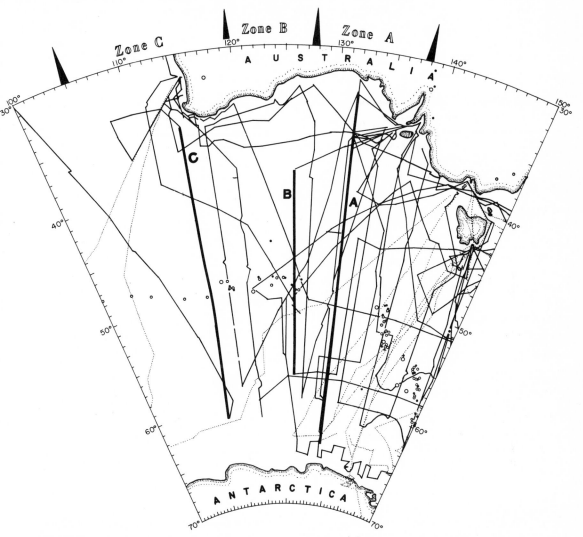

Figure 35-1
Index map of tracks along which geophysical data have been obtained. Relocated earthquake epicenters (Sykes, 1970c) are shown as circles, the smallest representing the most accurate determinations. Boundaries of the inferred zones of contrasting geophysical and morphological properties are indicated. Bold lines identify the profiles shown in Figure 35-2. Solid tracks are those of Lamont ships or programmes (r.v. Vema, Robert D. Conrad, Eltanin). Dotted tracks are those from all other sources. This is a stereographic conformal projection

except at their boundaries), though convincingly demonstrated on a global scale, may not strictly hold in a detailed analysis. The demonstration of systematic asymmetric spreading is the most important method to be dealt with in this article.

Zone A (138°E to 128°E) is almost aseismic, implying an absence of important trans-

form faults. Aseismic zones of similar extent also characterize segments of the East Pacific Rise (Barazangi and Dorman, 1969; Sykes, 1963). Zone B (128°E to about 120°E), however, shows a broad zone of seismic activity. Zone C also exhibits seismic activity—to a smaller extent than zone B—and is confined to a narrower zone (Sykes, 1970c). There is no linear grouping of epicentres in either zones B or C of the kind that might define large active transform faults. The presence of small offsets of the ridge crest is therefore inferred for both zones B and C.

Fig. 35-2,*A* shows a typical profile of topography and total intensity magnetic anomalies for zone A from the Antarctic continental slope to the Australian continental slope. Several key anomaly lineations (Pitman et al., 1968) can be identified. The classic morphological expression of a mid-oceanic ridge with an easily recognizable crestal zone is typical of zone A. The local topographic relief here is relatively low at about 100 to 200 m although a few large peaks are present on the southern flank. The magnetic lineations in this zone are well defined and the pattern can be traced both north and south from the axial zone at least to anomaly 13 and probably to anomaly 21. There are a number of "magnetic quiet zones" which begin just seaward of the continental slope and extend towards the land across the Antarctic and Australian continental margins.

Fig. 35-2,*B* shows representative profiles from zone B. The area is characterized by very high amplitude topographic relief (about 600 to 1,000 m with corresponding wavelengths of about 15 km). It is almost impossible to identify the ridge crest and so its location can only be inferred from the broad belt of epicentres extending through the zone. Several fracture zones are suggested by this data but individual zones cannot be correlated with the distribution of epicentres. About 500 km from the inferred ridge crest the relief is subdued and a sequence of magnetic anomalies from 6 to 13 is recognizable. It is important to note that anomalies older than anomaly 13 are continuous across the boundary between zone A and zone B. Magnetic anomalies within about 500 km of the epicentre belt are difficult to identify and no persistent pattern can be defined in this region.

Further to the west, zone C is characterized by a well developed ridge with local relief (generally less than 200 to 300 m) more similar to zone A than zone B (Fig. 35-2,*C*). The Diamantina Fracture Zone is shown near the northern edge of this profile and a minor fracture zone is indicated near the ridge crest.

There is no suggestion of seismically active, transform faults with large offsets. Magnetic anomaly lineations can be traced both north and south of the ridge crest to about anomaly 19.

If allowance is made for varying water depths, the magnetic anomaly amplitudes of zone A are consistently larger than those in zone C and zone B implying a contrast in magnetic properties of the rocks from zone to zone.

The amplitudes of anomalies in zone C older than anomaly 5 are larger over the north flank than over the south flank, and this relationship is contrary to that expected from inclination differences in the Earth's present field. The magnetic lineations and the morphology are symmetric about the ridge axis and the observed anomaly amplitude differences are, therefore, probably the result of decay (of unknown cause) of the remanent magnetization of the south flank.

The contrasts in seismic activity, inferred magnetic properties and ridge morphology demonstrate the occurrence of zones of markedly differing properties within two important plates generally considered to be internally uniform.

The locations of zone boundaries are clearly defined by the plot of "regional" depth against longitude as shown in Fig. 35-3. Depth variations with wavelengths < 100 km have been averaged to give the values quoted. Because the topographic "crest" for zone B is difficult to recognize, we define the "crest" as the minimum regional depth; the position of the "crest" coincides closely with the epicentre belt. There is a suggestion that the ridge crest within zone A shoals towards the east.

Fig. 35-4 shows the distance of a recognized magnetic anomaly (Pitman et al., 1968) from the ridge axis as a function of age taken from the geomagnetic reversal time scale (Heirtzler et al., 1968). Data from each zone were considered separately and different symbols used to distinguish lineations to the north of the axis from those to the south. There are several changes of slope common to all zones. If the time scale (Heirtzler et al., 1968) is correct, these changes reflect alterations in the rate or geometry of separation of the Antarctic and Indian plates. The time intervals between the inflexion points were examined

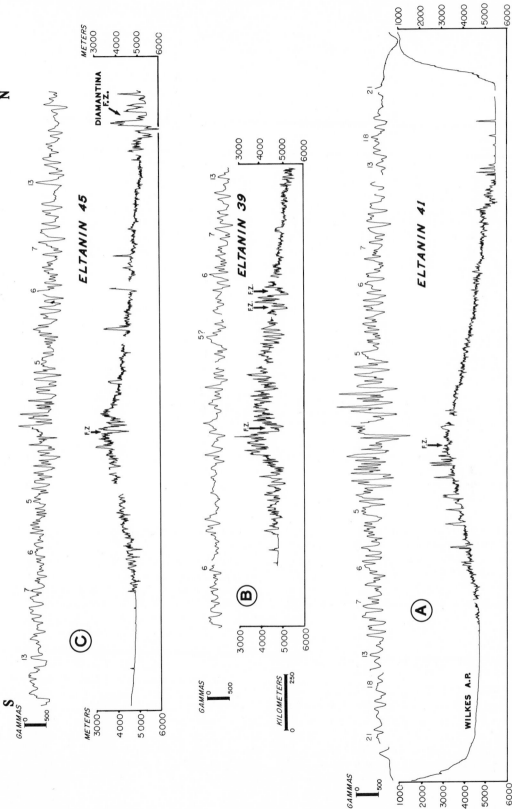

Figure 35-2
Representative magnetic and topographic profiles from each of the three zones. These are projected along a constant azimuth of average spreading direction determined from centers of rotation. Physiographic expressions of probable fracture zones are denoted by F.Z.

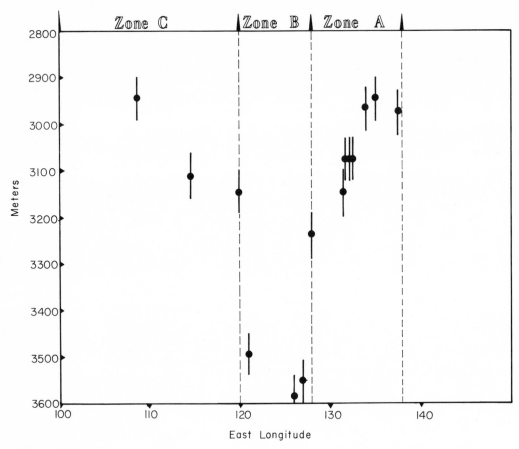

Figure 35-3
Variation of minimum regional ridge depth with longitude. The estimated confidence limits are shown by the vertical bars

as constant spreading rate intervals; the intervals chosen were 0 to 10 m.y., 10 to 20 m.y., 20 to 38 m.y., and 38 to 50 m.y. BP. Spreading rates (from each zone) were derived using a linear least squares analysis and are listed in Table 35–1. For the period 0 to 10 m.y., the crust in zone A was generated at the average half-rate of 3.69 cm yr^{-1} and at about the same rate for both the north and south flanks. The very limited data for the north flank of zone B indicate a relatively rapid half-rate of about 4.5 cm yr^{-1}. The spreading rate for the south flank for this period cannot be determined from these data. The spreading rate in zone C for this time interval was probably greater on the north flank than that on the south flank although the differences are only about 10%.

For the interval between 10 and 20 m.y. quite good evidence exists for asymmetric seafloor spreading in zone A. There is no indication whether or not spreading was symmetrical in zone B during this time interval. In zone C there is some evidence that asymmetric spreading occurred during this period; the rate for the north flank was greater (by about 10%) than that for the south. Constant spreading rates for this time interval are not well determined for any of the zones.

The interval between 20 and 38 m.y. shows significant and systematic asymmetric spreading for zone A. The half rates are 3.11 cm yr^{-1} on the north flank and 2.22 cm yr^{-1} on the south flank. This relationship has been defined on the basis of at least four tracks over

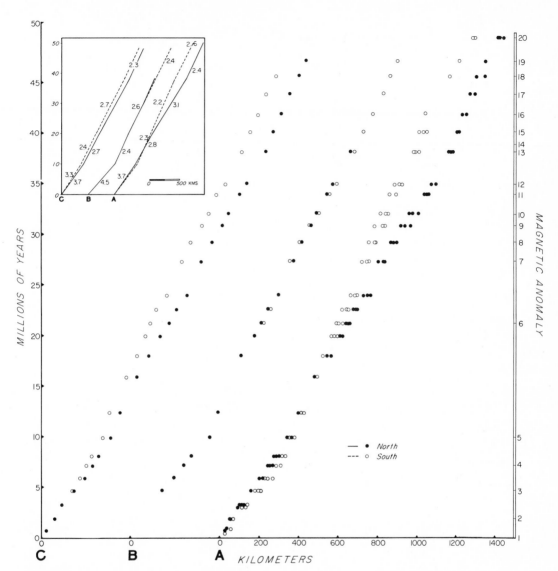

Figure 35-4
Distances to key magnetic anomalies from the ridge axis as a function of age for each of the three zones.
From left to right data for zone C, zone B, and zone A respectively are plotted. The distance scale is the
same for all profiles although the zeroes are displaced to avoid confusion. The numbers of the standard
magnetic anomalies (Pitman et al., 1968) are shown. These data were used to obtain a linear least squares
fit for four assumed constant spreading rate periods. The representative spreading rates (cm yr^{-1}) are
shown in the inset. The complete results of this analysis are given in Table 35-1. Solid circles and solid
lines, North. Open circles and dashed lines, South

each flank and is very well documented. During this time interval zone B and zone C show approximately symmetrical spreading. The total opening rates are 5.3 cm yr^{-1} for zone A, 5.0 to 5.4 cm yr^{-1} for zone B and about 5.5 cm yr^{-1} for zone C. The small differences observed probably relate to the variation in latitudes (total maximum range 35°) with respect to the pole of opening and to inherent errors in the spreading rate determinations.

Table 35–1
Calculated spreading rates for each of the three zones.

Zone	Time (m.y. BP)	North Half rates (cm yr⁻¹)	North Probable error	North No. of profiles	South Half rates (cm yr⁻¹)	South Probable error	South No. of profiles
A	0–10	3.60	0.05	2	3.78	0.08	2
	10–20	2.85	0.06	2	2.28	0.04	2
	20–38	3.11	0.04	2	2.20	0.03	2
	>38	2.35	0.08	2	2.57	0.16	1
	20–38	3.11	0.02	4–5	2.22	0.02	4–5
B	0–10	4.54?	0.07	2	—		
	10–20	2.35	0.06	1	—		
	20–38	2.63	0.02	1	2.54	0.05	2
	>38	—	—	—	2.43	0.07	1
C	0–10	3.71	0.07	1	3.29	0.05	2
	10–20	2.69	0.03	1	2.41	0.04	1
	20–38	2.82	0.03	1	2.70	0.03	1
	>38	2.38	0.02	1	2.35	0.08	1

Data for the period from approximately 40 to 50 m.y. give half-spreading rates of about 2.2 to 2.4 cm yr⁻¹ and seem to be generally comparable for all three zones.

A marked difference in the distribution of anomalies in narrow longitudinal zones and a general equivalence of total separation rates lead to an important conclusion about the evolution of this accreting plate boundary. The amount of crust generated on each flank (per unit length) as determined from the inferred spreading rates is shown in Table 35–2. The total formation of crust for each time interval is roughly comparable for each zone although complete calculations for zone B are not possible. In zone A there has been about 230 km more crust generated north of the ridge axis than south of it during the period from about 38 m.y. BP to 10 m.y. BP. The dis-

tance from anomaly 13 (about 38 m.y.) to the corresponding continental slopes of Australia and Antarctica is nearly the same (about 550 to 600 km), reaffirming that the asymmetric pattern cannot be extrapolated back to the time of initial separation. On the basis of the seismic pattern in zone B and the ridge crest morphology in zone A, we infer an offset of the ridge crest near the boundary of zones A and B of about 200 km which can be accounted for by the asymmetric spreading rates observed for zone A.

We visualize a situation in the earliest rifting history of Australia and Antarctica when there was no significant offset corresponding to the present boundary between zones A and B (Fig. 35–5). The present configuration of the spreading centre is partly a consequence of asymmetric spreading within zone A and

Table 35–2
Amount of crust generated per unit length of ridge crest (in kilometers).

	Time range (m.y.)	Zone A North	Zone A South	Zone A Total	Zone B North	Zone B South	Zone B Total	Zone C North	Zone C South	Zone C Total
	50–38	282	306	588	—	292	?	286	282	568
	38–20	560	400	960	473	457	930	507	486	993
	10–20	285	228	513	235	—	?	269	241	510
	0–10	360	378	738	454	—	?	371	329	700
Total	0–50	1,487	1,312	2,799				1,433	1,338	2,771

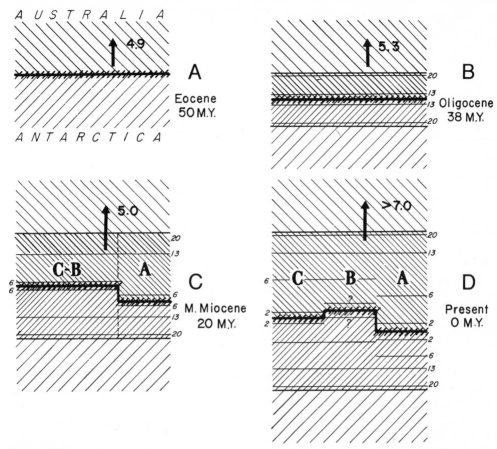

Figure 35-5
Schema of the evolution of the boundary between the Indian (upper) and Antarctic (lower) plates. Assuming Antarctica is fixed: (A) incipient fracturing of the continents with no offsets present along the rift; (B) following symmetric spreading for 12 m.y.; (C) asymmetric spreading has existed for 18 m.y. generating ridge offset as shown and differential offset of anomalies 13 to 6; (D) between 20 m.y. and 0 m.y. zone C experienced asymmetric spreading generating a slight offset at the boundary between zones C and B. Anomaly 2 as shown for zone B cannot be readily identified. Total spreading rate (cm yr^{-1}) is shown near the bold arrows

symmetric spreading in the adjacent zone B. The resulting differential ridge migration leads to a progressive offset of the spreading centre and provides an explanation for the observed differences in the offset of magnetic lineations across this fracture zone boundary. It is important to realize that the observed asymmetric pattern does not require any internal deformation within the principal crustal plates. The pattern of lineations can be explained and understood in terms of evolving boundaries for each individual zone. Only the free edge or accreting boundary of the plate is affected by this differential spreading. The only con-

dition for maintaining the continuity of the plates is that the total relative motion of the two plates is almost constant.

The contrast in the morphological properties, in the present seismicity and in the amplitudes of the magnetic anomalies—which lead us to the definition of individual zones—collectively demonstrate that these zones have evolved somewhat independently. This does not necessarily imply internal deformation of the plates. We have shown that the observed asymmetric seafloor spreading does not require internal deformation. Changes in the observed pattern of magnetic lineations, anomaly amplitudes, morphology, and seismicity are expected as a result of the motion of lithospheric plates on a sphere (Morgan, 1968b; Le Pichon, 1968; McKenzie and Parker, 1967). The zonal differences observed, however, seem to define discontinuous boundaries and the discontinuities suggest at least two feasible explanations.

(1) It may be that only the accreting plate boundaries evolve in a contrasting fashion. The mechanics of crustal accretion change along the boundary between separating plates, producing zones of rough and smooth morphology, high and low anomaly amplitudes, and symmetric and asymmetric spreading patterns. For this model, the present areal extent of these features is the result of the differing boundary mechanics and of seafloor spreading. (2) The ridge morphology and the magnetic anomalies may have been generated in a similar fashion for each zone. At some time after the generation of oceanic crust and its migration away from the ridge crest, second order deformation took place within the plate, altering the thermal remanent magnetization and fracturing the ridge flank morphology. The issue of deformation within the plate is a matter of scale. If there are no subduction or accreting zones within the dimensions of the plates, then by definition first order deformation is absent and the concepts of plate tectonics hold good. This does not preclude the possibility of tectonic activity within the plate on a somewhat smaller scale.

The results of recent studies clearly demontrate that the detailed motion of lithospheric plates about the Earth's surface requires a consideration of more than six plates (Le Pichon, 1968) or twenty plates (Morgan, 1968b). Recognition of active or ancient plate boundaries may be quite difficult if the contrast in the tectonic and magnetic fabrics is subtle.

The distribution of historical seismic activity and the observed pattern of magnetic lineations can be satisfied by our first explanation. Our second explanation best explains the absence of recognizable near crestal anomalies in zone B, the variation in the amplitudes of magnetic anomalies of the same age for all zones, and the highly fractured morphology of zone B. The inferred spreading rate here is fast (half rate $\sim$4.5 cm yr^{-1}) in contrast to the rate expected from the topographic roughness-spreading rate correlations (Menard, 1967b). The observed zonal differences in seafloor spreading between Australia and Antarctica probably result from a combination of both processes. The relative importance of each process has not yet been evaluated.

Zone A is the first large geographic area where systematic asymmetric spreading which persisted for a significant time ($>$20 m.y.) is well documented. There are many large areas of the world ocean in which only one limb of the inferred spreading system has been examined (for example, entire north-east Pacific, south-west Pacific). These areas must be reconsidered in the light of this important observation.

A detailed analysis of this entire area will be published by J. K. W. et al. Similarly, an analysis of asymmetric cooling at accreting zone boundaries is being considered (D. E. H., to be published) to explain the observed asymmetric spreading, and its implications for variable lineation offsets across fracture zones, generation of ridge crest offsets, and the abrupt disappearance of fracture zone morphology.

HEAT FLOW, GRAVITY, AND DRIVING MECHANISM

36

Introduction and Reading List

Search for a Driving Mechanism

In the articles included up to this point, motions of the earth's crust have, for the most part, been considered to be two dimensional. Plate tectonics describes the changing pattern of plates as they move over the earth's surface. However it doesn't explain *why* the plates are moving. Since the earth is a three-dimensional body, the question of driving mechanism inevitably involves tectonic processes occurring deep in the mantle. There can be little doubt that ultimately the plates are driven by thermal energy, so that the basic theoretical mechanism is one of convective overturn in a spherical shell of heterogeneous material in which the physical properties vary as a function of temperature. The task of gathering data to support the theoretical mechanism will be difficult in that it will require finding new knowledge about how rocks deform when they are subjected to high temperatures and extremely high pressures, and devising the apparatus to do the needed experiments. Moreover delineation of the mantle convection mechanism will require mathematical models more sophisticated than those employed heretofore. The research will undoubtedly occupy geophysicists for many years.

In this section several papers are presented in which the first steps have been taken to apply geophysical reasoning to learn something about processes occurring below the earth's surface.

Figure 36-1
David Griggs

Mechanics of Plate Motion

One of the main ways in which plate tectonics differs from earlier theories of mantle convection is in postulating the existence of lithospheric plates that are very long and very thin. The Pacific plate has a length of about 10,000 km and a thickness of no more than 100 km. Its length-to-width ratio is at least 100. Scaling considerations show that bodies with an aspect ratio this large tend to buckle rather easily; it's hard to push a long noodle, and if the noodle is long enough it's also hard to pull it without having it break. One of the first geophysicists to be concerned with the question of the mechanics of plates was Walter Elsasser (1967, 1969). Elsasser, a colleague of H. H. Hess, W. J. Morgan, and F. J. Vine at Princeton, found that a rigid lithosphere overlying a soft aesthenosphere could act as a stress guide, permitting stresses to be transmitted over great distance. In Elsasser's model, the cold, dense plate, as it sinks through the hot mantle beneath an island arc, effectively "pulls" the rest of the plate toward the trench. This is a special type of convection that does not depend on viscous coupling between a convecting mantle and the overlying lithosphere. Several objections have been raised to this model (Chapter 37). One is that plates exist that, although clearly moving, do not have a trench at any boundary. A second is that if Elsasser's model is correct, short plates should move faster because the driving force due to a trench would be the same as for a larger plate, whereas the resistance due to viscous drag at the base of the plate would be smaller. This does not agree with observation—some of the smaller plates are moving slowly. So the actual driving mechanism probably involves a more complex type of convection than that of Elsasser's model. However current thinking is still influenced by his demonstration that the lithosphere may transmit stress over great distances.

Figure 36–2
John Sclater (left) with the deep-sea
heat-flow probe

Since Elsasser's pioneering work, research has continued along two lines. One approach, taken by Dan McKenzie (Chapter 37) and David Griggs (Chapter 40), has been to start by assuming that a rigid slab is being thrust at uniform velocity into a mantle that is an ideal Newtonian viscous fluid, and to trace the thermal and mechanical consequences of this plate motion. A second approach has been to consider the classical convection problem, starting with a simple viscous layer heated from the bottom and introducing complexities such as (1) viscosity varying as a function of temperature, (2) more complex laws for stress versus rate of strain, (3) horizontal temperature gradients, (4) phase changes, and (5) internal heat sources. Oxborough and Turcotte have been the leaders in this research. (See reading list at the end of this chapter). The difficult task of producing mathematical models for mantle convection capable of explaining (and appropriately constrained by) geophysical and geologic observations is being pursued actively by numerous researchers.

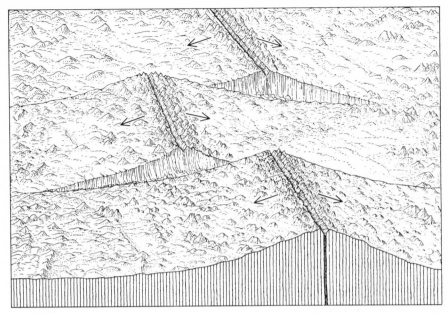

Figure 36-3
The average elevation of the sea floor is different on opposite sides of a transform fault for the following reason. As the lithosphere moves away from the ridge, it contracts and becomes denser as it slowly cools, sinking deeper and deeper into the soft aesthenosphere. (From "The Deep-Ocean Floor" by H. W. Menard. Copyright © 1969 by *Scientific American,* Inc. All rights reserved)

Heat Flow and the Topography of Ridges

Why are ridges higher than the adjacent ocean basins? Hess (1962) offered the qualitative answer that the lithosphere at the ridges, being hotter, was less dense and therefore would float higher on the soft aesthenosphere. The observation that the free-air gravity anomaly over the ridges was small (Chapter 39) supported this idea in a general way. The next step was to construct a theoretical heat-flow model for a spreading ridge to test this idea quantitatively. The results have proved unexpectedly fruitful and have lead to a method for estimating the age of the sea floor from its elevation.

The leader in this research has been John Sclater. Although a contemporary of Vine and McKenzie at Cambridge, initially he took neither the Vine-Matthews paper nor the Wilson paper on transform faults very seriously. While at Cambridge, Sclater, Wilson, and Hess had spent some time together looking for symmetry in magnetic profiles across the South Atlantic. At that time they were totally unsuccessful. As a result, Sclater became so skeptical about sea-floor spreading that he began to work on heat flow rather than magnetics, and continued this work after moving to Scripps Institution of Oceanography in California. Sclater and Dan McKenzie were roommates at La Jolla at the time the article included in this book as Chapter 7 was written and it was this study that finally convinced Sclater of the validity of plate tectonics. Sclater and McKenzie then began to work on the plate-tectonic history of the Indian Ocean using magnetic and topographic data. In 1969 Sclater and Jean Franchetau, then a Ph.D student at Scripps, began working on the article included here as Chapter 38. Their theoretical profiles of elevation as a function of distance from a spreading center turned out to give a surprisingly good fit to the observed bottom profiles measured with echo sounders. Features such as the change in elevation across fracture zones (Fig. 36–3, p. 443) are explained by their model not only qualitatively but quantitatively. Encouraged by this, Sclater et al. (1971) have proceeded to develop a standard curve that enables the age of rocks on the flank of a rise to be determined from their depth.

READING LIST

Mantle Convection and Sinking Slabs

Allan, D. W., Thompson, W. B., and Weiss, N. O., 1967, Convection in the earth's mantle, *in* Runcorn, S. K., ed., Mantles of the earth and terrestrial planets: New York, Wiley, p. 507–512.

Elsasser, W. M., 1969, Convection and stress propagation in the upper mantle, *in* Runcorn, S. K., ed., The application of modern physics to the earth and planetary interiors: New York, Wiley-Interscience, p. 223–246.

Elsasser, W. M., 1971, Sea-floor spreading as thermal convection: J. Geophys. Res., v. 76, p. 1101–1112.

Jacoby, W. R., 1970, Instability in the upper mantle and global plate movements: J. Geophys. Res., v. 75, p. 5671–5680.

Knopoff, L., 1964, The convection current hypothesis: Rev. Geophys., v. 2, p. 89–122.

Lliboutry, L., 1969, Sea-floor spreading, continental drift, and lithosphere sinking with an asthenosphere at melting point: J. Geophys. Res., v. 74, p. 6525–6540.

McKenzie, D. P., 1968a, The geophysical importance of high-temperature creep, *in* Phinney, R. A., ed., The history of the earth's crust: Princeton, N. J., Princeton Univ. Press, p. 28–44.

McKenzie, D. P., 1968b, The influence of the boundary conditions and rotation on convection in the earth's mantle: Roy. Astron. Soc., Geophys. J., v. 15, p. 457–500.

Minear, J. W., and Toksöz, M. N., 1970, Thermal regime of a downgoing slab and new global tectonics: J. Geophys. Res., v. 75, p. 1397–1419.

Rice, A. R., 1971, Mechanism of dissipation in mantle convection: J. Geophys. Res., v. 76, p. 1450–1459.

Rice, A. R., 1972, Some Benard convection experiments: their relationship to viscous dissipation and possible periodicity in sea-floor spreading: J. Geophys. Res., v. 77, p. 2514–2525.

Schubert, G., and Turcotte, D. L., 1972, One-dimensional model of shallow-mantle convection: J. Geophys. Res., v. 77, p. 945–951.

Schubert, G., Turcotte, D. L., and Oxburgh, E. R., 1970, Phase change instability in the mantle: Science, v. 169, p. 1075–1077.

Toksöz, M. N., Minear, J. W., and Julian, B. R., 1971, Temperature field and geophysical effects of a downgoing slab: J. Geophys. Res., v. 76, p. 1113–1138.

Tozer, D. C., 1967, Towards a theory of thermal convection in the earth's mantle, *in* Gaskell, T. F., ed., The earth's mantle: New York, Academic Press, p. 325–353.

Turcotte, D. L., and Oxburgh, E. R., 1967, Finite amplitude convection cells and continental drift: J. Fluid Mech., v. 28, p. 29–42.

Turcotte, D. L., and Oxburgh, E. R., 1968, A fluid theory for the deep structure of dip-slip fault zones: Phys. Earth Planet. Interiors, v. 1, p. 381–386.

Turcotte, D. L., and Oxburgh, E. R., 1969, Convection in a mantle with variable physical properties: J. Geophys. Res., v. 74, p. 1458–1474.

Turcotte, D. L., and Schubert, G., 1971, Structure of the olivine-spinel phase boundary in the descending lithosphere: J. Geophys. Res., v. 76, p. 7980–7987.

Heat Flow

Bullard, E. C., Maxwell, A. E., and Revelle, R., 1956, Heat flow through the deep sea floor: Adv. Geophys., v. 3, p. 153–181.

Burns, R. E., and Grim. P. J., 1967, Heat flow in the Pacific Ocean off central California: J. Geophys. Res., v. 72, p. 6239–6247.

Hamza, V. M., and Verma, R. K., 1969, The relationship of heat flow with age of basement rock: Bull. Volcanol., v. 33, p. 123–152.

Hasebe, K., Fujii, N., and Uyeda, S., 1970, Thermal processes under island arcs: Tectonophysics, v. 10, p. 335–355.

Herzen, R. von, 1959, Heat flow values from the southern Pacific: Nature, v. 183, p. 882–883.

Hsü, K. J., and Schlanger, S. O., 1968, Thermal history of the upper mantle and its relation to crustal history in the Pacific basin: Int. Geol. Congr., 23rd, Rep., sect. 1, p. 91–105.

Langseth, M. G., Le Pichon, X., and Ewing, M., 1966, Crustal structure of the mid-ocean ridges, 5, Heat flow through the Atlantic Ocean floor and convection currents: J. Geophys. Res., v. 71, p. 5321–5355.

Langseth, M. G., and von Herzen, R. P., 1970, Heat flow through the floor of the world oceans, *in* Maxwell, A. E., ed., The sea, v. 4, pt. 1: New York, Wiley-Interscience, p. 299–352.

Le Pichon, X., and Langseth, M. G., 1969, Heat flow from the mid-ocean ridges and sea-floor spreading: Tectonophysics, v. 8, p. 319–344.

Lister, C. R. B., 1972, On the thermal balance of a mid-ocean ridge: Roy. Astron. Soc., Geophys. J., v. 26, p. 515–535.

McKenzie, D. P., and Sclater, J. G., 1968, Heat flow inside the island arcs of the northwestern Pacific: J. Geophys. Res., v. 73, p. 3173–3179.

McKenzie, D. P., and Sclater, J. G., 1969, Heat flow in the eastern Pacific and sea floor spreading: Bull. Volcanol., v. 33, p. 101–118.

Oxburgh, E. R., and Turcotte, D. L., 1968, Problem of high heat flow and volcanism associated with zones of descending mantle convective flow: Nature, v. 218, p. 1041–1043.

Oxburgh, E. R., and Turcotte, D. L., 1969, Increased estimate for heat flow at oceanic ridges: Nature, v. 223, p. 1354–1355.

Oxburgh, E. R., and Turcotte, D. L., 1970, Thermal structure of island arcs: Geol. Soc. Amer., Bull., v. 81, p. 1665–1688.

Sclater, J. G., Anderson, R. N., and Bell, M. L., 1971, The elevation of ridges and the evolution of the central eastern Pacific: J. Geophys. Res., v. 76, p. 7888–7915.

Sclater, J. G., and Corry, C. E., 1967, Heat flow, Hawaiian area: J. Geophys. Res., v. 72, p. 3711–3715.

Sclater, J. G., and Harrison, C. G. A., 1971, Elevation of mid-ocean ridges and the evolution of the south-west Indian ridge: Nature, v. 230, p. 175–177.

Sclater, J. G., Mudie, J. D., and Harrison, C. G. A., 1970, Detailed geophysical studies on the Hawaiian arch near 24° 25′ N, 157° 40′ W: A closely spaced suite of heat flow stations: J. Geophys. Res., v. 75, p. 333–348.

Sleep, N. H., 1969b, Sensitivity of heat flow and gravity to mechanism of sea floor spreading: J. Geophys. Res., v. 74, p. 542–549.

Sleep, N. H., 1971, Thermal effects of formation of Atlantic continental margins by continental break-up: Roy. Astron. Soc., Geophys. J., v. 24, p. 325–350.

Smirnov, Ya. B., 1968, The relationship between the thermal field and the structure and development of the earth's crust and upper mantle: Geotectonics, p. 343–352.

Uyeda, S., and Vacquier, V., 1968, Geothermal and geomagnetic data in and around the island arc of Japan, in Knopoff, L., Drake, C. L., and Hart, P. J., eds., The crust and upper mantle of the Pacific area: Amer. Geophys. Union, Geophys. Monogr. 12, p. 349–366.

Vacquier, V., Sclater, J. G., and Corry, C. E., 1967, Studies of the thermal state of the earth. The 21st paper: Heat flow, eastern Pacific: Tokyo, Univ., Earthquake Res. Inst., Bull., v. 45, p. 375–393.

Vacquier, V., Uyeda, S., Yasui, M., Sclater, J., Corry, C., and Watanabe, T., 1966, Studies of the thermal state of the earth. The 19th paper: Heat-flow measurements in the northwestern Pacific: Tokyo, Univ., Earthquake Res. Inst., Bull., v. 44, p. 1519–1535.

Vacquier, V., and von Herzen, R. P., 1964, Evidence for connection between heat flow and the mid-Atlantic ridge magnetic anomaly: J. Geophys. Res., v. 69, p. 1093–1101.

Yasui, M., Kishii, T., Watanabe, T., and Uyeda, S., 1968, Heat flow in the Sea of Japan, in Knopoff, L., Drake, C. L., and Hart, P. J., eds., The crust and mantle of the Pacific area: Amer. Geophys. Union, Geophys. Monogr. 12, p. 3–16.

Gravity

Hess, H. H., 1938, Gravity anomalies and island arc structure with particular reference to the West Indies: Amer. Phil. Soc., Proc., v. 79, p. 71–96.

Hess. H. H., 1939, Island arcs, gravity anomalies and serpentinite inclusions: A contribution to the ophiolite problem: Int. Geol. Congr., 17th, Rep., v. 2, p. 263–283.

Kaula, W. M., 1969, A tectonic classification of the main features of the earth's gravitational field: J. Geophys. Res., v. 74, p. 4807–4826.

Kaula, W. M., 1970, Earth's gravity field: Relation to global tectonics: Science, v. 169, p. 982–984.

McKenzie, D. P., 1967, Some remarks on heat flow and gravity anomalies: J. Geophys. Res., v. 72, p. 6261–6273.

Morgan, W. J., 1965, Gravity anomalies and convection currents: J. Geophys. Res., v. 70, p. 6189–6204.

37

Speculations on the Consequences and Causes of Plate Motions

DAN P. McKENZIE
1969

From *Geophysical Journal of the Royal Astronomical Society*, v. 18, p. 1–32, 1969. Reprinted with permission of the author and the Royal Astronomical Society.

The original ideas of sea floor spreading (Hess, 1962; Dietz, 1961) were principally concerned with ridges, and with the creation of oceanic crust and upper mantle. They were first confirmed by the application of the Vine-Matthews hypothesis (Vine and Matthews, 1963; Pitman and Heirtzler, 1966; Vine, 1966) to account for oceanic magnetic lineations by production of normally and reversely magnetized ocean floor along the ridge axis. Sykes (1967) used the first motions of earthquakes as independent confirmation that fracture zones on ridges are transform faults (Wilson, 1965c) and require non-conservation of crust.

The remarkable success of these ideas concerning sea floor creation required either expansion of the Earth or destruction of the ocean floor away from ridges. The immediate difficulty all expansion hypotheses face is the rate required. The sea floor spreading velocities are an order of magnitude greater than had been expected, and therefore require catastrophic expansion starting in the Jurassic. This suggestion seems geologically unreasonable, and therefore oceanic crust and upper mantle must be destroyed somewhere. Vening Meinesz (1962, for instance) had maintained that trenches were the site of such destruction for many years, but until recently there was rather little evidence in favour of this belief. In particular the negative gravity anomalies in trenches were used by Vening Meinesz to support the theory of crustal contraction, and by Worzel (1966) to support that of crustal extension. Most of the surface features on both sides of the trench suggest normal faulting (Ludwig et al., 1966) or slumping into the trench from the arc side (Gates and Gibson, 1956; Brodie and Hatherton, 1958). Perhaps two guyots, one in the Aleutian (Menard and Dietz, 1951) and one in the Tonga trench (Raitt, Fisher, and Mason, 1955), were the best evidence that the crust under trenches was originally at a normal oceanic depth and had subsided after the formation of the guyots. Even if the subsidence is accepted, it could be caused by either extension or contraction. Also Hamilton and von Huene (1966) have now demonstrated that the guyot in the Aleutian trench is not in fact a guyot but a sea mount, since it does not possess a flat top.

The only clear evidence for crustal destruction in trenches has come from earthquake seismology. Detailed mapping of surface displacement in Alaska after the 1964 earthquake could only be explained by underthrusting of the ocean beneath the continent on an enormous scale (Plafker, 1965). Focal mechanism studies of the main shock and of many aftershocks confirmed this result (Stauder and Bollinger, 1966a). More evidence came from a careful study of the location of intermediate and deep focus earthquakes in the Tonga area (Sykes, 1966). This work demonstrated that the hook at the northern end of the Tonga Trench was matched by a corresponding feature in the intermediate and deep focus earthquake distribution. Such apparent mirroring of surface features at depths of 600 km is a remarkable result, and demonstrates that there is an intimate connection between surface and deep structures. There is now considerable evidence that this connection is the cold lithosphere moving down to great depths. Such a structure accounts for the propagation of high frequency P and S waves along the plane in this region (Oliver and Isacks, 1967) and in Japan (Utsu, 1967), because the oceanic lithosphere is known to have a high value of Q.

All these studies were concerned with the island arcs alone, and did not consider the problem of the conservation of surface area of a non-expanding earth. Nor did they discuss the motions of regions between ridges and trenches. Perhaps Wilson (1965c) and Bullard, Everett, and Smith (1965) were the first to realise the importance of rigidity of surface rocks in aseismic areas. Wilson stated the basic assumptions of plate theory, but made no further use of them. Bullard et al. fitted the Atlantic Continents together by a series of rotations about axes through the centre of the Earth. This procedure only succeeds because the continents have not deformed internally during their motion.

It is now clear that the major tectonic features of the Earth are produced by the relative rotation of few large aseismic plates, whose boundaries are the major seismic zones of the world. Since the seismic zones do not in general follow continental boundaries, the plates often contain both oceans and continents. For this reason 'continental drift' is a somewhat misleading name, and throughout this paper 'plate theory' is used instead. The relative motion between plates may be determined from the spreading velocities and the strike of transform faults on ridges (Morgan, 1968b; Le Pichon, 1968) or from the focal mechanisms of earthquakes (McKenzie and Parker, 1967; Isacks, Oliver, and Sykes, 1968). Le Pichon used the ridges to determine both the consumption rate and the direction of relative motion between the plates on either side of the trenches and island arcs, and both are in striking agreement with the seismic evidence (Brune, 1968; Isacks et al., 1968).

The remarkable consistency of these two independent methods of determining the motions is the principal evidence in favour of plate theory, and therefore for the destruction of the lithosphere beneath island arcs and trenches. The details of this process are, however, still not fully understood. In particular it is not yet clear exactly where the huge overthrust fault intersects the surface of the earth, or how the motion takes place without disturbing the sediments in the trench (Bunce, 1966; Shor, 1966; Scholl, von Huene, and Ridlon, 1968). These difficulties are principally caused by a lack of knowledge, and do not demonstrate that plate theory is wrong. Indeed there is a similar difficulty in relating the deatiled topography of ridges to the creation of oceanic crust on their axes (Atwater and Mudie, 1968).

It is therefore clear from this work on plate theory that the lithosphere is consumed asymmetrically by island arcs, and it is the purpose of this paper to discuss the consequences of this destruction. Certain features of island arcs are related in a general way to the consumption of lithosphere. The northern Pacific (McKenzie and Parker, 1967) demonstrates that active andesite volcanoes occur only where crust is destroyed (Fig. 37–1). The simplest explanation of this phenomenon is that oceanic crust

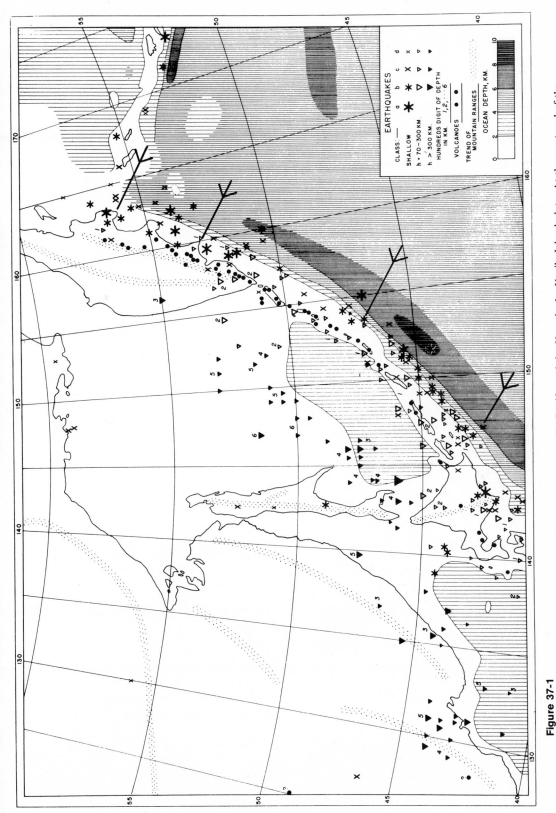

Figure 37-1
The arrows show the direction of the relative motion between the Pacific and the Kamchatka-Kurile Island plate. At the west end of the Aleutian Arc the motion is strike slip, and there are no active volcanoes or intermediate focus earthquakes. Both appear where the Pacific plate is being underthrust beneath Kamchatka and the Kurile Islands. (From Gutenberg and Richter, 1954)

is carried down into the mantle by the lithosphere. Partial melting at a depth of about 150 km then gives magmas of compositions between tholeite and andesite. Melting at different depths, and therefore pressures, may produce magmas of different compositions and hence account for Kuno's (1966) zones. This is the most obvious explanation of the observations, and appears to be in general agreement with high pressure experiments. Green and Ringwood (1966) partially melted various rocks of the calc-alkaline suite with different silica contents, and demonstrated that at 36 kb andesite had the lowest solidus and liquidus temperatures. Beneath the andesite volcanoes of island arcs the top of the sinking slab is at a depth of 100–150 km, or a pressure of 30–50 kb. Thus partial melting of the oceanic crust could produce the volcanics observed. This mechanism is simpler than that proposed by Ringwood and Green (1966), and can explain the remarkable correlation between active andesite volcanoes and consumption of the lithosphere (McKenzie and Parker, 1967). Further experiments are necessary, since small concentrations of elements may have a large effect on the composition of the melt.

Other phenomena related to plate destruction are the occurrence of a trench 3 to 4 km deeper than the surrounding abyssal ocean floor, of a large negative gravity anomaly and smaller positive one over the island arc, of intermediate and deep focus earthquakes and probably also of a large heat flow anomaly above the descending slab. Of these effects the trench and gravity anomalies are probably the result of the shallow angle overthrusting of one plate by another. The explanation is consistent with what little is known about the strength and thickness of the lithosphere (McKenzie, 1967).

Intermediate and deep focus earthquakes are of great interest, both because of the problem of their mechanism and also because they are a consequence of the three dimensional flow within the last 10 million years, though there are a few exceptions to this general rule (Isacks et al., 1968). Their focal mechanisms are not explosive, but are the usual double couple solutions generated by slip on a fault. The slip vectors demonstrate that these earthquakes are not caused by slip between the descending slab and the surrounding mantle. However, the internal fracturing of a homogeneous material does account for most focal mechanisms, at least in the Tonga-Fiji-Kermadec and Japan regions, if the axis of greatest principal stress is directed down the plane (Isacks et al., 1968; McKenzie 1969a). This mechanism probably requires larger stress differences than those observed in shallow earthquakes.

The high heat flow anomaly behind the island arcs of the western Pacific is now well established (Vacquier et al., 1966), but is not easy to explain. McKenzie and Sclater (1968) attempted to do so by producing the heat by viscous shearing on the plane containing the deep earthquakes. They were unable to transmit this heat to the surface at a geologically reasonable rate. Another attempt to account for the heat flow is made below, and avoids the time scale difficulties.

The major remaining problem in plate theory is the driving mechanism. Thermal convection in some form is the only source of sufficient energy, but agreement goes no further. A recent lengthy discussion of the convection problem (McKenzie, 1968b) merely demonstrated that there are no obvious objections to large-scale convection. This result is scarcely surprising because almost all surface effects previously believed to be related to mantle wide convection are more easily explained by creation and destruction of plates. Elsasser (1967) has suggested that the motion of the plates themselves is not caused by viscous coupling to the mantle beneath, but that the cold slabs beneath island arcs sink and pull the rest of the plates with them. This solution of the convection problem requires the lithosphere to act as a stress guide. The argument is very appealing because it is simple, and previous attempts at solving the convection problem (see McKenzie, 1968b, for a discussion of earlier work) were not. The mechanism is still a form of thermal convection because it is driven by the temperature induced density contrast between the cold slab and the hot mantle through which it sinks. Such convection is quite different from the Rayleigh-Benard problem of marginal stability because in Elsasser's problem convection of heat dominates the temperature distribution, and also because the flow is controlled by the extreme temperature dependence of the viscosity. Very little is known about convection of this type, and therefore the analysis below is rudimentary. It does, however, suggest that Elsasser's mechanism cannot maintain the motion of the major plates. This conclusion is supported by the observation that not all pairs of plates have a sinking slab attached to either of them. It is disappointing that neither observations nor theory support the idea, since the surface effects of all other types of flow are confused by the strength and rigidity of the lithosphere. Though the non-hydrostatic gravity field appears to be dominated by large-scale motions in the mantle, it cannot yet be used to determine the flow field because it is not known whether the geoid is elevated or depressed over a rising convection current (McKenzie, 1968b).

The analysis of the temperature within the slab and the flow and stress fields caused by its motion require some knowledge of the mechanical and thermal properties of mantle materials. The properties of the sinking slab are least in doubt, since it must be brittle enough to produce earthquakes by fracture, and sufficiently undeformable to maintain its shape at a depth of 600 km after passing through perhaps 1000 km of mantle (Sykes, 1966). Therefore for all purposes except that of generating earthquakes it behaves as a rigid plane slab. The mechanical properties of the mantle at depths of 100 km and greater, which are not part of the sinking slabs, have been the subject of fierce debate for many years. A recent collection of relevant experimental results (McKenzie 1968a) shows that pure ceramic oxides at high temperatures satisfy the viscous constitutional relationship at shearing stresses σ below $\sim 10^8$ dynes cm^{-2}. At greater stresses the creep rate $\dot{\varepsilon}$ obeys:

$$\dot{\varepsilon} = K\sigma^n \qquad (1)$$

where n and K are constants. Typical values for n are between 3 and 7. The mantle is not pure olivine, but a complex mixture of minerals. Creep of such a solid is more rapid than that of the pure material. Throughout the analysis below the mantle is assumed to obey a viscous constitutional relationship between stress and strain rate. This assumption is a

poor approximation in the region where the lithosphere bends, because the shearing stresses in the mantle outside the slab must be greater than 10^8 dynes cm^{-2}. Despite this limitation the solutions for the flow are believed to be similar to the real flow patterns, and permit a discussion of the stress distribution and viscous heating.

The thermal model for the mantle used below is considerably different from most of those commonly considered. The temperature gradient below the lithosphere is taken to be the adiabatic gradient everywhere, and all horizontal temperature variations outside the sinking slab are neglected. The lithosphere on top acts as a thermal boundary layer which can support large temperature and density gradients because of its finite strength. The effect of phase changes and of the adiabatic gradient within the mantle are contained in the analysis in the next section if T is defined to be the potential, rather than the ordinary, temperature. The arguments in favour of an adiabatic mantle below a mechanical and thermal boundary layer have been discussed previously (McKenzie, 1967, 1968b).

The following values of parameters are used through these calculations:

$$\begin{aligned}
C_p &= 0.25 \text{ cal g}^{-1}\,°C^{-1} \\
\kappa &= 0.01 \text{ cal cm}^{-1}\,°C^{-1}\,s^{-1} \\
T_1 &= 800°C \\
\alpha &= 4 \times 10^{-5}\,°C^{-1} \\
\rho &= 3 \text{ g cm}^{-3} \\
l &= 50 \text{ km} \\
\eta &= 3 \times 10^{21} \text{ poise} \\
g &= 10^3 \text{ cm s}^{-2}
\end{aligned} \qquad (2)$$

Only C_p, ρ and g are well determined. The thermal conductivity κ and its dependence on temperatre are both uncertain, and therefore the value chosen may be wrong by perhaps a factor of two. The thickness l of the lithosphere is a poorly defined quantity, since the change in mechanical properties with depth is a gradual process, and cannot produce a sharp boundary. The value chosen is probably within 30 km of the true effective thickness. T_1 is the temperature at the base of the lithosphere, and therefore the temperature through the mantle in the model. Since this temperature only enters the analysis as a scaling factor, an error in the value of T_1 produces an error in the temperature, but not in the shape, of the isotherms (see also McKenzie, 1967). The value of α the thermal expansion coefficient is that for pure olivine at high temperatures. Impurities are likely to increase this value, and there is some evidence (McKenzie and Sclater, 1969) that the value in equation (2) may be too small. The viscosity η is the least well known of all these parameters. The value within the mantle is probably very variable because it depends exponentially on the temperature. The value chosen is similar to those obtained from the uplift of Fennoscandia (McConnell, 1965) and Lake Bonneville (Crittenden, 1963), but could be in error by an order of magnitude.

THE THERMAL STRUCTURE OF THE SINKING SLAB

All the seismic evidence discussed in the previous section favours consumption of the lithosphere in island arcs by underthrusting. The underthrust plate then descends as a rigid slab deep into the mantle. Since the thermal conductivity of rocks is so small, the slab will retain its original temperature distributions even after it has sunk to considerable depths. The temperature structure within the slab may be obtained from the analysis used previously to discuss the heat flow anomalies associated with ridges (McKenzie, 1967). This method is successful only because the slab is rigid and two dimensional. The equation governing the temperature T within the descending lithosphere is:

$$\rho C_p \left(\frac{\partial T}{\partial t} + \mathbf{v} \cdot \nabla T \right) = \kappa \nabla^2 T + H \qquad (3)$$

where H is the rate of radioactive heat generation cm^{-3} and the other quantities are defined in the previous section. Provided the spreading velocity $\mathbf{v}$ has been constant during at least the last 10 million years all transient temperature distributions will now be unimportant. It is not yet possible to test this assumption against the spreading rate obtained from the magnetic lineations because the reversal time scale at present extends back only 5 million years. Thus a world-wide pause in spreading cannot be detected if it occurs over the same

period everywhere. Such an event has been suggested by Ewing and Ewing (1967) to explain sediment thicknesses on ridges. Seismic reflection records show that the thick sediments on the ridge flanks thin abruptly at anomaly 5. If the sedimentation rate has been constant, this observation requires a period of about 30 million years during which the plates did not move. Such an event is difficult to reconcile with any form of large-scale convection. Changes in convection will produce slow changes in the spreading rate over tens of millions years, and not sudden pauses: slow variations have been observed (Heirtzler et al., 1968) and do not occur simultaneously in different oceans. Thus it appears more likely that these observations are caused by large variations in the sedimentation rate, and not in the spreading rate. If this explanation is true, then $\partial T/\partial t$ may be neglected. H may also be ignored because of the small radioactivity of the lithosphere. Thus equation (3) becomes:

$$\rho C_p \mathbf{v} \cdot \nabla T = \kappa \nabla^2 T \qquad (4)$$

substitution of:

$$T = T_1 T', \qquad x = lx', \qquad z = lz'$$

where T_1 and l are defined in (2) gives:

$$\frac{\partial^2 T'}{\partial x'^2} - 2R \frac{\partial T'}{\partial x'} + \frac{\partial^2 T'}{\partial z'^2} = 0. \qquad (5)$$

The x axis is taken to be parallel to the dip of the slab, and the z axis to be normal to the plane of the slab. The origin of the co-ordinates is chosen to be on the lower boundary (Fig. 37–2). The Thermal Reynolds number R is:

$$R = \frac{\rho C_p v_x l}{2\kappa} \qquad (6)$$

R does not depend on v_y because the slab is assumed to be a two dimensional structure, and therefore $\partial T'/\partial y'$ is zero. The general solution to equation (5) is:

$$T' = A + Bz' + \sum_n C_n e^{\alpha_n x'} \sin k_n z' \qquad (7)$$

if:

$$\alpha_n = R - (R^2 + k_n^2)^{\frac{1}{2}} \qquad (8)$$

where A, B, C_n and k_n are constants. Since the temperature boundary conditions on both

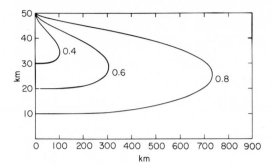

Figure 37-2
Isotherms within the sinking slab, × 10 vertical exaggeration. Temperatures are in dimensionless units. If the values of the parameters in Equation (2) are correct, they may be converted into °C by multiplying by 800. The ordinate is measured normal to the slab, the abscissa down its dip

the upper and lower surfaces of the slab descending through the mantle are $T' = 1$ everywhere, for sufficiently large values of x' the temperature T' must be 1. Thus $A = 1$ and $B = 0$, and:

$$1 = 1 + \sum_n C_n e^{\alpha_n x'} \sin k_n$$

$$k_n = n\pi. \qquad (9)$$

At $x' = 0$ the temperature distribution is the same as that in the lithosphere beneath the ocean (McKenzie, 1967), or:

$$T' = 1 - z'. \qquad (10)$$

Thus at $x' = 0$:

$$1 - z' = 1 + \sum_n C_n \sin n\pi z'$$

$$C_n = \frac{2(-1)^n}{n\pi}. \qquad (11)$$

Substitution of equations (8), (9), and (11) into equation (7) gives:

$$T' = 1 + 2 \sum_n \frac{(-1)^n}{n\pi}$$

$$\cdot \exp[(R - (R^2 + n^2\pi^2)^{\frac{1}{2}})x'] \sin n\pi z' \qquad (12)$$

Isotherms from equation (12) for $v = 10$ cm yr^{-1}, or $R = 62\cdot5$, in Figs. 37–2 and 37–3 show how they are convected deep into the mantle by the motion. An approximate expression for the temperature may be obtained by neglecting all terms in equation (12) except $n = 1$:

$$T' \simeq 1 - \frac{2}{\pi}$$

$$\exp[(R - (R^2 + \pi^2)^{\frac{1}{2}})x'] \sin \pi z' \qquad (13)$$

Since $R \gg \pi$ equation (13) gives:

$$T' \simeq 1 - \frac{2}{\pi} \exp\left(-\frac{\pi^2 x'}{2R}\right) \sin \pi z'. \qquad (14)$$

The greatest depth reached by any isotherm x_M' is now easily obtained, since at this depth $\partial T'/\partial z' = 0$. Thus:

$$-2 \exp\left(-\frac{\pi^2 x_M'}{2R}\right) \cos \pi z' = 0 \qquad (15)$$

or:

$$z' = \tfrac{1}{2}$$

since $0 < z' < 1$. Then equation (14) gives:

$$T' = 1 - \frac{2}{\pi} \exp\left(-\frac{\pi^2 x_M'}{2R}\right) \qquad (16)$$

or:

$$x_M' = \frac{2R \log_e [2/\pi(1 - T')]}{\pi^2}. \qquad (17)$$

If:

$$T' = 1 - \frac{1}{\pi} = 0\cdot682$$

$$x_M' = \frac{2R \log_e 2}{\pi^2}. \qquad (18)$$

Equation (18) is identical to the expression obtained by McKenzie (1967) for the half width of the heat flow anomaly on ridges. Thus there is a close relation between the temperature structure in the descending slab and that in the lithosphere near spreading ridges.

A remarkable relationship is now apparent between the depth of the deepest earthquakes and the temperature distribution within the

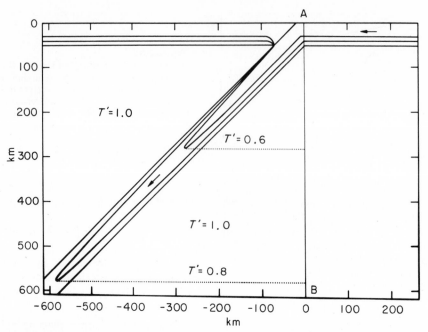

Figure 37-3
Temperature structure beneath island arcs with no vertical or horizontal exaggeration.
Temperatures as in Figure 37-2

sinking lithosphere. Fig. 37–4 shows a projection of the hypocentres of earthquakes and the isotherms beneath the Tonga-Fiji-Kermadec-New Zealand island arc onto a vertical plane approximately parallel to the arc. Only well located earthquakes were taken from Sykes (1966), and Hamilton and Gale (1968). The depths of the deepest point on the isotherms depends on both v_x and the angle ϕ between the dipping plane and the horizontal. v_x was obtained from plate theory, using 58°S, 168°E as the Australia-Pacific pole and an angular velocity of 12.3×10^{-7} degrees yr^{-1}. Le Pichon's (1968) pole at 52.2°S, 169.2°E was not used because it is not consistent with Sykes' (1967) fault plane solution on the Macquarie ridge. Substitution into equation (6) and (17) then gives x, the distance measured in the dipping plane, and hence the depth $x \sin \phi$. The correspondence between the deepest earthquakes and the isotherms is striking. Though the value of 680°C for the limiting temperature is probably inaccurate, the variation with depth of the isotherms is not. Thus Fig. 37–4 clearly demonstrates such

a limiting temperature exists if the present distribution of earthquakes is the steady state distribution. If the earthquakes are confined to those parts of the mantle below 680°C, then they should not occur on a plane but within a thin slab never thicker than 50 km. Fig. 37–3 shows the shape of the 640°C isotherm when $v_x = 10$ cm yr^{-1} and $\phi = 45°$ without any vertical exaggeration, and illustrates the great length of the plate in comparison with its thickness. If the temperature is the only controlling variable, earthquakes should occur throughout this volume. Sykes' (1966) locations are probably not sufficiently accurate to determine whether the hypocentres occur throughout the slab, or are confined to a smaller volume within it. There is, however, now some evidence (Mitronovas et al., 1968) that earthquakes are restricted to a layer about 25 km thick.

The temperature within the slab, equation (12), may also be used to determine the resultant force/unit length that the slab exerts on the lithosphere at the surface. If this calculation is to be valid the density difference

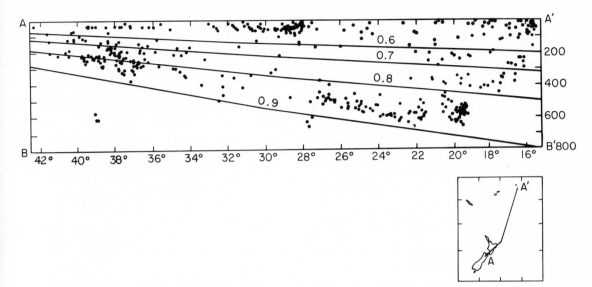

Figure 37-4
Projections of foci of earthquakes and isotherms onto a vertical plane along the line A′ in the Tonga Fiji Kermadec New Zealand region (see inset). Except for six anomalous deep earthquakes, intermediate and deep focus activity ceases at about $T' = 0.85$ or $T = 680°C$. Accurate hypocenters only are taken from Sykes (1966) and Hatherton and Gale (1968). Numbers on the x axis refer to the position of Sykes' sections

between the sinking lithosphere and the surrounding mantle must be caused by a difference in temperature and not in composition. This assumption is probably justified if the sinking slab was originally oceanic lithosphere, since all oceanic crust and upper mantle has been generated from the mantle below the plates. It is justified for continental lithosphere which must be considered separately. The effect of the oceanic and continental crust on the force is considered in detail later in this paper. With this assumption the force/unit length F is given by:

$$F = \int_0^l \int_0^\infty g(\rho(T_1) - \rho(T))\,dx\,dz. \quad (19)$$

Substitution from equation (12) then gives:

$$F = 2g\alpha\rho T_1 l^2 \int_0^1 \int_0^\infty \sum_{n=1}^\infty \frac{(-1)^n}{n\pi}$$

$$\cdot \exp\left[(R - (R^2 + n^2\pi^2)^{\frac{1}{2}})x'\right]$$

$$\cdot \sin n\pi z'\,dx'\,dz'. \quad (20)$$

Integration gives:

$$F = \frac{4g\alpha\rho T_1 l^2}{\pi^2}$$

$$\cdot \sum_{k=0}^\infty \frac{1}{(2k+1)^2\left[R - (R^2 + (2k+1)^2\pi^2)^{\frac{1}{2}}\right]}. \quad (21)$$

Since $R \gg (2k+1)\pi$ for small k equation (21) may be written:

$$F \simeq \frac{8g\alpha\rho T_1 l^2 R}{\pi^4} \sum_{k=0}^\infty \frac{1}{(2k+1)^4}. \quad (22)$$

But from Jolley (1961)

$$\sum_{k=0}^\infty \frac{1}{(2k+1)^4} = \frac{\pi^4}{4 \times 4!}. \quad (23)$$

Thus equation (22) gives:

$$F = \frac{g\alpha\rho T_1 l^2 R}{12}. \quad (24)$$

This force is the total resultant force on the slab due to its thermal structure. Only the component $F \sin \phi$ acts down the dip of the slab, and can cause plate motions. The other component, $F \cos \phi$, must be balanced by the pressure distribution within the mantle outside the slab. The corresponding force/unit area f acting on the lithosphere in the direction of the island arc is:

$$f = \frac{F}{l} = \frac{g \alpha \rho T_1 l R}{12} \sin \phi. \qquad (25)$$

This expression applies only if there is no resistance to the motion of the sinking slab. Equation (25) therefore gives the upper limit of f. Substitution from equation (2) with $v_x = 10$ cm yr^{-1} gives:

$$f = 2.5 \sin \phi \text{ kilobars.}$$

This value of f is surprisingly large, and since the numerical estimates of α and T_1 are probably too small, the true value may be even greater. Elsasser (1967) has suggested that this force maintains the observed surface motions, and this idea is discussed in detail in a later section. Since the lithosphere is at present moving at constant velocity, the convective force f must be opposed by an equal force either on the base of the surface plates or on the sinking slab. If Elsasser's suggestion is correct, then at least in the uppermost part of the slab the axis of least principal stress must be subparallel to the dip of the plane. If this condition is not satisfied the stress system within the slab will oppose the motion and will not provide a driving force. Elsasser's idea can therefore be tested by comparing the fault plane solutions of intermediate and deep earthquakes with the predicted stress field. A necessary but not sufficient condition for his hypotheses to be true is the occurrence of tensional fault plane solutions at shallow depths in all slabs. Few focal mechanisms of intermediate earthquakes have yet been determined. Those from deep earthquakes require the slab to be in compression, with the axis of greatest principal stress approximately parallel to the dip of the slab (Isacks et al., 1968). Such a stress field can arise only if the deepest part of the slab is being pushed into the mantle. Thus at depths greater than about 300 km the resistance to the motion is greater than the thermal driving forces. It is not yet known whether intermediate earthquakes are produced by a stress field like that of deep

earthquakes. If they are, then Elsasser's suggestion cannot be correct.

Though most intermediate and deep focus earthquakes occur within the cold sinking lithosphere, there exist also a few isolated earthquake sources which are less easily understood. Some sources, like those beneath Roumania and the Hindu Kush, produce a succession of earthquakes from a limited volume within the mantle. The most difficult sources to understand generate infrequent deep earthquakes, and perhaps the most striking of these is beneath eastern Spain. Only one earthquake, of magnitude 7 at a depth of 650 km, has ever been recorded from this source. Three deep earthquakes beneath North Island, New Zealand, in Fig. 37–4 are from another deep isolated source. All such special sources are beneath regions where crustal shortening has taken place in the Tertiary. It is therefore possible that the earthquakes are produced within relics of old slabs which are now no longer being regenerated by consumption of the lithosphere. The convective force, equation (25) generated by this density contrast is sufficient to produce earthquakes. If this explanation is correct, major changes in both the shape and relative motion between plates must have taken place in the second half of the Tertiary. Detailed geological studies of the deformation above isolated sources would therefore be of considerable interest. Such observations should provide a test of these ideas.

FLOW WITHIN THE MANTLE

The large plane slab which sinks beneath island arcs must set up stresses within the mantle through which it moves. Its motion will therefore govern the flow of the mantle if other forces are absent. Inside the earth there are also body forces, caused by horizontal density variations, which must strongly modify any flow produced by the sinking slab. Such thermal forces are neglected throughout this section, and all results must therefore be used with some caution. The other important assumption is that flow in the mantle outside the cold slab is governed by a viscous constitutional relationship, as already mentioned.

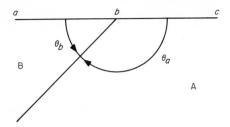

Figure 37-5

Since the stresses generated by the motion are less than 100 bars over large regions on both sides of the arc (Fig. 37–7), viscous flow governed by equation (26) probably resembles the real flow rather closely. It is much less likely that the temperature dependence of the viscosity can be neglected, as is done below. Rather important effects may well be governed by the strong dependence of viscosity on temperature. Thus several of these assumptions are unlikely to be satisfied by flow within the mantle, but without them non-linear terms dominate the equations and no analytic solutions can be obtained.

It is clear from these remarks that there is some doubt whether the resulting solutions apply to flow in the Earth. In this respect there is a considerable contrast between this section and the previous one. The motion of the slab is simple and also is rather well known from surface observations of spreading rates (Le Pichon, 1968) and from the location of deep earthquakes (Sykes, 1966). There is no similar method which can be used to observe the complicated three-dimensional flow within the mantle outside the slab. Even the geographic location of rising and sinking currents is in doubt. It is this extreme ignorance which makes the analysis below of at least some interest, however wrong the details may be.

Subject to these various assumptions the steady state velocity of an incompressible fluid must satisfy:

$$0 = \eta\nabla^2\mathbf{v} + \rho\nabla U - \nabla P \qquad (26)$$

$$0 = \nabla \cdot \mathbf{v} \qquad (27)$$

U is the gravitational potential and P the pressure. The curl of equation (26) gives:

$$\nabla^2\omega = 0 \qquad (28)$$

where $\omega(= \nabla \times \mathbf{v})$ is the vorticity. Most island arcs are approximately two-dimensional structures, and it is therefore convenient to use cylindrical co-ordinates with the z axis parallel to the arc. Fig. 37–5 shows a vertical section through an idealized arc, with the motions of the lithosphere and slab represented by the motion of planes. If the co-ordinate axes are fixed to ab, the lithosphere behind the arc, all boundary conditions are simple. Thus $\mathbf{v}$ may be written:

$$\mathbf{v} = \left(v_z, \frac{1}{r}\frac{\partial\psi}{\partial\theta}, -\frac{\partial\psi}{\partial r}\right) \qquad (29)$$

where ψ is the stream function.

The resulting equations are simple if $v_z = 0$ everywhere, a condition which requires the motion between ab and bc in Fig. 37–5 to be normal to the arc. This condition is not satisfied by all presently active island arcs (McKenzie and Parker, 1967; Le Pichon, 1968), and solutions to equation (29) can if necessary be obtained if $v_z \neq 0$. It is, however, doubtful if these general solutions would display any features which are not possessed by the special case discussed below. Equation (28) then becomes:

$$\nabla^4\psi = 0 \qquad (30)$$

Solutions to equation (30) are required which satisfy $\mathbf{v} = \mathbf{a}_r \times$ constant, where $\mathbf{a}_r$ is the radial unit vector, at specified values of θ (Fig. 37–5). Such solutions are easily obtained by substituting (Batchelor, 1967).

$$\psi = r\Theta(\theta) \qquad (31)$$

into equation (30) to give:

$$\frac{d^4\Theta}{d\theta^4} + 2\frac{d^2\Theta}{d\theta^2} + \Theta = 0. \qquad (32)$$

The general solution to equation (32) is:

$$\Theta = A \sin\theta + B \cos\theta + C\theta \sin\theta + D\theta \cos\theta \qquad (33)$$

where A, B, C and D are constants which must be obtained from the boundary conditions, and are not the same on each side of the island arc. The boundary conditions within region A (Fig. 37–5) are:

$$\mathbf{v} = -v\mathbf{a}_r \text{ on } \theta = 0$$
$$\mathbf{v} = \quad v\mathbf{a}_r \text{ on } \theta = \theta_a. \qquad (34)$$

These conditions are satisfied by:

$$\psi = \frac{-rv[(\theta_a - \theta) \sin \theta + \theta \sin(\theta_a - \theta)]}{\theta_a + \sin \theta_a}. \quad (35)$$

Similarly the flow within B must satisfy:

$$\mathbf{v} = 0 \quad \text{on} \quad \theta = 0$$
$$\mathbf{v} = v\mathbf{a}_r \quad \text{on} \quad \theta = \theta_b \quad (36)$$

giving:

$$\psi = \frac{rv[(\theta_b - \theta) \sin \theta_b \sin \theta - \theta_b \theta \sin(\theta_b - \theta)]}{\theta_b^2 - \sin^2 \theta_b}. \quad (37)$$

θ_a in equation (35) is measured clockwise from bc in Fig. 37–5, with bc as zero.

θ_b in equation (37) is measured anticlockwise from ab as zero. This sign convention simplifies the expressions and is permitted because it is possible to choose a right-handed set of co-ordinate axes in both cases.

Equations (34) and (36) are not rigorously satisfied in real island arcs because both the surface plates and the sinking slab are finite. This limitation is, however, unlikely to produce important differences in the flow close to the island arc.

The stream lines of constant ψ are easily obtained from equations (35) and (37) (Fig. 37–6). Throughout this discussion θ_a and θ_b are chosen to be 135° and 45° respectively. The stream lines demonstrate that the fluid is dragged down with the sinking slab, as indeed would be expected from rather simpler arguments. It is perhaps more surprising that fluid is swept upward in certain parts of region B.

The stresses which the fluid flow produces are also of considerable interest. Since the flow is two-dimensional, the stress tensor $\mathbf{S}$ may be written as:

$$\mathbf{S} = \begin{bmatrix} \sigma'_{rr} & \sigma'_{r\theta} \\ \sigma'_{r\theta} & \sigma'_{\theta\theta} \end{bmatrix} - P\mathbf{I} \quad (38)$$

where $\mathbf{I}$ is a 2×2 unit matrix and:

$$\sigma'_{rr} = 2\eta \frac{\partial v_r}{\partial r}$$
$$\sigma'_{\theta\theta} = 2\eta \left(\frac{1}{r} \frac{\partial v_\theta}{\partial \theta} + \frac{v_r}{r} \right) \quad (39)$$
$$\sigma'_{r\theta} = \eta \left(\frac{1}{r} \frac{\partial v_r}{\partial \theta} + \frac{\partial v_\theta}{\partial r} - \frac{v_\theta}{r} \right).$$

Equations (29) and (31) then give:

$$\sigma'_{rr} = \sigma'_{\theta\theta} = 0$$
$$\sigma'_{r\theta} = \frac{\eta}{r} \left(\frac{d^2\Theta}{d\theta^2} + \Theta \right). \quad (40)$$

Thus:

$$\mathbf{S} + P\mathbf{I} = \begin{bmatrix} 0 & \sigma'_{r\theta} \\ \sigma'_{r\theta} & 0 \end{bmatrix} = \mathbf{S}' \quad (41)$$

and forms a tensor $\mathbf{S}'$ without diagonal terms. Thus the greatest shearing stress is exerted parallel to the $r = $ const. and $\theta = $ const. surfaces, and is of magnitude $\sigma'_{r\theta}$. Substitution of equations (35) and (37) into equation (40) gives:

$$\sigma'_{r\theta} = \frac{2v\eta}{r(\theta_a + \sin \theta_a)} [\cos \theta_a + \cos(\theta_a - \theta)] \quad (42)$$

in A, and:

$$\sigma'_{r\theta} = \frac{2v\eta[\theta_b \cos(\theta_b - \theta) - \sin \theta_b \cos \theta]}{r(\theta_b^2 - \sin^2 \theta_b)} \quad (43)$$

in B. Both equations (42) and (43) are singular at the origin where $r = 0$. The singularities arise because the spreading lithosphere is required by the boundary conditions (34) and (36) to bend through an angle θ_b at the origin $r = 0$. Since the radius of curvature of the litho-

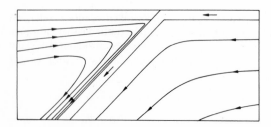

Figure 37-6
Stream lines for flow within the mantle. Motion is with respect to the plate behind the island arc, and is driven by the motion of the other plate and of the sinking slab. Thermal convection outside the slab is neglected, and the lithosphere is 50 km thick

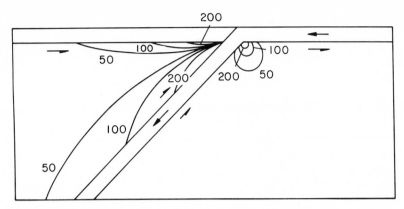

Figure 37-7
Shear stresses in bars caused by the flow in Figure 37-6. The half arrows show
the direction of the forces exerted by the fluid on the plates and slab. The
stresses on the plate behind the island arc exceed those on the plate in front.
The lithosphere is 50 km thick

sphere cannot in reality be zero, these singularities are not in practice possible. Thus equations (42) and (43) are probably good approximations to the stress field except perhaps within 50 km of $r = 0$ axis, and are sufficient for a qualitative discussion. Fig. 37–7 shows the contours of the shearing stress obtained from equations (42) and (43) with values for v and η taken from equation (2). The half arrows show the direction of the stresses exerted by the fluid on the rigid boundaries. The stresses in region B behind the arc are much greater than those in A. Within B there are two zones of high stress separated by a plane of zero stress. In both zones stresses exceed 100 bars and therefore the viscous constitutional relationship between stress and strain rate does not apply. The occurrence of shallow earthquakes at considerable distances inside the island arc may be related to the shallower of these zones (Hamilton and Gale, 1968). Earthquakes also occur within the deeper zone, though their frequency decreases rapidly with depth (Sykes, 1966). This observation agrees with the stress distribution in Fig. 37–7.

The stresses on both sides of the arc act in a direction which opposes the motion of the lithosphere. Thus work must be done by the mechanism which moves the plates against the viscous forces. The mechanical energy is then converted into heat by viscous dissipation within the mantle. Such stress heating is one source of energy which can maintain the large positive heat flow anomaly behind island arcs. McKenzie and Sclater (1968) attempted to produce the heat required from the shearing stresses between the sinking slab and the surrounding mantle. However, they were unable to transport the heat to the Earth's surface in a geologically reasonable time. This difficulty is avoided if the dissipation occurs within the shallow high stress zone in Fig. 37–7.

The expression for the stress heating H in a non-elastic material is:

$$H = S_{ij} \frac{\partial v_i}{\partial x_j}. \qquad (44)$$

S_{ij} is the stress tensor in equation (38), and summation over repeated indices is implied. Equation (44) may also be written:

$$H = \tfrac{1}{2}(S'_{ij} - P\delta_{ij})\left(\frac{\partial v_i}{\partial x_j} + \frac{\partial v_j}{\partial x_i}\right) \qquad (45)$$

where $\mathbf{S}'$ is given by equation (41) and $\delta_{ij} = 0$ if $i \neq j$, $= 1$ if $i = j$. Equations (44) and (45) give the heat generated by friction in any material. In this problem the mantle is taken to be a viscous incompressible fluid, and therefore:

$$S'_{ij} = \eta \left(\frac{\partial v_i}{\partial x_j} + \frac{\partial v_j}{\partial x_i}\right) \qquad (46)$$

and:

$$S'_{ii} = 0. \qquad (47)$$

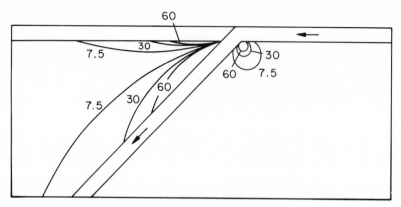

Figure 37-8
Stress heating caused by viscous dissipation within the flow in Figure 37-6 in units of 10^{-7} crg cm^{-3} s^{-1}. The heating within the mantle is more intense behind than in front of the arc

Equation (45) may be written:

$$H = \frac{1}{2\eta} (S'_{ij} - P\delta_{ij})S'_{ij}$$

$$= \frac{1}{2\eta} S'_{ij}S'_{ij}. \qquad (48)$$

$S'_{ij}S'_{ij}$ is an invariant of the stress tensor, and is therefore unaffected by the choice of axes. Thus equation (41) may be substituted directly into equation (48) to give:

$$H = \frac{1}{\eta} \sigma'^2_{r\theta}. \qquad (49)$$

The contours of the stress heating function are therefore the same as those of the shearing stress. The values on the contours in Fig. 37–8 are in units of 10^{-7} erg cm^{-3} s^{-1}, and are obtained from equations (2), (42), (43), and (49). A typical value for the radioactive heat generation rate for granites is 200×10^{-7} erg cm^{-3} s^{-1}, for basalts is 20×10^{-7} erg cm^{-3} s^{-1}, and for ultrabasic rocks is 2×10^{-7} erg cm^{-3} s^{-1}. The excess heat flow behind island arcs is ~ 1 μ cal cm^{-2} s^{-1}, and is therefore equivalent to a volume source of 80×10^{-7} erg cm^{-3} s^{-1} distributed through a depth of 50 km. This value is an order of magnitude greater than that calculated from equation (49). Since the viscosity is uncertain by perhaps an order of magnitude, the value of the stress heating could also be an order of magnitude different from that in Fig. 37–8. Thus stress heating could account for the observed high heat flow

behind island arcs. Since there is considerable heat generation at shallow depths, the heat can diffuse to the surface in a geologically reasonable time. Fig. 37–8 shows that heat is also generated by friction between the sinking slab and the surrounding mantle. McKenzie and Sclater (1968) attempted to explain the observations by conducting this heat upwards, but were unable to do so in less than about 300 million years. No such objection applies to the suggestion above because the heat is generated by shearing at shallow depths.

Stress heating can produce the observed surface heat flow only if the viscosity in equation (2) is too small. If the value of η used above is too large, viscous heating is quite unable to generate the heat required. Figs. 37–6 and 37–7 suggest two other possible explanations of the heat flow observations. The first depends on the shear stress on the base of the lithosphere above region B. The effective thickness of the lithosphere depends on the shearing stresses on its base; if these are large the lower part will be dragged away by the flow. Hot mantle material will be carried closer to the Earth's surface, and the heat flow will therefore be increased. Thus the temperature gradients in the thermal boundary layer at the Earth's surface must depend on the stress field.

A second effect can increase the temperature gradient at the base of the lithosphere. The fluid behind the island arc in Fig. 37–6

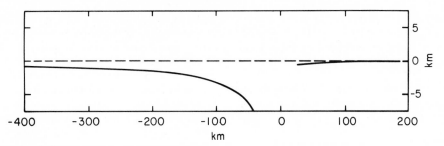

Figure 37-9
Surface deformation near trenches produced by stresses due to the flow and normal to the
earth's surface. × 10 vertical exaggeration. Horizontal distances measured from the
surface boundary between the plates, positive in front of the arc and negative behind (as
in Figure 37-3)

is carried towards the surface before being dragged down into the mantle by the plate motions. The flow therefore carries hot mantle material closer to the base of the lithosphere, and hence can produce a surface heat flow anomaly. All three possible explanations may well be involved since they each can produce an increase in the surface heat flow.

All the equations above were solved without any boundary conditions on the stress or displacement normal to the lithosphere. If the displacement is taken to be zero, the normal stress must be balanced by forces within the lithosphere. The alternative boundary condition is to require the normal stress to be zero at the deformed surface. The equations then determine the surface shape. The strength of the lithosphere probably cannot support the normal stress in this problem, because of the large horizontal extent of the forces (McKenzie, 1967). Therefore the deformation is calculated from the condition that the normal stress must vanish. The stress normal to any plane $r = \text{const.}$ is:

$$\sigma_{\theta\theta} = -P + 2\eta \left(\frac{1}{r} \frac{\partial v_\theta}{\partial \theta} + \frac{v_r}{r} \right). \qquad (50)$$

Equations (29) and (31) then give:

$$\sigma_{\theta\theta} = -P. \qquad (51)$$

If the fluid is at rest the hydrostatic pressure P_0 at any point is:

$$P_0 = \rho g r \sin \theta. \qquad (52)$$

The pressure P is perturbed by an amount P_1 from P_0 by the shear stresses of the flow.

P_1 may be obtained from equation (26) by perturbation theory:

$$P_1 = -\frac{\eta}{r} \left(\frac{d^3\Theta}{d\theta^3} + \frac{d\Theta}{d\theta} \right). \qquad (53)$$

If the normal stress is to vanish on the deformed surface:

$$\sigma_{\theta\theta} = 0 \qquad (54)$$

or:

$$\rho g r \sin \theta = \frac{\eta}{r} \left(\frac{d^3\Theta}{d\theta^3} + \frac{d\Theta}{d\theta} \right). \qquad (55)$$

Equation (55) determines $\theta = \theta(r)$ on the surface, provided the angle between the surface and the radius vector is small, and provided the fluid flow is not affected. Neither condition is satisfied close to the origin of co-ordinates. In front of the arc in region A equation (55) gives:

$$\tan \theta =$$

$$\frac{\sin \theta_a}{\left[1 + \cos \theta_a + \dfrac{\rho g}{2\eta v} r^2 (\theta_a + \sin \theta_a) \right]} \qquad (56)$$

and behind:

$$\tan \theta =$$

$$\frac{\theta_b \sin \theta_b}{\left[\theta_b \cos \theta_b - \sin \theta_b + \dfrac{\rho g}{2\eta v} r^2 (\theta_b{}^2 - \sin^2 \theta_b) \right]}. \qquad (57)$$

Fig. 37–9 shows the surface shape with a × 10 vertical exaggeration. The surface inside the island arc is depressed by several kilometres, whereas outside the arc, above region A, the deformation is much less. Therefore trenches are unlikely to be maintained by flow in the mantle, but are probably supported by the strength of the lithosphere (McKenzie, 1967). Depression within the arc has not been observed. If it does exist and is full of sediment it could be of considerable geological importance.

Though some of the results obtained in this section are probably of relevance to the flow beneath island arcs, the drastic approximations required to linearize the equations must not be forgotten.

THE CAUSES OF PLATE MOTION

Three causes of plate motions have been suggested. The oldest theory depends on large-scale convection throughout at least the upper mantle. Viscous forces are then required to couple the plates to the moving mantle below. The history of this idea is discussed by Holmes (1965), and a modified version is compatible with all relevant observations (McKenzie, 1968b). Another closely related theory has been put forward by Elsasser (1967). He suggested that the lithosphere may be considered as a stress guide, and that surface motions are maintained by the cold sinking slabs, as described earlier in this paper. At first sight it appears impossible to move the entire lithosphere by tensional forces because rocks have little strength in tension. However, this objection is not correct because materials fail in tension easily only if voids can form. Since void formation requires one of the principal stresses to be negative, this type of failure must be restricted to the rock within a few hundred metres of the surface, and does not affect the dynamics of the lithosphere. The energy source of Elsasser's mechanism is still thermal convection, but the force causing the motion is transmitted through the rigid lithosphere, and not through viscous coupling between the mantle and the plates.

A third mechanism which can maintain motions is the force exerted on the plate boundaries by neighbouring plates. In particular the motion of small plates is probably sustained by such forces. Two examples are the plate between the Juan de Fuca ridge and the continental margin of northwestern America, and that containing the Aegean Sea in the Mediterranean. This mechanism cannot produce the world-wide movements of large plates because it has no source of energy.

It is easy to demonstrate by energetic arguments that thermal convection must provide the driving force, and therefore that vertical movements of hot and cold material must be involved. The difference between Elsasser's model for thermal convection and those of previous authors is in the importance of the transmission of stress through the upper mantle. He suggested that the cold and therefore dense slab of lithosphere sinks into the mantle beneath island arcs and pulls with it the rest of the plate to which it is attached. Buoyancy forces throughout the mantle outside the slab are neglected. Thus this model is an extreme form of convection which takes place if the viscosity is a rapidly varying function of temperature. If the flow induced within the mantle by a sinking slab affects the motion of plates other than that to which it is attached, then Elsasser's model is too simple. Fortunately his assumptions have certain consequences which can be obtained from the analysis above and are not in agreement with observations.

The force/cm exerted by the sinking lithosphere on the surface plate is given by equations (6) and (24):

$$F = \frac{g\alpha\rho^2 T_1 C_p l^3}{24\kappa} v. \tag{58}$$

This force is directed towards the island arc and must overcome the frictional resistance between the sinking slab and the mantle, and also that between the surface plate and the mantle. The first of these cannot be obtained from equations (42) and (43) by integration because the stress is singular at $r = 0$. The singularity is not of physical importance, and therefore the resistive force/unit length of the arc R_1 may be written:

$$R_1 = \eta v f(\theta_b). \tag{59}$$

R_1 acts on the sinking slab in a direction opposite to its motion. The force resisting the motion of the surface plate depends on the variation of viscosity with depth. A simple model has an upper mantle of constant viscosity to a depth d, bounded by a rigid boundary below. The resistance/unit length R_2 acting on the plate is therefore:

$$R_2 = \eta v \frac{L}{d} \qquad (60)$$

where L is the length of the plate measured at right angles to the island arc. If the motion of the plates is driven by F in equation (58), then:

$$F > R_1 + R_2$$

or

$$\frac{g \alpha \rho T_1 C_p l^3}{\kappa v} > 24 \left(\frac{L}{d} + f(\theta_b) \right). \qquad (61)$$

The inequality (61) is very similar to Rayleigh's condition for convection in a fluid heated from below (see Saltzman, 1962). All parameters on the left of the inequality (61) and also d and θ_b ($\simeq 45°$) are the same for all plates. Thus only L varies between plates. Elsasser's suggestion could be correct if the inequality is satisfied by $L = 0$. There must then be some limiting value of L, L_c, perhaps very large, at which the inequality reverses. If the inequality (61) is satisfied, then v is determined only by the variation of viscosity with depth. Thus this type of convection will possess some rather remarkable features. All plates with $L < L_c$ and which have a sinking slab will spread at a rate limited only by the variation of viscosity within the mantle. Large plates cannot move faster than small ones, and plates with $L > L_c$ cannot move at all.

The motion of real plates is quite different. Various small plates with sinking slabs attached are known; an example is the one between the East Pacific Rise, the Galapagos Rift and the Middle America Trench. The consumption rate along the trench may be estimated from the depth of the intermediate earthquakes beneath Middle America (Gutenberg and Richter, 1954) to be about 3 cm yr⁻¹. This value is three times smaller than that of 9 cm yr⁻¹ for the Western Pacific arcs (Le Pichon, 1968). These arcs occur between three of the largest plates, containing India, America and the Pacific respectively. These observations suggest that the consumption rate increases with plate size. They do not demonstrate that the velocity depends only on the viscosity of the mantle, and therefore do not support Elsasser's suggestion.

It could be argued that these remarks are incorrect because they depend on the somewhat uncertain calculations of the previous sections. Though there is no reason to doubt the arguments already given, there is one other which does not depend on any analysis. If plate motions are maintained in the way Elsasser has suggested, spreading is not possible between two plates, neither of which possesses a sinking slab. The American and the African plate almost satisfy this condition. The only sinking parts of the American plate are beneath the Eastern Caribbean and the Scotia arc, both of which are small. Descending parts of the African plate are restricted to the Mediterranean, beneath the Aegean and Tyrrhenian Seas. The motion between these two plates is not small, reaching 3 cm yr⁻¹ in the South Atlantic. Elsasser's hypothesis can account for this result only if stress transmitted through the mantle can cause drift. This modification is of great importance, because the equations governing such flow are those of convection in a fluid of variable viscosity. Such equations have several important non-linear terms, and must be solved by numerical methods.

Convection currents in the mantle affect plate motions in two ways. They drag the lithosphere in the direction of the flow. They also distort the surface of the Earth, and therefore gravity acts to pull the plates downhill. If the lithosphere is sufficiently thick, the gravitational force is greater than the viscous stress on the base of the plate. Under these conditions the plate motions are dominated by gravity. In oceanic regions sea level is within about 2 m of the level surface (Stommel, 1966), and the structure of the plate remains the same over large regions. Thus the driving force on the plates can be obtained from the slope of the regional bathymetry, even if the bathymetry is itself maintained by convective motions within the mantle. In both the Pacific and the Indian Oceans the ocean floor slopes down towards the trenches, and therefore appears to support these arguments. Unfortunately the more detailed discussion below suggests that the viscous stress rather than gravity governs the motion of the largest plates, though in general both must be considered.

A useful model for convection in the mantle was suggested by Allan, Thompson, and Weiss (1967). In this model the flow is two-dimensional, and is driven by horizontal temperature differences applied to the upper surface of a semi-infinite viscous fluid. If convection of heat is neglected, analytic expressions may be obtained for the surface deformation and the shearing stress exerted on a rigid boundary. Though the model is undoubtedly too simple, it is probably sufficient for order of magnitude calculations. Values for the gravitational and viscous forces on a plate are easily obtained from the expressions derived by McKenzie (1968b). If the surface of the fluid is maintained at a temperature T:

$$T = T_0 \cos kx \qquad (62)$$

where T_0 is the amplitude and k the wavenumber of the applied temperature driving the flow. The difference z between the surface of the fluid and a level surface is given by:

$$z = \frac{3\alpha T_0}{4k} \cos kx. \qquad (63)$$

The surface slope is therefore:

$$\frac{dz}{dx} = -\frac{3\alpha T_0}{4} \sin kx. \qquad (64)$$

Provided $|dz/dx| \ll 1$ the horizontal force cm^{-3} f_x is:

$$f_x = \frac{3\alpha T_0}{4} \rho g \sin kx \qquad (65)$$

and the effective force cm^{-2} F_x acting on the lithosphere of thickness l is:

$$F_x = \frac{3\alpha T_0}{4} g\rho l \sin kx. \qquad (66)$$

The shearing stress σ_{xz} exerted by the flow is:

$$\sigma_{xz} = \frac{\alpha T_0}{4k} g\rho \sin kx. \qquad (67)$$

(In the equation given by McKenzie (1968b) the factor ρ was inadvertently omitted.)
Thus:

$$\frac{\sigma_{xz}}{F_x} = \frac{1}{3lk} = \frac{\lambda}{6\pi l}$$

where λ is the wavelength of the temperature disturbance. If $\lambda = 5000$ km, substitution from equation (2) gives:

$$\frac{\sigma_{xz}}{F_x} \simeq 7. \qquad (68)$$

Thus the viscous shearing stress cannot be neglected, and may well dominate the motion of large plates.

The arguments above all suggest that the forces driving plates can be understood only if solutions to the full convection equations are obtained. Solutions to simplified models appear to give only limited insight into the general problem, and also to be internally inconsistent. Numerical solutions to the equations governing convection must be obtained for a liquid whose viscosity varies rapidly with temperature. Flow patterns within the Earth can then probably be determined by combining these solutions with the non-hydrostatic gravity field, though the relation between the two is not simple. The rising and sinking parts of convection currents are unlikely to be reflected in the surface features (McKenzie, 1967). Thus there is no reason why elevations and depressions of the geoid should occur in the neighbourhood of trenches and ridges, as Runcorn (1965) has suggested.

Probably the most accurate determination of the external gravity field yet published is that of Gaposhkin (see Kaula, 1966). The corresponding geoid referred to the hydrostatic figure of the Earth (Fig. 37–10) shows that the non-hydrostatic equatorial bulge dominates other non-hydrostatic terms. At present there is no general agreement about how the non-hydrostatic bulge is maintained (MacDonald, 1963; McKenzie, 1966; Goldreich and Toomre, 1969), and therefore the geoid is commonly referred to a spheroid with the observed, rather than with the hydrostatic, ellipticity. Figs. 37–11 and 37–12 show such a geoid calculated from Gaposhkin's gravity field, and also the major ridges and island arcs. The ridges occur both where the geoid is elevated and where it is depressed. Therefore ridges must exist over both rising and sinking currents. This poor correlation is to be expected for the reasons discussed above. In contrast major trenches and island arcs round the Pacific occur in regions where the geoid is elevated. Unfortunately there are

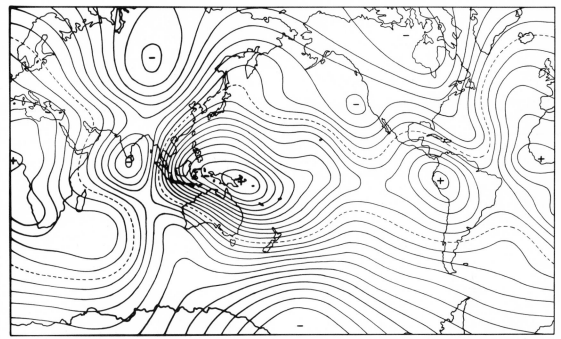

Figure 37-10
Contours at intervals of 10 m of the elevation of the geoid above the hydrostatic geoid given by Jeffreys (1963). The dashed contour is of zero elevation, positive and negative signs show elevations and depressions of the level surface. Mercator projection with the rotation axis as axis

rather few such island arcs, and therefore this correlation need not be significant. There is, however, a physical reason why island arcs, unlike ridges, might be closely related to flow deep within the mantle. The creation of lithosphere along ridges requires a large volume of hot mantle material to be intruded along the ridge axis. This rock presumably rises adiabatically from beneath the lithosphere, then cools as the plate moves away from the ridge. Since the temperature gradient in the mantle must be close to the adiabatic gradient, plate creation along ridges requires large volumes of mantle material, but, except within the lithosphere, convects little heat. Thus plate production produces little distortion of the isotherms within the mantle beneath the lithosphere. The opposite is true beneath island arcs. No change in the temperature structure within the lithosphere can take place until it starts to descend into the mantle. Thus the motion distorts the isotherms by hundreds of kilometres throughout the upper mantle. In the absence of surface deformation the cold slab will produce an elevation in the geoid of

about the observed magnitude. Thus the mantle below the lithosphere loses material beneath ridges, but loses heat beneath island arcs. Clearly this balance must be the exact opposite of that of the lithosphere. The worldwide heat loss due to plate creation on ridges (McKenzie and Sclater, 1969) is therefore equal to the rate of heat loss by the mantle beneath island arcs. This mechanism of heat transfer probably accounts for more than 15 per cent of the heat lost by the mantle beneath the lithosphere. The cold descending slab produces large horizontal, as well as vertical, temperature gradients, and should therefore govern the position of the descending limb of any convection cells. Control of convection by horizontal temperature gradients maintained by the boundary conditions has been widely studied in the Earth's oceans and atmosphere. This effect is not related to flow caused by the transmission of stress through the mantle (already discussed) though the direction of flow is the same.

There is a considerable difference between the convection hypotheses suggested here and

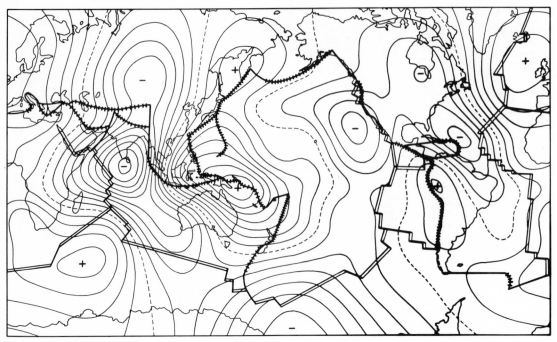

Figure 37-11
Contours of the height of the non-hydrostatic geoid with respect to a geoid with the observed value of
C_{20}, and not the hydrostatic value. Contours as in Figure 37-10. The approximate position and nature of
major plate boundaries are also shown. Major trenches all occur in regions where the geoid is above the
reference surface. The opposite is not true for major ridges. Mercator projection with the rotation axis
as axis. Hatched lines, trenches. Double lines, ridges. Single lines, tranform faults

those of Vening Meinesz (1962), Runcorn
(1965) and others. For reasons already discussed, it is believed that sinking currents
occur beneath island arcs because cold lithosphere is being thrust downwards into the
mantle. If the line where underthrusting takes
place moves, then so must the sinking current.
The arc is not the consequence but the cause
of the horizontal convergence beneath the
lithosphere. This separation between cause
and effect is of somewhat limited value, however, because the mantle and lithosphere are
in thermal and mechanical contact, and therefore affect each others' motions.

The discussion in this section suggests that
all detailed discussions of the convection problem so far published are too simple to apply
to the Earth. However it does appear that
there is now enough understanding of the
mechanical and thermal properties of the crust
and upper mantle for the convection problem
to be posed correctly, and perhaps even solved.

CONTINENTAL TECTONICS

All early theories of continental drift required
strong rigid continents to smash their way
through the oceans, in much the same way as
an ice-breaker moves through ice floes. This
is now known not to happen. The original
ideas have been greatly modified (see the beginning of this paper) and in plate theory there
is no difference between the motion of continents and oceans. There are, however, some
striking differences between the geological
and tectonic history of oceanic and continental
rocks. Oceanic crust is created and destroyed
rather rapidly, and the crust beneath most of
the world's oceans is probably less than 200
million years old. Plate boundaries are sharp
in oceanic areas, and the earthquake epicentres
are restricted to a narrow belt seldom wider
than the errors in location (Barazangi and
Dorman, 1969). Focal mechanisms of earthquakes on plate boundaries are determined

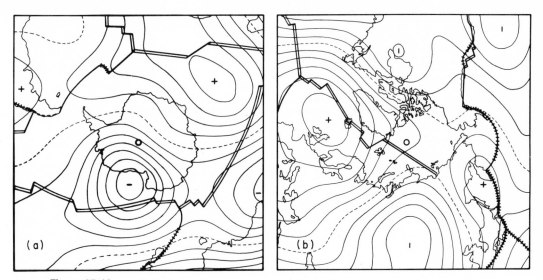

Figure 37-12
The south (a) and north (b) polar regions of Figure 37-11. The poles are shown as small circles. Azimuthal equidistant projections centered on the South (a) and North (b) Poles

by the motion of the plates alone, and are rarely complicated by the presence of small intervening plates. Intermediate and deep focus earthquakes occur only where plates are destroyed, and if the plate being consumed is oceanic there is a simple relationship between the consumption rate and the depth of the deepest earthquakes, as described earlier.

Continental crust behaves quite differently. Large areas were formed at least 1000 million years ago and have often been deformed by several orogenies. Plate boundaries within continents are rarely narrow, most commonly they are diffuse zones of seismic activity. Indeed there are few aseismic regions within the continents. Fault plane solutions for earthquakes reflect the complications and are not simply related to the motions of major plates because the boundaries consist of many small blocks in relative motion. The continental equivalents of island arcs are regions of active mountain building like Persia and the Himalaya (Le Pichon, 1968). In such regions deep earthquakes are absent and intermediate sources are infrequent and localized, like that beneath Roumania or the Hindu Kush. These sources were discussed earlier and could be relics of old oceanic lithosphere.

These differences are to be expected if the continental lithosphere cannot descend into the mantle, and also if it is more easily deformed than the oceanic lithosphere. Both properties are consequences of the nature and composition of continental rocks.

The buoyancy forces produced by the density contrast between the oceanic crust and the mantle were neglected in the section on the thermal structure of the sinking slab. The influence of such forces on the motion of the sinking slab is not easily determined because the composition of the oceanic crust is still uncertain, and because part of the crust will probably be converted to eclogite as it sinks. Since the density of eclogite is greater than that of peridotite, the presence of the oceanic crust could increase the downward force on the sinking slab. A simple calculation suggests that the same is not true for continental crust, which is probably sufficiently thick and light to prevent consumption. Before the lithosphere descends into the mantle the temperature within it is given by equation (10). Thus the maximum downward buoyancy force cm^{-2} that can be produced by the temperature difference between it and the surrounding mantle is:

$$g \int_0^l \rho (1 - \alpha T - (1 - \alpha T_1)) \, dz = \frac{\rho g \alpha T_1 l}{2}. \quad (69)$$

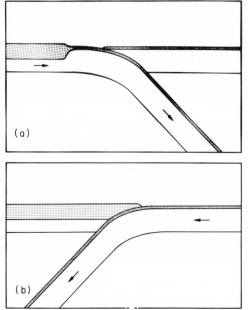

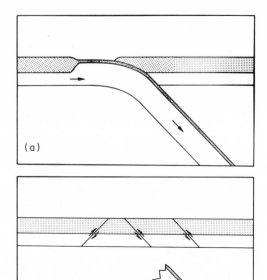

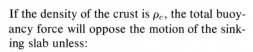

Figure 37-13
A trench which originally has oceanic crust on both sides (thin stippled surface layer) may attempt to consume a continent (thick stippled surface layer). Continental crust cannot sink, and therefore the direction of overthrusting changes (b) to consume the oceanic crust originally behind the island arc

Figure 37-14
If the trench in 37-13 (b) attempts to consume a continent (a), regeneration of the sinking slab ceases (b), and mountains are built over a wide zone by overthrusting

If the density of the crust is ρ_c, the total buoyancy force will oppose the motion of the sinking slab unless:

$$\frac{\rho \alpha T_1 l}{2} \geqslant (\rho - \rho_c)\, gd, \qquad (70)$$

where d is the crustal thickness. If $\rho_c = 2.7$ g cm^{-3}, substitution from equation (2) into (70) gives:

$$d \leqslant 4.5 \text{ km}. \qquad (71)$$

This value is comparable to the oceanic crustal thickness, but is much less than that of 30 km often observed beneath continents. Thus oceanic lithosphere can overcome the upward buoyancy due to its less dense crust, especially if there are phase changes to eclogite as the crust sinks. The same is not true of continental lithosphere. The upper granitic part is perhaps 15 km thick and has a density of ~ 2.5 g cm^{-3}. The lower part is probably gabbro, which has a density of 2.8 g cm^{-3}. Therefore if continental lithosphere is thrust into the mantle

buoyancy forces will oppose the motion. Thus continents once formed are very difficult to destroy. If an island arc attempts to consume a continent large stresses will be generated throughout the lithosphere on both sides. Changes in the boundaries of the plates, or in their motion, probably then take place. If the trench and island arc originally had oceanic crust on both sides, the island arc is likely to flip, and instead of attempting to consume the continent it will consume the oceanic crust originally behind the arc (Fig. 37–13). The same cannot happen if the trench was originally between an ocean and a continent, and was consuming oceanic lithosphere (Fig. 37–14). Either the plate motions must then thicken the crust and the lithosphere, or the plate boundaries must change to permit consumption of oceanic lithosphere. Such crustal thickening is at present taking place in the eastern part of the Alpide belt.

If the arguments above are correct the cold descending slab is only produced by consumption of oceanic crust. It is not produced by the continental equivalents of island arcs. Thus beneath continents horizontal temperature gradients are not generated by consumption of cold lithosphere, and the position of the descending limbs of convection currents are not controlled by plate motions. Thus the whole pattern of mantle convection must depend on the motion of the continents. However, changes in the geographical position of the sinking limb will not take place as soon as two continents meet. The mantle below an island arc is cold and will continue to sink. The motion can only change when this material has been convected elsewhere, and when the blanketing effect of continental radioactivity penetrates to the mantle. Thus the flow will continue to move two continents together after they have collided, and therefore can produce fold mountains by crustal shortening over large areas. Examples of such mountains are the Himalaya and the Persian ranges. Mountain ranges like the Andes between continents and oceans are much less broad, presumably because most of the relative motion is taken up by consumption of the oceanic plate. Perhaps the most important consequence of the collision of two continents is that the process provides a method of changing the convection pattern within the mantle. The generation of the cold sinking slab ceases when the continents meet, and therefore the control of the sinking limb of the cell by the trench concerned will also cease. Rearrangement of convection patterns can then occur. It should therefore be possible to relate world-wide slow changes in the motion of large plates to particular collisions. However, certain apparent changes in the directions of relative motion between plates are caused by changes in plate geometry, and not in their relative motion (McKenzie and Morgan, 1969). It is essential to separate these two effects, since only real changes of relative motion are connected with changes in convection patterns.

The other major difference between continental and oceanic plates is their seismicity. Throughout all continents except Antarctica some earthquakes occur which are not obviously related to the plate boundaries. The same is not true of oceanic regions. The difference in behaviour is caused by the contrast in the mechanical properties between the oceanic and continental rocks. The continental crust consists of about 15 km of granitic rocks and a similar thickness of gabbroic. The melting point of both rock types is less than about 1200°C, and together they form more than half of the continental lithosphere. Of the upper 30 km of the oceanic plate, only perhaps 5 km consists of rocks with low melting points, the other 25 km is peridotite strongly depleted in all low melting point phases, with a melting temperature of about 1800°C. The activation energy for high temperature creep is proportional to the melting temperature (Sherby, 1962). Thus oceanic plates are more resistant to deformation than continental ones, in contrast to the early ideas of drift.

The arguments in this section suggest that the differences between the deformation of oceans and continents is entirely due to the difference in chemical composition, and not to any differences in the convection cells beneath them. Continental plates are easy to deform but difficult to consume compared with oceanic ones, and much tectonic evolution of continents must be a consequence of these differences.

CONCLUSIONS

Perhaps the most important result obtained in this paper is the dominance of thermal boundary conditions and time constants in convection processes. The temperature structure of the sinking lithosphere is determined by the time constant of the slab and by the spreading rate, and in turn governs the distribution of intermediate and deep focus earthquakes. The sinking slab can also control the position of the descending limbs of convection cells.

The main difference between oceanic and continental plates are due to their different composition. Continental rocks are considerably less dense than the mantle, and cannot sink through it. Any attempt to consume continental rocks must therefore produce a change in the motion or the boundaries of the plates.

The only useful simplification of the convection problem appears to be the idea of plates. Attempts to understand the driving forces by similar methods show little promise. There is therefore no obvious alternative to numerical solution of the general non-linear convection equations.

38

The Implications of Terrestrial Heat Flow Observations on Current Tectonic and Geochemical Models of the Crust and Upper Mantle of the Earth

JOHN G. SCLATER
JEAN FRANCHETEAU
1970

From *Geophysical Journal of the Royal Astronomical Society*, v. 20, p. 509–542, 1970. Reprinted with permission of the authors and the Royal Astronomical Society.

Since the first reliable determinations of heat flow in England (Benfield, 1939; Anderson, 1939) and in South Africa (Bullard, 1939), the number of observation points has grown considerably in the last decade, chiefly due to measurements made at sea. Lee and Uyeda (1965), on the basis of 131 continental and 913 oceanic measurements, proposed the following arithmetic means for continental, oceanic and planetary heat flows—1.43, 1.60, and 1.58 μ cal cm $^{-2}$ s $^{-1}$ respectively. It is likely that the difference between the oceanic and continental means is not significant and is largely a consequence of over representation of the mid-ocean ridges which have high values and of the omission of volcanic areas on land. When the observations are averaged over five-degree squares, the means of the continental and oceanic squares are 1.49 and 1.47 μ cal cm $^{-2}$ s $^{-1}$ respectively (Lee and Uyeda, 1965).

It is usually supposed that most of the heat arriving at the surface is produced by radioactivity within the Earth. Because of the difference in radioactive content of the oceanic and continental crustal rocks, it was generally expected that the heat flow at sea would be less than that on land. The discovery that the two are equal poses a difficult problem (Bullard, 1952; Bullard, 1968).

In this paper, we attempt to answer this problem by analysing in more detail the distribution of heat flow values both on land and on the ocean floor. The explanation offered for the differing distribution of heat through continents and oceans reconciles terrestrial heat flow observations and plate tectonics and has strong implications for geochemical and thermal models of the Earth.

HEAT FLOW IN THE NORTH PACIFIC

In the sea-floor spreading model, new crust is created by the intrusion of hot material along the centre of spreading. The new material

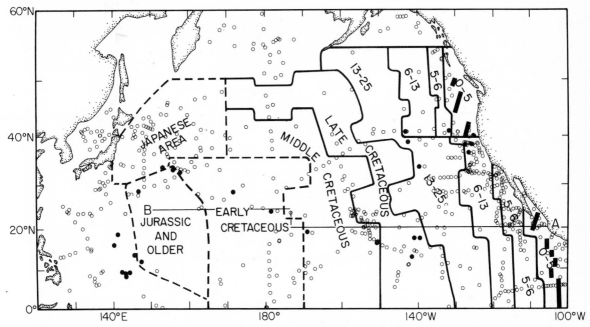

Figure 38-1
Provinces of the North Pacific. The boundaries of the provinces are determined from magnetic lineations (Atwater and Menard, 1970) and the age of sediment recovered by JOIDES deep sea drillings (McManus and Burns, 1969; Scientific Staff, Leg VI, 1969). The thick black line is the crest of the East Pacific Rise, Gorda and Juan de Fuca ridges. The open circles are the locations of the heat flow measurements and the filled-in circles are the locations of the JOIDES deep sea drillings. AB is the location of the topographic profile of Udintsev (1964)

cools and solidifies and moves away from the centre of spreading while more crust is intruded. McKenzie (1967) and Langseth and Von Herzen (in press) have shown that such a model is compatible with observed heat flow data. It explains the broad band of high values near the crest of the mid-ocean ridges and the general decrease of heat flow with increasing distance from the ridge crest. However, the plots of heat flow versus distance from the crest of ridges as shown by Lee and Uyeda (1965) are not suitable for detailed comparison with the theoretical models. These plots take no account of (a) the offsets of the ridge axis by transform faults, (b) the changes in velocity of spreading and (c) the increased scatter in the measurement, as the crest of the ridge is approached.

For comparison with a theoretical model, the North Pacific was separated into oceanic provinces of the same age and, to remove the scatter, the values within each province were averaged. The homogeneity and independence of the source data were investigated by comparing the distribution of heat flow within each province with a normal distribution.

The pattern of magnetic anomalies on the ocean floor provides a good estimate of the age of the oceanic crust (Vine, 1966; Heirtzler et al., 1968). Results from the JOIDES ocean floor sampling programme have established the validity of the magnetic interpretation (Maxwell, 1969; McManus and Burns, 1969). Further deep-sea drillings in the North Pacific have extended the ocean floor chronology to the upper Jurassic (Scientific Staff, JOIDES Leg VI, 1969).

The North Pacific Ocean was divided into regions of approximately similar age (Fig. 38–1). From the crest of the East Pacific Rise to anomaly 32, the age of the ocean floor is estimated from the magnetic anomalies. From anomaly 32 to the Marianas Trench, the results of the JOIDES drillings provided the age

Table 38–1
Statistics of the heat flow distribution in nine provinces of the North Pacific. The number of observations, their mean, standard error and standard deviation are listed for each province. The Kolmogorov-Smirnov $D_{0.1}$ statistic at the 10 per cent level and the largest observed difference, D_{max}, between sample and population cumulative percentages are also listed.

Province	n	$\bar{q}$	$\sigma_{\bar{q}}$	s	$D_{0.1}$	D_{max}
Jurassic and Older ⎰	14	1.44	0.095	0.357	–	–
⎱	11	1.28	0.059	0.187	44.0	10.0
Early Cretaceous	47	1.18	0.038	0.264	41.0	32.0
Japanese Area	47	1.11	0.032	0.218	47.0	6.0
Middle Cretaceous	60	1.38	0.055	0.427	41.0	20.0
Late Cretaceous	64	1.43	0.066	0.526	34.0	26.0
Anomaly 13–25	59	1.43	0.086	0.662	34.0	14.0
Anomaly 6–13	116	1.61	0.071	0.765	30.0	14.0
Anomaly 5–6	102	2.22	0.107	1.084	27.0	8.0
Anomaly 0–5	65	2.82	0.210	1.696	21.0	12.0

Note: n is the number of heat flow values; $\bar{q}$ is the mean value; $\sigma_{\bar{q}}$ is the standard error of the mean; s is the standard deviation; D is the Kolmogorov-Smirnov statistic.

estimate. The boundaries of the regions shown on Fig. 38–1 are: The crest of the East Pacific Rise, anomalies 5, 6, 13, and 25, the late/middle Cretaceous, the middle/early Cretaceous, the early Cretaceous/upper Jurassic. Magnetic lineations have been observed to the west of Japan (Uyeda et al., 1967) but could not be correlated with the known time scale. This area, which is definitely older than the late Cretaceous, is shown in Fig. 38–1 and is considered in this study.

Heat flow data were taken from Lee and Uyeda (1965), Vacquier et al. (1966), Vacquier et al. (1967), Sclater and Correy (1967), Burns and Grim (1967), Fisher, Von Herzen and Rhea (1967), unpublished data from M. Langseth and R. P. Von Herzen and Sclater et al. (1968). Only needle probe measurements showing full penetration or outrigger measurements of fair reliability (six or greater; Langseth, Le Pichon, and Ewing, 1966) were taken in averaging the data. All values less than 0.4 μ cal cm^{-2} s^{-1} were disregarded as Sclater, Mudie, and Harrison (1970) have shown that measurements this low are almost certainly caused by recent slumping of sediments from surrounding topographic highs. The distributions of heat flow data, satisfying these conditions, for each region of the North Pacific considered, are given in the histograms shown in Fig. 38–2. Table 38–1 gives the parameters of these distributions, the mean, standard devia-

tion, standard error, and the estimate of the goodness of fit of these distributions to a normal distribution function. The agreement of distributions in small sampling groups is best estimated by the Kolmogorov–Smirnov statistic (Massey, 1951). Fig. 38–3 presents a cumulative percentage plot of the heat flow values on normal probability paper and gives a graphical test for the null hypothesis that all the samples follow the normal distribution law. As can be seen from Table 38–1 and Fig. 38–3, the distribution of heat flow values in provinces of the same age does not contradict a normal distribution law at the 10 per cent level of confidence. Thus, in general, the heat flow values in each region of the same age are homogeneous. This suggests that our method of separation is statistically, as well as geologically, meaningful.

The Jurassic, early Cretaceous, and Japanese provinces have a mean and standard error of 1.44 ± 0.095, 1.18 ± 0.038, and 1.11 ± 0.032 μ cal cm^{-2} s^{-1} respectively. Though the Japanese and early Cretaceous provinces have a mean close to 1.15 and a small standard error, the Jurassic average of 14 values is much greater and has a significantly larger standard error. Three unpublished values of 1.93, 1.89, and 2.20 μ cal cm^{-2} s^{-1} contribute to this mean. As the average of the other 11 values is 1.28 μ cal cm^{-2} s^{-1} with a lower standard error and these three high values are

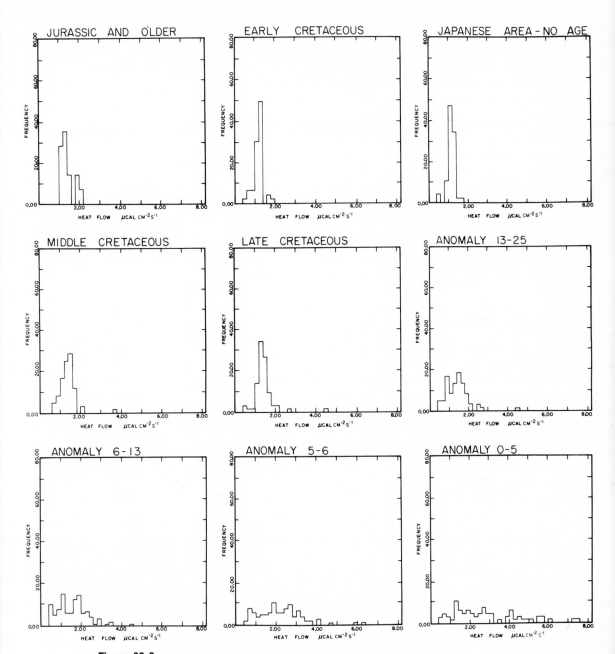

Figure 38-2
Histograms of the distribution of heat flow values in the nine North Pacific provinces

on the border of reliability (only one thermal gradient measured on an outrigger system) we consider the value of $1.28 \pm 0.059 \; \mu$ cal cm^{-2} s^{-1} to be closer to the actual mean heat flow through this area. The distribution of the heat flow in these three areas does not conflict with the hypotheses of normality at the 10 per cent level (Table 38–1 and Fig. 38–3).

The middle and late Cretaceous provinces and the region bordered by anomalies 25 and 13 have a mean and standard error of 1.38 ± 0.055, 1.43 ± 0.066, and $1.43 \pm 0.086 \; \mu$

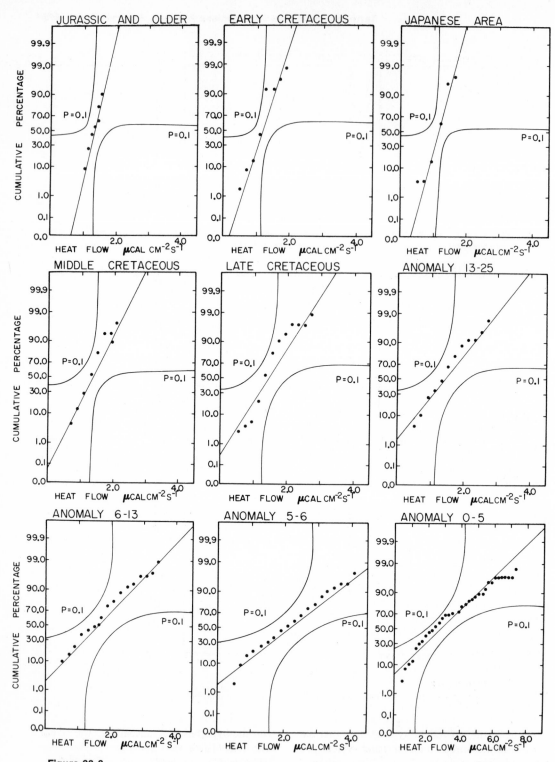

Figure 38-3
Graphical test of the hypothesis that the population of heat flow values in each of the nine North Pacific provinces is normal at the 10 per cent level. The cumulative relative frequency is plotted against heat flow value. The theoretical normal curve is indicated by the straight line. The band defines the maximum difference between sample and population cumulative percentages expected 10 per cent of the time (Kolmogorov-Smirnov statistic)

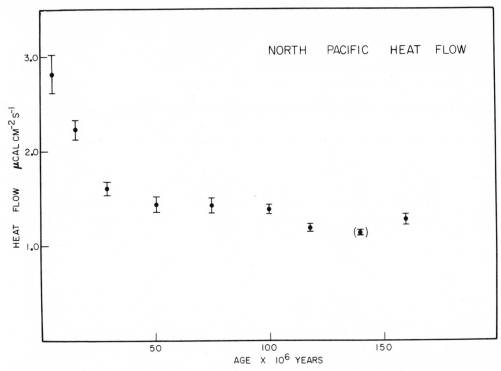

Figure 38-4
Plot of mean heat flow against age of province for the North Pacific. The length of the bar gives the magnitude of the respective standard error. The mean value for the youngest province has been plotted for a mean age of 5 m.y. in order to account for the paucity of observations in the crestal regions. An explanation for the lack of values near the crest of the ridge can be found in Oxburgh and Turcotte (1969)

cal cm^{-2} s^{-1} respectively. The Kolmogorov-Smirnov test indicates that the distribution of heat flow in these three areas does not differ significantly from a normal distribution.

The provinces extending from anomaly 13 to the crest of the East Pacific Rise have a mean and standard error of 1.61 ± 0.071, 2.22 ± 0.107, 2.82 ± 0.210 μ cal cm^{-2} s^{-1} respectively. The fit to a normal curve for these three provinces is not in conflict with the observed heat flow observations.

A plot of mean heat flow versus age for the nine regions considered shows a continuous and significant decrease of heat flow from the younger to the older crust (Fig. 38–4). The histograms of Fig. 38–2 and the length of the bars on Fig. 38–4 demonstrate that the scatter in the heat flow values increases from the older to the younger provinces. The continuous decrease in heat flow, away from the crest of the ridge (Fig. 38-4), is clearly compatible with a model in which hot material is intruded at the crest of the ridge and heat is lost by conduction, as this new crust moves away from the crest.

SOUTH ATLANTIC OCEANIC HEAT FLOW

Though reliable heat flow measurements are sparse in the South Atlantic and mostly concentrated around the ridge crest, it is tempting to investigate if the pattern of heat flow in the South Atlantic is similar to that in the North Pacific. The South Atlantic is the only other ocean where the age of the ocean floor is, at present, even partially understood. Deep drilling (Maxwell, 1969) in the South Atlantic has confirmed the age attributed to different parts of the ocean floor by the mapping of the magnetic anomalies (Dickson, Pitman, and Heirtzler, 1968).

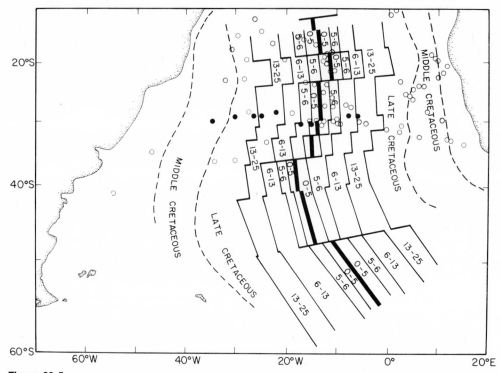

Figure 38-5
Provinces of the South Atlantic. The boundaries of the provinces are determined from the magnetic lineations (Dickson et al., 1968) and the age of the sediments recovered by JOIDES deep-sea drillings (Maxwell, 1969). The thick black line marks the crest of the mid-Atlantic ridge. The open circles are the location of the heat flow measurements and the filled-in circles are the locations of the JOIDES deep-sea drillings

The South Atlantic was separated into provinces of similar age (Fig. 38-5). The boundaries of these provinces have the same age as those chosen for the North Pacific. The crest of the mid-Atlantic ridge and anomalies 5, 6, 13, and 25 are well defined. The late, middle and early Cretaceous boundaries are determined by assuming a constant spreading rate of 2 cm yr^{-1} (Maxwell, 1969) from the present until the early Cretaceous.

The heat flow data for the South Atlantic were taken from Lee and Uyeda (1965) and Langseth et al. (1966). Excepting the low average heat flow value for the province extending from anomaly 6 to 13, the average heat flow decreases with increasing age of the oceanic crust (Fig. 38-6). This is similar to the pattern shown by the heat flow values in the North Pacific. However, the low number of values precludes a meaningful statistical study of the heat flow distribution within each prov-

ince. The anomalous value is the average of only five determinations (Table 38-2). It is not clear that this mean is representative of the

Table 38–2
Statistics of the heat flow distribution in seven provinces of the South Atlantic. The number of observations, their mean, standard error and standard deviation are listed for each province.

Province	n	$\bar{q}$	$\sigma_{\bar{q}}$*	s
Early Cretaceous	7	1.14	0.076	0.164
Middle Cretaceous	13	1.40	0.113	0.361
Late Cretaceous	12	1.42	0.174	0.530
13–25	6	1.26	0.428	0.820
6–13	5	0.69	0.102	0.172
5–6	10	1.75	0.334	0.912
0–5	16	2.15	0.557	2.009

Note: n is the number of heat flow values; $\bar{q}$ is the mean value; $\sigma_{\bar{q}}$ is the standard error of the mean; s is the standard deviation.

*The best estimate of the standard error of the mean of a small sample is given by the law of Student–Fisher.

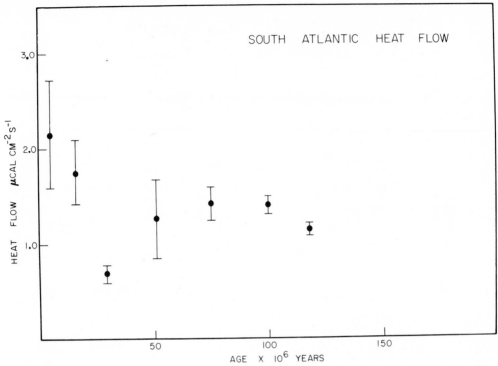

Figure 38-6
Plot of mean heat flow against age of province for the South Atlantic. The length of the bar gives the magnitude of the respective standard error. The mean value for the youngest province has been plotted for a mean age of 5 m.y. in order to account for the paucity of observations in the crestal regions. An explanation for the lack of values near the crest of the ridge can be found in Oxburgh and Turcotte (1969)

heat flow for this province and it was disregarded when fitting theoretical heat flow curves through the average heat flow values for the provinces. The similarity of the heat flow dependence on age for the North Pacific (Fig. 38–4) and the South Atlantic (Fig. 38–6) suggests that the heat flow decrease observed in the Pacific may be typical of all oceans.

CONTINENTAL HEAT FLOW

Polyak and Smirnov (1968) have analysed in considerable detail the relationship between terrestrial heat flow and the age of the orogenic province where the heat flow was measured. They demonstrate that the heat flow in each province is normally distributed and that the average value decreases with increasing age of the province.

Table 38-3 lists their data for several orogenic provinces and Fig. 38–7 presents a plot of the average heat flow values against the age of the corresponding province. The age of the Cenozoic, Mesozoic, Hercynian, and Caledonian orogenies are relatively well known. The Precambrian platforms have been given a rather arbitrary age of 700 My and the shield areas have been ascribed the mean age of the Grenville, of approximately 1100 My. Because a complex multipeak distribution is observed for heat flow values in the Cenozoic, the mean value, 1.78 μ cal cm^{-2} s^{-1}, may not be representative of the mean flow through this province (Polyak and Smirnov, 1966). The mean heat flow through the orogenic provinces decreases from 1.42 μ cal cm^{-2} s^{-1} for the Mesozoic to 0.90 μ cal cm^{-2} s^{-1} for the Precambrian shields. The small standard deviation of the values for the Precambrian and the exponential nature of the decrease on Fig. 38–7 suggest that, in this province, the heat flow may be close to an equilibrium value.

Table 38–3
Heat flow distribution in different continental tectonic provinces.

Tectonic provinces	n	q_{min}	q_{max}	$\bar{q}$	$\sigma_{\bar{q}}$	s	Distribution law
Provinces of Precambrian folding (undifferentiated)	88	0.53	1.33	0.93	0.02	0.17	Normal
Shields	69	0.61	1.32	0.90	0.02	0.15	Normal
Platforms	19	0.53	1.33	1.04	0.05	0.20	Normal
Provinces of Caledonian folding	17	0.68	1.71	1.11	0.07	0.28	Normal
Provinces of Hercynian folding	60	0.60	1.90	1.24	0.03	0.25	Normal
Provinces of Mesozoic folding and activation	26	1.00	2.12	1.42	0.06	0.31	Normal
Provinces of Cenozoic folding and activation							
Foredeeps and intermontane troughs	55	0.52	1.58	0.98	0.03	0.24	Normal
Folded mountain structures of miogeosynclinal zones	19	1.20	2.20	1.75	0.06	0.25	Normal
Zones of Cenozoic volcanism	55	1.20	3.49	2.20	0.06	0.42	Normal
Continental rift zones							
Nyasa	20	—	—	1.00	—	—	Unknown
Baikal	11	1.21	3.40	2.40	0.18	0.59	Unknown

Source: Polyak and Smirnov, 1968.
Note: n is the number of heat flow values; $\bar{q}$ is the mean value; $\sigma_{\bar{q}}$ is the standard error of the mean; s is the standard deviation.

There is some indication that the mean values of all heat flow measurements on land may have to be raised. Simmons and Nur (1968) compared the physical properties of granite *in situ* measured in two boreholes with those from laboratory measurements on dry samples. The correction for the effect of cracks and saturation with water on the thermal conductivity measurements could raise the average heat flow in all continental areas by as much as 10–15 per cent. When this correction is applied, the mean for the shield areas becomes 1.0 to 1.05 μ cal cm^{-2} s^{-1}. This value is very close to the average heat flow observed on the older portions of the oceanic crust.

EQUALITY OF CONTINENTAL AND OCEANIC HEAT FLOW

The mean heat flow through the deep-ocean crust is approximately equal to the mean flow through the continents (Lee and Uyeda, 1965; Polyak and Smirnov, 1968). Bullard (1952) proposed two alternative hypotheses in explanation: First, that the oceanic heat flow is due to convection; second, that, to a first approxi-

mation, the mean chemical composition of the mantle is the same over both continents and oceans when averaged down to depths of several hundred kilometres. The second hypothesis was the more popular and led to several geochemical models of the upper mantle (Ringwood, 1958; McDonald, 1963; Clark and Ringwood, 1964).

Several authors (Langseth et al., 1966; McKenzie, 1967; McKenzie and Sclater, 1969; Sleep, 1969) have demonstrated that, in a sea-floor spreading model, a considerable proportion of the heat lost by the oceans is due to the creation of oceanic lithosphere. Thus, there is no longer any necessity to assume that the mean radioactive composition beneath continents and oceans, when averaged down to considerable depths, are identical.

The mean heat flow in any continental or oceanic province is strongly related to the age of the province (Figs. 38–4 and 38–7). However, the time scales for the thermal decay of the heat flow for the two realms are an order of magnitude different. The continents take a billion years to reach a constant value of 1.00 μ cal cm^{-2} s^{-1} while the oceans reach a value of 1.1 μ cal cm^{-2} s^{-1} after only 100 My. The

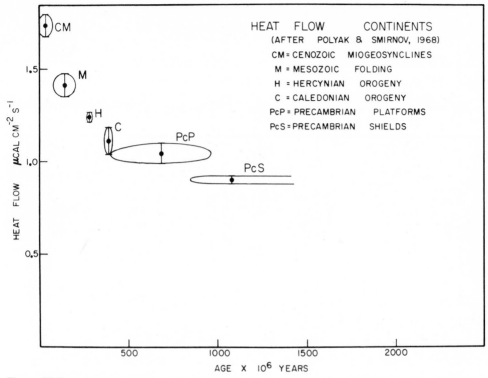

Figure 38-7
A plot of mean heat flow against age of orogenic province for continents. (After Polyak and Smirnov, 1968; Smirnov, 1968, Fig. 1)

continental and oceanic crustal rocks have a radically different chemical composition. This and the different time scales of the heat flow decay suggest that the processes responsible for the observed distribution of heat flow differ for continents and oceans. Although the mean heat flow through the two regions is the same, the strong dependence of heat flow on age (Figs. 38–4 and 38–7) suggest that this equality may be entirely fortuitous and need not necessarily be related to a deep-seated mantle process. Thus, a comparison of the gross means is not a meaningful method of evaluating the contribution from the deep interior of the Earth. It is more pertinent to examine the significance of the strong dependence of heat flow on age for both the oceanic and continental values and to pay special attention to the values in the older provinces. These values may be close to equilibrium and hence enable a better estimate to be made of the heat flow from the upper mantle in both cases.

The problems are then to account for (a) the dependence of heat flow on age, and (b) the approximate equality of the heat flow through the shield areas with that through the oldest portions of the deep oceans. We will consider the geochemical models of the crust and upper mantle of the continents proposed by Roy, Blackwell, and Birch (1968) and oceans proposed by Clark and Ringwood (1964) and Ringwood (1969; Fig. 5). Heat lost in the creation of oceanic lithosphere will play a significant role in the contribution to the oceanic heat flow (McKenzie, 1967; McKenzie and Sclater, 1969).

HEAT FLOW ON CONTINENTS

Continental heat flow values are a function of the age of the latest thermal event, the intensity of this event and the concentration of radioactive elements below the continents. An

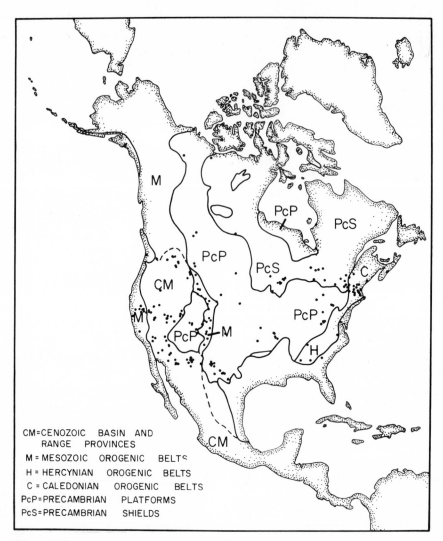

Figure 38-8
Tectonic map of North America, after King (1969). The black dots represent heat flow measurements

estimate of the vertical distribution of these elements has come from recent work by Birch, Roy, and Decker (1968), Roy, Blackwell, and Birch (1968) and Lachenbruch (1968). These authors have separated North America into five major thermal regions and have shown that within each region the heat flow values show a striking linear correlation with the concentration of radioactive elements in the surface rocks. To investigate whether the data for North America show a strong correlation with

age and to compare the thermal regions of Roy et al. (1968) with the tectonic provinces of Polyak and Smirnov (1968) we superimposed all available North American heat flow data (Lee and Uyeda, 1965; Roy et al., 1968; Blackwell, 1969) upon a tectonic map of North America (King, 1969; Fig. 5). The two youngest provinces were separated according to the most recent thermal event. The Cenozoic is the Basin and Range province of Blackwell (1969) and the Mesozoic is, for the most part,

Table 38–4
Heat flow distribution for the different continental tectonic provinces of North America.

Tectonic provinces	n	q_{min}	q_{max}	$\bar{q}$	$\sigma_{\bar{q}}$	s	$D_{0.1}$	D_{max}	Distribution law
Provinces of Precambrian folding									
Undifferentiated	63	0.44	1.90	1.09	0.032	0.253			
Shields	22	0.69	1.32	0.95	0.038	0.179	25.0	13.0	Normal
Platforms	41	0.44	1.90	1.17	0.040	0.254	19.1	10.1	Normal
Provinces of Caledonian folding	17	0.81	1.80	1.31	0.070	0.290	28.6	12.0	Normal
Provinces of Hercynian folding	6	0.73	1.40	1.09	0.093	0.229	47.0	14.0	Normal
Provinces of Mesozoic folding and activation	34	0.62	2.28	1.55	0.081	0.470	21.0	9.0	Normal
Provinces of Cenozoic folding and activation	63	1.10	6.89	2.18	0.100	0.795	15.4	16.5	Unknown

Note: n is the number of values; $\bar{q}$ is the mean value; $\sigma_{\bar{q}}$ is the standard error of the mean; s is the standard deviation; D is the Kolmogorov-Smirnov statistic; all values are in μ cal cm^{-2} s^{-1}.

the area of western North America folded in the Mesozoic and not affected by younger tectonic activity. Except for H (the mean of six values in the southern Appalachians) the data for North America (Fig. 38–9, Table 38–4) are remarkably similar to those presented by Polyak and Smirnov (1968; Fig. 7). Hamza and Verma (1969) have compiled radioactive age determinations and heat flow measurements for Africa, Australia and India. They suggest that the decrease of heat flow with age is true for all continents.

The two oldest thermal regions of Roy et al. (1968), the Central Stable region and the New England area, correspond to the Precambrian Platforms (PcP) and Caledonian (C) provinces respectively (Fig. 38–8). These authors have shown that the heat flow and surface radioactivity data for these two provinces plot on the same straight line (Fig. 38–10). Roy (personal communication) has further shown that two published values (Hyndman et al. 1968) and one unpublished value for the Australian shield, and three unpublished values for the Canadian shield also fall on this line (Fig. 38–10). This linear relationship affords a simple explanation of the heat flow in the Caledonian and older provinces. The slope of the line gives a measure of the thickness of the radioactive layer and the intercept gives the heat flow from below. For these three provinces the variation of heat flow can be ex-

plained by assuming an 8-km thick radioactive layer with a uniform heat flow of 0.8 μ cal cm^{-2} s^{-1} at its base. The decrease in heat flow from the Caledonian to the shields can be explained by the reduction of radioactivity in the surface layer. There is no need to invoke any thermal energy due to intrusion. Thus, the intercept heat flow may represent the equilibrium flux from the lower crust and upper mantle.

It is supposed that the continental rocks have differentiated from the underlying mantle and that, in the course of this differentiation, the uranium, thorium and potassium are selectively removed from the mantle and are concentrated at high levels in the crust by magmatism. For the Cenozoic province, it is likely that the thermal energy due to magmatism is responsible for a large fraction of the surface heat flow. The decay of this heat flow is a function of the age of emplacement, the size of the magmatic body, the degree of metamorphism and the impoverishment of the radioactive elements. From the explanation of Roy et al. (1968) for Fig. 38–10 one can infer that all energy of intrusion has been lost after 400 million years. If radioactive studies on rocks of Mesozoic age yield the same high radioactive concentrations as those reported by Birch et al. (1968) for the White Mountains, the rest of the Mesozoic values would plot on the line of Roy et al. (1968; Fig. 38–10) and would

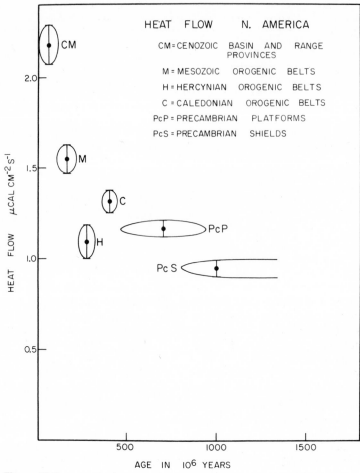

Figure 38-9
A plot of mean heat flow against age of last thermal event for North
America

imply that the time scale of the decay of the intrusive energy could be as short as 200 million years. However, there are few reported measurements to substantiate these high concentrations. Lachenbruch (1968) has shown that a model where the radioactive concentrations of the crustal rocks decrease logarithmically with depth can explain the linear relationship of heat flow to radioactivity found for the provinces older than the Caledonian. Such a model might reflect a continuous history of fractional melting in the mantle resulting in a strong upward differentiation of the lighter elements (Ringwood, 1969). One could then qualitively account for

the decrease of the mean surface radioactivity with increasing age by two main factors: (a) The decay of the radioactive elements themselves, particularly potassium; and (b) the erosion of the logarithmic layer.

For a comparison with the heat flow through the old ocean basins we will consider the regional flow through the shield areas as this value is close to the equilibrium and hence sets a lower limit to temperatures in the mantle. Our model which is compatible with the data of Roy et al. (1968) consists of a two-layer crust. The upper layer, where measurements of radioactivity have been made, is 8 km thick and contributes 0.25 μ cal cm^{-2} s^{-1} to

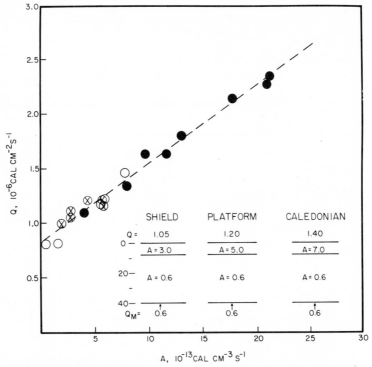

Figure 38-10
Heat flow and heat productivity data for plutons in the New England area
(solid circles), the Central Stable region (open circles), and the Australian
and Canadian shields (open circles with crosses). The sections represent
steady state models of the crust compatible with the data (Roy et al., 1968;
Hyndman et al., 1968; Roy, personal communication). The three highest
values are from the White Mountains, of Mesozoic age, and have not been
considered in evaluating a mean radioactivity for the Caledonian surface layer

the surface heat flow; the lower layer is assumed to have a radioactive concentration of
0.6×10^{-13} cal cm^{-3} s^{-1} and contributes only
0.2 μ cal cm^{-2} s^{-1} to the surface flux. Below
this crust, Ringwood's (1969) model consists
of a chemically zoned upper mantle with a
refractory dunite-peridotite layer overlying
primitive pyrolite (Fig. 38–11). Brune and
Dorman (1963) have suggested that there may
exist a low velocity zone at a depth of approximately 200 km beneath the shield areas. If the
lower ultrabasic layer indeed consists of
dunite-peridotite with a mean radioactive heat
production of 0.02×10^{-13} cal cm^{-3} s^{-1}
(Holmes, 1965; p. 1002), then the total heat
contribution of this layer is 0.03 μ cal cm^{-2}
s^{-1}. Thus, in the model we have chosen, the
lithosphere comprising the Precambrian

shields contributes 0.48 μ cal cm^{-2} s^{-1} to the
surface heat flow. This leaves a heat flow of
approximately 0.6 μ cal cm^{-2} s^{-1} coming from
below the lithosphere. *(20 mW/m² as in notes)*

OCEANIC HEAT FLOW

Heat flow observations in the North Pacific
and South Atlantic show a systematic decrease
of mean heat flow with increasing age of the
ocean floor. Such a decrease is expected in the
sea-floor spreading hypothesis where all hot
material is being intruded at the crest of the
ridges. Langseth et al. (1966), McKenzie
(1967) and Langseth and Von Herzen (in
press) have computed the heat flow resulting
from the cooling of a slab of lithosphere moving away from the ridge axis. Models with a

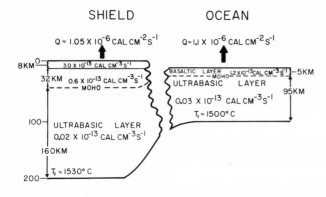

SHIELD   OCEAN

$Q \approx 1.05 \times 10^{-6}$ CAL CM^{-2}S^{-1} $Q \approx 1.1 \times 10^{-6}$ CAL CM^{-2}S^{-1}

8KM — 0 —— 3.0×10^{-13} CAL CM^{-3}S^{-1} ——
32KM — 0.6×10^{-13} CAL CM^{-3}S^{-1} —
— — — MOHO — — —

BASALTIC LAYER 1.2×10^{-13}CAL CM^{-3}S^{-1} — 5KM
—— MOHO ——
ULTRABASIC LAYER
95KM
0.03×10^{-13} CAL CM^{-3}S^{-1}
$T_i \approx 1500°$ C

100 — ULTRABASIC LAYER
0.02×10^{-13} CAL CM^{-3}S^{-1}
160KM

200 — $T_i \approx 1530°$ C

PYROLITE II   PYROLITE II

Figure 38-11
Geochemical model of continental
shield and oceanic lithosphere (after
Fig. 5 of Ringwood, 1969) assuming
convection maintains a constant
temperature at the base of the two
lithospheres. A mean conductivity of
7.1×10^{-3} cal cm^{-1} s^{-1} °C^{-1} and
adiabatic gradient of 0.3 °C km^{-1}
have been assumed

slab thickness of 50–100 km and a basal temperature of 550°–1200°C have given reasonable fits to the observed heat flow data. The problem of preserving the equality of the heat flow through shields and older ocean basins when oceanic and continental plates are permitted to move has not been considered in these models (Bullard, 1968). To explain this equality, within the framework of plate tectonics, it is necessary to examine the lower boundary condition and the possible heat contribution of the oceanic lithosphere. The lower boundary condition can be set on either temperature or heat flow.

As the temperature structure of the mantle is not very different from adiabatic, it is reasonable to assume that the low velocity layer is nearly isothermal. Such a state is easily maintained if convective processes are efficient. Thus, the temperatures at the base of the oceanic and continental lithospheres can be considered roughly equal. If this is so, and heat sources contribute 0.48 μ cal cm^{-2} s^{-1} to the heat flow through the shields, then the approximate equality of heat flow through shields and old ocean basins is a consequence of the lithosphere being twice as thick beneath the shields (Fig. 38–11). In this model the oceanic lithosphere is depleted in radioactive elements. The low radioactivity is consistent with measurements made on lherzolites, rocks recovered from the ocean floor (Wakita et al., 1967; Table 38–5). As the base of both lithospheres have nearly the same temperature, no difficulty

is encountered in preserving the heat flow equality while moving the plates. In this model, the composition of the mantle is the same under both continents and oceans.

An alternative way to explain the equality of heat flow when allowing for plate movements is to assume that the heat flowing through the base of the two lithospheres is approximately equal. This demands that the oceanic lithosphere carry a significant component of the surface heat flow, roughly 0.4 μ cal cm^{-2} s^{-1}, along with it. The material comprising the lithosphere would then be considerably more radioactive than common ultrabasic rocks. In the geochemical models of Ringwood (1969) and Clark and Ringwood (1964), the parental material is pyrolite from which are derived the basaltic and ultrabasic rocks. If the basaltic layer of the oceanic crust is 5 km thick (layer 2 and layer 3), then the underlying ultrabasics (lherzolite ?) may be as much as 15 km thick. A mixture of one part basalt with three parts ultrabasic gives pyrolite with a radioactive heat generation of 0.32×10^{-13} cal cm^{-3} s^{-1} (Table 38–5). To obtain a heat flow of 0.4 μ cal cm^{-2} s^{-1} from the oceanic lithosphere, one has to choose a plate thickness of 120 km comprised of the following layers: Five kilometres of basalt with a radioactive heat production of 1.2×10^{-13} cal cm^{-3} s^{-1}, 15 km of ultrabasic (lherzolite) with a heat production of 0.03×10^{-13} cal cm^{-3} s^{-1} and 100 km of pyrolite with a heat production of 0.32×10^{-13} cal cm^{-3} s^{-1}. This

Table 38–5
Average abundances of heat-producing elements and heat production from radioactivity in possible rocks of the oceanic lithosphere.

Rock groups	Average abundances of the radioactive elements in grams per 10^6 grams of rock			Heat production from radioactivity					
				In calories per year					Total in calories per second per cm^3
				per 10^6 grams of rock				Total per 10^6 cm^3	
	U	Th	K^{40}	U	Th	K^{40}	$Total$		
Basaltic rocks[a]	0.7	3.0	1.1	0.5	0.6	0.2	1.30	3.80	1.20×10^{-13}
Ultrabasic rocks[a]	0.013	0.05	0.001	0.01	0.01	0.0002	0.02	0.07	0.02×10^{-13}
Lherzolites[b]	0.019	0.05	0.007	0.02	0.01	0.0014	0.03	0.10	0.03×10^{-13}
Pyrolite I	0.189	0.79	0.28	–	–	–	–	–	0.32×10^{-13}
Pyrolite II[c]	0.059	0.25	0.09	–	–	–	–	–	0.10×10^{-13}

[a]Holmes, 1965.
[b]Wakita et al., 1967.
[c]Ringwood, 1958.

plate rests on a more primitive pyrolite with a mean heat production of 0.1×10^{-13} cal cm^{-3} s^{-1} down to a depth of 400 km (Fig. 38–12). These values for the heat production are consistent with Clark and Ringwood's (1964) proposal that under the oceans there is a uniform gradient of heat generation for pyrolite from 0.35×10^{-13} cal cm^{-3} s^{-1} at the base of the Moho to 0.007×10^{-13} cal cm^{-3} s^{-1} at a depth of 400 km.

This model of an ocean basin adjoining a shield area is shown in Fig. 38–12. The contribution of the oceanic lithosphere to the surface heat flow is 0.4 μ cal cm^{-2} s^{-1}. This leaves 0.7 μ cal cm^{-2} s^{-1} coming from below the lithosphere. At a depth of 200 km, the heat flow from below is approximately 0.6 μ cal cm^{-2} s^{-1}. This value is similar to that at a depth of 200 km under the shields. As in the first model the composition of the mantle at the base of the lithosphere under both continents and oceans is the same and plate movements will not alter the heat flow.

Talwani et al. (1965a) have observed essentially a zero free air gravity anomaly over oceanic ridges. This suggests that the lithosphere is isostatically compensated and that the relief of the ridge may be accounted for by thermal expansion of the lithosphere (McKenzie and Sclater, 1969). The questions remain as to whether (a) the production of oceanic lithosphere can explain satisfactorily both the observed heat flow and topographic relief of ridges, and (b) the observations give clear support for one or other of the geochemical models.

PROPERTIES OF THE MODELS

The model of the oceanic lithosphere used here is similar to the one described by McKenzie (1967). A plate of constant thickness is produced on a vertical boundary at the ridge axis and moves away at a constant velocity. The solution for the temperature structure within the lithosphere, the surface heat flow and the elevation can be found in Appendix I. The radioactive heat production term was kept in the heat flow equation and an adiabatic gradient of temperature, 0.3°C km^{-1}, was specified for the vertical boundary beneath the ridge axis. At the base of the lithosphere, the temperature is that of incipient melting of pyrolite at that depth and is obtained from the solidus/liquidus curve of pyrolite (Green and Ringwood 1967; Fig. 38–12). This curve has a gradient of approximately 10°C kb^{-1}. Once the thickness of the slab is assumed, the temperature is then completely determined. The thermal conductivity is the least well known of all the parameters controlling the temperature of the lithosphere. For the models considered in this paper, once the thickness (and hence

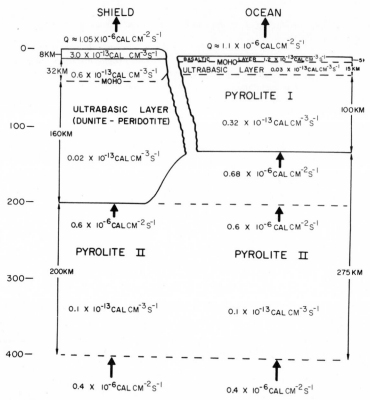

Figure 38-12
Models of continental shield and oceanic lithosphere (based on Clark and Ringwood, 1964) assuming a constant heat flux at a depth of 200 km. The thick arrows indicate the heat flow at different levels

the temperature) and radioactive heat production are assumed, the mean conductivity of the lithosphere is defined uniquely by the condition of a uniform heat flow of 1.1 μ cal cm^{-2} s^{-1} observed at great distances from the crest of the ridge. If a different temperature were assigned to the base of a slab of the same thickness, the conductivity would vary so as to retain a constant value for $K(T_1 - T_0)$. With these boundary conditions a small variable radioactive heat production term has no significant effect on either the heat flow anomaly or the elevation (Fig. 38–13). This arises from fixing the temperature at the base of the slab and demanding that the conductivity be lowered when the radioactivity is increased. Because of this lack of sensitivity to the radioactive content, one is free to choose any internal heat generation term in the range 0−0.32 × 10^{-13} cal cm^{-3} s^{-1}. In the models used to

compare theoretical curves with the observed heat flow and topography, a value of 0.1 × 10^{-13} cal cm^{-3} s^{-1} has been assumed. Such a value is consistent with a model of oceanic lithosphere where a basaltic layer overlies a thick ultrabasic lid (Fig. 38–12 and Ringwood, 1969, Fig. 5). The isotherms within an oceanic lithosphere 100 km thick moving at 5 cm yr^{-1} are shown in Fig. 38–14.

OCEANIC HEAT FLOW AND THE TOPOGRAPHY OF RIDGES

The North Pacific is the only ocean where there are enough heat flow observations to enable a meaningful comparison between the theoretical profiles and the observed heat flow and topography. The theoretical heat flow is strongly dependent on the assumed velocity of

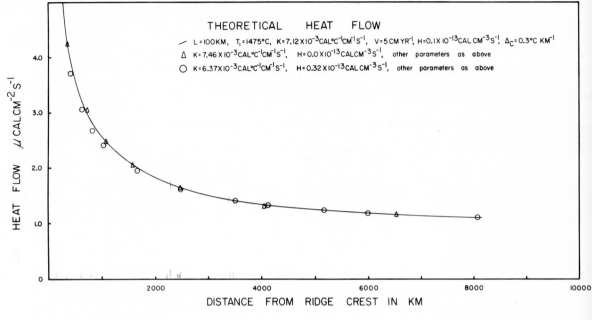

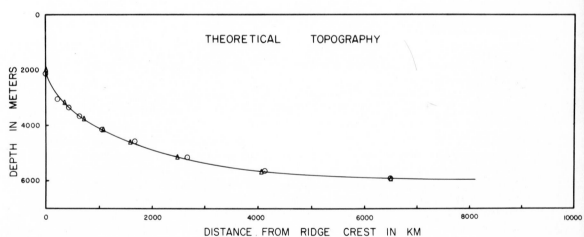

Figure 38-13
Comparison of theoretical models of heat flow and topography showing the effect of varying the radio-active content of the lithosphere.

/ $L = 100$ km, $T_1 = 1475$ °C, $K = 7.12 \times 10^{-3}$ cal °C^{-1} cm^{-1} s^{-1},
$V = 5$ cm yr^{-1}, $H = 0.1 \times 10^{-13}$ cal cm^{-3} s^{-1}, $\Delta_C = 0.3$ °C km^{-1}
△ $K = 7.46 \times 10^{-3}$ cal °C^{-1} cm^{-1} s^{-1}, $H = 0.0 \times 10^{-13}$ cal cm^{-3} s^{-1}, other parameters as above
○ K / 6.37×10^{-3} cal °C^{-1} cm^{-1} s^{-1}, $H = 0.32 \times 10^{-13}$ cal cm^{-3} s^{-1}, other parameters as above

spreading. Examination of linear magnetic anomalies in the North Pacific reveals that the spreading rate varies from 4 cm yr^{-1} for the two provinces near the ridge crest to 5 cm yr^{-1} for the provinces extending from anomaly 6 to 32. The age of the sediment recovered in re-

cent deep drillings in the Northwest Pacific (Scientific Staff, Leg VI, 1969) suggests that the spreading rate could have varied from 7 cm yr^{-1} in the late Cretaceous to 9 cm yr^{-1} in the Jurassic (Fig. 38–1). Because of this large change in spreading rate one should not expect

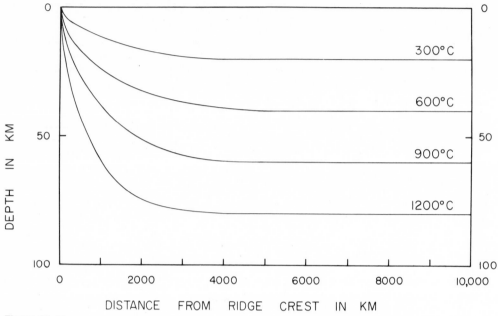

Figure 38-14
Isotherms under the North Pacific. The lithosphere of thickness 100 km is moving at 5 cm yr^{-1} to the right

to be able to match the observed heat flow with a curve based on a single spreading rate.

The heat flow and topography due to slabs of lithosphere 75, 100 and 125 km thick, moving at a velocity of 5 cm yr^{-1} are compared with observed heat flow averages in the North Pacific and an east — west topographic profile at 20° N (Udintsev, 1964) (Fig. 38–15).

In regions of rough topography the heat flow average may be biased on the low side. Sclater, Mudie, and Harrison (1970) have shown that, in a small region of high topographic relief on the Hawaiian Arch, the preferential position-ing of heat flow measurements in sediment ponds yielded an average heat flow for the small region that was 30 per cent lower than the mean for the whole Arch. A systematic heat flow survey at right angles to the Gala-pagos rift zone yielded a mean of 5.0 μ cal cm^{-2} s^{-1} for 18 heat flow observations within a 100 km wide zone about the spreading centre (Piquero 7, unpublished). These two results suggest that the average heat flow in regions of rough topography may have to be raised. To correct for this possible bias, the average for the younger provinces has been increased by a conservative 15 per cent and shown as open circles in Fig. 38–15.

The theoretical elevations for 75- and 125-km thick lithospheres effectively bracket the observed topography (Fig. 38–15). A slab of lithosphere 100 km thick with a basal tempera-ture of 1475°C spreading at 5 cm yr^{-1} gives a reasonable fit to the topographic section along 20° N. There is strong experimental (Sclater et al., 1970) and theoretical (Oxburgh and Turcotte, 1969) evidence that we have under-estimated the bias in heat flow for the crestal zone. If the lowering of the heat flow is as much as 50 per cent for the provinces close to the crest, a slab, 100 km thick, moving at 4 cm yr^{-1} would give a theoretical anomaly con-sistent with the heat flow observations near the crest of the ridge. Further from the ridge, the curve for a spreading rate of 5 cm yr^{-1} would adequately account for the data. A combina-tion of the two spreading rates for a lithosphere 100 km thick would then give a reasonable fit to the observed heat flow. As the theoretical topographic profile is less sensitive to changes in spreading rate, this model would also fit the topographic data. Differences in spreading rate

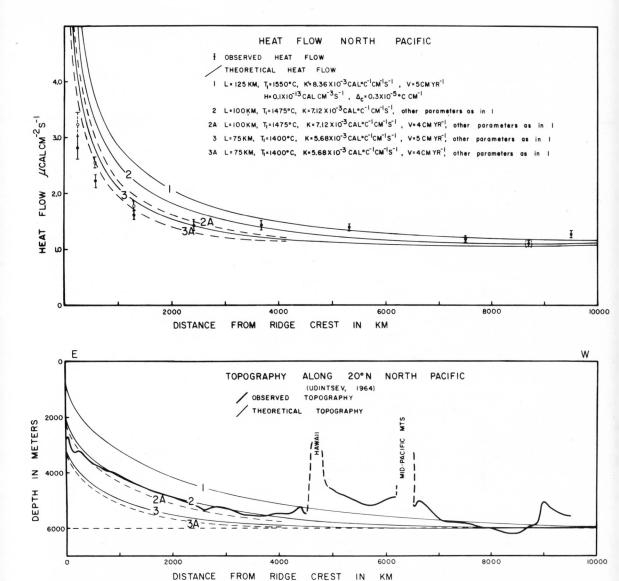

Figure 38-15

(a) Comparison of observed heat flow averages in the North Pacific with five theoretical profiles. (b) Comparison of observed topography (Udintsev 1964) with five theoretical profiles. L is the thickness of the lithosphere, T_1 is the basal temperature, K is the thermal conductivity, v is the half spreading rate of the slab, H is the rate of internal heat generation and Δ_c is the adiabatic temperature gradient. The open circles represent an increase of 15 per cent in the heat flow for the three provinces close to the crest of the ridge.

1   $L = 125$ km, $T_1 = 1550$ °C, $K = 8.36 \times 10^{-3}$ cal °C^{-1} cm^{-1} s^{-1},
     $V = 5$ cm yr^{-1}, $H = 0.1 \times 10^{-13}$ cal cm^{-3} s^{-1}, $\Delta_c = 0.3 \times 10^{-3}$ °C cm^{-1}
2   $L = 100$ km, $T_1 = 1475$ °C, $K = 7.12 \times 10^{-3}$ cal °C^{-1} cm^{-1} s^{-1}, other parameters as in 1
2A $L = 100$ km $T_1 = 1475$ °C, $K = 7.12 \times 10^{-3}$ cal °C^{-1} cm^{-1} s^{-1},
     $V = 4$ cm yr^{-1}, other parameters as in 1
3   $L = 75$ km, $T_1 = 1400$ °C, $K = 5.68 \times 10^{-3}$ cal °C^{-1} cm^{-1} s^{-1},
     $V = 5$ cm yr^{-1}, other parameters as in 1
3A $L = 75$ km, $T_1 = 1400$ °C, $K = 5.68 \times 10^{-3}$ cal °C^{-1} cm^{-1} s^{-1},
     $V = 4$ cm yr^{-1}, other parameters as in 1

have little effect on the tail of the theoretical profiles (Fig. 38–15). Thus, it is not necessary to compute a curve of the expected heat flow and topography for the faster spreading in the older regions. A point of interest is that in the region comprising the Hawaiian chain and the mid-Pacific mountains the anomalous topography is accompanied by a heat flow slightly higher than that in the surrounding deep ocean basins.

An attempt is made to explain the heat flow and topographic relief in the South Atlantic (Fig. 38–16). In this ocean, the spreading rate has been roughly constant since the early Cretaceous (Maxwell, 1969). A model of oceanic lithosphere 100 km thick would adequately explain the topographic profile V-22 (Dickson et al., 1968). This profile was chosen because it is long, south of the heavily sedimented Argentine Basin and it does not cross either the Walvis or Rio Grande Ridges. The scatter in the observed heat flow in each province and the possible bias of the heat flow average in the regions of rough topography prevent us from selecting a model of lithosphere from the heat flow measurements. It is encouraging that a good fit to the ridge topography is obtained by taking the same model of the lithosphere as that for the North Pacific. Further, the theoretical topography has the same basal depth of 6000 m in both oceans. This suggests that the model for the Pacific may be valid for all oceans.

DISCUSSION OF THE OCEANIC MODEL

Because of the uncertainty in the thermal conductivity, a model of lithosphere in which a small amount of radioactive heat generation is assumed cannot be separated from one which is totally depleted in heat producing elements. However, the two models which were suggested to explain the equality of heat flow through the shields and old oceans would have different chemical compositions. A lithosphere which contributes significantly to the surface heat flow is likely to have a composition close to that of pyrolite with a thin lid of ultrabasic and basaltic rocks. Within the framework of plate tectonics, where the lithosphere is envisaged as the lowest melting fraction of the upper mantle, this model implies a two-stage differentiation process with increasing enrichment in the lighter elements. Pyrolite II is the parent of pyrolite I which in turn gives rise to the basaltic and ultrabasic layers. A depleted lithosphere consisting mainly in ultrabasics can be derived directly from the more primitive pyrolite II. The differentiation history of such a model is simpler.

In essence, the depleted model assumes an efficient convective process in the upper mantle which maintains a constant temperature at the bottom of the plate. The enriched model assumes a constant supply of heat from the deep mantle and requires an efficient differentiation process. These two contrasting models are essentially the same as those proposed by Bullard (1952) rephrased to fit modern ideas about the oceans. In spite of the difficulty in selecting a preferred geochemical model, the previous analysis has shown that the heat flow anomaly and topography of ridges can be accounted for satisfactorily by the creation of oceanic lithosphere. The thickness of the lithosphere is of the order of 100 km and the base of the lithosphere may have the temperature of incipient melting of pyrolite.

However, the basis for the previous analysis was an upper mantle with no phase changes. Though relevant experimental data on the sub-solidus $P-T$ stability fields is sparse, high pressure laboratory work in the past decade has shown that several phases are likely to be present in any model of the upper mantle. It is important to evaluate the effect of these phase changes on the surface elevation. To compute the increase in elevation resulting from the presence of several phases in a 75-km thick lithosphere, it is necessary, if we assume the bulk of the lithosphere to have the composition of pyrolite (Fig. 38–12), to consider a 'wet' upper mantle. In the presence of water, the pyrolite solidus is lowered and the temperature at the base of the lithosphere would be correspondingly decreased. The 'wet' pyrolite sub-solidus stability fields of Green (1969; Fig. 4) were plotted on the temperature structure of a 75-km thick oceanic lithosphere with a basal temperature of 1300°C. This temperature is the incipient melting of 'wet' pyrolite at that depth. As shown in Appendix I the effect on the surface elevation of density variations due

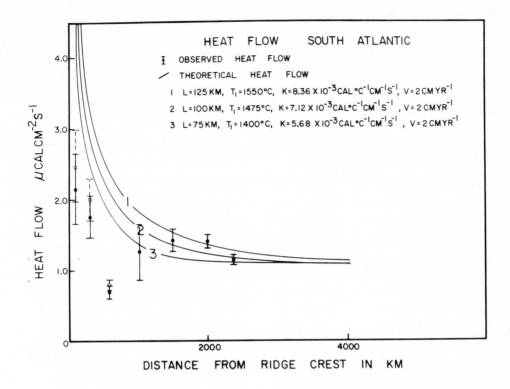

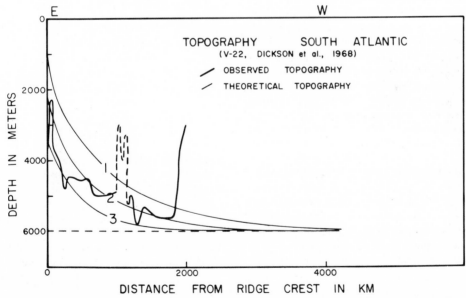

Figure 38-16
(a) Comparison of observed heat flow averages in the South Atlantic and three theoretical profiles. (b) Comparison of observed topography in the South Atlantic V-22 (Dickson et al., 1968) and three theoretical profiles. Symbols are defined in Appendix I and the caption of Figure 38-15

1 $L = 125$ km, $T_1 = 1550$ °C, $K = 8.36 \times 10^{-3}$ cal °C^{-1} cm^{-1} s^{-1},
  $V = 2$ cm yr^{-1}
2 $L = 100$ km, $T_1 = 1475$ °C, $K = 7.12 \times 10^{-3}$ cal °C^{-1} cm^{-1} s^{-1},
  $V = 2$ cm yr^{-1}
3 $L = 75$ km, $T_1 = 1400$ °C, $K = 5.68 \times 10^{-3}$ cal °C^{-1} cm^{-1} s^{-1},
  $V = 2$ cm yr^{-1}

to thermal expansion and phase changes can be added. Because of the lack of information on the thermal expansion of mantle materials, we have assumed the basalt, the ultrabasics and the pyrolite, all to have the same coefficient of thermal expansion. Rather arbitrarily, α was taken to be the coefficient of thermal expansion of pure olivine at high temperatures (Skinner, 1966). The increase of elevation due to phase changes is of the order of 1 km at the crest of the ridge. Were the lithosphere 100 km thick, the additional elevation would be even greater and the theoretical profile would be considerably above the observed topography. Though the theoretical elevation for the 75-km thick lithosphere does not match the observed topography, a small variation in the high pressure and temperature fields for the plagioclase, pyroxene or garnet assemblages could easily yield a distribution of density contrast that would give a better fit. Further the stability fields at low temperatures and pressures are not well established. For example, it is possible that amphibole pyrolite be stable below the lid of ultrabasics. Considering these uncertainties, the difference between the theoretical and observed profiles (Fig. 38–17b) is not large enough to rule out a 75-km thick lithosphere. A slab of this thickness moving at 5 cm yr^{-1} gives a theoretical heat flow curve which lies close to the observed profile at distances greater than 2000 km from the ridge crest. A theoretical profile was not computed for the lower spreading rate near the axis. It can be seen from Fig. 38–15a that a decrease in spreading rate shifts the theoretical heat flow curve closer to the ridge axis. With a rate of 4 cm yr^{-1}, a fit to the heat flow near the crest could be obtained by assuming a bias in the observed values of no more than 30 per cent. Besides giving a better fit to the observed heat flow and topographic data, a 75-km thick litho-

sphere has a more realistic composition and is preferred from other lines of evidence such as surface wave dispersion studies (Kanamori and Abe, 1968) and highly precise group velocity data (Press and Kanamori, 1970).

This analysis was carried out for the composition of the enriched model where pyrolite is the major constituent of the lithosphere. MacGregor's (1968) P−T stability fields for plagioclase, spinel and garnet peridotite plotted on the temperature structure of a 75-km thick lithosphere would yield an increase in elevation comparable to that produced by the pyrolite assemblages of Green (1969). Thus, phase changes are just as important in a depleted as in an enriched model.

The rate of heat loss from the Earth due to the production of lithosphere can be computed once the temperature structure in the slab is known (McKenzie and Sclater, 1969). The derivation for the heat loss is given in Appendix I. If:

$$\rho = 3.3 \text{ g cm}^{-3}$$
$$L = 75 \times 10^5 \text{ cm}$$
$$V = 3 \text{ cm yr}^{-1}$$
$$H = 0.10 \times 10^{-13} \text{ cal cm}^{-3} \text{ s}^{-1}$$
$$K = 0.066 \text{ cal } °\text{C}^{-1} \text{ cm}^{-1} \text{ s}^{-1}$$
$$C_p = 0.25 \text{ cal g}^{-1} °\text{C}^{-1}$$
$$d = 5 \times 10^9 \text{ cm}$$
$$\alpha = 4 \times 10^{-5} °\text{C}^{-1}$$
$$T_1 - T_0 = 1300 °\text{C}$$
$$\Delta_c = 3.0 \times 10^{-6} °\text{C cm}^{-1}$$

then equation (23) of Appendix I yields a heat loss, for the production of lithosphere by all ridges, of

$$Q = 3.6 \times 10^{12} \text{ cal s}^{-1}.$$

The total terrestrial heat loss is 7.7×10^{12} cal s^{-1} (Lee and Uyeda, 1965). The excess heat

Figure 38-17 *(facing page)*
(a) Comparison of observed heat flow averages in the North Pacific with the theoretical profile for a 75 km thick lithosphere. (b) Comparison of observed topography (solid line) (Udintsev, 1964) with two theoretical profiles. The upper dashed curve is the profile expected from the thermal expansion and the phases of the model shown in (c). The lower dashed curve is the profile assuming thermal expansion of a lithosphere of uniform density. (c) Isotherms and chemical zoning of a 75 km thick lithosphere moving at 5 cm yr^{-1} to the right. The parameters are $T_1 = 1300°\text{C}$, $K = 6.15 \times 10^{-3}$ cal$°\text{C}^{-1}$ cm^{-1} s^{-1}, $H = 0.1 \times 10^{-13}$ cal cm^{-3} s^{-1}, $\Delta_c = 0.3 \times 10^{-5} °\text{C}$ cm^{-1} (see Figure 38-12 for explanation). The densities at $0°\text{C}$ for plagioclase, pyroxene and garnet pyrolite are 3.25, 3.33 and 3.38 g cm^{-3} (Ringwood, 1969). $\bar{\rho}_0$ (see Appendix I) is taken as 3.3 g cm^{-3}. The phase boundaries are for a "wet" pyrolite (Green, 1969; Fig. 4)

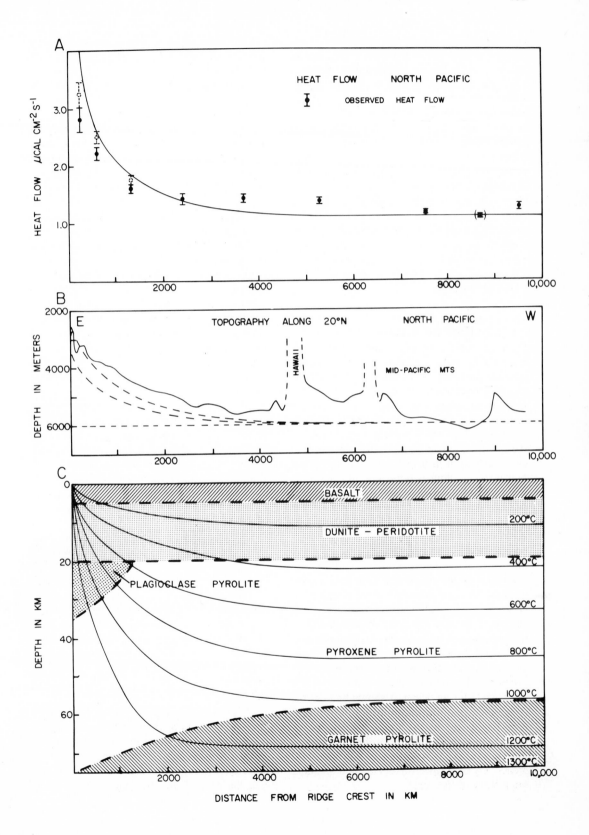

A

HEAT FLOW NORTH PACIFIC

● OBSERVED HEAT FLOW

HEAT FLOW μCAL CM^{-2}S^{-1}

3.0

2.0

1.0

2000   4000   6000   8000   10,000

B

2000

E   TOPOGRAPHY ALONG 20°N   NORTH PACIFIC   W

DEPTH IN METERS

4000

HAWAII

MID-PACIFIC MTS

6000

2000   4000   6000   8000   10,000

C

0

BASALT

200°C

DUNITE – PERIDOTITE

20

400°C

PLAGIOCLASE PYROLITE

600°C

DEPTH IN KM

40

PYROXENE PYROLITE

800°C

1000°C

60

GARNET PYROLITE

1200°C

1300°C

2000   4000   6000   8000   10,000

DISTANCE FROM RIDGE CREST IN KM

produced by creation of the lithosphere is approximately 45 per cent of the total average heat loss of the Earth. The only way this estimate can be reduced significantly is to decrease either the thickness of the lithosphere or the temperature at its base or both. At present there seem to be no strong arguments to reduce these parameters. It is likely that the creation of oceanic lithosphere has occurred during a significant portion of geologic history. The loss of heat by this process is large and may represent a major mode by which the Earth has cooled. It must be considered in any attempt to unravel the thermal history of the Earth.

CONCLUSIONS

The decrease of heat flow with age is well established for the continents (Polyak and Smirnov, 1968; Smirnov, 1968). The currently accepted hypothesis of plate tectonics predicts a similar decrease for the oceans but on a much shorter time scale (Langseth et al., 1966; McKenzie, 1967). The heat flow observations in the North Pacific support this prediction. More heat flow measurements are needed in the other oceans to test whether the continuous decrease in heat flow with the age of the North Pacific floor is typical of all oceans. In these oceans more emphasis should be placed on measurements away from the crest of ridges and each region of different age should be given equal weight. The oldest provinces appear to be the most revealing in evaluating the flow of heat through the upper mantle. Measurements in the provinces close to the crest may be biased by the necessity of taking these measurements in sediment ponds. In the Pacific more detailed surveys should be made in these provinces to establish: (a) The effect of environment on the measurement of heat flow; and (b) the actual flow of heat through these provinces. If a reliable heat flow curve can be established for an ocean, it should be possible to tell the age of the ocean floor from the heat flow and also to test whether one model of the lithosphere can explain the heat flow decrease in all oceans. Thus, oceanic heat flow measurements may be a powerful tool in understanding the history of the oceans.

The two models which explain the near equality of heat flow through the continental shields and the old ocean basins within the framework of plate tectonics require that the base of the oceanic plate be at a depth of between 75 and 100 km and that the continental plate be almost twice as thick. In both models, at depths of approximately 100 km, the temperatures are significantly higher under the oceans. This temperature distribution should be checked from induction studies. It would be of interest to compare transient variations in the magnetic field on the Australian shield with the same studies on the sea floor to the south.

The fit of the theoretical profiles to the observed heat flow and topography in the North Pacific depend critically on our knowledge of many crucial parameters. The computation of elevation is very sensitive to the choice of composition, the coefficient of thermal expansion, the basal temperature, and the $P - T$ stability fields of subsolidus mineral assemblages. None of these are well known. The largest uncertainty of all is perhaps the chemical nature of the lower part of the lithosphere and the upper mantle. The presence of water has also been recognized as a very important factor for, as well as lowering the solidus temperature, it would also alter the stability fields of the phases present in the lithosphere. Considering these uncertainties; a 75-km thick oceanic lithosphere may well be found consistent with heat flow and topography of the North Pacific.

The difficulty in separating models with varying radioactive content arises through the uncertainty in the thermal conductivity K. However, there are several reasons for preferring the depleted model. A lithosphere with a low radioactivity will have a higher thermal conductivity. Recent work on a forsterite-rich olivine by Fukao (1969) has shown that, for the temperatures and pressures likely to exist in the lower part of the lithosphere, the thermal conductivity of ultrabasic rocks may be as high as 10.0×10^{-3} cal $°C^{-1}$ cm^{-1} s^{-1}. The observed heat flow and elevation across the East Pacific Rise are better fit by theoretical models assuming a lithosphere with a thickness of 75 km if several phases are present. The surface wave dispersion data in the North Pacific also imply that the base of the lithosphere be at a depth

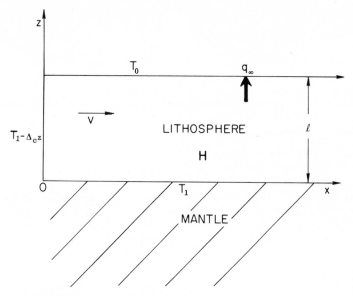

Figure 38-18
Simple intrusive model (after McKenzie, 1967)

close to 75 km. An enriched model of this thickness would have a heat production of 0.45×10^{-13} cal cm^{-3} s^{-1} for pyrolite II. Though not impossible such a high value seems unlikely. The depleted model is more appealing because it has a simpler differentiation history and its parameters lie comfortably within the allowed ranges rather than at the extreme ends.

Rocks dredged from the fracture zones that cut deeply through the oceanic crust (Engel and Fisher, 1969) may yield information concerning the nature of the oceanic lithosphere. Careful dredging of these fracture zones followed by extensive chemical, petrological, trace element, and heat production studies are essential to set constraints on possible models of the oceanic lithosphere.

APPENDIX I

Temperature and Heat Flow

The general equation for the temperature within a moving material is

$$\rho C_p \left[\frac{\partial T}{\partial t} + \mathbf{v} \cdot \nabla T \right] = K \nabla^2 T + H \quad (1)$$

where $\mathbf{v}$ is the velocity of the material, ρ its density, C_p its specific heat at constant pressure, K the thermal conductivity and H the rate of internal heat generation. For the case shown in Fig. 38–18 where $\mathbf{v}$ is directed along the x-axis equation (1) becomes:

$$\rho C_p \, v \, \frac{\partial T}{\partial x} = K \left(\frac{\partial^2 T}{\partial x^2} + \frac{\partial^2 T}{\partial z^2} \right) + H . \quad (2)$$

The time dependent term $\partial T/\partial t$ has been neglected for two reasons: It simplifies the solutions and, for the oceans considered, an approximately steady state has been reached. Following the notation of McKenzie (1967) we introduce the following non-dimensional variables:

$$\left. \begin{array}{l} x = l x' \\ z = l z' \\ T = (T_1 - T_0) T' + T_0 \\ H = A_1 A' \end{array} \right\} (3)$$

substitution into (2) gives:

$$\frac{\partial^2 T'}{\partial x'^2} - 2R \frac{\partial T'}{\partial x'} + \frac{\partial^2 T'}{\partial z'^2} - B_1 A' = 0 \quad (4)$$

where R is the thermal Reynolds number.

$$R = \rho C_p \frac{vl}{2K}. \quad (5)$$

The boundary conditions for this problem are those of McKenzie (1967) except that at $x' = 0$:

$$
\left. \begin{array}{lll}
z' = 0 & T' = 1 & \\
z' = 1 & T' = 0 & \text{for all } x' \\
x' = 0 & T' = 1 - \dfrac{\Delta_c\, lz'}{T_1 - T_0} = 1 - Dz'
\end{array} \right\} (6)
$$

Δ_c, 0.3°C km^{-1}, is the adiabatic temperature gradient. The rate of internal heat generation, H, is assumed constant throughout the plate. The solution of equation (4) under these conditions is:

$$
T' = 1 + \frac{B_1 A'}{2} z'^2 - \left(1 + \frac{B_1 A'}{2}\right) z'
$$

$$
+ \sum_{n=1}^{\infty} A_n \exp\{[R - \sqrt{(R^2 + n^2\pi^2)}]x'\} \sin n\pi z'
$$

where

$$
A_n = \frac{2(-1)^{n+1}}{n\pi}[1 - D]
$$

$$
+ \frac{2B_1 A'}{n^3\pi^3}[1 - (-1)^n] \tag{7}
$$

and

$$
-\left(\frac{\partial T'}{\partial z'}\right)_{z'=1} = 1 - \frac{B_1 A'}{2} + 2 \cdot \tag{8}
$$

$$
\left\{ \begin{array}{l}
\displaystyle\sum_{n=1}^{\infty}\left(1 - D + \frac{2B_1 A'}{n^2\pi^2}\right)\exp\{[R - \sqrt{(R^2 + n^2\pi^2)}]x'\} \\
n \text{ odd} \\[6pt]
\displaystyle\sum_{n=2}^{\infty}(1 - D)\exp\{[R - \sqrt{(R^2 + n^2\pi^2)}]x'\}. \\
n \text{ even}
\end{array} \right.
$$

If we neglect all terms in the sum except $n = 1$ and remark that $R^2 \gg \pi^2$ we have:

$$
-\left(\frac{\partial T'}{\partial z'}\right)_{z'=1} \simeq 1 - \frac{B_1 A'}{2}
$$

$$
+ 2\left(1 - D + \frac{2B_1 A'}{\pi^2}\right)\exp\left(\frac{-\pi^2 x'}{2R}\right). \tag{9}
$$

If q_∞ is the heat flow measured at the surface of the plate at great distance from the crest of the ridge (9) yields:

$$
q_\infty \simeq \frac{K(T_1 - T_0)}{l} + \frac{Hl}{2} \tag{10}
$$

q_∞ is measured and H and l are assumed in the geochemical model chosen for the oceanic model. T_1 is the temperature of incipient melting of pyrolite at depth l. Under these conditions the thermal conductivity has to be equal to:

$$
K = \frac{l}{T_1 - T_0}\left(q_\infty - \frac{Hl}{2}\right). \tag{11}
$$

If

$$
\begin{aligned}
T_0 &= 0°C \\
l &= 1.0 \times 10^7 \text{ cm} \\
H &= 0.1 \times 10^{-13} \text{ cal cm}^{-3}\text{ s}^{-1} \\
q_\infty &= 1.1 \times 10^{-6} \text{ cal cm}^{-2}\text{ s}^{-1}
\end{aligned}
$$

then

$$
T_1 = 1475°C
$$

and

$$
K = 7.1 \times 10^{-3} \text{ cal°C}^{-1}\text{ cm}^{-1}\text{ s}^{-1}.
$$

If the half-width $\delta_{\frac{1}{2}}'$ of the heat flow anomaly is the value of x' such that:

$$
-\left(\frac{\partial T'}{\partial z'}\right)_{z'=1} = 2
$$

then:

$$
\delta_{\frac{1}{2}}' = \frac{2R}{\pi^2}\log_e\left[\frac{4[5K(T_1 - T_0 - l\Delta_c) - Hl^2]}{5[2K(T_1 - T_0) - Hl^2]}\right]. \tag{12}
$$

But:

$$
\delta_{\frac{1}{2}}' = \frac{\delta_{\frac{1}{2}}}{l} = \frac{vt_{\frac{1}{2}}}{l}
$$

so that the heat flow should be

$$
q_{\frac{1}{2}} = \frac{2K(T_1 - T_0)}{l}
$$

at a time

$$
t_{\frac{1}{2}} = \frac{\rho C_p l^2}{K\pi^2}\log_e\left[\frac{4[5K(T_1 - T_0 - l\Delta_c) - Hl^2]}{5[2K(T_1 - T_0) - Hl^2]}\right) \tag{13}
$$

after the time of intrusion at the axis of the ridge of the corresponding material. This provides a means of checking directly if this radioactive model agrees with the curve of average heat flow within a province versus the age of the province.

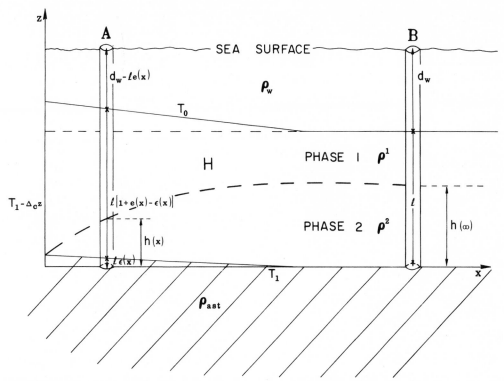

Figure 38-19
Isostatic model of the oceanic lithosphere. A and B are two columns of unit area for the base and equal mass. When no phase changes are considered the density of the lithosphere is ρ_{slo}

Topography

If the elevation of the ridges above the surrounding ocean basins is due chiefly to thermal expansion of the lithosphere, this elevation can be computed (McKenzie and Sclater, 1969). If isostatic compensation takes place at the bottom of the slab of lithosphere at great distance from the crest of the ridge and if the mass of the lithosphere is conserved the shape of the slab must be as shown in Fig. 38–19. The mass of the two columns A and B can be equated:

$$\rho_w d_w + \int_0^l \rho_{slo}[1 - \alpha T_\infty(z)] \, dz$$

$$= \rho_w[d_w - le(x)]$$

$$+ \int_{l\varepsilon(x)}^{l[1+e(x)]} \rho_{slo}[1 - \alpha T_x(z)] \, dz + \rho_{ast} l\varepsilon(x) \quad (14)$$

where ρ_w is the density of seawater, ρ_{slo} the density of the lithosphere at 0°C, ρ_{ast} the density of the asthenosphere, α the thermal expansion coefficient, d_w the normal depth of an ocean basin far from a ridge crest, $le(x)$ the elevation of the ridge above the basin and $l\varepsilon(x)$ the elevation of the bottom of the lithosphere above the level of isostatic compensation.

Substitution of $T_\infty(z)$ and $T_x(z)$ from (7) into (14) and integration with respect to z yields:

$$e(x) = \frac{\alpha \rho_{slo}(T_1 - T_0)}{(\rho_{slo} - \rho_w) - \alpha \rho_{slo} T_0 + \alpha \rho_w T_1}$$

$$\cdot \sum_{k=0}^{\infty} \left[\frac{4(1-D)}{(2k+1)^2 \pi^2} + \frac{8B_1 A'}{(2k+1)^4 \pi^4} \right]$$

$$\cdot \exp\left\{ [R - \sqrt{(R^2 + (2k+1)^2 \pi^2)}] \frac{x}{l} \right\} \quad (15)$$

if terms in $e^2(x)$ are neglected and $\rho_{ast}\varepsilon(x)$ $= \rho_w e(x)$ with $\rho_{ast} \simeq \rho_{slo}$. The elevation $le(0)$ at the crest of the ridge is:

$$le(0) \simeq \frac{\alpha l \rho_{slo}(T_1 - T_0)}{\rho_{slo} - \rho_w}\left[\tfrac{1}{2}(1 - D) + \frac{8B_1A'}{100}\right]$$

(16)

since

$$\sum_{k=0}^{\infty} \frac{1}{(2k+1)^2} \simeq 1.23, \qquad \sum_{k=0}^{\infty} \frac{1}{(2k+1)^4} \simeq 1.01$$

and

$$\rho_{slo} - \rho_w \gg \alpha \rho_{slo} T_0.$$

Substitution of:

$$\rho_{slo} = 3.3 \text{ g cm}^{-3}$$
$$\rho_w = 1 \text{ g cm}^{-3}$$
$$\alpha = 4 \times 10^{-5}°\text{C}^{-1}$$
$$T_1 = 1475°\text{C}$$
$$T_0 = 0°\text{C}$$
$$l = 1.0 \times 10^7 \text{ cm}$$
$$H = 0.1 \times 10^{-13} \text{ cal cm}^{-3} \text{ s}^{-1}$$
$$K = 7.1 \times 10^{-3} \text{ cal}°\text{C}^{-1} \text{ cm}^{-1} \text{ s}^{-1}$$
$$\Delta_c = 0.3 \times 10^{-5}°\text{C cm}^{-1}$$

gives:

$$B_1A' = \frac{-Hl^2}{K(T_1 - T_0)} = -0.095$$

$$D = \frac{l\Delta_c}{T_1 - T_0} = 0.022$$

so that:

$$le(0) = 4.0 \times 10^5 \text{ cm} = 4000 \text{ m}.$$

Topography with Phase Changes

If we assume a chemical composition for the lithosphere, the effect of phase changes on the topography can be computed. The sub-solidus P–T stability fields are plotted on a cross section of the lithosphere. For simplicity we will consider the case where only two phases are stable in the lithosphere. The two phases are assumed to have the same coefficient of thermal expansion. Their densities are:

$$\left.\begin{array}{l} \rho_0{}^1 = \bar\rho_0 + \delta\rho_1 \\ \rho_0{}^2 = \bar\rho_0 + \delta\rho_2 \end{array}\right\}$$

(17)

where $\delta\rho_1$ and $\delta\rho_2$ are small compared with $\bar\rho_0$, the mean density at 0°C. Again assuming isostatic compensation takes place at the bottom of the lithosphere (Fig. 38–19) we can write:

$$\rho_w d_w + \int_0^l (\bar\rho_0 + \delta\rho)[1 - \alpha T_\infty(x)]\, dz$$
$$= \rho_w[d_w - le_1(x)] + \rho_{ast}le_1(x)$$
$$+ \int_{le_1(x)}^{l[1+e_1(x)]} (\bar\rho_0 + \delta\rho)[1 - \alpha T_x(z)]\, dz. \quad (18)$$

Evaluating the integrals for intervals where the density $\bar\rho_0 + \delta\rho$ remains constant gives:

$$e_1(x) = \frac{\alpha\bar\rho_0(T_1 - T_0)}{\bar\rho_0 - \rho_w - \sigma\bar\rho_0 T_0 + \delta\rho_1}$$
$$\cdot \left\{\sum_{k=0}^{\infty}\left[\frac{4(1 - D)}{(2k+1)^2\pi^2} + \frac{8B_1A'}{(2k+1)^4\pi^4}\right]\right.$$
$$\left.\cdot \exp\left\{[R - \sqrt{(R^2 + (2k+1)^2\pi^2)}]\frac{x}{l}\right\}\right\}$$
$$+ \frac{(\delta\rho_2 - \delta\rho_1)(h(\infty) - h(x))}{(\bar\rho_0 - \rho_w - \alpha\bar\rho_0 T_0 + \delta\rho_1)l}$$

(19)

if all terms in $\alpha\delta\rho$ are ignored. In this expression $h(x)$ is the height above the level of isostatic compensation at which the phase transformation takes place.

If we write $e_1(x) = d(x) + \delta e(x)$ equation (19) gives:

$$\delta e(x) = \frac{(\delta\rho_2 - \delta\rho_1)}{(\bar\rho_0 - \rho_w - \alpha\bar\rho_0 T_0 + \delta\rho_1)}\frac{(h(\infty) - h(x))}{l}$$

(20)

and $e(x)$ has the same form as the expression, equation (15), found for thermal expansion of a medium of density $\bar\rho_0$ at 0°C since $\delta\rho_1$ is small in comparison with $\bar\rho_0 - \rho_w$.

This indicates that the elevation $le_1(x)$ due to phase changes and thermal expansion of the lithosphere can be calculated by adding the elevation $l\delta e(x)$ due to phase changes alone for a uniform temperature to the elevation $le(x)$ due to thermal expansion alone for a medium of uniform density. This result is valid only to first order and requires that the phases vary little in density and have the same coefficient of thermal expansion.

Heat Loss Due to Production of Lithosphere

McKenzie and Sclater (1969) have shown that the rate of heat loss due to creation of lithosphere can be computed by considering the change in heat content of the lithosphere as it moves from the axis to the flanks of the ridge. When the lithosphere is produced it contains:

$$2\left\{\rho_{slo}C_p l\left[T_1(1-\alpha T_1)\right.\right.$$

$$\left.\left.+\Delta_c\left(\alpha l T_1 - \frac{l}{2} - \frac{\alpha\Delta_c l^2}{3}\right)\right]+Hl\right\}$$

$$\text{cal unit area}^{-1} \quad (21)$$

At great distance from the ridge the heat content, neglecting terms in $\alpha B_1 A'$, is:

$$2\left\{\rho_{slo}C_p l\left[\frac{T_1+T_0}{2} - \frac{B_1 A'}{12}(T_1-T_0)\right.\right.$$

$$\left.\left.-\frac{\alpha}{3}(T_1^2+T_0^2+T_1 T_0)\right]+Hl\right\}$$

$$\text{cal unit area}^{-1}. \quad (22)$$

The heat lost by plate creation for a total length of ridges of d cm and a mean spreading velocity of v cm yr^{-1} is:

$$2\rho_{slo}C_p ldv\left[\frac{T_1-T_0}{2} + \frac{B_1 A'}{12}(T_1-T_0)\right.$$

$$-\frac{\alpha}{3}(2T_1^2-T_0^2-T_1 T_0)$$

$$\left.+\Delta_c\left(\alpha l T_1 - \frac{l}{2} - \frac{\alpha\Delta_c l^2}{3}\right)\right]\text{cal s}^{-1}.$$

$$(23)$$

39

Global Gravity and Tectonics

WILLIAM M. KAULA
1972

From *The Nature of the Solid Earth,* ed. by E. C. Robertson, J. F. Hays, and L. Knopoff, p. 385–405, 1972. Reprinted with permission of the author and McGraw-Hill, Inc.

This paper is a continuation of Kaula (1969), which attempted a tectonic classification of the main features of the earth's gravitational field. On the basis of magnitude and extent of mean gravity anomalies for 5° squares, 19 areas on the earth were selected as markedly positive, 14 as markedly negative, 10 as exceptionally mild. Other geological and geophysical data for each of these 43 areas were examined. On the basis of certain patterns of correlation, 11 types of areas were defined. It was found that, where characteristics of different types appeared, certain characteristics were dominant over others. In general, characteristics associated with positive anomalies were dominant over those associated with mild or negative anomalies, and characteristics associated with recent tectonics dominant over those associated with ancient. The 11 types in order of dominance, with sign and a leading example of each given in parentheses, were: trench and island arc (+, Indonesia-Philippines), Cenozoic oceanic flood basalts (+, Iceland-North Atlantic), Cenozoic orogeny with Quaternary extrusives (+, Caucasus), Quaternary glaciation (−, Canadian Shield), vigorous ocean rise (0, southeast Pacific), current orogeny without extrusives (−, Himalayas), ocean basin (−, Somali-Arabian), continental basin (−, Parnaiba Basin), pre-Cenozoic orogeny (0, eastern United States), continental shield (0, Brazilian Shield), pre-Cenozoic ocean flood basalts (0, Darwin Rise). The strongest correlation found was between positive gravity anomalies and Quaternary volcanism. Positive correlation of gravity anomalies with topography residual to a fifth-degree figure was almost universal. The extent to which the different area types relate to the global tectonics inferred from paleomagnetic and seismic data varied from strong (trench and island arc, vigorous ocean rise) to negligible (Cenozoic oceanic flood basalts, Quaternary glaciation). The lack of systematic correlation between temperature indicators (heat flow, P_n velocities, seismic station delays) and gravity anomalies indicated that horizontal variations in petrology are significant.

Since Kaula (1969), there has been a major improvement in the determination of the global gravity field by Gaposhkin and Lambeck (1971). In this paper, we first examine this improved determination, and attempt to transform it so as to be most useful for geophysical interpretation. Then we hypothesize mechanisms by which the main features of the gravitational field are maintained.

THE DATA

Figure 39–1 of this paper differs from Fig. 1 of Kaula (1969) in four significant respects:

1. The new determination of the gravity field by Gaposchkin and Lambeck (1971) is used. This analysis is based primarily on the orbits of 21 artificial satellites and secondarily on mean gravity anomalies for 5° by 5° squares covering 56% of the earth.

2. The gravimetry is the same as that used by Kaula (1966), but the manner of combination of data in effect gives higher weight to the satellites than in the 1966 analysis.

3. The results are given in the form of spherical harmonic coefficients of the potential, complete through the 16th degree, (plus 33 coefficients of higher degree to which satellite orbits are sensitive), rather than area means. Hence, the resolution, or shortest half wavelength represented, is about 11° or 1200 km.

4. The free air anomalies in Fig. 39–1 are referred to the figure of hydrostatic equilibrium, an ellipsoid of flattening 1/299.8, in accord with the explanation of Goldreich and Toomre (1969) for the excess oblateness.

There are two major effects of these changes:

1. The improved resolution results in the breakup of the two largest features in the southern oceans. The large area of mild anomaly in the South Pacific is now resolved into two negative areas with a positive area between, the former over basins and the latter along the East Pacific Rise. In the area between Africa and Antarctica, a single large positive feature centered in the "vee" between

the two rises is now divided into two positive features over the rises and an area of mild anomaly between. In general, most of the ocean rises are now positives, rather than "mild" features.

2. The use of the hydrostatic flattening results in the intensification of the negative anomalies in the glaciated areas near the poles: at the South Pole to an extent that is much greater than can be imputed to glacial loading.

Lesser effects are the appearance of the highest Himalayas as a small positive belt; the reduction of an overlap of the southeast Indian Ocean rise by the south Australian basin negative anomaly; the emphasis of the positive belt from the Carpathians to Iran; the reduction of the East Mediterranean negative; and the reduction or removal of positive features in areas of slight recent tectonic activity in northeast USSR, the Central Pacific, and Australia.

Figure 39–2 is the corresponding isostatic anomaly map, using the spherical harmonic expansion of the Airy-Heiskanen 30-km crust isostatic correction calculated by Uotila (1962). As usual, oceanic maxima and continental minima are enhanced in the isostatic map, but significant change in the pattern would occur only if the compensation were placed at, or below, asthenospheric depths.

As previously pointed out (Kaula, 1967), the correlation of gravity with topography is poor for the fifth and lower degrees. On the other hand, Hide and Malin (1970) have recently shown that the low degree harmonics of the gravity field have a high correlation with the corresponding harmonics of the magnetic field, provided that the latter is rotated 160° eastward. The obvious application of these facts for the purpose of interpreting upper mantle and crustal phenomena is to use a residual field. Figure 39–3 is the free air anomaly field calculated from spherical harmonic coefficients of degrees 6 through 16, and Fig. 39–4 is the corresponding isostatic anomaly field. In the four successive representations of Figs. 39–1 through 39–4, the correlation of ocean rises with positive anomalies appears more and more emphasized.

As discussed in Kaula (1969), it seems appropriate to analyze the gravity field in terms of reasonably contiguous blocks of anomaly x area, since, by the half-space application of

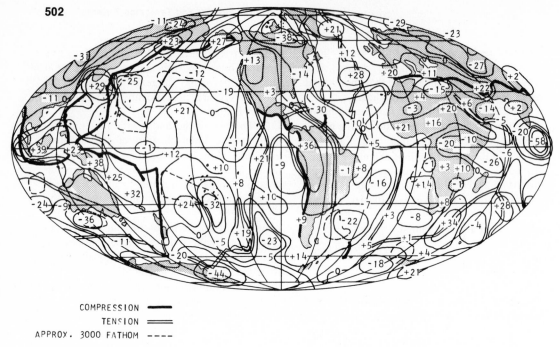

COMPRESSION ▬▬
TENSION ▭▭
APPROX. 3000 FATHOM ----

Figure 39-1
Free air anomalies in milligals referred to an ellipsoid of flattening 1/299.8. Calculated from the spherical harmonic coefficients of the gravitational field of degrees 2 through 16 of Gaposchkin and Lambeck (1971). (Non-zero contours enclosing only one value have been omitted on all figures.) Global tectonic lines of compression and tension from Isacks et al. (1968), and major basins indicated by approximate 3000 fathom line on all figures.

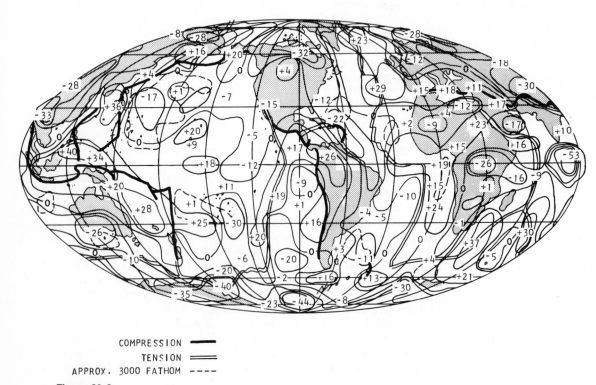

COMPRESSION ▬▬
TENSION ▭▭
APPROX. 3000 FATHOM ----

Figure 39-2
Isostatic anomalies in milligals referred to an ellipsoid of flattening 1/299.8. Airy-Heiskanen compensation with nominal crustal thickness of 30 km. Calculated from Figure 39-1, less the spherical harmonic coefficients for the isostatic correction of degrees 2 through 16 of Uotila (1962)

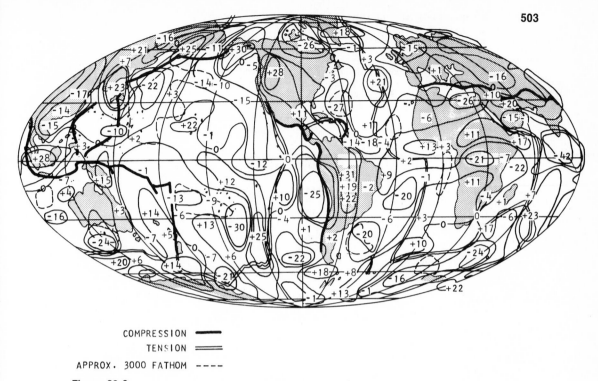

COMPRESSION ━━━
TENSION ═══
APPROX. 3000 FATHOM ----

Figure 39-3
Free air anomalies in milligals referred to a fifth-degree figure. Calculated from the spherical harmonic coefficients of the gravitational field of degrees 6 through 16 of Gaposchkin and Lambeck (1971)

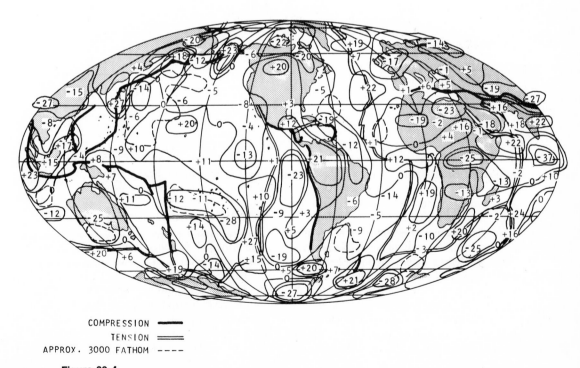

COMPRESSION ━━━
TENSION ═══
APPROX. 3000 FATHOM ----

Figure 39-4
Isostatic anomalies in milligals referred to a fifth-degree figure. From Figure 39-3, less the spherical harmonic coefficients for the isostatic correction of degrees 6 through 16 of Uotila (1962)

Table 39–1
Areas of exceptional gravity anomaly, defined as having an area × free air anomaly referred to the hydrostatic figure more than 50 mgal × 10^6 km^2 (1.17×10^{21} gm) in absolute magnitude, and absolute anomaly of more than 10 mgal throughout the area.

General location	Area (10^6 km^2)	Free air anomaly × area (mgal × 10^6 km^2)	Mean free air anomaly (mgal)	Type
POSITIVE FEATURES				
Sumatra–Philippines–Solomons	18.9	+461	+24	Arc
Andes–W. Amazon Basin	9.4	206	22	Arc–Orogenic
Solomons–Tonga–Kermadec	10.4	178	17	Arc
Mid-Indian Rise–Indian Antarctic Rise	9.4	175	19	Rise
Crozet Plateau–S. Madagascar Rise	6.3	117	19	Rise
Mexico–N. W. Colombia	7.3	107	15	Arc
Carpathians–Turkey–Iran	7.2	103	14	Orogenic
Hawaii	6.1	96	16	Shield
Azores Plateau	5.8	95	16	Rise
Japan–Bonins	4.6	92	20	Arc
Atlas–Iberia–W. Mediterranean	5.5	86	16	Arc–Orogenic
West Africa–Guinea Basin	6.2	85	14	Orogenic
Ahaggar–Tibesti–Nigeria	4.7	77	16	Orogenic
Greenland–Iceland–Norwegian Sea	4.6	76	16	Rise
East Pacific Rise, N. of Easter Island	4.2	64	15	Rise
Walvis Rise–S. W. Africa	4.5	+ 58	+13	Rise
NEGATIVE FEATURES				
Antarctica	22.4	−511	−23	Glac.–Basin?
Siberian Platform–Turkestan–Himalayas	15.1	289	19	Glac.–Orogenic
North Canada	10.1	218	22	Glaciated
N. American–Guiana Basins	11.7	212	18	Basin
Somali Basin–Central Indian Ocean	6.2	193	31	Basin
N. Pacific Basin–N. E. Pacific Slope	12.3	175	14	Basin
W. Australian Shield–S. Australian Basin	4.6	121	26	Basin
N. W. Pacific Basin	5.3	116	22	Basin
Wharton Basin	5.6	98	17	Basin
N. W. Siberia–Aleutian Basin	5.6	87	16	Glac.–Basin
Society Islands–S. W. Pacific Basin	3.2	69	22	Basin
Congo–Kenya	5.0	60	12	Basin–Rift
Argentine Basin	3.6	54	15	Basin
Chile Rise–Pacific Antarctic Basin	2.8	− 50	−18	Basin

Gauss's theorem, this quantity is directly proportional to excess mass, which in turn is a primary measure of the stresses required. Table 39–1 gives the 30 largest blocks in terms of free air anomalies referred to the hydrostatic figure, while Table 39–2 gives the 25 largest blocks in terms of isostatic anomalies referred to the fifth-degree figure. Of the 12 question marks in Table 4 of Kaula (1969), about 10 seem to be resolved. The greatest question remaining is the great negative over Antarctica; it is too large by more than a factor of 3 to be attributable to the loss of ice in recent geologic time (O'Connell, 1971).

The types given in Tables 39–1 and 39–2 are those used in Kaula (1969), with some obvious modifications.

INTERPRETATION

The principal inference from the large areas of postglacial uplift is, of course, the existence of an asthenosphere: a relatively plastic layer in the upper mantle, 80 to 400 km or more deep. Of the seven or eight major feature types, the glaciated areas are alone in being transient, with a decay time on the order of a few thousand years (O'Connell, 1971).

Table 39–2
Areas of exceptional gravity anomaly, defined as having an area × isostatic anomaly referred to a fifth-degree figure more than 50 mgal × 10^6 km² (1.17×10^{21} gm) in absolute magnitude, and absolute anomaly of more than 10 mgal throughout the area.

General location	Area $(10^6\ km^2)$	Isostatic anomaly × area $(mgal \times 10^6\ km^2)$	Mean isostatic anomaly $(mgal)$	Type
	POSITIVE FEATURES			
Southeast Pacific Rise	6.6	+135	+20	Rise
Mid-Indian–Amsterdam–Naturaliste Ridge	6.1	100	16	Rise
Borneo–Sumatra–Cocos	5.3	98	18	Arc
Indian Peninsula–Bay of Bengal	4.7	82	17	Sediment?
S. E. Indian and MacQuarrie Rises	5.1	76	15	Rise
North Atlantic–Arctic Ocean	7.1	73	10	Glaciated(?) Rise
Azores Plateau	4.7	73	16	Rise
Indian–Antarctic, Gaussberg Ridges	4.5	72	16	Rise
N. Andes–W. Amazon Basin	5.0	72	14	Arc–Orogenic
N. E. Georgia–S. Sandwiches–Mid-Atlantic	4.3	68	16	Rise–Arc
Japan–Bonins	4.0	66	16	Arc
Carlsberg Ridge–Gulf of Aden	3.7	65	18	Rise
South Alaska	3.6	65	18	Arc–Orogenic
Walvis Rise	4.5	+ 61	+14	Rise
	NEGATIVE FEATURES			
Himalayas–China	8.0	−141	−18	Orogenic
Antarctica	8.1	140	17	Glaciated Basin?
Laccadives–Ceylon–Mid-Indian Ocean	5.0	130	26	Basin?
N. Canada–Greenland	7.5	123	16	Glaciated
Australian Shield–S. Australian Basin	7.7	113	15	Basin
N. Amer.–Guiana–Parnaiba Basins	7.3	96	13	Basin
Galapagos–Peru Basin	6.1	96	16	Basin
Society–Tuamotu–Austral Seamount	6.8	96	14	Basin?
Congo–Kenya	3.7	74	20	Basin–Rift
E. Crozet Basin–Kerguelen	2.6	53	20	Basin
N. W. Europe	3.5	− 51	−15	Glaciated

Since the lithosphere is not capable of supporting elastically the necessary stresses for features thousands of kilometers in extent (McKenzie, 1967), the other broad departures in the earth from equilibrium must entail flow in the asthenosphere. The asthenosphere is stiff enough, however, and the thermal conductivity of the earth is poor enough (the Prandtl number is large), that it is generally agreed that the flow is essentially *steady state* (Turcotte and Oxburgh, 1967, 1969). As indicated by magnetic reversal patterns, the present pattern of tectonic motion has persisted for about 10 m.y. (Heirtzler et al., 1968).

In a steady-state flow system, to maintain a mass excess in a particular region, there must be effectively a Lagrangian deceleration of matter entering the region and an acceleration of matter leaving it; the converse must apply to a region of mass deficiency. Mathematically, this condition requires that for the volume containing the mass excess, the surface integral

$$- \int_{\text{surface}} \rho \frac{\partial \mathbf{v}}{\partial t} \cdot d\mathbf{n} = \int_{\text{surface}} \rho \mathbf{v} \cdot \nabla \mathbf{v} \cdot d\mathbf{n} > 0 \qquad (1)$$

where ρ is density, $\mathbf{v}$ is velocity, and $\mathbf{n}$ is the outward drawn unit vector normal to the surface. The uppermost boundary of this volume for purposes of gravity anomaly interpretation should be a segment of a bounding equipotential, such as the geoid. Since the compressibility of upper mantle material is slight, these "decelerations" must be accomplished by: (1) the piling up of material at the surface; (2) the replacement of less dense by more dense material at an interior interface; (3) thermal contraction; (4) transition to a denser phase; or (5) petrological fractionation in which a less dense component is left behind. The reverse of one or more of these processes is needed to accomplish an "acceleration." It is to be emphasized that mechanisms such as (3), (4), and (5) do not affect the gravity field directly by increasing the density, but rather by inducing mass transfers in accord with Eq. (1).

If the asthenosphere is a relatively thin layer, then the obvious direction to transfer matter so as to affect the external gravitational field is lateral. Vertical transfers, however, are not to be ruled out: an upward displacement of material making the density higher than the average at a shallow level, balanced by a mass deficiency at considerable depth (below about 200 km) could account for a gravity excess. From the formula for the potential arising from a spherical harmonic surface distribution of mass (Kaula, 1968, p. 67), we have for a mass excess of $\Delta\rho h$ of width L compensated at depth D:

$$\Delta g \approx 2\pi^2 G \frac{D}{L} \Delta\rho h \qquad (2)$$

Then to say the data are satisfied by isostatic compensation at great depth, however, is to beg the question as to the response of the asthenosphere to the stresses that must necessarily exist at the intervening levels.

The greater effectiveness of lateral transfer also suggests that stagnation points (regions where the flow changes direction, such as ocean rises or trench and island arcs) will tend to be regions of gravity excess, while regions dominated by horizontal flow will tend to be regions of gravity deficiency.

The relationship of gravity anomalies to the flow system depends considerably on the boundary conditions. In a system of thermal convection, if the boundary is rigid, then an upcurrent is associated with a negative anomaly, because of its lower density (Runcorn, 1965). If the upper boundary is free, however, while the lower boundary remains fixed, then an upcurrent is associated with a positive

anomaly because the effect of the mass pushed up at the surface outweighs the density effect (McKenzie, 1968b; Pekeris, 1935).

In the case of the real earth, the question becomes to what extent the lithosphere (the layer of relative strength) and the crust (the lower density, uppermost layer of the lithosphere) act as a part of the convective flow, and to what extent they act as a restraining boundary to the flow. Manifestly, they act both roles to differing degrees in different parts of the earth. The lithosphere can even be simultaneously a rigid boundary for horizontal forces in being able to act as a rigid plate in tectonic motions and a free boundary for vertical forces in not resisting convective upthrusts. The extent to which a particular portion of the lithosphere acts as a free or rigid boundary depends on its temperature, size of feature, rate of motion of material into and out of a feature, and composition, particularly its water content. The situation may be further complicated by steady surface transfers of matter: erosion and sedimentation.

How a boundary acts in the range between perfectly free and perfectly rigid depends on both (1) its elastic properties (its rigidity and thickness) and (2) its plastic properties (most simply expressed as a decay time in response to a transient loading, dependent on dimensions of the loading and stress as well as creep properties of the material). Under small stresses, the decay time of the lithosphere is very long: it is effectively acting as an elastic layer in areas of post glacial uplift. Under greater stress, however, such as in the major areas of mass excess, the effective decay time may be much shorter because of the nonlinear dependence of strain rate on stress (Weertman, 1970), as evidenced by the seismicity of these regions. Qualitatively, for both elastic and plastic behavior, we should expect that the thicker, the colder, the less hydrous the lithosphere is in a particular region, the more it will behave like a rigid boundary. Quantitatively, however, we should expect that in some cases it may be difficult even to infer the correct sign of the gravity anomaly.

The flow system for a body that has boundaries that are partly rigid, partly free, would be a difficult problem to treat rigorously. However, we might expect that usually the nature of the local boundary conditions would predominate in determining the characteristics of a particular region. We shall apply this assumption in the analysis of feature types.

Of the 11 gravity anomaly area types proposed in Kaula (1969), 6 appear to be associated with current internal activity in the earth. We shall discuss these 6 (somewhat modified) in an order suggested by their apparent relationship to the global tectonic pattern: (1) active ocean rises, (2) oceanic shield basalts still active in Quaternary, (3) basins, (4) trench and island arcs currently active, (5) current orogeny without extrusives, and (6) Cenozoic orogeny with extrusives in Quaternary.

Active Ocean Rises

The indication from the new data that these areas are generally of positive gravity anomaly is consistent with their being free boundaries over upcurrents in a convective system. Their well-known characteristics of high heat flow, shallow depth in the ocean, thin sediments, large scale volcanism, frequent moderate earthquakes, lack of a distinct Moho, and prevalence of intermediate seismic primary velocities in the range 7.2 to 7.7 km/sec are generally taken to indicate that the rises are the sites of upwelling and spreading out in a convective cycle. The intensity and uniformity of heating is apparently sufficient to prevent this mass imbalance from being large. The small temperature gradients entailed are the expected consequence of a strong temperature dependence of viscosity (Tozer, 1967; Turcotte and Oxburgh, 1969).

Oceanic Shield Basalts

With the improved data, all major oceanic positive areas appear to be associated with spreading centers except one: Hawaii. Hawaii appears to be the buildup of an appreciable mass excess by extrusive activity off the rise. This buildup is in spite of a sinking of the crust, as pointed out by Menard (1969). An approach to isostatic adjustment is also suggested by depths to the Moho somewhat greater than the oceanic average (Drake and Nafe, 1968). Apparently the lithosphere has cooled sufficiently to cause a lag in the attainment of equilibrium. This notion is corroborated by the relatively low heat flow. The existence of such a feature indicates that the asthenospheric flows that generate the required pressure do not necessarily have a simple and direct relationship to the lithospheric plate motions (McKenzie, 1969b).

Hawaii is unique in that it falls midway in the 10,000-km stretch from the East Pacific Rise to the trench and island arc along the west Pacific margin: the only region in the world where the rise-continent distance exceeds 5000 km. It is also the only region in the world where the rise-basin distance exceeds 3500 km.

Basins

This commonest of the major features always occurs somewhere to the flanks of ocean rises. Landward of the basin, however, may be a trench and island arc, an orogenic belt, or a relatively quiescent continent on the same tectonic plate. This suggests that the nature of the flows associated with basins depends more on where the material came from than where it is going.

The direct source of the negative isostatic anomaly is most likely that the crust carried along in the sea floor spreading is thicker than compatible with the depth of the basin; a Moho deeper by less than a kilometer is adequate to account for the average isostatic anomaly of -14 mgal.

The underlying cause, of course, is asthenospheric withdrawal, which results in the 3-km topographic drop from the rise to the basin. Such a drop could be caused by either (1) a horizontal acceleration in the asthenosphere or (2) a downflow of a denser component of the asthenospheric material.

A horizontal acceleration in the asthenosphere at a distance from the rise equal to the typical rise-basin distance (1500 to 3500 km) seems implausible, given that the velocity at the upper boundary is maintained constant by the lithosphere. If we assume a two-dimensional flow, neglect the effect of temperature gradients on the flow, then adopting the customary stream function form (Batchelor, 1967, p. 76)

$$u = \frac{\partial \psi}{\partial y}, \qquad v = -\frac{\partial \psi}{\partial x} \qquad (3)$$

where (x, y) and (u, v) are, respectively, the horizontal and vertical position and velocity (positive downward); we obtain the biharmonic equation

$$\nabla^4 \psi = 0 \qquad (4)$$

for which a solution is (Batchelor, 1967, p. 225)

$$\psi(r, \theta) = r f(\theta)$$
$$f(\theta) = A \sin \theta + B \sin \theta + C\theta \sin \theta \qquad (5)$$
$$+ D\theta \cos \theta$$

where r is distance from the origin and θ is the angle from the x axis. Make the boundary conditions

$$u = 0, \qquad v = -v_0, \qquad x = 0$$
$$u = u_0, \qquad v = 0, \qquad y = 0 \qquad (6)$$

i.e., the x axis is the asthenospheric-lithospheric boundary, u_0 is the spreading velocity, and the y axis is the vertical center of the plume of constant velocity v_0. The solution then is

$$u = u_0[1 - f(\xi)] + v_0 g(\xi)$$
$$v = v_0[c(\xi) - 1] - u_0 d(\xi) \qquad (7)$$

where

$$\xi = \frac{y}{x} = \tan \theta \qquad (8)$$

and

$$f(\xi) = \frac{\pi}{2} q \left(\frac{\xi}{1 + \xi^2} + \tan^{-1} \xi \right) - q \frac{\xi^2}{1 + \xi^2}$$

$$g(\xi) = q \left(\frac{\xi}{1 + \xi^2} + \tan^{-1} \xi \right) - \frac{\pi}{2} q \frac{\xi^2}{1 + \xi^2}$$
$$(9)$$

$$c(\xi) = \frac{\pi}{2} q \left(\frac{\xi}{1 + \xi^2} - \tan^{-1} \xi \right) + q \frac{\xi^2}{1 + \xi^2} + 1$$

$$d(\xi) = q \left(\frac{\xi}{1 + \xi^2} - \tan^{-1} \xi \right) + \frac{\pi}{2} q \frac{\xi^2}{1 + \xi^2}$$

$$q = \frac{4}{\pi^2 - 4} = 0.678 \ldots \qquad (10)$$

The condition imposed by Eq. (1) necessary to make the rise a mass excess and the basin a mass deficiency is that

$$\frac{\partial u}{\partial x} > 0 \qquad (11)$$

or

$$u_0 \frac{\partial f}{\partial \xi} - v_0 \frac{\partial q}{\partial \xi} > 0 \qquad (12)$$

Developing f and g in series of ξ, for small ξ:

$$v_0 < \frac{\pi}{2} \left(1 + \frac{4 - \pi^2}{\pi} \xi + \cdots \right) u_0 \qquad (13)$$

In words, the process of lithospheric formation has to consume more than a certain fraction of the material brought up by the vertical flow v_0 if the lithosphere of velocity u_0 is to cause an acceleration in the asthenosphere by dragging the asthenosphere along with it.

The settling out of a denser component would be expected in a multicomponent laterally moving flow that was cooling. A 3-km drop requires much more than thermal contraction, however. If there is an appreciable negative gravity anomaly as well, then Eq. (2) indicates that the settling out cannot just be immediately below the basin lithosphere, but must be either (1) at several 100 km depth below the basin or (2) between the ocean rise and the basin. The process (1) could be induced by the phase transitions of olivine and pyroxene, which occur at depths of 300 to 600 km, while the process (2) might be facilitated by gabbro-to-eclogite transitions at shallower depths.

The sharpness of the crust-mantle boundary, with the 7.2 to 7.7 km/sec gap in velocities (Drake and Nafe, 1968), makes it impossible for the crust to be directly involved in causing the negative anomaly. If the crust is being carried passively along (as suggested by the lack of seismicity, volcanism, or disturbance of the sea floor), then it is hard to understand why it is thicker under the basins than on the flanks of the rises, as emphasized by Le Pichon (1969). Could it be that consolidated sediments are mistaken for basement rock? Sedimentation itself is a secondary process in explaining the gravity anomaly pattern, more a result than a cause: if the thick sediments were the driving force, then the isostatic anomalies in ocean basins would be positive, rather than negative.

Possibly includable in the category of basins caused by behavior of the lithosphere as a free boundary over flows with horizontal accelerations or settling out of denser components are two land features, Antarctica and the Congo Basin. Antarctica is an extremely large feature —large enough to require an unique explanation.

Trench and Island Arcs

The now generally accepted model of McKenzie (1969b) and others of a colder, denser oceanic lithospheric slab being thrust down under a less dense but stiffer continental margin fits a simple notion of the gravity pattern: the dominant feature is the broad positive anomaly associated with the denser downthrust slab, while the secondary feature is the narrow negative belt associated with the trench caused by tensile cracking along the downward breaking line. This simple picture is based on the assumption that the applicable boundary condition of the convective flow is more "rigid" than "free": in other words, the time scale of the process is short enough that the strength of the continental lithosphere (and perhaps the oceanic lithosphere as well) significantly resists being pulled down by the downcurrent. This general idea that resistance to flow combined with densification creates positive gravity anomalies applies not only to the boundary layer, but also to deeper strata: the downthrust slab could in part be supported by stiffer matter below the asthenosphere, as suggested by Isacks and Molnar (1969) and others from seismic data.

The association of the downthrust slab with positive anomalies also suggests that the driving cause is a push from above rather than withdrawal from below. Whether this "push" is the gravitationally caused sinking of the denser oceanic lithosphere, the pressure of the spreading sea floor behind it, or the viscous drag by the sublithospheric flow, does not seem resolvable from the gravity data.

Current Orogeny Without Extrusives

The hypothesis of McKenzie (1969b) that purely continent versus continent compression results in folding rather than downthrust because of the excessive bouyancy of the thicker crust is appealing as an explanation for the strongly negative gravity anomalies associated with the Asian part of the Alpide belt. The resulting pileup of lower density material results in a mass deficiency in the short run because the stiffness of the lithosphere containing low density crust enables it to push out of the way higher density asthenospheric material. In the longer run, however, the trend from "rigid" to "free" boundaries is expressed by the forcing upward of the lithosphere; geologic and geodetic indications are that the Himalayas-Turkestan complex is currently rising (Artyushkov and Mescherikov, 1969; Gansser, 1964).

The thick layers of sedimentary and metamorphic rocks constituting the upper part of the Himalayas have existed since Precambrian times. The resulting excess of radioactive material combined with low thermal conductivity will lower crustal densities. Furthermore, there may be a contribution to the negative anomaly by erosion, as corroborated by the positive features over the corresponding sedimentation basins, the Bay of Bengal and the Arabian Sea, which appear in Fig. 39–4.

It is possible that the foregoing suggested mechanisms are all quantitatively insufficient and that an asthenospheric withdrawal is necessary.

Cenozoic Orogeny with Extrusives

These mountain-building areas, listed as orogenic among the positive features in Table 39–1 are of more limited extent and closer to isostatic equilibrium. Most are associated with compressive belts of the global tectonic system, but this is not entirely so. The reason why they differ from the Himalayas-Turkestan complex in being positive may be (1) the lack of the pre-existing great thicknesses of sedimentary and metamorphic rocks or (2) the presence in the eastern Mediterranean of oceanic crust that can be "consumed" or "subducted" (McKenzie, 1970). They may also be the continental equivalents of Hawaii to some extent: the coincidence of weak features in the lithosphere with regions of excess pressure and heat in a convective system that is not directly related to surface features. Most of these areas have positive seismic delay residuals, suggesting high temperatures to considerable depth.

DISCUSSION AND CONCLUSIONS

The gravity data now appear to be quite reconcilable with the dependence of plate tectonics on mantle convection inferred from other phenomena associated with the mid-ocean rises and the compressive belts (Isacks et al., 1968). Gravity still suffers, however, from its traditional ambiguity in being insufficient to infer the exact mechanism, such as whether the ocean basin negatives are caused by horizontal acceleration or downflow of a denser component.

The greatest feature not readily related to the global tectonic system is the Antarctic negative, much too large to be explained by glacial melting. Antarctica is five-sixths surrounded by ocean rises. Hence, either the spreading rises are migrating away from Antarctica, or Antarctica is a sink for lithospheric material. The latter seems ruled out by the complete absence of the seismicity expected with the destruction or folding of lithosphere. Given that the rises are migrating away, Antarctica must be a mass deficiency because

there is not an asthenospheric flow to match the lithospheric spread: i.e., the condition of Eq. (13) applies over most of the rises around Antarctica.

Other features not well explained by the global tectonic pattern are the gravity excesses associated with extrusive flows that occur away from the ocean rises and trench and island arcs, in both oceanic and continental areas. These features appear to require higher temperatures in the asthenosphere generating excess pressures, together with weaknesses in the lithosphere allowing the extrusions. It is, however, difficult to choose whether the resulting net mass excess is a consequence of sufficient overall strength in the lithosphere to support the extruded load or of behavior as a free boundary over a horizontal deceleration or an upcurrent, the reverse of the processes that appear necessary to account for the ocean basins.

Anticipated properties of the mantle convective system that need to be better related to the gravity field are the stress- and temperature-dependence of the effective viscosity, the horizontal temperature gradients arising from variations in radiogenic heating and the contributions to driving the system by fractionations and phase transitions. All these properties are important, of course, to solution of the entire global tectonic problem.

40

The Sinking Lithosphere and the Focal Mechanism of Deep Earthquakes

DAVID T. GRIGGS
1972

From *The Nature of the Solid Earth,* ed. by E. C. Robertson, J. F. Hays, and L. Knopoff, p. 361–384, 1972. Reprinted with permission of the author and McGraw-Hill, Inc.

The origin of deep earthquakes seems to be paradoxical. They occur only where there is reason to believe that the lithosphere is plunging into the mantle. When the rate of this motion was first realized, it became clear that such known processes as dehydration embrittlement and weakening could cause shear fractures at even the maximum earthquake depths. Such fractures were expected to occur roughly parallel to the surface of the sinking plate, as in the underthrust earthquakes occurring in the Aleutians. Recent focal plane solutions for intermediate and deep earthquakes seem to point to shear surfaces inclined $\sim 45°$ to the sinking plate, with the stress axis of greatest compression (or occasionally greatest extension) parallel to the dip of the slab. If these quakes are of simple mechanical origin, they would seem to imply that the colder center of the slab serves as a stress guide (Elsasser, 1967), though the origin of the compressive stresses is uncertain.

The origin of shallow earthquakes will not be pursued in this paper for two reasons: (1) shallow earthquakes are being investigated in detail by many seismologists, and it would be beyond the scope of this paper to do justice to this extensive work, and (2) Crustal rock motions seem everywhere consistent with mechanisms of conventional fracture and stick-slip behavior facilitated at depth by high fluid pressure. There thus seems to be no problem of mechanisms, although the details of these processes will only become clear as more intensive studies are made.

Since deep earthquakes seem to be intimately related to processes within the sinking lithospheric slab, the first part of this paper examines in an exploratory way the character of the slab. The second part comments on the earthquake mechanics.

NATURE OF THE SINKING LITHOSPHERE

Introduction

It has been dimly perceived for decades that the crust and upper mantle are underthrust at the oceanic trenches, and the discovery of the deep earthquake zones extending downward from the trenches lent credence to this idea. Sykes' detailed study (1966) of the Tonga and Kermadec zones first showed that these earthquakes occur in a very narrow zone (<20 km thick), which is for the most part nearly planar and inclined at roughly 45° to the horizontal.

Oliver and Isacks (1967) showed that these earthquake zones have anomalously low attenuation of S waves and 1 to 2% higher P and S wave velocities. From these facts, they deduced that the upper part of the mantle under the ocean (the lithosphere) sinks or is thrust down at the trenches and extends in recognizable form at least as far down as the deepest earthquakes.

Isacks, Oliver, and Sykes (1968), hereafter called IOS, argued that this behavior is characteristic of trenches and showed that the downdip length along the earthquake zone is roughly proportional to the sinking velocity derived for that trench from the rigid plate model of the earth's surface motion (LePichon, 1968).

This author has reasoned (Griggs, 1968, 1969) (1) that the lithosphere is similar to the underlying asthenosphere except that it is cooler, which accounts for its higher seismic velocity and lower attenuation, and (2) that the limiting depth of earthquakes is that point at which the lithosphere, which is heated as it sinks, becomes too hot for seismic instability (fracture?) to occur. Below this depth, all deformation occurs by flow.

Related Work

Two recent papers have dealt with the nature of the sinking lithospheric plates (McKenzie, 1969b; Minear and Toksöz, 1970). McKenzie extended his analysis of the heat flow anomalies associated with the ridges (McKenzie, 1967) to the solution of the temperature distribution in the sinking lithosphere. He also derived the streamlines, the stresses, and the strain heating in the mantle above and below the sinking plate. He then considered the broad problem of mantle convection and the role played in this process by the sinking plates.

Minear and Toksöz did a numerical two-dimensional heat flow calculation of the sinking lithosphere and the surrounding mantle, taking into account adiabatic compression, radioactivity, and phase changes. Estimates of the effect of strain heating in the slab boundary, surface heat flux, gravity anomalies, and seismic travel time in the vicinity of the slab are presented.

Because of uncertainties in the values of the large number of parameters and the mathematical difficulties in handling the nonlinear real-earth processes, these analyses (and mine also) can only hope to throw additional light on the probable trend and the relative importance of key factors, not to solve the problem. They may, however, point the way toward future geophysical exploration of significance and toward refinements to be sought in the mathematical and calculational techniques.

McKenzie's analysis (1969b) is made tractable by assuming that the temperature gradient in the mantle below the lithosphere is adiabatic everywhere except in the slab and that mantle flow is characterized by a constant Newtonian viscosity. The latter assumption is probably apt for the upper mantle below the low-velocity zone because it is very hot (near the melting point) and the stresses are low. I believe the former assumption is grossly in error and that the mantle more nearly follows the melting point gradient below the low-velocity zone. The thermal history of the slab is not sensitive to the assumed geotherm, however, and McKenzie's analysis gives a very good qualitative picture of the nature of the temperature distribution in the slab and most of the gross effects that he considers.

Minear and Toksöz (1970) used a 20- by 20-km grid and translated the boundary temperatures 60 km with respect to the slab in each step of the calculation. They adopted MacDonald's chondritic mantle radioactivity and temperature gradient (1963). Adiabatic heating and heating resulting from passing through phase transitions are introduced.

Strain heating in the mantle at the slab boundaries is treated parametrically. They

Table 40–1
Six sets of parameters used for the olivine-spinel transition in the calculations.

Transition	ρ_{sp}/ρ_{ol}	dT/dP ($°C/kb$)	T_0 ($P = 0$)	ΔT
A	1.10	30.0	−1747°C	35°C
B	1.10	15.7	−2	70
C	1.10	20.0	−527	50
D	1.05	15.7	−2	35
E	1.05	20.0	−527	25
F	1.05	30.0	−1747	17.5

apparently assume constant shear stress over the whole length of the slab rather than conform to McKenzie's model, which has a maximum near the top and diminshes several-fold downward. Typical shear stresses assumed (2500 bars in Figure 11) are very much higher than McKenzie's average stress of about 100 bars. The strain heating in Minear and Toksöz' model is thus very much greater and even causes inversion in the slab. The slab at depth is hotter than the surrounding mantle.

I believe McKenzie's values of strain heating in the mantle are about right. Certainly, in the model of the earth which I use, neither the drag stresses nor the strain heating could approach Minear and Toksöz' values for reasons that will become clear later.

Calculational Model of the Sinking Lithosphere

Attention is restricted to the interior of the downgoing slab. For simplicity, I employ a one-dimensional heat flow calculation of the most rudimentary sort (Carslaw and Jaeger, 1959, p. 470). Because of the fact that the transverse temperature gradient is vastly greater than the longitudinal, one-dimensional heat flow is a good approximation except at very low sinking velocities. This may be qualitatively verified by comparing my isotherms with those of McKenzie (1969b).

The basic difference equation is:

$$T_j = R_j + w(R_{j-1} - 2R_j + R_{j+1}) \quad (1)$$

where T is the new temperature, R the old, the subscript j denotes the grid point number across the slab, and $w = \kappa \Delta t/\Delta x^2$. κ is the diffusivity, Δt the time interval between each set of calculations and Δx the distance between grid points across the slab. For calculational stability, w must be less than 0.5, and is given the value 0.316. A grid of 27 points across the slab (25 active cells) suffices to reveal with the requisite fidelity the thermal structure and also the fine structure in the resultant gravity anomaly. Duplicate calculations with 50 active cells and hence with a time interval reduced fourfold show very small differences in the thermal structure. The κ is taken as 0.01 cm²/ sec. Radiative conductivity is of no consequence except very near the borders of the slab at depth and then only if the opacity is 10 cm⁻¹ or less. It is neglected in these calculations, but it may not be negligible for calculations on the mantle.

Adiabatic heating is added at each calculational step as follows (Jeffreys, 1929, p. 139):

$$\frac{dT}{dZ} = g \frac{\alpha T}{c_p} \quad (2)$$

where Z = depth, α = coefficient of thermal expansion, and c_p = specific heat. The value of α is of first-order importance for calculating the gravity anomalies and the stresses resulting from the excess density of the slab. Values of α and α/c_p versus depth are shown in Fig. 40–1. These are taken from Birch (1952) and seem as good as any in the later literature.

Only a single phase change was explored in this model—the olivine-spinel transition presumed by Anderson (1967b) to occur at a depth of 365 km. As will be seen, the P-T slope of this transition has a pronounced effect on the density structure of the slab. Table 40–1 shows the transition parameters that were used in the calculations. Transition B is consistent with Akimoto and Fujisawa (1968). Its slope is so nearly parallel to the geotherm

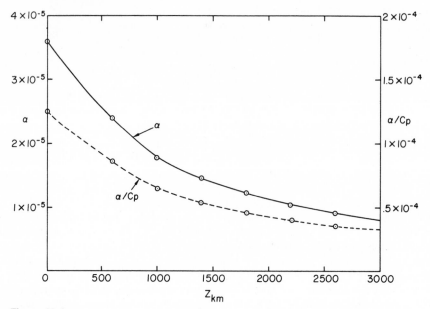

Figure 40-1
Thermal expansion coefficient α/c_p versus depth in the earth (from Birch, 1952)

that the transition occurs at shallow depth in the center of the cold slab. For this reason, other slopes were considered. Transitions D, E, and F correspond to a mantle composition of 50% olivine.

The melting temperature within the mantle is approximated by:

$$T_{mp} = 1273 \left(1 + \frac{P}{15}\right)^{0.25} \tag{3}$$

where P is the pressure in kilobars. This is shown in Fig. 40-2 faired to 1100°C at $P = 0$. At low pressure, this corresponds to the solidus of peridotite (lherzolite) as determined by Ito and Kennedy (1967) and Kushiro, Syono, and Akimoto (1968). At higher pressures, it is adjusted by a Simon-type fit to a melting temperature of 4000°K at the core boundary. While this is functionally wrong, the result corresponds to the best guesses of my colleagues (principally that of G. C. Kennedy, 1970, oral communication) as to the temperature of first melting in the mantle. The importance of melting temperature in these calculations is twofold: (1) it is used to find β (the ratio of temperature in the slab to the melting temperature at that pressure) and (2) it sets an upper limit for the rise of temperature resulting from strain heating, etc.

The principal factor governing the choice of the undisturbed mantle geotherm is my belief that the mantle temperature must closely approach or equal the temperature of first melting in the low-velocity zone. A wide variety of geotherms were explored, and it was found that the results of the calculations were not very sensitive to the shape of the geotherm, provided it approached the melting point at depths comparable to that of the low-velocity zone. The geotherm used in the calculations reported here was:

$$T = 1340\left[1 - \exp\left(-\frac{Z}{45}\right)\right] + 1.7Z$$
$$0 < Z < 215 \text{ km}$$
$$T = T_{mp} - \frac{(Z - 215)}{4} \qquad Z \geq 215 \text{ km} \tag{4}$$

where Z is the depth in kilometers and T_{mp} from Eq. 3 is here used in °C. This is shown in Fig. 40-2, together with the adiabat from 215-km depth. This geotherm would give a surface heat flow of 1.6 μcal/cm²sec if the surface conductivity were 0.005 cal/cm deg C sec. It will be noticed that this geotherm corresponds to a Lachenbruch (1968) steady-state model with radioactivity decreasing exponentially from a surface value of 3×10^{-13}

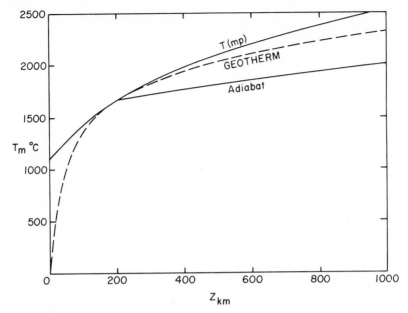

Figure 40-2
Solidus melting temperature, T(mp), geotherm, and adiabat used in most calculations, versus depth

cal/cm³sec and an *e*-folding length of 45 km. Actual upper mantle temperatures, however, must be largely determined by their cooling history as they spread away from the ridges. the geotherm chosen is intended to approximate such cooling after transport from ridge to trench.

Pressure and density in the mantle were calculated as follows:

$$P = g \int \rho dZ$$
$$\rho = \rho_0 \left[1 + \int (a - 2bP)dP - \int adT \right], \tag{5}$$

where $a = 8 \times 10^{-4}$ and $b = 1 \times 10^{-6}$, from the two-constant fit for the compressibility of olivine (Clark, 1966, p. 131). Density versus depth is shown in Fig. 40–3 for the two types of olivine-spinel transitions (Table 40–1), each of which is presumed to be gradational over a depth range of 35 km. At depths greater than 200 km, these densities are lower than those commonly estimated for the outer mantle, suggesting that the compressibility used in these calculations is too low. Calculations with other equations of state have shown, however, that the effects to be discussed below are not sensitive to the magnitude of the compressibility, but only to the general shape of the ρ versus Z curve.

We are now ready to describe the calculational procedure. It is assumed that the slab slips down through undisturbed mantle. This is clearly not correct, but is akin to McKenzie's assumption (1969b) of an adiabatic mantle. As he points out, there is no obvious alternative to numerical solution of the general nonlinear convection equations, which is beyond my capability.

Values are chosen for sinking velocity (parallel to the slab), V, angle of descent (assumed constant throughout the sinking), ψ, slab thickness, D, and type of transition. The details of the bending of the lithosphere are ignored, since it was found that these have little effect on the thermal structure in the slab except at very low velocity. The transverse grid points are initially assigned the temperature of the mantle at their depths. The grid is then translated parallel to the slab a distance ΔY—the distance the slab would have traveled in time Δt ($\Delta Y = Vw\Delta x^2/\kappa = V\Delta x^2/10$ km, where V is in cm/yr). The upper and lower boundary temperatures are set to the corresponding mantle temperatures. The heat flow calculation is done by Eq. (1), and adiabatic heating is added

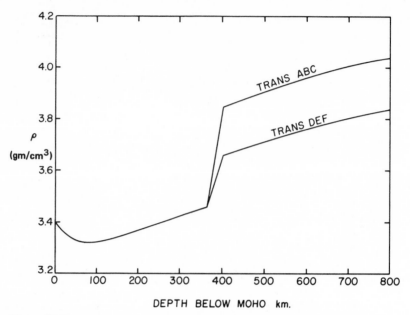

Figure 40-3
Density versus depth for the six sets of olivine-spinel transitions (Table 40-1)

for each grid point. Radioactive heating is negligible except at very low velocities, and so it is neglected. Pressure and density are calculated for each grid point and each equivalent point in the undisturbed mantle. The contribution of the density difference in each cell to the gravity anomaly is calculated. The contribution of these density differences to the down-slab extensile stress is also calculated. The minimum temperature in the slab is found, and $B = T/T_{mp}$ is calculated for that point. The calculation is then iterated and the behavior of the slab is observed to some chosen depth. Temperature and density at each grid point (if 25 cells, or every other grid point, if 50 cells are used) are printed out, so that the details of the process may be observed. Cumulative gravity anomalies, extensile stress, and diagnostic data are printed out every five time steps.

At each time step, a test is made to see if any point in the slab has passed the threshold for the transition and the density, pressure, and temperature are appropriately incremented if this condition is met, until the transition is complete. This is done separately, of course, for the slab grid points and the corresponding mantle points, since the transition occurs earlier in the cooler parts of the slab.

Isotherms and Transition Elevation in the Slab

The thermal structure in the slab for a typical calculation is shown in Fig. 40–4. Similarity between this and McKenzie's calculation (1969b, Fig. 3) is readily apparent. The upward extension of the transition within the slab is also shown for the two extreme transition slopes considered. The early history of the slab is dominated by the slowness of thermal diffusion. The time required for changes in the boundary temperatures to be felt appreciably at the center of a slab is (Carslaw and Jaeger, 1959, p. 98, Fig. 10a):

$$\tau \simeq \frac{D^2}{30\kappa} \simeq \frac{D^2}{1000} \tag{6}$$

where D is the total slab thickness in km and τ is in millions of years. IOS show that the sinking time to the deepest earthquake is about 10 m.y., so if the thickness is greater than 100 km, the center of the slab will be nearly as cold as it was initially, except for adiabatic heating and transition heating. In a 50-km-thick slab, however, the central temperature will have increased to about half the elevation of the boundary temperature.

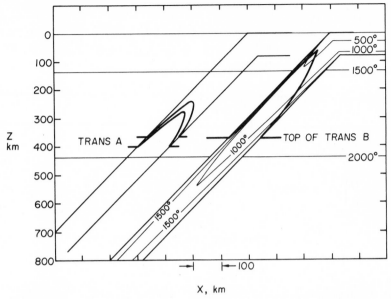

Figure 40-4
Isotherms and transition boundaries for typical calculations. Lithosphere
thickness $(D) = 80$ km, $\psi = 45°$, $v = 8$ cm/yr, $\rho c_p = 1.0$, adiabatic heating and
transition heating included

It follows that the maximum elevation of an olivine-spinel transition such as B or D within the slab is insensitive to slab thickness because the sinking time to the transition boundary is small compared to the thermal time constant. Values for transitions A and B (Table 40–1) are given in Table 40–2. Figure 40–5 shows temperature versus pressure profiles through a slab and the intersection with the various transitions for various lengths, L, down the slab.

I have dwelt on this elevation of the olivine-spinel transition in more detail than Minear and Toksöz (1970) because this seems to be an effect capable of exploration by seismic methods. Since Anderson (1967b) and others now have strong evidence that such a transition exists, it seems appropriate to mount special studies to determine the shape of the high-density, high-velocity region within the sinking slab.

Maximum Depth of Earthquakes

We turn next to the limiting depth of earthquakes. McKenzie (1969b) assumes that this limit occurs when the minimum temperature in the slab exceeds some critical value. Testing this against the variations of sinking velocity along the Tonga-Kermadec trench zone, he shows that the limit corresponds to heating of the center of the 50-km-thick slab in his model to 85% of his adiabatic mantle temperature (McKenzie, 1969b, Fig. 4). I believe it is physically more reasonable to assume that the limiting depth is that point at which the minimum slab temperature exceeds a critical fraction β_c of the melting temperature.

IOS (their Fig. 16) give the length down the slab to the deepest earthquake versus sinking velocity for trenches all over the world. This seems to be the best data to use for a test of the limiting β hypothesis. Figure 40–6 shows the results from 30 calculations at three different angles of sinking and 10 velocities from 1 to

Table 40–2
Depth to uppermost point of olivine-spinel transition in slab at $v = 8$ cm/yr.

Transition	Slab thickness, kilometers			
	50	75	100	150
B	47	47	47	47
A	225	220	218	216

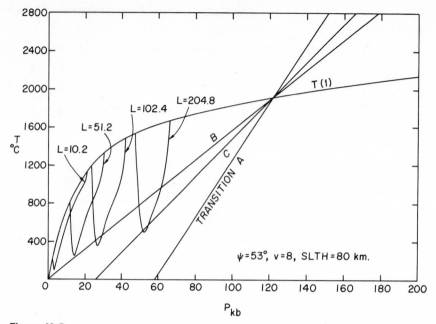

Figure 40-5
Thermal sections through lithosphere at different values of length down the slab (L) in km, plotted as a function of ambient pressure. Intersection with transition boundaries illustrates degree of elevation of transition within the sinking lithosphere. T(1) is temperature at slab boundary, ψ is angle of descent, v is velocity of sinking in cm/yr, and SLTH is slab thickness

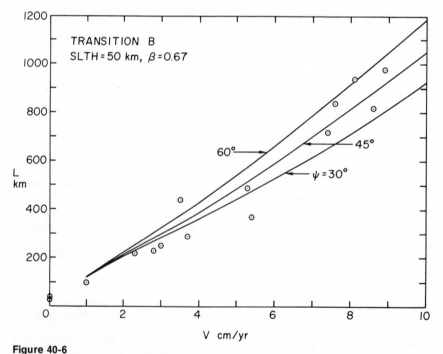

Figure 40-6
Length down-dip in slab to the point where the minimum temperature in the slab reaches $\beta = 0.67$ versus velocity of sinking, for three different angles of dip. Points are from IOS (Fig. 16) excluding points of widest scatter. This shows that the critical β hypothesis of the limiting depth of earthquakes fits the observations within the scatter of the data, and that it fits better than the 10 m.y. straight line of IOS, which extends from $L = 0$, $v = 0$, to $L = 1000$, $v = 10$

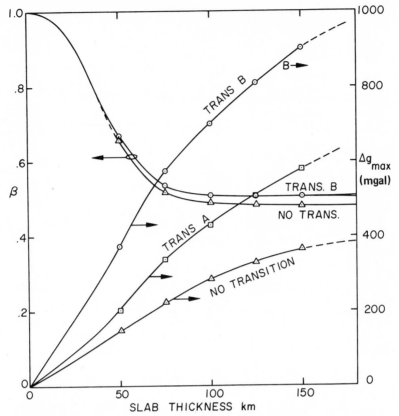

Figure 40-7

Ratio of minimum temperature in slab to the melting temperature (β) and the maximum value of the gravity anomaly component attributable to the slab alone (Δg_{max}) versus slab thickness. Arrows indicate the appropriate ordinate scale. $L = 820$ km, $v = 8$ cm/yr, $\psi = 45°$. Illustrates increase in gravity anomaly attributable to elevation of transition in slab. As slab thickness increases, thermal conduction asymptotically approaches zero, in which case β is determined solely by the initial temperature plus adiabatic and transition heating

10 cm/yr, for $\beta_c = 2/3$, and the IOS data with points of widest scatter omitted. It is seen that the 45° curve fits the data somewhat better than the 10-m.y. straight line of IOS. I do not have sufficient data on the dip of the zones to check whether the quality of the fit would be improved or not by correcting each point for dip. Unfortunately, a wide variety of model parameters will give about equally good fit to the IOS data, so this can not serve as a discriminant of model validity.

The variation of β with slab thickness is shown in Fig. 40–7, together with the amplitude of the gravity anomaly produced by the slab. The nature of the transition affects β very little, but drastically affects the magnitude of the gravity anomaly, because of the elevation of the transition in the slab. Density contours for a typical calculation illustrate this effect (Fig. 40–8).

Gravity Anomalies

The characteristic shape of the gravity anomaly due to the slab is shown in Fig. 40–9 (see also Minear and Toksöz, 1970, Fig. 14). This anomaly is partly compensated by the characteristic negative anomaly associated with the trenches caused by the sinking lithosphere.

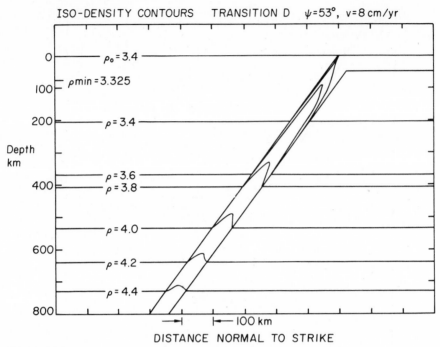

Figure 40-8
Distribution of density, ρ, in gm/cm^3, in a typical calculation

Taking Tonga as the model, a symmetric negative anomaly is created by mirroring the oceanic-side bathymetry about the trench axis and assuming a density contrast of 1.8 (Talwani et al., 1961b). Sykes (1966, Fig. 9) shows that the trench is offset about 50 km from the top of the sinking slab as revealed by earthquakes. This symmetric anomaly is shown in Fig. 40–9 with the 50-km offset. The sum of this and the slab anomaly is a broad positive anomaly of large magnitude, which must be regionally compensated.

The problem of regional compensation is very complicated, since it involves time-dependent strength and non-Newtonian plastic flow of the crust and lithosphere plus flow of the asthenosphere (McConnell, 1968). Based on McConnell's and other comparable analyses of isostatic adjustment, it seems that a fairly good approximation of regional compensation in this case would be achieved by a $\cos^2(\pi X/\lambda)$ deviation from the geoid, with $\lambda \cong 2000$ km, where X is now the horizontal distance from the center of effective mass of the sum of slab and trench anomalies. Such a

regional compensation anomaly is shown in Fig. 40–9, adjusted so that the net regional anomaly is zero. The sum of these three anomaly components yields my best estimate of the observable free-air anomaly attributable to the sinking lithosphere and the trench alone.

Net anomalies derived in this way are shown in Fig. 40–10 for three different types of transition, several lithosphere thicknesses, and three angles of dip of the sinking slab. If there were no transition, or a transition with large dT/dP, a lithospheric thickness of 100 km would fit the observed anomaly well. If there were a transition with the slope of Akimoto and Fujisawa (1968), then the slab thickness must be less, but also the effect of dip becomes more important (Fig. 40–10d). The dip of the Benioff zone at 20 to 25°S, where the anomalies were measured, is about 53° (Sykes, 1966, Fig. 9). The net anomalies with this dip are shown in Fig. 40–11 for transitions B and D. Of these, transition D with a lithosphere thickness of 50 km best fits the observed data. Thus, it seems possible to explain the island arc anomalies without resort

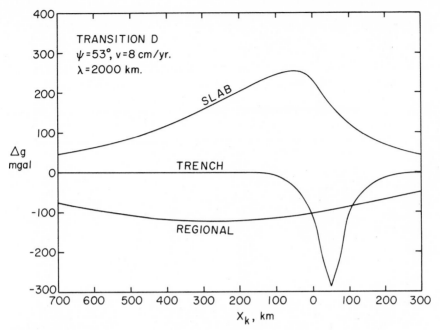

Figure 40-9
Gravity anomaly components attributable to slab, symmetric trench, and regional compensation. Surface distance X_K measured normal to trench axis from the point at which the projection of the upper surface of the slab intersects the surface ($X_K = 0$)

to the assumed tapered transition zone adduced by Talwani et al. (1961b).

Kanamori and Press (1970) conclude from seismic evidence that the thickness of the lithosphere is close to 70 km, and that the bottom of the lithosphere is probably at the solidus, because of the sharp decrease in rigidity observed. If the lithosphere differs from the underlying upper mantle material only in being cooler from surface conduction, as presumed here, then the thickness of the lithosphere should increase as it moves from the ridges to the trench. The magnitude of this effect will bear investigation, both from seismic observations and from theory of mantle currents. As it stands, the finding of Kanamori and Press is consonant with the model presented here.

Strain Heating in the Slab

Strain heating can at most raise the boundary temperature to the melting point. Because the geotherm is so near the melting point at depth

in my model, this has little effect on the temperature in the interior of the slab. A sample calculation showed that the central temperature in the slab is only raised from 1410 to 1433°C (0.668 to 0.677 β) when the boundary temperature was assumed to be at melting throughout ($D = 50$ km, $v = 8$ cm/yr; transition B, $L = 800$ km). This is consistent with the strain heating calculated by McKenzie, (1969b), but far less than that assumed by Minear and Toksöz (1970).

Since heating by boundary drag is unimportant in our model, let us estimate the maximum strain heating that could be produced *within* the slab. The flow law applicable to the interior of the slab must be of the Weertman form:

$$\dot{\gamma} = a\tau^n \exp\left(-\frac{E}{RT}\right) = a'\tau^n \exp\left(-\frac{G}{\beta}\right) \quad (7)$$

where $\dot{\gamma}$ is strain rate, τ shear stress, T temperature, β is T/T_{mp}, and a, n, E, R, G are constants.

The integration of this over the length of the slab proved very difficult, and an upper limit

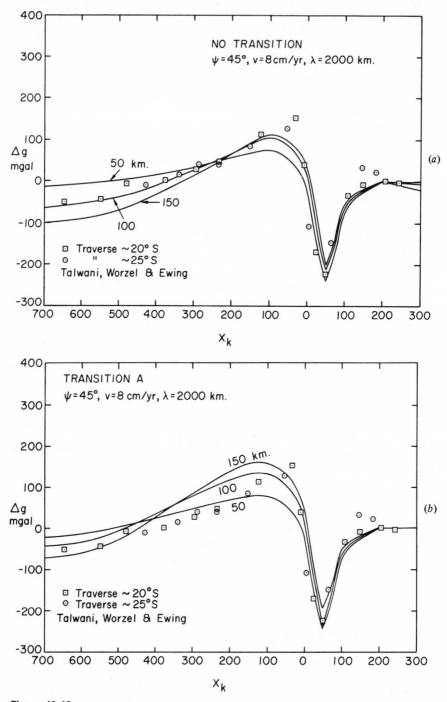

Figure 40-10
Net anomalies attributable to slab plus trench plus regional compensation compared with observed gravity anomalies (Talwani et al., 1961b). Effects of transitions, lithosphere thickness, and dip angle are illustrated

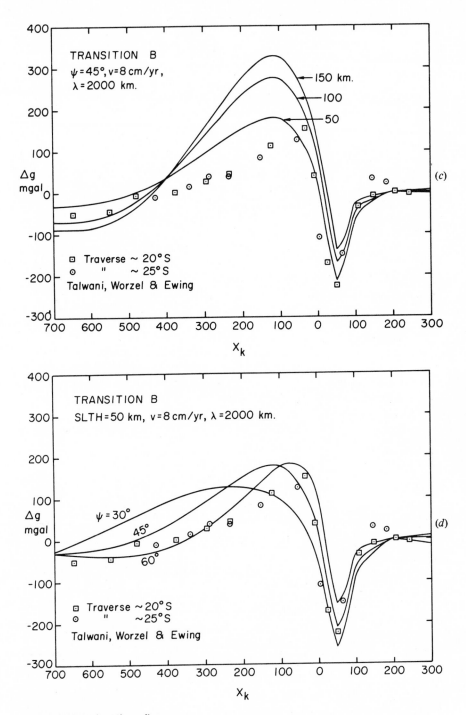

Figure 40-10 (continued)

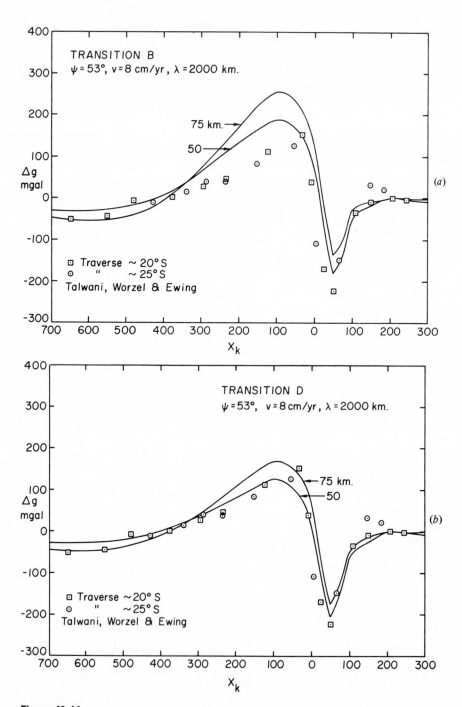

Figure 40-11
Net anomalies with transitions having the parameters of Akimoto and Fujisawa (1968)
and dip angle of Sykes (1966) compared with observed anomalies (Talwani, et al., 1961b).
(a) Mantle 100% olivine composition. (b) Mantle 50% olivine

estimate is made by assuming that all the work done by the sinking lithosphere is converted to internal heat in the slab. The work done is σDL where σ is the extensile stress described above and L is the length of the slab. The maximum slab heating is:

$$\Delta T = \frac{\sigma DL}{DL\rho c_p} = 23.9\sigma \,(^{\circ}C)$$

For a typical case ($D = 50$ km, $v = 8$ cm/yr $\psi = 53^{\circ}$, Transition D), $\sigma = 9.14$ kb, $\Delta T = 218^{\circ}$ C. This is to be compared with the total heating of the slab by conduction: $\Delta T' = 1466^{\circ}C$. Hence the maximum strain heating of the slab is only 15% of the conduction heating.

This is a substantial overestimate of the actual strain heating, since much of the work must go to pulling the lithosphere horizontally ($\sigma_l \sim 1$–3 kb) some goes into the flow field (σ_f from McKenzie ~ 2 kb) and much of the remainder will be lost by conduction. It is concluded that strain heating is not of great importance in this model of lithosphere sinking.

It may be noticed that the boundary shear stress of 2.5 kb assumed by Minear and Toksöz (1970, Figure 11, $v = 8$ cm/yr, $D = 160$ km) is 25 times McKenzie's calculated average shear stress and 5 times the maximum value available from my calculation above if all the sinking energy went into heating the slab.

EARTHQUAKE MECHANISMS

If deep earthquakes were of the shear fracture type, as seems indicated by all studies of the focal mechanisms, I expected that they would reflect the shearing flow associated with the mantle currents (the shear surfaces would be predominantly coincident with the upper surfaces of the sinking slab). Discovery by IOS, Isacks, Sykes, and Oliver (1969), Isacks and Molnar (1969), and Fitch and Molnar (1970) that focal mechanisms involve predominantly an axis of maximum compressive stress or occasionally maximum extensile stress down-dip in the sinking slabs requires some other mechanism. Elsasser's (1967) conception of the lithosphere as a stress guide has been appealed to by most of these authors. It is suggested by them that the slab hits a more resistant zone at depth, causing down-dip compression at the bottom, which may grade upward into extension if the resistance is sufficiently small. The extension is sometimes believed sufficient to pull apart the lithosphere, accounting for seismic gaps in some Benioff zones.

Raleigh and Kirby (1970) advance persuasive arguments that Weertman power law flow characterizes behavior of the lithosphere. Hot creep experiments on dunite by Post and Blacic (unpublished data) in our laboratory yield a preliminary flow law as follows:

$$\dot{\epsilon} = 13.2\sigma^{5.9} \exp\left(-\frac{66700}{RT}\right)$$

where $\dot{\epsilon}$ and σ are compressive strain rate in $\sec^{-1}$ and stress in kb, respectively. Converting to shear strain and shear stress and noting the Weertman observation (1970) that the effects of pressure may be included by expressing temperature in units of β (Raleigh and Kirby, 1970); we obtain:

$$\dot{\gamma} = 788\tau^{5.9} \exp\left(-\frac{66700}{RT}\right) \qquad (8)$$
$$\dot{\gamma} = 788\tau^{5.9} \exp\left(-\frac{24.5}{\beta}\right)$$

where $\dot{\gamma}$ is shear strain rate in $\sec^{-1}$ and τ is shear stress in kb.

This flow law allows us to calculate the shear stress at any point in the cooler interior of the slab corresponding to a chosen shear strain rate. It does not apply to the mantle below about 100 km nor to the hot outer regions of the slab where Nabarro-Herring flow is to be expected. A strain rate of 10^{-14} $\sec^{-1}$ seems appropriate for an illustrative calculation. At this strain rate, the slab could, for instance be pulled apart in about 3 m.y. Figure 40–12 shows stress contours in a typical slab calculation with this strain rate and the flow law of Eq. (8). These represent the effective strength distribution within the slab for this strain rate.

It is seen that, viewed as a stress guide, the sinking lithosphere resembles a tongue getting thinner and weaker as it dips into the mantle. We now face the question: how can such a thin tongue if subject to resistance in the nether regions yield so as to exhibit shear fractures of predominantly horizontal strike and dipping at

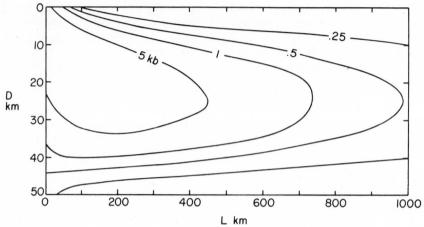

Figure 40-12
"Strength" distribution within sinking lithosphere. Note exaggerated transverse scale
(D). Values plotted are stress (calculated from preliminary laboratory flow law from hot
creep of dunite at a strain rate of 10^{-4}/sec and appropriate values of β) calculated
throughout the sinking slab. Transition D, 50 km slab, $\psi = 53°$, $v = 8$ cm/yr. Note
progressive weakening and/or mechanical thinning of slab as it sinks from surface (left)
to depth of deepest earthquakes (L $\simeq$ 800 km)

angles of about 45° to the slab trend? Further,
how can such fractures occur over most of its
length from near surface to the limiting depth
of earthquakes?

One might expect buckling in such a thin
tongue subject to resistance near its tip. This
could cause short wavelength folding, with
maximum shear on the limbs of the folds. This
is one farfetched solution to the enigma of the
observed focal mechanisms. If this were the
case, one would expect to see evidence for
these folds in earthquake distributions and in
the distribution of focal mechanisms. There is
no trace of it in Sykes (1966) earthquake distributions in the Tonga-Kermadec region,
though the apparent thickening toward the
bottom might suggest some such mechanism.
As yet, good focal mechanism solutions are
too sparse to check this hypothesis.

Another problem is that the effective length
of the strong "tongue" is roughly proportional
to the velocity of sinking. "Resistance" in the
mantle must occur at roughly equal depth
everywhere, so that such resistance should be
felt much more by long, high-velocity tongues
than by short slow tongues. There seems to be
no evidence of such an effect.

Other related problems are associated with
the finding of extension axes down-dip in the

slabs at odd places and the apparent separation
in zones of no earthquakes. Clearly, the slab
can be subject to extension whenever the resistance to downward motion is sufficiently small.
The slab can readily be pulled apart by gradual
necking with increased heating.

With an assumed flow law, the total mechanical behavior of the slab can, in principle, be
numerically calculated for any set of assumed
boundary conditions.

I believe that shear instabilities already
known in the laboratory may suffice to explain
deep earthquake phenomena once the puzzling
geometry of the sources is understood. The
slab center is sufficiently cool at the depths of
the deepest earthquakes so that dehydration
embrittlement and fracture (Raleigh and Paterson, 1965) may occur, so this is one possible
mechanism.

Another type of mechanism is suggested by
our hot creep experiments. In the extensive
work on dunite deformation done in our laboratory first by Blacic and now by Post (unpublished data), shear instabilities akin to hot
creep fracture are common. Characteristically,
these seem to develop from zones of distributed shear that progressively become narrower. At the same time, the creep rate (at
constant stress) rises, so that the local strain

rate is increasing roughly exponentially with time, until there is a sudden stress drop. Our understanding of the phenomena involved is at this stage inadequate to know whether these hot-creep instabilities have the characteristics requisite for deep focus earthquakes, but this phenomenon appears to be a promising candidate.

One wonders what the role of transitions is in deep earthquakes. We have done hundreds of deformation experiments in which we transgressed the stability fields of two reconstructive transitions (quartz-coesite and calcite-aragonite) and have noted no mechanical effects equivalent to shear fracture or any other indication of sudden stress drop resulting from the transitions. We cannot, however, exclude the possibility that some such phenomenon exists and may play a role in deep earthquakes.

CONCLUSIONS

The thermal structure in the sinking lithosphere according to my calculations is similar to that of McKenzie and unlike Minear and Toksöz' calculations (1970) with strain energy added. I believe Minear and Toksöz have greatly overestimated the magnitude of strain heating in the boundary layer of the sinking slab. The effects of adiabatic heating, radioactive heating, latent heat of transitions, and radiative conductivity are second order compared to the effects resulting from variations in the assumed thickness of the lithosphere and the rate of descent. (This is true in both Minear and Toksöz' calculations (1970) and in mine.)

The hypothesis is advanced that the limiting depth of earthquakes is that depth at which the minimum temperature in the slab exceeds a critical value of $\beta(T/T_{mp})$. This agrees with the data on maximum depth of earthquakes within the rather wide scatter of the observations. This hypothesis seems physically more reasonable than McKenzie's assumption of a critical temperature.

The loci of mantle transitions within the sinking lithosphere is sensitive to the dT/dP of the transition. An olivine-spinel transition with the Akimoto and Fujisawa slope (1968) would rise within a rapidly descending lithosphere to a level above the ambient low-velocity zone.

Since the density change is large (10%), the actual location of the transition within the sinking slab should be determinable by appropriate studies of body wave transmission from deep earthquakes if an adequate network of local seismometers can be established. This would provide important additional information as to the nature of mantle transitions above 600 to 700 km depth.

The gravity anomaly comprised only of that calculated for the slab, the contribution of a *symmetric* trench and plausible regional compensation, is remarkably close to the observed anomaly across the Tonga trench. If one could solve the problem of determining the nature of regional compensation, then using the known bathymetry and crustal structure, it should be possible to reduce greatly the uncertainty as to lithosphere thickness and the loci of transitions within the sinking slab. It might even be possible to derive outer mantle temperatures with some precision. I favor McKenzie's idea of counter-flow on the landward side of the trench, which would aid in explaining the high heat flow in this region and would also explain the net east-west gravity asymmetry at Tonga, when the known crust and bathymetry are included with the compensated slab anomaly.

The thickness of the lithosphere cannot be uniquely determined from present calculations. It can vary from perhaps 50 to 100 km depending on the nature of the actual transitions in the mantle.

Deep focus earthquakes can be explained by either or both of two mechanisms now known in the laboratory: (a) the dehydration-embrittlement and weakening of Raleigh and Paterson (1965) and (b) hot creep instability as found by Blacic and Post.

The geometry required by recent focal mechanism solutions for deep earthquakes (down-dip compression) presents problems. It is shown that the stress-guide concept has difficulties because strengthwise the sinking slab is a long thin tongue, thinning and weakening from the surface down. Further, if the compression were a result of resistance at some depth, there should be a highly nonlinear relationship between maxim depth of earthquakes and sinking velocity rather than the linear relationship that is observed.

The role of transitions, if any, in deep earthquakes is unknown.

The energy released by the sinking lithosphere is of the order of 10% of the energy given off by the total surface heat flow of the earth, according to my calculations. Hence, this must be an important part of the driving potential for worldwide mantle currents and plate tectonics. It seems paradoxical, however, that focal mechanism solutions of earthquakes in these sinking slabs indicate predominantly down-dip compression. If the sinking slabs were driving the mantle and plate motions, I would expect down-dip extension to predominate.

PLATE TECTONICS AND GEOLOGY

41

Introduction and Reading List

The paradigm of plate tectonics, although immensely stimulating to the geologist, has not provided a universal framework that accommodates all tectonic problems. It is a common experience, in fact, for the observations of a field geologist mapping a given quadrangle on a continent to be so dissimilar to what the marine geologist tells him about the plate tectonics of the adjacent sea floor that the two scientists might well be mapping on different planets. The discrepancy may in part be a matter of scale. Generalizations drawn from plate tectonics are commonly more useful on a global scale than on a detailed local scale. Also, the continents are thicker than the oceanic crust and more heterogeneous, reflecting the longer and more complex histories of the continents. As a result, recognizing many of the geologic implications of plate tectonics has been a slow process.

Geology of the Ocean Basins

It was early research into the geology of the ocean basins that originally suggested the idea of sea-floor spreading, as has already been discussed in Section II. Some important later discoveries about the geology of the ocean basins are presented in this section.

Earth scientists have long dreamed of doing deep-sea drilling to recover complete sedimentary sections of the ocean floor. A decade ago the Mohole Project was started for the purpose of drilling all the way through the oceanic crust to the mantle. This project was terminated short of completion for economic and political reasons. However interest in drilling the sea floor continued and eventually a consortium named the Joint Oceanographic Institutions for Deep Earth Sampling (JOIDES) was formed under the sponsorhsip of the National Science Foundation. Its objective was to drill and sample the upper part of the oceanic crust. This research has turned out to be one of the most important experiments in the history of earth science.

Drilling for the JOIDES program was done by the *Glomar Challenger,* a ship equipped with oil well drilling equipment modified to permit collection of cores up to one kilometer long in the deep ocean. The first scientific problem attacked by JOIDES was sea-floor spreading in the Atlantic — that is, does the geology of the Atlantic sea floor support the idea that the sea floor is growing symmetrically at a steady rate? Negative evidence in the form of the discovery of very old sediments in the center of the Atlantic ocean basin would have rung the death knell for the theory of sea-floor spreading and plate tectonics.

The initial 18-month cruise of the *Glomar Challenger* in the Atlantic, culminating in Leg 3 from Dakar, Senegal, to Rio de Janeiro, Brazil, hit scientific pay dirt. The results are described in Chapter 44 by Chief Scientists Art Maxwell and Richard Von Herzen and by the other specialists who constituted the shipboard scientific party.

This cruise confirmed beyond any reasonable doubt that the Atlantic had formed by sea-floor spreading. Pre-Mesozoic sediments were absent; the sediments were systematically older away from the ridge; and the age of the sediments lying immediately on the basalt basement were consistent with the time scale of magnetic reversals derived from magnetic anomalies in the South Atlantic (Chapter 25). A more direct and compelling confirmation of sea-floor spreading at a nearly constant rate cannot be imagined.

Another important recent development has been the discovery by Frank Press (Chapter 42) that the suboceanic lithospheric layer is denser than the rocks in the mantle beneath it. Press found the variation of density with depth from seismic surface waves traveling across ocean basins, combining this information with other geophysical data to obtain an internally consistent earth model. A more complete account of this research is given by Forsyth and Press (1971). In the article included in this book as Chapter 42, Press proposed that the dense lithospheric layer was about half eclogite, as had been suggested by geologists back to the time of Arthur Holmes (Chapter 3). The result is a gravitationally unstable system in which mantle convection is highly probable.

Island Arcs, Benioff Zones, and Mountain Belts

Island arcs have long challenged the imagination of geologists. As early as 1962, for example, Robert Coats realized that the Aleutian Arc had been formed by under-thrusting of the oceanic crust. As evidence for this he cited the focal-plane

Figure 41-1
Extruding a deep-sea core drilled in the South Atlantic from the *Glomar Challenger*. Research results from the cruise during which this photograph was taken established unambiguously that sea-floor spreading had occurred in the South Atlantic at a nearly constant rate. Left: Jim Dean, Cruise Operations Manager. Center and right: Art Maxwell and Dick Von Herzen, chief scientists on this cruise. (Courtesy of Scripps Institution of Oceanography)

solutions known at that time for earthquakes beneath the arc, and he presented geochemical data to show that the magma of andesite volcanoes was formed by the mixing of basaltic magma melted from the mantle with sediments and serpentine dragged down along the downward thrusting oceanic plate (Fig. 41–2).

There is no science of paleoseismology. Where earthquakes were formerly generated along a Benioff zone, there are no "fossil seismograms" that may be analyzed to determine the location and depth of the prehistoric earthquakes. However, William Dickinson, a geologist, and Trevor Hatherton, a geophysicist, have developed an approach that uses the geochemistry of andesite volcanoes to determine indirectly the depth of a Benioff Zone (Chapter 43). Their collaborative work

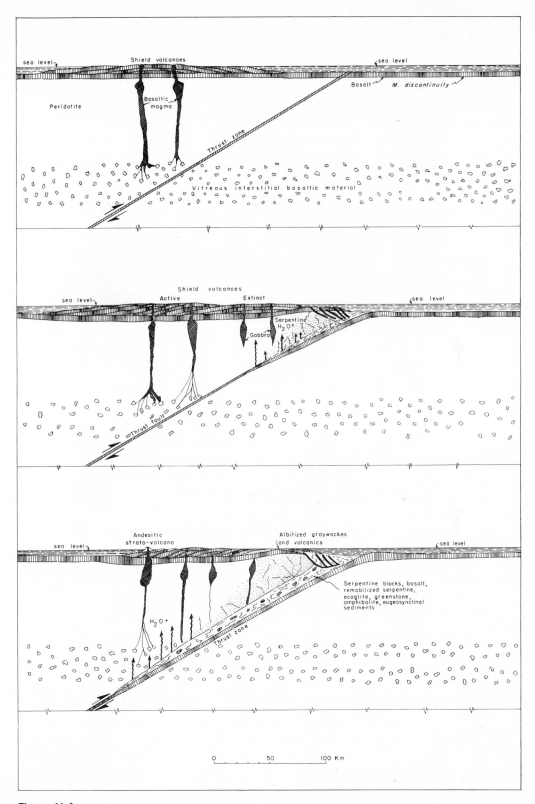

Figure 41-2
A model suggested by Robert Coats (1962) to account for the initial formation of island arcs such as the Aleutians. The andesitic volcanic rocks are formed by the addition of water and hyperfusible material from eugeosynclinal deposits to basaltic material in the mantle. (From R. R. Coats, Geophysical Monograph Number 6: *The Crust of the Pacific Basin,* 1962. With permission of the author and the American Geophysical Union)

Figure 41-3
Seiya Uyeda aboard the *Mayflower* at Plymouth

began late in 1966 when Hatherton was visiting Stanford University. Dickinson, on the staff at Stanford, was trying to make some sense of volcanic rocks from Fiji and had compiled a vast array of chemical analyses of volcanic rocks from around and within the Pacific basin. The geologist and the geophysicist began to worry jointly about the connection between volcanic activity above island arcs and seismic activity beneath the arcs. Dickinson noted that the amount of potassium increased with distance from the trench. Hatherton was intrigued with the observation that andesite volcanoes occur along linear trends parallel to the arcs. He realized that geometrically a line can be the locus of either a moving point (e.g., hot spot in the mantle) or the intersection of two planes (e.g., Benioff zones and isothermal surfaces). Putting their ideas and data together, they found, as Kuno (1966) had found in Japan, that magma sources at different depths along the Benioff zone produce lavas with different compositions. This relationship permits the geologist to delineate ancient Benioff zones from the geochemistry of ancient volcanic rocks.

New insight into some of the larger scale features of island arcs has been provided by the collaboration of another geologist-geophysicist pair, Tokihiko Matsuda and Seiya Uyeda (Chapter 47). They define a Pacific-type orogenic zone consisting of a pair of orogenic belts: an outer eugeosynclinal belt and an inner belt containing granitic intrusions. Starting with the immense body of geologic information available for Japan, they developed a model that explains the broad geologic pattern of metamorphism and intrusion in terms of the thermal effects of thrusting a lithospheric plate beneath the Japan Sea. Their study represents an excellent synthesis of geologic and geophysical data from a complex area.

An even broader look at the geology of plate tectonics is taken by John Dewey and John Bird in Chapter 46. They envisage mountain belts originating through a complex sequence of events beginning with the opening of a new ocean basin such as the Atlantic. All of the deposits seen at present along a profile from the continental shelf to the oceanic ridge are eventually compressed to form a metamorphic mountain belt when a new trench develops near the continental margin. The evolution of magma at different stages in this process is described by Gilluly (Chapter 48), who, while acknowledging that he is now a believer in plate tectonics, points out that all tectonic activity and all magma generation can be attributed to plate tectonics only as an act of faith.

North America-Pacific Plate Boundary

North Americans have an active plate boundary in their own backyard: the North American-Pacific boundary. Along the coast of California this boundary is the San Andreas fault. Unravelling the history of this complex zone has been one of the great achievements of plate-tectonic theory. The first surprise in applying plate tectonics to this area came with the discovery that along magnetic profiles extending westward from California, the anomalies became *older* with increasing distance from the continent, not younger as they do on the east coast. Vine (1966) realized that this required the former existence of a rise in the eastern Pacific separated from North America by a trench. Neither the rise nor the trench exist off the coast of California today, but their former presence is clearly indicated by the magnetic anomalies.

The sequence of tectonic events that transpired as the ancient East Pacific rise migrated toward North America is complex. Contributions toward understanding this history were made by McKenzie and Parker (Chapter 7), Morgan (Chapter 8), Pitman and Hayes (1968), and numerous others. A masterful synthesis tying together most of the relevant geophysics and much of the geology of western North America is given in Chapter 45. This research was done by Tanya Atwater while she was a student at Scripps Institution of Oceanography. Describing how she got caught up in this research (which was not part of her thesis research), she writes (personal communication):

> From the moment the plate concept was introduced, the geometry of the San Andreas system was an obviously interesting example. The night Dan McKenzie and Bob Parker told me the idea, a bunch of us were drinking beer at the Little Bavaria in La Jolla. Dan sketched it on a napkin. "Aha!" said I, "but what about the Mendocino trend?" "Easy!" and he showed me three plates. As simple as that! The simplicity and power of the geometry of those three plates captured my mind that night and has never let go since.
>
> It is a wondrous thing to have the random facts in one's head suddenly fall into the slots of an orderly framework. It is like an explosion inside. That is what happened to me that night and that is what I often felt happen to me and to others as I was working out (and talking out) the geometry of the western U.S. I took my ideas to

John Crowell one Thanksgiving day. I crept in feeling very self-conscious and embarrassed that I was trying to tell him about land geology starting from ocean geology, using paper and scissors. He was very patient with my long bumbling, but near the end he got terribly excited and I could feel the explosion in his head. He suddenly stopped me and rushed into the other room to show me a map of when and where he had evidence of activity on the San Andreas system. The predicted pattern was all right there. We just stood and stared, stunned.

The best part of the plate business is that it has made us all start communicating. People who squeeze rocks and people who identify deep ocean nannofossils and people who map faults in Montana suddenly all care about each others' work. I think I spend half my time just talking and listening to people from many fields, searching together for how it might all fit together. And when something does fall into place, there is that mental explosion and the wondrous excitement. I think the human brain must love order.

Plumes: an Epilogue

Much of the current research in plate tectonics is concerned with testing the idea of Wilson (1955a) that chains of islands like the Hawaiian Islands are produced as a crustal plate advances over a "hot spot," or plume, rising from deep in the mantle. The plume is envisaged as a convective upwelling similar to a thunderhead in the earth's atmosphere. Chapter 49 by Morgan presents the basic elements of this idea. A reading list is given below.

READING LIST

Structure of Upper Mantle and Crust

Anderson, D. L., 1965, Recent evidence concerning the structure and composition of the earth's mantle: Phys. Chem. Earth, v. 6, p. 1–131.

Anderson, D. L., 1967a, Latest information from seismic observations, in Gaskell, T. F., ed., The earth's mantle: New York, Academic Press, p. 355–420.

Anderson, D. L., 1967b, Phase changes in the upper mantle: Science, v. 157, p. 1165–1173.

Anderson, D. L., Sammis, C., and Jordan, T., 1971, Composition and evolution of the mantle and core: Science, v. 171, p. 1103–1112.

Birch, F., 1951, Remarks on the structure of the mantle and its bearing upon the possibility of convection currents: Amer. Geophys. Union, Trans., v. 32, p. 533–534.

Dickinson, W. R., and Luth, W. C., 1971, A model for plate tectonic evolution of mantle layers: Science, v. 174, p. 400–404.

Forsyth, D. W., and Press, F., 1971, Geophysical tests of petrological models of the spreading lithosphere: J. Geophys. Res., v. 76, p. 7963–7979.

Kanamori, H., 1970, Velocity and Q of mantle waves: Phys. Earth Planet. Interiors, v. 2, p. 259–275.

Kanamori, H., and Press, F., 1970, How thick is the lithosphere?: Nature, v. 226, p. 330–331.

Le Pichon, X., 1969, Models and structure of the oceanic crust: Tectonophysics, v. 7, p. 385–401.

Macdonald, G. J. F., 1964, The deep structure of continents: Science, v. 143, p. 921–929.

Menard, H. W., 1967c, Transitional types of crust under small ocean basins: J. Geophys. Res., v. 72, p. 3061–3073.

Pekeris, C. L., 1966, The internal constitution of the earth: Roy. Astron. Soc., Geophys. J., v. 11, p. 85–132.

Press, F., 1968, Earth models obtained by Monte Carlo inversion: J. Geophys. Res., v. 73, p. 5223–5234.

Press, F., 1970, Earth models consistent with geophysical data: Phys. Earth Planet. Interiors, v. 3, p. 3–22.

Press, F., and Kanamori, H., 1970, How thick is the lithosphere (abs.): Amer. Geophys. Union, Trans., v. 51, p. 363.

Thompson, G. A., and Talwani, M., 1964, Crustal structure from Pacific basin to central Nevada: J. Geophys. Res., v. 69, p. 4813–4837.

Geochemistry, Petrology, and Plate Tectonics

Dickinson, W. R., 1968, Circum-Pacific andesite types: J. Geophys. Res., v. 73, p. 2261–2269.

Dickinson, W. R., 1969, Evolution of calc-alkaline rocks in the geosynclinal system of California and Oregon, in McBirney, A. R., ed., Proceedings of Andesite Conference: Oreg., Dept. Geol. Miner. Ind., Bull., no. 65, p. 151–156.

Dickinson, W. R., 1970, Relation of andesites, granites, and derivative sandstones to arc-trench tectonics: Rev. Geophys., v. 8, p. 813–860.

Dickinson, W. R., and Hatherton, T., Andesitic volcanism and seismicity around the Pacific: Science, v. 157, p. 801–803.

Engel, A. E. J., Engle, C. G., and Havens, R. G., 1965, Chemical characteristics of oceanic basalts and the upper mantle: Geol. Soc. Amer., Bull., v. 76, p. 719–734.

Green, D. H., 1969, The origin of basaltic and nephelinitic magmas in the earth's mantle: Tectonophysics, v. 7, p. 409–422.

Green, D. H., 1971, Composition of basaltic magmas as indicators of conditions of origin: application to oceanic volcanism: Roy. Soc. London, Phil. Trans., ser. A, v. 268, p. 707–725.

Green, D. H., and Ringwood, A. E., 1967, The genesis of basaltic magmas: Contrib. Mineral. Petrology — Beitr. Mineral. Petrologie, v. 15, p. 103–190.

Green, T. H., Green, D. H., and Ringwood, A. E., 1967, The origin of high-alumina basalts and their relationships to quartz tholeiites and alkali basalts: Earth Planet. Sci. Lett., v. 2, p. 41–51.

Green, T. H., and Ringwood, A. E., 1968, Genesis of the calc-alkaline igneous rock suite: Contrib. Mineral. Petrology — Beitr. Mineral. Petrologie, v. 18, p. 105–162.

Hatherton, T., 1969a, The geophysical significance of calc-alkaline andesites in New Zealand: N. Z. J. Geol. Geophys., v. 12, p. 436–459.

Hatherton, T., and Dickinson, W. R., 1968, Andesite volcanism and seismicity in New Zealand: J. Geophys. Res., v. 73, p. 4615–4619.

Hatherton, T., and Dickinson, W. R., 1969, The relationship between andesitic volcanism and seismicity in Indonesia, the Lesser Antilles, and other island arcs: J. Geophys. Res., v. 74, p. 5301–5310.

Kay, R., Hubbard, N. J., and Gast, P. W., 1970, Chemical characteristics and origin of oceanic ridge volcanic rocks: J. Geophys. Res., v. 75, p. 1585–1613.

Kuno, H., 1966, Lateral variation of basalt magma across continental margins and island arcs, *in* Poole, W. H., ed., Continental margins and island arcs: Canada, Geol Surv., Pap. 66–15, p. 317–336.

Kuno, H., 1967, Volcanological and petrological evidence regarding the nature of the upper mantle, *in* Gaskell, T. F., ed., The earth's mantle: New York, Academic Press, p. 89–110.

McBirney, A. R., ed., 1969, Proceedings of Andesite Conference: Oreg., Dept. Geol. Mineral Ind., Bull., no. 65, 193 p.

Rises and Rifts

Bodvarsson, G., and Walker, G. P. L., 1964, Crustal drift in Iceland: Roy. Astron. Soc., Geophys. J., v. 8, 285–300.

Chase, G. G., Menard, H. W., Larson, R. L., Sharman, G. F., III, and Smith, S. M., 1970, History of sea-floor spreading west of Baja California: Geol. Soc. Amer., Bull., v. 81, p. 491–498.

Deffeyes, K. S., 1970, The axial valley: a steady-state feature of the terrain, *in* Johnson, H., and Smith, B. L., eds., The megatectonics of continents and oceans; New Brunswick, N. J., Rutgers Univ. Press, p. 194–222.

Demenitskaya, R. M., and Karasik, A. M., 1969, The active rift system of the Arctic Ocean: Tectonophysics, v. 8, p. 345–351.

Dietz, R. S., 1966, Passive continents, spreading sea floors, and collapsing continental rises: Amer. J. Sci., v. 264, p. 177–193.

Engel, A. E. J., and Engel, C. G., 1964, Composition of basalts from the mid-Atlantic ridge: Science, v. 144, p. 1330–1333.

Engel, C. G., and Fisher, R. L., 1969, Lherzolite, anorthosite, gabbro, and basalt dredged from the mid-Indian Ocean ridge: Science, v. 166, p. 1136–1141.

Herron, E. M., and Hayes, D. E., 1969, A geophysical study of the Chile ridge: Earth Planet. Sci. Lett., v. 6, p. 77–83.

Krause, D. C., and Watkins, N. D., 1970, North Atlantic crustal genesis in the vicinity of the Azores: Roy. Astron, Soc., Geophys. J., v. 19, p. 261–283.

Le Pichon, X., Houtz, R. E., Drake, C. L., and Nafe, J. E., 1965, Crustal structure of the mid-ocean ridges: 1. Seismic refraction measurements: J. Geophys. Res., v. 70, p. 319–339.

Le Pichon, X., Hyndman, R. D., and Pautot, G., 1971b, Geophysical study of the opening of the Labrador Sea: J. Geophys. Res., v. 76, p. 4724–4743.

Loncarevic, B. D., Mason, C. S., and Matthews, D. H., 1966, Mid-Atlantic ridge near 45° north, I. The median valley: Can. J. Earth Sci., v. 3, p. 327–349.

Matthews, D. H., Vine, F. J., and Cann, J. R., 1965, Geology of an area of the Carlsberg ridge, Indian Ocean: Geol. Soc. Amer., Bull., v. 76, p. 675–682.

Melson, W. G., Thompson, G., and Van Andel, T. H., 1968, Volcanism and metamorphism in the mid-Atlantic ridge, 22° N lat.: J. Geophys. Res., v. 73, p. 5925–5941.

Melson, W. G., and Van Andel, T. H., 1968, Metamorphism in the mid-Atlantic: Mar. Geol., v. 4, p. 165–186.

Menard, H. W., 1969a, Elevation and subsidence of oceanic crust: Earth Planet. Sci. Lett., v. 6, p. 275–284.

Menard, H. W., 1969b, Growth of drifting volcanoes: J. Geophys. Res., v. 74, p. 4827–4837.

Phillips, J. D., Thompson, G., Von Herzen, R. P., and Bowen, V. T., 1969, Mid-Atlantic ridge near 43° N: J. Geophys. Res., v. 74, p. 3069–3081.

Sclater, J. G., Hawkins, J. W., Mammerickx, J., and Chase, C. G., 1972, Crustal extension between the Tonga and Lau ridges: Petrologic and geophysical evidence: Geol. Soc. Amer., Bull., v. 83, p. 505–518.

Van Andel, T. H., and Bowin, C. P., 1968, Mid-Atlantic ridge between 22° and 23° north latitude and the tectonics of mid-ocean rises: J. Geophys. Res., v. 73, p. 1279–1298.

Van Andel, T. H., Corliss, J. B., and Bowen, V. T., 1967, The intersection between the mid-Atlantic ridge and the Vema fracture zone (abs.): Amer. Geophys. Union, Trans., v. 48, p. 133.

Van der Linden, W. J. M., 1969, Extinct mid-ocean ridges in the Tasman Sea and in the western Pacific: Earth Planet. Sci. Lett., v. 6, p. 483–490.

Vogt, P. R., 1971, Asthenosphere motion recorded by the ocean floor south of Iceland: Earth Planet. Sci. Lett., v. 13, p. 153–160.

Subduction Zones

Bunce, E. T., 1966, The Puerto Rico trench, *in* Poole, W. H., ed., Continental margins and island arcs: Canada, Geol. Surv., Pap. 66–15, p. 165–176.

Chase, C. G., 1971, Tectonic history of the Fiji plateau: Geol. Soc. Amer., Bull., v. 82, p. 3087–3110.

Chase, R. L., and Bunce, E. T., 1969, Underthrusting of the eastern margin of the Antilles by the floor of the western North Atlantic Ocean, and origin of the Barbados ridge: J. Geophys. Res., v. 74, p. 1413–1420.

Coats, R. R., 1962, Magma type and crustal structure in the Aleutian arc, *in* Macdonald, G. A., and Kuno, H., eds., Crust of the Pacific basin: Amer. Geophys. Union, Geophys. Monogr. 6, p. 92–109.

Dickinson, W. R., 1971c, Plate tectonic models of geosynclines: Earth Planet. Sci. Lett., v. 10, p. 165–174.

Dietz, R. S., and Holden, J. C., 1966, Miogeosynclines in space and time: J. Geol., v. 74, p. 566–583.

Fisher, R. L., 1961, Middle America trench: topography and structure: Geol. Soc. Amer., Bull., v. 72, p. 703–720.

Hayes, D. E., 1966, A geophysical investigation of the Peru-Chile trench: Mar. Geol., v. 4, p. 309–351.

James, D. E., 1971, Plate tectonic model for the evolution of the central Andes: Geol. Soc. Amer., Bull., v. 82, p. 3325–3346.

Jones, J. G., 1971, Aleutian enigma: a clue to transformation in time: Nature, v. 229, p. 400–403.

Karig, D. E., 1970, Ridges and trenches of the Tonga-Kermadec island arc system: J. Geophys. Res., v. 75, p. 239–254.

Karig, D. E., 1971, Origin and development of marginal basins in the western Pacific: J. Geophys. Res., v. 76, p. 2542–2561.

Ludwig, W. J., Ewing, J., Ewing, M., Murauchi, S., Den, N., Asano, S., Hotta, H., Haya-kawa, M., Asanuma, T., Ichikawa, K., and Noguchi, I., 1966, Sediments and structure of the Japan trench: J. Geophys. Res., v. 71, p. 2121–2135.

Ludwig, W. J., Hayes, D. E., and Ewing, J., 1967, The Manila trench and West Luzon trough, 1. Bathymetry and sediment distribution: Deep-Sea Res., v. 14, p. 533–544.

Miyashiro, A., 1967, Orogeny, regional metamorphism and magmatism in the Japanese islands: Dansk Geol. Foren., Medd., v. 17, p. 390–446.

Scholl, D. W., and Christensen, M. N., Von Huene, R., and Marlow, M. S., 1970, Peru-Chile trench sediments and sea-floor spreading: Geol. Soc. Amer., Bull., v. 81, p. 1339–1360.

Scholl, D. W., Von Huene, R., and Ridlon, J. B., 1968b, Spreading of the ocean floor: Undeformed sediments in the Peru-Chile trench: Science, v. 159, p. 869–871.

Stoiber, R. E., and Carr, M. J., 1971, Lithospheric plates, Benioff zones, and volcanoes: Geol. Soc. Amer., Bull., v. 82, p. 515–522.

Worzel, J. L., 1966, Structure of continental margins and the development of ocean trenches, *in* Poole, W. H., ed., Continental margins and island arcs: Canada, Geol. Surv., Pap. 66–15, p. 357–375.

Transforms and Fracture Zones

Burk, C. A., and Morres, E. M., 1968, Problems of major faulting at continental margins, with special reference to the San Andreas fault system, *in* Dickinson, W. R., and Grantz, A., eds., Proceedings of conference on geologic problems of San Andreas fault system: Stanford Univ., Publ. Geol. Sci., v. 11, p. 358–374.

Crowell, J. C., 1968, Movement histories of faults in the Transverse ranges and speculations on the tectonic history of California, *in* Dickinson, W. R., and Grantz, A., eds., Proceedings of the conference on geologic problems of San Andreas fault system: Standford Univ., Publ. Geol. Sci., v. 11, p. 323–341.

Dickinson, W. R., and Grantz, A., 1968a, Indicated cumulative offsets along the San Andreas fault in the California coast ranges, *in* Dickinson, W. R., and Grantz, A., eds., Proceedings of the conference on geologic problems of San Andreas fault system: Stanford Univ., Publ. Geol. Sci., v. 11, p. 117–120.

Hayes, D. E., and Ewing, M., 1971, The Louisville ridge—A possible extension of the Eltanin fracture zone: Amer. Geophys. Union, Antarctic Res. Ser., v. 15, p. 223–228.

Hill, M. L., and Hobson, H. D., 1968, Possible post-Cretaceous slip on the San Andreas fault zone, *in* Dickinson, W. R., and Grantz, A., eds., Proceedings of the conference on geologic problems of San Andreas fault system: Stanford Univ., Publ. Geol. Sci., v. 11, p. 123–129.

Johnson, G. L., 1967, North Atlantic fracture zone near 53° N.: Earth Planet. Sci. Lett., v. 2, p. 445–448.

Van Andel, T. H., Phillips, J. D., and Von Herzen, R. P., 1969, Rifting origin for the Vema fracture in the North Atlantic: Earth Planet. Sci. Lett., v. 5, p. 296–300.

Wellman, H. W., 1971, Reference lines, fault classification, transform systems, and ocean-floor spreading: Tectonophysics, v. 12, p. 199–210.

Mountain Belts, Orogeny, and Plate Tectonics

Bird, J. M., and Dewey, J. F., 1970, Lithosphere plate-continental margin tectonics and the evolution of Appalachian orogen: Geol. Soc. Amer., Bull., v. 81, p. 1031–1060.

Dewey, J. F., and Horsfield, B., 1970, Plate tectonics, orogeny and continental growth: Nature, v. 225, p. 521–525.

Dickinson, W. R., 1971a, Plate tectonic models for orogeny at continental margins: Nature, v. 232, p. 41–42.

Dickinson, W. R., 1971b, Plate tectonics in geologic history: Science, v. 174, p. 107–113.

Dietz, R. S., 1963, Collapsing continental rises: an actualistic concept of geosynclines and mountain building: J. Geol., v. 71, p. 314–333.

Dott, R. H., Jr., 1969, Circum-Pacific Late Cenozoic structural rejuvenation: Implications for sea floor spreading: Science, v. 166, p. 874–876.

Ernest, W. G., 1970, Tectonic contact between the Franciscan melange and the Great Valley sequence — Crustal expression of a late Mesozoic Benioff zone: J. Geophys. Res., v. 78, p. 886–902.

Malfait, B. T., and Dinkelman, M. G., 1972, Circum-Caribbean tectonic and igneous activity and the evolution of the Caribbean plate: Geol. Soc. Amer., Bull., v. 83, p. 251–272.

Marine Sediments and Plate Tectonics

Burckle, L. H., Ewing, J., Saito, T., and Leyden, R., 1967, Tertiary sediment from the East Pacific rise: Science, v. 157, p. 537–540.

Ewing, J., and Ewing, M., 1967, Sediment distribution on the mid-ocean ridges with respect to spreading of the sea floor: Science, v. 156, p. 1590–1592.

Ewing, J., Ewing, M., Aitken, T., and Ludwig, W. J., 1968, North Pacific sediment layers measured by seismic profiling, *in* Knopoff, L., Drake, C. L., and Hart, P. J., eds., The crust and upper mantle of the Pacific area: Amer. Geophys. Union, Geophys. Monogr. 12, p. 147–173.

Ewing, J., Worzel, J. L., and Ewing, M., 1962, Sediments and oceanic structural history of the Gulf of Mexico: J. Geophys. Res., v. 67, p. 2509–2527.

Ewing, M., and Ewing, J., 1964, Distribution of oceanic sediments, *in* Yoshida, K., ed., Studies on oceanography: Tokyo, Univ. of Tokyo Press, p. 525–537.

Ewing, M., Ewing, J., and Talwani, M., 1964, Sediment distribution in the oceans: The mid-Atlantic ridge: Geol. Soc. Amer., Bull., v. 75, p. 17–35.

Gilluly, J., 1969, Oceanic sediment volumes and continental drift: Science, v. 166, p. 992–994.

Gilluly, J., Reed, J. C., Jr., and Cady, W. M., 1970, Sedimentary volumes and their significance: Geol. Soc. Amer., Bull., v. 81, p. 353–376.

Maxwell, A. E., Von Herzen, R. P., Andrews, J. E., Boyce, R. E., Milow, E. D., Hsü, K. J., Percival, S. F., and Saito, T., 1970a, Initial reports of the Deep Sea Drilling Project, v. 3: Dakar, Senegal to Rio de Janeiro, Brazil, December 1968 to January 1969: Washington, D. C., Gov. Print. Off., 806 p.

Maxwell, A. E., Von Herzen, R. P., Hsü, K. J., Andrews, J. E., Saito, T., Percival, S. F., Milow, E. D., and Boyce, R. E., 1970b, Deep-sea drilling in the South Atlantic: Science, v. 168, p. 1047–1059.

Plate Tectonics of Western North America and the North Pacific

Christiansen, R. L., and Lipman, P. W., 1972, Cenozoic volcanism and plate tectonic evolution of the western United States, II. Late Cenozoic: Roy. Soc. London, Phil. Trans., ser. A, v. 271, p. 249–284.

Churkin, M., Jr., 1972, Western boundary of the North American continental plate in Asia: Geol. Soc. Amer., Bull., v. 83, p. 1027–1036.

Coney, P. J., 1971, Cordilleran tectonic transitions and motion of the North American plate: Nature, v. 233, p. 462–465.

Gilluly, J., 1970, Crustal deformation in the western United States, in Johnson, H., and Smith, B. L., eds., The megatectonics of continents and oceans: New Brunswick, N. J., Rutgers Univ. Press, p. 47–73.

Grow, J. A., and Atwater, T., 1970, Mid-Tertiary tectonic transition in the Aleutian arc: Geol. Soc. Amer., Bull., v. 81, p. 3715–3722.

Hamilton, W., 1961, Origin of the Gulf of California: Geol. Soc. Amer., Bull., v. 72, p. 1307–1318.

Hamilton, W., 1969a, Mesozoic California and the underflow of Pacific mantle: Geol. Soc. Amer., Bull., v. 80, p. 2409–2430.

Hamilton, W., and Myers, W. B., 1966, Cenozoic tectonics of the western United States: Rev. Geophys., v. 4, p. 509–549.

Larson, R. L., Menard, H. W., and Smith, S. M., 1968, Gulf of California, a result of ocean-floor spreading and transform faulting: Science, v. 161, p. 781–784.

Lipman, P. W., Prostka, H. J., and Christiansen, R. L., 1972, Cenozoic volcanism and plate tectonic evolution of the western United States, I. Early and Middle Cenozoic: Roy. Soc. London, Phil. Trans., ser. A., v. 271, p. 217–248.

Page, B. M., 1972, Oceanic crust and mantle fragment in subduction complex near San Luis Obispo, California: Geol. Soc. Amer., Bull., v. 83, p. 957–972.

Scholz, C. H., Barazangi, M., and Sbar, M. L., 1971, Late Cenozoic evolution of the Great Basin, western United States, as an ensialic inter-arc basin: Geol. Soc. Amer., Bull., v. 82, p. 2979–2990.

Silver, E. A., 1969a, Late Cenozoic underthrusting of the continental margin off northernmost California: Science, v. 166, p. 1265–1266.

Silver, E. A., 1971a, Small plate tectonics in the northeastern Pacific: Geol. Soc. Amer., Bull., v. 82, p. 3491–3496.

Silver, E. A., 1971b, Tectonics of the Mendocino triple junction: Geol. Soc. Amer. Bull., v. 82, p. 2965–2978.

Silver, E. A., 1971c, Transitional tectonics and Late Cenozoic structure of the continental margin off northernmost California: Geol. Soc. Amer., Bull., v. 82, p. 1–22.

Wilson, J. T., 1965b, Transform faults, oceanic ridges and magnetic anomalies southwest of Vancouver Island: Science, v. 150, p. 482–485.

Large Scale Plate Movements (Continental Drift)

Allard, G. O., and Hurst, V. J., 1969, Brazil-Gabon geologic link supports continental drift: Science, v. 163, p. 528.

Briden, J. C., 1967, Recurrent continental drift of Gondwanaland: Nature, v. 215, p. 1334–1339.

Bullard, E. C., 1964, Continental drift: Geol. Soc. London, Quart. J., v. 120, p. 1–34.

Bullard, E. C., Evertt, J. E., and Smith, A. G., 1965, Fit of continents around Atlantic, *in* Blackett, P. M. S., Bullard, E. C., and Runcorn, S. K., eds., A symposium on continental drift: Roy. Soc. London, Phil. Trans., ser. A, v. 258, p. 41–75.

Creer, K. M., 1965, Paleomagnetic data from the Gondwanic continents, *in* Blackett, P. M. S., Bullard, E., and Runcorn, S. K., eds., A symposium on continental drift: Roy. Soc. London, Phil. Trans., ser. A, v. 258, p. 27–40.

Dietz, R. S., and Holden, J. C., 1970, Reconstruction of Pangaea: Breakup and dispersion of continents, Permian to present: J. Geophys. Res., v. 75, p. 4939–4955.

Fox, P. J., Pitman, W. C., III, and Shepard, F., 1969, Crustal plates in the central Atlantic: Evidence for at least two poles of rotation: Science, v. 165, p. 487–489.

Francheteau, J., Harrison, C. G. A., Sclater, J. G., and Richards, M., 1970, Magnetization of Pacific seamounts: A preliminary polar curve for the northeastern Pacific: J. Geophys. Res., v. 75, p. 2035–2062.

Funnell, B. M., and Smith, A. G., 1968, Opening of the Atlantic Ocean: Nature, v. 219, p. 1328–1332.

Gough, D. I., Opdyke, N. D., and McElhinny, M. W., 1964, The significance of paleomagnetic results from Africa: J. Geophys. Res., v. 69, p. 2509–2519.

Herron, E. M., 1971, Crustal plates and sea floor in the southeastern Pacific: Amer. Geophys. Union, Antarctic Res. Ser., v. 15, p. 229–237.

Herron, E. M., 1972, Sea-floor spreading and the Cenozoic history of the east-central Pacific: Geol. Soc. Amer., Bull., v. 83, p. 1671–1692.

Kawai, N., Hirooka, K., and Nakajima, T., 1969, Paleomagnetic and potassium-argon age informations supporting Cretaceous-Tertiary hypothetic bend of the main island Japan: Palaeogeogr. Palaeoclimatol. Palaeoecol., v. 6, p. 277–282.

Kawai, N., Nakajima, T., and Hirooka, K., 1971, The evolution of the island arc of Japan and the formation of granites in the circum-Pacific belt: J. Geomag. Geoelec., v. 23, p. 267–294.

Larson, R. L., and Chase, C. G., 1970, Relative velocities of the Pacific, North America and Cocos plates in the middle America region: Earth Planet. Sci. Lett., v. 7, p. 425–428.

Laughton, A. S., 1971, South Labrador Sea and the evolution of the North Atlantic: Nature, v. 232, p. 612–617.

Le Pichon, X., and Fox, P. J., 1971, Marginal offsets, fracture zones, and the early opening of the North Atlantic: J. Geophys. Res., v. 76, p. 6294–6308.

Le Pichon, X., and Hayes, D. E., 1971, Marginal offsets, fracture zones, and the early opening of the South Atlantic: J. Geophys. Res., v. 76, p. 6283–6293.

McKenzie, D. P., 1970, Plate tectonics of the Mediterranean region: Nature, v. 226, p. 239–243.

Phillips, J. D., and Forsyth, D., 1972, Plate tectonics, paleomagnetism, and the opening of the Atlantic: Geol. Soc. Amer., Bull., v. 83, p. 1579–1600.

Phillips, J. D., and Luyendyk, B. P., 1970, Central North Atlantic plate motions over the last 40 million years: Science, v. 170, p. 727–729.

Pitman, W. C., III, and Talwani, M., 1972, Sea floor spreading in the North Atlantic: Geol. Soc. Amer., Bull., v. 83, p. 619–646.

Roy, J. L., 1972, A pattern of rupture of the eastern North American-western European paleoblock: Earth Planet. Sci. Lett., v. 14, p. 103–114.

Smith, A. G., and Hallam, A., 1970, The fit of the southern continents: Nature, v. 225, p. 139–144.

Williams, C. A., and McKenzie, D., 1971, The evolution of the North-East Atlantic: Nature, v. 232, p. 168–173.

Wilson, J. T., 1966, Did the Atlantic close and then reopen?: Nature, v. 211, p. 676–681.

Plumes

Jackson, E. D., Silver, E. A., and Dalrymple, G. B., 1972, Hawaiian-Emperor chain and its relation to Cenozoic circumpacific tectonics: Geol. Soc. Amer., Bull., v. 83, p. 601–618.

Johnson, G. L., and Lowrie, A., 1972, Cocos and Carnegie ridges result of the Galapagos "hot spot"?: Earth Planet. Sci. Lett., v. 14, p. 279–280.

McDougall, I., 1971, Volcanic island chains and sea floor spreading: Nature, Phys. Sci., v. 231, p. 141–144.

Morgan, W. J., 1972, Plate motions and deep mantle convection: Geol. Soc. Amer., Mem. 132 (in press)

Morgan, W. J., in prep., in Shagam, R., ed., Hess memorial volume: Geol. Soc. Amer., Mem. (not verified)

Wilson, J. T., 1963a, Evidence from islands on the spreading of the ocean floor: Nature, v. 197, p. 536–538.

Wilson, J. T., 1963c, A possible origin of the Hawaiian Islands: Can. J. Phys., v. 41, p. 863–870.

Wilson, J. T., 1965a, Evidence from ocean islands suggesting movement in the earth, *in* Blackett, P. M. S., Bullard, E., and Runcorn, S. K., eds., A symposium on continental drift: Roy. Soc. London, Phil. Trans., ser. A, v. 258, p. 145–167.

42

The Suboceanic Mantle

FRANK PRESS
1969

The composition, state, and mechanical properties of the crust and mantle beneath the sea floor are highly pertinent to the mechanism of sea floor spreading and continental drift and are key to understanding the origin of basic and ultrabasic rocks. In this report I give the first independent determination of the density in a portion of the suboceanic upper mantle. Previously only relative densities could be inferred from lateral variations in gravity, or very approximate indications could be obtained by using seismic velocities and empirical or theoretical equations of state. The latter results were rough not only because of the uncertainty in the seismic velocity distribution below the very top of the mantle but also because of the dependency of the equation of state on composition and temperature.

The following geophysical data have been inverted to obtain density and shear velocity models representative of a section from the sea surface to the center of the earth: spheroidal oscillations of the earth $_0S_0$, $_0S_2$ through $_0S_{22}$, $_1S_2$, $_1S_3$, $_1S_5$, $_1S_6$, $_1S_8$, $_1S_{12}$, $_2S_4$, $_2S_6$, $_2S_{10}$; toroidal oscillations[1] $_0T_3$ to $_0T_{21}$; Rayleigh wave phase velocities (Ben-Menahem, 1965) for predominantly oceanic paths, in the period range 125 to 325 seconds; Love wave phase velocity for oceanic paths (Toksöz and Anderson, 1965), period range 80 to 340 seconds; shear velocity distribution in the lower mantle restricted to a narrow range below 800 km determined from apparent shear wave velocities at the Large Aperture Seismic Array in Montana and concomitant shear wave travel times in the range

From *Science*, v. 165, p. 174–176, 1969. Reprinted with permission of the author and the American Association for the Advancement of Science. Copyright 1969 by the AAAS.

[1] I have used the experimental eigenperiods as reviewed and summarized by J. S. Derr (1970) except as follows: $_0S_2$, 3229.0 seconds: $_0S_{11}$, 537.5 seconds.

30° to 100° (Fairborn, 1968); a fixed compressional wave velocity distribution[2]; mass and moment of inertia of the earth[3]. The uncertainty in the eigenperiod and dispersion data was taken to be ± 0.4 percent, which should allow for asphericity and experimental errors (Dahlen, 1968); the shear wave travel times were required to fit to within ± 5 seconds, which represents the scatter in the observations.

The surface wave data derived from oceanic paths primarily provide the resolving power for the structure of the upper mantle under the oceans. These data merge smoothly into the eigenperiod data at a long period, as would be expected if gross lateral variations do not persist below the asthenosphere. The shear wave data primarily constrain the mantle below 800 km. Although the emphasis in this report is on the upper mantle, all of the data must be used to obtain self-consistent models and absolute rather than relative values of upper-mantle density. This procedure allows us to use nearly homogeneous data where most needed, that is, in deducing the structure of the suboceanic upper mantle.

A Monte Carlo procedure was used to find earth models consistent with the preceding data. In comparison with that reported previously (Press, 1968) the program can find a larger number of successful models from among the millions of randomly generated models. The successful models fit a more extensive suite of new data with better precision than was achievable in the earlier study. The eigenperiods, for example, typically fit the data to ± 0.2 percent, although ± 0.4 percent was acceptable.

Complete details of the successful models will be published elsewhere (Press, 1970). Figure 42–1 shows density distributions to a depth of 400 km for 18 successful models. Heavy solid lines show bounds of permissible

solutions, and ticks on these lines indicate depths at which density and velocity were varied randomly (linear gradients being assumed between these points). At the very top of the mantle where the models begin, the densities occupy the entire permissible range, indicating that the data are insufficient to constrain the models to narrow (and geophysically interesting) bounds. However, in the vicinity of 100 km the densities fall in the narrow band between 3.5 and 3.6 g/cm[3], which lies in the upper part of the permissible range. To confirm this result, biased searches were made without success to find acceptable models with densities falling below 3.4 g/cm[3] in this depth range. In the depth range 250 to 400 km, control of density deteriorates, with the models filling more than half the permissible range.

Also plotted in Fig. 42–1 are densities computed by Clark and Ringwood (1964) for petrologic models of the mantle composed of pyrolite and eclogite. Their eclogite model alone is consistent with my results between 80 and 150 km. Either model is acceptable above this region, and the pyrolite model is weakly favored in the region near 300 km. I shall return to this result later.

The successful models differ because of errors in the data and nonuniqueness of the inversion due to lack of a complete data set. With regard to the latter, Backus and Gilbert (1968) have provided a powerful method (δ-ness criterion) for drawing conclusions about earth structure from a given set of earth data. They showed that properly chosen data, in certain cases, can yield local values of density and velocity obtained from a single model by averaging these parameters over a restricted depth range. These local averages are stable even though details of successful models differ as in Fig. 42–1. In computing the averages, weighting functions are used which are determined directly from the data. If the weighting functions are concentrated over narrow depth intervals (that is, short resolving lengths) then geophysically meaningful resolution is possible. Weighting functions pertinent to my data were calculated by Wiggins.[4] These

---

[2]The fixed compressional velocity in the mantle is very close to the array-determined models of L. R. Johnson (personal communication) and J. Fairborn (1968). Numerical tests show that use of a variable compressional velocity would not alter the conclusions regarding density. The velocity in the core is fixed and based on the results of E. S. Husebye and M. N. Toksöz.

[3]I have used 5.976 × 10[27] g and 0.3308 for the mass and dimensionless moment of inertia of the earth.

[4]These weighting functions extend the numerical work of Backus and Gilbert (1968) in considering velocity and density. I am grateful to Dr. Ralph Wiggins for allowing me to use his results in advance of publication.

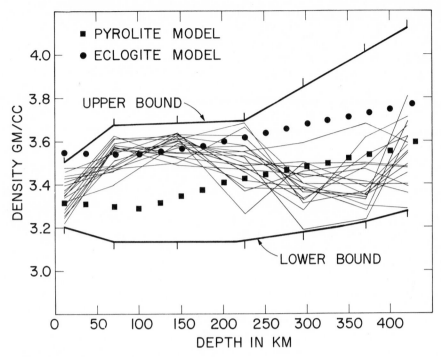

Figure 42-1
Successful density models for the suboceanic upper mantle. Bounds define range permitted in Monte Carlo selection. Points show density values according to Clark and Ringwood (1964) for pyrolite and eclogite mantles

functions together with my density models yield average densities which all fall in the range 3.5 to 3.6 g/cm³ for the depth interval 75 to 125 km. The agreement to 0.1 g/cm³ in the local average density for each of 27 successful models verifies numerically the stability predicted by Backus and Gilbert for weighted averages obtained according to their δ-ness criterion. Presumably the average density near 100 km is uniquely determined and any model computed from our data set should give the same value. The resolution deteriorates below 100 km. For example, at 300 km the resolving length is 200 km.

All successful models show a low velocity zone for shear waves with a lid about 100 km thick. The center of the low velocity zone occurs at depths between 150 and 250 km. The low shear velocity below 150 km supports, on physical grounds, models with low densities below 150 km, implying a density reversal from 3.5 to 3.6 g/cm³ at 100 km to 3.3 to 3.5 g/cm³ at 300 km. No model without a low ve-

locity zone was found, despite a special search in which only monotonic models were generated. Possibly the monotonic model reported by Haddon and Bullen (1969) resulted from a lack of resolution in their work because only modes through $n = 44$ were used.

More complex models for the upper mantle were also found, involving two low-velocity or low-density zones. These models also indicate high density in the upper mantle.

Recent data in support of sea floor spreading and continental drift also imply that the suboceanic mantle-crust system consists of a lithosphere about 100 km thick which behaves mechanically like a rigid plate (Le Pichon, 1968; Isacks et al. 1968). It is underlain by the asthenosphere which is associated with the low velocity, low Q zone and presumably is a region of low strength. These properties of the asthenosphere probably result from partial melting, forming basaltic magma and peridotite or dunite residue (Yoder and Tilley, 1962; Green and Ringwood, 1967; Vinogradov,

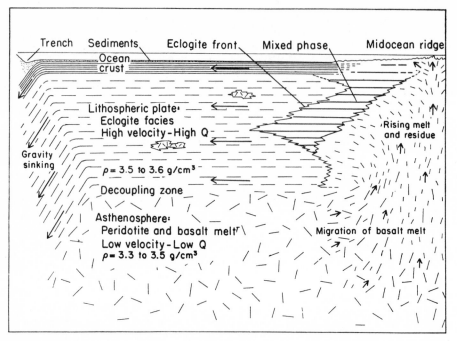

Figure 42-2
Role of eclogite fractionation in the creation of the suboceanic lithospheric plate (not to scale)

1961). The lithospheric plate is produced beneath midocean ridges from which it spreads away, cooling in the process.[5] The asthenosphere serves to decouple the plate from the underlying mantle.

My results uniquely associate high densities of 3.5 to 3.6 g/cm³ with at least the lower half of the lithosphere and suggest reduced densities (3.3 to 3.5 g/cm³) in the asthenosphere. The high density for the lower part of the lithosphere indicates that there it is predominantly of eclogitic composition.[6] I propose that fractionation of eclogite is a key element in the synthesis of the lithosphere.[7] The mechanism, depicted in Fig. 42–2 involves the basalt-

eclogite phase transition (Ringwood and Green, 1966). Basaltic melt rises buoyantly under midocean ridges together with residual dunite or peridotite which moves upward as part of a large-scale advective process involving the asthenosphere and lithosphere. The basalt is extruded to form the ridges, but large amounts are also intruded within the mantle to depths around 100 km. The material cools as it moves laterally away from the ridges, the basaltic melt cooling as a closed system. When critical temperatures are reached at various depths, a zone is defined in which the material solidifies and transforms to an eclogitic facies.

The shape and width of the transformation zone depend on the details of the isotherms in the spreading lithosphere and the nature of the boundary between the basalt and eclogite stability field. The former depends on the spreading rate and the lithosphere thickness[5] and the latter on the composition of the melt (Ringwood and Green, 1966). With so many parameters to adjust, only general statements can be made. The transformation front can be complex in shape and may extend laterally 1000 km or more from the ridge, for the higher spreading rates. An extensive intermediate

---

[5]Isotherms for a spreading lithosphere have been computed under various assumptions by N. H. Sleep (1969); M. G. Langseth et al. (1966); D. P. McKenzie (1967).

[6]The association of high density with eclogite is not unique and is only justified if the choice is made from the current (petrologic) hypotheses for the composition of the upper mantle.

[7]The fractionation of eclogite from a more primitive rock (for example garnet peridotite) has been discussed by several investigators, for example, Yoder and Tilley (1962); O'Hara (1968); O'Hara and Yoder (1967).

mixed phase region may occur, analogous to the pyroxene and garnet granulite facies discussed by Ringwood and Green (1966). Below about 70 km the melt would transform directly to eclogite.

A minimum thickness of about 35 km of eclogite facies has a density 3.5 g/cm³, is 70 km thick, and contains 50 percent eclogite. The melting of large amounts of basalt is implied by the proposed mechanism. Conceivably this could be derived from an asthenosphere (without fully depleting it) about 300 km thick if the level of partial melting is about 15 percent. Otherwise, a source in the deeper mantle would be needed. Inasmuch as our results do not specify the density at the very top of the mantle, the suboceanic M-discontinuity could represent a composition or phase change or hydrous metamorphism.

The dense lithosphere overlying a less dense asthenosphere is gravitationally unstable. Gravitational sliding, which has been suggested as the driving mechanism for the spreading sea floor, would be enhanced by the density inversion (Hales, 1958).

The proposed mechanism is similar in some respects to several earlier concepts. A shell of eclogite around the earth was proposed by Birch (1957), by Lovering, and by Kennedy (1959), among others. Birch argued, as I do, that the conventional density of 3.3 g/cm³ for the top of the mantle was too poorly founded to eliminate eclogite from consideration on this basis. Ringwood and Green (1966) proposed that the transformation of small pockets of basalt to eclogite in the crustal segment of the lithospheric plate drags the crust down near continental margins or island arcs or both. They did not envisage the large-scale transformation to eclogite proposed here. Talwani et al. (1965a) used gravity profiles together with seismic refraction data to infer the existence of a wedge of low-density "anomalous mantle" below the normal mantle and extending as far as 1000 km from the mid-Atlantic ridge and even farther from the east Pacific rise. They suggested that the anomalous mantle was transformed from normal mantle by an unspecified phase change. Simple changes in density and the shape of their anomalous mantle can reconcile their hypothesis with mine and relate their proposed phase change to the basalt-eclogite transformation. The fit to the gravity and seismic refraction data remains unaltered. I take this as substantial support for my concept of the role of eclogite fractionation in the origin of the suboceanic lithosphere.

The preceding conclusion requires the use of all of the experimental data cited earlier, but rests particularly on the phase velocity of surface waves for oceanic paths. Numerical tests indicate that an accuracy of 1 percent in these data is required to establish the high density for the lithosphere. An error analysis of the experiment indicates that this precision was achieved for Love and Rayleigh waves. Comparison with phase velocities for other oceanic paths verifies this.[8] One source of uncertainty is the reduction of suboceanic surface wave phase velocities due to the effect of dissipation or to passage across midocean rises.[9] However, correction for these factors would tend to raise the densities to slightly higher values than I found.

---

[8]For example, comparison of phase velocities in Love and Rayleigh waves of Ben-Menahem (1965) and A. Dziewonski and M. Landisman (personal communication) for entirely different (mostly oceanic) paths, sources, and procedures reveals agreement to 0.5 to 0.7 percent. This argues against the possibility that systematic or other errors in data could vitiate my conclusions.

[9]D. Davies, *Geophys. J.* 13, 421 (1967). Davies overestimates the effect of dissipation by a factor of about two, since I use 15-second shear waves rather than the 1-second shear waves of his calculation.

43

The Relationship Between Andesitic Volcanism and Seismicity in Indonesia, the Lesser Antilles, and Other Island Arcs

TREVOR HATHERTON
WILLIAM R. DICKINSON
1969

From *Journal of Geophysical Research*, p. 5301–5310, 1969. Reprinted with permission of the authors and the American Geophysical Union. Copyright 1969.

We have previously shown (Dickinson and Hatherton, 1967) that in the circum-Pacific island arcs and active continental margins there appears to be a relationship between the potash content of an andesite volcano and the depth of the seismic (Benioff) zone below that volcano. We obtained curves of K_2O contents at 55 and 60% SiO_2, respectively, against the depth h of the Benioff zone (hereafter called the K-h curves). The range of h above which andesitic rocks appear is 80 to 290 km, an interval corresponding approximately to the low-velocity layer of the upper mantle. The simplest model we can devise to explain the correlation is shown by Figure 43–1. Andesitic magmas or their parents are generated at the Benioff zones and rise to the ground surface without undergoing enough contamination to mar the pattern of chemical variations established at sites of partial melting along the Benioff zone.

Our original curves (Dickinson and Hatherton, 1967) were based on a relatively small number of samples, particularly at the higher potash levels; the limiting factor in our statistics was the small number of areas where seismic cross sections had previously been published. We are now endeavouring (*a*) to test our K-h relationship in other areas (Hatherton and Dickinson, 1968) and (*b*) to improve our curves by the addition of other data, especially for volcanoes erupting lavas with high potash contents. In the present study, data from Indonesia, where many of the lavas are high in potash, and from the Lesser Antilles and New Zealand are added to our previous results from circum-Pacific arcs. Together with our original data they allow us to make a statistical assessment of the correlation between K and h.

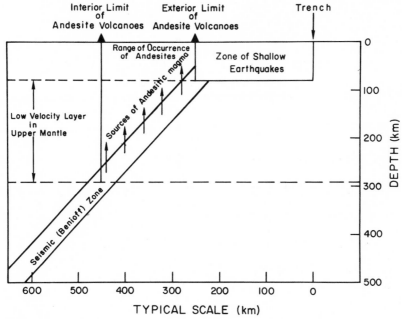

Figure 43-1
Schematic diagram of model proposed for andesite production and distribution

INDONESIA

Seismic sections

Unfortunately no detailed modern studies have been made of the seismicity of this region and to derive depths to the Benioff zones beneath the volcanoes we have had to construct our own seismic sections.

We have examined the monthly Seismological Bulletins of the U. S. Coast and Geodetic Survey from January 1961 to April 1966 and the Catalogue of Epicenters in the International Seismological Summary for 1959 and 1960 and have plotted the epicenters and depths of earthquakes in the region bounded by 6°N, 12°S, 95°E, and 134°E. We have followed Berlage (1937) in drawing, though for limited areas only, lines of equal depth. The positions of the epicenters (of which about 1000 were used), the equal depth lines, and the locations of all active or recently active volcanoes of this area are shown in Figure 43-2. All volcanoes, with the exceptions of Batu Tara in the Flores Sea and Gunung Api in the

Banda Sea, lie between the 100 and 300 km contours, consistent with our model of Figure 43-1.

The positions where seismic sections can be drawn for *K-h* investigations are determined primarily by the availability of chemical analyses for rocks from the various volcanoes. The chemical analyses used in this investigation have been taken from Neuman van Padang (1951), who gives isolated analyses of rocks from many of the volcanoes, but for only a few are there sufficient analyses to draw Harker variation diagrams. Even so, some of the potash contents quoted later are based on as few as four analyses.

The volcanoes for which sufficient analyses for good or approximate K_{55} or K_{60} (percentage $-K_2O$ at 55% SiO_2 or 60% SiO_2) determinations can be made are noted in Figure 43-2. Cross sections AA', BB', and CC' have been drawn in Figures 43-3, 43-4, and 43-5 by projection onto these lines of the epicenters in the enclosed neighboring regions.

Cross section AA' (Figure 43-3) through Java lies close to the volcanoes Ungaran,

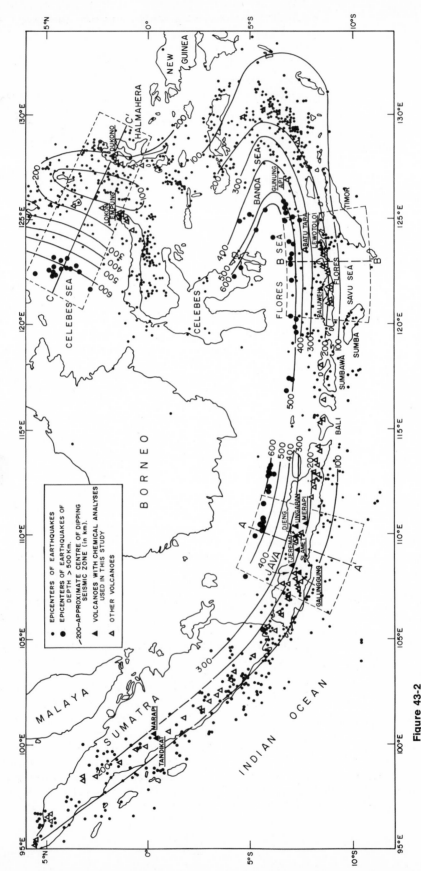

Figure 43-2
Seismic isobaths and volcanoes in Indonesian region with epicenters on which isobaths and sections (Figures 43-3, 43-4, and 43-5) are based

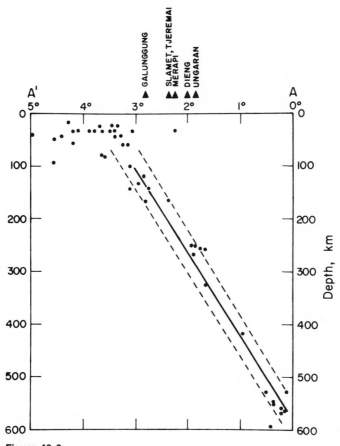

Figure 43-3
Seismic section AA' across Java

Dieng, Merapi, Tjeremai, Slamet, and Galung-gung. This seismic cross section is similar to several that have been published by Sykes (1966) and Hamilton and Gale (1968) for other regions that have a simple trench-andesite arc-continent asymmetry. The thickness of the envelope of the Benioff zone is about 50 km, again similar to the thicknesses obtained by the above authors, and the zone dips beneath the arc from near the inner margin of a region of relatively shallow seismicity between the arc and the trench.

Cross section BB' across Flores and close to the volcanoes Paluweh and Lewotolo lies in a more complex morphological situation. Deep water lies to the northwest of the cross section in the Flores Sea and to the southeast of the section in the Savu Sea, though neither of

these could be considered to be trench-like features. The islands of Sumba and Timor strike obliquely to Flores and intervene between the Sunda trench to the west and the trench to the east of the Banda Sea. Cross section BB' shows a fairly typical Benioff zone but differs from cross section AA' in showing some relatively shallow seismicity both north and south of the upper extension of the inclined zone. The Benioff zone dips away from the region to the south of Flores, and the seismic isobaths of Figure 43-2 show that the deep earthquakes lie to the north of Flores. The volcanoes Lewotolo and Paluweh are in similar positions on the section about 240 km above the center of the zone. Batu Tara lies about 380 km above the center of the Benioff zone but is not an island arc volcano; silica

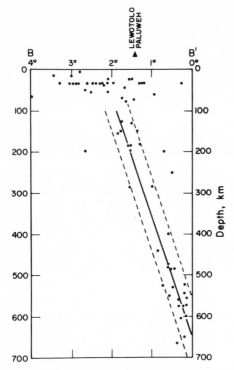

Figure 43-4
Seismic section BB' across Flores

Depths to the centers h of the Benioff zones below the volcanoes have been measured from the seismic sections, with the exceptions of the depths below Tandikat and Merapi that have been estimated from the seismic isobaths. The values of h for the various volcanoes are given in Table 43–1.

Chemical Analyses

Harker partial variation diagrams for K_2O-SiO_2 for the volcanoes discussed above have been plotted from the analyses in Neuman van Padang (1951). Analyses of ash have been rejected and so have all analyses with more than 2% water because of the suspicion of alteration. All other analyses have been corrected to 100% volatile-free. These include basalts and dacites as well as andesites. The K_{55} and K_{60} values for each volcano have been derived after Dickinson and Hatherton (1967); they are listed in Table 43–1, together with the Benioff zone depths beneath the volcanoes.

LESSER ANTILLES

The seismic data of Sykes and Ewing (1965) allow us to construct two transverse seismic sections across the Lesser Antillean arc in places where petrologic data adequate to apply tests of the K-h correlation are available. The two sections are: (1) the more northerly section crossing the arc at Nevis Peak to correspond with petrologic data for Nevis (C. O. Hutton, p. 196 in Weyl, 1966), for nearby Montserrat (MacGregor, 1938, p. 74), and for St. Kitts (Baker, 1968, p. 136–137); and (2) the more southerly section crossing the arc at Mont Pelee to correspond with petrologic data for Martinique (Lacroix, 1924, p. 394–395) and nearby St. Lucia (Robson and Tomblin, 1961, p. 43). The seismic depths h and the potash contents K_{55} and K_{60} for the Lesser Antilles are given in Table 43–1.

NEW ZEALAND

We have previously tested the K-h relationship for some volcanoes in New Zealand (Hatherton and Dickinson, 1968) by using the seis-

contents of Batu Tara are in the range 47–58% and potash contents are between 3.3 and 5.2% in analyses of the lavas.

Cross section CC' lies across another complex situation. The section runs from continental West New Guinea through Halmahera across the Molucca Passage, with a narrow trench separating the floor of the passage from the Northern Celebes; to the northwest of the Northern Celebes is the deep Celebes Sea, one of several such features separating the island arcs from the coast of Asia. Thus we have two opposite-facing systems, which are dissimilar at their extremes, Halmahera-New Guinea being continental and the Celebes Sea presumably being oceanic. These contrasts are reflected in the seismicity: the Celebes system displays a relatively conventional seismic zone dipping west-northwest to depths of 650 km, whereas the Halmahera system dips more gently east-southeast to depths of about 240 km. Although a depth can clearly be assigned to the seismic zone of the Celebes below Lokon-Empung, it is difficult to make such a choice below Dukono.

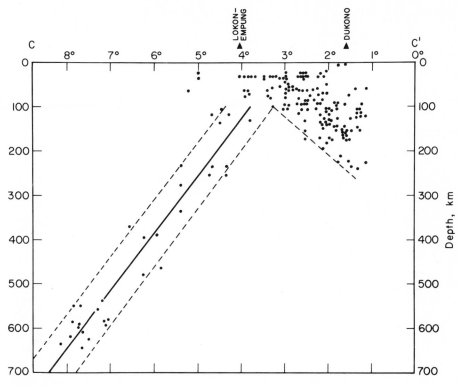

Figure 43-5
Seismic section CC′ across Celebes Sea, Molucca Passage, and Halmahera

mic sections of Hamilton and Gale (1968) and the analyses in Steiner (1958). Recently (Hatherton, 1969), the analyses of rocks from Mt. Egmont, the westernmost andesite volcano in New Zealand, which is high in potash, have been gathered together and K_{55} and K_{60} determined. The depths to the Benioff zone beneath Egmont are also available from Hamilton and Gale (1968), section CC′. All the New Zealand data are also included in Table 43–1.

THE *K-h* PLOTS AND THEIR STATISTICAL SIGNIFICANCE

Figure 43–6 shows the results of our previous study (Dickinson and Hatherton, 1967, Table 1) together with the results reported here, plotted as K_{55}-h and K_{60}-h diagrams. The straight lines of best fit obtained by least squares are also shown in Figure 43–6 and the coefficients of correlation are $r = +0.86$ for Figure 43–6a (55% SiO_2 level) and $r = +0.80$

for Figure 43–6b (60% SiO_2 level. Our previous results with little control at the high potash end appeared to show a quadratic relationship between the two variables; although polynomials might fit the data in Figure 43–6 (particularly at the 60% SiO_2 level) slightly better than straight lines, there appears to be no real justification at present for trying to develop a polynomial relationship between K and h.

The angles of dip of the various Benioff zones we have examined range from 20° to 70°. If the true thicknesses of the zones are similar, the vertical distance between the top and bottom faces increases with the angle of dip. Thus, $h = h_{min} + (t/2 \cos \alpha)$ where h_{min} is the distance from the ground surface to the upper surface of the Benioff zone, t is the true thickness of the zone, and α the dip of the zone. As the difference between h and h_{min} is not constant, we have plotted K_{55} and K_{60} against h_{min} also, to determine if there is a more significant correlation between K and h_{min}. The correlation coefficients are respectively $+0.82$

Table 43–1
Relevant chemistry and geometry of active volcanoes in Indonesia, Lesser Antilles, and New Zealand.

Volcano[a]	No.[b]	Intercept K_{55}[c]	Intercept K_{60}[d]	h[e], km	d[f], km
Indonesia					
(21) Marapi	6, 1–14	1.7	2.5	160	320
(22) Tandikat	6, 1–15	–	1.85	130	300
(23) Galunggung	6, 3–14	0.6	–	130	270
(24) Tjeremai	6, 3–17	1.15	1.95	210	320
(25) Slamet	6, 3–18	1.7	–	210	–
(26) Dieng	6, 3–20	1.9	2.65	260	330
(27) Ungaran	6, 3–23	3.2	–	290	–
(28) Merapi	6, 3–25	2.0	–	225	–
(29) Paluweh	6, 4–15	1.85	2.5	240	–
(30) Lewotolo	6, 4–23	3.0	3.95	240	–
(31) Lokon-Empung	6, 6–10	0.9	–	130	–
(32) Dukono	6, 8–1	1.5	2.5	180	–
Lesser Antilles					
(33) Mt. Misery	16–3	0.6	0.7	120	250
(34) Nevis Peak	16–4	–	1.2	120	250
(35) Montserrat	16–5	0.8	0.9	120	250
(36) Mont Pelee	16–12	0.7	1.0	130	–
(37) Qualibou, St. Lucia	16–14	0.8	1.3	130	–
New Zealand					
(38) Ohakune-Edgecumbe	–	0.9	1.45	140	300
(39) White Island	–	–	2.1	210	300
(40) Egmont	–	2.1	2.5	280	380

[a] For volcanoes 1–20, see *Dickinson and Hatherton* (1967, Table 1).
[b] Number of volcano in *Catalogue of Active Volcanoes of the World* (1951, 1961).
[c] K_{55}, % K_2O at 55% SiO_2.
[d] K_{60}, % K_2O at 60% SiO_2.
[e] Depth to center of Benioff zone below volcano.
[f] Distance to volcano from center of trench, where present.

and +0.72 for K_{55} and K_{60} against h_{min}. If the results of volcano 30, Lewotolo, are ignored, the correlation coefficients improve to +0.88 and +0.83, respectively. These are not significantly different from the coefficients of correlation for K_{55} and K_{60} against h, which include the Lewotolo data.

The significance of the relationship between the potash content of an andesite and its height above the Benioff zone may be compared with the relationship between potash content and other parameters of asymmetry in an island arc, such as the distance d of the volcano from the center of the trench associated with the arc. Previously we have only commented that 'in the gross picture the relationship of d with percentage-K_2O is not as significant as that between h and percentage-K_2O' (Dickinson and Hatherton, 1967, p. 803). In Figure 43–7 we plot K_{55} against d for volcanoes in arcs with well developed trenches. Our K-h relationship implies that in an individual arc K must also increase regularly with d (see Figure 43–1 and linear relationship for Kamchatka data in Figure 43–7). Despite this, the coefficient of correlation of K_{55} and d in Figure 43–7 is only +0.60, and the much greater significance of the relationship between potash content and depth to Benioff zone, with correlation coefficients in the range 0.82 to 0.88, is apparent.

PETROLOGICAL SIGNIFICANCE OF *K-h* PLOTS

If the mantle beneath the arcs is assumed initially to be chemically homogeneous, our model implies that andesitic volcanism effects

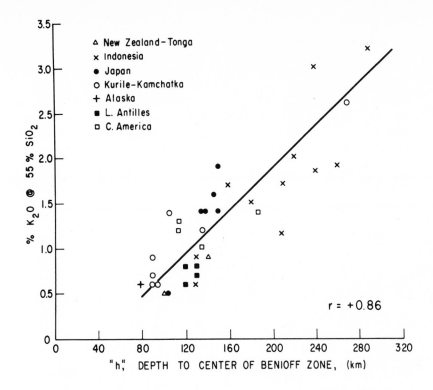

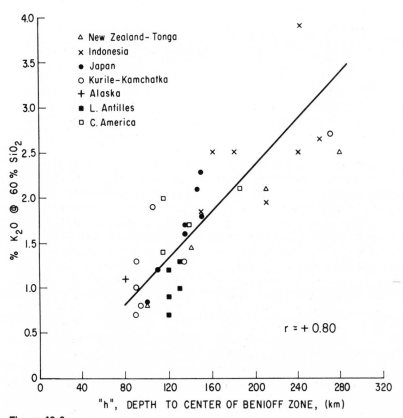

Figure 43-6
Variation in K_2O content of andesites (K_{55} and K_{60}) with depth to center of Benioff zone h beneath volcanoes. Straight lines are least-squares best fits; r is coefficient of correlation

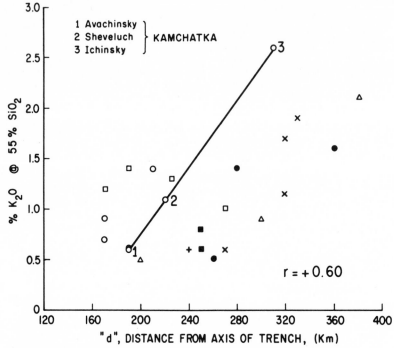

Figure 43-7
Variation in K_2O content at 55% SiO_2 (K_{55}) with distance of volcano from center
of trench d. Straight line shows correlation of K_{55} and d for a single arc,
Kamchatka, resulting from the K-h relationship, r is coefficient of correlation.
For explanation of symbols see Figure 43-6

a primary and irreversible geochemical evolution of the upper mantle (Dickinson, 1968). However, Oliver and Isacks (1967) have suggested that the Benioff seismic zones beneath the arcs may be at the upper surfaces of thick slabs of oceanic crust and upper mantle carried beneath the arcs as part of a fundamental movement plan related to ocean-floor spreading and continental drift. If so, the magmas produced at the seismic zones may gain part of their substance and volatile content from the material descending beneath the seismic zones (Coats, 1962). Green and Ringwood (1968) have shown qualitatively that the low melting fraction of quartz eclogite, to which basalt would invert if carried deep into the mantle, is the andesitic composition in the pressure region corresponding to a depth of 100–150 km. Kushiro (1969) has stated, on the basis of experimentation with hydrous systems, that andesitic magmas could be generated directly from the mantle down to depths of at least 70

km, provided a little water is present. In either case, the ascent of the andesitic magmas may be a means of forming new continental crust in the arcs (Markhinin, 1968), whether wholly from primary fractionation of mantle within the low-velocity zone or partly from secondary fractionation of shallower mantle and oceanic crust carried tectonically to the deeper levels.

The explanation for the successively higher potash contents in magmas arriving at the surface from correspondingly deeper sites of partial melting along the seismic zones remains uncertain in the absence of adequate experimental data relevant to the process. One of us (Dickinson, 1968) has suggested that the equilibrated distribution of potash between a melt phase and appropriate solid phases may shift with changing conditions of temperature and pressure at the melting point at different depths along the seismic zones. As the *K-h* correlation implies not only consistently varying depths of origin for the magmas but also consistently varying path lengths of travel to the surface, it may be that some systematic history of magmatic evolution during ascent can equally well account for the K-h correlation, even if the magmas have nearly the same compositions at all levels of partial melting along the seismic zones. The recent suggestion by Stanton (1967) that the evolution of andesitic magmas may be induced by internal gas transfer in open conduit systems may lead to fruitful lines of thought in this regard. Also of interest is the suggestion of Green et al. (1967) that mantle diapirs may be activated at varying depths along the inclined seismic zones. As such diapirs rose into lower pressure regions, induced partial melting might lead to eventual magma segregation and differentiation at shal-

lower levels. The levels of magma generation would be controlled by the amount of potential superheat in the diapirs, and this would be related to the varying initial diapir temperatures governed by the depth of initiation of upward rise from the seismic zone. Although these and other inferential schemes might explain the K-h correlation, we continue to prefer, at this stage of investigation, the simple and direct hypothesis that relative levels of potash content in the different lava suites are set by partial melting at the Benioff zone. Kuno (1968a) in a thorough review of the andesite problem, has concluded that andesite magmas probably form by crystal fractionation of basalt magmas derived from the mantle and not by contamination.

44

Deep Sea Drilling in the South Atlantic

ARTHUR E. MAXWELL
RICHARD P. VON HERZEN
K. JINGHWA HSÜ
JAMES E. ANDREWS
TSUNEMASA SAITO
STEPHEN F. PERCIVAL, JR.
E. DEAN MILOW
ROBERT E. BOYCE
1970

From *Science*, v. 168, p. 1047–1059, 1970. Reprinted with permission of the authors and the American Association for the Advancement of Science. Copyright 1970 by the AAAS.

For many years ocean scientists have dreamed of recovering complete sedimentary sections of the ocean floor in order to understand more thoroughly the earth's history. The epic 18-month cruise of the *Glomar Challenger* under the aegis of the Joint Oceanographic Institutions for Deep Earth Sampling (JOIDES) has made the dream a reality. Results from the early legs indicate that the sediments so far recovered constitute scientific pay dirt. Leg 3, which covered the South Atlantic between Dakar, Senegal, and Rio de Janeiro, Brazil, has been particularly rewarding because it has provided evidence to support two current global geological hypotheses: sea-floor spreading and continental drift.

A prime objective of leg 3 was to investigate sea-floor spreading with a series of long cores transecting the axis of the Mid-Atlantic Ridge. Figure 44–1 shows the locations of the seven sites on the Mid-Atlantic Ridge flanks and two sites on the Rio Grande Rise. Discussion of site 13 on the Sierra Leone Rise is not included in this paper but has been described in detail elsewhere (Maxwell et al., 1970a).

Each site in this series of seven holes was located within the relatively well-defined magnetic anomaly pattern near 30°S (Dickson et al., 1968). Since the age of basement rock in this area had been predicted from the lineal sequence of magnetic anomalies on the assumption of a constant rate of sea-floor spreading (Heirtzler et al., 1968) the ages of the sediment or rock samples obtained from these sites, as determined by paleontological studies or potassium-argon dating techniques, would provide a test both of the concept of magnetic stratigraphy and of the hypothesis of sea-floor spreading.

From the *Glomar Challenger*, a specialized, deep sea drilling vessel, sediment samples were recovered with standard oil well drilling and coring techniques. The vessel is capable of drilling through sediment thicknesses of about

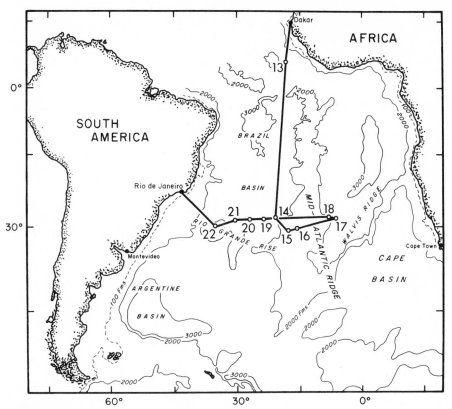

Figure 44-1
Geographic location of drilling sites on leg 3 and general topography of the South Atlantic.
The site numbers relate to the sequence in which the holes were drilled

1 kilometer in water depths greater than 5 kilometers. Operation of the ship was handled by Global Marine, Inc., under subcontract to the Scripps Institution of Oceanography, which is managing the Deep Sea Drilling Project as the operating institution for JOIDES.

PHYSICAL DESCRIPTION OF SITES

Of the nine sites drilled in the South Atlantic, more than one hole was drilled and cored at four sites (two each at sites 13 and 21, three at site 17, and four at site 20). Table 44–1 summarizes the physical parameters at each site that are relevant to this discussion. Most of the sites have been correlated with a previously established numbering system for magnetic anomalies in the sea-floor spreading hypothesis (Heirtzler et al., 1968; Pitman and Heirtz-

ler, 1966). Several of the sites were surveyed before drilling by the Lamont-Doherty Geological Observatory vessel *Vema*.

PALEONTOLOGY

Most of the sediments cored contained an abundance of calcareous microfossils — the calcareous nannoplankton and the planktonic Foraminifera. Siliceous microfossils, such as Radiolaria and diatoms, were also found sporadically at sites 17, 18, and 22. Ages assigned to the sediments in the cores are based mostly on the calcareous microfossils.

An almost continuous composite stratigraphic section has been recovered; it ranges in age from Campanian to Late Pleistocene. The stratigraphic interval recovered from each

Table 44–1
Physical characteristics of South Atlantic drilling sites. The numbering of marine magnetic anomalies (column 6) follows the previously established system (Heirtzler et al., 1968; Pitman and Heirtzler. 1966). Sites 21 and 22 on the Rio Grande Rise were not correlated within the magnetic anomaly numbering scheme. The acoustic reflection time referred to in columns 7 and 8 is the round-trip travel time below the bottom, measured with a 10-cubic-inch air gun, which is part of the seismic profiling equipment.

Site No.	Lati- tude (S)	Longi- tude (W)	Water depth (m)	Maximum sub- bottom pene- tration (m)	Mag- netic anomaly (No.)	Acoustic reflec- tion time (sec)	Remarks
14	28°20′	20°56′	4343	107	Negative 25 km west of 13	0.13	Upper flank of small hill in topography of 40- to 200-m amplitude; acoustic reflection time, 0 to 0.15 sec
15	30°53′	17°59′	3927	142	6	0.15	Lineated N-S topography, sediment thickness, and magnetic anomalies; 40- to 200 m topographic amplitude; acoustic reflection time, 0.10 to 0.15 sec
16	30°20′	15°43′	3527	192	5	0.15	Eastern flank of 1 km high, N-S lineated, 10 km wide topographic rise; acoustic reflection time, 0 to 0.30 sec. Site on small hill 50 m high
17	28°03′	6°36′	4360	127	West of 13	0.18	Lineated N-S topography, sediment thickness, and magnetic anomalies; hills locally 10- to 200-m amplitude. Site at base of hill
18	27°59′	8°01′	4018	178	Between 6 and 13	0.16–0.32	Hills locally 20- to 80-m amplitude; 4800-m deep depression, 50 to 60 km wide in E-W direction, about 100 km east of site
19	28°32′	23°41′	4677	145	21		On east flank of N-S trending ridge, 400-m high; locally irregular topography with diffuse bottom reflections (12 kilohertz)
20	28°31′	26°51′	4506	72	30	0.15(?)	East side of a valley 10 km wide, 4850 m deep. Site on slope extending 600 m above valley floor with uniform sediment thickness
21	28°35′	30°36′	2113	131		0.20–0.25	Northeast slope of Rio Grande Rise. Strong acoustic reflector at 0.1 sec, probably a middle Eocene chalk layer
22	30°01′	35°15′	2134	242		0.52	Small hills (20- to 40-m amplitude); intermediate acoustic reflector at 0.31 and 0.52 sec; higher reflector probably *Braarudosphaera* chalk layer (see Fig. 44–5)

site is shown in Fig. 44–2. Within this geological time interval, there has been continuous coring of numerous stage and series boundaries, including the Pliocene/Pleistocene boundary, the Miocene/Pleistocene boundary, the Oligocene/Miocene boundary, the Eocene/Oligocene boundary, and the Paleocene/Eocene boundary. Many Cretaceous stage boundaries have also been cored, some repeatedly at the various sites, and the Cretaceous/Tertiary boundary has been cored in holes 20C and 21A.

Cretaceous and Early to Middle Cenozoic floras and planktonic Foraminifera faunas are diverse and bear a close resemblance to those reported from tropical regions of the world. However, temperate planktonic Foraminifera faunas dominate the Upper Miocene through Upper Pleistocene sections in the southernmost sites (sites 15 and 16). These assemblages are markedly different from the highly diverse tropical ones in that they are represented by few species. Because many of the diagnostic tropical species currently used for age determinations are absent, determining the age of these assemblages was difficult. However, the calcareous nannoplankton in this same interval is apparently more eurytopic and very similar to that reported from tropical areas.

One of the more important findings is the presence of several layers of *Braarudosphaera* chalk which occur most frequently in the Late Upper Oligocene but which sometimes occur

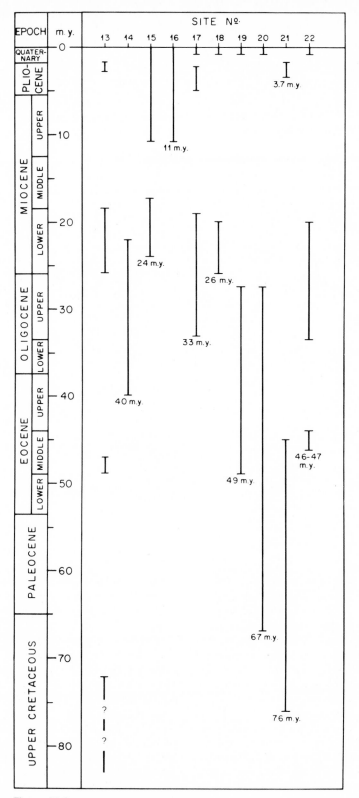

Figure 44-2
Stratigraphic interval recovered at each site plotted to illustrate the almost complete coverage of samples dating from Upper Cretaceous to Late Pleistocene. Only Middle Miocene microfossils were not recovered (m.y., million years)

in the Lower Oligocene. The chalk consists almost exclusively of *Braarudosphaera rosa* Levin and Joeger, as complete specimens and floods of isolated fragments, with a few other species of calcareous nannoplankton. In the Late Oligocene, these chalks usually occur near the boundary of the *Globorotalia opima opima* and *Globigerina ampliapertura* planktonic foraminiferal Zones of Bolli (1957); in the Early Oligocene, in the interval equivalent to the *Globigerina sellii/Pseudoahstigerina barbadiensis* planktonic foraminiferal Zone of Blow (1969). Although studies of modern *Braarudosphaera bigelowi* suggest that it is abundant in sediments originating at very shallow (approximately 10-meter) depths, all the floral and faunal evidence associated with the Oligocene *Braarudosphaera* indicates bathyal depths. This type of chalk is found widely distributed in the South Atlantic; it occurs at sites 14, 17, 19, 20, and 22. There are two hypotheses that may explain its origin. Either unusual oceanographic conditions must have prevailed in this region for short intervals of geologic time to cause the "bloom" of *Braarudosphaera rosa,* or currents carried these shallow water deposits into deep water. A modern analogy of such a bloom would be the so-called "red tide" bloom of certain dinoflagellates. Whereas the "red tide" bloom is of short duration (several weeks), these Oligocene nannoplankton blooms might have lasted for several hundred or several thousand years to produce the chalk.

With minor exceptions, the paleontologic studies showed that at each site the sediment increased in age with increasing depth, implying relatively undisturbed sedimentation conditions. One of the most useful results of these paleontologic studies is the comparison of the paleontologic ages of the sediments immediately overlying the basalt basement at the sites on the Mid-Atlantic Ridge (sites 14 to 20) with the sea-floor spreading hypothesis (Hess, 1962; Dietz, 1961) and with the magnetic stratigraphy concept (Heirtzler et al., 1968). By using the stratigraphic correlation chart of Berggren (1969b), radiometric age dates were determined for the sediments immediately overlying the basalt basement. The planktonic zone correlation and equivalent radiometric ages of the recovered basal sediments are presented in Table 44–2. The implications of these results for the sea-floor spreading concept are discussed more fully in a later section.

STRATIGRAPHIC NOMENCLATURE AND PRACTICES

In order to communicate lengthy and repetitive core descriptions in an effective manner, submarine lithologic units that can be traced from one drill site to another have been designated as submarine subsurface formations. A sequence of nine formations, ranging from Upper Cretaceous to Holocene, has been recognized in the Mid-Atlantic Ridge province (Andrews and Hsü, N.V.). Rather than designating submarine subsurface formations by geographic names, which are relatively rare in the central part of the South Atlantic, we have used the names of historic exploratory vessels. Such a practice employs already familiar names and has an added advantage that the formations at any region could be named in an alphabetical order—for example, from the youngest to the oldest—as a flexible mnemonic device.

MID-ATLANTIC RIDGE LITHOLOGY

The sediments of the Mid-Atlantic Ridge province are almost exclusively pelagic. No turbidity-current deposits or mass slide debris were found, with the exception of a Cretaceous slump block in the Paleocene at site 20. Nonetheless, the lithological changes in time and in space permitted us to identify nine formations. They are, in order of increasing depth, Albatross Ooze, Blake Ooze, Challenger Ooze, Discovery Clay, Endeavor Ooze, Fram Ooze, Gazelle Ooze, Grampus Ooze, and Hirondelle Ooze. The formations have been recognized mainly on the basis of lithological changes with time. Initially, the only criteria selected were (i) variations in Foraminifera content and (ii) differences in noncarbonate content (with corresponding color changes) as determined by visual and smear slide examinations on shipboard, supplemented by shorebased calcium carbonate and grain size analyses. Later, a third criterion—namely, the presence of pink

Table 44–2
Correlation of planktonic foraminiferal zones and calcareous nannoplankton zones from basal Mid-Atlantic Ridge sediments with equivalent radiometric ages (m.y., million years).

Site No.	Planktonic foraminiferal zone	Calcareous nannoplankton zone	Equivalent radiometric age (m.y.)
14	Cribrohantkenina inflata[b]	Helicopontosphaera reticulata[d]	40
15	Globergerinita dissimilis[c]	Triquetrorhabdulus carinatus[d]	24
16[a]	Globorotalia acostaensis[b]		11
	Globorotalia merotumida[b]		
17	Globigerina ampliapertura[b]	Sphenolithus predistentus[d]	33
18	Globorotalia kugleri[c]	Sphenolithus ciperoensis[d]	26
19	Hantkenina aragonensis[c]	Chiphragmolithus quadratus[e]	49
20	Abathomphalus mayaroensis[c]		67

[a]Lowermost sediments recovered were 13.4 m above basalt.
[b]Blow, 1969.
[c]Bolli, 1957.
[d]Bramlette and Wilcoxon, 1967.
[e]Hay, 1967.

oozes — was chosen to establish the Hirondelle as a formation. The properties of the nine formations are summarized in Table 44–3 by use of these criteria.

The identification of the lithologic formations appears to be supported by measurements of natural gamma radiation and, to a lesser extent, by the mass physical properties of the sediments, such as wet-bulk density,

porosity, sediment sound velocity, and a thermal conductivity. Since the natural gamma radiation is probably the least disturbed physical property that was measured on the cores, it seems reasonable that it would provide the most reliable identification.

In the measurement apparatus on shipboard, natural gamma radiation indicated the presence of clay minerals, zeolites, phosphates,

Table 44–3
Summary of diagnostic characteristics used to identify Mid-Atlantic Ridge sedimentary formations.

Formation	Foraminifera (approx. %)	Terrigenous material (approx. %)	Natural gamma radiation[a]	Remarks
Albatross	>10	<10	0->2000	Natural gamma count decreased with depth
Blake	<10	<10	100–300	
Challenger	<10	10–20	200–400	Darker color than Albatross or Blake
Discovery	0	55–100	1100–1700	Red clay
Endeavor	1–8	20–60	250–1600	
Fram	1–3	20–30	250–500	
Gazelle	<3	40–60	500–1200	
Grampus	10–15	20–30	200–400	Darker color toward base of formation
Hirondelle	<10	~30	400	Pink color component

[a]Units are counts per 1.25 minutes, scanning a 7.62-cm core segment.

and possibly hematite, dolomite, and some opaque minerals. Some distinct natural gamma radiation signatures of the lithologic formations are shown in Fig. 44–3 and are listed in Table 44–3.

A composite stratigraphic section of the Mid-Atlantic Ridge formations has been constructed and is shown with the time range of the formations in Fig. 44–4. The type section for each of the formations in the composite section (Fig. 44–4) has been chosen with the following considerations: (i) preferably, the top and bottom contacts of the unit fell within cored intervals; and (ii) lithology of the type section is fairly representative of the formation. In general, the type section is the most completely developed (thickest) section, but not necessarily.

RIO GRANDE RISE LITHOLOGY

Sites 21 and 22 lie on the Rio Grande Rise at a water depth of about 2 kilometers. A composite stratigraphic section of the two sites is as follows: (i) Pleistocene and Pliocene white foraminiferal chalk oozes; (ii) Lower Miocene to Maestrichtian, very pale brown and pink nannofossil chalk oozes, partly recrystallized; (iii) Maestrichtian and Campanian nannofossil chalk oozes, recrystallized, with *Inoceramus* fragments at several horizons; (iv) Campanian or older coquina, cemented by sparry calcite. Noteworthy is the presence of a major unconformity within the Cenozoic. Oligocene and Miocene sediments are missing at site 21, and Pliocene directly overlies Lower Miocene at site 22. The Rio Grande Rise sediments include significant amounts of Foraminifera in calcareous oozes of all ages. Red clays are absent.

LITHOLOGICAL SUMMARY

The distribution of the various Mid-Atlantic Ridge formations in time and in depth is summarized in Fig. 44–5. The stratigraphic columns of sites 21 and 22 on the Rio Grande Rise are also shown for comparison.

Several obvious conclusions may be deduced from Fig. 44–5.

1) Basalt basement has been reached at all the seven sites on the Mid-Atlantic Ridge.

2) The age of the sediments above the basalt bears a direct relation to the distance from the ridge axis: older sediments were found farther from the axis.

3) The sediments increase in age with increasing depth below the sea floor at each site.

4) The thickness of the sediments bears no simple relation to the age of the sediments. Paradoxically, the sites with the youngest basement ages (Miocene in sites 16 and 18) have the greatest sediment thickness (176 and 179 meters, respectively), whereas the site with the oldest basement age (Upper Cretaceous at site 20) has the thinnest sedimentary sequence (72 meters). Local variability in sediment thickness and nonrandom site selection probably account for this pattern.

5) The nature of the topmost formation is related to the distance from the ridge axis and to the present depth of the drill sites. The Albatross Ooze is present at sites 15, 16, 17, and 18, at distances ranging from 221 to 718 kilometers from the ridge axis at depths ranging from 3527 to 4265 meters. The Endeavor Ooze is present at site 14, at 745 kilometers from the ridge axis and at a depth of 4343 meters. The Discovery Clay (covered by thin veneers of local units of similar lithology) is present at sites 19 and 20, 1010 and 1303 kilometers from the ridge axis, at depths of 4677 and 4506 meters, respectively.

6) The nature of the sediments below the topmost formation bears no simple relation to the present depth of the drill site, which indicates that there were changes in depositional environment at each site with time.

7) The sediments at sites less than 500 kilometers from the ridge axis (sites 15, 16, and 18) are mainly Neogene, whereas the sediments farther out are mainly Paleogene (sites 14, 17, 19, and 20).

8) The formations on the whole are not isochronous. The bases of the formations older than Middle Miocene (Hirondelle, Grampus, Gazelle, Fram, Endeavor, and Discovery) tend to become older at drill sites farther from the ridge axis. On the other hand, the trend is reversed for younger formations (Challenger, Blake, and Albatross), the bases of which are either nearly synchronous or younger at drill sites farther from the ridge axis.

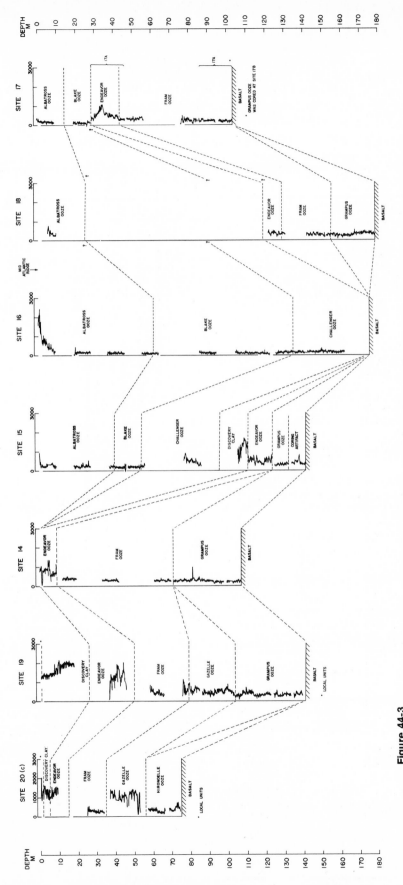

Figure 44-3

Natural gamma radiation signatures versus depth at Mid-Atlantic Ridge sites. Sites are arranged with respect to location from the Mid-Atlantic Ridge axis. Units in counts per 7.62 centimeters of core per 1.25 minutes

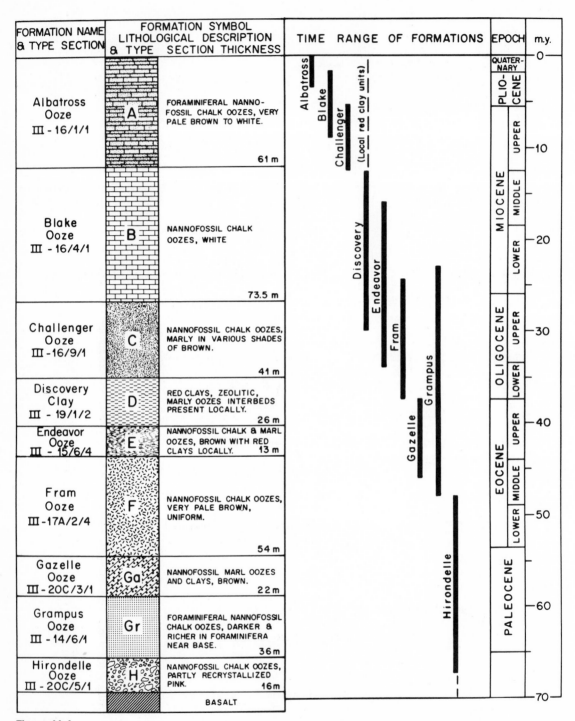

FORMATION NAME & TYPE SECTION	FORMATION SYMBOL & TYPE SECTION	LITHOLOGICAL DESCRIPTION SECTION THICKNESS
Albatross Ooze III - 16/1/1	A	FORAMINIFERAL NANNO-FOSSIL CHALK OOZES, VERY PALE BROWN TO WHITE. 61 m
Blake Ooze III - 16/4/1	B	NANNOFOSSIL CHALK OOZES, WHITE 73.5 m
Challenger Ooze III - 16/9/1	C	NANNOFOSSIL CHALK OOZES, MARLY IN VARIOUS SHADES OF BROWN. 41 m
Discovery Clay III - 19/1/2	D	RED CLAYS, ZEOLITIC, MARLY OOZES INTERBEDS PRESENT LOCALLY. 26 m
Endeavor Ooze III - 15/6/4	E	NANNOFOSSIL CHALK & MARL OOZES, BROWN WITH RED CLAYS LOCALLY. 13 m
Fram Ooze III - 17A/2/4	F	NANNOFOSSIL CHALK OOZES, VERY PALE BROWN, UNIFORM. 54 m
Gazelle Ooze III - 20C/3/1	Ga	NANNOFOSSIL MARL OOZES AND CLAYS, BROWN. 22 m
Grampus Ooze III - 14/6/1	Gr	FORAMINIFERAL NANNOFOSSIL CHALK OOZES, DARKER & RICHER IN FORAMINIFERA NEAR BASE. 36 m
Hirondelle Ooze III - 20C/5/1	H	NANNOFOSSIL CHALK OOZES, PARTLY RECRYSTALLIZED PINK. 16 m
		BASALT

Figure 44-4

A composite stratigraphic section of the Mid-Atlantic Ridge formations showing formation type, relative thickness of formation, and range of age of each formation. The numbers identify the core and the location in it at which the lithologic unit was first encountered (m.y., million years)

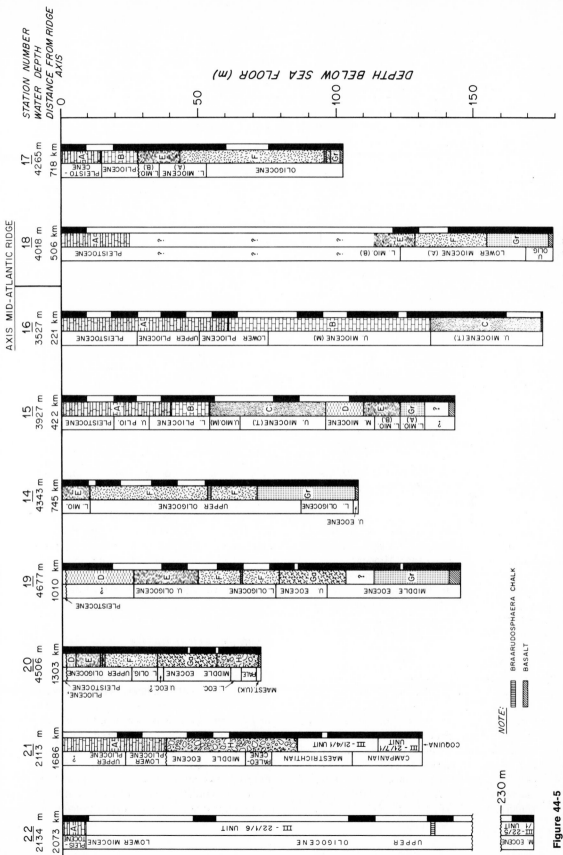

Figure 44-5
Summary of formations, ages, and cores recovered from the Mid-Atlantic Ridge and Rio Grande Rise sites. Water depth and distance from the ridge axis have been added, although they are not shown to scale on the plots

Table 44–4

Percentage of calcium carbonate measured in the various formations at each site on leg 3. Measurements were made by Anthony C. Pimm, Scripps Institution of Oceanography (Maxwell et al., 1970). The number of samples analyzed are in parentheses. Where no value is given, the formation was absent at the site. A question mark indicates that the value is unknown, not sampled, or insufficiently sampled.

| Formation | Site No. | | | | | | | Average of sites |
	16	15	18	17	14	19	20	
Albatross	90.7 (16)	76.8 (10)	? (1)	86.2 (5)				84.6
Blake	90.7 (25)	91.1 (10)	?	87.6 (6)				89.9
Challenger	89.8 (17)	80.1 (6)	?					85.0
Discovery		46.9 (3)	?			0.0	?	23.5
Endeavor		80.0 (8)	78.9 (5)	76.9 (10)	57.6 (6)	42.8 (6)	37.4 (7)	62.3
Fram			83.8 (9)	84.9 (25)	76.8 (15)	74.5 (8)	72.8 (6)	78.6
Gazelle						67.9 (14)	34.1 (10)	50.1
Grampus		83.8	82.6	? (2)	? (0)	78.6 (23)		81.1
Hirondelle							74.3	74.3

9) The Oligocene *Braarudosphaera* chalk is a remarkable time-stratigraphic marker that has been recognized at drill sites some 2800 kilometers apart (sites 17 to 22). This marker is absent in the sites on the ridge crest (sites 15, 16, and 18) because the age of the basalt basement is not older than Miocene.

10) The stratigraphy has a symmetry about the ridge axis, so that the sedimentary sequence at a drill site is more similar to that at a site nearly equidistant from the axis on the other flank of the ridge than to that of its immediate neighbor (compare sites 14 and 17, sites 15 and 18).

The graphical summary, which shows formations as stratigraphic units, cannot express the lateral variations within each of those units. These variations can be examined through the statistical summaries of shore-based carbon/carbonate analyses. Table 44–4 lists these summaries as percent $CaCO_3$ by sites that have been arranged in the order of increasing distance from the ridge axis— namely, 16, 15, 18, 17, 14, 19, 20. From this table the variation of the $CaCO_3$ content of each formation is found, in general, to decrease with the distance from the ridge axis; the same formation is least calcareous at the sites farthest away. A similar decrease in the Foraminifera content with the distance from the ridge axis has been measured; therefore, the same formation also contains, in general, less Foraminifera at the sites farther from the ridge axis.

SEDIMENTATION RATES

The sedimentation rates of the different formations vary with lithology. The rates vary from about 1.8 centimeters per thousand years for a foraminiferal chalk ooze unit to 0.02 centimeter per thousand years for a red clay unit— a difference of some two orders of magnitude. To evaluate the changes resulting from production and dissolution of calcareous planktons on net accumulation rate, we have computed both total and noncarbonate depositional rates

Table 44–5

Summary of total and noncarbonate component sedimentation rates in various formations at each site on leg 3. Rates are in centimeters per thousand years, and the noncarbonate component is given in parentheses. Where no value is given, the formation was absent; a question mark indicates an uncertain rate.

Formation	Site No.							Average of sites
	16	15	18	17	14	19	20	
Albatross	1.8 (0.17)	1.2 (0.28)	?	0.5 (0.07)				1.2 (0.17)
Blake	1.2 (0.11)	0.6 (0.05)	?	0.45 (0.06)				0.8 (0.07)
Challenger	1.0 (0.10)	0.6 (0.12)	?					0.8 (0.11)
Discovery		0.3 (0.16)	?			?	0.02 (0.02)	0.16 (0.09)
Endeavor		?	?	0.3 (0.07)	?	0.35 (0.20)	0.15 (0.09)	0.27 (0.12)
Fram			2 (0.4)	0.75 (0.18)	1.0 (0.4)	0.6 (0.15)	0.4 (0.11)	0.96 (0.25)
Gazelle						0.3 (0.10)	0.2 (0.13)	0.25 (0.12)
Grampus		?	2 (0.4)	?	0.45 (?)	1.2 (0.26)		1.2 (0.33)
Hirondelle							0.1 (0.03)	0.1 (0.03)
Average								0.64 (0.16)

of each formation. As Table 44–5 shows, the average values of noncarbonate components still vary from 0.07 to 0.33 centimeter per thousand years (with the exception of Hirondelle). The variation is probably real, reflecting changes in terrigenous influx, although the errors resulting from assigning absolute ages to paleontologically determined stages might have exaggerated the difference. The average rate of noncarbonate deposition is 0.16 centimeter per thousand years, approximately one-tenth that of a chalk ooze that contains some 10 percent noncarbonate impurities (for instance, Albatross Ooze at site 16).

CALCIUM CARBONATE DISSOLUTION

A synthesis of the facts so far presented has led us to adopt a hypothesis that dissolution during deposition played an important role in determining the lithology of the ridge sedi-ments and their net rates of accumulation. It seems reasonable that the red clays represent the insoluble residues of chalk oozes, especially when the depositional rate of a sediment is compared with its calcium carbonate content. On the other hand, whether the rarity or absence of Foraminifera in a nannofossil chalk ooze could be attributed to dissolution is a debatable question.

Bramlette (1958) emphasized the role of production in determining the composition of calcareous plankton in a pelagic ooze. Production rates do, no doubt, need to be considered. However, several lines of evidence suggest that calcareous Foraminifera are more readily soluble in ocean water than are calcareous nannofossils, so that the paucity of Foraminifera in the nannofossil sediments of the South Atlantic could be attributed largely to differential dissolution. The following indications are noted:

1) The parallel trend in variations of the Foraminifera and $CaCO_3$ content of the ridge

formations suggests that dissolution that increases the noncarbonate impurities of a sediment also tends to decrease its Foraminifera content.

2) The calcareous planktons present in siliceous oozes commonly are exclusively nannofossils, not Foraminifera.

3) The more soluble Foraminifera species are now being readily dissolved at 3000-meter oceanic depth, considerably above the carbonate-compensation depth, so that deeper pelagic oozes include more resistant forms (Berger, 1967; 1968). The fact that the only Foraminifera found in some more marly nannofossil oozes of the ridge province are species resistant to dissolution (for instance, *Globorotalia index* and *G. suteri* in Gazelle Ooze, site 19) suggests that all but a trace of the original Foraminifera fauna has been removed by dissolution from such nannofossil sediments.

Although these arguments may not be definitive, they seem sufficient to us to justify the adoption of a working hypothesis to the effect that the formations of the Mid-Atlantic Ridge province originally represent primarily calcareous sediments that have undergone different degrees of dissolution during deposition. Accordingly, we have postulated five different dissolution facies and have described their physical properties as exemplified by the South Atlantic sediments:

a) Chalk oozes with no evidence of being dissolved. The Quaternary Foraminifera oozes of the Rio Grande Rise may belong to this facies.

b) Chalk oozes with signs of initial dissolution. The ridge sediments with 10 percent terrigenous matter or less but more than 10 percent Foraminifera may belong to this facies.

c) Chalk oozes with signs of moderate dissolution, particularly the dissolution of Foraminifera. The ridge sediments with 10 to 30 percent terrigenous matter and less than 10 percent Foraminifera may belong to this facies.

d) Marl oozes with signs of considerable dissolution; only nannofossils and the more resistant Foraminifera species are preserved. The ridge sediments with 30 to 70 percent terrigenous matter and less than 3 percent Foraminifera may belong to this facies.

e) Red clays; all calcareous planktons have been dissolved.

With this outline, we have interpreted the genesis of the formations as follows: Albatross Ooze, (b) slight loss of carbonate; Blake Ooze, (c) moderate loss of carbonate; Challenger Ooze, (c) moderate loss of carbonate but more terrigenous than Blake; Discovery Clay, (d) considerable loss of carbonate at a crestal site (site 15), but otherwise (e) total loss of carbonate; Endeavor Formation, (d) considerable to (c) moderate loss of carbonate; Fram Ooze, (c) moderate loss of carbonate; Gazelle Ooze, (d) considerable loss of carbonate; Grampus Ooze, (c) moderate loss of carbonate, slight loss at more crestal sites; and Hirondelle Ooze, (d) considerable loss of carbonate but only moderate loss (c) at base of hole 20C.

These interpretations permit us to derive two simple generalizations on the sedimentary history of the Mid-Atlantic Ridge province:

1) Each time marker is represented by a more carbonate-rich formation toward the ridge crest and by a more carbonate-free formation farther out.

2) The Tertiary succession undergoes a cyclic change from carbonate-rich facies to carbonate-free facies, followed by a return to carbonate-rich facies, although the second half of the cycle was developed at the upper flank sites only.

To us, the different dissolution facies appear to be an expression of the varying depths of the accumulation sites as related to the carbonate-compensation depth. The present compensation depth for calcite is approximately 4500 meters in the Pacific (Bramlette, 1958; 1961) and is about the same in the South Atlantic at 30°S, as indicated by the distribution of the modern red clay sediments there. This depth is related to an increase in the rate of calcite dissolution, not to a boundary of equilibrium solubility (Peterson, 1966; Hudson, 1966); the boundary of equilibrium solubility may be only several hundred meters deep (Berger, 1967; Hudson, 1966; Berner, 1965). The level of compensation depth during the past has been a matter of much speculation. There is evidence that ocean waters were warmer during the Tertiary and Cretaceous than they are at present (Emiliani and Edward, 1953; Lowenstarn, 1964). However, whether the compensation level should be deeper or shallower for those warmer oceans is a debatable point.

Hudson (1966) emphasized the effect of temperature on kinetics and postulated a shallower compensation depth for the warmer Cretaceous Chalk Sea. This unorthodox approach did not take into consideration the effect of the degrees of undersaturation on kinetics. That the calcite was being rapidly dissolved in the cold waters below 4500 meters but not at a more nearly saturated warmer level is an evidence that departure from equilibrium exerts the predominant control on the calcite dissolution by ocean waters.

Various authors (Bramlette, 1958; Arrhenius, 1952; Funnell, 1964) have postulated that the calcite-compensation depth may have been considerably greater during the Tertiary. Bramlette (1958) suggested a 6700-meter compensation depth for a Tertiary bottom-water temperature of 12°C. This figure is probably too high. On the assumption of a nonsubsiding ocean-floor, Heath (1969) found an apparent maximum compensation depth of 5200 meters some 35 million years ago (Oligocene). He recognized, however, that this depth may reflect the postdepositional subsidence of sea floor; there may have been very little change in the actual compensation depth.

A further complicating factor is the effect of cold, CO_2-rich bottom currents. Berger (1967) suggested the term "lysocline" for the level at which the solution rate increases drastically, a level that is 500 meters or more above the compensation depth. He cited evidence to show that this surface is not horizontal but is inclined in the South Atlantic, because the Antarctic Bottom Water is responsible for the pronounced abyssal $CaCO_3$ dissolution. As this water at 30°S is confined to a path west of the Mid-Atlantic Ridge on its way northward, the lysocline is probably hundreds of meters shallower on the west side than on the east side of the ridge. Also, since depth differences between successive dissolution facies are probably of the order of some hundreds of meters, it would be unwise to make interpretations in terms of absolute depth. Therefore, a relative bathymetric scale for the dissolution facies with reference to the calcite-compensation

depth (CCD) is suggested, taking into consideration the distribution of Holocene sediments at 30°S. It is (i) no dissolution: some 1500 meters above CCD; (ii) slight dissolution: 500 to 1500 meters above CCD; (iii) moderate dissolution: 200 to 500 meters above CCD (just below lysocline); (iv) considerable dissolution: 200 meters or less above CCD; and (v) complete dissolution below CCD.

The rapid facies changes near the lysocline surface express the fact that dissolution rate changes rapidly there.

As the Tertiary facies changes of the Mid-Atlantic Ridge sediments at 30°S involve changes in relative depth of 1000 meters or more, it does not appear warranted to attribute such changes to past fluctuations in the absolute depth of calcite compensation. Furthermore, even if Heath's interpretations are accepted, a depression of actual compensation depth during the early Tertiary should result in a succession of increasingly carbonate-rich facies if there had been no crustal subsidence. Instead, a progression from high to low carbonate sediments has been found. Consequently, these facies changes are interpreted in terms of the chronologically changing depth at each depositional site.

SEA-FLOOR SPREADING

As has been noted (Fig. 44–5), the age of the sediments recovered immediately above basalt basement at each Mid-Atlantic Ridge site increases with distance from the ridge axis. To decipher the history of sea-floor spreading from these data, the distance of each site from its point of origin at the ridge axis must be determined. The procedure to obtain this distance is not always straightforward, because the Mid-Atlantic Ridge axis has numerous offsets and changes in strike along its length (Heezen and Tharp, 1965).

Near 30°S, the ridge has not been surveyed in detail, although there are a few ship tracks along which geophysical data have been recorded as profiles across the ridge. Location of the ridge axis along these tracks (Fig. 44–6) has been determined from the prominent magnetic anomaly associated with it, or from the

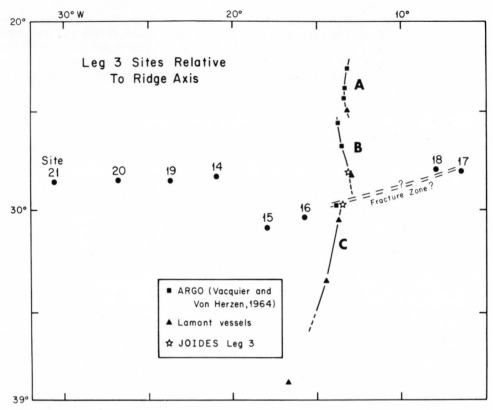

Figure 44-6
Trend of the axis of the Mid-Atlantic Range in the South Atlantic as determined from geophysical evidence obtained at the locations indicated, plus earthquake epicenter data. The drilling sites are shown relative to the ridge axis. (Sources of data: squares from Vacquier and Von Herzen, 1964; stars, from Maxwell et al., 1970a; triangles, from Lamont Geological Observatory data)

characteristic crestal topography (Vacquier and Von Herzen, 1964). Additional data for the trend of the axis are provided from the locations of earthquake epicenters (Stover, 1968). From Fig. 44–6 it is possible to measure the distance from any site to the nearest ridge axis. If only distances from the axis are considered, then the displacement near 29½°S suggests that sites 15 and 16 are associated with ridge segment C, sites 17 and 18 with B, and all others are nearly equidistant from either B or C. Linear distances of sites from the ridge axis are summarized in Table 44–6, in which the errors in distances represent a subjective estimate of the uncertainty in axis location.

Other values of distances from the ridge axis are obtained from a model of rigid rotation of crustal plates on a sphere. This model requires the trace of the motion of any part of the surface to describe a small circle about its axis of rotation. Thus, the distance traversed by any site on the spreading ocean floor, which is rotating away from the Mid-Atlantic Ridge axis, would be slightly different from that obtained simply by measuring the linear distance to the nearest ridge axis. More importantly, the original location of any site along the axis will be different in the two cases. From a hypothesized fitting of the continents of Africa and South America, it has been shown (Bullard et al., 1965) that the present positions of these continents can be explained as a rotation about an axis at 44°N, 30.6°W. Other authors (Morgan, 1968b; Le Pichon, 1968) have refined the location of rotation axes farther north on the basis

Table 44–6
Magnetic anomaly ages, paleontologic ages, and distances of Mid-Atlantic Ridge sites from the axis. The number of the magnetic anomaly and its age has been taken from Heirtzler et al. (1968). The location of sites 18 and 17 within the characteristic magnetic anomaly pattern is uncertain. Basement rock was not reached at site 21. Magnetic ages (column 2) are based on the magnetic anomaly numbers listed in Table 44–1; m.y., million years.

Site No.	Magnetic age of basement (m.y.)	Paleontological age sediment above basement (m.y.)	Distance from ridge axis (km)	
			Linear	Rotation at 62°N, 36°W
16	9	11 ± 1	191 ± 5	221 ± 20
15	21	24 ± 1	380 ± 10	422 ± 20
18		26 ± 1	506 ± 20	506 ± 20
17	34–38	33 ± 2	643 ± 20	718 ± 20
14	38–39	40 ± 1.5	727 ± 10	745 ± 10
19	53	49 ± 1	990 ± 10	1010 ± 10
20	70–72	67 ± 1	1270 ± 20	1303 ± 10
21		>76	1617 ± 20	1686 ± 10

of the trends of linear magnetic anomalies and fracture zones and have inferred differential rates of spreading along the ridge axis.

The last column of Table 44–6 shows the distances of drilling sites from the ridge axis for the rotational spreading hypothesis, obtained by using the rotational axis at 62°N, 36°W proposed by Morgan (1968). The major difference in distances of sites from the ridge axis is for site 17, which for the case of rotation is associated with segment C (Fig. 44–6) of the ridge axis. Site 18, however, remains associated with segment B, so that the difference in distances of sites 17 and 18 from the ridge axis is greater than their actual geographic separation.

A fracture zone that displaces the ridge crest, perhaps extending eastward between sites 17 and 18, may support this interpretation. Unusually large depths were recorded between sites 17 and 18; such depths are sometimes characteristic of fracture zones (Heezen and Tharp, 1965; Menard, 1964). If the fracture zones of the South Atlantic trend somewhat north of due east, a direction consistent with the proposed axes of rotation, then this fracture zone may be the same as that displacing the ridge crest at 29½°S between segments B and C (Fig. 44–6).

Estimated uncertainties in the paleontologic ages have been determined from the stratigraphic correlation by Berggren (1969b). However, a more recent compilation (Berggren, N.V.), based on radiometric dates within stratigraphic sequences on the continents, gives somewhat different ages for some paleontologic stages, particularly during the mid-Tertiary. Implications of some of these uncertainties will be discussed.

The values in Table 44–6 show an overall linear relationship of age versus distance from the ridge axis. If the nearest distance to the ridge axis is selected, the average spreading rate back to 66 million years ago for these drilling sites appears slightly less than 2 centimeters per year. For a rotation axis in the North Atlantic, the best-fitting spreading rate appears to be nearly equal to 2 centimeters per year (see Fig. 44–7). If the more recent paleontologic ages of Berggren (N.V.) are accepted, the average rates may be slightly increased in both cases but are still near 2 centimeters per year.

Sites 17 and 18 were selected on the east side of the ridge to test symmetry of the spreading pattern. Whereas the sediment ages at these sites seem to fit well with the pattern for the western flank for linear distances, site 17 appears to be somewhat younger by the

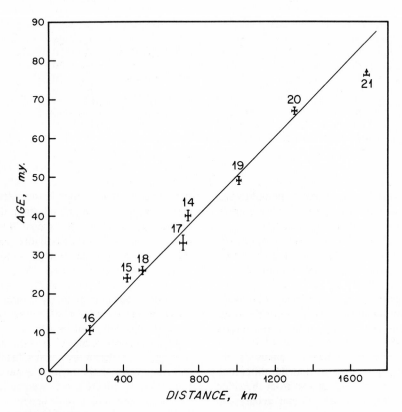

Figure 44-7
The age of the sediment immediately above basalt basement is plotted as
a function of distance from the ridge axis (m.y., million years)

rotation axis interpretation (Fig. 44–7). The lack of fit for site 17 in this case may result from some east-west asymmetry in the spreading pattern, or from fracture zone displacements that may occur between the ridge crest and sites 17 and 18. Unfortunately, the magnetic anomaly pattern on the east flank of the ridge is not well enough determined to be of much help in revolving this minor problem.

The scatter of points in Fig. 44–7 may be interpreted to indicate past changes in the sea-floor spreading rate (Langseth et al., 1966; Ewing and Ewing, 1967). For short-term constant spreading, the range of spreading rates corresponding to variations in slope between points of Fig. 44–7 is from 0.4 to 4 centimeters per year. Similar ranges result if the alternate paleontologic dates of Berggren (in preparation) are used. Because of the uncertainty of the data, it does not appear warranted to conclude that sea-floor spreading has actually varied that much over periods of 5 to 10 million years. Further, there seems to be little support in the data for cessation of sea-floor spreading for periods of more than 10 million years, as has been proposed for the Miocene (5.5 to 22.5 million years) and the Paleocene (53 to 65 million years) (Le Pichon, 1968). Greater ranges in spreading rates than have been observed would be needed if long periods of low or no spreading are postulated.

Another possible explanation for some of the scatter in the data may be shifts in the axis of injection of new rocks in the sea floor, perhaps resulting in a normal distribution about a geographic axis (Matthews and Bath, 1967; Harrison, 1968b). For some clearly defined magnetic profiles across ridges, the standard deviation of the normal distribution appears to be as small as 3 kilometers. For the South Atlantic, where the pattern appears relatively regular, the standard deviation is probably not more than 5 kilometers. In this case, 95 percent of the sea floor has an origin within 10 kilometers of its position predicted from ideal sea-floor spreading (±2 standard deviations, or ±10 kilometers, to either side of the ridge axis); or, in other words, there is only a 5 percent probablility that any part of the sea floor originated more than 10 kilometers away from its most probable position within the sea-floor spreading scheme, if it is assumed that the

process has remained constant throughout the time required to form the sea floor. This additional uncertainty of 10 kilometers might explain some of the data scatter in Fig. 44–7.

Relative displacements of the ocean floor between drilling sites due to fracture zones may also contribute to the scatter of the data. The relatively small scatter implies that any such displacements have been less than 50 to 100 kilometers.

A South Atlantic spreading rate of 2 centimeters per year has been deduced earlier from studies of the marine magnetic field anomalies (Dickson et al., 1968). The magnetic anomaly ages predicted for the drilling sites are listed in Table 44–6. Reasonably good agreement exists between predicted magnetic ages and ages determined from paleontology, especially when the uncertainty in locating the site in relation to the center of the magnetic anomaly is considered. Perhaps the only significant departures from the predicted ages are at sites 19 and 20. Both sites had ages somewhat younger than predicted, which suggests that the proposed geomagnetic time scale (Heirtzler et al., 1968) may require some minor revision for ages of 50 million years or greater, in the same direction but probably not so large as suggested by Le Pichon (1968).

It is concluded that the sea floor of the South Atlantic has been spreading at an essentially constant rate for the past 67 million years. Further, the spreading half-rate of 2 centimeters per year (determined from the drilling) is in agreement with and has provided a critical test for both sea-floor spreading and magnetic stratigraphy.

HISTORY OF THE SOUTH ATLANTIC BASIN

From the data presented above, an interpretation of the geologic history of the South Atlantic evolves from a model of nearly constant sea-floor spreading with significant vertical movements of the crestal region. Changes in the ridge elevation may be related to small but significant changes in the spreading rate. The idea of a lapse in sea-floor spreading has been suggested earlier by Ewing and Ewing (1967) on the basis of interpretations of the seismically determined sediment thickness. In addition, the narrowness of the high heat flow at the axial region of the Mid-Atlantic Ridge led Langseth et al. (1966) to suggest an increase in the rate of spreading during the last 10 million years. Schneider and Vogt (1968) also postulated nonsteady Atlantic spreading, on the basis of additional evidence furnished by discontinuities in ridge topography and in amplitudes of magnetic anomalies. However, the approximately constant spreading rate of 2 centimeters per year since the Cretaceous (determined from the drilling) almost precludes large changes or long interruptions in the spreading rate. Nonetheless, some variations in spreading rate have not been excluded. As we have noted, the data could be interpreted to include short-term changes in spreading rates of nearly one order of magnitude, although we believe it is probable that the spreading has been much more constant.

We have discussed the strong evidence furnished by the sedimentary record of a nonsteady sea-floor depth, which may be correlated with spreading rate variations. For, if there had been steady spreading and flank subsidence only, the geologic column at each site should be a regular succession of gradually deepening sediments. The superposition of moderate to rich carbonate sediments, of Upper Miocene age and younger, over sediments that had been accumulated in much deeper waters (at sites 15 and 17) indicates a Late Cenozoic rejuvenation of the Mid-Atlantic Ridge.

The cause of crestal relief is no doubt related to the abnormally light mantle that now underlies the ridge (Le Pichon, et al., 1965; Talwani et al., 1965a). This light mantle density is, in turn, related to the abnormally high heat flows measured on the axial portion of the ridge (Vacquier and Von Herzen, 1964; Langseth et al., 1966). As a newly created crust was conveyed aside to flank regions underlain by a mantle of normal density, or as the mantle was altered (thermally or chemically) to a more normal structure, subsidence must have occurred. The crestal uplift and the flank subsidence are probably not simply the result of thermal expansion alone. Mantle-density variations caused by phase changes or other processes, as well as isostatic subsidence, must be taken into consideration to explain the relief of ocean bottoms (Schneider and Vogt, 1968; Hsu, 1965; Hsu and Schlanger, 1968).

The drilling results clearly support the evidence that the sea floor of the South Atlantic has been spreading apart since Cretaceous at an average half-rate of 2 centimeters per year, with new basalt crust being continually created at the crestal region. We adopt the premise that the South Atlantic lithostratigraphic formations represent dissolution facies, the depths of which are related to the calcite-compensation depth. In the present discussion, the dissolution facies are attributed primarily to absolute variations in depth of deposition — that is, to variations in ridge elevations — rather than to relative variations of the depth of the lysocline.

As the elevation difference between the crest and flank of the Mid-Atlantic Ridge is some 3000 meters, a crustal segment has probably subsided this amount as it was moved from the crest to the lower flank. The early Tertiary chalk ooze to red clay succession provides evidence for a subsiding conveyor-belt model of sedimentation. Sediments showing slight to moderate dissolution were deposited on a newly created basalt basement in the crestal areas of an ancestral Mid-Atlantic Ridge, which stood hundreds or a thousand meters above the calcite-compensation depth. Sediments of more nearly red clay facies were accumulated as the sea floor was conveyed to the deeper outer flanks by spreading. The sea floor at site 20 reached the compensation depth during Late Oligocene, but the sea floor at site 15 reached it only during Middle Miocene. The carbonate-rich Neogene sediments were deposited in the crestal areas, which remained high above the calcite-compensation depth. This interpretative model best explains the observed facies pattern. The fact that each time marker is represented by a more carbonate-rich formation toward the ridge crest and by a more nearly carbonate-free formation farther out can be considered an expression of the topography of the ancestral Mid-Atlantic Ridge.

A graphic presentation of this interpretation is shown in Fig. 44–8. In this figure a series of stratigraphic sections has been reconstructed for five reference dates: (a) end of Eocene (37 million years ago), (b) end of Oligocene (26 million years ago), (c) end of Lower Miocene Aquitanian (23 million years ago), (d) end of Miocene (6 million years ago), and (e) present. The following procedures were used:

1) For each reference data the sites were displaced toward the ridge axis a distance corresponding to the amount the sea floor had spread during the period between the preceding reference date and the new one.

2) At the new locations all sediments younger than the reference date were removed from the sections.

3) The bathymetry of each site was then redetermined by examining the newly exposed formation and relating its depositional depth to the calcite-compensation depth, as was discussed earlier.

4) The elevation of the ridge axis at each reference date was extrapolated, primarily on the basis of the interpreted bathymetry of the nearby sites.

5) The lithostratigraphy has been shown by a columnar section for each site, and time lines have been drawn between sections to indicate age of the formations.

Although the average linear spreading rate was 2.0 centimeters per year, in some instances this rate was modified slightly to accommodate stratigraphy. For example, site 15, which is now located 422 kilometers from the axis, includes Aquitanian sediments; therefore, a spreading rate of 1.5 centimeters per year had to be used for the reference date of 23 million years.

Figure 44–8 shows that the sedimentary regime in the South Atlantic basin has not changed radically during the last 80 million years. A relatively thin blanket of chalk oozes, with some marl oozes and red clays, has covered the ridge. No turbidity current deposit managed to find its way here, probably because it was trapped in basins near the continents (such as the Brasilian and Angola basins). The formations of the ridge province have been distinguished mainly by their different degrees of dissolution. In contrast, production rate and current erosion may play a more important role in the Rio Grande Rise province. Whereas unconformities between the ridge sediments appear to have resulted largely from dissolution, the hiatus in the rise sedi-

ments has almost certainly been the work of mechanical removal. Data from site 21 suggest that the rise remained high, while the ridge widened and, at times, deepened during the Tertiary.

With the assumption of an average spreading rate of 2 centimeters per year, it is possible to estimate that the separation of South America and Africa began some 130 million years ago, or during the Early Cretaceous. This is consistent with a geological comparison of Brazil and Gabon, which has shown a striking similarity of the Aptian and older rocks in these countries, and their separation was thus dated as Middle Cretaceous, or some 120 million years old (Allard and Hurst, 1969). Unfortunately, the results of the leg 3 drilling cannot give a definitive answer on the age of separation, since the oldest dated sediment in the South Atlantic is Campanian (<80 million years). Yet the evidence does indicate that the South Atlantic came into being either during Late Jurassic or during Early Cretaceous, and the ocean was already some 3000 kilometers wide as the Campanian (oldest cored sediment) at site 21 was deposited.

The oldest Cretaceous sediments of the Rio Grande Rise province were deposited in relatively shallow waters, either on a fragment of a foundered continent or on a group of subsiding guyots. The oldest cored sediments of the ridge province is Maestrichtian, or some 70 million years old. By the end of Eocene (Fig. 44–8a), the ridge was already a distinct topographic feature, although the Atlantic was not yet sufficiently deep for red clays to have been deposited. The very slow rate of deposition of the Maestrichtian, Paleocene, and Lower Eocene sediments in the ridge province and the existence of disconformities indicate, however, considerable dissolution of $CaCO_3$, particularly of planktonic Foraminifera; thus, the Hirondelle Formation at site 20 is considerably thinner and is much poorer in Foraminifera than its counterpart on the Rio Grande Rise.

The Grampus Ooze first appeared during the Middle Eocene at site 19. From then until Early Miocene this unit of moderate and locally high carbonate content invariably includes the first sediments deposited on the newly formed basalt crust (see Fig. 44–8). The

580

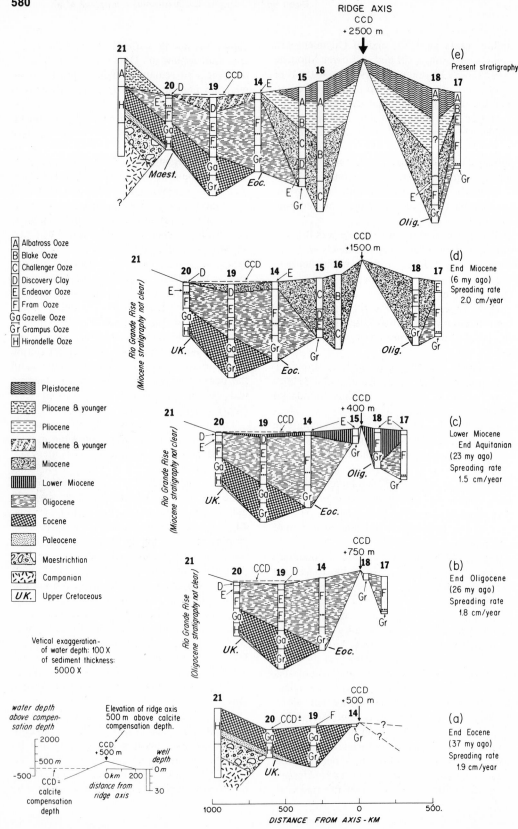

RIDGE AXIS
CCD
+2500 m

(e) Present stratigraphy

CCD
+1500 m

(d) End Miocene
(6 my ago)
Spreading rate
2.0 cm/year

CCD
+400 m

(c) Lower Miocene
End Aquitanian
(23 my ago)
Spreading rate
1.5 cm/year

CCD
+750 m

(b) End Oligocene
(26 my ago)
Spreading rate
1.8 cm/year

CCD
+500 m

(a) End Eocene
(37 my ago)
Spreading rate
1.9 cm/year

|A| Albatross Ooze
|B| Blake Ooze
|C| Challenger Ooze
|D| Discovery Clay
|E| Endeavor Ooze
|F| Fram Ooze
|Ga| Gazelle Ooze
|Gr| Grampus Ooze
|H| Hirondelle Ooze

Pleistocene
Pliocene & younger
Pliocene
Miocene & younger
Miocene
Lower Miocene
Oligocene
Eocene
Paleocene
Maestrichtian
Campanian
U.K. Upper Cretaceous

Vetical exaggeration-
of water depth: 100 X
of sediment thickness:
5000 X

*water depth
above compen-
sation depth*

Elevation of ridge axis
500 m above calcite
compensation depth.

CCD
+500 m

*well
depth*
0 m

— 2000

— 500 *m*

—500 -

CCD =
calcite
compensation
depth

0 km   200
*distance from
ridge axis*

30

1000   500   0   500.

DISTANCE FROM AXIS - KM

Grampus Ooze can thus be considered the crestal deposit on the ancestral Mid-Atlantic Ridge. The Middle and Upper Eocene flank deposits constitute the low carbonate Gazelle Formation. Site 20 must have deepened to the compensation depth during the Late Eocene time, so that the Upper Eocene deposits are almost entirely missing there, probably having been removed by dissolution.

There may have been a slight uplift of the ocean bottom, or a slight depression of the absolute depth of calcite compensation during the Early Oligocene. In any case, the Fram Ooze of moderate $CaCO_3$ content was superposed on the flank above the low carbonate Gazelle. The rates of the Fram sedimentation are also significantly higher. By the end of the Oligocene, however, the flank sites (sites 19 and 20) dropped below the calcite-compensation depth, whereas the crestal area stood some 1000 meters higher to permit high carbonate sedimentation (Grampus of site 18).

A noteworthy sedimentary event took place during the Middle Oligocene, when the *Braarudosphaera* chalk was deposited. This chalk unit extended across the entire basin at this time and is thickest on the Rio Grande Rise. The reason for this impressive bloom of *Braarudosphaera rosa* and the exclusion of other calcareous planktons is unknown. The condition was apparently recurrent, because more than one such chalk layer was found at site 20.

A general subsidence must have commenced during Early Miocene. The Lower Miocene sediment at site 15, which then lay close to the ridge axis, is the Endeavor Ooze of moderate $CaCO_3$ content, suggestive of a deeper ridge axis. Accordingly, a height for the crest of the calcite-compensation depth plus 400 meters has been postulated as 4100 meters by assuming the present calcite-compensation depth, or as 4600 meters by assuming Heath's

(1969) calcite-compensation depth. The subsidence led also to an encroachment of the compensation depth laterally toward the ridge axis. The subsidence continued during the Middle Miocene, when the Endeavor was superposed by a carbonate-poor Discovery unit at site 15. Solution rates on this deep Mid-Atlantic Ridge must have been rapid. The Middle Miocene is represented either by a red clay barren of calcareous planktons, or by a solution unconformity at the ridge sites farther out than site 15.

The condition of red clay sedimentation in the Middle Miocene was not restricted to the South Atlantic. The Miocene at site 10 on the flank of the northern Mid-Atlantic Ridge showed largely zeolitic red clays (Peterson et al., 1970). Mid-Tertiary red clays barren of fossils at site 12 in the Cape Verde Basin, (Peterson et al., 1970) and at site 13 on the Sierra Leone Rise (Maxwell et al., 1970a) were also deposited during a time span that included the Middle Miocene.

The great depth of the Middle Miocene Atlantic may have coincided with a period of lower crustal heat flow and of denser subcrestal material, which produced a slower rate of sea-floor spreading. This theory is supported for the period from the end of Aquitanian to end of Miocene by the sedimentation pattern, which suggests a rate markedly slower than 2.0 centimeters per year.

A rejuvenation of the Mid-Atlantic Ridge and an increase in spreading rate took place during the Late Miocene. By the end of the Miocene (Fig. 44–8), the new crestal area stood at about 1500 meters above the calcite-compensation depth (at a depth of 3000 meters, if the present calcite-compensation depth is assumed). At site 16, located near the crest, the Challenger Ooze is high in $CaCO_3$, though poorer at site 15 some 200 kilometers to the

Figure 44-8 *(facing page)*
Reconstruction of the history of the South Atlantic, with sea-floor spreading, paleontologic ages, and lithologic formations taken into consideration. Details of the procedures used are described in the text. The distances between sites, the bathymetry, and the stratigraphical thickness are shown at different scales for effective illustration. The top line connecting each section represents the past topography; the bottom line illustrates the sediment-basalt contact. (a) End of Eocene (37 million years ago); spreading rate, 1.9 centimeters per year. (b) End of Oligocene (26 million years ago); spreading rate, 1.8 centimeters per year. (c) Lower Miocene to end of Aquitanian (23 million years ago); spreading rate, 1.5 centimeters per year. (d) End of Miocene (6 million years ago); spreading rate, 2.0 centimeters per year. (e) Present stratigraphy

west. At sites 14 and 19, the sea floor remained below the compensation depth, where the Upper Miocene is represented by a red clay and by unconformity.

The uplift of the ridge may have continued during the Plio-Pleistocene. The present crestal elevation at a depth of 2 kilometers has probably never been exceeded. The Albatross Ooze with moderate to high $CaCO_3$ content is now being deposited to a distance of some 650 kilometers from the ridge axis (site 17). Farther out, at sites 14, 19, and 20, the ocean floor has remained largely below the compensation depth since Late Oligocene or Early Miocene. Younger calcareous plankton deposits are either absent or are present only as a very thin veneer.

SUMMARY

A series of nine holes drilled in profile across the Mid-Atlantic Ridge in the South Atlantic has provided samples from which a sedimentary history of the South Atlantic can be reconstructed back to Early Cretaceous. Paleontologic studies of the age of the sediment immediately above the basalt basement indicate that the sea floor has, overall, spread uniformly at an average half-rate of 2 centimeters per year during this period. Nine separate sediment formations were identified from their lithologies and physical properties. By assuming a hypothesis of calcium carbonate dissolution during deposition, we have interpreted the distribution of these formations as an indication of major changes in the elevation of the Mid-Atlantic Ridge. The changes in elevation may be associated with short-term changes in the rate of sea-floor spreading. For example, the ridge was depressed in elevation during the Middle Miocene, at a time when the sea floor may have spread more slowly as compared with earlier and more recent periods.

The overall rate of sea-floor spreading determined from the drilling has been found to be nearly identical with the rate determined from the study of magnetic anomalies. This agreement provides strong support for the hypothesis of sea-floor spreading and the concept of magnetic stratigraphy. An extrapolation of the sea-floor spreading rate to the continental shelves in the South Atlantic suggests that the ocean in this area was formed during the Early Cretaceous Period.

45

Implications of Plate Tectonics for the Cenozoic Tectonic Evolution of Western North America

TANYA ATWATER
1970

From *Bulletin of the Geological Socity of America*, v. 81, p. 3513–3536, 1970. Reprinted with permission of the author and the Geological Society of America.

The theories of sea-floor spreading and plate tectonics as described by Vine (1966), Morgan (1968b), and Isacks and others (1968), have had such great success in predicting large-scale phenomena in the oceans that they are almost universally accepted by marine earth scientists. On the other hand, their implications for continental work are more obscure and complex and are only beginning to be developed. It is likely that many of the larger scale features of continental geology are related to plate motions, the energy for tectonic activity being derived from the interactions of plates. Many modern examples exist to support this view: the uplift and volcanism in the Andes appear to be related to the destruction of an oceanic plate, the uplift and deformation of the Himalayas to the collision of two continental plates, the opening of the African rift valley to the breaking of a plate, and so on. Although these continental expressions of plate interactions are less straightforward than the oceanic expressions, they hold the key to the past. While the sea-floor spreading process has left ample evidence in the oceanic crust to reveal Cenozoic plate motions, the subduction of oceanic crust at the trenches has annihilated similar evidence for any plate motions that might have occurred before a few hundred million years ago. Thus, the main possibility for unraveling plate motions during most of geologic time lies in understanding the relationship of continental geology to plate interactions. It behooves us to study known Cenozoic examples of such interactions where oceanic evidence still exists to indicate the nature of the motions.

Of the presently active continental tectonic zones, western North America seems especially promising for study since a great deal is known about both onshore and offshore features, and discernible relationships presently exist between the 2 regimes. The purpose of this paper is to outline what is known about plate motions from marine geophysical data in

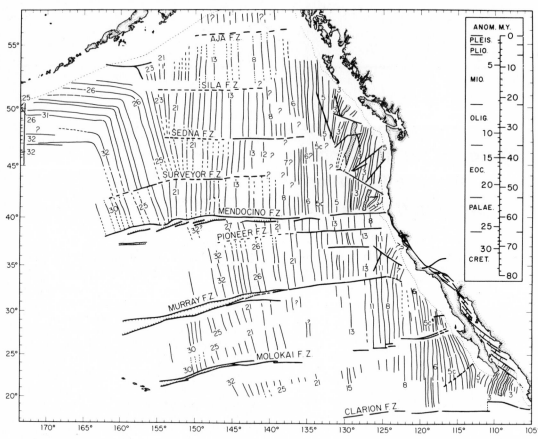

Figure 45-1
Magnetic anomalies in the northeast Pacific from Atwater and Menard (1970). Numbering of anomalies and their ages shown in the scale follow Heirtzler et al. (1968); ages of geologic epochs follow Berggren (1969a)

this region and to discuss the possible relationships of these motions to continental tectonics. Although many of the conclusions reached have been previously expressed or implied by Vine (1966), Morgan (1968b), and McKenzie and Morgan (1969), their geologic implications have not been stressed.

IMPLICATIONS DRAWN FROM THE MAGNETIC ANOMALIES IN THE NORTHEAST PACIFIC OCEAN

Figure 45-1 shows the pattern of magnetic anomalies in the northeast Pacific. The numbering of the anomalies follows the time scale set up by Heirtzler and others (1968, Fig. 3). Although most of the ages in this time scale

were highly speculative when they were proposed, dates from the Deep Sea Drilling Project and from an abyssal hill dredge haul indicate that they are approximately correct (Maxwell and others, 1970b; McManus and Burns, 1969; Luyendyk and Fisher, 1969). All anomaly ages used in this paper were taken from the Heirtzler time scale, and the correlation of ages to geologic epochs follows Berggren (1969a).

If we use the plate model for sea-floor spreading and continental drift developed by Morgan (1968b) and McKenzie and Parker (1967), and assume that the central Pacific Ocean floor acted as a rigid plate during the Cenozoic, then the locations of the anomalies and the fracture zones indicate the configurations through time of the ancient spreading

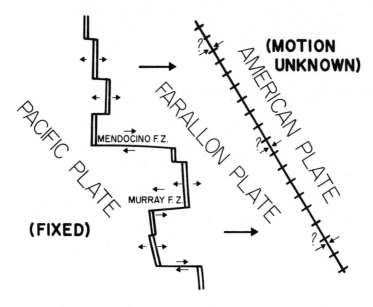

Figure 45-2
Configuration of plate boundaries at the time of anomaly 21, about 53 m.y.
ago, as deduced from Figure 45-1. Magnetic anomalies and fracture zones are
being created by the ridges and transform faults between 2 rigid oceanic
plates. In this and subsequent figures, symbols follow McKenzie and Parker
(1967): single lines are transform faults, double lines are spreading centers,
and hatched lines are zones of subduction (usually trenches). Large arrows
show motions of plates with respect to the Pacific plate which is arbitrarily
held fixed. Small arrows show relative motions at points along plate boundaries

ridge and its transform faults. For example,
Figure 45–2 shows the configuration of the
spreading system at the time when anomaly 21
was being created, the early Eocene.

A Mid-Cenozoic Trench off
Western North America

A striking feature of Figure 45–1 is that the
anomalies present represent only the western
half of the symmetrical pattern expected. All
presently known ridges spread approximately
symmetrically, and so we might expect to find
the other halves of these anomalies somewhere. In fact, the eastern halves and the ridge
itself are for the most part missing. This geometry indicates that there once was another
plate lying to the east of the ridge, the "Farallon plate" of McKenzie and Morgan (1969),

which contained the missing anomalies, and
since most of this plate no longer exists, there
must have been a trench which consumed it at
its boundary with the American plate. This
trench apparently consumed the Farallon plate
at a rate that was, on the average, faster than
the rate at which it was being created at the
ridge (5 cm/yr), so that eventually the ridge
itself was overrun by the trench. At any point
along the margin, the age of the youngest
anomaly present in the oceanic plate indicates
a time at which the ridge was still viable and
spreading. Furthermore, since the symmetrical
eastern anomaly is missing, the trench must
also have been viable and consuming crust at
least until that time.[1]

---

[1]It is geometrically possible that the consumption
of the Farallon plate occurred entirely after the formation of all anomalies, but this possibility seems
highly unlikely.

The numbers along the continental margin in Figure 45–3 show the ages of the youngest anomalies recognizable. They indicate that the trench was active until at least 29 m.y. ago just south of the Mendocino fracture zone. Between the Pioneer and Murray fracture zones, the Farallon plate broke up, and spreading slowed about 32 m.y. ago, so that rapid creation and consumption of crust ceased for this entire segment at that time. The trench continued to consume crust more slowly until at least as recently as 29 to 24 m.y. ago. South of the Murray fracture zone, the easternmost anomalies become younger southward from 20 m.y. Near Guadalupe Island, they lack distinctive character and thus remain unidentified except that they probably belong to the anomaly 5B-5A sequence (16 to 12 m.y.). South of the Shirley trough (the eastward extension of the Molokai fracture zone), the anomalies changed strike so that they are approximately parallel to the coast; thus anomaly 5A (11 m.y.) is the easternmost one along much of southern Baja California (Chase and others, 1970).

All of these ages supply only upper limits for the time when the trench ceased activity, for it is possible that younger anomalies once existed offshore and subsequently were overridden by the continent. It should also be emphasized that these anomalies lie within the Pacific plate and move with it. The geometrical relationship between the events they date and events on the continent cannot be deciphered until the relative motion of the North American plate is determined. Nevertheless, the anomalies do indicate that a trench existed along the western United states and Mexico in middle Tertiary times.

The character of the trench that consumed the Farallon plate may be guessed from the rate at which it consumed crust. This rate was, on the average, greater than 5 cm/yr, since the trench was able to overtake the ridge; the most likely models, discussed below, suggest consumption rates of 7 to 10 cm/yr. Thus, the trench may have been approximately equivalent to the more active trenches known today (Kurile, Japan, Tonga; *see* Le Pichon, 1968), having a well-developed Benioff zone, calc-alkaline volcanism, and associated vertical tectonics.

Limits on the Age of the San Andreas Fault

Some conclusions can also be drawn concerning the history of the San Andreas fault system. According to the plate tectonics model (Fig. 45–4) this fault system is simply part of the present boundary between the North American and Pacific plates and expresses their relative motion (Morgan, 1968b; McKenzie and Parker, 1967). In these terms, the San Andreas can exist only where the 2 plates are in direct contact. As long as the Farallon plate lay between the American and Pacific plates, the San Andreas system could not function as a strike-slip boundary. Thus, no part of the San Andreas system began movement in the present sense and rate before 30 m.y. ago, and the ages of the offshore anomalies discussed above give age limits not only for the ceasing of trench activity but also for the beginning of the various segments of the San Andreas system.

This young age for the San Andreas system contradicts data which suggest a considerable amount of early and middle Tertiary slip: the post-Oligocene offset is about 350 km (Huffman, 1970), while the post-Cretaceous offset of the basement is about 500 km (Hill and Hobson, 1968). It is possible that a pre-San Andreas fault may have existed behind the middle Tertiary trench as a trench-related, strike-slip fault; such relationships are not uncommon behind presently active trenches (Allen, 1962; Burk and Moores, 1968). Another possibility, that older offsets on the San Andreas fault were related to a distinct, early Tertiary episode of oblique motion, is discussed in the last section of this paper.

CONCERNING THE MOTION OF NORTH AMERICA

To understand how the Eocene plate configuration (Fig. 45–2) evolved into the present one (Fig. 45–4), we must establish how the North American plate moved with respect to the oceanic plates during the intervening time span. As discussed below, evidence suggests that North America has been moving with respect to the Pacific plate at a rate of 6 cm/yr

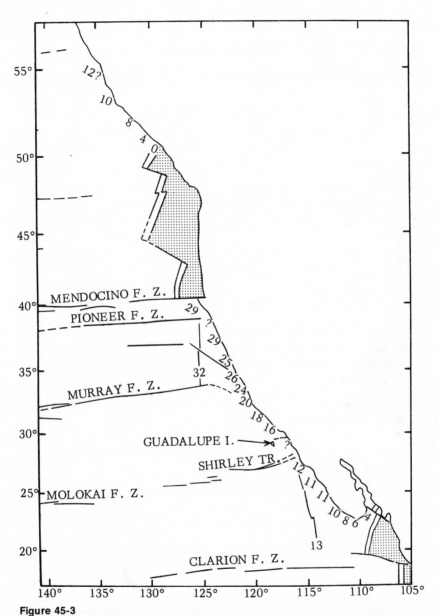

Figure 45-3
Ages (in millions of years before present) of easternmost recognizable anomalies in
the Pacific plate. These ages indicate the earliest possible time that the ridge and
trench collided and the Farallon plate was destroyed in a given region. Ceasing of
trench activity and the starting of motion related to the San Andreas fault may be
indicated by these ages. The coastline has been omitted to emphasize that the position
of the North American plate during most of the collisions is unknown

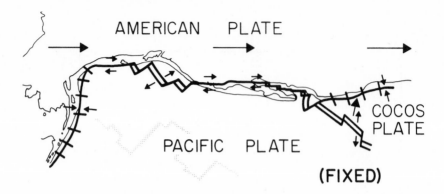

Figure 45-4
Present configuration of plate boundaries in the northeast Pacific and western North America, after Morgan (1968b). The map is a Mercator projection about the pole of relative motion between the Pacific and American plates, 53°N., 53°W., Morgan (1968b). Transform faults between the two plates lie on small circles about the pole of relative motion; thus, in this projection they form horizontal lines. Gray line marks the location of anomaly 21, used in Figure 45-2. Boundary symbols and arrows are as described in Figure 45-2

for at least the last 4 m.y. For the relative motions prior to 4 m.y. ago, 2 distinct models have been previously proposed and still others are possible. In the following sections, I shall first present the proposed models and then examine the data which concerns them.

A Model with Constant Relative Motion

If the Pacific-American motion is constant,[2] if all of it is taken up by a single transform fault, and if the coastline is parallel to the motion, then the simplified evolution described by McKenzie and Morgan (1969) results (Fig. 45–5). The ages of the easternmost anomalies in Figure 45–3 just predate the passage of a ridge-trench-transform triple junction (that is, a point where the ridge and trench collide and a

transform fault emerges). Figure 45–6 is a plot of time versus distance along the coast of North America. It shows how the triple junctions migrated up and down the coast and which plate was interacting with North America at any point along the coast at any time. For example, it shows that off San Francisco, the Farallon plate was impinging on North America at a trench until 6 m.y. ago, when the Mendocino fault migrated past and the North America-Pacific interactions (San Andreas fault) began in this region. As another example, it shows that 17 m.y. ago, a trench bordered North America from southern British Columbia to Los Angeles and south from central Baja California, while San Andreas-type interactions were taking place off southern California, northern Baja California, and northern British Columbia.

Models with Changing Motions

The interactions represented in Figure 45–5 and 45–6 are derived from the assumption of constant relative motions. If we assume instead that the relative motions changed, many other models are possible. One such model (Morgan, 1968a; Vine and Hess, 1970) is that the North American and Pacific plates were

---

[2]For this and all models of "constant motion" considered below, relative motion between the Pacific and North American plates is a rotation about a pole (fixed with respect to both plates) which presently lies at 53° N., 53° W. The rate of rotation used amounts to 6 cm/yr at the mouth of the Gulf of California. This geometry was derived by Morgan (1968b) from trends of faults in the San Andreas and Fairweather systems. A similar geometry was independently derived by McKenzie and Parker (1967) using the first motions of earthquakes on the San Andreas fault and in the Aleutian arc.

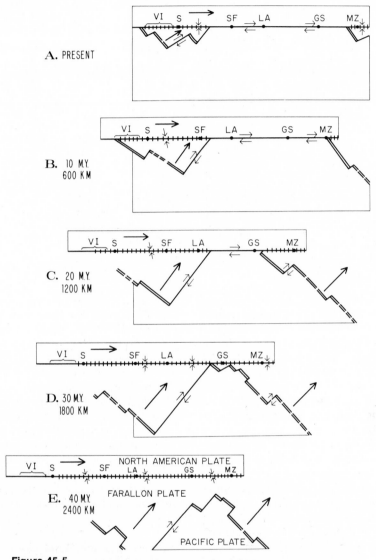

Figure 45-5
Schematic model of plate interactions assuming the the North American
and Pacific plates moved with a constant relative motion of 6 cm/yr parallel
to the San Andreas fault. The coast is approximated as parallel to the
San Andreas. Farallon-Pacific plate motions are approximated from anomalies
in Figure 45-1. Initials represent cities listed in Figure 45-6 and Vancouver
Island. Boundaries and arrows are as in Figure 45-2, the Pacific plate being
held fixed. Captions show the time represented by each sketch in millions
of years before present and the distance that the North American plate must
subsequently be displaced to reach its present position with respect to the
Pacific plate

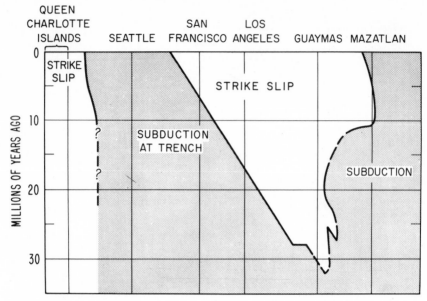

Figure 45-6
Evolution with time of boundary regimes along the coast of North America assuming
the model of constant American-Pacific motion described in Figure 45-5. White areas
denote times and coastal locations where the Pacific plate was touching North America
and interacting with it. Gray area shows times and coastal locations where the
Farallon plate lay offshore and was underthrusting North America

fixed with respect to one another until about
5 m.y. ago, at which time they broke and began
to move past one another at a rate of 6 cm/yr.
Figure 45–7 illustrates this possibility, and
Figure 45–8 shows the evolution of boundary
conditions for the North American plate. Figure 45–8 shows that off San Francisco in this
model, the trench ceased its activity 30 m.y.
ago, after which there were no plate interactions until the San Andreas began, 5 m.y. ago.

Other models postulating changes of motion
are possible. However, models that include
significant overthrusting of the Pacific by the
North American plate are discounted by evidence (discussed below) that the coast was
nearby at the time of the formation of anomalies that are now nearshore, and models that
include significant southwest spreading of the
Pacific away from the North American plate
are ruled out by the lack of symmetrical anomalies near the coast. Thus, the other likely
models lie between the two presented: prior
to 4 m.y. ago, the rate of motion may have
varied considerably, but the direction re-

mained approximately parallel to the San
Andreas and the coast.

The considerable differences between Figures 45–6 and 45–8 demonstrate how important it is to decipher the true relative motions
between the Pacific and North American
plates. Although this cannot be done unambiguously at the present time, considerable data
exist which bear on the problem, and appropriate tests can be suggested.

Spreading at the Mouth of the Gulf of California

The magnetic anomalies at the mouth of the
Gulf of California (Fig. 45–9A) indicate
spreading at a half-rate of 3 cm/yr (6 cm/yr
total) in a direction approximately parallel to
the San Andreas system for the last 4 m.y.
(Larson and others, 1968). According to Figure 45–4, Baja California is presently part of
the Pacific plate and Mexico is part of the
North American plate, and so spreading in the

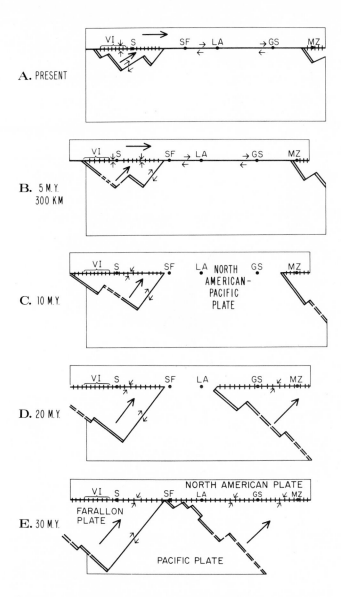

A. PRESENT

B. 5 M.Y.
300 KM

C. 10 M.Y.

D. 20 M.Y.

E. 30 M.Y.

FARALLON PLATE

PACIFIC PLATE

NORTH AMERICAN PLATE

NORTH AMERICAN-PACIFIC PLATE

Figure 45-7
Schematic model of plate interactions assuming that the North American and Pacific plates were fixed to one another until 5 m.y. ago, at which time they broke apart and began to move at a rate of 6 cm/yr. Other assumptions and symbols are similar to those in Figure 45-5

Gulf is a manifestation of the relative motion of these two plates, and the Gulf anomalies may be used to deduce the rate of this relative motion. Thus, it is crucial to discover how realistic Figure 45-4 is as a model for this region.

Seismicity provides one test. The lack of earthquakes west of the ridge (Fig. 45-9A) shows that it is probably safe to assume that Baja California moves with the Pacific plate. On the other hand, the continuation of the mid-America trench north of the Rivera fracture zone, the epicenters near the coast of Mexico in that region, and the distinct orientation of the Rivera fracture zone may indicate that the ocean floor east of the ridge acts as an independent plate (the Rivera plate, Fig. 45-9B). This plate was once part of the Cocos plate (McKenzie and Morgan, 1969, p. 131), and although it now appears to be moving with the American plate, it may not be entirely coupled to it yet. A complicated motion for this small

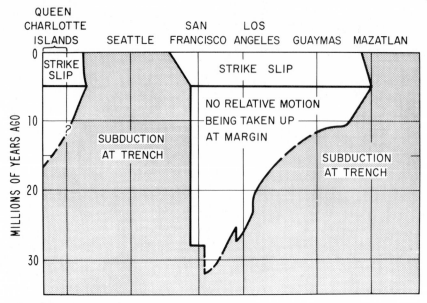

Figure 45-8
Evolution with time of boundary regimes along the coast of North America assuming the model of changing motions described in Figure 45-7. White areas denote times and locations of Pacific-North American interactions. Prior to 5 m.y. ago, these plates were locked so that no plate-related tectonics should be detected in the regions and during the times indicated by the white spaces marked "no relative motion . . ." White areas from zero to 5 m.y. ago indicate a regime of strike slip. Gray areas show times and places where the Farallon plate lay offshore and was underthrusting North America

plate would not be surprising, since it is caught in a triple junction (East Pacific Rise, Mid-America Trench, and Gulf of California spreading system) and shares boundaries with 3 large plates (North American, Pacific, and Cocos). Unfortunately, most of the magnetic anomalies known in the Gulf were created at the boundary between the Pacific plate and this small wayward one, so that their relationship to the American plate is uncertain. On the other hand, 2 magnetic profiles northeast of the Tamayo fracture zone show rates of 2.9 cm/yr (5.8 cm/yr full rate) for the last 2 m.y. (Larson and others, 1968; R. L. Larson, 1969, personal commun.). These anomalies are not involved with the Rivera plate and appear to be truly a part of the Gulf spreading system. The similarity of rates across the Tamayo fracture zone suggests that the Rivera plate is nearly coupled to the American one. The spreading rate northeast of the Tamayo fracture zone is probably the strongest evidence we have concerning the Pacific-North American rate of relative mo-

tion. It is the principal source for the 6 cm/yr rate used in the models described above.

The onset of spreading of the Gulf 4 or perhaps 6 m.y. ago (Larson and others, 1968; Larson, 1970) may be interpreted as the onset of Pacific-North America motion in the model of changing motions, or it can be attributed to the southward migration of a triple junction in the model of constant motions. This second possibility will be discussed below.

Spreading at the Juan de Fuca Ridge

The magnetic anomalies at the Juan de Fuca Ridge (Fig. 45–10A) show that spreading is occurring there at a half-rate of about 2.9 cm/yr (5.8 cm/yr total), presumably in the direction parallel to the Blanco fracture zone. The northern Gorda Ridge is also spreading at this rate and direction. (Activity on the southern part of the Gorda Ridge and on the Mendocino fault may be disregarded for our

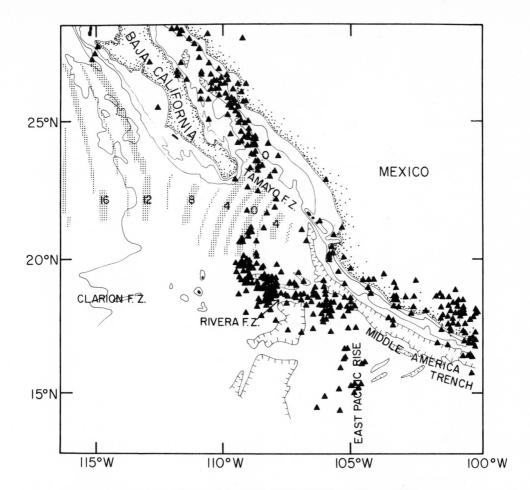

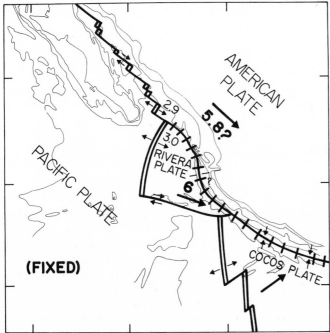

Figure 45-9A
Earthquake epicenters and sea-floor age near the mouth of the Gulf of California. Earthquakes (triangles) include all preliminary determinations of hypocenters of the USCGS, ESSA, between 1961 and 1967. Sea-floor ages (in millions of years before present) were deduced from anomalies as described in Larson et al. (1968) and Chase et al. (1970). Contours of 200, 1000, and 2000 fathoms are from Chase and Menard (1965)

Figure 45-9B
Present plate configurations and motions near the mouth of the Gulf of California. Spreading rates and ridge orientation indicate that the Rivera plate may be moving approximately with the American plate; however, earthquakes, the topographic trench, and the difference in orientation of the fracture zones may be indications of compression between these two plates

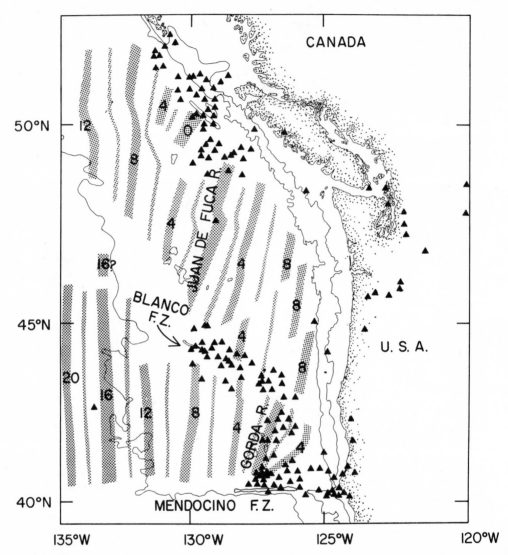

Figure 45-10A
Earthquake epicenters and sea-floor age near the Juan de Fuca ridge. Earthquakes (triangles) are from
Tobin and Sykes (1968). Sea-floor ages (in millions of years before present) are deduced from anomalies
of Raff and Mason (1961) and Atwater and Menard (1970). Contours of 200, 1000, and 2000 fathoms
are from McManus (1967)

purposes, since distorted anomalies and scattered earthquakes indicate that the sea floor is
not acting as a rigid plate in this region.) The
Juan de Fuca Ridge and Blanco fracture zone
have been considered to be a continuation of
the San Andreas system by Vine and Wilson
(1965), Bolt and others (1968), and others. As
such, it would be an indicator of the relative
rate of motion between the American and

Pacific plates. However, Figure 45–4 emphasizes a serious problem in this interpretation.
The figure is a Mercator projection which uses
the North American-Pacific pole of relative
motion as its north pole (similar to that of Mc-
Kenzie and Parker, 1967, Fig. 1). In this projection, any fault which is acting as a transform
fault between the Pacific and American plates
will appear as a horizontal line. The strike of

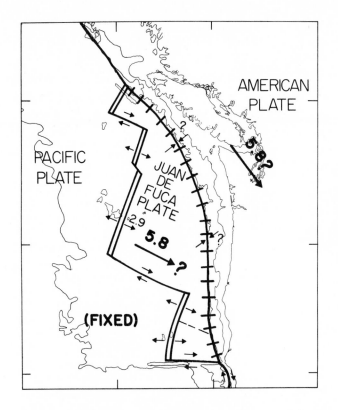

Figure 45-10B
Present plate configuration and
motions near the Juan de Fuca ridge.
The probability that the Juan de Fuca
plate is underthrusting the North
American plate makes spreading at
the Juan de Fuca ridge unreliable as
an indicator of North American-
Pacific motion

the Blanco fracture zone differs from this trend
by about 25°. Either the Blanco is absorbing a
large amount of compression, or else the ocean
floor between the ridge and the continent is
acting as a separate plate (the Juan de Fuca
plate, Fig. 45–10B).

If the Juan de Fuca plate is a separate plate
and is moving parallel to the Blanco fracture
zone at 5.8 cm/yr while the American plate is
moving parallel to the San Andreas fault at
5.8 cm/yr, Figure 45–11A shows that the resulting motion between these 2 plates is a
north-northeastward compression of about 2.5
cm/yr. If Oregon and Washington are moving
at a rate less than 5.8 cm/yr (as in Fig. 45–16),
then the compression is faster and is in a more
easterly direction (Figure 45–11B). There is a
considerable amount of evidence that the
margin of North America is being underthrust
by the ocean floor in this region (Silver, 1969a,
1969b; Byrne and others, 1966), and the
andesitic volcanism of the Cascade mountains
independently suggests the existence of a
down-welling oceanic plate. The lack of a clear

Benioff zone of earthquakes may be considered as evidence against downwelling; however, a few earthquakes of intermediate depth
do occur (Tobin and Sykes, 1968), and boundaries where slow compression is predicted are
noted for their low seismicity (southern Chile
and the Macquarie ridge are examples; see
Le Pichon, 1968). Furthermore, the Juan de
Fuca plate is young and thus not very thick
and cool, and so it may reheat very quickly,
producing few deep earthquakes. The possibility of decoupling of the Juan de Fuca and
American plates renders the spreading rate at
the Juan de Fuca ridge unreliable as a Pacific-
American rate indicator.

A short discussion of the history of the Juan
de Fuca ridge as shown by magnetic anomalies is in order. As stated above, the older
anomalies were formed by spreading between
the Pacific and Farallon plates (Fig. 45–2), and
a trench existed which was consuming the
eastern edge of the Farallon plate. After the
ridge and trench collided south of the Mendocino fracture zone, a large piece of the Farallon

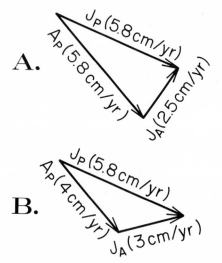

Figure 45-11
Vector diagrams for deducing relative motion between the Juan de Fuca and North American plates. Curved line shows trend of North American coast line. In both cases, the motion of the Juan de Fuca plate with respect to the Pacific plate (J_P) is assumed to be 5.8 cm/yr parallel to the Blanco fracture zone. A, The motion of the North American with respect to the Pacific plate (A_P) is assumed to be 5.8 cm/yr parallel to the San Andreas fault. The resultant motion of the Juan de Fuca with respect to the North American plate (J_A) is seen to be a compression in a north-northeast direction of 2.5 cm/yr. B, If the North American-Pacific rate of motion is assumed to be 4 cm/yr, then the Juan de Fuca-American motion (J_A) is seen to be an eastward compression of 3 cm/yr

plate remained to the north and continued to move eastward from the Pacific plate. Even though the spreading rate and direction changed (Vine, 1966; Menard and Atwater, 1968), and the manner by which the present spreading configuration evolved from the previous one is not understood, the Juan de Fuca plate is clearly a direct descendent of the Farallon plate. The fact that the anomaly sequence is complete just north of the Mendocino fracture zone and is very nearly complete near the Sila fracture zone shows that spreading has been continuous. If the Juan de Fuca plate is presently moving with the American plate, this coupling is relatively recent.

The direction of spreading changed at the Juan de Fuca ridge between 7 and 4 m.y. ago. This may be related to the onset of motion in the model of changing motions. However, it can be as easily explained using the model of constant motions by noting that the Farallon plate was steadily diminishing in size. Perhaps about 7 m.y. ago, it became too small to maintain its motion, and so it became partially coupled to the American plate.

Evidence from the Sea Floor off Central California

Figure 45–12 shows the magnetic anomalies west of central California. The complicated broken geometry and the slowing of the spreading rate may be interpreted as an indication that the Farallon plate broke up about 32 m.y. ago, as the ridge neared the trench and this part of the plate became very narrow (McKenzie and Morgan, 1969; Atwater and Menard, 1970). If this is correct, the broken pattern indicates that some part of the continent was nearby. Further support for this conclusion comes from Deep Sea Drilling Project operations off San Francisco. Late Miocene abyssal fan deposits were recovered near the bottom of the section (more than 400 km from the coast; see McManus and Burns, 1969), indicating that the continent was nearby and that the trench had been destroyed or overfilled before this time.

These lines of evidence support plate models in which North America either was moving approximately parallel to its own coastline or was fixed with respect to the Pacific for the last 30 m.y. In such models, the ages of near-coast anomalies of Figure 45–3 represent the actual times of the ridge-trench collisions. Both models presented above are of this type.

Evidence from the Aleutian Abyssal Plain and Island Arc

Grim and Naugler (1969) have reported a headless submarine channel on the Aleutian plain (Fig. 45–13), and E. L. Hamilton (1967) found that the plain consists of old turbidites

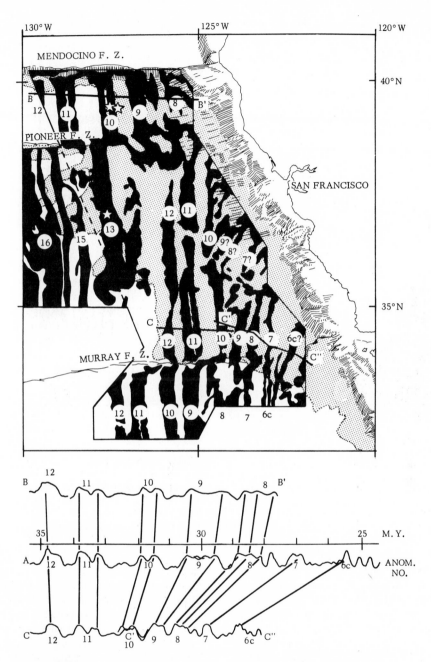

Figure 45-12
Magnetic anomalies and sedimentation off central California. Positive magnetic anomalies are shown in black (after Mason and Raff, 1961; Bassinger et al., 1969. Distribution of fan deposits (gray) and topographic features are from Menard (1964; physiographic diagram). White stars show locations of Deep Sea Drilling holes 32 (southern star), 33 and 34 which bottomed in late Miocene fan deposits. Magnetic anomaly profiles show the basis for identification of certain anomalies. Profile A, for comparison, is from north of the Mendocino fracture zone. Profile C-C'-C" shows slowing of spreading about 32.5 m.y. ago, presumably associated with disruption of the Farallon plate.

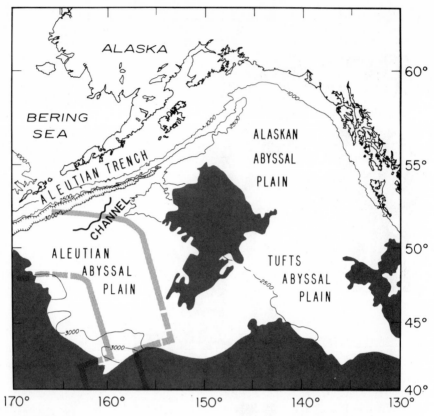

Figure 45-13
Location of sourceless sediments and beheaded deep-sea channel on the Aleutian abyssal
plain, after Hamilton (1967) and Gram and Naugler (1969). Dark gray shows areas of
abyssal hills and sea-mounts. Gray lines show location of anomalies 25 and 32 in the
Great Magnetic Bight of Elvers et al. (1967a). Contours of 1000, 2500, and 3000 fathoms
are generalized from U.S. Navy Hydrographic Office chart H. O. Misc. 15, 254-6 (1961).
Ages of starting and ceasing of turbidite sedimentation may provide clues concerning the
life time of the ridges which made the anomalies and the location of this plain on the
Pacific plate with respect to possible source areas which lay in the North American plate

from a northern source, overlain by a blanket
of pelagic sediments. The underlying crust is
part of the magnetic bight. After this part of the
crust was created, ridges lay north and east of
it, probably forming a barrier for turbidites.
Thus the starting of turbidite sedimentation
may mark the demise of these ridges, about 25
m.y. ago according to constant motions (described below; see Fig. 45–18), or about 60
m.y. ago according to the changing-motion
model (Pitman and Hayes, 1968). The cessation of turbidite deposition marks the cutting
of pathways leading to the source area either
by the encroachment of the Aleutian trench or
by formation of the trench, depending upon
which model is assumed. Both possibilities

require younger ages than those estimated
from pelagic sediment thickness by Hamilton.
Future deep-sea drilling will supply dates
which may distinguish between the models.

The Aleutian trench and island arc are the
result of the Pacific plate underthrusting the
North American plate, so that clues about the
history of motions between these plates might
be found from a study of the continuity of
activity in the Aleutian chain. The constant-
motion model predicts a definite history for the
Aleutian arc which includes a change in the
rate and direction of underthrusting in the mid-
Tertiary, and data available appear to be
compatible with such a model (Grow and Atwater, 1970).

Separation of Cretaceous
Paleomagnetic Poles

Another method of measuring the motion of large plates is to compare their paleomagnetic polar wandering curves. Paleo-pole positions for the Pacific are known from the magnetization of seamounts (Francheteau and others, 1970) and from comparative magnetization of the limbs of the magnetic bight (Vine, 1968; Vine and Hess, 1970). These data indicate that the Pacific plate has moved northward a considerable distance since Cretaceous (at least 25° in the last 70 m.y.; see Vine, 1968). The pole position for North America during the Cretaceous is fairly well known (Gromme and others, 1967). The discrepancy between it and the seamount pole is 50° to 60°. Farrell (1968) shows that these poles can be brought together by assuming the Pacific and American plates moved with a constant relative motion parallel to the present San Andreas and at a rate equivalent to about 6 cm/yr at the San Andreas since Cretaceous time. This solution is not unique, but the fit is encouraging for the constant-motion model. The model with changing motions presented above predicts only a few degrees of total relative motion, so that it does not fit unless an older period of extremely rapid motion is postulated.

CONTINENTAL PHENOMENA
RELATED TO PLATE BOUNDARIES

Evidence concerning the evolution of the plate boundaries and triple junctions should be found within the continental geology of North America. Two active boundary regimes were predicted above for the edge of the North American plate: a trench and a strike-slip dominated regime.

Igneous Activity Related to
an Offshore Trench

Probably the most easily detected continental indication of the existence of a presently active trench and associated Benioff zone is the eruption and intrusion of calc-alkaline magmas of predominantly intermediate and silicic compositions. Such activity should have existed above formerly active Benioff zones and should have ceased when subduction ceased. Thus, once the Pacific-North American motion is known, a figure similar to Figures 45–6 and 45–8 can be constructed and used to predict the type of igneous activity expected. Conversely, the distribution of igneous rocks may help to determine what the motions were. Lipman and others (1970) and Christiansen and Lipman (1970) report that intermediate volcanism was prevalent throughout the western United States in middle Cenozoic times and appears to have ceased in the southwest between 20 and 10 m.y. ago. This pattern is similar to that predicted by the models. Establishment of these relationships in time and space may provide the detail needed to choose the correct model of relative motion from among the possible ones described above.

The San Andreas System as
a Transform Fault

The present boundary between the North American and Pacific plates is usually drawn as a single break following the San Andreas fault and Gulf of California rift system (Vine and Wilson, 1965; Morgan, 1968b; McKenzie and Parker, 1967). If this is a realistic description, a study of the history of offsets across it will yield a history of the Pacific-North American motions.

Geodetic measurements across the fault zone indicate average slips of 5 or 6 cm/yr in the San Francisco area (Meade and Small, 1966) and about 8 cm/yr in the Imperial Valley region (Whitten, 1955), and an estimate of the seismic moment for the San Andreas system since 1800 indicates an average slip rate of 6.6 cm/yr (Brune, 1968). These measurements may indicate that the San Andreas system is taking up all the motion, but any long-range conclusion is highly uncertain because of the short time span measured. For geologic time spans, offsets of the San Andreas can be measured by correlating and dating distinctive units which bridged the fault and since have been offset by it. Such studies have been summarized by Dickinson and Grantz (1968a). They tend to favor a rate of offset of about 1.3 cm/yr in central California; however, the scatter in the points is great, and few of the correlations are sufficiently unique and well studied to stand alone. Three intensively studied ties suggest that while an offset of 350 km has occurred since the Oligocene (post-23.5 m.y.), much of it (about 275 km) is post-Miocene (Huffman, 1970; Turner and others, 1970).

These ties may be compatible with a model of changing motions like the one discussed above in which the San Andreas fault broke relatively recently and moved quickly, taking up all of the Pacific-North American motion (300 km in 5 m.y.), no offset occurring earlier. If the San Andreas fault took up all the American-Pacific motion, the constant motion model predicts that it would show 1400 km of offset since 23 m.y. ago. This obviously does not fit any of the measurements of offset. However, the constant model may fit reasonably well into a picture of broader deformation.

Western United States as a Broad Transform Fault Zone

The idea that the San Andreas fault constitutes a simple boundary between 2 large, perfectly rigid plates is almost certainly too simplistic a view. Other active and inactive faults in California lie parallel to the San Andreas and probably have taken up some motion, and the folding of the California Coast Ranges may be drag folding that has taken up some of it. We might take a still broader view and include the late Tertiary deformation of Oregon, Washington, Idaho, and the Basin and Range province, since this deformation has been described as a megashear in the San Andreas direction and sense. In other words, we might consider western North America to be a very wide, soft boundary between 2 rigid, moving plates. These unifying ideas for western U.S.A. tectonics have been suggested by Carey (1958), Wise (1963), and Hamilton and Myers (1966). It is especially enticing to include them here because it allows most of the late Tertiary tectonic activity of western North America to derive its driving energy from the interactions of moving plates. The concept of a wide, soft boundary is also suggested by the broad, diffuse seismicity pattern of the western United States (Barazangi and Dorman, 1968).

This pattern of earthquakes is entirely different from the narrow, linear belt of activity associated with oceanic transform faults.

Figure 45–14 contains some of the more prominent tectonic features from King (1969), plotted on the Mercator projection of Figure 45–4. The relationship of the Pacific-North American motion to features within the boundary can be visualized by moving the top of the map horizontally to the right. Faults running horizontally across the figure (San Andreas) show pure strike slip. Faults at other angles show opening (Basin and Range) and closing (Transverse Ranges). Rotations of small blocks will complicate the picture somewhat.

Relations of the Onset of Continental Deformation to Ridge-Trench Collisions

Before we can discuss the age of onset and the total amount of deformation expected within the continent, we must consider what occurs at a migrating ridge-trench-transform triple junction. Figure 45–15 is a sketch of plates in cross section as the ridge and trench approach. Figure 45–15A was constructed from Mc-Kenzie's (1967, 1969) models of ridges and trenches. The rigid plate has zero thickness at the ridge center and thickens as it moves away from the ridge and cools. At the trench, the down-going slab thins again as it is warmed by the surrounding mantle. In Figure 45–15B, the ridge is so near the trench that the plate is still very thin when it begins to descend into the mantle and to be thinned by heating. By the time the trench overruns the ridge, Figure 45–15C, the Farallon plate has practically ceased to exist.

In plate models, a spreading ridge is considered to be just the weakest place between 2 diverging plates; it is not especially related to a convective up-welling zone in the mantle. Thus, when the Farallon plate ceases to exist, the ridge also ceases. This idea is incompatible with hypotheses which relate continued activity of the overrun East Pacific Rise to rifting in the Gulf of California and the Basin and Range province. Those features are here regarded as weak places in the continental crust which broke and thereby became parts of the boundary between the obliquely diverging North American and Pacific plates.

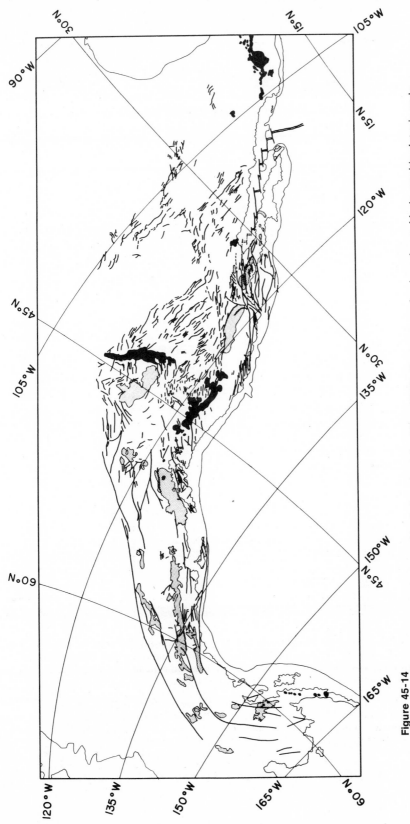

Figure 45-14
Some major tectonic features of western North America (after King, 1969). Quaternary volcanic rocks are black; granitic plutonic rocks are gray; most thrust faults have been omitted. Map projection is that used in Figure 45-4, so that deformation related to the motion between the American and Pacific plates can be imagined by keeping the ocean floor rigid and moving the rigid part of North America horizontally to the right. Horizontal faults experience pure strike slip while oblique faults have components of rifting or compression. A large-scale version of this projection is available from the author.

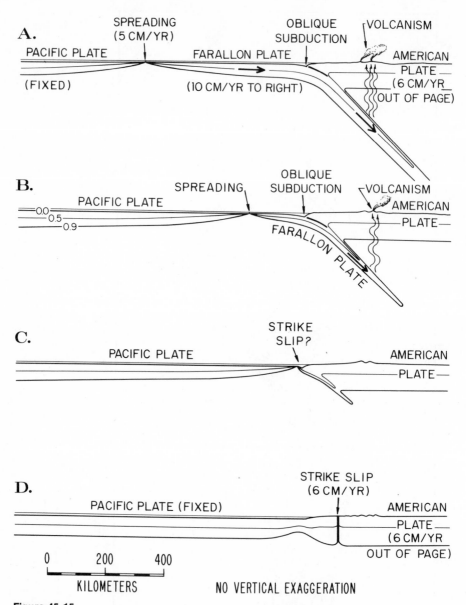

Figure 45-15

Sketch of plate cross sections during the collision of a ridge and trench. The Pacific plate is held fixed. The spreading center spreads at 5 cm/yr half-rate, accreting material onto both plates so that it moves to the right at 5 cm/yr and the Farallon plate moves to the right at 10 cm/yr. The American plate moves out of the page at 6 cm/yr. Consumption at the trench is oblique. As the Farallon plate moves away from the ridge, it thickens by cooling; as it moves down the Benioff zone, it is thinned by being heated. Plate thickness in A is sketched following the 0.9 isotherm of McKenzie (1967, 1969b) (where 1.0 is the normalized potential temperature of material in the mantle beneath the plates and intruding at the ridge-crest). (A) represents the plates in early Tertiary. (C) represents the situation when the ridge and trench collided, and (D) represents the plate configuration in central California at the present time. The time which elapses during the evolution of the situation in (C) to that in (D) must be considered when origin times for the San Andreas fault are predicted

Figure 45–15C is a sketch of the situation when the trench meets the ridge and the Pacific and North American plates first come into contact. The Pacific plate was still thin and hot at the juncture. In Figure 45–15D, the present situation, a piece of the continent has become attached to the Pacific plate and motion is taken up inland. This can occur only after the lithosphere at the juncture in Figure 45–15C has cooled and thickened. Thus, a cooling time must be introduced between the time that the triple junction passes a given point and the time when deformation related to the new boundary regime might be expected to be felt within the continent. For much of southern Baja California, there appears to have been a cooling time of 5 to 7 m.y., since the last anomaly offshore is 11 m.y. old, while spreading began inside the mouth of the Gulf 6 to 4 m.y. ago (Chase and others, 1970; Larson, 1970). During the cooling time, all American-Pacific motion was apparently taken up along the continental margin. If the margin is parallel to the motion, deformation will be pure strike slip, difficult to detect at a later time. If the margin lies at an angle to the motion, oblique spreading or compression will result. Much of the borderland rifting off southern California and Baja California may have occurred during these times before deformation jumped inland, and may be simply related to the slight nonalignment of the coast with the San Andreas trend. This possibility will be used in the reconstruction below (Fig. 45–16).

The fact that motion is presently taken up by deformation within the continent indicates that the continental lithosphere is very weak, weaker even than the continent-ocean interface, even after that interface was continually sheared and rifted during the cooling time.

A Reconstruction of Middle and Late Cenozoic Interactions

Using the concepts just discussed, a slightly more realistic version of Figure 45–5 can be constructed. In Figure 45–16, the North American plate is arbitrarily held fixed and, going backward in time, the Pacific plate is progressively moved horizontally to the right by the amount specified. The map projection is such that pieces may be moved horizontally across the page without changing size or shape. The offshore anomalies are used to delineate the Farallon plate. From zero to 20 m.y. ago, 2 continental boundaries are used. The wide, gray, inland zone schematically represents deformation in the Basin and Range province, while the coastal boundary represents offset on the San Andreas and other nearshore faults. Deformation is divided arbitrarily: one-third on the inland boundary, two-thirds on the coastal one.

The black regions in Figures 45–16D, 45–16E, and 45–16F are overlaps of continental and oceanic crust which arise in the construction because the coast is not perfectly parallel to the direction of motion. These overlaps indicate some basic error in the assumptions, either the sea floor has been grossly misdated, or Mexico deformed much more than the amount accounted for by the rifting of the Gulf and borderland, or else the American-Pacific motion before 4 m.y. ago was not constant but changed and was more nearly parallel to the coastline. Minor, rather sudden changes in direction and rate of motion have been noted between other plates (Menard and Atwater, 1968; Heirtzler and others, 1968, Fig. 3) so that a small change would not be surprising. A change in direction 10 m.y. ago of about 20° would be adequate to avoid the overlap, for instance. Evidence cited above for the proximity of the coast to near-coast anomalies during their formation suggests that the change in direction was small. Assuming that the direction changed enough to resolve this conflict while the rate was about constant, a time-distance plot, Figure 45–17, can be constructed and the following history might be described.

About 32 m.y. ago, the Farallon plate broke up off Baja California between the Pioneer and Murray fracture zones, and thereafter, pieces of the ridge began colliding with the trench. By 24 m.y. ago, the Farallon plate between the Mendocino and Murray fracture zones had disappeared, and American-Pacific motion was being taken up at the young, hot, continent-ocean boundary. By 20 m.y. ago, this section lay off southern California and northern Baja California. Apparently the margin had cooled

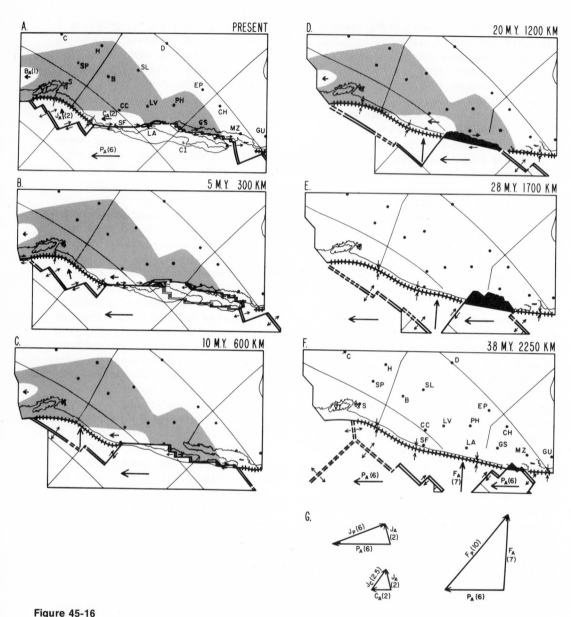

Figure 45-16
Reconstruction of plate evolution and deformation related to late Cenozoic interaction of the North American and Pacific plates. Initials are cities listed in Figure 45-17 and symbols and arrows follow Figure 45-2 except that North America is now arbitrarily held fixed and large arrows show motion relative to it. Diagrams in G show the derivation of various vectors. Captions give time in millions of years before present and amount of offset which must subsequently occur to bring the Pacific and inner North American plates back to their present relative positions. Pacific-North American motion (P_A) is assumed constant at 6 cm/yr in a horizontal direction (map projection as in Figures 45-14 and 45-4). For the last 20 m.y., 4 cm/yr is assumed to be taken up on near-coast faults, while 2 cm/yr is accommodated by inland faults (gray region). Thus, California was moving northwest at 2 cm/yr (C_A). City locations and coordinate lines are deformed accordingly. Prior to 20 m.y. ago, all motion is presumed to be taken up at the coast. Black regions are unacceptable overlaps of oceanic and continental crust, showing that the direction of Pacific-North American motion probably suffered at least a minor change sometime between 20 and 4 m.y. ago

and strengthened sufficiently that American-Pacific motion began to be felt inland, on the San Andreas (Crowell, 1968), and perhaps in the Basin and Range province. Between 20 and 10 m.y. ago, more southerly sections of the ridge collided with the trench while the Mendocino continued to move northward so that the Pacific-American boundary was greatly lengthened. By 10 m.y. ago (Fig. 45–16C), the Farallon plate had disappeared along the full length of the present San Andreas-Gulf of California system. Opening of the lower Gulf of California does not appear to have begun until 4 or 5 m.y. ago. Apparently between 11 and 5 m.y. ago, the continent-ocean juncture off Baja California was still hot and weak so that motion was being taken up in the margin and borderland, perhaps causing the buried deformation of the margin noted by Normark and others (1969). In California, the ocean-continent coupling appears to have grown stronger so that more motion moved inland, accelerating the slip rate of the San Andreas (Huffman, 1970). Between 20 and 5 m.y. ago, the San Andreas and Basin-Range systems are assumed to have extended coastward to connect into the Baja margin system. The San Andreas apparently passed through the transverse Ranges while the Basin-Range deformation extended across southern California. Both systems pass into the borderland and southeastward along the coast. The trends of these extensions are such that movement in the San Andreas direction will cause them to open obliquely. In Figure 45–16C, this is schematically drawn as a rift-transform fault system, but a more realistic description would be a zone of northwest-southeast stretching of the crust, manifested by the formation of numerous basins and rifts. The timing of subsidence of the Los Angeles basin (Yerkes and others, 1965) fits this model well.

The configuration shown in 45–16C lasted until about 5 m.y. ago, when the margin apparently became stronger than an inland zone which broke to take up the motion (45–16B), opening the Gulf of California. In southern California, the San Andreas had to break its way inland to connect into the new Baja California boundary. The bend was in such a direction that oblique compression began in the

Transverse Ranges (Crowell, 1968). The two-stage development of the San Andreas system with motion first taken up outboard and then inboard of the Baja peninsula has been suggested by Crowell (Dec. 1969, personal commun.) and developed by Suppe (1970a); although the suggested timing is different.

AN EXTRAPOLATION INTO THE EARLY CENOZOIC

When the model of constant motions is extended into the early Cenozoic, yet another triple junction is seen to have migrated along the coast of North America. This is the junction between the Farallon, North American, and Kula plates (Fig. 45–18).

The late Mesozoic existence of the Kula plate (Grow and Atwater, 1970) has been postulated to account for the formation of the east-west–trending magnetic anomalies which lie south of the Aleutian trench (Pitman and Hayes, 1968; Grim and Erickson, 1969). The rate and direction of spreading of the Kula and Farallon plates away from the Pacific plate can be determined from anomaly spacing and fracture zone trends near the magnetic bight. The 3 plates were all diverging so that a third ridge lay between the Kula and Farallon plates (Pitman and Hayes, 1968). The direction and rate of spreading of the third ridge can be deduced for the period from 75 to 63 m.y. ago, but the Aleutian trench has destroyed data concerning more recent motions of the Kula plate and spreading of the third ridge. The configuration of the third ridge and its transform faults cannot be determined, for the anomalies created by its spreading lay in the Kula and Farallon plates which have both subsequently been destroyed.

If we take the Kula-Pacific motion of the late Mesozoic and the North American-Pacific motion of the last 4 m.y. and extrapolate them both throughout the Cenozoic, and if we assume that the third ridge had no transform faults, then the history shown in Figure 45–18 results. This extrapolation is very tenuous. The Kula-Pacific motion since 60 m.y. ago is unknown and no compelling reasons exist to

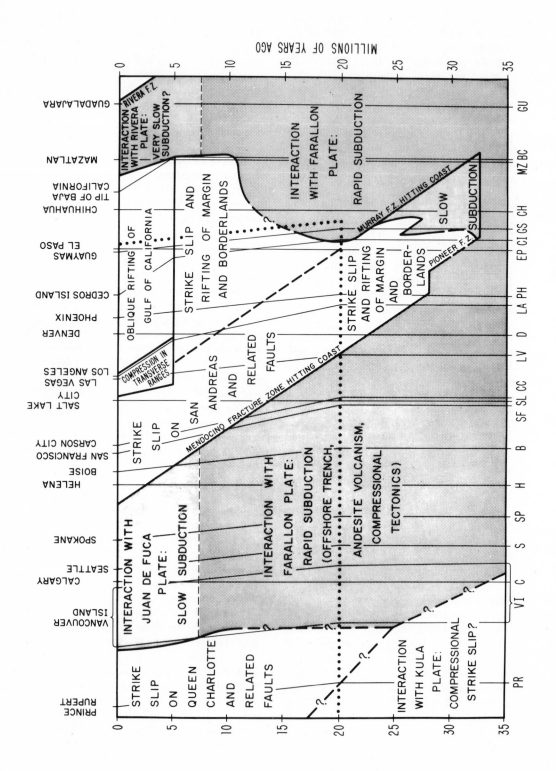

suggest that it was constant. The North American-Pacific motion amounted to a large offset in the present direction (from paleomagnetic evidence), but it is almost certain that it suffered at least minor changes in direction and rate along the way. The crustal overlaps in Figures 45–16E and 45–16F seem to require such a change between 20 and 4 m.y. ago. The Farallon-Pacific motion is known to have had a minor change about 58 m.y. ago (Menard and Atwater, 1968; Atwater and Menard, 1970). In Figure 45–18, this was assumed to reflect a change only in the Farallon plate motion, but the Pacific plate motion may have changed as well. Furthermore, McKenzie and Morgan (1969, p. 131) show that 3 plates cannot all maintain constant relative motion in the sense that it is used here (rotation about a pole which is fixed with respect to the plates whose motions it describes); constant minor readjustment is required.

Another problem with the reconstruction in Figure 45–18 is that the pole for Kula-Pacific relative motion is difficult to establish since anomalies and fracture zones concerning it occupy such a small area. For the extrapolation, the pole is assumed to be far away so that rate and direction of motion is uniform over the entire Kula plate. This uncertainty makes the Kula-North American vector, K_A, particularly unreliable. Yet another problem is that other oceanic plates may have once existed north and east of the Kula and Farallon plates. This could drastically change the predicted North American boundary regimes.

Despite these many uncertainties, the gross geometry of Figure 45–18 is probably correct. It is clear that at least 3 ridges existed in late Mesozoic and early Cenozoic times, and that their triple junction lay farther south than the present location of the magnetic bight. Unless other small plates intervened, the third ridge intersected North America, and this triple junction moved up the coast, the Farallon-North American trench developing south of it.

This geometry suggests possible relationships to some geologic problems. Although the Kula-North American vector is poorly known, it appears to have had a significant component of right lateral strike slip. A late Mesozoic-early Cenozoic era of Kula-North American interaction is indicated off the western United States. This suggests a possible early episode of slip on the San Andreas which could account for the pre-late-Oligocene offset of Mesozoic terrains discussed above. A late Mesozoic episode of San Andreas slip is also suggested by Wentworth (1968) to explain contrasting source areas and trough-like depositional characteristics for the Gualala basin. The geometry also shows that the mid-Tertiary trench discussed near the beginning of this paper can begin only after the passage northward of the third ridge triple junction. Dating of igneous rocks suggests the existence of a short-lived mid-Tertiary trench off California (McKee and others, 1970); however, it appears to have started in middle or late Eocene, somewhat later than predicted. Another implication of Figure 45–18 (and Fig. 45–17) is that the Farallon plate never extended farther north than Vancouver Island. Central British Columbia and southeast Alaska are presently interacting with the Pacific plate. The Kula and other unknown plates lay offshore in earlier times. No definite predictions can be made concerning plate interactions in this region.

Figure 45-17 *(facing page)*
Location of plate boundary regimes with respect to points in western North America, assuming motions and deformations described for Figure 45-16. Inland cities are raised above the line. They have been projected to the coast vertically in the projection of Figure 45-16, roughly parallel to the direction of Farallon-North American underthrusting. Fine lines trace the shifting locations of the cities as the continent deforms. Gray areas show times and places where tectonic and igneous activity related to subduction are predicted. White areas show times and places where North America was in contact with the Pacific plate (or with the Kula plate, to be discussed below). The probable near-coast manifestations of these interactions are stated. The dotted line encloses the time and space included in the inland deformation zone of Figure 45-16. Although it is part of the North American-Pacific interaction, this zone overlaps the Farallon-Pacific field, so that effects of the 2 regimes may be superimposed. For example, around Carson City, andesite volcanism is predicted through the middle Tertiary until 11 m.y. ago, while stiike-slip and basin-range rifting is predicted to have started 20 m.y. ago, lasting to the present day

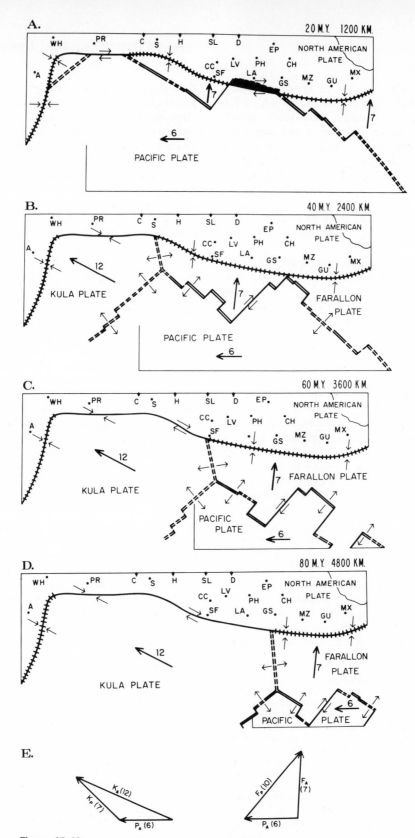

Figure 45-18
Plate relationships derived from extrapolation of Pacific-North American and Pacific-Kula motions through the Cenozoic. Conventions and assumptions are as in Figure 45-16

A Mesozoic trench is often associated with Franciscan rocks (Dietz, 1963; Ernst, 1965, 1970; Hamilton, 1969a; and others) and with the emplacement of the Sierra Nevada batholith (Gilluly, 1969; Hamilton, 1969b). It appears to have been active until about 80 m.y. ago (Evernden and Kistler, 1970; Suppe, 1970b). According to Figure 45–18, the Kula or some other northern plate lay off western North America during the Mesozoic. The trench had nothing to do with the Darwin rise nor with the Pacific or Farallon plate. If the Tertiary motion of the Kula plate was approximately as shown, its relative motion must have changed about 80 m.y. ago from a more compressional motion to the nearly strike-slip one shown. Alternatively, another unknown plate may have lain between the Kula and American plates. A major change in motion seems likely since extrapolation of Farallon-Kula-Pacific motions backward in time shows that the ridge-ridge-ridge triple junction represented by the magnetic bight intersected the Surveyor and Mendocino fracture zones about 100 and 115 m.y. ago (about 1300 and 1950 km southwest of anomaly 31). Consideration of ridge-ridge-ridge triple junctions shows that the offsets on fracture zones must come into existence at or after this intersection time (Atwater and Menard, 1970). Thus, the huge offset of the Mendocino fracture zone indicates some major change between 115 and 77 m.y. ago. The variable spacing and trend of anomalies 32 to 31 south of the Aleutian trench (Hayes and Heirtzler, 1968; Grim and Erickson, 1969) may be an indication that until about 72 m.y. ago, the Kula-Pacific ridge was still getting adjusted to a large change or to its original formation.

Probably the most important contribution of Figure 45–18 is that it emphasizes the inconstant nature of plate tectonic boundary effects. The figure was constructed assuming constant relative motions of the plates, and yet, 3 different successive boundary regimes are predicted at many points along the coast.

DISCUSSION

The conclusions in the 3 main sections of this paper represent 3 different levels of uncertainty. All assume that the basic principles of sea-floor spreading and plate tectonics are approximately correct. Given this assumption, the conclusions in the first section—that a mid-Tertiary trench lay off western North America and that the San Andreas fault began activity in its present role not earlier than 30 m.y. ago—are nearly inescapable. The conclusions in the second section depend upon which model is assumed for the history of motions of the North American plate with respect to the oceanic plates. Most of the discussions and reconstructions (Figs. 45–5, 45–6, 45–16, and 45–17) in this section assume that the motions were approximately constant during the late Tertiary. Although this model appears to be the most probable one, it cannot be definitely established. The third section deals with an outrageous extrapolation of the constant motion model. Its value lies mainly in that it presents the most straightforward model for early Tertiary plate motions, and thus may serve as a starting point for discussions of plate reconstructions of that era.

46

Mountain Belts and the New Global Tectonics

JOHN F. DEWEY
JOHN M. BIRD
1970

Any attempt to explain the development of mountain belts must account for the following:

1. They are long linear/arcuate features.
2. They have distinctive zones of sedimentary, deformational, and thermal patterns that are, in general, parallel to the belt.
3. They have complex internal geometry, with extensive thrusting and mass transport that juxtaposes very dissimilar rock sequences, so that original relationships have been obscured or destroyed.
4. They have extreme stratal shortening features and, often, extensive crustal shortening features.
5. They have asymmetric deformational and metamorphic patterns.
6. They are belts of marked sedimentary composition and thickness changes that are normal to the trend of the belt.
7. The bulk of the sediments involved are marine.
8. The basement beneath mountain belts is dominantly continental, but mountain belts have zones in which basic and ultrabasic (ophiolite suite) rocks occur as basement and as upthrust slivers.
9. They have some sedimentary sequences that were deposited during very long intervals when volcanicity was completely absent in the regions of sedimentation.
10. Intense deformation and metamorphism is comparatively short-lived when compared with the time during which much of the sedimentary rock of mountain belts was deposited.

We propose that understanding of the evolution of mountain belts should be sought by examining processes currently active within the oceans for the following reasons:

1. The emergence of the theory of lithosphere plate tectonics, the new global tectonics (McKenzie and Parker, 1967; Le Pichon, 1968; Morgan, 1968b; Isacks et al., 1968), for the first time provides a unifying worldwide explanation for tectonic processes. Plates

are generated at oceanic ridges. Mechanical and thermal energy processes occur in zones associated with oceanic trenches, adjacent to continental margins or island arcs, where plates return into the asthenosphere as rigid slabs. Plate tectonics is essentially ocean-based, i.e., the bulk of intense seismic and thermal activity is associated with plate margins within the oceans and at continental margins.

2. Young mountain belts and modern island arcs are associated with the highly seismic and volcanic belts adjacent to consuming plate margins. This association occurs at continent/ocean boundaries (e.g., the Andes), in island arcs within oceans (the New Hebrides), and where continental collision has occurred (the Himalayas) or is impending (the eastern Mediterranean). The close similarity of the features of these younger orogenic belts with older ones strongly indicates that similar processes have been involved in their development.

We take the view that plate tectonics is too powerful and viable a mechanism in explaining modern mountain belts to be disregarded in favour of ad hoc, nonactualistic models for ancient mountain belts, and that understanding of mountain belts can only come from a full integration of their features with observed sedimentary, volcanic, and tectonic processes of modern oceans and continental margins. In this paper we attempt to apply the new global tectonics to the problem of the development of mountain belts following the fundamental work of the proponents of plate tectonics and the perceptive preceding ideas of Wilson (1963a, 1965a,b, 1966) and Dietz and Holden (1967). We believe that the viability of the new global tectonics in terms of analyzing an active orogenic belt has been amply demonstrated by Hamilton (1969).

PLATE, OCEAN, AND CONTINENT RELATIONSHIPS

Plates are generated at oceanic ridges (accreting plate margins) and may be regarded as boundary conduction layers that cool, thicken, and subside as they move away from the seismically and thermally active 'mountains' of the oceanic ridges. Continental drift is merely a corollary of plate motion in that continents as much as 70 km thick are superficial passengers on plates as much as 150 km thick. Where continents ride on the same plate with adjacent oceanic crust, such as the Americas and the western Atlantic oceanic crust on the American plate (Figure 46–1), the ocean expands with continental margins that are essentially aseismic. The Red Sea is an example of a juvenile expanding ocean, and the Baikal suture may be part of a plate margin, separating a European/Asian plate from a Southeast Asian plate (Figure 46–1) and representing the embryonic stage of oceanic opening. In the Pacific Ocean, plates are being consumed in marginal trenches. The western boundary of the American plate is the western boundary of the American continents, where a complex ridge-transform system approximately marks the western boundary of the Cordilleran Mountains, and a trench marks the western boundary of the Andean Mountains (Figure 46–1; Figures 46–2B and 46–2C). In the western Pacific Ocean, plates are being consumed in extremely complex trench-transform systems associated with island arcs that are mostly within the ocean. The trenches consume lithosphere mainly from the Pacific Ocean side, thereby trapping small ocean basins (Menard, 1967; Karig, 1970), such as the Sea of Okhotsk, the Sea of Japan and the South Fiji Basin, between island arc and continent. However, complications exist, e.g., in the Philippine Sea, where a small ocean plate is bounded by trenches (Figure 46–2C) and where the junction between the Japan and the Marianas arc is migrating southwestward. The Marianas arc is therefore moving towards and being consumed under the Japan arc, although the position of the Manila trench west of the northern part of the Philippine arc indicates that a small segment of the Philippine plate (Figure 46–1) is being pushed westward into the South China Sea. This indicates that the oceanic crust of the South China Sea is being consumed and that the northern Philippine arc is approaching the Asian continent just as the Australian continent is approaching the New Hebrides (Figure 46–2E). Thus, where a trench is on the continental side of an island arc, and hence where marginal small ocean basins are contracting, island arcs may

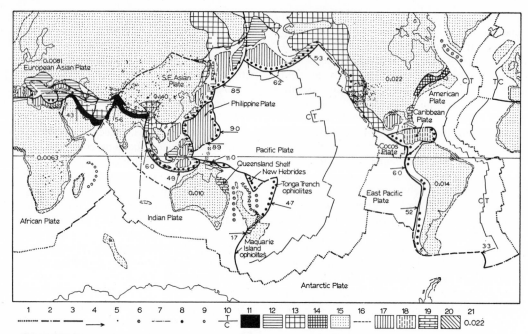

Figure 46-1
Global lithosphere plates. Key to ornament: (1) accreting plate margins; (2) active transform fault; (3) consuming plate margin; (4) slip vector across consuming plate margin in cm/yr (Le Pichon, 1968); (6) extinct accreting plate margins; (7) inactive transform fault; (8) volcanic arc; (9) extinct volcanic arc; (10) boundary of Cretaceous/Tertiary oceanic crust (Dr. W. C. Pitman, personal communication, 1969); (11) Tertiary collisional mountain belt; (12) Tertiary folding and metamorphism; (13) Mesozoic mountain belt; (14) northern Appalachian Orogen; (15) continental crust; (16) continental margin; (17) small ocean basin; (18) intermediate crust; (19) extensive carbonate platform on continental shelf; (20) thick wedges and cones of sediment on continental margin at major river mouth; (21) rate of continental erosion in cm/yr. (Dr. P. F. Friend, personal communication, 1969)

eventually collide with continental margins. Similarly, island arcs may approach one another and collide. This suggests that although island arcs may develop within an ocean, they may accumulate at continental margins.

Although continents are passive passengers on plates, they place significant restraints on plate motion in that their buoyancy prevents any significant consumption of continental crust along consuming plate margins (McKenzine, 1969). Peninsular India has collided with Asia to produce the Himalayas and the wide zone of shallow seismicity across Tibet (Figures 46–1 and 46–2G) is probably the result of the dissemination of contraction by the splintering of lithosphere over an extensive area. Sykes (1970b) suggests that the diffuse belt of shallow seismicity from the southern tip of India to Australia (Figures 46–1 and 46–2G) represents the incipient development

of a trench, and that this zone may be the result of the collision of India with Asia. Before this collision the Himalaya trench was the site of plate consumption. Plate consumption ceased at collision, and the Indus suture subsequently represented the Himalayan trench. Because plate accretion is continuing in the Indian Ocean, a new trench is required for plate consumption, and this zone of shallow seismicity may represent the beginning of a new consuming plate margin south of the Indian continent (Sykes, 1970b). Where a continent collides with an island arc, a new trench may form on the oceanic side of the arc, resulting in the change of the plate underthrusting direction (McKenzie, 1969). This has probably happened in northern New Guinea following the collision of the Australian continent with the Bismarck arc. Thus mountain belts on continental margins, where the oce-

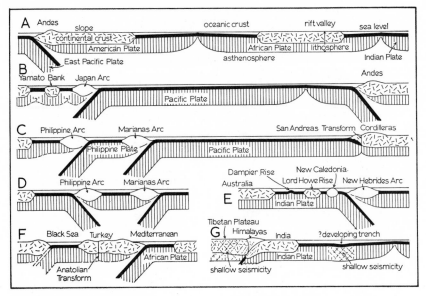

Figure 46-2
Schematic sections showing plate, ocean, continent, island arc relationship. The ?
developing trench indicated in (G) is from Sykes (1970b)

anic lithosphere underthrusts the continent, may be extremely complex accumulations of island arcs that have been driven against the continent. Although new trenches may form following collisions of continents with continents or island arcs, it seems possible that trenches may also form due to the sinking of lithosphere where it is thick, cold and dense. This is likely to happen near continental margins of Atlantic type after a long period of oceanic expansion when the oceanic lithosphere has moved a considerable distance from its accreting margin. Thus continental margins of Atlantic type may be converted into margins of Pacific type by the decoupling of the continental lithosphere from the oceanic lithosphere.

Perpetual combination and recombination of all these geometric relationships, at varying distances from rotation poles, in a complex yet continuously evolving global system, yields a great variety of possible plate, ocean, and continent combinations. In spite of the complexity, however, the continent/ocean interface is the essential locus of mountain building because it is the site of continent/continent and continent/island arc collision and is the likely site for the development of

trenches. This forms the basis for the ensuing discussion of mountain building.

The complex patterns of older mountain belts found on the continents are probably the result of perpetual rupture, oceanic opening, and continental collision for, we believe (Bird and Dewey, 1970), at least the past 10^9 years, notwithstanding strong arguments to the contrary (Hurley and Rand, 1969).

EVOLUTION OF ATLANTIC-TYPE OCEANS AND CONTINENTAL MARGINS

The Gulf of Suez/Red Sea/Gulf of Aden system illustrates the pattern of embryonic and juvenile ocean expansion. The Gulf of Suez is a 50-km-wide fault trough (graben) with at least 4 km of Neogene and Quaternary sediment (Figure 46–3E). Farther south, the Red Sea graben is wider (Figure 46–3F) where a thick marginal sedimentary sequence lies on faulted and distended continental crust. Figures 46–3A to 46–3D show a sequence of schematic sections showing the probable general nature of events during the rupture of a continental mass and the development of an ocean. Initially (Figure 46–3A), the continental

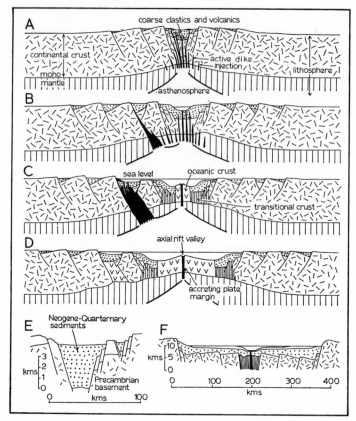

Figure 46-3
Schematic sections A-D illustrate a possible sequence of events in the rupturing of a continent and the early expansion of an ocean. E represents a section across the Gulf of Suez (Picard, 1966); F shows a section across the Red Sea (Girdler 1966)

crust is distended, and graben form and widen (Figure 46–3B) to accumulate coarse clastic sediments from intervening horsts. The largest and deepest graben will be along the central site of rupture, where the thinned and faulted continental crust is invaded by dike swarms that supply volcanos at the surface and sills injected into the sediments. Flanking graben are likely to be asymmetric, with the largest faults developing on the side away from the central site of distension. Eventually, after considerable distension of the continental crust, the central graben is torn apart, and oceanic crust is produced at a distinct accreting plate margin. Thus the central half-graben and flanking graben with thick clastic and volcanic sequences, developed during initial continental

rupture, are preserved on continental margins that move away from a mid-oceanic ridge.

New oceanic crust is generated at accreting plate margins, from which the plates move away as cooling, thickening, and subsiding slabs (Figure 46–4A). Pillowed tholeiites are erupted in the axial rift valley above a distending zone of dike injection and move outward to be uplifted and segmented by flanking faults parallel to the ridge axis. We adopt a model (Figure 46–4C) for the oceanic crust, based on refraction seismology (Raitt, 1963; Talwani et al., 1965b) and dredge haul (Aumento, 1967; Melson et al., 1968) evidence, of pillowed basalts for layer 2, below which dike complexes form the upper part of layer 3 and layered gabbros form the dominant component

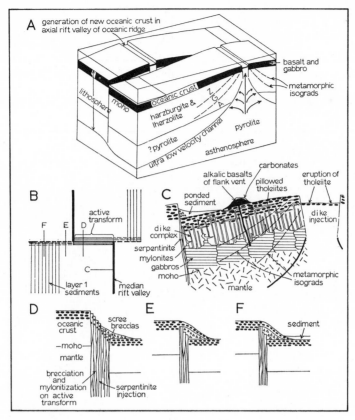

Figure 46-4
(A) Generation of lithosphere at an accreting plate margin, Z is zeolite
facies, G is green schist facies, A is amphibolite facies; (C) model for
the oceanic crust; (D-F) development of structures along a transform
fault, position of sections shown in (B)

of the lower parts of layer 3. The Moho is a
narrow transition zone from layer 3 into underlying ultrabasic rocks (probably harzburgite
and lherzolite) of the mantle. A minor component of ridge volcanicity is the alkalic basalt
of ridge-flank volcanos that form the founda-
tions of sea mounts. Recent dredge hauls of
deformed and metamorphosed rocks (Cann
and Vine, 1966; Van Andel and Bowin, 1968;
Melson and Van Andel, 1968) from the ridge
indicate that ridges are not merely sites of
passive plate accretion. Zeolitization, chlori-
tization and amphibolitization of basalt and
gabbro is probably related to the attitude of
critical isotherms for these metamorphic reactions (Figures 46-4A and 46-4C). Ridge-
flanking faults are probably sites of cataclasis

and serpentinite injection from the mantle;
transform offsets may be even more important
for deformation. In Figures 46-4D, 46-4E,
and 46-4F, we illustrate a possible sequence
of events on a transform fault. On the active
segment of a transform (Figure 46-4D), a
wide zone of cataclasis probably affects the
crust and higher parts of the mantle; it is believed that this zone may be a site of serpentinite injection. Scree breccias of broken pillow
lavas and gabbros are likely to accumulate
against the transform scarp. As the cataclastic
zone moves past the offset end of the ridge
(Figure 46-4E), fresh tholeiite may flow onto
the earlier scree deposits and sheared rocks.
Eventually, as the inactive transform moves
down the ridge flank, the whole complex will

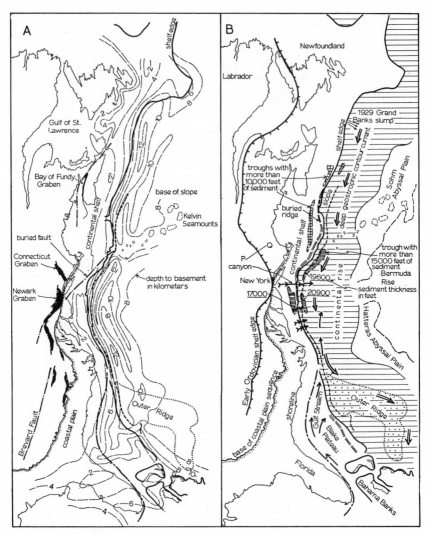

Figure 46-5
Morphology and structure of the continental margin of eastern North America. Data
from Drake (1966), Drake et al. (1968), Drake et al. (1959), Drake and Nafe (1968),
Ewing et al. (1966a), Heezen (1968), and Heezen and Drake (1964)

be draped by oceanic sediments. Thus local
complex unconformities are likely to be developed along transform faults. We believe
that ophiolite complexes (e.g., the Troodos
complex of Cyprus [Gass, 1968]), caught up
as upthrust wedges and slices in mountain
belts, are oceanic crust and mantle, originally
produced at accreting plate margins in the
manner described above.

The Atlantic continental margin of North
America (Figure 46–5) is a margin initially
developed in late Triassic times (Dr. W. C.

Pitman, personal communication, 1969) by
rifting from Africa. The coastal plain and
continental shelf are underlain by a prism of
sediment, lying on continental basement, that
generally thickens oceanward (Figure 46–6A
and 46–6B), locally into deep troughs with
no present-day surface expression and contains as much as 17,000 ft. of sediment. Cretaceous sediments of the feather edge of the
coastal plain sequence, from Georgia northward, are dominantly littoral sands and silts
of a marine transgression. Southward from

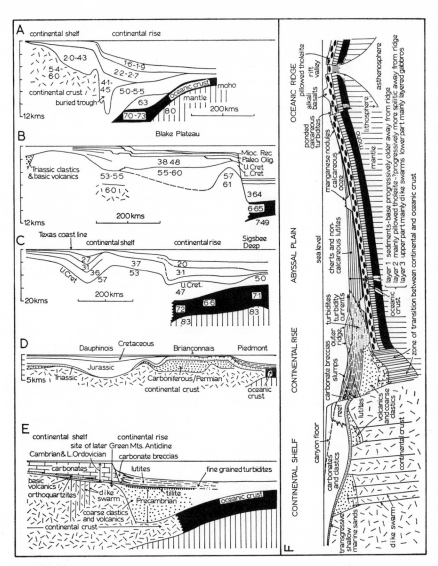

Figure 46-6
(A) Section of Atlantic margin of North America east of Cape May, New Jersey
(Drake et al., 1968). (B) Section across the Blake Plateau (Drake et al., 1968; Ewing
et al., 1966a). (C) Section across the continental shelf and rise of the Texas Coast of
the Gulf of Mexico (Antoine and Ewing, 1963; Ewing et al., 1960). Seismic velocities
are indicated. (D) Reconstruction of Late Cretaceous relationships in the French Alps
(Ramsay, 1963). (E) Reconstruction of Early Ordovician relationships in the western
margin of the northern Appalachians (Bird and Dewey, 1970). (F) Schematic section
illustrating relationships between continental crust, oceanic crust and sediments in an
Atlantic-type ocean and continental margin

Georgia these deposits grade laterally into
carbonates. Present-day carbonate deposition
on the Bahama banks changes northward on
the Blake plateau (Figure 46–5B and 46–6B)
into a shallow sea, the floor of which is swept

by the northward-flowing Gulf Stream. Further north, sand, silt, and lutite are being deposited on the shelf, which, north of Cape
Hatteras, is cut by oceanward-deepening canyons. Large segments of the outer margin of

the shelf are characterized by a buried ridge. This could be a rise in the continental basement but may be in part, or even mostly, a buried outer reef complex along the shelf edge.

The Atlantic continental slope and upper continental rise are regions of bypass for coarse, land-derived sediment. The upper rise is characterized by gravitational instability that leads to slumping (e.g., the Grand Banks slump, triggered by an earthquake in 1929). The continental rise is a region dominated by lutite deposition. Terrigenous sediments, supplied through the submarine canyons, are distributed onto the lower rise as cones and onto the deep-ocean floor as turbidites. Major sediment accumulations, building the shelf edge outward, are only found in the delta of the Mississippi River, but even there sediments are spread in a cone onto the ocean floor. The extent to which sediment is transported southward by geostrophic currents at depth and parallel to the continental margin is indicated by the red lutites of the outer ridge of the Blake plateau. The provenance of this red lutite is in the Gulf of St. Lawrence (Stanley, 1969). Thus, the Atlantic continental rise is not a region of rapid sediment accumulation; it is shaped mainly by contour-following geostrophic currents (Heezen, 1968). However, the upper continental rise is underlain in many areas by deeply buried troughs, some of which contain more than 20,000 ft. of sediment over continental crust (Figures 46–5B/46–6A). East of New York, the crystalline basement beneath the upper continental rise is over 12 km below sea level (Figure 46–5A). These thick sediment accumulations cannot have been produced by the type of sedimentation currently active on the upper continental rise. It is suggested that much of the locally thick accumulation of sediment was deposited on graben that formed during the earliest stages of continental rifting during late Triassic times. This is supported by the presence of late Triassic volcanics and coarse red clastics in the Bay of Fundy, Connecticut, and the Newark, graben.

Sheridan (1969) has indicated that the North American continental margin has undergone considerable subsidence. We suggest that this subsidence was related to the over-all subsidence of the American plate during its movement away from the mid-Atlantic ridge. The sea floor has an increasing depth away from ridges and the subsidence rate is related to plate thickness and density. Subsidence is probably initially fast and subsequently slower as the continental margin moves farther from the ridge. The accumulation of salt deposits is an important feature of the early stages of continental separation because salt deposition is likely to be an integral feature of sedimentation in juvenile oceans where circulation is poor. Salt deposits with associated later diapirism are frequently associated with continental margins, for example, the Gulf of Mexico (Ewing and Antoine, 1966; Ewing et al., 1962) and the African continental margins (Ayme, 1965).

In Figure 46–6F, a schematic section of an Atlantic-type continental margin and half-ocean illustrates relationships between continental crust, ocean crust, and sediments. *Drake et al.*, (1959) have pointed out striking analogies between the morphology and crustal structure of the North Atlantic continental margin and *Kay*'s (1951) Ordovician miogeosyncline/eugeosyncline couple, reconstructed for the Northern Appalachians. A modified version (*Bird and Dewey*, 1970) of Kay's reconstruction of the early Ordovician, preorogenic, continental margin of North America (Appalachian Atlantic) is shown in Figure 46–6E, and the position of the outer edge of the early Ordovician continental shelf (miogeosyncline) is indicated in Figure 46–5B. Cambrian-Ordovician carbonates are underlain by a westwardly-transgressive orthoquartzite lying on the Precambrian Grenville basement. In Newfoundland, basic dike swarms are related to flood basalts below the orthoquartzite. The shelf carbonates overlying the orthoquartzite change into an eastern carbonate breccia, which interfingers with a lutite sequence. The lutite sequence overlies a great thickness of Precambrian clastics and volcanics. In the Alpine fold belt of France, preorogenic reconstructions (Figure 46–6D) of sedimentary facies relations (Ramsay, 1963) involve shelf carbonate-lutite sequences (Dauphinois and Briançonnais) passing eastward into lutite sequences of the Piedmont. Below the thin Mesozoic carbonate sequence of the Briançonnais, thick Carboniferous and

Permian clastic and volcanic sequences probably represent the distension of continental crust along the northern margin of the Tethyan Ocean. In 1959 Drake, Ewing, and Sutton incisively summarized the problems of mountain belt evolution in the following series of statements:

> If the continental margin of eastern North America is accepted as a future mountain system, it would be interesting to outline the major requirements necessary to convert it to such a system, . . . the single most important process in an orogenic cycle must be the one which increases the crustal thickness in this area to continental proportions, . . . the key to orogenesis lies in the heat problem. Once the sources of heat are determined and the method by which it is focused on small areas is established, the nature of orogenic movements should be revealed.

We believe that plate tectonics provides answers to these problems, and in the following sections we suggest mechanisms by which an Atlantic-type continental margin is converted into a mountain belt. We consider crustal, mechanical, and thermal relationships associated with consuming plate margins, first, where oceanic lithosphere underthrusts island arcs; second, where oceanic lithosphere underthrusts a continental margin; third, where a continent collides with an island arc; and last, where a continent collides with a continent. Some geophysical and petrologic data relevant to the structure and origin of the Japan arc are assembled on a section from Primorye across the Sea of Japan, northern Honshu, and the Japan trench, to the Pacific Ocean (Figure 46–7A). The Pacific Ocean lithosphere underthrusts the arc at a rate of between 8.5 and 8.7 cm/yr (Le Pichon, 1968). The partial melting (Ringwood, 1969), under high load pressures and shear stresses (Oxburgh and Turcotte, 1968), of oceanic crust on the descending plate provides a likely source of calc-alkaline magma erupted on the arc as andesite and dacite (Figure 46–7A). This probably involves (Ringwood, 1969) the partial melting of quartz-eclogite (transformed lower, 'dry', basaltic oceanic crust), leaving a refractory residue of

dense eclogite, and of amphibolite (transformed upper, 'wet' oceanic crust), leaving residual eclogite. Green and Ringwood (1968) have suggested that the origin of the basalt series (tholeiitic, high alumina, alkalic) of island arcs (Kuno, 1966) lies in convective instability triggered by the rise of calc-alkaline magmas, resulting in partial melting of pyrolite in the wedge of asthenosphere between the descending plate and the arc. The experimental work of Green and Ringwood (1963) suggests that quartz tholeiites and high alumina basalts result from the fractional melting of pyrolite at depths shallower than 35 km, and that alkalic basalts originate at depths between 35 km and 70 km. Active volcanicity in northern Honshu is entirely restricted to an area west of a line (volcanic front) 130 km from the trench, suggesting that the beginning of partial melting of amphibolite and quartz-eclogite does not occur until the descending plate has reached a depth of about 120 km (Figure 46–7E). If the depth at which the eclogite and amphibolite begin to partially melt is dominantly controlled by load pressure and temperature, the position of the volcanic front will be related to the inclination of the descending plate.

The region between the volcanic front and the trench has a dominantly submarine, oceanward thickening, prism of sediment derived from the volcanic portion of the arc. Sugimura et al. (1963) have shown that the largest volumes of Quaternary volcanic rock coincide with the region of tholeiitic basalt immediately behind the volcanic front, and that these volumes decrease rapidly as they approach the region characterized by high alumina basalt and alkalic basalt toward the Sea of Japan (Figure 46–7D). There is a fairly good correlation of crustal thickness with this volume variation, suggesting that the crust of Honshu may have been built largely by volcanic accumulation. The accumulation of a thick pile of basalt and andesite to form the crust of Japan would explain the thick crust characterized by positive Bouguer gravity anomalies up to 150 mgal. The Japan trench is characterized by low gravity and heat-flow values. The inner trench wall is steep and terraced (Figure 46–8B) and, judging by similar structures in the Manila trench (Figure 46–8A), has perched sediment accumulations on the terraces. The outer

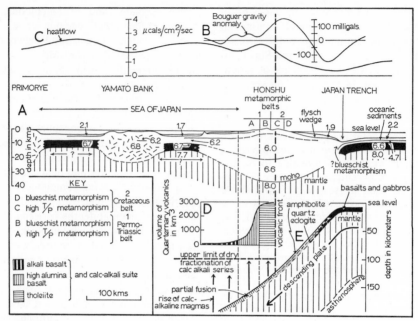

Figure 46-7
(A) Section from Primorye to the Pacific Ocean across the Japan Sea and northern Honshu (Miyashiro, 1961, 1967; Matsuda et al., 1967; Rikitake et al., 1968). Seismic velocities are indicated. (B) Bouguer gravity anomaly (Hagiwara, 1967). (C) Heat flow (Uyeda and Vacquier, 1968). (D) Volumes of Quaternary volcanics (Sugimura et al., 1963). (E) Morphology (Sykes, 1966) of the descending plate and probable composition (Ringwood, 1969) of the descending oceanic crust

trench wall is gently inclined, and the sediments of layer 1 are affected by steep faulting, probably resulting from tension due to bending of the plate (Isacks et al., 1968) before it descends.

Oceanic trenches (e.g., the Puerto Rico trench; see Chase and Bunce, 1969) commonly display a complete lack of crumpling and folding of oceanic sediments as they are carried into the trench by the oceanic plate but show extreme deformation at the foot of the inner trench wall. We suggest that although some oceanic sediment is carried down into the asthenosphere by the descending slab (possibly to contribute to the calc-alkaline suite), a significant component of sediment is scraped from the plate and plastered onto the inner trench wall (Figure 46–8C). There may be a thick prism of intensely deformed sediment (chert and argillite; carbonates and alkalic basalt of seamounts) beneath and behind the trench. This region of high pressure and low

geothermal gradient may be a site of present-day blueschist (glaucophane, lawsonite, aragonite, jadeite assemblages) metamorphism (Takeuchi and Uyeda, 1965). The oceanic trenches of consuming plate margins are apparently the only regions where the high pressure/low temperature conditions required for the formation of blueschist assemblages occur (Figure 46–9).

Japan has a history of deformation and metamorphism dating back to Triassic times. Honshu is characterized by two pairs of metamorphic belts (Miyashiro, 1961; 1967). The inner pair (Figure 46–7A), of Triassic age, has a blueschist zone on the outer, oceanward side and a high-temperature zone with andalusite, sillimanite, cordierite assemblages on the inner side. The outer pair, of Cretaceous age, consists of the outer Sanbagawa blueschist zone and the inner Ryoke high-temperature zone. The likely pressure/temperature fields of the Sanbagawa and Ryoke zones are indi-

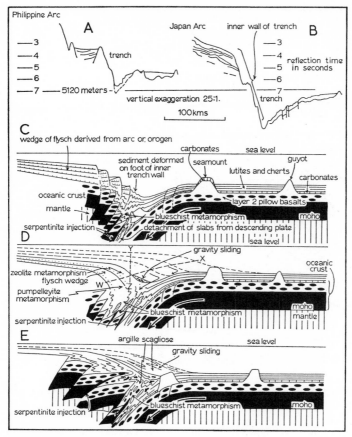

Figure 46-8
Structures associated with trenches. (A) Section of the Manila Trench
(Ludwig et al., 1967). (B) Section of the Japan Trench (Ludwig et al.,
1966). (C-E) Various relationships between trench structures, sediments,
and metamorphism discussed in text

cated in Figure 46–9. Miyashiro (1967) has
argued that each blueschist zone lies on the
site of an old trench, and that each high-temperature metamorphic zone corresponds with
the high heat-flow zone behind the volcanic
front. Although this interpretation is consistent
with the present over-all distribution of geothermal gradients, it seems unlikely that the
Triassic and Cretaceous blueschist zones
represent trenches in their present relative
positions because the inner high-temperature
Cretaceous zone is too narrow to represent the
whole width of an arc. We suggest that two
arcs have been driven together by the consumption of oceanic crust between the Triassic
and Cretaceous belts. This may have been

similar to the approach of the Japan and
Marianas arcs of the present day by the contraction of the Philippine plate (Figure 46–1).
A major problem exists in accounting for the
origin of the Sea of Japan (Figure 46–7A), a
small ocean basin (Menard, 1967). It has high
heat-flow values and oceanic crust with, in
places, 2 km of sediment. The Yamato bank
(Figure 46–7A) is a microcontinent located
approximately in the center of the Sea and
has lower heat-flow values. The Sea of Japan
could be a very old part of the Pacific Ocean
crust, trapped between Asia and Japan by the
growth of the Japan arc, and forming a sediment trap since Triassic times. Alternatively, it
may have been produced, at least in part, by a

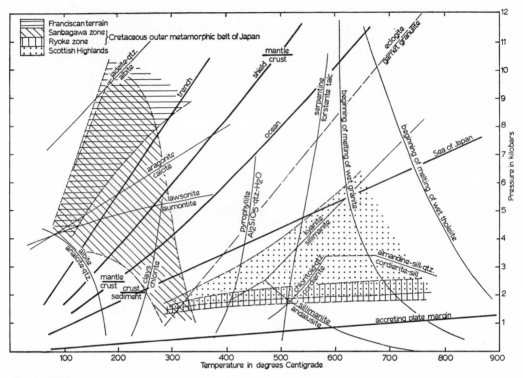

Figure 46-9
Pressure-temperature plot of various metamorphic reactions and metamorphic assemblages (Miyashiro, 1961, 1967; Chinner, 1966; Ernst and Seki, 1967; Turner. 1968)

more recent detachment of Japan from the Asian continent (Dr. K. Nakamura, personal communication, 1969). If so, the Sea of Japan may have fairly young oceanic crust, formed by a slow extension, with alkalic composition because it is at present entirely within the alkali basalt field. It is probable that the Lord Howe rise/New Zealand/Campbell plateau microcontinent-arc system moved away from Australia in post-Permian times by the creation of the Tasman Sea (van der Linden, 1969). Sea-floor spreading behind the western Pacific arcs may be a major factor in their present position within the ocean. Although Japan is probably the result of a complex history of arc/arc collision, of change in trench position, and of separation from the Asian continent, we believe that many arcs, such as the Tonga/Kermadec arc are the result of the growth of a volcanic pile in the ocean above the descending plate.

We suggest that the sequence of events in the growth of a single arc might be as follows:

1. Development of a trench and the beginning of plate descent. This stage will be accomplished by the intricate thrusting of wedges of oceanic crust and mantle to form a ridge such as the Macquarie ridge (Figure 46–1). During this early thrusting, oceanic chert, argillite, and carbonates may become involved in large submarine gravity slides into the new trench (Figure 46–8E). These gravity slides are likely to carry blocks of basic and ultra-basic rocks derived from the displaced thrust wedges. These slides will be carried down to some extent into the deformed sedimentary wedge beneath and behind the trench and, together with chert, argillite, and seamounts scraped from the descending plate, may undergo strong deformation in the blueschist field.

2. When the descending plate reaches depths of over 100 km, the amphibolite and quartz-eclogite crust begins to partially melt, and calc-alkaline magmas undergo fractional crystallization as they rise. Basalt magmas are generated at shallow levels, and accumula-

tion of a volcanic pile begins at the surface. These rocks may undergo high-temperature metamorphism and deformation if the lithosphere becomes thinned and weakened by the heat flux. As the crust thickens by the growth and deformation of the volcanic pile, an oceanward-thickening wedge of sediment (flysch), derived from the volcanic belt, develops between the volcanic front and the trench. Thus a volcanic arc with a paired metamorphic zonation is developed.

In many mountain belts there are chaotic mélange terrains often affected by blueschist metamorphism. These are important because they are the products of trench activity and probably in most cases indicate the position of an earlier trench. The term 'argille scagliose,' from the Mesozoic-Tertiary mélanges of the Appennines, is often applied to these rocks. Judging by the heterogeneity of rock types and structures common to mélange terrains such as the Mesozoic Franciscan zone of California (Bailey et al., 1964; Blake et al., 1967; Hsü, 1968), considerable variations in trench structures, sediment and morphology seem likely. These variations are probably related to the rate of plate consumption and the rate of growth of the sedimentary (flysch) wedge behind the trench. If the rate of plate consumption is slow and the rate of oceanward building of the sediment wedge is fast, the flysch will build across the trench onto oceanic sediments (Figure 46–8D). In this way, enormous apparent thicknesses of flysch may accumulate (thickness measured along WX of Figure 46–8D), but when measured vertically (YZ of Figure 46–8D), will be considerably thinner. The 'burial metamorphism' of the sediments of the South Island syncline of New Zealand (Coombs, 1961) may be related to the building of such a flysch wedge. Rafts of semilithified flysch may be dragged from the base of the wedge and incorporated in the mélange (Hsü, 1968).

During periods of fast plate consumption, blueschist metamorphism is likely to be active with the development of jadeite assemblages. During periods of slow plate consumption, when considerable amounts of flysch may advance across the trench, blueschist mélanges may rise isostatically. The elevation of blueschist rocks may also be related to a more continuous upward motion of the leading plate edge behind the trench caused by the addition of deformed oceanic sediments and slices of oceanic crust and mantle torn from the descending plate. These slices of oceanic crust and mantle are amply represented by ophiolite (chert, pillow lava, basic dike swarms, gabbro, peridotite) rafts in mélange terrains. In addition to basaltic rocks, serpentinite sheets and lenses, and blocks of amphibolite and blocks of eclogite are common in mélange environments, suggesting that oceanic crust and mantle are ripped from the downgoing plate at considerable depths and injected into the mélanges. Thus trenches and their associated sedimentary wedges are likely to be extremely complicated terrains, with intricate geometric and time relationships between oceanic crust, mélange, and flysch sediments.

CORDILLERAN-TYPE MOUNTAIN BELTS

We suggest that a mountain belt marginal to a continent where oceanic lithosphere underthrusts a continental margin, such as the Cordilleras or the Andes, develops where a trench originates at, or very near to, a continental margin of Atlantic-type. A sequence of events by which an Atlantic-type continental margin is converted into a cordilleran-type mountain belt is schematically illustrated in Figure 46–10. These sequences are based on an analysis (Bird and Dewey, 1970) of northwestern parts of the Appalachians for Ordovician times and on the general structure of the Mesozoic Cordilleran system along the western continental margin of North America. During and after the formation of a consuming plate margin near the continental margin, the sequence of events in and immediately behind the newly-formed trench (Figures 46–10A–46–10E) is likely to be broadly similar to sequences of events already discussed for island-arc trenches. The sequence of events will involve early, oceanward-driven wedges of oceanic crust and mantle, flysch accumulations thickening toward the trench, and blueschist mélanges. As the oceanic plate descends beneath the continental rise to depths greater than 100 km, submarine volcanics are erupted behind the volcanic front (Figure 46–10A). As the heat flux generated

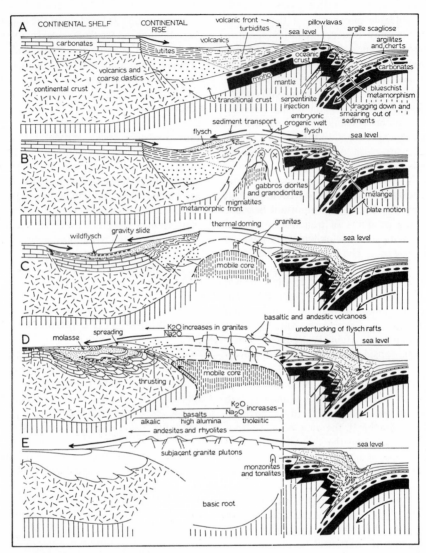

Figure 46-10
Schematic sequence of sections illustrating a model for the evolution of a cordilleran-type mountain belt developed by the underthrusting of a continent by an oceanic plate

by the rise of basaltic and calc-alkaline magmas increases, an embryonic orogenic welt (Figure 46–10B) rises above an expanding dome, the core of which is occupied by rising gabbroic and granodioritic magmas. High-temperature deformation and metamorphism begins to affect the sedimentary pile of the lower continental rise as the mobile core (Figure 46–10C) expands and grows toward the continent. The coarse sediments and volcanics that accumu-

lated in graben formed during the early stages of continental rupture are affected by the high-temperature deformation and metamorphism. When the growing welt rises above sea level, it forms an axis of sedimentary polarity (Figure 46–10B). Sediment is transported toward the ocean, forming a flysch wedge between the volcanic front and the trench, and also toward the continent, progressively filling the trough between the continental margin and the orogenic

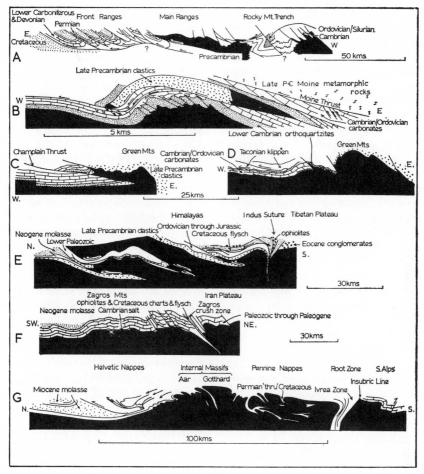

Figure 46-11
(A) Section across the Rocky Mountains (King, 1959). (B) Section across the
northwestern margin of the Lower Paleozoic Orogen of the Scottish Highlands.
(C-D) Sections across the western margin of the Appalachians in Vermont (Doll et al.,
1961). (E) Section across the Himalayas (Gansser. 1964). (F) Section across the Zagros
Mountains (Stocklin, 1968). (G) Section across the Swiss Alps (Argand, 1911; Heim,
1921, 1922, Staub, 1924)

welt. Eventually, the deformation wave, accompanied by flysch migrating toward the continent, arrives at the continental margin. At about that time the continental shelf subsides (Figure 46–10C) to form a trough into which turbidites (flysch), shale, and massive gravity slides from the welt accumulate. The gravity slides (e.g., Taconian of the Appalachians, Figure 46–11D) overlie chaotic zones (wild-flysch) produced by the erosion products and bulldozing effect of the advancing slides. The regime of gravity-based deformation mechan-

isms progressively changes to the emplacement of metamorphosed sheets being driven toward the continent. Sometimes the continental basement is involved in the thrusting (Figure 46–11C). As the movement of material toward the continent changes from a flysch-wildflysch-gravity slide regime to hard-rock thrust emplacement, the exterior troughs begin to accumulate thick fluviatile sequences (molasse; see Figure 46–10D). A major factor in the emplacement of the thrust sheets, in addition to the rise and continentward-extrusion of the mobile thermal core (Figure 46–10D), may be lateral extrusion by the spreading of the orogenic welt. This mechanism to some extent obviates the problem of transmitting lateral compressive stresses at high levels through the marginal thrusts. Eventually, the whole orogenic system becomes resistent to further intense contractional strains, and post-kinematic granites are emplaced at high levels in the orogenic welt, below a block-faulted terrain of basaltic and calc-alkaline volcanoes (Hamilton, 1969).

The mountain belt thus developed has a symmetry in terms of thrust direction and sediment distribution from the orogenic welt. However, this symmetry is only one of gross morphology. The oceanward thrusting, on the oceanic side of the volcanic front, is early and is related largely to the *underthrusting* of the oceanic plate. The diachronous slides and thrusts driven toward the continent are more superficial and are related to later progressive *overthrusting* of rocks from the orogenic welt. The polarity axis, from which the sense of thrusting and sediment distribution diverges, coincides with the axis of a high-temperature metamorphic terrain. King (1959) emphasizes that the Jurassic Nevadan deformation, involving the Franciscan mélange terrain and the Klamath Mountain ophiolite thrust sheets of western California (Davis, 1968), was the beginning of continentward-migrating orogeny which culminated in the late Mesozoic Laramide deformation of the Rocky Mountain front ranges. The time of the appearance of clastic sediments transported toward the continent interior is of great importance because it coincides with the growth of the embryonic orogenic welt above sea level.

The marginal overthrusts sometimes involve the transport of deformed and metamorphosed rocks toward the continent (Figure 46–11B), but in the Cordilleran belt the large thrust sheets driven toward the continent often lie east of internal rigid massifs, such as the Colorado plateau, underlain by continental basement and a deep Moho. This internal resistance to deformation may be a function of a descending plate having a shallow dip under the continent, so that such heat as is generated by the rise of magmas is blanketed by the continental crust. This is strongly indicated by the San

Juan volcanic field (Doe et al., 1969; Dickinson, 1969).

In terms of Kay's (1951) analysis involving a paired miogeosynclinal (nonvolcanic-carbonate) – eugeosynclinal (clastic-volcanic) couple in the Appalachians, the continental shelf edge, subsequently a site of major thrusting, approximately delineates the boundary between miogeosynclinal and eugeosynclinal assemblages. Kay's miogeosyncline is the preorogenic continental shelf orthoquartzite-carbonate assemblage, followed by an exogeosynclinal shale-flysch-molasse sequence (Kay, 1951). Kay's eugeosyncline involves all the preorogenic, synorogenic, and postorogenic clastic and volcanic sequences between the continental margin and the outer wall of the trench.

It seems likely, at least in theory, that all gradations may exist between cordilleran-type orogens and island arcs. If the trench forms at some considerable distance from the continental margin so that an island arc grows within the ocean, a small ocean basin forms a sediment trap (Menard, 1967) between arc and continent margin. If the trench forms adjacent to the continental margin, a cordilleran-type mountain belt develops.

CONTINENT/ISLAND ARC COLLISION

An alternative method of developing a volcanic (eugeosynclinal) and nonvolcanic (miogeosynclinal) couple is by colliding a continental margin of Atlantic-type with an island arc (Figure 46–12,A-D), an event that probably happened in northern New Guinea in Miocene times. The volcanic arc with its flysch wedge and trench blueschist-mélange rocks, and the continent with its continental margin assemblage of sediments, approach by the consumption of an intervening ocean (Figure 46–12A). Flysch fills the small ocean basin remaining just before collision. When the continental margin is driven into the region of the trench (Figure 46–12, B-C), the buoyancy of the continental rocks prevents much underthrusting of the plate (McKenzie, 1969). Thrust slices moving toward the continent develop while the flysch and the blueschist rocks, together with slices of oceanic crust, are driven onto the deforming rocks of the continental rise and shelf. This may

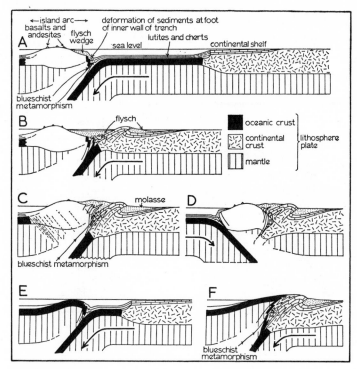

Figure 46-12
(A-D) Schematic sequence of sections illustrating the collision of a
continental margin of Atlantic type with an island arc, followed by
change in the direction of plate descent. (E-F) Proposed mechanism for
thrusting oceanic crust and mantle onto continental crust

be followed by the development of a new trench on the oceanic side of the arc (Figure 46–12D). The underthrusting of an upthrust ridge of oceanic crust and mantle such as the Macquarie ridge, followed by the collision of a continent against the ridge, is a probable mechanism (Figure 46–12, E-F) by which oceanic crust may be thrust onto continental crust.

This method of developing an orogenic belt along a continental margin may bear superficial resemblance to cordilleran-type mountain building but differs in the following:

1. The earliest deformation is in the ophiolite-flysch-blueschist environment, in the suture zone, followed by collisional deformation producing the continentward-driven thrust sheets. The outer blueschist deformation in the new trench, following change in the direction of plate underthrusting (McKenzie, 1969) (Figure 46–12D), will be younger than the previous deformation. This sequence of events is in di-

rect contrast to the sequence of Nevadan early deformation on the western side of the Cordilleras, continuous with younger Laramide deformation on the continental side.

2. The high-temperature metamorphic and volcanic belt will be separated from the nonvolcanic belt by a suture zone of blueschist, ophiolite, and flysch.

3. The tholeiitic to alkalic polarity of the basalt series will be oceanward for the earlier volcanics. It will be related to the dip of the plate under the arc before collision and, after change of the underthrust direction, will be continentward for the later volcanics.

CONTINENT/CONTINENT COLLISION

A schematic sequence of events in the approach and collision of an Atlantic-type continental margin and another continent with a

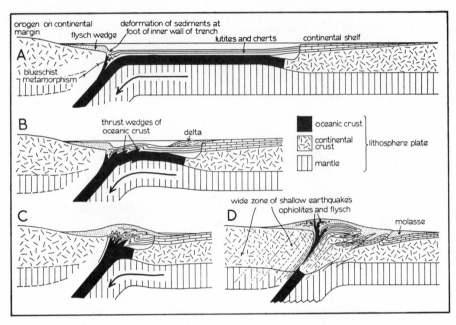

Figure 46-13
Schematic sequence of sections illustrating the collision of two continents.

marginal trench is shown in Figure 46–13. The trench-bearing margin may be associated with an existing or developing cordilleran-type orogen or with a margin resulting from an island arc/continent collision. In the Alpine-Mediterranean system (Figure 46–14) the situation is even more complex. The African plate is at present being consumed in a trench system south of the Aegean arc, and a collision of North Africa with Greece and Turkey seems inevitable. The Alpine fold belt, between the Ionian trench and the European shield-platform, is a complex maze of ophiolite-flysch-blueschist sutures representing the sites of old trenches. The massifs between these sutures were probably microcontinents and island arcs accumulated by collisions on the northern margin of Tethys. In the collision model shown in Figure 46–13, we illustrate a continental margin of indeterminate type with a flysch wedge and trench, and structures and relationships described in previous sections. The structures developed as the Atlantic-type margin is driven onto the trench are likely to be initially similar to those already described for an arc/continent collision, which involve the splintering and thrusting of continental basement to form cores of nappes. Oceanic crust, chert and

lutite, and flysch, are squeezed and thrust over lower thrust sheets. Eventually, however, the buoyancy of the underthrust continental rocks prevents further destruction, and the descending plate may break off and sink into the asthenosphere (McKenzie, 1969). At that time a single trench zone of plate consumption will be replaced by cracking and splintering of the lithosphere over a wide area (Figure 46–13D). Eventually, a new trench is likely to form near the Atlantic-type trailing edge of the collided continent. The Himalayas (Figure 46–11E) are taken as the type example of this kind of mountain belt. The kinds of structures developed by collision may depend to a large extent on the nature of the sedimentary sequence developed on the Atlantic-type margin before collision. The Zagros Mountains of Iran began to develop after the collision of the Arabian shield with the Iran plateau. The Zagros crush zone (Figure 46–11F) is a suture zone of present-day underthrusting (Wells, 1970) of the Arabian shield and has ophiolite, chert, flysch, and mélange sequences of Mesozoic Age (Stöcklin, 1968) marking the site of a pre-collision trench along the southern margin of the Iran plateau. The Zagros stratigraphic sequence is a thick carbonate-shale assemblage of

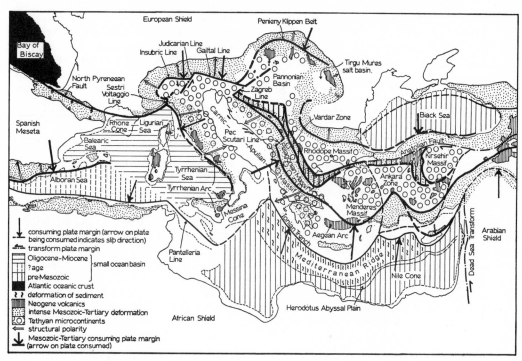

Figure 46-14
Outline of the structure of the Alpine fold belt and the Mediterranean.

Lower Paleozoic to Tertiary age lying on Precambrian basement. The basal part of the sequence consists of late Precambrian-Cambrian salt deposits. The salt seems to have acted as a lubricant-detachment horizon, thereby allowing the overlying rocks to deform to some extent independently of the basement and with structural styles that are simple and that lack large thrust sheets. The main arcuate loop of the French, Swiss, and Austrian Alps probably resulted from the collision of an Atlantic-type continental margin (Figure 46–6D) with a Tethyan microcontinent (Carnic-Apulian massif; Figure 46–14). The site of a Mesozoic precollision trench is now marked by the Insubric-Judicarian-Gailtal line (Figure 46–14). The dominant sense of thrusting (structural polarity) in the Swiss Alps is toward the north (Figure 46–11G), with the Pennine nappes of the south involving thorough basement remobilization and the Helvetic nappes of the north largely detached from their root zone (the internal Aar and Gotthard basement massifs) and emplaced dominantly by gravity sliding.

The degree of suturing accompanying continental collision and the consequent degree and timing of deformation will depend on the shape of the colliding margins (Wilson, 1965a). Where margins are very irregular, the first segments of the margins to collide will exhibit the earliest and most intense deformation. In these regions the trench zone will be reduced to a narrow suture from which ophiolite, oceanic sediments, and flysch are extruded. Also, these regions are likely to have extensive fluviatile clastic deposits (molasse) in exterior troughs. Indented portions of the continental margins may never collide and may thus preserve extensive regions of oceanic crust within the orogenic belt. These indented regions will show less deformation and thin (or absent) exterior molasse deposits.

CONCLUSIONS

We propose that mountain building occurs in two basic ways. One type, the island arc/cordilleran, is for the most part thermally driven and develops on leading plate edges above a descending plate. The other type, the result of continent/island arc or continent/continent collisions, is for the most part mechanically driven. In Figure 46–15 we show schematic time-space models for the development of cordilleran-type and collision-type mountain belts, together with a series of associated typical stratigraphic (geosynclinal) sequences. The main differences between the two types of mountain belt may be summarized as follows:

1. The symmetry and sense of thrusting in a cordilleran mountain belt is divergent, involving oceanward-driven wedges of oceanic crust related to the underthrusting of the oceanic lithosphere plate and more superficial continentward-driven sheets of sediment and continental basement. A collision mountain belt is characterized by a single, dominant, sense of thrusting onto the consumed plate. This often involves the complete remobilization of basement near the site of collision, and gravity slides further onto the site of the old continental shelf.

2. In cordilleran-type mountain belts, gravity slides predate hard-rock thrusting, whereas in collision-type belts the opposite happens.

3. A paired metamorphic zonation is developed in cordilleran mountain belts with an earlier blueschist belt on the ocean side and a later, high-temperature, metamorphic belt on the continental side. An important point is that the high-temperature metamorphism affects rocks, which, although usually involved in continentward-driven thrust sheets, were originally arranged adjacent to the continental shelf as thin continental-rise sediments overlying thick accumulations of coarse clastics and volcanics (Figure 46–6E). These sediments and volcanics cannot represent an island arc driven into the continental margin because of their continuity of sedimentary and chronologic relationships with the shelf sediments, and because of the absence of a possible collisional suture zone marked by blueschist-ophiolite mélanges. By contrast, a collision type does not have a paired metamorphic zonation, and the dominant metamorphic assemblages are in the blueschist facies (e.g., the French Alps; see Ramsay, 1963).

4. Cordilleran types have a paired miogeosynclinal (continental shelf)/eugeosynclinal (re-gion between continental shelf edge and trench) relationship. Collision types are the result of the juxtaposition of a continental shelf and rise-sediment assemblage and a trench-volcanic arc assemblage, originally separated by an ocean. It is important to note that the preorogenic continental margin sediment assemblages of both orogenic types are basically the same as indicated in Figure 46–15.

5. The flysch of cordilleran types has a dual polarity, continentward and oceanward, and is dominantly derived from the high-temperature metamorphic axis region of the mountain belt. The flysch of collision types has, essentially, a single polarity and is of two distinct origins. The early flysch (e.g., Cretaceous Helminthoid flysch of the Embrunais nappe of the Alps) is a trench and small ocean-basin accumulation and during collision is driven as high-level thrust sheets onto the continent. The later flysch is related to the continentward-movement of nappes.

6. Cordilleran types have a dense, basic root (Thompson and Talwani, 1964), probably related to the emplacement of basic intrusions beneath the volcanic high-temperature metamorphic axis. The root of collision mountain belts is sialic and probably results from continental underthrusting and thickening.

The ophiolite suite is of fundamental importance in evaluating the stages of development of mountain belts. We suggest that the term ophiolite be restricted to a full sequence of ultrabasic rocks, gabbro, dike complexes, pillow lava, and chert, a sequence that, if fully developed, almost certainly represents oceanic mantle and crust. Ophiolites are well developed in cordilleran mountain belts, for example, in the Klamath Mountains of northern California (Davis, 1968); they form extensive upthrust regions behind the blueschist trench terrains. Ophiolites also occur as smaller, detached rafts in the mélanges of trenches, and these occurrences may represent blocks that have slid down the inner trench wall, slices of oceanic crust or mantle, or both, and seamounts torn from the descending plate. In collision mountain belts, ophiolite blocks are extruded from the trench during collision and lie in flysch-mélange suture zones that mark the collision 'join lines.' The composition of ophiolite pillow basalts may be a criterion for distinguishing

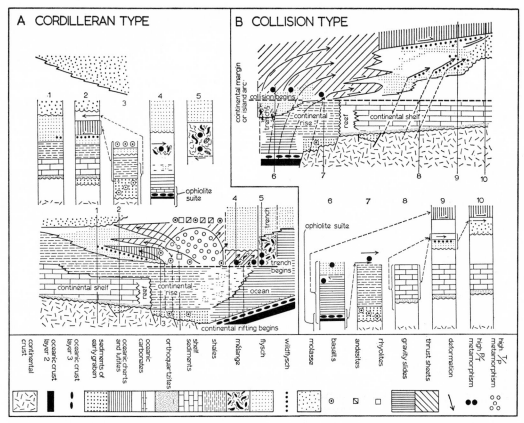

Figure 46-15
Schematic time (vertical coordinate)/space (horizontal coordinate) relationships of sedimentation, volcanicity, deformation, and metamorphism in cordilleran-type, and collision-type mountain belts

between the crust of the main oceans (tholeiite and spilite) and the crust of small ocean basins (alkalic) if the latter are produced by the separation of arcs from continents.

Although the cordilleran/island arc and collision mechanisms are probably the fundamental ways by which mountain building occurs, mountain belts are generally the result of complex combinations of these mechanisms. The evolution of the Appalachian orogen (Bird and Dewey, 1970) involved Ordovician cordilleran and island arc mechanisms, and Devonian continental collision. The Alpine-Himalayan system has been developing since early Mesozoic times by multiple collision

resulting from the sweeping of microcontinents and island arcs across the Tethyan-Indian Ocean. Mountain belts such as the Urals, lying at present within continents, may be complex combinations of cordilleran belts, microcontinents, and volcanic arcs, of widely different ages, that become juxtaposed by the driving out of a major ocean basin. Furthermore, the possibility of expanding and contracting transform offsets of consuming plate margins raises the likelihood of distinctive belts of volcanism, deformation, and metamorphism coming to an abrupt termination along the strike of a mountain belt.

47

On the Pacific-type Orogeny and its Model: Extension of the Paired Belts Concept and Possible Origin of Marginal Seas

TYKIHIKO MATSUDA
SEIYA UYEDA
1971

From *Tectonophysics*, v. 11, p. 5–27, 1971. Reprinted with permission of the authors and Elsevier Publishing Co.

The concept of paired metamorphic belts, originally proposed by Miyashiro (1961) from the study of the Late Mesozoic orogeny in Japan and other circum-Pacific areas, has been developed along with the view that the modern island arc features are nothing but a present manifestation of the paired orogeny (Matsuda, 1964; Takeuchi and Uyeda, 1965; Miyashiro, 1967; Matsuda et al., 1967; Sugimura and Uyeda, in preparation). Miyashiro (1961) pointed out that the Mesozoic metamorphic belt in the circum-Pacific region consists of outer and inner belts having contrasting natures in metamorphism and in associated magmatism. He further pointed out that the outer metamorphic belt, characterized by high P/T type metamorphism, lies always on the oceanic side, which has no sialic basement. Matsuda (1964) noted that the two types of regional metamorphism, i.e. the high P/T type and low P/T type may be taking place at present under the trench and inner belts of the island arc respectively. Later, Takeuchi and Uyeda (1965) drew attention to the remarkable coincidence of the P/T conditions under the trench and inner belts, estimated from heat-flow data (Uyeda and Horai, 1964), with the P/T conditions for the two types of metamorphism, estimated from mineralogical grounds (Miyashiro, 1961). Miyashiro (1967) and Sugimura and Uyeda (in preparation) proposed a model for the origin of the paired zone putting the prime importance on the mantle convection sinking along a deep-seismic plane beneath the island arc system.

In this paper we describe the general nature of the paired orogeny which we propose to call the Pacific-type orogeny, and discuss its possible mechanism. This type of orogeny may be only a local variety of the classical type of orogeny established in the European or American continents (e.g., Kay, 1951; Aubouin, 1965), but the present authors think otherwise. There seem to be many fundamental differ-

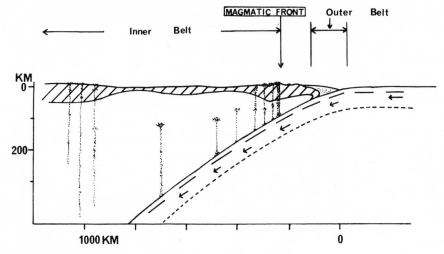

Figure 47-1
A schematic profile of the Pacific-type orogenic belt

ences between the two types of orogeny. For example, in the generally accepted pattern of orogenic evolution of a eugeosyncline, the "geosynclinal phase" is followed by "orogenic phase" with granitic magmatism in the same belt. In an orogenic zone with paired belts, however, the eugeosynclinal environment and the granitic magmas are active almost simultaneously in the juxtaposed outer and inner belts.

FUNDAMENTAL FRAMEWORK OF THE PACIFIC-TYPE OROGENIC BELT

The framework of the Pacific-type orogenic belt is asymmetric and paired, consisting of the outer and inner belts as schematically shown in Fig. 47–1. The outer belt is the place of regional subsidence, sedimentation and high P/T type metamorphism, as suggested by the negative isostatic gravity anomaly and low heat flow in active island arcs. On the continental side, on the other hand, is the inner belt, characterized by various types of magmatism and high heat flow. The oceanward front of the inner belt, called the *magmatic front* in this paper, is defined by the volcanic front (Sugimura, 1960) or a frontal line limiting the distribution of granites. The features of the inner belt extend more than 1,000 km from

the magmatic front to the interior of the continent. The inner belt may or may not include a marginal sea behind the magmatic front. The inner belt of the present writers includes the inner metamorphic belt described by Miyashiro (1961), but it is much wider than the latter. The outer and inner belts are separated, as a rule, by a non-volcanic geanticlinal ridge. The narrow outer belt on the oceanic side, the wide inner belt on the continental side, and the median geanticlinal belt form the basic framework of the Pacific-type orogenic belt.

A brief explanation follows of the two main orogenies of the Japanese Islands, i.e. the Late Cenzoic orogeny and the Late Mesozoic– Early Tertiary one from the viewpoint of the paired orogeny as defined above.

LATE CENOZOIC EXAMPLE OF A PAIRED BELT IN NORTHEASTERN JAPAN

The Late Tertiary–Quaternary pair of the Pacific type orogeny has persisted till today to form the present island arcs. Figure 47–2 is a cross-section of northeastern Japan, which is one of the most typical and active island arcs. Approaching the arc from the Pacific Basin, the order of the main features in the

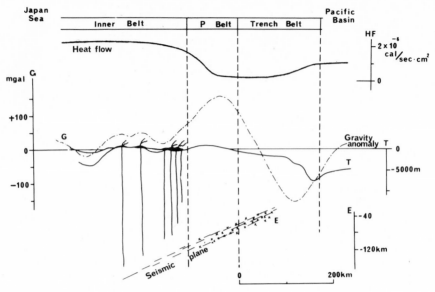

Figure 47-2
A schematic profile across northeastern Japan (Matsuda et al., 1967, revised)

zonal arrangements is as follows: (1) Japan Trench; (2) negative gravity anomaly belt; (3) positive gravity anomaly belt on the non-volcanic outer arc; (4) front of volcanoes (Sugimura, 1960); (5) volcanoes; and (6) Japan Sea. Among these, the outer (1), (2) and (3) zones have low heat flow and the inner (4), (5), and (6) zones have high heat flow (Uyeda and Vacquier, 1968). Beneath these features, there is a deep-seismic plane dipping toward the Asiatic continent. The above-cited belts were classified previously by Matsuda (1964) into the trench belt (1,2), P belt (3), and inner belt (4–6) as shown in Fig. 47–2.

The inner belt of northeastern Japan, often called the green tuff region of Uetsu-Fossa Magna idiogeosyncline (Sugimura and Uyeda, in preparation), was described previously (Minato et al., 1965; Matsuda et al., 1967). Since the Early Miocene, the belt has been subjected to active volcanism, which began suddenly after a long period of denudation that had lasted since the end of the Paleozoic. Various volcanic products, mostly acidic to intermediate in composition, were laid down successively in either subaerial, lacustrine or submarine conditions, interfingering with normal clastic sediments. Granitic intrusions

accompanied them. The Quaternary volcanism in the inner belt is a continuation of the Late Tertiary one. Although Quaternary volcanoes are concentrated in a fairly narrow zone just west of the volcanic front, the Late Cenozoic and Quaternary magnetism is known to have been active over a much wider area, extending to the eastern margin of the Asiatic continent (Kuno, 1952; Ministry of Geology, U.S.S.R., 1965). Heat flow is persistently high, not only in the volcanic zone on the main islands of Japan, but all over the Japan Sea (Yasui et al., 1968). Thus, it appears reasonable to consider the inner belt in the Late Cenozoic of northeastern Japan to be more than 1,000 km wide and to extend from the volcanic front to the interior of the Asiatic continent beyond the Japan Sea (Fig. 47–3).

East of the magmatic front, lies a median non-volcanic geanticlinal ridge of positive gravity anomaly (P belt) which has been rising gradually throughout the Cenozoic. This geanticlinal belt separates the inner belt or the green tuff region from the offshore outer or trench belt.

The outer belt is represented, in the present island arc, by the trench belt, where the geosynclinal sedimentation and the high P/T

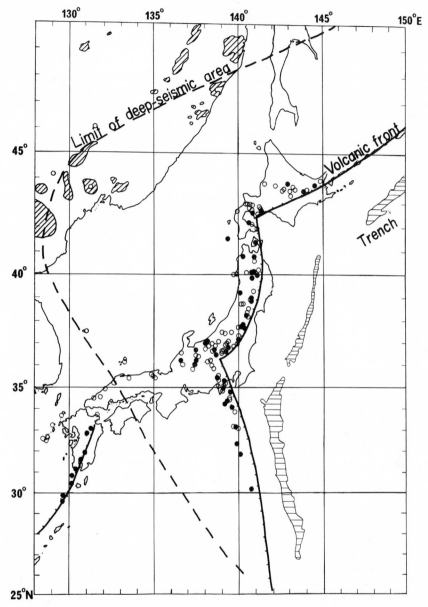

Figure 47-3
Trench, volcanic front, volcanoes and deep seismic area around the Japanese islands.
Solid circles: active volcanoes; open circles: Quaternary volcanoes; dotted areas: Late
Tertiary volcanic rocks; hatched areas: Pliocene–Quaternary regional basalts (after
Kuno, 1968a and Geologic Map of U.S.S.R., Ministry of Geology. U.S.S.R., 1965);
and volcanic front and limit of deep-seismic area: after Sugimura (1960)

type of regional metamorphism are inferred
to occur (Miyashiro, 1967; Matsuda et al.,
1967). A post-Pliocene large-scale subsidence
and tectonism of this belt has been suggested
(Iijima and Kagami, 1961; Ludwig et al.,

1966). It is inferred that the belt came into
existence at least by the Early Miocene, possibly in the Early Tertiary or latest Cretaceous
as a counterpart of the inner belt of the Late
Cenozoic pair.

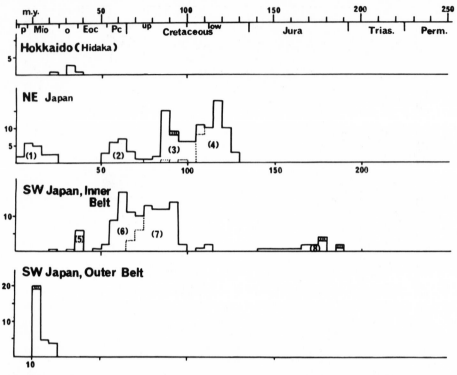

Figure 47-4
Potassium/argon age of granitic rocks from Japan (after Kawano and Ueda, 1967; Nozawa, 1968). (1) = Joshin-Kobu area; (2) = Uetsu area; (3) = Abukuma-Sekiryo area; (4) = Kitakami-southwest Hokkaido; (5) = San'in area; (6) = San'in central Japan; (7) = N. Kyushu-Seto I.S.-Kinki; and (8) = Hida area. Data for southwest Japan-outer belt include effusive acidic rocks

LATE MESOZOIC EXAMPLE OF PAIRED BELTS IN SOUTHWESTERN JAPAN

Cretaceous – Early Tertiary granitic plutons constitute 90% of the area of exposure of all the granitic rocks in Japan (Ono and Isomi, 1967). As shown in Fig. 47–4, their age ranges from 130 m.y. (million years) to about 50 m.y. according to K/Ar age determinations (Kawano and Ueda, 1967). Particularly in southwestern Japan, granitic plutons and associated acidic volcanic rocks of Cretaceous – Early Tertiary age (mostly 95–50 m.y.) are distributed widely in the inner or Japan Sea side of the area (Fig. 47–5). The metamorphic belt (the Ryoke metamorphic belt) of the low P/T type (andalusite – sillimanite type) lies along the southern margin of this wide belt of granitic plutons. On the other hand, in the outer or the Pacific Ocean side of southwestern

Japan, no granitic rocks of the same age are present. Instead, a geosynclinal sedimentary belt (Shimanto belt) parallels the contemporaneous magmatic front to the north.

The Inner Granitic Belt

Intrusions and extrusions of granitic magma took place repeatedly from Cretaceous to Early Tertiary times in the inner belt, where a non-geosynclinal tectonic state had preceded, and local temporary sedimentary basins of either marine or lacustrine environments were present (Research group for Late Mesozoic igneous activity of southwest Japan, 1967; Ichikawa et al., 1968). Most of the extrusions, especially the younger ones were subaerial, yielding a tremendous amount of pyroclastic rhyolite and andesite, estimated at ca. $6.7 \cdot 10^4$ km^3. Before and after these eruptions, granitic

Figure 47-5
Cretaceous-Early Tertiary granites and the contemporaneous outer sedimentary terrain
(Shimanto zone) in Japanese islands and the vicinity (after Saito, 1940; Geological
Survey of Korea and Geological Society of Korea, 1956; Ministry of Geology. U.S.S.R.,
1965; Geological Survey of Japan, 1968)

magma intruded many times into the Paleozoic
or volcanic formations, which were formed by
the extrusion of co-genetic magmas. As already pointed out by many authors (e.g.,
Ichikawa et al., 1968; Matsumoto, 1968), it
is notable that these large-scale magmatic
activities were not immediately preceded by
eugeosynclinal subsidence. Before this Cretaceous igneous activity began, most of the
area had been undergoing erosion ever since
the end of the Permian when the Paleozoic
geosyncline was converted into land. The
southern limit of the inner belt is defined
sharply by a magmatic front which approximately corresponds to the present position
of the median tectonic line.

The Cretaceous — Early Tertiary granitic
activities were not restricted to the Japanese
Islands, but extended further inland, beyond
the Japan Sea to the Korean Peninsula and
other marginal areas of the Asiatic continent
(Fig. 47–5). The general nature of the granitic
activity and the tectonic situation of these
regions in Late Mesozoic — Early Tertiary
times were quite similar to those of the
Japanese Islands (e.g., Ustiyev, 1963).
Welded rhyolitic pyroclastic flows like those
of the Late Mesozoic in southwestern Japan
were formed also at submarine highs, Yamato
Bank and other banks in the southwestern
Japan Sea (Sato and Ono, 1964). These facts
indicate that part of the southwestern Japan

Sea area, as well as the main Asiatic continent, was a land area having inner belt features probably in Late Mesozoic — Early Tertiary times.

The Outer Sedimentary Belt (the Shimanto Belt)

Contemporaneously with the granitic magmatism in the inner belt, a geosynclinal sedimentary belt (the Shimanto belt) persisted in the outer zone of southwest Japan. This outer belt is underlain mainly by a thick sequence of turbidite graywacke and shale, interbedded irregularly with conglomerate, chert and submarine basaltic lavas and tuffs. Ultramafic intrusions are also found. These rocks are regionally tightly folded to form almost isoclinal folds dipping north at high angles (e.g., Harada, 1964; Hashimoto, 1966; Kimura, 1967; Yamada et al., 1969), and they are regionally metamorphosed up to the greenschist facies through the pumpellyite — prehnite metagraywacke facies (Matsuda and Kuriyagawa, 1965). Neither contemporaneous acidic volcanism nor granitic intrusions are known in this belt.

Although the rocks of the Shimanto belt contain few fossils, available paleontological data indicate that most of the sequence is Middle Cretaceous — Early Tertiary in age and was laid down in open sea. Jurassic and Triassic formations may be included in the lower part of the sequence (Takai et al., 1963; Minato et al., 1965). The basal part of this zone is not exposed, but it contacts tectonically with the Triassic and Paleozoic terrains to the north. The upper part of the sedimentary sequence of the Shimanto belt includes the Lower Miocene rocks which are overlain locally by Middle Miocene sediments with a remarkable unconformity.

Median Geanticlinal Ridge

The Sanbagawa belt and the Paleozoic Chichibu zone, both to the north of the Shimanto belt, were in the stage of gradual rising in Cretaceous and Early Tertiary times as a geanticlinal ridge separating the inner

granitic belt from the outer Shimanto belt. Thus the Sanbagawa belt is considered to have been a belt similar, tectonically, to the geanticlinal non-volcanic outer arc of positive gravity anomaly (P belt) in the present island arcs.

The K/Ar dates of the Middle Cretaceous for the Sanbagawa metamorphic rocks (Miller et al., 1961) probably represent the time of the beginning of the uprise of the belt (Hurley et al., 1962). This view is different from those of Miyashiro (1961, 1967) and others, who regard the Sanbagawa belt as the outer counterpart to the Ryoke metamorphic belt in the Mesozoic pair of orogenies.

The Late Mesozoic — Early Cenozoic orogenic activity mentioned above had apparently terminated by the Middle Tertiary, although in part of southwestern Japan it continued, with a lesser magnitude up to Late Cenozoic time. Instead, a new paired orogenic feature came into existence in the northeastern Japan as described in the section "Late Cenozoic eaxmple of a paired belt in northeastern Japan."

SYNTHESIS OF THE PACIFIC-TYPE OROGENY

As illustrated in the preceding sections, the major features of the orogenies in the Japanese area are well described by pairs of contemporaneous outer and inner belts having contrasting characteristics. Their geologic characteristics are summarized in Table 47–1. There is little need for further explanations of the items in this table. It may be worthwhile, however, to emphasize that the outer belt is eugeosynclinal, with the Steinmann trinity (Holmes, 1965) and, unless the mechanism suggested later comes into play, is free from granitic intrusions. The inner belt, on the other hand, is granitic, not necessarily accompanied by geosynclinal subsidence. A spatial separation of geosynclinal environment and granitic intrusions is the most important feature of the Pacific-type orogeny. Thus, the classical evolutionary pattern of a eugeosyncline, starting with geosynclinal magmatism followed by the granitic magmatism of orogenic and subsequent stages and ending with post-orogenic volcanism, is applicable to neither

Table 47–1
Pacific-type orogenic belt.

Example	Inner belt	Median ridge	Outer belt
Present Island arc	Volcanic arc and marginal sea behind, with high heat flow	Geanticlinal non-volcanic arc of positive gravity anomaly	Trench and belt of negative gravity anomaly
Late Mesozoic SW Japan	Ryoke metamorphic belt and circum-Japan Sea area		Shimanto belt

	Inner orogenic features	Outer orogenic features
Tectonism	I_1 = Block movement, with local sedimentary basins	0_1 = Geosynclinal sinking, sedimentation and regional folding
Magmatism	I_2 = Acidic-intermediate magmatism (basalts may be also accompanied)	0_2 = Basic to ultrabasic magmatism ("ophiolites")
Metamorphism	I_3 = Low P/T metamorphism (andalusite-sillimanite type, etc.)	0_3 = High P/T metamorphism (glaucophanitic metamorphism, etc.)

the inner nor the outer belt of the Pacific-type orogeny. Instead, the belts of the Pacific-type orogeny have their own evolutionary patterns. Table 47–1 lists only the main events in each belt.

The age and duration of the high P/T type metamorphism in the outer belt are essentially the same as those of the sedimentation in this belt; both are involved in the single process of geosynclinal evolution. This contrasts with the inner belt, where geosynclinal subsidence does not necessarily occur and granitic magmas intrude or extrude through a pre-existing continental crust of various geologic ages. For example the Cretaceous granites intruded into both the Precambrian orogenic belts in the interior of the continent and the Late Paleozoic belts in the Japanese Islands. These granites are exotic to the surrounding terrains, having no genetic relationships with the sediments that they intrude. Likewise, the age of the low P/T type regional metamorphism of this belt has no definite relationship with that of the formations metamorphosed.

Granites are largely of "post-orogenic" discordant type. Concordant gneissose granodiorite and low P/T type regional metamorphism formed only along the magmatic front in the Japanese Late Mesozoic orogeny (Ryoke

metamorphic belt). Of the volcanic rocks, it can be shown in the Japanese Islands that the volcanic products during the Quaternary Period increase in amount toward the volcanic front (Sugimura et al., 1963). These facts suggest that the thermal activity beneath the inner belt is asymmetric, fading continentward from the magmatic front. It has also been shown that there is a systematic zonal areal variation in the chemical composition of volcanic rocks, which arranges parallel to the magmatic front; the nearer the magmatic front, the less the alkali contents (or tholeiitic) (Kuno, 1960; Sugimura, 1960).

The outer and inner belts also contrast in tectonism. While the outer belt is characterized by regional subsidence, the inner belt undergoes block movements or German-type tectonism. Differential vertical crustal movements in the inner belt result in basin and range structures, whose basins undergo sedimentation. "Back deep" or "idiogeosyncline" (Umbgrove, 1947) is an example of such sedimentary basins. The amount of local vertical displacement during the Late Cenozoic exceeded 5,000 m in the northeastern Japan inner belt (Matsuda et al., 1967, fig. 5).

These Japanese examples suggest that the initiation of the inner belt activity may be

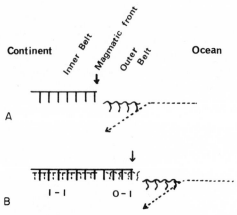

Figure 47-6
Overlapping of the inner and outer belts (portion indicated as 0-1), as magmatic front migrates oceanward from stage A to stage B

later than that of the outer belt. However, the epoch of the maximum activity (magmatism) of the inner belt seems roughly contemporary with the maximum sedimentation of the outer belt.

It is presumed (see later section "Possible mechanism of the Pacific-type orogeny and the origin of marginal seas") that the Pacific-type orogeny formed on a boundary zone between the continental and oceanic lithosphere by the oblique downgoing of the latter beneath the former (Fig. 47–1). If this is accepted, it appears that the orthodox evolutionary pattern of eugeosynclines in the European and American continents (e.g., Kay, 1951; Aubouin, 1965) would form when the lithosphere descends vertically, so that the cross section of the orogenic belt is symmetrical and the granitic magmatism takes place subsequently in the same belt. Matsumoto (1967) noticed an evolutionary change in the type of tectonic framework of orogeny through geologic ages in the circum-Pacific mobile belt, i.e., a change from the Paleozoic eu- and miogeosynclinal framework to the Late Cenozoic island-arc type through a transitional Mesozoic-type. In fact, Paleozoic orogeny in the Japanese Islands seems to be more complex than the Mesozoic – Cenozoic orogeny. Recently Dewey and Horsfield (1970) presented a valuable discussion of types of orogeny, based on the plate tectonic theory. Further investigation is desired on this subject.

MIGRATION OF MAGMATIC FRONT AND OVERLAPPING OF THE INNER AND OUTER BELTS

In the Pacific-type orogeny, the outer belt is always free from the magmatism active in the inner belt. There are, however, cases where a geosynclinal belt is later subjected to granitic magmatism as in the case of the classical type orogeny. These are interpreted as results of migration of magmatic fronts in the Pacific-type orogeny. As shown in Fig. 47–6, the outer belt at one time would become the site of the inner belt at a later time, if the magmatic front shifts oceanward, resulting in the overlapping of the inner belt features on the former outer belt. The overlapping portions would have an apparent sequence in which granitic intrusions took place where geosynclinal sinking with a Steinmann trinity had previously gone on. In such a case, the high P/T type metamorphic belt in the former outer belt would be affected by the subsequent low P/T type metamorphism (O−I type polymetamorphism). Polymetamorphism of repeated low P/T type (I−I type polymetamorphism) is expected in inner regions.

Accepting the model of a downgoing lithosphere beneath the island arc as discussed in the following section, a migration of the magmatic front is thought to be caused by a lateral migration of the Benioff zone. The change in the inclination angle or the slip-rate of the downgoing lithosphere may also cause minor and temporal shifting of the magmatic front. The position and trend of the magmatic front may, thus, be good indicators of the downgoing lithosphere and hence, of the migration of orogeny. The migration of the magmatic front on the Japanese islands since the Mesozoic is shown in Fig. 47–7 and Fig. 47–8. The Mesozoic and Early Tertiary magmatic fronts are inferred from the results of the radiometric age determinations of granites (Kawano and Ueda, 1967). The trend of the pre-Early Tertiary magmatic fronts has a break in central Japan as is shown in Fig. 47–7. This break may suggest that northeastern Japan moved eastward relative to southwestern Japan and rotated anticlockwise between 50 m.y. and 20 m.y. ago. This inferred movement of northeastern Japan is in harmony with the contention of the eastward shift of the northeastern

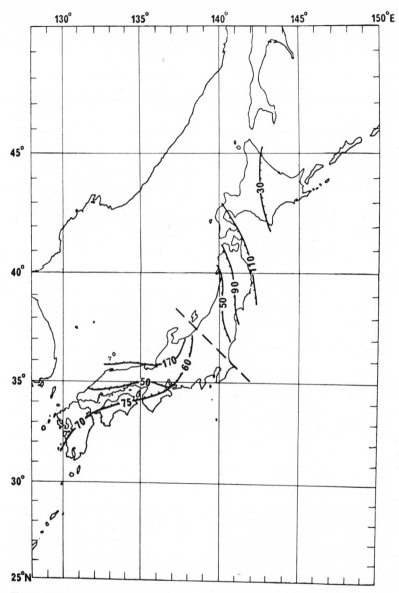

Figure 47-7
Migration of magmatic fronts during Mesozoic and Early Tertiary times.
Numerals: million years

Japan postulated by Kobayashi (1941) and with the hypothesis of the bending of the Japanese islands suggested by Kawai et al. (1961, 1969). A well-known discordance between orogenic trends of Late Mesozoic – Early Tertiary time and of Late Cenozoic time in northeastern Japan can be explained by this rotational movement. In contrast to northeastern Japan, magmatic fronts in southwestern Japan lie nearly parallel with each other through the Late Mesozoic to Late Cenozoic time, although there were considerable migrations to the north and south. It is a task for the future to understand fully the tectonic significance of these migrations and changes of magmatic fronts.

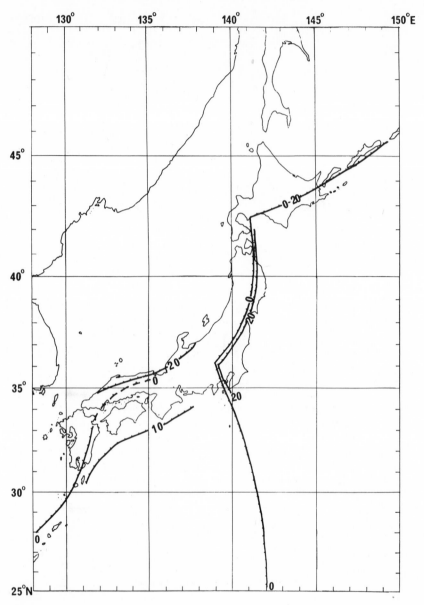

Figure 47-8
Migration of magmatic fronts since Early Miocene. Numerals: million years

POSSIBLE MECHANISM OF THE PACIFIC-TYPE OROGENY AND THE ORIGIN OF MARGINAL SEAS

Taking the above characteristic features into consideration, we propose a model for the pro-

cess of the Pacific-type orogeny as shown in Fig. 47–9. It is based on the hypothesis of the downgoing mantle flow under the island arc, and may be considered to be consistent with the plate tectonics, in that the downgoing part represents a slab of lithosphere. In the outer belt, the sea floor is dragged down to

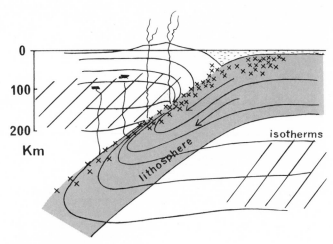

Figure 47-9
Cartoon describing the cross section under the Japanese Arc
(Sugimura and Uyeda, 1972)

form the trench where geosynclinal sedimentation proceeds. Low heat flow and shallow earthquakes are expected there. As has been proposed (Sugimura and Uyeda, in preparation), we assume that the deep earthquakes along the inclined seismic plane are caused by shear between the sliding lithosphere and the mantle above it, and the same shear generates heat along the seismic plane giving rise to partial melting. The molten material is the parent of the inner belt magma. The front of volcanoes or the magmatic front is the oceanward limit of magma production whose position depends on many factors such as the relative velocity of the slab and overlying mantle, material properties of the mantle and slab, etc. Basic or ultrabasic intrusions or ophiolites found in the outer belt may be difficult to explain in the present model. But they may be interpreted as produced in the mid-oceanic area when the plate was formed to be carried to the trench and there scraped off.

Recently, Hasebe et al. (1970) developed numerical experiments on the model proposed by Sugimura and Uyeda (in preparation) which is outlined in Fig. 47-9. Hasebe et al. solved numerically the equation of heat conduction in which the motion of the slab, heat generation due to radioactive sources and to shear are involved. The computation starts from the initial state which is considered as the normal oceanic and follows the variations in the thermal field. When the temperature exceeds the solidus temperature of water-saturated basalts, partial melting is presumed and heat transfer due to the rise of magma comes into play. This effect is taken care of by an increased thermal conductivity in the vertical direction. In their model, which was meant to simulate northeastern Japan, a 100 km thick slab is assumed to descend with an inclination 1:3 at a rate of 3 cm/year. Other parameters used in computations were the depth below which shear heating is operative, increment of thermal conductivity by the rise of magma and the rate of shear heating.

Computations were designed to seek the best combination of these parameters which results in the surface heat flow pattern in, say 10^8 years after the initiation of the process, as observed. The results showed that the heat generation some five times as high as the mean heat flow (1.5 H.F.U.) at the upper surface of the slab and effective heat conduction ten times that of normal phonon conduction are required to account for the observed high heat in the Japan Sea area. An example from their results is shown in Fig. 47–10. These values of heat generation and effective thermal conductivity may appear much higher than one would expect from ordinary physical consideration on viscous heating (Turcotte and Oxburgh, 1968; McKenzie and Sclater, 1968; Minear and Toksöz, 1970), but it seems that these large

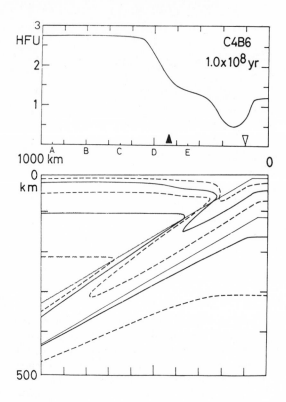

Figure 47-10
Heat-flow and temperature distribution at 10^8 years after commencement of downgoing of lithosphere (Hasebe et al., 1970). Shaded area represents the partially molten region. The solid triangle represents volcanic front, the open inverted triangle, trench. The solid lines are 500°C, 1,000°C and 1,500°C isotherms. The broken lines show 250°C, 750°C and 1,250°C isotherms

figures are needed as long as the model of a descending slab is assumed. Readers are referred to the paper cited above(Hasebe et al., 1970).

One interesting point to be noted is that the material flux needed to provide effective vertical thermal conductivity ten times greater than normal would be some $0.6 - 0.1$ cm/year. If we take the median value 0.3 cm/year, the total thickness of the material ascended during 10^8 years would become 300 km. This figure is not inconsistent with the idea that the Japan Sea was formed by opening (e.g., Kobayashi, 1956; Murauchi, 1966; Karig, 1969): The present Japanese islands were the eastern marginal parts of the Asiatic continent originally, but after the commencement of the process envisaged here, magma has continued to rise so that the Japanese islands had to drift southeastward to provide room for the magma (Fig. 47–11). Present considerations, thus, call for a revival of the seaward drift of the Japanese islands (Tokuda, 1926; Terada, 1934).

REPRODUCTION OF PLATES IN MARGINAL SEAS AND THE DEVELOPMENT OF ARCUATE FORM OF ISLAND FESTOONS

If the model described in the preceding section is applicable to other active island arcs, it may be postulated that other marginal seas are also expanding by intrusions of magma from the seismic plane. In plate tectonics or the new global tectonic, trench-arc systems are considered to be the area where plates are consumed. In our case, however, at least part of the descending lithosphere would rise again to form the upper mantle of the marginal seas. We propose here that, in addition to the production of plates over mid-oceanic ridges and their consumption at trenches, the above-mentioned process of reproduction of plates behind island arcs should also be regarded as an integral element of the scheme of plate tectonics. In the process, the opening of marginal seas or the spreading of the plates of marginal seas is caused by the ascent of magmas

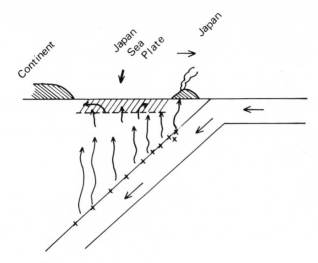

Figure 47-11
Formation of Japan Sea by
reproduction of marginal sea plate

almost everywhere as indicated in Fig. 47–11.
Most of the magmas are supposed to solidify
before reaching the surface. This type of
spreading differs from that at mid-oceanic
ridges which produces the regular symmetric
magnetic lineations by the Vine-Matthews
mechanism (1963). In the case of the spreading of marginal seas, there may be many short-
lived microspreading centers in the basin, so
that the pattern of magnetic anomalies may
accordingly be different from those under big
spreading oceans. The geomagnetic total force
anomaly patterns of the Japan Sea reported by
Yasui et al. (1967) seem to favor the above
conjecture. However, it appears that their
ships' tracks were still too sparse to delineate
the small-scale anomalies satisfactorily. A
more detailed magnetic survey in the Japan
Sea is underway.

Spreading of marginal seas will separate
the island arcs from the main continent. If,
for some reason, the movements of both ends
of a festoon of islands are constrained, the
parting will shape the festoon into an arc convex toward the ocean. As pointed out recently
by Kawai and Nakajima (1970), the possible
bending of Japan during Cenozoic time, deduced from paleomagnetic studies (Kawai et
al., 1961, 1969), may be due to a mechanism
related to the situations dealt with here: the
bending was caused by a compressive force
longitudinal to the axis of Honshu Island.

Fig. 47–12 illustrates that such a longitudinal
compression may be a geometrical consequence of relative plate motions. In this figure,
the oceanic plate is moving in a direction normal to the central arc A. Seismic studies have
shown that the relative motions between the
oceanic plate and the plates containing arcs
and marginal seas are essentially normal to

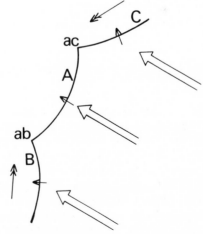

Figure 47-12
Axial compression in a chain of arcs.
White arrows: oceanic plate motion;
small arrows: relative motion between
oceanic and marginal sea plates; double
arrows: motion of marginal sea plates C
and B, relative to A

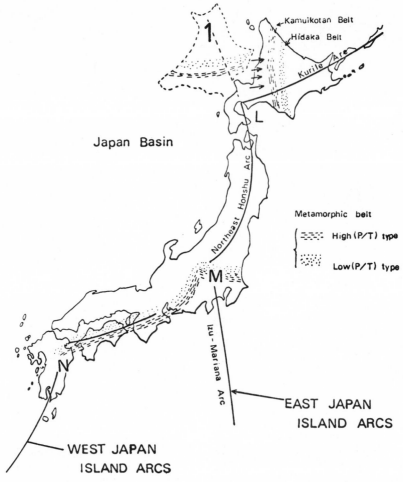

Figure 47-13
Two possible explanations of the reversal of metamorphic belts in Hokkaido

the arcs as shown by small arrows. If we as- sume that plate *A* is fixed to the reference frame, plates *B* and *C*, of necessity, are moving towards *A* to meet the above conditions. Thus, once a chain of arcs starts to form, it will con- tinue to accentuate its arcuate shape with angled intersections. In this scheme, inter- sections between arcs represent an area of intense compression. Pronounced negative gravity anomalies have long been noted at such intersections as noted by ab and ac in Fig. 47–12 (Hess, 1948). It is the interpreta- tion of the authors that such a negative anom- aly represents the effect of undiscovered downwarping of light crustal material at the intersection.

Another interesting feature to be noted is concerned with the possible reversal of the zonal arrangement of the metamorphic belts in Hokkaido. As shown schematically in Fig. 47–13, the Hidaka metamorphic belt (low P/T type) and Kamuikotan metamorphic belt (high P/T type) in Hokkaido are arranged in the opposite way to those in other parts of island arcs (Miyashiro, 1961). The reversal might be interpreted in two ways. One is that at the time of Hidaka orogeny the area west of the belts, i.e. the present Japan Basin was underthrusting eastward. This interpretation might harmonize with the notion that the Japan Basin plate was developing in Late Mesozoic times by the mechanism mentioned already:

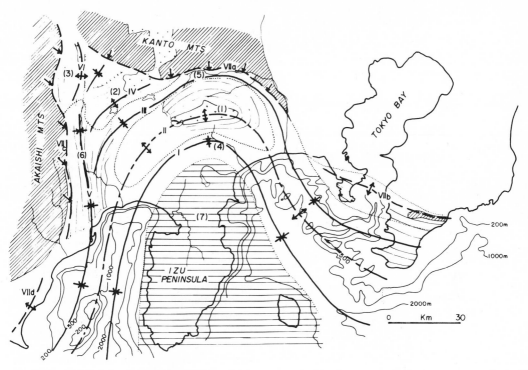

Figure 47-14
Tectonic map of the south Fossa Magna (Matsuda, 1962)

the developing Japan Basin plate caused the southeastward drift of the main island of Japan (Honshu) but it underthrusted beneath Hokkaido, thereby producing an eastdipping Benioff zone locally (east-pointing arrows in Fig. 47-13).

Another interpretation is related to the development of the arcuate form and angled intersection of arcs just mentioned: Hokkaido had originally been located in the straight extension of the axis of northeastern Japan as schematically shown by *1* in Fig. 47-13 and then was rotated clockwise relative to northeastern Honshu during the development of the intersection *L* between the northeast Honshu Arc and the Kurile Arc to take the present orientation. It must be noted, of course, that the argument here applies only to the *relative orientation* of Honshu and Hokkaido and does not mean that Hokkaido was actually located at the position *1* of Fig. 47-13. The same argument may be applied to the bend of the Mesozoic metamorphic belts at the central part of Honshu, that is, the intersection *M* of the northeastern Hon-

shu Arc and the Izu–Mariana Arc (Fig. 47-13). The direction of the bend is concave to the ocean and reverse to that of Kawai's general bending of Honshu Island. The development of the intersections *L* and *M* is considered to be the product of the activity of the East Japan Island Arcs which started in Cenozoic time. Perhaps Kawai's general bending of Honshu Island convex toward the ocean was caused by the Late Mesozoic orogeny and, therefore, predated the bending mentioned here. Fig. 47-14 shows the geologic structure of the southern part of the bending in central Honshu. The age of sedimentary strata which have the curved fold axes as indicated in this figure is Miocene. Therefore, if the curvature of fold axes was caused by post- or syn-depositional bending, it might be explained by the Late Cenozoic development of the intersection *M* described here. Verification of this hypothesis by, for example, paleomagnetic investigations is highly desirable. (The argument of this section, however, is highly speculative and held by only one (S.U.) of the authors).

48

Plate Tectonics and Magmatic Evolution

JAMES GILLULY
1971

From *Bulletin of the Geological Society of America,*
v. 82, p. 2382–2396, 1971. Reprinted with permission of
the author and the Geological Society of America.

So far as I know, no one has yet suggested a model for the generation of plate motion that is acceptable to anyone else. Nevertheless, the arguments from magnetic strips, from the distribution of blue schists and ophiolite belts, from sedimentary volumes, from the mutual relations of volcanic and plutonic belts, ocean ridges and deeps, and from the JOIDES drill cores, are cumulatively so compelling that the reality of plate tectonics seems about as well demonstrated as anything ever is in geology. Of course these phenomena do not demonstrate that plate motion is the only kind nor even the generating engine of all tectonics; in fact, many tectonic and magmatic features are so situated that any connection with plate tectonics is so tenuous as to be visible only to the eye of faith.

In this paper I first discuss the magmatic features that seem most clearly related to plate motion, then pass to some more doubtfully related, and finally to those magmatic developments that seem definitely independent of plate motion.

MAGMAS ALONG THE OCEANIC RIDGES

Magmas are developed at both ends of the mvoing-sidewalk-like tread of a migrating plate, both where new crust is being formed and along the Benioff zone, where crust is apparently being consumed. Most of the new crust developed along the active oceanic ridges is tholeiitic basalt, as Engel and Engel (1964), Engel and others (1965), and Nichols (1965) have suggested, but it must not be forgotten that dredge hauls have brought up several kinds of alkalic olivine basalt and quantities of ultramafic rocks from some scarps along the Mid-Atlantic Ridge, as Tilley (1947), Muir (1965), Muir and Tilley (1964), and Wiseman (1965) have reported. The ultramafics may be residual from the partial melting of pyrolite at greater depths while the tholeiite was being formed in shallower zones, as suggested by Kuno (1959, 1960, 1962, 1966), Yoder and Tilley (1962), Kushiro and Kuno (1963), and Kushiro (1965). Or they might be fragments of the mantle that rode up on injected sills of basalt.

McBirney and Gass (1967) have compiled data to show that Bouvet and Jan Mayen Islands, both on the crest of the Mid-Atlantic Ridge, are largely composed of alkali basalt, with very minor differentiates of trachyte and even rhyolite. The volumes of these siliceous differentiates are so small that they may readily be accounted for by differentiation in a high-level chamber within the volcanic edifice.

Iceland is much more aberrant; along the 40,000 mi length of the oceanic ridges, it is the locus of the only considerable stocks of granite and the only voluminous rhyolites, with the possible exception of Ascension Island (Mitchell-Thomé, 1970). The virtual uniqueness of this occurrence and the restriction to the continents and so-called microcontinents (Seychelles) of all the other granite masses large enough to be called stocks, make it very doubtful that the granites of eastern Iceland are direct differentiates of the mantle. Even though Deffeyes (1970) has shown how very improbable would be the persistence of sialic rocks astride a spreading ridge, I cannot help suspecting that one such block remains trapped beneath Iceland, there to contaminate a basaltic magma to produce the granite. The masses of granite seem much too large to have been formed by crystallization differentiation within the volcanic edifice in the way that might satisfactorily account for the much smaller volumes of siliceous rock found on Jan Mayen and Bouvet. There are not enough intermediate rocks such as would be expected if crystallization differentiation of basaltic magma were the only process involved in granite production. Sigurdssen (1968) has recorded granitic xenoliths from more than a dozen vents in central and western Iceland and from Surtsey, at considerable distances from the granites of the eastern coast. Cargill and others (1928), Walker (1963, 1966), and Thorarinssen (1967) have all suggested that the Icelandic basement is granitic.

McBirney and Gass (1967) have also shown that the volcanic islands tend generally to be more alkalic the farther they are from the oceanic ridges, and they attribute this to deeper sources of magma in the areas of generally lower heat flow away from the active ridges. The magmatism of the scattered volcanic islands, then, is not related to any plate margin as was once postulated by Tuzo Wilson (1963a). I return to the problem of island volcanism later and turn now to the supposed extension of the oceanic ridge system into Africa.

AFRICAN RIFT MAGMATISM

If the African Rifts are destined to become mid-oceanic features from which the fragments of the continent are to drift in a new pattern of plate tectonics, nothing in the chemistry of the volcanics of the region suggests any similarity to those of the mid-ocean ridges. The magmas of East Africa differ decisively from the potash-poor tholeiites of the oceanic ridges; they are among the most strongly alkaline rocks of the earth, many with $Na + K$ more than 10 percent, according to the studies of Mohr (1970a, 1970b) in the Afar triangle, McCall (1957, 1964), Baker (1967), Saggerson (1963), Williams (1970), and King (1966) in Kenya and Uganda, and of Dawson (1970) in Tanzania. Some are potash rich, others are highly sodic, and many carbonatite pipes exist, some of which are still active. Most of these rocks average rather basic but some of the younger lavas are salic and, according to Baker and Wohlenberg (1971), perhaps owe their peculiar chemistry to contamination by fenitized crustal rocks.

Whatever the mechanism creating the African Rifts, it is wholly different from that operating in the crust-creating margins of the oceanic ridges. The rifts range in date of beginning from Permian on the south (McConnell, 1970), through Miocene in East Africa (Baker, 1967), to Pliocene in Ethiopia (Mohr, 1970b), a time-spread twice as long as was involved in the opening of the Atlantic. The Rift mechanism is also very different from that operating in the Basin and Range province, which several workers have suggested overlies an enfeebled residuum of a former extension of the East Pacific Rise (Menard, 1964; Cook, 1966; Armstrong, 1968). The rates of distention of the Basin and Range province and of the African Rift System are orders of magnitude slower than those of the ocean ridges and the sprawling geometric arrangement of their faults is wholly different from the closely parallel pattern of the faults of the mid-ocean ridges. These differences may be partly attributed to the more heterogeneous structure of the continental as compared to the oceanic crust, but it seems there are other more fundamental differences. Certainly any generalization yet made as to the magmatic associations of the oceanic ridges is entirely inapplicable to the continental rifts so far studied.

MAGMA EVOLUTION ALONG SUBDUCTION ZONES

We turn now to the distal ends of the plates, where the crust is being consumed along subduction zones. There are three varieties of such plate junctions: oceanic against continental, as with the Pacific and South American plates; oceanic against oceanic, as with the Pacific and the western two-thirds of the Aleutian arc; and continenal against continental, as with the Himalayas, and formerly with the Alps and Apennines.

Oceanic-Continental Plate Junctions

At first glance it might appear that the introduction of a cold surficial slab into the mantle would hinder the development of magma, a subject to which Griggs devoted considerable discussion at the Birch symposium of 1970 (Chapter 40). But the consistent relations between volcanic belts and the oceanic deeps where the subduction zones crop out show that the chilling effect is overcome within a few tens of kilometers. It seems that Oxburgh and Turcotte (1968) must be right in their contention that frictional heat soon overcomes the original temperature difference between mantle and oceanic plate. Whatever the validity of the analysis, the volcanic belts do exist to prove that magma is indeed developed along the subduction zones, a relation first emphasized by Harry Hess in 1937

(published in Hess, 1939), before anyone else thought that the Benioff zones (not yet identified even by Benioff) are zones of consistent motion. Hess called them the limbs of orogens, but had the same general picture as is now generally accepted.

Kuno showed in 1959 that there is a consistent relation between magma composition and the depth to the underlying Benioff zone, the magmas becoming more alkalic as the depth of their origin increases. In 1962 the experimental work of Yoder and Tilley showed how these differences come about. Kuno (1960, 1962), Kushiro and Kuno (1963), Dickinson and Hatherton (1967), and Hatherton and Dickinson (1969) have examined these relations in several parts of the Pacific ring of fire and found them consistent; the deeper the Benioff zone the more potassic the magma, thereby rationalizing the existence of Moore's Quartz Diorite Line (1959). The work of Stoiber and Carr (1971) shows that the subduction zone is not a single surface but is 50 to 75 km thick.

Most experimentalists who have worked with the problem of magma genesis have started with some kind of pyrolite. But it should be noted that oceanic crust on the average includes about 5 or 6 km of potassium-poor tholeiitic basalt beneath about 1 km of sediment. And neither dredge nor JOIDES drill has yet brought up sedimentary rock older than Jurassic from any ocean floor. Unless the re-entry drilling now beginning reveals revolutionary surprises, the pre-Jurassic and many younger sediments shed off the leading edges of the drifting continents have all disappeared down the subduction zones, along with most of the 5 or 6 km of tholeiitic basalt.

It has been argued that these lighter rocks — the basalts and sediments — could not have been dragged far down a subduction zone into the much denser mantle. But many arguments seem to show that huge volumes of both sediment and basalt are so dragged. First, and perhaps less convincing, is the widespread occurrence along subduction zones of blue schist, eclogite, and aragonite marbles, associated in mélanges with such low-density rocks as chert and shale. The mineral paragenesis indicates deep burial, but both dense blue schist and light chert have gone down to depths of several tens of kilometers and been returned to the surface. The common occurrence of ophiolites high in the sialic crust of many ranges also demonstrates that not all rocks are disposed in the earth's gravity field in strict accordance with their density.

But a much stronger argument, and one that seems to me unassailable, if the plate tectonics model is accepted, is this: if the westward drift of the Americas has really gone on, the American plates must have overridden those to the west for fully 1,500 km. Indeed, if the absence of any subduction zone between the Mid-Atlantic Ridge and the continents to the east means what several of the plate advocates think it does, that is, that the ridge has pushed itself westward, away from Europe and Africa — the American plates have overridden those to the west by a distance equal to the full width of the Atlantic — 3,000 km, not merely the half width.

The oceanic crust averages about 1 km or so of sediment overlying 5 or 6 km of tholeiitic basalt. Accordingly, somewhere between 7,500 and 15,000 km^3 of basalt and about 1,500 to 3,000 km^3 of sediment have been overridden for *each km* length of continent-ocean interface. These are gigantic numbers. But that is not all that must be accounted for. Very considerable masses of Paleozoic and even Precambrian strata lay immediately off the west coast of North America, judging from the cutoff edges of such rocks along the southeastern Alaskan, Juan de Fucan, Klamath, and Transverse Range segments of coast. What happened to all these rocks? The sediments now offshore the Pacific Coast of the United States are less than a sixth as voluminous as those off the Atlantic, despite the roughly comparable areas tributary to the two coasts (Gilluly and others, 1970). The greater volume of Mesozoic and Cenozoic rocks on the west of the continent by no means equalizes the discrepancy.

Obviously, if the sediments had been scraped off the downgoing slab, we should find them piled up against the west edge of the continent. According to Pakiser and Steinhart (1964), the average thickness of the crust under the Coast Range in California is about 20 km. If only 1,500 cu km of sedi-

ment had been scraped off the diving slab, we ought to find a 75 km average width of Paleozoic, Mesozoic, and Cenozoic rocks piled up against the continent; if the drift did indeed equal the whole width of the Atlantic, the pile should be 150 km wide. But the Franciscan mélange is a mere fraction of this volume. It seems to me undeniable that most of the 1,500 to 3,000 cu km of sediment and virtually all of the 7,500 to 15,000 cu km of basalt per km of latitude have disappeared down the subduction zone — this despite the fact that when they started down, at least, they were much less dense than the surrounding material. How far down do they go? Even though under pressure there would doubtless be phase changes, with production of such minerals as kyanite in the sediments and with the basalts converting to eclogite, I doubt that they go all the way to the bottom of the zone. Probably they get heated up and drift off to form the magma pockets that many seismologists think characteristic of the low velocity zone — somewhere between 80 or 90 km to, say, 200 to 300 km.

I therefore think that when we hypothesize about magma genesis along a subduction zone we are not restricted to pyrolite alone as the parental material — in fact, we are *compelled* to think otherwise. Perhaps it is the principal starting material where the magmas produced are the tholeiite-high alumina basalt-alkalic basalt sequence as in northern Japan and northern California and Oregon. But where the magmas produced are dominantly andesitic, as along the Mariana arc, the Bonins, and the coast of both Americas, perhaps the starting material was considerably more siliceous — the tholeiitic basalt of course converted to eclogite in the downgoing oceanic slab. And perhaps, where the volcanics of the coast zone are still more siliceous, we are dealing with a palingenic magma derived from the remelting of sediment and tholeiite — a parental material already a long way along the differentiation trend when compared to a suite derived from partial melting of pyrolite. One need not go so far as Tom Barth (1961), who considered that the differentiation of the rocks we recognize as igneous has been largely brought about by sedimentation and erosion, but it certainly seems inescapable that a non-negligible contribution to the evolution of igneous rocks has been by this process of sedimentary differentiation. A lot of the rocks that went down the subduction zone were sandstone, chert, and radiolarian oozes. We must consider the implications of the missing sediment and basalt when discussing magma generation along the subduction zones; partial melting of a pyrolite is not the answer.

Such a secondary derivation might account for much of the great volume of andesitic magmas of the circum-Pacific. It is impossible to find the gigantic volumes of intermediate rocks that would have had to be formed if crystal differentiation of a primary basaltic magma were the dominant process in producing the huge granitic batholiths of the Andes, the Sierra Nevada, the Idaho, and British Columbia batholiths.

The Sr^{87}/Sr^{86} ratios of most andesites are such as to suggest derivation directly from the mantle, without notable contamination by crustal material. But if the material recycled were mainly the low potassium tholeiitic basalt of the ocean floor we could expect the same low strontium ratios in the magma.

Ocean to Ocean Contacts

Some plate contacts are ocean plate against ocean plate as in the western two-thirds of the Aleutian arc. This arc is anomalous; the eastern part involves continental crust above a north-dipping Benioff zone, while the western two-thirds has oceanic crust both above and below. Ten years ago, Coats (1962) suggested that the predominantly andesitic volcanics of the western part represent partial remelting of eugeosynclinal sediments and basalts as they were dragged down the thrust zone. There are a few albite granites in the island arc and very minor amounts of rhyolite and olivine basalt; to the east, in the Alaskan peninsular part of the arc where the volcanoes overlie sialic crust, stocks of quartz diorite and quartz monzonite abound. It is hard to avoid the inference that the parental magmas of this segment of the arc were either contaminated by the incorporation of large volumes of sialic crust or else were derived by the incorporation of considerable volumes of subducted sialic sediment, as inferred for the coast farther southeast. The ocean to ocean plate contact, on the other hand, yielded only trivial amounts of magmas more salic than andesite—amounts that could readily be accounted for by sodic metasomatism or by crystallization differentiation of small masses of magma within the volcanoes.

The Aleutian arc is unusual in still another way. As Grow and Atwater (1970) and Jones (1971) have pointed out, the younger magnetic anomalies lie close inshore along the Kenai and Aleutian arcs, the older ones offshore, showing that the north boundary of the Pacific plate has changed from a source of crustal material (the Kula Ridge of Grow and Atwater), to a subduction zone along which crust is being destroyed. The conversion probably took place in early Tertiary time. It might be said that the Kula Ridge was first destroyed, along with a wide belt of oceanic crust, by the Aleutian Trench subduction zone before it began to consume the Pacific plate. Along the eastern half of the arc, which is still volcanically active, the Pacific plate is still being consumed, but to the west it is moving parallel with the boundary of the North American plate; there is only strike-slip motion, no subduction, and no volcanic activity. Along the eastern segment there does not appear to have been any change in the magmatism at the time of the destruction of the Kula Ridge; strangely enough, both the Kula Ridge and the Aleutian Trench subduction zone seem to have produced similar magmas.

In the same way, the East Pacific Rise seems to have disappeared down a now fossil subduction zone along southern California, leaving younger magnetic bands in the Pacific plate nearer the continent than offshore (Atwater, 1970).

This overwhelming of the East Pacific Rise is another of the many differences between it and the Mid-Atlantic Ridge. Although both have narrow zones of high heat flow near their crests, the Mid-Atlantic Ridge is cut by innumerable normal faults, subparallel to its trend, which create a nearly continuous central graben. Such faults are all but completely missing from the East Pacific Rise and there are no recognized grabens. Also, the Mid-Atlantic Ridge, by creating new crust that is not consumed on any subduction zone along the European-African continental border, has receded from the border as though it were pushing itself relatively westward. But the East Pacific Rise, though once powerful enough to resist the westward advance of the American plate sufficiently to dispose of great quantities of crust down a former subduction zone, has now become so feeble north of the Clarion fracture zone as to allow long segments to be destroyed by the westward motion of America (Atwater, 1970). Where this has happened, as between Cape San Lucas and Cape Mendocino, volcanism onshore has virtually, if not completely, ceased. The connection between volcanism and active subduction here seems very direct, though we shall see elsewhere no comparable connection.

Continent-Continent Junctions

The third type of plate junction—continent against continent—is best illustrated by the Himalayan Range. Here the continental plate of India is actively underriding the continental plate of Asia. According to Gansser (1964), the Himalayan Range is not geosynclinal. The main range is largely composed of crystalline rocks identical with those of peninsular India, and indeed relics of some Precambrian structural trends of the peninsula can still be recognized in the thrust sheets. The Tethyan geosynclinal rocks lie well to the north of the main range. The most southerly thrusts of the range carry Precambrian crystallines relatively southward over Neogene sedimentary rocks several kilometers thick (Berthelsen, 1951). Unlike the molasse of the Alpine foreland, these sedimentary rocks were not derived from the advancing thrust sheets but from the foreland itself. Overlying the crystalline thrust sheets are thrusts involving the Tethys geosynclinal sedimentary rocks. These rise from a narrow root zone of vertical shears along which the Indus and Tsang-po Rivers have carved their valleys.

The Tsang-po line is followed by a string of ophiolites and a mélange of huge breccia blocks that range widely in stratigraphic position. The ophiolites may be a part of the old Tethyan sea floor, squeezed up as the Indian crustal block dove northward beneath Asia along a subduction zone. The reality of the subduction zone is signalled by the many deep-focus earthquakes beneath the Tibetan Plateau (Gutenberg and Richter, 1949). The Tibetan Plateau, the highest in the world, has probably been elevated by the doubling of the sialic crust beneath it—a process still going on, as is indicated by many physiographic features (Wager, 1937), as well as by the seismic activity.

Despite this being the region of greatest tectonic activity on earth, there has been no sign of magmatic activity along the Himalayas; the quiescence has been attributed by Gansser to the persistent tangential orogenic pressure being so great that magma is unable to work its way upward. It seems most likely, however, that the tectonic activity indicated by the deep-focus earthquakes far beneath that great thickness of sial will eventually result in considerable siliceous magmatism.

Both to the east, along the meridional Arakan-Yoma line in Burma, and to the west, along the Quetta line in Baluchistan, volcanism was active in the Tertiary, and some persists to the present. These lines are the lateral boundaries of the Indian plate as it moves relatively northward; they are the continental continuations of the Ninety East Ridge and the Owen fracture zone–Murray Ridge, respectively, and seemingly the only recognizable continental prolongations of oceanic fracture zones, with the unlikely exception of the Murray fracture zone of the eastern Pacific which may doubtfully control the Transverse Ranges of southern California (Gilluly, 1970; von Huene, 1971). The movement along these bounding shear zones has been chiefly strike-slip (Gansser, 1966), but there has been some east-west overthrusting along the Burma line, presumably caused by the subduction of the Indian plate as it crowds northward into the Asian continent (Reed, 1949; Krishnan, 1956).

Unlike the western part of the Aleutian arc, where volcanism died out when this segment became a strike-slip feature, volcanism has continued along both the Arakan-Yoma and Quetta lines, spoiling a too facile generalization about the connection between magmatism and plate motion. The Quetta line is andesitic (Reed, 1949; Krishnan, 1956); the volcanism on the Arakan-Yoma line has been considerably more diversified, yielding both olivine-bearing and olivine-free basalt, augite andesite, dacite, rhyolite, and even a little nepheline basalt (Washington, 1924; Reed, 1949).

The Himalayas are the youngest of the Tethyan ranges and the Alps the oldest, but all represent junctions of continental plates. The Tethyan Sea, though surely several hundred kilometers across in the longitude of the Betic Cordillera and the Alps, was nevertheless largely underlain by sialic crust, with only minor basins of oceanic crust, some of which are now represented by masses of ophiolite. The massifs of the Alps, from which the Helvetian nappes were derived—Mont Blanc, Mercantour, Gotthard, Aar, and the others, are all sialic. So, too, are the narrow antiforms of the root zone of the Penninic nappes and the gneiss zones along the Insubric line, from which the sialic gneisses of the Austro-Alpine nappes were sheared off (Heim, 1922–1924; Staub, 1924).

Here, as in the Himalayas, the ancient Tethys Sea has been completely destroyed; the Mediterranean is clearly a new sea, not a relic of Tethys, as shown both by the paleogeography (Kraus, 1960, 1962; Dürr and others, 1962; Colom and Escandell, 1962; Bourcart, 1962; Egeler and de Booy, 1962; de Booy, 1969; Glangeaud and others, 1966; Pannekoek, 1969) and by seismic studies (Ritsema, 1969). Staub (1924) contended that the crystalline nappes of the Austro-Alpine group once underlay the northern continental shelf of Africa; they now constitute the highest structural units of the European Alps. The Tethyan sediments are represented in the Pennine, and to a lesser extent in the Austro-Alpine nappes, but vast quantities of them have disappeared down one or more subduction zones as the great Austrian structural geologist Ampferer demonstrated as long ago as 1906. His thesis is now accepted by many students of the Alps (Amstutz, 1955; Trümpy, 1960; Kraus, 1951, Gansser, 1970, oral commun.).

It is now generally conceded that the Insubric line along the southern side of the Alps is such a subduction zone, bordered by the ultramafic massif of Ivrea—perhaps a piece of mantle trapped above the descending plate (Gansser, 1968; Giese, 1968; Kaminski and Menzel, 1968). Unlike the Himalayan area, where no magmatism is apparent along the subduction zone, and western America and Japan, where a gap of 100 to 200 km intervenes between the outcrop of the subduction zone and the zone of igneous activity, the rather small Alpine intrusive masses of tonalite, quartz monzonite, and related rocks lie directly along the Insubric line. Nevertheless, the magmas may have formed at depths equivalent to those of the American and Japanese magmas, for the Insubric zone is now essentially vertical, perhaps because of the late back-tilting of the southern Pennine nappes.

Except for the Ivrea mass, the ophiolite bodies of the western Alps are all small, and though their contacts are commonly somewhat sheared, they appear not to be oceanic basement but intrusive into the Pennine eugeosynclinal sediments (Trümpy, 1960). Those of the Apennines (Steinmann, 1926), Dinarides (Hiessleitner, 1951–1952; Brunn, 1960; Moores, 1969), Crete (Bear, 1966), and Anatolia (Hiessleitner, 1951–1952; Bailey and McCallien, 1950a, 1954) may represent deep ocean floor, at least in part. Some of these seem to fit readily into a scheme of subduction zones, but the great width of the band of ultramafic masses in Anatolia seems difficult to relate to a single plate boundary; if they are all along subduction zones several small plates must have been involved. Bailey and McCallien (1954), however, thought that the pillow lavas and intrusive ultramafic masses were both derived from ultramafic magmas; they were not cumulates.

SUMMARY AS TO MESOZOIC AND YOUNGER PLATE BOUNDARY MAGMAS

The discussion thus far has had to do with plate boundaries active during the Mesozoic and Cenozoic; if we can find consistent relations between plate motion and magmatic character for these younger activities, we may be able better to interpret some of the older tectonic and magmatic features. What generalizations can we reasonably make?

It seems safe to say that tholeiitic basalt, with minor amounts of alkaline and olivine-rich basalts, are the principal volcanics associated with spreading ridges. Locally, small quantities of more siliceous volcanics are found but these are generally in trivial volume and reasonably attributed to crystal differentiation within the volcanic piles. The two ridge islands with considerable exposures of siliceous rock, Iceland and Ascension, may have sialic basements; this is more doubtful in the case of Ascension but highly probable with Iceland. In both islands some of the basic flows carry xenoliths of granite.

Along converging plate boundaries things are a good deal more complex. As a generalization with many exceptions, where the boundary is between downgoing oceanic plate and overriding continent, the subduction zone is marked by ophiolitic blocks and there is a spread of 100 to 200 km between the trench and the first volcanic and plutonic zone inland. This has been called the arc-trench gap by Dickinson (lecture, February 1971). The volcanic belt itself may be zoned, with tholeiite, high-alumina basalt and alkali basalt in succession inland, as in northern Japan, northern California, and southern Oregon, or it may be simply andesitic as in part of western America, southern Japan, and South America. In these areas the volcanic zone may be succeeded inland by more K-rich rocks at Moore's (1959) quartz-diorite line.

Where the boundary is between two oceanic plates the magmas produced seem to be dominantly andesitic, although more differentiated rocks such as dacite may be represented; local masses of albite granite may be of metasomatic origin.

At continent-continent plate contacts, andesitic volcanics, such as are represented in the Taveyanez sandstone of the Alps, may be formed, but the later developments is of quartz diorite and quartz monzonite. Perhaps because of the essentially vertical dip of the Alpine subduction zone, there is no lateral spread of more potassic magmas inland from less potassic magmas, such as we find in Japan, Indonesia, California, and New Caledonia.

These generalizations, weak as they are when looked at rigorously, immediately recall the ideas of petrographic provinces, very popular a generation or two ago. Suess (1885) called attention to the contrasting tectonics of the Atlantic and Pacific shores and named those shores that generally accord with the strike of the folds "Pacific" and the ria coasts "Atlantic." Judd (1886) pointed out some petrological characters of various igneous rocks that seemed to him to be associated with each of these coastal structures, and soon Bertrand (1888), Iddings (1892), Becke (1903), Harker (1909), and Washington (1906) were all involved in attempting to characterize various petrologic associations as belonging either to an Atlantic or a Pacific petrographic province. Washington (1923) showed that some, at least, of the Atlantic coastal rocks fail to conform with Judd's generalization, and indeed, that some fail to confirm the drift hypothesis, although they do not firmly negate it, either.

Some petrographic provinces do indeed present striking contrasts, such as the strongly alkalic Tertiary volcanics in the foreland of the Carpathians in Bohemia with the siliceous calc-alkaline rocks of the same age in the upper plate of the Carpathian thrusts in Hungary (Becke, 1903). Note that this is opposite to the relation in subduction zones, where the more alkaline rocks are in the hinterland of the thrusts. Similar chemical contrasts between the calc-alkaline rocks of the Yellowstone and the Boulder batholith with the highly alkaline rocks farther east in Montana, again are in the foreland of the thrusts. The highly leucitic rocks of Washington's "Roman comagmatic region," are readily distinguished from those of other areas. According to the classification of all these men, the Pacific suite, subject to a lot of minor exceptions, would be the one closely associated with most of our presently recognized subduction zones. Their Atlantic suite would not, however, embrace most of the rocks of the Mid-Atlantic Ridge, although it would include those of many of the other Atlantic islands (McBirney and Gass, 1967).

About 50 yrs ago, Niggli (1921) added a Mediterranean suite to take care of the strongly potassic rocks of Washington's Roman region, the Lipari, Aeolian Islands, Vesuvius, and others. A fuller compilation was made by Murri (1927). These rocks are hard to relate to any subduction zones identifiable in the region, but all except Pantelleria are on the hinterland sides of the Apennine and Silician thrusts, which have sporadic ophiolites along them; presumably these overlie subduction zones at considerable depths, appropriate to the production of K-rich magmas. But Pantelleria lies in the foreland of all known subduction zones and is equally alkalic. Evidently Nature does not always cooperate with our models.

Loewinson-Lessing (1923) showed quite conclusively that, although there are many marked regions of high consanguinity of magmas, hardly any considerable area of either Atlantic (alkaline) or Pacific (calc-alkaline) affinities is free from some rocks belonging to the other suite. Certainly if one follows Harker (1909) in saying that the Pacific suite is found in areas of folded rocks and the Atlantic in areas of non-folding and normal faulting, there are innumerable exceptions. On the Colorado Plateau, the Tertiary teschenite and analcite syenite of the San Rafael Swell (Gilluly, 1927) are within 20 mi of the calc-alkaline diorite porphyry of Mount Ellen (Hunt and others, 1953) of similar age and tectonic setting. The volcanic pile of K-rich quartz latite overlying the Bingham stock in the Oquirrh Range at the east side of the Basin and Range province contains minor flows of nepheline basalt (Gilluly, 1932). Hundreds of other examples could be given.

MAGMAS OF UNCERTAIN TECTONIC ASSOCIATION

Thus far we have considered magmatic phenomena that seem readily related to plate motion; we turn now to some that are not so clearly related. First we must consider how far a particular belt of magmatism may be from a plate boundary and still be reasonably attributed to plate motion. Believing, as I do, that the best guide to the past lies in the present, let us see how closely magmatic activity is associated with currently active plate boundaries.

Benioff (1949) traced a zone of earthquake foci from the Chilean Trench to a depth of 700 km beneath Argentina; although deformation of the crustal rocks extends locally as much as 700 km east of the trench and averages about 350 km from it, none of the Andean volcanics

are more than 600 km from the trench and most are within 400 km. The volcanics inland from the Middle American Trench lie within 500 km to the northeast. The volcanics associated with the Aleutian Trench are within 400 km of it, although some Jurassic plutons are as much as 550 km away. The volcanic belts of the Marianas, Japan, and the Kurile Islands lie 200 to 300 km from th trenches (Hess, 1948).

Benioff (1955) showed that several zones of earthquake foci steepen downward, but on the average most dip at about 45°. Both the Tonga-Kermadec and the fossil Insubric zone dip much more steeply, and Plafker (1965) has attributed the Prince William Sound earthquake of 1964 to an underthrust related to the Aleutian Trench that dips only 7°. There is no magmatic activity associated with the Aleutian Trench this far east, however. At a dip of 45° a Benioff zone would reach the low-velocity zone at about 100 km; probably this accounts for the usual gap between deep and volcanic arc. It would reach the bottom of the zone about 400 to 500 km inland. One would hardly expect magma generation very far beneath the low-velocity zone, and magmas that would be formed at such depths should be those such as kimberlite, rather than the basaltic and andesitic varieties associated with the currently active zones.

For this reason I find it impossible to accept the proposal of my friends and colleagues Lipman, Prostka, and Christiansen (in press) that all the Eogene volcanics of the western United States, even as far east as the San Juan province in southern Colorado, 1,550 km from the Pacific, are related to active subduction zones; one cropping out at the Pacific shore and the other en échelon with it but only extending downward from the low-velocity zone and nowhere cropping out. Their basis for the easterly occult zone is that they identify two belts of easterly increasing potash content in the volcanics, the first related to the coastal Benioff zone and the second, the more easterly one, they think explicable by the same mechanism. They leave the mechanism for the origin of the eastern subcontinental zone unexplained, but this objection may not be important considering our ignorance as to the formation of the demonstrable Benioff zones.

Nevertheless, it is noteworthy that, if the coastal zone were destroyed in Miocene time, as contended by Atwater (1970), this did not stop the production of magmas, for volcanism has continued as far east as the Yellowstone in Pliocene and even Pleistocene time. These gentlemen all realize this but contend that the magmas of the younger fields are very different from the earlier ones, being bimodal, either basalt or high-silica rhyolite rather than such intermediate lavas as latite and quartz latite.

This difference is probably significant of a different mechanism of differentiation, but even granting that it is, the very presence of the volcanics demonstrates that no subduction zone is needed to produce magmas. If it is not necessary for the bimodal volcanics, why is it required for the others?

Moore's (1959) quartz diorite line is locally as much as 700 km inland, but this is a segment in southern Washington and northern Oregon where an embayment of oceanic crust may once have projected into what is now continent. The subduction zone may have had a parallel course; elsewhere the line is no more than 300 km inland. If, at this distance from the plate boundary the Benioff zone had been at depths appropriate for the generation of K-rich magmas, it is hard to see how it could remain at similar depth for another 1,200 km farther inland. Or if we grant the second en échelon zone, it seems fair to ask why we should not have a third one still farther east, to yield the Eogene rhyolites of the Black Hills (Noble, 1948; Jaggar, 1901), the alkalic province of central Montana (Weed and Pirsson, 1895, 1896) and the other Tertiary igneous rocks of West Texas, the Cretaceous tuffs and plutons of Arkansas, the Cretaceous intrusives of the Monteregian Hills of Quebec, and the minor hypabyssal Tertiary intrusives of Virginia (Fullagar and Bottino, 1969; Dennison and Johnson, 1971). I think that to attribute magmatism more than 700 km from the outcrop of a subduction zone to activity along that zone is simply to assert that all magmatism of intermediate chemistry is related to plate tectonics and to prejudge the whole problem.

Having in mind the distribution of tectonic and magmatic activity related to the currently active subduction zones, I feel extremely doubtful about attributing any tectonism or magmatism farther inland than the Idaho batholith or western Nevada to the influence of the formerly active zone at the boundary of the American plate. Menard (1969a) has demonstrated intraplate tectonism and McBirney and Gass (1967) intraplate magmatism in oceanic plates; I think the Charleston and Reelfoot Lake earthquakes, both almost surely greater than anything known from California, show that we have intraplate tectonism in the continental plates as well. So too the igneous activity from the Rockies to Virginia seems of intraplate origin. Think, also, of the late Tertiary orogenesis of the Kuen Lun and northern ranges of central Asia, thousands of kilometers from any recognized plate boundary.

Hess (1948) has shown, by tracing zones of peridotite, that the Mesozoic and Cenozoic magmatic rocks of Korea and Japan, which form a belt 1,200 km across, are related not to a single subduction zone but to three distinct ones, successively younger toward the Pacific. (Hess did not call them subduction zones, but his cross sections show that that is what he considered them.) There are no comparable peridotite zones in the western United States to mark former subduction zones other than the one along the Pacific coast. It seems to me entirely gratuitous to attribute to a single subduction zone all the tectonic and magmatic activity in western America. The belt involved here is even wider than the Korea-Japan belt, which was related to three different zones.

I think, then, that many igneous rocks can be quite satisfactorily related to plate tectonics — after all, the earthquake zones now active are generally closely associated with the zones of active volcanoes. Others, however, are less surely related and still others are so remote from plate boundaries that any association with them must be based on faith rather than evidence.

Among the unrelated volcanics are those of most of the oceanic islands that are not sited on an active ridge, as McBirney and Gass (1967) have shown. The more alkalic character of most islands distant from the ridges suggests that their lavas are derived from deeper sources in the mantle. This accords with the fact that they are in areas of much lower heat flow than the active ridges. Clearly most of them have nothing whatever to do with plate boundaries.

The Hawaiian Island volcanics are especially instructive in this regard. The islands have formed in succession from northwest to southeast. Each began as a tholeiitic volcano and later changed to an alkaline basalt before dying out. As several have commented, this sequence suggests a downward migration of the source area with time. Although Tuzo Wilson (1963a) and W. J. Morgan (1971) have suggested the possibility that the ocean crust is drifting northwestward over a more or less stationary "hot spot" in the mantle, thus forming the sequence, the behavior of the successive volcanoes seems to me inconsistent with this idea. If the magma source descended toward the close of the activity of one volcano in order to furnish the terminal alkaline magmas, it is hard to see why it should rise again as the site of the next-to-be-born volcano drifted over the hot spot—and especially why it should do it every time.

Much of the volcanism of the ocean basins thus seems independent of plate boundaries, as with much of the continental magmatism. It is difficult to relate the Tertiary dike swarms of Britain (Richey, 1948), the subaerial flood basalts of Greenland (Wager and Deer, 1938), the Faeroes (Rasmussen and Noe-Nygaard, 1970), Columbia River and Snake River, and Siberia with any mechanism of plate tectonics, yet all of these were formed after the present pattern of plate motion had been established. The volcanics of the Rhineland, the Plateau Central, and of the Pribiloff Islands are other examples.

In summary, it seems clear that much Mesozoic and younger magmatism is directly associated with plate tectonics, but there remains a goodly number of magmatic occurrences that cannot reasonably be attributed to plate motion. Tectonism is assuredly not all related to plate motion and neither is magmatism.

PLATE JUNCTIONS OF THE GEOLOGIC PAST

Almost 25 yrs ago, Harry Hess, following up the orogen idea of Vening-Meinesz, thought to trace former mountain chains and to date the onset of their deformation by following the belts of Alpine-type ultramafic rocks (Hess, 1939). He compiled a series of maps which show these belts. The orogen idea as originally proposed has been displaced by plate tectonics; the features Vening-Meinesz and Hess called orogens we now call subduction zones, but the maps are virtually identical.

49

Convection Plumes in the Lower Mantle

W. JASON MORGAN
1971

From *Nature*, v. 230, p. 42, 1971. Reprinted with permission of the author and Macmillan Journals, Ltd.

The concept of crustal plate motion over mantle hotspots has been advanced (Wilson, 1965a) to explain the origin of the Hawaiian and other island chains and the origin of the Walvis, Iceland–Farroe and other aseismic ridges. More recently the pattern of the aseismic ridges has been used in formulating continental reconstructions (Dietz, 1970). I have shown (Morgan, in preparation) that the Hawaiian–Emperor, Tuamotu–Line and Austral–Gilbert–Marshall island chains can be generated by the motion of a rigid Pacific plate rotating over three fixed hotspots. The motion deduced for the Pacific plate agrees with the palaeomagnetic studies of seamounts (Francheteau et al., 1970). It has also been found that the relative plate motions deduced from fault strikes and spreading rates agree with the concept of rigid plates moving over fixed hotspots. Fig. 49–1 shows the absolute motion of the plates over the mantle, a synthesis which satisfies the relative motion data and quite accurately predicts the trends of the island chains and aseismic ridges away from hotspots.

I now propose that these hotspots are manifestations of convection in the lower mantle which provides the motive force for continental drift. In my model there are about twenty deep mantle plumes bringing heat and relatively primordial material up to the asthenosphere and horizontal currents in the asthenosphere flow radially away from each of these plumes. The points of upwelling will have unique petrological and kinematic properties but I assume that there are no corresponding unique points of downwelling, the return flow being uniformly distributed throughout the mantle. Elsasser has argued privately that highly unstable fluids would yield a thunderhead pattern of flow rather than the roll or convection cell pattern calculated from linear viscous equations. The currents in the asthenosphere spreading radially away from each upwelling will produce stresses on

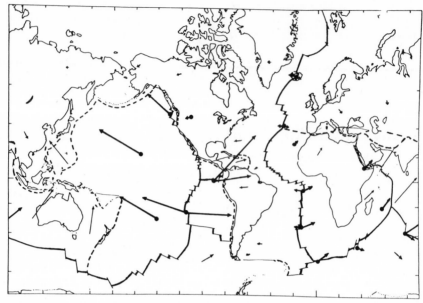

Figure 49-1
The arrows show the direction and speed of the plates over the mantle: the heavier arrows show the plate motion at hotspots. This synthesis was based on relative plate motion data (fault strikes and spreading rates) and predicts the directions of the aseismic ridges/island chains emanating from the hotspots

the bottoms of the lithospheric plates which, together with the stresses generated by the plate to plate interactions at rises, faults and trenches, will determine the direction in which each plate moves.

Evidently the interactions between plates are important in determining the net force on a plate, for the existing rises, faults and trenches have a self-perpetuating tendency. The plates are apparently quite tough and resistant to major changes, because rise crests do not commonly die out and jump to new locations and points of deep upwelling do not always coincide with ridge crests. (For example, the Galapagos and Réunion upwellings are near triple junctions in the Pacific and Indian Oceans. Asthenosphere motion radially away from these hotspots would help to drive the plates from the triple junctions, but there is considerable displacement between the "pipes to the deep mantle" and the lines of weakness in the lithosphere which enable the plates to move apart.) Also, a large isolated hotspot such as Hawaii can exist without splitting a plate in two. I believe it is possible to construct a simple dynamic model of plate motion by

making assumptions about the magnitude of the flow away from each hotspot and assumptions about the stress/strain rate relations at rises, faults and trenches. Such a model has many possibilities to account for past plate motions; hotspots may come and go and plate migration may radically change the plate to plate interactions. But the hotspots would leave visible markers of their past activity on the seafloor and on continents.

This model is compatible with the observation that there is a difference between oceanic island and oceanic ridge basalts (Engel et al., 1965; Gast, 1968). It suggests a definite chain of events to form the island type basalt found on Hawaii and parts of Iceland. Relatively primordial material from deep in the mantle rises adiabatically up to asthenosphere depths. This partially fractionates into a liquid and solid residual, the liquid rising through vents to form the tholeiitic part of the island. The latter alkaline "cap rocks" would be generated in the lithosphere vent after plate motion had displaced the vent from the "pipe to the deep mantle". In contrast, the ridge basalts would come entirely from the asthenosphere, passively rising to fill the void created as plates are pulled apart by the stresses acting on them. The differences in potassium and in rare earth pattern for island type and ridge type basalts may be explained by this model. Moreover, the 2 billion year "holding age" advocated by Gast (Oversby and Gast, 1970) to explain lead isotope data of Gough, Tristan da Cunha, St. Helena and Ascension Islands may reflect how long the material was stored in the lower mantle without change prior to the hotspot activity.

My claim that the hotspots provide the driving force for plate motions is based on the following observations to be discussed below. (1) Almost all of the hotspots are near rise crests and there is a hotspot near each of the ridge triple junctions, agreeing with the notion that asthenosphere currents are pushing the plates away from the rises. (2) There is evidence that hotspots become active before continents split apart. (3) The gravity pattern and regionally high topography around each hotspot suggest that more than just surface volcanism is involved at each hotspot. (4) Neither rises nor trenches seem capable of driving the plates.

The symmetric magnetic pattern and the "mid-ocean" position of the rises indicate that the rises are passive. If two plates are pulled apart, they split along some line of weakness and in response asthenosphere rises to fill the void. With further pulling of the plates, the laws of heat conduction and the temperature dependence of strength dictate that future cracks appear down the centre of the previous "dike" injection. If the two plates are displaced equally in opposite directions or if only one plate is moved and the other held fixed, perfect symmetry of the magnetic pattern will be generated. The axis of the ridge must be free to migrate (as shown by the near closure of rises around Africa and Antarctica). If the "dikes" on the ridge axis are required to push the plates apart, it is not clear how the symmetric character of the rises could be maintained. The best argument against the sinking lithospheric plates providing the main motive force is that small trench-bounded plates such as the Cocos plate do not move faster than the large Pacific plate (McKenzie, 1969b). Also, the slow compressive systems would not appear to have the ability to pull other plates away from other units. The pull of the sinking plate is needed to explain the gravity minimum and topographic deep locally associated with the trench system (Morgan, 1965), but I do not wish to invoke this pull as the principal tectonic stress. This leaves sublithospheric currents in the mantle and the question now is: are these currents great rolls (mirrors of the rise and trench systems), or are they localized upwellings (that is, hotspots)?

A recent world gravity map (Kaula, 1970) computed for spherical harmonics up to order 16 shows isolated gravity highs over Iceland, Hawaii, and most of the other hotspots. Such gravity highs are symptomatic of rising currents in the mantle. Even if the gravity measurements are inaccurate (different authors have very different gravity maps), the fact remains that the hotspots are associated with abnormally shallow parts of the oceans. For example, note the depth of the million square kilometres surrounding the Iceland, Juan de Fuca, Galapagos, and Prince Edward hotspots. The magnitude of the gravity and topographic effect should measure the size of the mantle flow at each hotspot.

There is evidence of continental expression of hotspot activity in the lands bordering the Atlantic: the Jurassic volcanics in Patagonia (formed by the present day Bouvet Island plume), the ring dike complex of South-west Africa and flood basalts in the Parana Basin (Tristan da Cunha plume), the White Mountain Magma series in New Hampshire (the same hotspot that made the New England Seamount Chain (Azores plume?), the Skaegaard and the Scottish Tertiary Volcanic Province (Iceland plume) and perhaps others. I claim this line-up of hotspots produced currents in the asthenosphere which caused the continental break-up leading to the formation of the Atlantic. Likewise the Deccan Traps (Reunion plume) were symptomatic of the forthcoming Indian Ocean rifting. A search should be made for such continental activity, particularly in East Africa and the western United States (the Snake River basalts?) as an explanation for the rift features found there. There is a paucity of continental hotspots in Fig. 49–1; perhaps this is a bias due to continental complexity versus oceanic simplicity, but the model presented here predicts that most hotspots will be near a spreading rise.

Bibliography

This bibliography contains all of the references cited in the chapters. Where citations in the original article were in an abbreviated format, we have attempted to complete the reference by adding the title and inclusive page numbers. For those articles originally cited as being "in press" that have now been published, we have completed the references. To avoid perpetuating incomplete or erroneous bibliographic material, references that we could not verify after expending a reasonable amount of effort are indicated by the notation (not verified).

The bibliography contains, in addition to the articles referenced in the individual chapters, several hundred more recent articles on plate tectonics. To aid the student who would like to read further but finds himself lost in a bibliography this large, the newer references, together with selected references from the older literature, are listed by subject in reading lists at the end of the introductory chapters. An index to the reading lists is given at the end of Chapter 1.

Adams, R. D., 1963, Source characteristics of some deep New Zealand earthquakes: N. Z. J. Geol. Geophys., v. 6, p. 209–220.

Adams, R. D., and Christoffel, D. A., 1962, Total magnetic field surveys between New Zealand and the Ross Sea: J. Geophys. Res., v. 67, p. 805–813.

Ade-Hall, J. M., 1964, The magnetic properties of some submarine oceanic lavas: Roy. Astron. Soc., Geophys. J., v. 9, p. 85–91.

Aki, K., 1966, Earthquake generating stress in Japan for the years 1961 to 1963 obtained by smoothing the first motion radiation patterns: Tokyo, Univ., Earthquake Res. Inst., Bull., v. 44, p. 447–471.

Akimoto, S., and Fujisawa, H., 1968, Olivine-spinel solid solution equilibria in the system $Mg_2SiO_4 - Fe_2SiO_4$: J. Geophys. Res., v. 73, p. 1467–1479.

Algermissen, S. T., 1964, Address presented on 19 November 1964 at the annual meeting of the Geological Society of America (not verified).

Allan, D., 1958, Reversals of the earth's magnetic field: Nature, v. 182, p. 469–470.

Allan, D. W., 1962, On the behaviour of systems of coupled dynamos: Cambridge Phil. Soc., Proc., v. 58, p. 671–693.

Allan, D. W., Thompson, W. B., and Weiss, N. O., 1967, Convection in the earth's mantle, in Runcorn, S. K., ed., Mantles of the earth and terrestrial planets: New York, Wiley, p. 507–512.

Allan, T. D., Charnock, H., and Morelli, C., 1964, Magnetic gravity and depth surveys in the Mediterranean and Red Sea: Nature, v. 204, p. 1245–1248.

Allard, G. O. and Hurst, V. J., 1969, Brazil-Gabon geologic link supports continental drift: Science, v. 163, p. 528.

Allen, C. R., 1962, Circum-Pacific faulting in the Philippines-Taiwan region: J. Geophys. Res., v. 67, p. 4795–4812.

Allen, C. R., 1965, Transcurrent faults in continental areas, in Blackett, P. M. S., Bullard, E., and Runcorn, S. K., eds., A symposium on continental drift: Royal Soc. London, Phil. Trans., ser. A, v. 258, p. 82–89.

Allen, C. R., St. Amand, P., Richter, C. F., and Nordquist, J. M., 1965, Relationship between seismicity and geologic structure in the southern California region: Seismol. Soc. Amer., Bull., v. 55, p. 753–797.

Ampferer, O., 1906, Über das Bewegungsbild von Faltengebirge: Austria, Geol. Bundesanst., Jahrb., v. 56, p. 539–622.

Ampferer, O., and Hammer, W., 1911, Geologischer Querschnitt durch die Ostalpen vom Allgau zum Gardasee: Autstria, Geol. Bundesanst., Jahrb., v. 61, p. 531–710.

Amstutz, A., 1955, Structures alpines, subductions successives dans l'Ossola: Acad. Sci. Paris, C. R., ser. D, v. 241, p. 967–969.

Anderson, D. L., 1965, Recent evidence concerning the structure and composition of the earth's mantle: Phys. Chem. Earth, v. 6, p. 1–131.

Anderson, D. L., 1967a, Latest information from seismic observations, in Gaskell, T. F., ed., The earth's mantle: New York, Academic Press, p. 355–420.

Anderson, D. L., 1967b, Phase changes in the upper mantle: Science, v. 157, p. 1165–1173.

Anderson, D. L., Sammis, C., and Jordan, T., 1971, Composition and evolution of the mantle and core: Science, v. 171, p. 1103–1112.

Anderson, E. M., 1939–1940, The loss of heat by conduction from the earth's crust in Britain: Roy. Soc. Edinburgh, Proc., v. 60, p. 192–209.

Andrews, J. E., and Hsü, K. J., 1970, American Stratigraphic Commission Note 38: Tulsa, American Association of Petroleum Geologists (in press) (not verified).

Antoine, J., and Ewing, J., 1963, Seismic refraction measurements on the margins of the Gulf of Mexico: J. Geophys. Res., v. 68, p. 1975–1996.

Argand, E., 1911, Les nappes de recouvrement des Alpes pennines et leurs prolongement structuraux: Beitr. zur Geol. Karte Schweiz, n.s. 31, p. 1–26.

Armstrong, R. L., 1968, Sevier orogenic belt in Nevada and Utah: Geol. Soc. Amer., Bull., v. 79, p. 429–458.

Arrhenius, G. O. S., 1952, Sediment cores from the east Pacific: Swedish Deep-Sea Exped., 1947–48, Rep., v. 5, f. 1, 227 p.

Artyushkov, E. V., and Mescherikov, Y. A., 1969, Recent movements of the earth's crust and isostatic compensation, *in* Hart, P. J., ed., The earth's crust and upper mantle: Amer. Geophys. Union, Geophys. Monogr. 13, p. 379–390.

Attix, F. H., and Roesch, W. C., 1966, Radiation dosimetry, v. 2: New York, Academic Press, 462 p.

Atwater, T., 1970, Implications of plate tectonics for the Cenozoic tectonic evolution of western North America: Geol. Soc. Amer., Bull., v. 81, p. 3513–3536.

Atwater, T., and Menard, H. W., 1970, Magnetic lineations in the north-east Pacific: Earth Planet. Sci. Lett., v. 7, p. 445–450.

Atwater, T., and Mudie, J. D., 1968, Block faulting on the Gorda rise: Science, v. 159, p. 729–731.

Aubouin, J., 1965, Geosynclines: Amsterdam, Elsevier, 335 p.

Aumento, F., 1967, The mid-Atlantic rise near 45° N. II., Basalts from the area of Confederation Peak: Can. J. Earth Sci., v. 5, p. 1–21.

Avery, O. E., 1963, Geomagnetic and bathymetric profiles across the North Atlantic Ocean: U.S. Naval Oceanographic Office, Tech. Rep. TR-161, 74 p.

Avery, O. E., Burton, G. D., and Heirtzler, J. R., 1968, An aeromagnetic survey of the Norwegian Sea: J. Geophys. Res., v. 73, p. 4583–4600.

Ayme, J. M., 1965, The Senegal salt basin, *in* Salt basins around Africa: London, Institute of Petroleum, p. 83–90.

Backus, G. E., 1964, Magnetic anomalies over oceanic ridges: Nature, v. 201, p. 591–592.

Backus, G. E., and Gilbert, F., 1968, The resolving power of gross earth data: Roy. Astron. Soc., Geophys. J., v. 16, p. 169–205.

Badgley, P. C., 1965, Structural and tectonic principles: New York, Harper and Row, 521 p.

Bailey, E. B., and McCallien, W. J., 1950, The Ankara melange and the Anatolian thrust: Nature, v. 166, p. 938–940.

Bailey, E. B., and McCallien, W. J., 1954, Serpentine lavas, the Ankara melange and the Anatolian thrust: Roy. Soc. Edinburgh, Trans., v. 62, p. 403–441.

Bailey, E. H. ed., 1966, Geology of northern California: Calif., Div. Mines Geol., Bull. 190, 508 p.

Bailey E. H., Irwin, W. P., and Jones, D. L., 1964, Franciscan and related rocks and their significance in the geology of western California: Calif., Div. Mines Geol., Bull. 183, 177 p.

Baker, B. H., 1967, Geology of the Mount Kenya area: Kenya, Geol. Surv., Rep., no. 79, 78 p.

Baker, B. H., and Wohlenberg, J., 1971, Structure and evolution of the Kenya Rift Valley: Nature, v. 229, p. 538–542.

Baker, P. E., 1968, Petrology of Mt. Misery volcano, St. Kitts, West Indies: Lithos, v. 1, p. 124–150.

Balakina, L. M., 1962, General regularities in the directions of the principal stresses effective in the earthquake foci of the seismic belt of the Pacific Ocean: Acad. Sci. USSR, Bull., Geophys. Ser., no. 11, p. 918–926.

Balakina, L. M., Shirokova, H. I., and Vvedenskaya, A. V., 1960, Study of stresses and ruptures in earthquake foci with the help of dislocation theory: Canada, Dominion Observ., Publ., v. 24, p. 321–327.

Balsley, J. R., and Buddington, A. F., 1954, Correlation of reverse remanent magnetism and negative anomalies with certain minerals: J. Geomag. Geoelec., v. 6, p. 176–181.

Banghar, A., and Sykes, L. R., 1969, Focal mechanism of earthquakes in the Indian Ocean and adjacent areas: J. Geophys. Res., v. 74, p. 632–649.

Barazangi, M., and Dorman, J., 1969, World seismicity map compiled from ESSA Coast and Geodetic Survey epicenter data, 1961–1967: Seismol. Soc. Amer., Bull., v. 59, p. 369–380.

Barazangi, M., and Isacks, B., 1971, Lateral variations of seismic wave attenuation in the upper mantle above the inclined earthquake zone of the Tonga island arc: deep anomaly in the upper mantle: J. Geophys. Res., v. 76, p. 8493–8516.

Barazangi, M., Isacks, B., and Oliver, J., 1972, Propagation of seismic waves through and beneath the lithosphere that descends under the Tonga island arc: J. Geophys. Res., v. 77, p. 952–958.

Barth, T. F. W., 1961, Ideas on the interrelation of igneous and sedimentary rock: Finl., Comm. Géol., Bull., no. 196, p. 321–326.

Bassinger, B. G., DeWald, D. E., and Peter, G., 1969, Interpretations of magnetic anomalies off central California: J. Geophys. Res., v. 74, p. 1484–1487.

Batchelor, G. K., 1967, An introduction to fluid dynamics: Cambridge, Cambridge Univ. Press, 615 p.

Bateman, P. C., 1956, Economic geology of the Bishop tungsten district, California: Calif., Div. Mines Geol., Spec. Rep. 47, 87 p.

Bateman, P. C., and Eaton, J. P., 1967, Sierra Nevada batholith: Science, v. 158, p. 1407–1417.

Bear, L. M., 1966, The evolution and petrogenesis of the Troodos complex: Cyprus, Geol. Surv., Ann. Rep., p. 26–37.

Becke, F., 1903, Die Eruptivgebiete des böhmischen Mittelgebirge und der amerikanischen Andes: Atlantische und pazifische Sippe der Eruptiv-gesteine: Tschermaks Mineral. Petrogr. Mitt., ser. 2, v. 22, p. 209–265.

Ben-Menahem, A., 1965, Observed attenuation and Q values of seismic surface waves in the upper mantle: J. Geophys. Res., v. 70, p. 4641–4651.

Ben-Menahem, A., Jarosch, H., and Rosenman, M., 1968, Large scale processing of seismic data in search of regional and global stress patterns: Seismol. Soc. Amer., Bull., v. 58, p. 1899–1932.

Benfield, A. E., 1939, Terrestrial heat flow in Great Britain: Roy. Soc. London, Proc., ser. A, v. 173, p. 428–450.

Benioff, H., 1949, Seismic evidence for the fault origin of oceanic deeps: Geol. Soc. Amer., Bull., v. 60, p. 1837–1856.

Benioff, H., 1951, Earthquakes and rock creep, part 1: Creep characteristics of rocks and the origin of aftershocks: Seismol. Soc. Amer., Bull., v. 41, p. 31–62.

Benioff, H., 1954, Orogenesis and deep crustal structure: Additional evidence from seismology: Geol. Soc. Amer., Bull., v. 65, p. 385–400.

Benioff, H., 1955, Seismic evidence for crustal structure and tectonic activity, *in* Poldervaart, A., ed., Crust of the earth—a symposium: Geol. Soc. Amer., Spec. Pap. 62, p. 61–75.

Benioff, H., ed., 1958, Contributions in geophysics in honor of Beno Gutenberg: New York, Pergamon, 244 p.

Benioff, H., 1962, Movements on major transcurrent faults, *in* Runcorn, S. K., ed., Continental drift: New York, Academic Press, p. 103–134.

Benioff, H., 1964, Earthquake source mechanisms: Science, v. 143, p. 1399–1406.

Berckhemer, H., and Jacob, K. H., 1968, Investigation of the dynamical process in earthquake foci by analyzing the pulse shape of body waves: Frankfurt am Main, Institut für Meteorologie und Geophysik (not verified).

Berg, E., 1964, The Alaskan earthquake of March 1964: Alaska, Univ., Geophys. Inst., Ann. Rep., 1963–64, p. 69–82.

Berg, J. W., Jr., and Baker, C. D., 1963, Oregon earthquakes, 1841 through 1958: Seismol. Soc. Amer., Bull., v. 53, p. 95–108.

Berger, W. H., 1968, Foraminiferal ooze: solution at depths: Science, v. 156, p. 383–385.

Berger, W. H., 1968, Planktonic foraminifera: selective solution and paleoclimatic interpretation: Deep-Sea Res., v. 15, p. 31–43.

Berggren, W. A., 1969a, Cenozoic chronostratigraphy, planktonic foraminiferal zonation and the radiometric time scale: Nature, v. 224, p. 1072–1075.

Berggren, W. A., 1969b, Rates of evolution in some Cenozoic planktonic foraminifera: Micropaleontology, v. 15, p. 351–365.

Berggren, W. A., Phillips, J. D., Bertels, A., and Wall, D., 1967, Late Pliocene-Pleistocene stratigraphy in deep sea cores from the south-central North Atlantic: Nature, v. 216, p. 253–255.

Berlage, H. P., 1937, A provisional catalogue of deep-focus earthquakes in the Netherlands East Indies, 1918–36: Gerlands Beitr. Geophys., v. 50, p. 7–17.

Berner, R. A., 1965, Activity coefficients of bicarbonate, carbonate and calcium ions in sea water: Geochem. Cosmochim. Acta, v. 29, p. 947–965.

Berthelsen, A., 1951, A geological section through the Himalayas; a preliminary report: Dansk Geol. Foren., Medd., v. 12, p. 102–104.

Bertrand, M., 1888, Sur la distribution géographique des roches eruptives en Europe: Soc. Géol. France., Bull., ser. 3, v. 16, p. 593–617.

Biehler, S., Kovach, R. L., and Allen, C. R., 1964, Geophysical framework of northern end of Gulf of California structural province, *in* Van Andel, T. H., and Shor, G. G., eds., Marine geology of the Gulf of California: Tulsa, American Association of Petroleum Geologists, p. 126–143. (Amer. Ass. Petrol. Geol., Mem. 3.)

Birch, F., 1951, Remarks on the structure of the mantle and its bearing upon the possibility of convection currents: Amer. Geophys. Union, Trans., v. 32, p. 533–534.

Birch, F., 1957, Elasticity and constitution of the earth's interior: J. Geophys. Res., v. 57, p. 227–286.

Birch, F., Roy, R. F., and Decker, E. R., 1968, Heat flow and thermal history in New England and New York, *in* Zen, E., White, W. S., Hadley, J. B., and Thompson, J. B., eds., Studies of Appalachian geology: Northern and maritime: New York, Interscience, p. 437–451.

Bird, J. M., and Dewey, J. F., 1970, Lithosphere plate-continental margin tectonics and the evolution of Appalachian orogen: Geol. Soc. Amer., Bull., v. 81, p. 1031–1060.

Bishop, W. W., and Miller, J. A., eds., 1972, Calibration of hominid evolution: Edinburgh, Scottish Academic Press, 487 p.

Black, D. I., 1967, Cosmic ray effects and faunal extinctions at geomagnetic field reversals: Earth Planet. Sci. Lett., v. 3, p. 225–236.

Blackett, P. M. S., Bullard, E., and Runcorn, S. K., eds., 1965, A symposium on continental drift: Roy. Soc. London, Phil. Trans., ser. A, v. 258, 323 p.

Blackman, L. C., Saunders, G., and Ubbelohde, A. R., 1961, Defect structure and properties of pyrolytic carbons: Roy. Soc. London, Proc., ser. A, v. 264, p. 19–40.

Blackwell, D. D., 1969, Heat flow determinations in the northwestern United States: J. Geophys. Res., v. 74, p. 992–1007.

Blake, M. C., Irwin, W. P., and Coleman, R. G., 1967, Upside down metamorphic zonation, blueschist facies, along a regional thrust in California and Oregon: U.S. Geol. Surv., Prof. Pap. 575-C, 9 p.

Blanc, A. C., 1958, Discurso tenuto da A. C. Blanc all' Assemblea Generale di chiusura del V. Congresso Internazionale delle Science Preistoriche e Proto-istoriche, in Amburgo, il 30 Agosto 1958: Quaternaria, v. 5, p. 399–400.

Bloomfield, K., 1966, A major east-north-east dislocation zone in central Malawi: Nature, v. 211, p. 612–614.

Blow, W. H., 1969, Late middle Eocene to Recent planktonic foraminiferal biostratigraphy, *in* Brönniman, P., and Renz, H. H., eds., Proceedings of the 1st International Conference on Planktonic Microfossils, v. 1: Leiden, E. J. Brill, p. 199–422.

Bodvarsson, G., and Walker, G. P. L., 1964, Crustal drift in Iceland: Roy. Astron. Soc., Geophys. J., v. 8, p. 285–300.

Bolli, H. M., 1957, Planktonic foraminifera from the Oligocene-Miocene Cipero and Lengua formations of Trinidad, B. W. I.: U.S. Nat. Mus., Bull. 215, p. 97–123.

Bolt, B. A., 1964, Seismic air waves from the great 1964 Alaskan earthquake: Nature, v. 202, p. 1095–1096.

Bolt, B. A., Lomnitz, C., and McKevilly, T. V., 1968, Seismological evidence on the tectonics of central and northern California and the Mendocino Escarpement: Seismol. Soc. Amer., Bull., v. 58, p. 1725–1767.

Bolt, B. A., and Nuttli, O. W., 1966, P wave residuals as a function of azimuth: J. Geophys. Res., v. 71, p. 5977.

Bonhommet, N., and Babkine, J., 1967, Sur là presence d'aimentation inversées dans la Chaîne des Puys: Acad. Sci. Paris, C. R., ser. B, v. 264, p. 92–94.

Bonhommet, N., and Zahringer, J., 1969, Paleomagnetic and potassium-argon age determination of the Laschamp event: Earth Planet. Sci. Lett., v. 6, p. 41–46.

Bott, M. H. P., 1967a, Solution of the linear inverse problem in magnetic interpretation with application to oceanic magnetic anomalies: Roy. Astron. Soc., Geophys. J., v. 13, p. 313–323.

Bott, M. H. P., 1967b, Terrestrial heat flow and the mantle convection hypothesis: Roy. Astron. Soc., Geophys. J., v. 14, p. 413–428.

Bourcart, J., 1962, La Méditerranée et la revolution du Pliocene: Livre à la mémoire du Prof. Paul Fallot: Soc. Géol. Franc, Bull., v. 1, p. 103–116.

Bramlette, M. N., 1958, Significance of coccolithophorids in calcium-carbonate deposition: Geol. Soc. Amer., Bull., v. 69, p. 121–128.

Bramlette, M. N., 1961, Pelagic sediments, *in* Sears, M., ed., Oceanography: Washington, D.C., American Association for the Advancement of Science, p. 345–366.

Bramlette, M. N., and Wilcoxon, J. A., 1967, Middle Tertiary calcareous nannoplankton of the Cipero section, Trinidad, W. I.: Tulane Stud. Geol., v. 5, p. 93–131.

Briden, J. C., 1967, Recurrent continental drift of Gondwanaland: Nature, v. 215, p. 1334–1339.

Brodie, J. W., 1964, Bathymetry of the New Zealand region: N. Z. Oceanogr. Inst., Mem. 11, 55 p.

Brodie, J. W., and Dawson, E. W., 1965, Morphology of the north Macquarie ridge: Nature, v. 207, p. 844–845.

Brodie, J. W., and Hatherton, T., 1958, The morphology of Kermadec and Hikurangi trenches: Deep-Sea Res., v. 5, p. 18–28.

Brönniman, P., and Renz, H. H., eds., 1969, Proceedings of the 1st International Conference on Planktonic Microfossils, Geneva, 1967: Leiden, E. J. Brill, 2 v.

Brune, J. N., 1968, Seismic moment, seismicity, and rate of slip along major fault zones: J. Geophys. Res., v. 73, p. 777–784.

Brune, J. N., and Allen, C. R., 1967, A low-stress-drop, low-magnitude earthquake with surface faulting: the Imperial California earthquake of March 4, 1966: Seismol. Soc. Amer., Bull., v. 57, p. 501–514.

Brune, J. N., and Dorman, J., 1963, Seismic waves and earth structure in the Canadian shield: Seismol. Soc. Amer., Bull., v. 53, p. 167–209.

Brunhes, B., 1906, Recherches sur la direction d'aimentation des roches volcaniques (1): J. Physique, 4e sér., v. 5, p. 705–724.

Brunn, J. H., 1960, Mise en place et differenciation de l'association pluto-volcanique du cortege ophiolitique: Rev. Géogr. Phys. Géol. Dyn., sér. 2, v. 3, p. 115–132.

Bucha, V., 1970, Geomagnetic reversals in Quaternary revealed from a paleomagnetic investigation of sedimentary rocks: J. Geomag. Geoelec., v. 22, p. 253–272.

Bucher, W. H., 1933, The deformation of the earth's crust: Princeton, N.J., Princeton Univ. Press, 518 p.

Budinger, T. F., and Enbysk, B. J., 1967, Late Tertiary data from the East Pacific rise: J. Geophys. Res., v. 72, p. 2271–2274.

Bullard, E. C., 1939, Heat flow in South Africa: Roy. Soc. London, Proc., ser. A, v. 173, p. 474–502.

Bullard, E. C., 1952, Discussion of paper by R. Revelle and A. E. Maxwell, Heat flow through the floor of the eastern North Pacific Ocean: Nature, v. 170, p. 200.

Bullard E. C., 1955, The stability of a homopolar dynamo: Cambridge Phil. Soc., Proc., v. 51, p. 744–760.

Bullard, E. C., 1964, Continental drift: Geol. Soc. London, Quart. J., v. 120, p. 1–34.

Bullard, E. C., 1968, Reversals of the earth's magnetic field: The Bakerian Lecture, 1967: Roy. Soc. London, Phil. Trans., ser. A, v. 263, p. 481–524.

Bullard, E. C., Everett, J. E., and Smith, A. G., 1965, Fit of continents around Atlantic, in Blackett, P. M. S., Bullard, E. C., and Runcorn, S. K., eds., A symposium on continental drift: Roy. Soc. London, Phil. Trans., ser. A, v. 258, p. 41–75.

Bullard, E. C., and Mason, R. G., 1963, The magnetic field over the oceans, in Hill, M. N., ed., The sea, v. 3: New York, Wiley-Interscience, p. 175–217.

Bullard, E. C., Maxwell, A. E., and Revelle, R., 1956, Heat flow through the deep sea floor: Adv. Geophys., v. 3, p. 153–181.

Bunce, E. T., 1966, The Puerto Rico trench, in Poole, W. H., ed., Continental margins and island arcs: Canada, Geol. Surv., Pap. 66-15, p. 165–176.

Burkle, L. H., Ewing, J., Saito, T., and Leyden, R., 1967, Tertiary sediment from the East Pacific rise: Science, v. 157, p. 537–540.

Burek, P. J., 1970, Magnetic reversals: their application to stratigraphic problems: Amer. Ass. Petrol. Geol., Bull., v. 54, p. 1120–1139.

Burk, C. A., 1965, Geology of the Alaska peninsula-island arc and continental margin: Geol. Soc. Amer., Mem. 99, pt. 1–3, 250 p.

Burk, C. A., and Morres, E. M., 1968, Problems of major faulting at continental margins, with special reference to the San Andreas fault system, in Dickinson, W. R., and Grantz, A., eds., Proceedings of conference on geologic problems of San Andreas fault system: Stanford Univ., Publ. Geol. Sci., v. 11, p. 358–374.

Burns, R. E., and Grim, P. J., 1967, Heat flow in the Pacific Ocean off central California: J. Geophys. Res., v. 72, p. 6239–6247.

Burri, C., 1927, Kritische Zusammenfassung unserer Kenntnisse über die differentiationstypen postmesozoischen Vulkangebiete: Schweiz. Mineral. Petrog. Mitt., v. 7, p. 254–310.

Burridge, R., and Knopoff, L., 1964, Body force equivalents for seismic dislocations: Seismol. Soc. Amer., Bull., v. 54, p. 1875–1888.

Byerly, P., 1926, The Montana earthquake of June 28, 1925, G.M.C.T.: Seismol. Soc. Amer., Bull., v. 16, p. 209–265.

Byerly, P., 1928, The nature of the first motion in the Chilean earthquake of November 11, 1922: Amer. J. Sci., v. 216, p. 232–236.

Byerly, P., and DeNoyer, J., 1958, Energy in earthquakes as computed from geodetic observations, in Benioff, H., ed., Contributions in geophysics in honor of Beno Gutenberg: New York, Pergamon Press, p. 17–35.

Byerly, P., and Stauder, W. V., 1957, Motion at the source of an earthquake: Canada, Dominion Observ., Publ., v. 20, p. 255–261.

Byrne, J. V., Fowler, G. A., and Maloney, N. J., 1966, Uplift of the continental margin and possible continental accretion off Oregon: Science, v. 154, p. 1654–1656.

Cain, J. C., Daniels, W. E., Hendricks, S. J., and Jensen, D. C., 1965, An evaluation of the main geomagnetic field, 1940–1962: J. Geophys. Res., v. 70, p. 3647–3674.

Cain, J. C., and Hendricks, S. J., 1968, The geomagnetic secular variation 1900–1965: NASA Tech. Note, TN D-4527, 221 p.

Cann, J. R., and Vine, F. J., 1966, An area on the crest of the Carlsberg ridge—petrology and magnetic survey: Roy. Soc. London, Phil. Trans., ser. A, v. 259, p. 198–217.

Carder, D. S., 1945, Seismic investigations in the Boulder Dam area, 1940–1944, and the influence of reservoir loading on local earthquake activity: Seismol. Soc. Amer., Bull., v. 35, p. 175–192.

Carder, D. S., Tocher, D., Bufe, C., Stewart, S. W., Eisler, J., and Berg, E., 1967, Seismic wave arrivals from Longshot, 0° to 27°: Seismol. Soc. Amer., Bull., v. 57, p. 573–590.

Carey, S. W., 1955, The orocline concept in geotectonics, part 1: Roy. Soc. Tasmania, Pap. Proc., v. 89, p. 255–288.

Carey, S. W., ed., 1958a, Continental drift; a symposium: Hobart, Univ. of Tasmania, Geol. Dept., 375 p.

Carey, S. W., 1958b, The tectonic approach to continental drift, in Carey, S. W., ed., Continental drift; a symposium: Hobart, Univ. of Tasmania, Geol. Dept., p. 177–358.

Carey, S. W., ed., 1964, Syntaphral tectonics and diagenesis—a symposium: Hobart, Univ. of Tasmania, Geol. Dept., 1 v. (v.p.)

Cargill, H. K., Hawkes, L., and Ledeboer, J. A., 1928, The major intrusions of south-eastern Iceland: Geol. Soc. London, Quart. J., v. 84, p. 534–537.

Carmichael, C. M., 1961, The magnetic properties of ilmenite-haematite crystals: Roy. Soc. London, Proc., ser. A, v. 263, p. 508–530.

Carslaw, H. S., and Jaeger, J. C., 1959, Conduction of heat in solids, ed. 2: Oxford, Clarendon Press, 510 p.

Centre National de la Recherche Scientifique, 1962, Oceanographie géologique et géophysique de la Méditerranée Occidentale; Colloque nationale, Centre National de la Recherche Scientifique, Villefranche,

Avril 1961: Paris, Centre National de la Recherche Scientifique, 256 p.

Chamalaun, F. H., and McDougall, I., 1966, Dating geomagnetic polarity epochs in Réunion: Nature, v. 210, p. 1212–1214.

Chander, R., and Brune, J. N., 1965, Radiation pattern of mantle Rayleigh waves and the source mechanism of the Hindu Kush earthquake of July 6, 1962: Seismol. Soc. Amer., Bull., v. 55, p. 805–819.

Chandrasekhar, S., 1953, The onset of convection by thermal instability in spherical shells: Phil. Mag., ser. 7, v. 44, p. 233–241.

Chase, C. G., 1971, Tectonic history of the Fiji plateau: Geol. Soc. Amer., Bull., v. 82, p. 3087–3110.

Chase, C. G., Menard, H. W., Larson, R. L., Sharman, G. F., III, and Smith, S. M., 1970, History of sea-floor spreading west of Baja California: Geol. Soc. Amer., Bull., v. 81, p. 491–498.

Chase, R. L., and Bunce, E. T., 1969, Underthrusting of the eastern margin of the Antilles by the floor of the western North Atlantic Ocean, and origin of the Barbados ridge: J. Geophys. Res., v. 74, p. 1413–1420.

Chase, T. E., and Menard, H. W., 1963, Topographic chart no. 13: U.S. Bureau of Commercial Fisheries and University of California Institute of Marine Resources.

Chase, T. E., and Menard, H. W., 1965, Topographic charts 3, 4a, 5 and 6: La Jolla, Calif., U.S. Bureau of Commercial Fisheries.

Chevallier, R., 1925, L'Aimentation des lavas de Etna, chap. VII: Résultats, première partie – Uniformité d'aimentation. Rôle des chocs: Ann. Physique, ser 10, v. 4, p. 131–162.

Childs, O. E., and Beebe, B. W., eds., 1963, Backbone of the Americas: tectonic history from pole to pole: Tulsa, American Association of Petroleum Geologists, 320 p. (Amer. Ass. Petrol. Geol., Mem. 2.)

Chinner, G. A., 1966, The distribution of pressure and temperature during Dalradian metamorphism: Geol. Soc. London, Quart. J., v. 122, p. 159–186.

Chinnery, M. A., 1961, The deformation of the ground around surface faults: Seismol. Soc. Amer., Bull., v. 51, p. 355–372.

Christiansen, R. L., and Lipman, P. W., 1970, Cenozoic volcanism and tectonism in the western United States and adjacent parts of the spreading ocean floor, pt. 2, Late Cenozoic (abs.): Geol. Soc. Amer., Abstr., v. 2, p. 81.

Christiansen, R. L., and Lipman, P. W., 1972, Cenozoic volcanism and plate tectonic evolution of the western United States, II. Late Cenozoic: Roy. Soc. London, Phil. Trans., ser. A, v. 271, p. 249–284.

Christoffel, D. A., and Ross, D. I., 1965, Magnetic anomalies south of the New Zealand plateau: J. Geophys. Res., v. 70, p. 2857–2861.

Churkin, M., Jr., 1972, Western boundary of the North American continental plate in Asia: Geol. Soc. Amer., Bull., v. 83, p. 1027–1036.

Clark, D. L., 1970, Magnetic reversals and sedimentation rates in the Arctic Ocean: Geol. Soc. Amer., Bull., v. 81, p. 3129–3134.

Clark, S. P., ed., 1966, Handbook of physical constants: Geol. Soc. Amer., Mem. 97, 587 p.

Clark, S. P., and Ringwood, A. E., 1964, Density distribution and constitution of the mantle: Rev. Geophsy., v. 2, p. 35–88.

Cleary, J., 1967, Azimuthal variation of the Longshot source term: Earth Planet. Sci. Lett., v. 3, p. 29–37.

Cleary, J., and Hales, A. L., 1966, An analysis of the travel times of P waves to North American stations, in the distance range 32°–100°: Seismol. Soc. Amer., Bull., v. 56, p. 467–489.

Closs, H., 1967, Geophysical results and problems on a cross-section from the Tyrrhenian Sea to Iceland (abs.): IUGG Upper Mantle Symposium, Zurich, 1967 (not verified).

Coats, R. R., 1961, Magma type and crustal structure in the Aleutian arc (abs.): Pac. Sci. Congr., 10th, Abstr. Symp. Pap., p. 390–391.

Coats, R. R., 1962, Magma type and crustal structure in the Aleutian arc, in MacDonald, G. A., and Kuno, H., eds., Crust of the Pacific basin: Amer. Geophys. Union, Geophys. Monogr. 6, p. 92–109.

Coe, R. S., 1967a, The determination of paleo-intensities of the earth's magnetic field with emphasis on mechanisms which could cause nonideal behavior in Thellier's method: J. Geomag. Geoelec., v. 19, p. 157–179.

Coe, R. S., 1967b, Paleo-intensities of the earth's magnetic field determined from Tertiary and Quaternary rocks: J. Geophys. Res., v. 72, p. 3247–3262.

Cogné, J., Gèze, B., Goguel, J., Grolier, J., Letourneur, J., Pellet, J., Rothé, J., and Sittler, C., 1966, Les "rifts" et les failles de décrochement en France: Rev. Géogr. Phys. Géol. Dyn., sér. 2, v. 8, p. 123–131.

Coleman, J. M., and Smith, W. G., 1964, Late Recent rise of sea level: Geol. Soc. Amer., Bull., v. 75, p. 833–840.

Colom, G., 1955, Jurassic-Cretaceous pelagic sediments of the western Mediterranean zone and the Atlantic area: Micropaleontology, v. 1, p. 109–124.

Colom, G., and Escandell, B., 1962, L'Evolution du géosynclinal Baleare: Livre à la mémoire du Prof. Paul Fallot: Soc. Géol. France, Bull., v. 1, p. 125–136.

Condon, W. H., and Cass, J. T., 1958, Map of a part of the Prince William Sound area, Alaska, showing linear geologic features as shown on aerial photographs: U.S. Geol. Surv., Misc. Geol. Investig. Map I-273.

Coney, P. J., 1971, Cordilleran tectonic transitions and motion of the North American plate: Nature, v. 233, p. 462–465.

Cook, K. L., 1966, Rift system in the Basin and Range province, in Irvine, T. N., ed., The world rift system: Canada, Geol. Surv., Pap. 66-14, p. 246–279.

Cooke, R. J. S., 1966, Some seismological features of the north Macquarie ridge: Nature, v. 211, p. 953–954.

Coombs, D. S., 1961, Some recent work on the lower grades of metamorphism: Austral. J. Sci., v. 24, p. 203–215.

Cox, A., 1959, The remanent magnetization of some Cenozoic rocks: Ph.D thesis, Univ. of California, Berkeley, 193 p.

Cox, A., 1961, Anomalous remanent magnetization of basalt: U.S. Geol. Surv., Bull. 1083-E, p. 131–160.

Cox, A., 1968, Lengths of geomagnetic polarity intervals: J. Geophys. Res., v. 73, p. 3247–3260.

Cox, A., 1969, Geomagnetic reversals: Science, v. 163, p. 237–245.

Cox, A., 1970, Reconciliation of statistical models for reversals: J. Geophys. Res., v. 75, p. 7501–7503.

Cox, A., and Cain, J. C., 1972, International conference on the core-mantle interface: EOS (Amer. Geophys. Union, Trans.), v. 53, p. 591–623.

Cox, A., and Dalrymple, G. B., 1967a, Geomagnetic polarity epochs – Nunivak Island, Alaska: Earth Planet. Sci. Lett., v. 3, p. 173–177.

Cox, A., and Dalrymple, G. B., 1967b, Statistical analysis of geomagnetic reversal data and the precision of potassium-argon dating: J. Geophys. Res., v. 72, p. 2603–2614.

Cox, A., Dalrymple, G. B., and Doell, R. R., 1967, Reversals of the earth's magnetic field: Sci. Amer., v. 216, p. 44–54.

Cox, A., and Doell, R. R., 1960, Review of paleomagnetism: Geol. Soc. Amer., Bull., v. 71, p. 645–768.

Cox, A., and Doell, R. R., 1961, Paleomagnetic evidence

relevant to a change in the earth's radius: Nature, v. 190, p. 36–37.

Cox, A., and Doell, R. R., 1962, Magnetic properties of basalt in hole EM7, Mohole Project: J. Geophys. Res., v. 67, p. 3997–4004.

Cox, A., and Doell, R. R., 1964, Long period variations of the geomagnetic field: Seismol. Soc. Amer., Bull., v. 54, p. 2243–2270.

Cox, A., and Doell, R. R., 1968, Paleomagnetism and Quaternary correlation, in Morrison, R. B., and Wright, H. E., Jr., eds., Means of correlation of Quaternary successions: Int. Ass. Quater. Res., 7th Congr. (1965), Proc., v. 8, p. 253–265.

Cox, A., Doell, R. R., and Dalrymple, G. B., 1963a, Geomagnetic polarity epochs and Pleistocene geochronometry: Nature, v. 198, p. 1049–1051.

Cox, A., Doell, R. R., and Dalrymple, G. B., 1963b, Geomagnetic polarity epochs: Sierra Nevada II: Science, v. 142, p. 382–385.

Cox, A., Doell, R. R., and Dalrymple, G. B., 1964a, Geomagnetic polarity epochs: Science, v. 143, p. 351–352.

Cox, A., Doell, R. R., and Dalrymple, G. B., 1964b, Reversals of the earth's magnetic field: Science, v. 144, p. 1537–1543.

Cox, A., Doell, R. R., and Dalrymple, G. B., 1965, Quaternary paleomagnetic stratigraphy, in Wright, H. E., Jr., and Frey, D. G., eds., The Quaternary of the United States: Princeton, N.J., Princeton Univ. Press, p. 817–830.

Cox, A., Doell, R. R., and Dalrymple, G. B., 1968, Radiometric time scale for geomagnetic reversals: Geol. Soc. London, Quart, J., v. 124, p. 53–66.

Cox, A., Doell, R. R., and Thompson, G., 1964c, Magnetic properties of serpentinite from Mayaguez, Puerto Rico, in A Study of serpentinite, the AMSOC core hole near Mayaguez, Puerto Rico: Nat. Acad. Sci.–Nat. Res. Council, Publ., no. 1188, p. 49–60.

Cox, A., Hopkins, D. M., and Dalrymple, G. B., 1966, Geomagnetic polarity epochs: Pribilof Islands, Alaska: Geol. Soc. Amer., Bull., v. 77, p. 883–910.

Crain, I. K., and Crain, P. L., 1970, New stochastic model for geomagnetic reversals: Nature, v. 228, p. 39–41.

Crain, I. K., Crain, P. L., and Plaut, M. G., 1969, Long period Fourier spectrum of geomagnetic reversals: Nature, v. 223, p. 283.

Crary, A. P., Gould, L. M., Hurlburt, E. O., Odishaw, H., and Smith, E. E., eds., 1956, Antarctica in the International Geophysical Year: Amer. Geophys. Union, Geophys. Monogr. 1, 133 p.

Creer, K. M., 1965, Paleomagnetic data from the Gondwanic continents, in Blackett, P. M. S., Bullard, E., and Runcorn, S. K., eds., A symposium on continental drift: Roy. Soc. London, Phil. Trans., ser. A, v. 258, p. 27–40.

Creer, K. M., 1971, Mesozoic paleomagnetic reversal column: Nature, v. 233, p. 545–546.

Creer, K. M., and Ispir, Y., 1970, An interpretation of the behavior of the geomagnetic field during polarity transitions: Phys. Earth Planet. Interiors, v. 2, p. 283–293.

Crittenden, M. D., Jr., 1963, Effective viscosity of the earth derived from isostatic loading of Pleistocene Lake Bonneville: J. Geophys. Res., v. 68, p. 5517–5530.

Crowell, J. C., 1962, Displacement along the San Andreas fault, California: Geol. Soc. Amer., Spec. Pap. 71, 58 p.

Crowell, J. C., 1968, Movement histories of faults in the Transverse ranges and speculations on the tectonic history of California, in Dickinson, W. R., and Grantz, A., eds., Proceedings of the conference on geologic problems of San Andreas fault system: Stanford Univ., Publ. Geol. Sci., v. 11, p. 323–341.

Curtis, G. H., and Hay, R. L., 1972, Further geologic studies and K-Ar dating of Olduvai Gorge and Ngorongoro Crater, in Bishop, W. W., and Miller, J. A., eds., Calibration of hominid evolution: Edinburgh, Scottish Academic Press, p. 289–302.

Dagley, P., and Wilson, R. L., 1971, Geomagnetic field reversals—a link between strength and orientation of a dipole source: Nature, Phys. Sci., v. 232, p. 16–18.

Dagley, P., Wilson, R. L., Ade-Hall, J. M., Walker. G. P L., Haggerty, S. E., Sigurgeirsson, T., Watkins, N. D., Smith, P. J., Edwards, J., and Grasty, R. L., 1967, Geomagnetic polarity zones for Icelandic lavas: Nature, v. 216, p. 25–29.

Dahlen, F. A., 1968, The normal modes of a rotating elliptical earth: Roy. Astron. Soc., Geophys. J., v. 16, p. 329–367.

Dalrymple, G. B., 1963, Potassium-argon dates of some Cenozoic volcanic rocks of the Sierra Nevada, California: Geol. Soc. Amer., Bull., v. 74, p. 379–390.

Dalrymple, G. B., 1972, Potassium-argon dating of geomagnetic reversals and North American glaciations, in Bishop, W. W., and Miller, J. A., eds., Calibration of hominid evolution: Edinburgh, Scottish Academic Press, p. 107–134.

Dalrymple, G. B., Cox, A., and Doell, R. R., 1965, Potassium-argon age and paleomagnetism of the Bishop tuff, California: Geol. Soc. Amer., Bull., v. 76, p. 665–673.

Dalrymple, G. B., Cox, A., Doell, R. R., and Grommé, C. S., 1967, Pliocene geomagnetic polarity epochs: Earth Planet. Sci. Lett., v. 2, p. 163–173.

Dalrymple, G. B., and Hirooka, K., 1965, Variation of potassium, argon, and calculated age in a Late Cenozoic basalt: J. Geophys. Res., v. 70, p. 5291–5296.

Daly, R. A., 1940, Strength and structure of the earth: Englewood Cliffs, N. J., Prentice-Hall, 434 p.

Davies, D., 1967, On the problem of compatibility of surface wave data, Q and body wave travel times: Roy. Astron. Soc., Geophys. J., v. 13, p. 421–424.

Davies, D., and McKenzie, D. P., 1969, Seismic traveltime residuals and plates: Roy. Astron. Soc., Geophys. J., v. 18, p. 51–63.

Davies, K. A., 1951, The Uganda section of the western rift: Geol. Mag., v. 88, p. 377–385.

Davis, G. A., 1968, Westward thrusting in the south-central Klamath Mtns., Calif.: Geol. Soc. Amer., Bull., v. 79, p. 911–933.

Davis, T. N., and Sanders, N. K., 1960, Alaska earthquake of July 10, 1958; intensity distribution and field investigation of northern epicentral region: Seismol. Soc. Amer., Bull., v. 50, p. 221–252.

Dawson, J. B., 1970, Neogene rift tectonics and volcanism in northern Tanzania: Geol. Soc. London, Proc., no. 1663, p. 151–153.

DeBooy, T., 1969, Repeated disappearance of continental crust during the geological development of the western Mediterranean area: Ned. Geol. Mijnbouwk., Genoot., Verh., v. 26, p. 79–103.

Deffeyes, K. S., 1970, The axial valley: a steady-state feature of the terrain, in Johnson, H., and Smith, B. L., eds., The megatectonics of continents and oceans; New Brunswick, N. J., Rutgers Univ. Press, p. 194–222.

Demenitskaya, R. M., and Karasik, A. M., 1969, The active rift system of the Arctic Ocean: Tectonophysics, v. 8, p. 345–351.

Denham, C. R., and Cox, A., 1971, Evidence that the Laschamp polarity event did not occur 13,300–30,400 years ago: Earth Planet. Sci. Lett., v. 13, p. 181–190.

Dennison, J. M., and Johnson, R. W., Jr., 1971, Tertiary intrusions and associated phenomena near the thirty-eighth parallel fracture zone in Virginia and West Virginia: Geol. Soc. Amer., Bull., v. 82, p. 501–508.

Derr, J. S., 1970, Discrimination of earthquakes and explosions by the Rayleigh-wave spectral ratio: Seismol. Soc. Amer., Bull., v. 60, p. 697–716.

Dewey, J. F., and Bird, F. M., 1970, Mountain belts and the new global tectonics: J. Geophys. Res., v. 75, p. 2625–2647.

Dewey, J. F., and Horsfield, B., 1970, Plate tectonics, orogeny and continental growth: Nature, v. 225, p. 521–525.

Dibblee, T. W., Jr., 1966, Evidence for cumulative offset on the San Andreas fault in central and northern California, in Bailey, E. H., ed., Geology of northern California: Calif., Div. Mines Geol., Bull. 190, p. 375–384.

Dickinson, W. R., 1968, Circum-Pacific andesite types: J. Geophys. Res., v. 73, p. 2261–2269.

Dickinson, W. R., 1969, Evolution of calc-alkaline rocks in the geosynclinal system of California and Oregon, in McBirney, A. R., ed., Proceedings of Andesite Conference: Oreg., Dept. Geol. Miner. Ind., Bull., no. 65, p. 151–156.

Dickinson, W. R., 1970, Relation of andesites, granites, and derivative sandstones to arc-trench tectonics: Rev. Geophys., v. 8, p. 813–860.

Dickinson, W. R., 1971a, Plate tectonic models for orogeny at continental margins: Nature, v. 232, p. 41–42.

Dickinson, W. R., 1971b, Plate tectonics in geologic history: Science, v. 174, p. 107–113.

Dickinson, W. R., 1971c, Plate tectonic models of geosynclines: Earth Planet. Sci. Lett., v. 10, p. 165–174.

Dickinson, W. R., and Grantz, A., 1968a, Indicated cumulative offsets along the San Andreas fault in the California coast ranges, in Dickinson, W. R., and Grantz, A., eds., Proceedings of the conference on geologic problems of San Andreas fault system: Stanford Univ., Publ. Geol. Sci., v. 11, p. 117–120.

Dickinson, W. R., and Grantz, A., eds., 1968b, Proceedings of the conference on geologic problems of San Andreas fault system: Stanford Univ., Publ. Geol. Sci., v. 11, 374 p.

Dickinson, W. R., and Hatherton, T., 1967, Andesitic volcanism and seismicity around the Pacific: Science, v. 157, p. 801–803.

Dickinson, W. R., and Luth, W. C., 1971, A model for plate tectonic evolution of mantle layers: Science, v. 174, p. 400–404.

Dickson, G. O., and Foster, J. H., 1966, The magnetic stratigraphy of a deep-sea core from the North Pacific Ocean: Earth Planet. Sci. Lett., v. 1, p. 458–462.

Dickson, G. O., Pitman, W. C., III, and Heirtzler, J. R., 1968, Magnetic anomalies in the South Atlantic and ocean floor spreading: J. Geophys. Res., v. 73, p. 2087–2100.

Dietz, R. S., 1961, Continent and ocean basin evolution by spreading of the sea floor; Nature, v. 190, p. 854–857.

Dietz, R. S., 1963, Collapsing continental rises: an actualistic concept of geosynclines and mountain building: J. Geol., v. 71, p. 314–333.

Dietz, R. S., 1966, Passive continents, spreading sea floors, and collapsing continental rises: Amer. J. Sci., v. 264, p. 177–193.

Dietz, R. S., 1968, Reply: J. Geophys. Res., v. 73, p. 6567.

Dietz, R. S., and Holden, J. C., 1966, Miogeosynclines in space and time: J. Geol., v. 74, p. 566–583.

Dietz, R. S., and Holden, J. C., 1970, Reconstruction of Pangaea: Breakup and dispersion of continents, Permian to present: J. Geophys. Res., v. 75, p. 4939–4955.

Doe, B. R., Lipman, P. W., and Hedge, C. E., 1969, Radiogenic traces and the source of continental andesites; a beginning at the San Juan volcanic field, Colorado,

in McBirney, A. R., ed., Proceedings of Andesite Conference: Oreg., Dept. Geol. Mineral Ind., Bull., no. 65, p. 143–149.

Doell, R. R., and Cox, A., 1961, Paleomagnetism: Adv. Geophys. v. 8, p. 221–313.

Doell, R. R., and Dalrymple, G. B., 1966, Geomagnetic polarity epochs: A new polarity event and the age of the Brunhes-Matuyama boundary: Science, v. 152, p. 1060–1061.

Doell, R. R., Dalrymple, G. B., and Cox, A., 1966, Geomagnetic polarity epochs: Sierra Nevada data, 3: J. Geophys. Res., v. 71, p. 531–541.

Doell, R. R., Dalrymple, G. B., Smith, R. L., and Bailey, R. A., 1968, Paleomagnetism, potassium-argon ages, and geology of rhyolites and associated rocks of the Valles Caldera, New Mexico: Geol. Soc. Amer., Mem. 116, p. 211–248.

Doll, C. G., Cady, W. M., Thompson, J. B., Jr., and Billings, M. P., 1961, Geologic map of Vermont, scale 1:250,000: Montpelier, Vermont Geological Survey.

Domen, S. R., 1969, A heat loss compensated calorimeter and related theorems: U.S. Nat. Bur. Standards, J. Res.: C, v. 73C, p. 17–20.

Dorman, J., Ewing, M., and Oliver, J., 1960, Study of shear velocity distribution by mantle Rayleigh waves: Seismol. Soc. Amer., Bull., v. 50, p. 87–115.

Dott, R. H., Jr., 1969, Circum-Pacific Late Cenozoic structural rejuvenation: Implications for sea floor spreading: Science, v. 166, p. 874–876.

Doty, M. S., 1957, Rocky intertidal surfaces, in Hedgpeth, J. W., ed., Treatise on marine ecology and paleoecology, v. 1, Ecology: Geol. Soc. Amer., Mem. 67, v. 1, p. 535–585.

Drake, C. L., 1966, Recent investigations on the continental margin of the eastern United States, in Poole, W. H., ed., Continental margins and island arcs: Canada, Geol. Surv., Pap. 66-15, p. 33–47.

Drake, C. L., Campbell, N. J., Sander, G., and Nafe, J. E., 1963, A mid-Labrador sea ridge: Nature, v. 200, p. 1085–1086.

Drake, C. L., Ewing, J., and Stockard, H., 1968, The continental margin of the eastern United States: Can. J. Earth Sci., v. 5, p. 993–1010.

Drake, C. L., Ewing, M., and Sutton, J., 1959, Continental margins and geosynclines: The east coast of North America north of Cape Hatteras: Phys. Chem. Earth, v. 5, p. 110–198.

Drake, C. L., and Girdler, R. W., 1964, A geophysical study of the Red Sea: Roy. Astron. Soc., Geophys. J., v. 8, p. 473–495.

Drake, C. L., and Nafe, J. E., 1967, Geophysics of the North Atlantic region, paper presented at the UNESCO-IUGS Symposium on Continental Drift, Montevideo, 1967. (Available on microfilm from American Geophysical Union.)

Drake, C. L., and Nafe, J. E., 1968, The transition from ocean to continent from seismic refraction data, in Knopoff, L., Drake, C. L., and Hart, P. J., eds., The crust and upper mantle of the Pacific area: Amer. Geophys. Union, Geophys. Monogr. 12, p. 174–186.

Drake, C. L., and Woodward, H. P., 1963, Appalachian curvature, wrench faulting, and offshore structures: N. Y. Acad. Sci., Trans., ser. 2, v. 26, p. 48–63.

Durr, S., Hoeppener, R., Hoppe, P., and Kockel, F., 1962, Geologie des montagnes entre le Rio Guadalhorce et Campo de Gibraltar (Espagne meridional): Livre à mémoire du Prof. Paul Fallot: Soc. Géol. France, Bull., v. 1, p. 209–227.

Du Toit, A. L., 1937, Our wandering continents: Edinburgh, Oliver and Boyd, 366 p.

Dutro, J. T., Jr., and Payne, T. C., 1957, Geologic map of Alaska, scale 1:2,500,000: Washington, D.C., U.S.

Geological Survey.

Dymond, J., 1969, Age determinations of deep-sea sediments: a comparison of three methods: Earth Planet. Sci. Lett., v. 6, p. 9–14.

Dymond, J., and Deffeyes, K., 1968, K-Ar ages of deep-sea rocks and their relation to sea floor spreading (abs.): Amer. Geophys. Union, Trans., v. 49, p. 364.

Dymond, J., Watkins, N. D., and Nayudu, Y. R., 1968, Age of the Cobb seamount: J. Geophys. Res., v. 73, p. 3977–3979.

Eaton, J. P., 1962, Crustal structure and volcanism in Hawaii, in Macdonald, G. A., and Kuno, H., eds., Crust of the Pacific basin: Amer. Geophys. Union, Geophys. Monogr. 6, p. 13–29.

Eaton, J. P., 1967, Instrumental seismic studies in the Parkfield-Cholame, California, earthquakes of June-August 1966 — surface geologic effects, water-resources aspects and preliminary seismic data: U.S. Geol. Surv., Prof. Pap. 579, p. 57–65.

Edgar, N. T., 1968, Seismic refraction and reflection in the Caribbean Sea: Ph.D. thesis, Columbia University.

Egeler, C. G., and DeBooy, T., 1962, Signification tectonique de la presence d'élèments du Bètique de Malaga dans le sud-est de Cordillères Betiques avec quelques remarques sur les rapports entre le Betique de Malaga et le Subbetique: Livre à la mémoire du Prof. Paul Fallot: Soc. Géol. France, Bull., v. 1, p. 155–162.

Egyed, L., 1956, The change of the earth's dimensions determined from paleogeographical data: Pure Appl. Geophys., v. 33, p. 42–48.

Egyed, L., 1957, A new dynamic conception of the internal constitution of the earth: Geol. Rundschau, v. 46, p. 101–121.

Elsasser, W. M., 1955, Hydromagnetism, I: A review: Amer. J. Phys., v. 23, p. 590–609.

Elsasser, W. M., 1967, Convection and stress propagation in the upper mantle: Princeton Univ., Tech. Rep. 5, 65 p. (not verified).

Elsasser, W. M., 1969, Convection and stress propagation in the upper mantle, in Runcorn, S. K., ed., The application of modern physics to the earth and planetary interiors: New York, Wiley-Interscience, p. 223–246.

Elsasser, W. M., 1971, Sea-floor spreading as thermal convection: J. Geophys. Res., v. 76, p. 1101–1112.

Elvers, D. J., Mathewson, C. C., Kohler, R. E., and Moses, R. L., 1967a, Systematic ocean surveys by the U. S. C. and G. S. S. Pioneer, 1961–1963: U.S. Coast Geodetic Surv., Oper. Data Rep. C, GSDR-1, 19 p.

Elvers, D. J., Peter, G., and Moses, R., 1967b, Analysis of magnetic lineations in the North Pacific (abs.): Amer. Geophys. Union, Trans. v. 41, p. 89.

Emery, K. O., Uchupi, E., Phillips, J. D., Bowin, C. O., Bunce, E. T., and Knott, S. T., 1970, Continental rise off eastern North America: Amer. Ass. Petrol. Geol., Bull., v. 54, p. 44–108.

Emilia, D. A., and Heinrichs, D. F., 1969, Ocean floor spreading: Olduvai and Gilsa events in the Matuyama epoch: Science, v. 166, p. 1267–1269.

Emiliani, C., and Edwards, G., 1953, Tertiary ocean bottom temperatures: Nature, v. 171, p. 887–888.

Engdahl, E. R., 1971, Explosion effects and earthquakes in the Amchitka Island region: Science, v. 173, p. 1232–1238.

Engdahl, E. R., and Flinn, E. A., 1969, Seismic waves reflected from discontinuities within the earth's upper mantle: Science, v. 163, p. 177–179.

Engel, A. E. J., and Engel, C. G., 1964, Composition of basalts from the mid-Atlantic ridge: Science, v. 144, p. 1330–1333.

Engel, A. E. J., Engle, C. G., and Havens, R. G., 1965, Chemical characteristics of oceanic basalts and the upper mantle: Geol. Soc. Amer., Bull., v. 76, p. 719–734.

Engel, A. E. J., James, H. L., and Leonard, B. F., eds., 1962, Petrologic studies: a volume in honor of A. F. Buddington: Geological Society of America, 660 p.

Engel, C. G., and Fisher, R. L., 1969, Lherzolite, anorthosite, gabbro, and basalt dredged from the mid-Indian Ocean ridge: Science, v. 166, p. 1136–1141.

Environmental Science Services Administration, 1966, ESSA symposium on earthquake prediction; proceedings: Washington, D. C., Gov. Print. Off., 167 p.

Erickson, B. H., and Grim, P. J., 1969, Profiles of magnetic anomalies south of the Aleutian island arc: Geol. Soc. Amer., Bull., v. 80, p. 1387–1390.

Ernst, W. G., 1965, Mineral parageneses in Franciscan metamorphic rocks, Panoche Pass, California: Geol. Soc. Amer., Bull., v. 76, p. 879–914.

Ernst, W. G., 1970, Tectonic contact between the Franciscan melange and the Great Valley sequence — Crustal expression of a late Mesozoic Benioff zone: J. Geophys. Res., v. 78, p. 886–902.

Ernst, W. G., and Seki, Y., 1967, Petrologic comparison of the Franciscan and Sanbagawa metamorphic terrains: Tectonophysics, v. 4, p. 463–478.

Evans, A. L., 1970, Geomagnetic polarity reversals in a late Tertiary lava sequence from the Akaroa volcano, New Zealand: Roy. Astron. Soc., Geophys. J., v. 21, p. 163–183.

Everitt, C. W. F., 1962, Self-reversal of magnetization in a shale containing pyrrhotite: Phil. Mag., v. 7, p. 831–842.

Evernden, J. F., and Curtis, G. H., 1965, The potassium-argon dating of Late Cenozoic rocks in East Africa and Italy: Curr. Anthropol., v. 6, p. 343–385.

Evernden, J. F., Curtis, G. H., and Lipson, J. L., 1957, Potassium-argon dating of igneous rocks: Amer. Ass. Petrol. Geol., Bull., v. 41, p. 2120–2147.

Evernden, J. F., and Kistler, R. W., 1970, Chronology of emplacement of Mesozoic batholithic complexes in California and western Nevada: U. S. Geol. Surv., Prof. Pap. 623, 42 p.

Evernden, J. F., Savage, D. E., Curtis, G. H., and James, G. T., 1964, Potassium-argon dates and the Cenozoic mammalian chronology of North America: Amer. J. Sci., v. 262, p. 145–198.

Ewing, J., Antoine, J., and Ewing, M., 1960, Geophysical measurements in the western Caribbean Sea and the Gulf of Mexico: J. Geophys. Res., v. 65, p. 4087–4126.

Ewing, J., and Ewing, M., 1959, Seismic-refraction profiles in the Atlantic Ocean basins, in the Mediterranean Sea, on the mid-Atlantic ridge and in the Norwegian Sea: Geol. Soc. Amer., Bull., v. 70, p. 291–318.

Ewing, J., and Ewing, M., 1967, Sediment distribution on the mid-ocean ridges with respect to spreading of the sea floor: Science, v. 156, p. 1590–1592.

Ewing, J., Ewing, M., Aitken, T., and Ludwig, W. J., 1968, North Pacific sediment layers measured by seismic profiling, in Knopoff, L., Drake, C. L., and Hart, P. J., eds., The crust and upper mantle of the Pacific area: Amer. Geophys. Union, Geophys. Monogr. 12, p. 147–173.

Ewing, J., Ewing, M., and Leyden, R., 1966a, Seismic-profiler survey of Blake plateau: Amer. Ass. Petrol. Geol., Bull., v. 50, p. 1948–1971.

Ewing, J., Worzel, J. L., and Ewing, M., 1962, Sediments and oceanic structural history of the Gulf of Mexico: J. Geophys. Res., v. 67, p. 2509–2527.

Ewing, J., Worzel, J. L., Ewing, M., and Windisch, C., 1966b, Ages of horizon A and the oldest Atlantic sediments: Science, v. 154, p. 1125–1132.

Ewing, M., 1963, Sediments of ocean basins, in Higginbotham, S. W., ed., Man, science, learning, and education: Houston, Rice Univ. Press, p. 41–59.

Ewing, M., and Antoine, J., 1966, New seismic data concerning sediments and diapirie structures in Sigsbee deep and upper continental slope, Gulf of Mexico: Amer. Ass. Petrol. Geol., Bull., v. 50, p. 479–504.

Ewing, M., Eittreim, S. L., Ewing, J., and Le Pichon, X., 1971, Sediment transport and distribution in the Argentine basin; 3, Nepheloid layer and processes of sedimentation: Phys. Chem. Earth, v. 8, p. 49–77.

Ewing, M., and Ewing, J., 1964, Distribution of oceanic sediments, in Yoshida, K., ed., Studies on oceanography: Tokyo, Univ. of Tokyo Press, p. 525–537.

Ewing, M., Ewing, J., Leyden, R., and Aitken, T., 1968, Seismic reflection profiler results in the Pacific (not verified).

Ewing, M., Ewing, J., and Talwani, M., 1964, Sediment distribution in the oceans: The mid-Atlantic ridge: Geol. Soc. Amer., Bull., v. 75, p. 17–35.

Ewing, M., and Heezen, B. C., 1956, Some problems of antarctic submarine geology, in Crary, A., Gould, L. M., Hurlburt, E. O., Odishaw, H., and Smith, W. E., eds., Antarctica in the International Geophysical Year: Amer. Geophys. Union, Geophys. Monogr. 1, p. 75–81.

Ewing, M., Hirshman, J., and Heezen, B. C., 1959, Magnetic anomalies of the mid-ocean ridge system (abs.); in Sears, M., ed., International Oceanographic Congress preprints: Washington, D. C., American Association for the Advancement of Science, p. 24.

Ewing, M., Le Pichon, X., and Ewing, J., 1966, Crustal structure of the mid-Atlantic ridge: J. Geophys. Res., v. 71, p. 1611–1636.

Ewing, M., Ludwig, W. J., and Ewing, J., 1965, Oceanic structural history of the Bering Sea: J. Geophys. Res., v. 70, p. 4593–4600.

Fairborn, J. W., 1968, Mantle P and S wave velocity distributions from DT/D triangle measurements; Ph.D thesis, Massachusetts Institute of Technology.

Farrell, W. E., 1968, Has the Pacific basin moved as a rigid plate since the Cretaceous?: Nature, v. 217, p. 1034–1035.

Fedotov, S. M., Bagdasarova, A. M., Kuzin, I. P., Tarakanov, R. Z., and Shmidt, O. Yu., 1963, Seismicity and deep structure of the southern part of the Kurile Island arc: Acad. Sci. USSR, Dokl., Earth Sci. Sect., no. 153, p. 71–73.

Fisher, R. L., 1961, Middle America trench; topography and structure: Geol. Soc. Amer., Bull., v. 72, p. 703–720.

Fisher, R. L., Von Herzen, R. P., and Rhea, K. P., 1967, Topographic heat flow studies in seven Micronesian and Melanesian trenches (abs.): Amer. Geophys. Union, Trans., v. 48, p. 218.

Fitch, T. J., and Molnar, P., 1970, Focal mechanisms along inclined earthquake zones in the Indonesia-Philippines region: J. Geophys. Res., v. 75, p. 1431–1444.

Förstemann, F. C., 1859, Ueber den Magnetismus der Gesteine: Ann. Physik, ser. 4, v. 106, p. 106–136.

Forsyth, D. W., and Press, F., 1971, Geophysical tests of petrological models of the spreading lithosphere: J. Geophys. Res., v. 76, p. 7963–7979.

Foster, J. H., and Opdyke, N. D., 1970, Upper Miocene to Recent magnetic stratigraphy in deep-sea sediments: J. Geophys. Res., v. 75, p. 4465–4473.

Fox, P. J., Pitman, W. C., III, and Shepard, F., 1969, Crustal plates in the central Atlantic: Evidence for at least two poles of rotation: Science, v. 165, p. 487–489.

Francheteau, J., Harrison, C. G. A., Sclater, J. G., and Richards, M., 1970, Magnetization of Pacific seamounts: A preliminary polar curve for the northeastern Pacific: J. Geophys. Res., v. 75, p. 2035–2062.

Friedman, M., 1964, Petrofabric techniques for the determination of principal stress directions in rocks, in Judd, W. R., ed., State of stress in the earth's crust: New York, American Elsevier, p. 451–552.

Fukao, Y., 1969, On the radiative heat transfer and the thermal conductivity in the upper mantle: Tokyo, Univ., Earthquake Res. Inst., Bull., v. 47, p. 549–569.

Fullagar, P. D., and Bottino, M. L., 1969, Tertiary felsite intrusions in the Valley and Ridge province, Virginia: Geol. Soc. Amer., Bull, v. 80, p. 1853–1858.

Funnell, B. M., and Riedel, W. R., eds., 1971, Micropaleontology of oceans: Cambridge, Cambridge Univ. Press, 828 p.

Funnell, B. M., and Smith, A. G., 1968, Opening of the Atlantic Ocean: Nature, v. 219, p. 1328–1332.

Gansser, A., 1964, Geology of the Himalayas: New York, Wiley-Interscience, 289 p.

Gansser, A., 1966, The Indian Ocean and the Himalayas; a geological interpretation: Eclogae Geol. Helv., v. 59, p. 831–848.

Gansser, A., 1968, The Insubric line, a major geotectonic problem: Schweiz. Mineral. Petrog. Mitt., v. 48, p. 123–163.

Garland, G. D., ed., 1966, Continental drift: Toronto, Univ. of Toronto Press, 140 p. (Roy. Soc. Canada, Spec. Pub. no. 9.)

Gaskell, T. F., ed., 1967, The earth's mantle: New York, Academic Press, 509 p.

Gaposchkin, E. M., and Lambeck, K., 1971, Earth's gravity field to the sixteenth degree and station coordinates from satellite and terrestrial data: J. Geophys. Res., v. 76, p. 4855–4883.

Gass, I. G., 1968, Is the Troodos Massif of Cyprus a fragment of Mesozoic ocean floor?: Nature, v. 220, p. 39–42.

Gast, P. W., 1968, Trace element fractionation and the origin of tholeiitic and alkaline magma types: Geochim. Cosmochim. Acta, v. 32, p. 1057–1086.

Gates, G. O., and Gryc, G., 1963, Structure and tectonic history of Alaska, in Childs, O. E., and Beebe, B. W., eds., Backbone of the Americas; Tectonic history from pole to pole: Tulsa, American Association of Petroleum Geologists, p. 264–277. (Amer. Ass. Petrol. Geol., Mem. 2.)

Gates, O., and Gibson, W., 1956, Interpretation of the configuration of the Aleutian ridge; Geol. Soc. Amer., Bull., v. 67, p. 127–146.

Gibson, W. M., 1960, Submarine topography in the Gulf of Alaska: Geol. Soc. Amer., Bull., v. 71, p. 1087–1108.

Giese, P., 1968, Die Struktur der Erdkruste in Bereich der Ivrea-zones: Ein Vergleich verschiedener seismischer Interpretationen und der Versuch einer petrographisch-geologischen Deutung: Schweiz. Mineral. Petrog. Mitt., v. 48, p. 261–284.

Gilluly, J., 1927, Analcite diabase and related alkaline syenite from Utah: Amer. J. Sci., v. 214, p. 199–211.

Gilluly, J., 1932, Geology and ore deposits of the Stockton and Fairfield quadrangles, Utah: U. S. Geol. Surv., Prof. Pap. 173, 171 p.

Gilluly, J., 1954, Geologic contrasts between continents and ocean basins, in Poldervaart, A., ed., Crust of the earth—a symposium: Geol. Soc. Amer., Spec. Pap. 62, p. 7–18.

Gilluly, J., 1969, Oceanic sediment volumes and continental drift: Science, v. 166, p. 992–994.

Gilluly, J., 1970, Crustal deformation in the western United States, in Johnson, H., and Smith, B. L., eds., The megatectonics of continents and oceans: New Brunswick, N. J., Rutgers Univ. Press, p. 47–73.

Gilluly, J., 1971, Plate tectonics and magmatic evolution:

Geol. Soc. Amer., Bull., v. 82, p. 2383–2396.

Gilluly, J., Reed, J. C., Jr., and Cady, W. M., 1970, Sedimentary volumes and their significance: Geol. Soc. Amer., Bull., v. 81, p. 353–376.

Girdler, R. W., 1962, Initiation of continental drift: Nature, v. 194, p. 521–524.

Girdler, R. W., 1964, Research note—how genuine is the Circum-Pacific Belt?: Roy. Astron. Soc., Geophys. J., v. 8, p. 537–540.

Girdler, R. W., 1966, The role of translational and rotational movements in the formation of the Red Sea and the Gulf of Aden, in Irvine, T. N., ed., The world rift system: Canada, Geol. Surv., Pap. 66–14, p. 65–75.

Glangeaud, L., 1962, Paléogéographie dynamique de la Méditerranée et de ses bordures. Le role des phases Ponto-Plio-Quaterenaires, in Oceanographie géologique et géophysique de la Mediterranée Occidentale: Paris, Centre National de la Recherche Scientifique, p. 125–265.

Glangeaud, L., Alinat, J., Polveche, J., Guillaume, A., and Leenhardt, O., 1966, Grandes structures de la mer Ligure, leur evolution et leurs relations avec les chaines continentales: Soc. Géol. France., Bull., ser. 7, v. 8, p. 921–937.

Glass, B., Ericson, D. B., Heezen, B. C., Opdyke, N. D., and Glass, J. A., 1967, Geomagnetic reversals and Pleistocene chronology: Nature, v. 216, p. 437–442.

Godby, E. A., Baker, R. C., Bower, M. E., and Hood, P. J., 1966, Aeromagnetic reconnaissance of the Labrador Sea: J. Geophys. Res., v. 71, p. 511–517.

Goldreich, P., and Toomre, A., 1969, Some remarks on polar wandering: J. Geophys. Res., v. 74, p. 2555–2567.

Goldstein, H., 1950, Classical mechanics: Cambridge, Mass., Addison-Weslery, 399 p.

Goodell, H. G., and Watkins, N. D., 1968, The paleomagnetic stratigraphy of the Southern Ocean: 20° West to 160° East longitude: Deep-Sea Res., v. 15, p. 89–112.

Gorter, E. W., and Schulkes, J. A., 1953, Reversal of spontaneous magnetization as a function of temperature in Li Fe Cr spinels: Phys. Rev., v. 90, p. 487–488.

Gough, D. I., and Gough, W. I., 1962, Geophysical investigations at Lake Kariga, in Meisser, O., ed., I. Internationales Symposium über rezente Erdkrustenbewegungen: Leipzig, Akademie-Verlag, p. 441–447.

Gough, D. I., Opdyke, N. D., and McElhinny, M. W., 1964, The significance of paleomagnetic results from Africa: J. Geophys. Res., v. 69, p. 2509–2519.

Graham, K. W. T., 1961, The re-magnetization of a surface outcrop by lightning currents: Roy. Astron. Soc., Geophys. J., v. 6, p. 85–102.

Grantz, A., Plafker, G., and Kachadoorian, R., 1964, Alaska's Good Friday earthquake, March 27, 1964— a preliminary geologic evaluation: U. S. Geol. Surv., Circ. 491, 35 p.

Grantz, A., and Zietz, I., 1960, Possible significance of broad magnetic highs over belts of moderately deformed sedimentary rocks in Alaska and California: U. S. Geol. Surv., Prof. Pap. 400-B, p. B342–B347.

Great Britain, Admiralty, 1963, Bathymetric magnetic and gravity investigations, H. M. S. Owen, 1961–62: Admiralty Mar. Sci. Publ., no. 4, pt. 1–2.

Green, D. H., 1969, The origin of basaltic and nephelinitic magmas in the earth's mantle: Tectonophysics, v. 7, p. 409–422.

Green, D. H., 1971, Composition of basaltic magmas as indicators of conditions of origin: application to oceanic volcanism: Roy. Soc. London, Phil. Trans., ser. A, v. 268, p. 707–725.

Green, D. H., and Ringwood, A. E., 1963, Mineral assemblages in a model mantle composition: J. Geophys. Res., v. 68, p. 937–945.

Green, D. H., and Ringwood, A. E., 1966, Origin of the calc-alkaline igneous rock suite: Earth Planet. Sci. Lett., v. 1, p. 307–316.

Green, D. H., and Ringwood, A. E., 1967, The genesis of basaltic magmas: Contrib. Mineral. Petrology—Beitr. Mineral. Petrologie, v. 15, p. 103–190.

Green, T. H., Green, D. H., and Ringwood, A. E., 1967, The origin of high-alumina basalts and their relationships to quartz tholeiites and alkali basalts: Earth Planet. Sci. Lett., v. 2, p. 41–51.

Green, T. H., and Ringwood, A. E., 1968, Genesis of the calc-alkaline igneous rock suite: Contrib. Mineral. Petrology-Beitr. Mineral. Petrologie, v. 18, p. 105–162.

Gregory, J. W., 1920, The African rift valleys: Geogr. J., v. 56, p. 13–47, 327–328.

Griggs, D. T., 1939, A theory of mountain building: Amer. J. Sci., v. 237, p. 611–650.

Griggs, D. T., 1954, Discussion, Verhoogen, 1954: Amer. Geophys. Union, Trans., v. 35, p. 93–96.

Griggs, D. T., 1966, Reflections on the earthquake mechanism, in Page, R., ed. Proceedings of the Second United States–Japan Conference on Research Related to Earthquake Prediction Problems: National Science Foundation–Japan Society for Promotion of Science, p. 63–64.

Griggs, D. T., 1968–1969, Invited lectures (no abstracts), Nat. Acad. Sci. fall meeting 1968, and Amer. Geophys. Union, 1969 (not verified).

Griggs, D. T., 1972, The sinking lithosphere and the focal mechanisms of deep earthquakes, in Robertson, E. C., ed., The nature of the solid earth: New York, McGraw-Hill, p. 361–384.

Griggs, D. T., and Baker, D. W., 1969, The origin of deep-focus earthquakes, in Mark, H., and Fernback, S., eds., Properties of matter under unusual conditions; in honor of Edward Teller's 60th birthday: New York, Wiley, p. 23–42.

Grim, P. J., and Erickson, B. H., 1968, Marine magnetic anomalies and fracture zones south of the Aleutian trench (abs.): Geol. Soc. Amer., Spec. Pap. 115, p. 84.

Grim, P. J., and Erickson, B. H., 1969, Fracture zones and magnetic anomalies south of the Aleutian trench: J. Geophys. Res., v. 74, p. 1488–1494.

Grim, P. J., and Naugler, F. P., 1969, Fossil deep-sea channel on the Aleutian abyssal plain: Science, v. 163, p. 383–386.

Griscom, A., 1966, Magnetic data and regional structure in northern California, in Bailey, E. H., ed., Geology of northern California: Calif., Div. Mines Geol., Bull. 190, p. 407–417.

Grommé, C. S., and Hay, R. L., 1963, Magnetizations of basalt of Bed I, Olduvai Gorge, Tanganyika: Nature, v. 200, p. 560–561.

Grommé, C. S., and Hay, R. L., 1967, Geomagnetic polarity epochs—new data from Olduvai Gorge, Tanganyika: Earth Planet. Sci. Lett., v. 2, p. 111–115.

Grommé, C. S., and Hay, R. L., 1971, Geomagnetic polarity epochs: age and duration of the Olduvai normal polarity event: Earth Planet. Sci. Lett., v. 10, p. 179–185.

Grommé, C. S., Merrill, R. T., and Verhoogen, J., 1967, Paleomagnetism of Jurassic and Cretaceous plutonic rocks in the Sierra Nevada, and its significance for polar wandering and continental drift: J. Geophys. Res., v. 72, p. 5661–5684.

Grow, J. A., and Atwater, T., 1970, Mid-Tertiary tectonic transition in the Aleutian arc: Geol. Soc. Amer., Bull., v. 81, p. 3715–3722.

Gunn, R., 1937, A quantitative study of mountain building on an unsymmetrical earth: J. Franklin Inst., v. 224, p. 19–53.

Gunn, R., 1947, Quantitative aspects of juxtaposed ocean deeps, mountain chains, and volcanic ranges: Geophysics, v. 12, p. 238–255.

Gutenberg, B., 1939, Tsunamis and earthquakes: Seismol. Soc. Amer., Bull., v. 29, p. 517–526.

Gutenberg, B., ed., 1959, Physics of the earth's interior: New York, Academic Press, 240 p.

Gutenberg, B., and Richter, C. F., 1954, Seismicity of the earth and associated phenomena, ed. 2: Princeton, N. J., Princeton Univ. Press, 310 p.

Hackel, O., 1966, Summary of the geology of the Great Valley, in Bailey, E. H., ed., Geology of northern California: Calif., Div. Mines Geol., Bull. 190, p. 217–238.

Haddon, R. A. W., and Bullen, K. E., 1969, An earth model incorporating free earth oscillation data: Phys. Earth Planet. Interiors, v. 2, p. 35–49.

Hagiwara, Y., 1967, Analyses of gravity values in Japan: Tokyo, Univ., Earthquakes Res. Inst., Bull., v. 45, p. 1091–1228.

Hales, A. L., 1969, Does continental drift arise as a result of gravitational sliding (abs.): Geol. Soc. Amer., Spec. Pap. 121, p. 124–125.

Hallam, A., 1967, The bearing of certain paleozoogeographic data on continental drift: Palaeogeogr. Palaeoclimatol. Palaeoecol., v. 3, p. 201–241.

Hamilton, E. L., 1956, Sunken islands of the mid-Pacific mountains: Geol. Soc. Amer., Mem. 64, 98 p.

Hamilton, E. L., 1965, Isotopic composition of strontium in a variety of rocks from Réunion Island: Nature, v. 207, p. 1188.

Hamilton, E. L., 1967, Marine geology of abyssal plains in the Gulf of Alaska: J. Geophys. Res., v. 72, p. 4189–4214.

Hamilton, E. L., and Von Huene, R. E., 1966, Kodiak seamount not flat-topped: Science, v. 154, p. 1323–1325.

Hamilton, R. M., and Evison, F. F., 1967, Earthquakes at intermediate depths in south-west New Zealand: N. Z. J. Geol. Geophys., v. 10, p. 1319–1329.

Hamilton, R. M., and Gale, A. W., 1968, Seismicity and structure of the North Island, New Zealand: J. Geophys. Res., v. 73, p. 3859–3876.

Hamilton, W., 1961, Origin of the Gulf of California: Geol. Soc. Amer., Bull., v. 72, p. 1307–1318.

Hamilton, W., 1969a, Mesozoic California and the underflow of Pacific mantle: Geol. Soc. Amer., Bull., v. 80, p. 2409–2430.

Hamilton, W., 1969b, The volcanic central Andes — A modern model for the Cretaceous batholiths and tectonics of western North America, in McBirney, A. R., ed., Proceedings of Andesite Conference: Oreg., Dept. Geol. Mineral Ind., Bull., no. 65, p. 175–184.

Hamilton, W., and Myers, W. B., 1966, Cenozoic tectonics of the western United States: Rev. Geophys., v. 4, p. 509–549.

Hamza, V. M., and Verma, R. K., 1969, The relationship of heat flow with age of basement rock: Bull. Volcanol., v. 33, p. 123–152.

Hanna, G. D., 1964, Biological effects of an earthquake: Pac. Disc., v. 17, p. 24–26.

Harada, T., 1964, The Muro group in the Kii Peninsula, southwest Japan: Kyoto, Univ., Coll. Sci., Mem., ser. B, v. 31, p. 71–94.

Harding, S. T., and Rinehart, W., 1966, Preliminary seismological report, in U. S. Coast and Geodetic Survey, The Parkfield, California, earthquake of June 27, 1966: Washington, D. C., Gov. Print. Off., p. 1–16.

Harker, A., 1909, The natural history of the igneous rocks: London, Methuen, 384 p.

Harland, W. B., 1961, An outline structural history of Spitzbergen, in Raasch, G. O., ed., Geology of the Arctic, v. 1: Toronto, Univ. of Toronto Press, p. 68–132.

Harland, W. B., Smith, A. G., and Wilcock, B., eds., 1964, The phanerozoic time scale: London, Geological Society of London, 458 p.

Harris, P. G., and Rowell, J. A., 1960, Some geochemical aspects of the Mohorovicic discontinuity: J. Geophys. Res., v. 65, p. 2443–2460.

Harrison, C. G. A., 1966, The paleomagnetism of deep sea sediments: J. Geophys. Res., v. 71, p. 3033–3043.

Harrison, C. G. A., 1968a, Evolutionary processes and reversals of the earth's magnetic field: Nature, v. 217, p. 46–47.

Harrison, C. G. A., 1968b, Formation of magnetic anomaly patterns by Dyke injection: J. Geophys. Res., v. 73, p. 2137–2142.

Harrison, C. G. A., 1969, What is the true rate of reversals of the earth's magnetic field?: Earth Planet. Sci. Lett., v. 6, p. 186–188.

Harrison, C. G. A., and Somayajulu, B. L. K., 1966, Behaviour of the earth's magnetic field during a reversal: Nature, v. 212, p. 1193–1195.

Hart, P. J., ed., 1969, The earth's crust and upper mantle: Amer. Geophys. Union, Geophys. Monogr. 13, 735 p.

Hasebe, K., Fujii, N., and Uyeda, S., 1970, Thermal processes under island arcs: Tectonophysics, v. 10, p. 335–355.

Hashimoto, I., 1966, The Mesozoic strata of uncertain ages, north of Saeki, Oita Prefecture: Kyushu Univ., Earth Sci. Dept., Rep., v. 13, p. 15–24.

Hatherton, T., 1969a, The geophysical significance of calc-alkaline andesites in New Zealand: N. Z. J. Geol. Geophys., v. 12, p. 436–459.

Hatherton, T., 1969b, Gravity and seismicity of asymmetric active regions: Nature, v. 221, p. 353–355.

Hatherton, T., and Dickinson, W. R., 1968, Andesite volcanism and seismicity in New Zealand: J. Geophys. Res., v. 73, p. 4615–4619.

Hatherton, T., and Dickinson, W. R., 1969, The relationship between andesitic volcanism and seismicity in Indonesia, the Lesser Antilles, and other island arcs: J. Geophys. Res., v. 74, p. 5301–5310.

Hay, R. L., 1963, Stratigraphy of beds I through IV, Olduvai Gorge, Tanganyika: Science, v. 139, p. 829–831.

Hay, W. W., 1967, Zonation of the Middle-Upper Eocene interval: Gulf Coast Ass. Geol. Soc., Trans., v. 17, p. 438–439.

Hayes, D. E., 1966, A geophysical investigation of the Peru-Chile trench: Mar. Geol., v. 4, p. 309–351.

Hayes, D. E., and Ewing, M., 1971, The Louisville ridge — A possible extension of the Eltanin fracture zone: Amer. Geophys. Union, Antarctic Res. Ser., v. 15, p. 223–228.

Hayes, D. E., and Heirtzler, J. R., 1968, Magnetic anomalies and their relation to the Aleutian island arc: J. Geophys. Res., v. 73, p. 4637–4646.

Hayes, D. E., Pimm, A. C., Benson, W. E., Berger, W. H., Von Rad, V., Supko, P. R., Beckman, J. P., Roth, P. H., and Musich, L. F., 1971, Deep sea drilling project leg Ir: Geotimes, v. 16, m. 2, p. 14–17.

Hayes, D. E., and Pitman, W. C., III, 1970, Magnetic lineations in the North Pacific, in Hays, J. D., ed., Geological investigations of the North Pacific: Geol. Soc. Amer., Mem. 126, p. 291–314.

Hays, J. D., ed., 1970a, Geological investigations of the North Pacific: Geol. Soc. Amer., Mem. 126, 323 p.

Hays, J. D., 1970b, The stratigraphy and evolutionary trends of radiolaria in North Pacific deep-sea sediments, *in* Hays, J. D., ed., Geological investigations of the North Pacific: Geol. Soc. Amer., Mem. 126, p. 185–218.

Hays, J. D., 1971, Faunal extinctions and reversals of the earth's magnetic field: Geol. Soc. Amer., Bull., v. 82, p. 2433–2447.

Hays. J. D., and Berggren, W. A., 1971, Quaternary boundaries and correlations, *in* Funnell, B. M., and Riedel, W. R., eds., Micropaleontology of oceans: Cambridge, Cambridge Univ. Press, p. 669–691.

Hays, J. D., and Ninkovitch, D., 1970, North Pacific deep-sea ash – a chronology and age of present Aleutian underthrusting, *in* Hays, J. D., ed., Geological investigations of the North Pacific: Geol. Soc. Amer., Mem. 126, p. 263–290.

Hays, J. D., and Opdyke, N. D., 1967, Antarctic radiolaria, magnetic reversals, and climatic change: Science, v. 158, p. 1001–1010.

Hays, J. D., Saito, T., Opdyke, N. D., and Burckle, L. H., 1969, Pliocene-Pleistocene sediments of the equatorial Pacific: their paleomagnetic, biostratigraphic, and climatic record: Geol. Soc. Amer., Bull., v. 80, p. 1481–1514.

Heath, G. R., 1969, Carbonate sedimentation in the abyssal equatorial Pacific during the past 50 million years: Geol. Soc. Amer., Bull., v. 80, p. 689–694.

Heck, N. H., 1947, List of seismic sea waves: Seismol. Soc. Amer., Bull., v. 37, p. 269–286.

Hedgpeth, J. W., ed., 1957, Treatise on marine ecology and paleoecology: Geol. Soc. Amer., Mem. 67, v. 1–2.

Heezen, B. C., 1960, The rift in the ocean floor: Sci. Amer., v. 203, p. 98–110.

Heezen, B. C., 1962, The deep-sea floor, *in* Runcorn, S. K., ed., Continental drift: New York, Academic Press, p. 235–288.

Heezen, B. C., 1968, The Atlantic continental margin: Univ. Missouri, Rolla, J., v. 1, p. 5–25.

Heezen, B. C., Bunce, E. T., Hersey, J. B., and Tharp, M., 1964a, Chain and Romanche fracture zones: Deep-Sea Res., v. 11, p. 11–33.

Heezen, B. C., and Drake, C. L., 1964, Gravity tectonics, turbidity currents and geosynclinal accumulations in continental margin of eastern North America, *in* Carey, S. W., ed., Syntaphral tectonics and diagenesis – a symposium: Hobart, Univ. of Tasmania, Geol. Dept., p. D1–D10.

Heezen, B. C., and Ewing, M., 1961, The mid-oceanic ridge and its extension through the Arctic basin, *in* Raasch, G. O., ed., Geology of the Arctic, v. 1: Toronto, Univ. of Toronto Press, p. 622–642.

Heezen, B. C., and Ewing, M., 1963, The mid-oceanic ridge, *in* Hill, M. N., ed., The sea, v. 3: New York, Wiley-Interscience, p. 388–410.

Heezen, B. C., Ewing, M., and Miller, E. T., 1953, Trans-Atlantic profile of total magnetic intensity and topography, Dakar to Barbados: Deep-Sea Res., v. 1, p. 25–33.

Heezen, B. C., Gerard, R. D., and Tharp, M., 1964b, The Vema fracture zone in the equatorial Atlantic: J. Geophys. Res., v. 69, p. 733–739.

Heezen, B. C., and Sheridan, R. E., 1966, Lower Cretaceous rocks (Neocomian-Albian) dredged from Blake escarpment: Science, v. 154, p. 1644–1647.

Heezen, B. C., and Tharp, M., 1964, Physiographic diagram of the Indian Ocean, with explanatory sheet: Geological Society of America.

Heezen, B. C., and Tharp, M., 1965, Tectonic fabric of Atlantic and Indian Oceans and continental drift, *in* Blackett, P. M. S., Bullard, E. C., and Runcorn, S. K., eds., Symposium on continental drift: Roy. Soc. London,

Phil. Trans., ser. A., v. 258, p. 90–108.

Heezen, B. C., Tharp, M., and Ewing, M., 1959, The floors of the oceans, 1, The North Atlantic: Geol. Soc. Amer., Spec. Pap. 65, 122 p.

Heim, A., 1921, Geologie der Schweiz, 2: Leipzig, Tauchnitz, 476 p.

Heim, A., 1922, Geologie der Schweiz, 3: Leipzig, Tauchnitz, 1641 p.

Heirtzler, J. R., 1961, Vema cruise no. 16 magnetic measurements: Columbia Univ., Lamont Geol. Observ., Tech Rep. 2, contract CU-3-61 Nonr Geology.

Heirtzler, J. R., 1968, Evidence for ocean floor spreading across the ocean basins, *in* Phinney, R. A., ed., The history of the earth's crust: Princeton, N. J., Princeton Univ. Press, p. 90–100.

Heirtzler, J. R., Dickson, G. O., Herron, E. M., Pitman, W. C., III, and Le Pichon, X., 1968, Marine magnetic anomalies, geomagnetic field reversals, and motions of the ocean floor and continents: J. Geophys. Res., v. 73, p. 2119–2136.

Heirtzler, J. R., and Hayes, D. E., 1967, Magnetic boundaries in the North Atlantic Ocean: Science, v. 157, p. 185–187.

Heirtzler, J. R., and Le Pichon, X., 1965, Crustal structure of the mid-ocean ridges, 3. Magnetic anomalies over the mid-Atlantic ridge: J. Geophys. Res., v. 70, p. 4013–4033.

Heirtzler, J. R., Le Pichon, X., and Baron, J. G., 1966, Magnetic anomalies over the Reykjanes ridge: Deep-Sea Res., v. 13, p. 427–443.

Helsley, C. E., 1969, Magnetic reversal stratigraphy of the Lower Triassic Moenkopi formation of western Colorado: Geol. Soc. Amer., Bull., v. 80, p. 2431–2450.

Helsley, C. E., and Steiner, M. B., 1969, Evidence for long intervals of normal polarity during the Cretaceous period: Earth Planet. Sci. Lett., v. 5, p. 325–332.

Herrin, E., 1966, Travel-time anomalies and structure of the upper mantle (abs.): Amer. Geophys. Union, Trans., v. 47, p. 44.

Herrin, E., and Taggart, J., 1966, Epicenter determinations for Longshot (abs.): Amer. Geophys. Union, Trans., v. 47, p. 164.

Herron, E. M., 1971, Crustal plates and sea floor in the southeastern Pacific: Amer. Geophys. Union, Antarctic Res. Ser., v. 15, p. 229–237.

Herron, E. M., 1972, Sea-floor spreading and the Cenozoic history of the east-central Pacific: Geol. Soc. Amer., Bull., v. 83, p. 1671–1692.

Herron, E. M., and Hayes, D. E., 1969, A geophysical study of the Chile ridge: Earth Planet. Sci. Lett., v. 6, p. 77–83.

Herron, E. M., and Heirtzler, J. R., 1967, Sea-floor spreading near the Galapagos: Science, v. 158, p. 775–780.

Herzen, R. von, 1959, Heat flow values from the southern Pacific: Nature, v. 183, p. 882–883.

Hess, H. H., 1938, Gravity anomalies and island arc structure with particular reference to the West Indies: Amer. Phil. Soc., Proc., v. 79, p. 71–96.

Hess, H. H., 1939, Island arcs, gravity anomalies and serpentinite inclusions: A contribution to the ophiolite problem: Int. Geol. Congr., 17th, Rep., v. 2, p. 263–283.

Hess, H. H., 1946, Drowned ancient islands of the Pacific basin: Amer. J. Sci., v. 244, p. 772–791.

Hess, H. H., 1948, Major structural features of the western North Pacific: an interpretation of H. O. 5485, Bathymetric Chart, Korea to New Guinea: Geol. Soc. Amer., Bull., v. 59, p. 417–446.

Hess, H. H., 1954, Serpentines, orogeny and epeirogeny, *in* Poldervaart, A., ed., Crust of the earth-a symposium:

Geol. Soc. Amer., Spec. Pap. 62, p. 391–408.

Hess, H. H., 1955, The oceanic crust: J. Mar. Res, v. 14, p. 423–439.

Hess, H. H., 1959a, The AMSOC hole to the earth's mantle: Amer. Geophys. Union, Trans., v. 40, p. 340–345.

Hess, H. H., 1959b, Nature of the great oceanic ridges, in Sears, M., ed., International Oceanographic Congress preprints: Washington, D. C., American Association for the Advancement of Science, p. 33–34.

Hess, H. H., 1962, History of ocean basins, in Engel, A. E. J., James, H. L., and Leonard, B. F., eds., Petrologic studies: a volume in honor of A. F. Buddington: Geological Society of America, p. 599–620.

Hess, H. H., 1965, Mid-oceanic ridges and tectonics of the sea floor, in Whittard, W. F., and Bradshaw, R., eds., Submarine geology and geophysics: London, Butterworth, p. 317–332.

Hess, H. H., 1968, Reply: J. Geophys. Res., v. 73, p. 6569.

Hide, R., 1967, Motions of the earth's core and mantle, and variations of the main geomagnetic field: Science, v. 157, p. 55–56.

Hide, R., and Malin, S. R. C., 1970, Novel correlations between global features of the earth's gravitational and magnetic fields: Nature, v. 225, p. 605–609.

Hide, R., and Roberts, P. H., 1961, The origin of the main geomagnetic field: Phys. Chem. Earth, v. 4, p. 27–98.

Hiessleitner, G., 1951–1952, Serpentin- und Chromerz-Geologie der Balkanhalbinsel und eines Teiles von Kleinasien: Austria, Geol. Bundesanst., Jahrb., Sonderb. 1, pt. 1–2, 668 p.

Higginbotham, S. W., ed., 1963, Man, science, learning and education: Houston, Rice Univ. Press, 254 p.

Hill, M. L., and Dibblee, T. W., Jr., 1953, San Andreas, Garlock, and Big Pine faults, California—A study of the character, history, and tectonic significance of their displacements: Geol. Soc. Amer., Bull., v. 64, p. 443–458.

Hill, M. L., and Hobson, H. D., 1968, Possible post-Cretaceous slip on the San Andreas fault zone, in Dickinson, W. R., and Grantz, A., eds., Proceedings of the conference on geologic problems of San Andreas fault system: Stanford Univ., Publ. Geol. Sci., v. 11, p. 123–129.

Hill, M. N., ed., 1963, The sea, v. 3: The earth beneath the sea—history: New York, Wiley-Interscience, 963 p.

Hirasawa, T., 1965, Source mechanism of the Niigata earthquake of June 16, 1964, as derived from body waves: J. Phys. Earth, v. 13, p. 35–66.

Hirasawa, T., 1966, A least-square method for the focal mechanism determinations from S wave data, 2: Tokyo, Univ., Earthquake Res. Inst., Bull., v. 44, p. 919–938.

Hoare, J. M., Condon, W. H., Cox, A., and Dalrymple, G. B., 1968, Geology, paleomagnetism, and potassium-argon ages of volcanic rocks from Nunivak Island, Alaska: Geol. Soc. Amer., Mem. 116, p. 377–413.

Hodgson, J. H., 1957a, Current status of fault-plane studies—a summing up: Canada, Dominion Observ., Publ., v. 20, p. 413–418.

Hodgson, J. H., 1957b, Nature of faulting in large earthquakes: Geol. Soc. Amer., Bull., v. 68, p. 611–644.

Hodgson, J. H., 1964, Earthquakes and earth structure: Englewood Cliffs, N. J., Prentice-Hall, 166 p.

Hodgson, J. H., and Adams, W. M., 1958, A study of inconsistent observations in the fault plane project: Seismol. Soc. Amer., Bull., v. 48, p. 17–31.

Hodgson, J. H., and Allen, J. F. J., 1954, Tables of extended distances for PKP and PcP: Canada, Dominion Observ., Publ., v. 16, p. 329–348.

Hodgson, J. H., and Cock, J. I., 1956, Direction of faulting in the deep focus Spanish earthquake of March 29, 1954: Tellus, v. 8, p. 321–328.

Hodgson, J. H., and Milne, W., 1951, Direction of faulting in certain earthquakes of the North Pacific: Seismol. Soc. Amer., Bull., v. 41, p. 221–242.

Hodgson, J. H., and Stevens, A. E., 1964, Seismicity and earthquake mechanism, in Odishaw, H., ed., Research in geophysics, v. 2: Cambridge, Mass., M. I. T. Press, p. 27–50.

Hodgson, J. H., and Storey, R. S., 1953, Tables extending Byerly's fault-plane technique to earthquakes of any focal depth: Seismol. Soc. Amer., Bull., v. 43, p. 49–61.

Hodgson, J. H., Storey, R. S., and Bremner, P. C., 1952, Progress report on fault plane project (abs.): Geol. Soc. Amer., Bull., v. 63, p. 1354.

Holmes, A., Radioactivity and earth movements: Geol. Soc. Glasgow, Trans., v. 18, p. 559–606.

Holmes, A., 1933, The thermal history of the earth: Wash. Acad. Sci., J., v. 23, p. 169–195.

Holmes, A., 1965, Principles of physical geology: New York, Ronald Press, 1288 p.

Holmes, M. L., Von Huene, R., and McManus, D. A., 1972, Seismic reflection evidence supporting underthrusting beneath the Aleutian arc near Amchitka Island: J. Geophys. Res., v. 77, p. 959–964.

Honda, H., 1962, Earthquake mechanism and seismic waves: J. Phys. Earth, v. 10, p. 1–97.

Honda, H., and Masatsuka, A., 1952, On the mechanism of the earthquakes and the stresses producing them in Japan and its vicinity: Tohoku Univ., Sci. Rep., ser. 5, Geophys., v. 4, p. 42–60.

Honda, H., Masatsuka, A., and Emura, K., 1957, On the mechanism of the earthquakes and the stresses producing them in Japan and its vicinity (second paper): Tohoku Univ., Sci. Rep., ser. 5, Geophys., v. 8, p. 186–205.

Hopkins, W., 1839, Researches in physical geology: Roy. Soc. London, Phil. Trans., v. 129, p. 381–385.

Hospers, J., 1951, Remanent magnetism of rocks and the history of the geomagnetic field: Nature, v. 168, p. 1111–1112.

Hospers, J., 1953, Reversals of the main geomagnetic field: Ned. Akad. Wetensch., Proc., ser. B, v. 56, p. 467–477.

Hospers, J., 1954, Magnetic correlation in volcanic districts: Geol. Mag., v. 91, p. 352–360.

Hospers, J., and Van Andel, S. I., 1967, Paleomagnetism and the hypothesis of an expanding earth: Tectonophysics, v. 5, p. 5–24.

Houtz, R., Ewing, J., and Le Pichon, X., 1968, Velocities of deep-sea sediments from sonobuoy data: J. Geophys. Res., v. 73, p. 2615–2641.

Hsü, K. J., 1965, Isostasy, crustal thinning, mantle changes, and the disappearance of ancient land masses: Amer. J. Sci., v. 263, p. 97–109.

Hsü, K. J., 1968, Principles of melanges and their bearing on the Franciscan-Knoxville paradox: Geol. Soc. Amer., Bull., v. 79, p. 1063–1074.

Hsü, K. J., and Schlanger, S. O., 1968, Thermal history of the upper mantle and its relation to crustal history in the Pacific basin: Int. Geol. Congr., 23rd, Rep., sect. 1, p. 91–105.

Hubbert, M. K., 1951, Mechanical basis for certain familiar geologic structures: Geol. Soc. Amer., Bull., v. 62, p. 355–372.

Hudson, J. D., 1966, Speculations on the depth relations of calcium carbonate solution in Recent and ancient seas: Mar. Geol., v. 5, p. 473–480.

Huffman, O. F., 1970, Miocene and post-Miocene offset on the San Andreas fault in central California (abs.): Geol. Soc. Amer., Abstr., v. 2, p. 104–105.

Hunt, C. B., Averitt, P., and Miller, R. L., 1953, Geology and geography of the Henry Mountains region: U. S. Geol. Surv., Prof. Pap. 228, 234 p.

Hunting Survey Corp., Ltd., 1960, Reconnaissance geology of part of West Pakistan: Toronto, Queen's Printer, 550, 172 p.

Hurley, P. M., ed., 1966, Advances in earth science: Cambridge, Mass., M. I. T. Press. 502 p.

Hurley, P. M., Hughes, H., Pison, W. H., Jr., and Fairbairn, H. W., 1962, Radiogenic argon and strontium diffusion parameters in biotite at low temperatures obtained from alpine fault uplift in New Zealand: Geochim. Cosmochim. Acta, v. 26, p. 67–80.

Hurley, P. M., and Rand, J. R., 1969, Pre-drift continental nucleii: Science, v. 164, p. 1229–1242.

Hurwitz, L., Knapp, D. G., Nelson, J. H., and Watson, D. E., 1966, Mathematical model of the geomagnetic field for 1965: J. Geophys. Res., v. 71, p. 2373–2383.

Husebye, E. S., and Toksöz, M. N., 1969, J. Geophys. Res. (in press) (not verified).

Hyndman, R. D., Lambert, I. B., Heier, K. S., Jaeger, J. C., and Ringwood, A. E., 1968, Heat flow and surface radioactivity measurements in the Precambrian shield of Western Australia: Phys. Earth Planet. Interiors, v. 1, p. 129–135.

Ibrahim, A., and Nuttli, O. W., 1957, Travel-time curves and upper-mantle structure from long period S waves: Seismol. Soc., Bull., v. 57, p. 1063–1092.

Ichikawa, K., Murakami, N., Hase, A., and Wadatsumi, K., 1968, Late Mesozoic igneous activity in the inner side of southwest Japan: Pac. Geol., v. 1, p. 97–118.

Ichikawa, M., 1961, On the mechanism of earthquakes in and near Japan during the period from 1950 to 1957: Geophys. Mag., v. 30, p. 355–403.

Ichikawa, M., 1966, Mechanism of earthquakes in and near Japan, 1950–1962: Pap. Meteorol. Geophys., v. 16, p. 201–229.

Iddings, J. P., 1892, The origin of igneous rocks: Phil. Soc. Wash., Bull., v. 12, p. 128–130.

Iida, K., 1963, Magnitude, energy, and generation mechanisms of tsunamis and a catalogue of earthquakes associated with tsunamis: Int. Union Geod. Geophys., Monogr., no. 24, p. 7–18.

Iijima, A., and Kagami, H., 1961, Cenozoic tectonic development of the continental slope, northeast of Japan: Geol. Soc. Japan, J., v. 67, p. 561–577.

Institute of Petroleum, 1965, Salt basins around Africa; proceedings of a joint meeting of the Institute of Petroleum and the Geological Society: London, Institute of Petroleum, 122 p.

International Commission on Radiation Units and Measurements, 1969, Radiation dosimetry: x rays and gamma rays with maximum photon energies between 0.6 and 50 MeV: Int. Comm. Radiation Units and Measurements, ICRU Rep., no. 14, 30 p.

Irvine, T. N., ed., 1966, The world rift system: Canada, Geol. Surv., Pap. 66–14, 471 p.

Irving, E., 1956, Palaeomagnetic and palaeoclimatological aspects of polar wandering: Pure Appl. Geophys., v. 33, p. 23–41.

Irving, E., 1959, Paleomagnetic pole positions: Roy. Astron. Soc., Geophys. J., v. 2, p. 51–77.

Irving, E., 1964, Paleomagnetism and its application to geological and geophysical problems: New York, John Wiley, 399 p.

Irving, E., 1966a, The great Paleozoic reversal of the geomagnetic field (abs.): Amer. Geophys. Union, Trans., v. 47, p. 78.

Irving, E., 1966b, Paleomagnetism of some carboniferous rocks from New South Wales and its relation to geological events: J. Geophys. Res., v. 71, p. 6025–6051.

Irving, E., 1971, Nomenclature in magnetic stratigraphy: Roy. Astron. Soc., Geophys. J., v. 24, p. 529–531.

Irving, E., Stott, P. M., and Ward, M. A., 1961, Demagnetization of igneous rocks by alternating magnetic fields: Phil. Mag., v. 6, p. 225–241.

Isacks, B., and Molnar, P., Mantle earthquake mechanisms and the sinking of the lithosphere: Nature, v. 223, p. 1121–1124.

Isacks, B., and Molnar, P., 1971, Distribution of stresses in the descending lithosphere from a global survey of focal-mechanism solutions of mantle earthquakes: Rev. Geophys. Space Phys., v. 9, p. 103–174.

Isacks, B., Oliver, J., and Sykes, L. R., 1968, Seismology and the new global tectonics: J. Geophys. Res., v. 73, p. 5855–5899.

Isacks, B., Sykes, L. R., and Oliver, J., 1969, Focal mechanisms of deep and shallow earthquakes in the Tonga-Kermadec region and the tectonics of island arcs: Geol. Soc. Amer., Bull., v. 80, p. 1443–1470.

Ishikawa, Y., and Syono, Y., 1963, Order-disorder transformation and reverse thermo-remanent magnetism in the $FeTiO_3$-Fe_2O_3 system: Phys. Chem. Solids, v. 24, p. 517–528.

Ito, H., 1970, Polarity transitions of the geomagnetic field deduced from the natural remanent magnetization of Tertiary and Quaternary rocks in southwest Japan: J. Geomag. Geoelec., v. 22, p. 273–290.

Ito, K., and Kennedy, G. C., 1967, Melting and phase relations in a natural peridotite to 40 kilobars: Amer. J. Sci., v. 265, p. 519–538.

Jackson, E. D., Silver, E. A., and Dalrymple, G. B., 1972, Hawaiian-Emperor chain and its relation to Cenozoic circumpacific tectonics: Geol. Soc. Amer., Bull., v. 83, p. 601–618.

Jacobs, J. A., 1963, The earth's core and geomagnetism: New York, Pergamon Press, 137 p.

Jacoby, W. R., 1970, Instability in the upper mantle and global plate movements: J. Geophys. Res., v. 75, p. 5671–5680.

Jaggar, T. A., Jr., 1901, The laccoliths of the Black Hills: U. S. Geol. Surv., Ann. Rep., 21st, pt. 3, p. 163–290.

James, D. E., 1971, Plate tectonic model for the evolution of the central Andes: Geol. Soc. Amer., Bull., v. 82, p. 3325–3346.

Japan, Geological Survey, 1968, Geological map of Japan, scale 1:2,000,000: Tokyo, Japan Geological Survey.

Jeffreys, H., 1929, The earth, ed. 2: Cambridge, Cambridge Univ. Press, 346 p.

Jeffreys, H., 1963, On the hydrostatic theory of the figure of the earth: Roy. Astron. Soc., Geophys. J., v. 8, p. 196–202.

Johnson, A. H., Nairn, A. E. M., and Peterson, D. N., 1972, Mesocoic reversal stratigraphy: Nature, v. 237, p. 9–10.

Johnson, G. L., 1967, North Atlantic fracture zone near 53° N.: Earth Planet. Sci. Lett., v. 2, p. 445–448.

Johnson, G. L., and Lowrie, A., 1972, Cocos and Carnegie ridges result of the Galapagos "hot spot"?: Earth Planet. Sci. Lett., v. 14, p. 279–280.

Johnson, H., Smith, B. L., eds., 1970, The megatectonics of continents and oceans: New Brunswick, N. J., Rutgers Univ. Press, 282 p.

Johnson, L. R., 1967, Array measurements of P velocities in the upper mantle: J. Geophys. Res., v. 72, p. 6309–6325.

Jolley, L. B. W., 1961, Summation of series: New York, Dover, 251 p.

Jones, J. G., 1971, Aleutian enigma: a clue to transformation in time: Nature, v. 229, p. 400–403.

Judd, W. J., 1886, On the gabbros, dolerites and basalts of Tertiary age in Scotland and Ireland: Geol. Soc. London, Quart. J., v. 42, p. 49–97.

Judd, W. R., 1964, State of stress in the earth's crust: New York, American Elsevier, 732 p.

Kaminski, W., and Menzel, H., 1968, Zur Deutung der Schwereanomalie des Ivrea-Körpers: Schweiz. Mineral. Petrog. Mitt., v. 48, p. 255–260.

Kanamori, H., 1970, Velocity and Q of mantle waves: Phys. Earth Planet. Interiors, v. 2, p. 259–275.

Kanamori, H., 1971a, Great earthquakes at island arcs and the lithosphere: Tectonophysics, v. 12, p. 187–198.

Kanamori, H., 1971b, Seismological evidence for a lithospheric normal faulting: the Sanriku earthquake of 1933: Phys. Earth Planet. Interiors, v. 4, p. 289–300.

Kanamori, H., and Abe, K., 1968, Deep structure of island arcs as revealed by surface waves: Tokyo, Univ., Earthquake Res. Inst., Bull., v. 46, p. 1001–1025.

Kanamori, H., and Press, F., 1970, How thick is the lithosphere?: Nature, v. 226, p. 330–331.

Karig, D. E., 1969, Extensional zones in island arc systems (abs.): Amer. Geophys. Union, Trans., v. 50, p. 182.

Karig, D. E., 1970, Ridges and trenches of the Tonga-Kermadec island arc system: J. Geophys. Res., v. 75, p. 239–254.

Karig, D. E., 1971, Origin and development of marginal basins in the western Pacific: J. Geophys. Res., v. 76, p. 2542–2561.

Karlstrom, T. N. V., 1964, Quaternary geology of the Kenai lowland and glacial history of the Cook Inlet region, Alaska: U. S. Geol. Surv., Prof. Pap. 443, 69 p.

Kasahara, K., 1957, The nature of seismic origins as inferred from seismological and geodetic observations, 1: Tokyo, Univ., Earthquake Res. Inst., Bull., v. 35, p. 473–532.

Kasameyer, P. W., Von Herzen, R. P., and Simmons, G., 1972, Layers of high thermal conductivity in the North Atlantic: J. Geophys. Res., v. 77, p. 3162–3167.

Katsumata, M., 1967, Seismic activities in and near Japan, 3: Seismic activities versus depth (in Japanese with English summary): Seismol. Soc. Japan, J., v. 20, p. 75–84.

Katsumata, M., and Sykes, L. R., 1969, Seismicity and tectonics of the western Pacific; Izu-Mariana-Caroline and Ryukyu-Taiwan regions: J. Geophys. Res., v. 74, p. 5923–5948.

Kaula, W. M., 1966, Test and combination of satellite determinations of the gravity field with gravimetry: J. Geophys. Res., v. 71, p. 5303–5314.

Kaula, W. M., 1967, Geophysical implications of satellite determinations of the earth's gravitational field: Space Sci. Rev., v. 7, p. 769–794.

Kaula, W. M., 1968, An introduction to planetary physics: the terrestrial planets: New York, John Wiley, 490 p.

Kaula, W. M., 1969, A tectonic classification of the main features of the earth's gravitational field: J. Geophys. Res., v. 74, p. 4807–4826.

Kaula, W. M., 1970, Earth's gravity field: Relation to global tectonics: Science, v. 169, p. 982–984.

Kaula, W. M., 1972, Global gravity and tectonics, in Robertson, E. C., ed., The nature of the solid earth: New York, McGraw-Hill, p. 385–405.

Kawai, N., Hirooka, K., and Nakajima, T., 1969, Paleomagnetic and potassium-argon age informations supporting Cretaceous-Tertiary hypothetic bend of the main island Japan: Palaeogeogr. Palaeoclimatol. Palaeoecol., v. 6, p. 277–282.

Kawai, N., Ito, H., and Kume, S., 1961, Deformation of the Japanese Islands as inferred from rock-magnetism: Roy. Astron. Soc., Geophys. J., v. 6, p. 124–130.

Kawai, N., Nakajima, T., and Hirooka, K., 1971, The evolution of the island arc of Japan and the formation of granites in the circum-Pacific belt: J. Geomag. Geoelec., v. 23, p. 267–294.

Kawano, Y., and Ueda, Y., 1967, Periods of the igneous activities of the granitic rocks in Japan by K-Ar dating method: Tectonophysics, v. 4, p. 523–530.

Kay, M., 1951, North American geosynclines: Geol. Soc. Amer., Mem. 48, 143 p.

Kay, R., Hubbard, N. J., and Gast, P. W., 1970, Chemical characteristics and origin of oceanic ridge volcanic rocks: J. Geophys. Res., v. 75, p. 1585–1613.

Kaye, C. A., 1964, The upper limit of barnacles as an index of sea-level change on the New England coast during the past 100 years: J. Geol., v. 72, p. 580–600.

Keen, M. J., 1963, Magnetic anomalies over the mid-Atlantic ridge: Nature, v. 197, p. 888–890.

Kelly, T. E., 1963, Geology and hydrocarbons in Cook Inlet basin, Alaska, in Childs, O. E., and Beebe, B. W., eds., Backbone of the Americas; tectonic history from pole to pole: Tulsa, American Association of Petroleum Geologists, p. 278–296. (Amer. Ass. Petrol. Geol., Mem. 2.)

Kennedy, G. C., 1959, The origin of continents, mountain ranges and ocean basins: Amer. Scientist, v. 47, p. 491–504.

Kennet, J. P., and Watkins, N. D., 1970, Geomagnetic polarity change, volcanic maxima and faunal extinction in the southern ocean: Nature, v. 227, p. 930–934.

Kennet, J. P., Watkins, N. D., and Vella, P., 1971, Paleomagnetic chronology of Pliocene-Early Pleistocene climates and the Plio-Pleistocene boundary in New Zealand: Science, v. 171, p. 276–279.

Khattri, K. N., 1969, Focal mechanism of the Brazil deep focus earthquake of Nov. 3, 1965, from the amplitude spectra of isolated P waves: Seismol. Soc. Amer., Bull., v. 59, p. 691–704.

Khramov, A. N., 1957, Paleomagnetism—the basis of a new method of correlation and subdivision of sedimentary strata: Acad. Sci. USSR, Dokl., Earth Sci. Sect., v. 112, p. 129–132.

Khramov, A. N., 1958, Paleomagnetism and stratigraphic correlation: Leningrad, Gostoptechizdat, 204 p. (English transl., A. J. Lojkine, Australian National University, Canberra, 1960) (not verified).

Khramov, A. N., Rodionov, V. P., and Komissarova, R. A., 1965, Novyye dannyye o paleozoyskoy istorii zamnogo magnitnogo polya na territorii SSR (New data on the Palezoic history of the geomagnetic field in the territory of the USSR) in Nastoyashchee i proshloye magnitnogo polya zemli: Moscow, Akademiia Nauk SSR, Institut Fiziki Zemli, p. 206–213 (English transl., E. R. Hope, Directorate Scientific Information Services, Canada).

Kimura, T., 1967, Geologic structure of the Shimanto Group in the southern Oigawa district, central Japan, analyzed by the minor geologic structures, in Commem. Publ. of Prof. Y. Sasa, p. 21–38 (not verified).

King, B. C., 1966, Volcanism in eastern Africa and its structural setting: Geol. Soc. London, Proc., no. 1629, p. 16–19.

King, E. R., Zietz, I., and Alldredge, L. R., 1966, Magnetic data on the structure of the central Arctic region: Geol. Soc. Amer., Bull, v. 77, p. 619–646.

King, L. C., 1962, The morphology of the earth: New York, Hafner, 577 p.

King, P. B., 1959, The evolution of North America: Princeton, N. J., Princeton Univ. Press, 189 p.

King, P. B., 1969, Tectonic map of North America: Washington, D. C., U. S. Geological Survey.

Klein, C. A., 1962, Electrical properties of pyrolytic graphites: Rev. Mod. Phys., v. 34, p. 56–79.

Knopoff, L., 1964, The convection current hypothesis: Rev. Geophys., v. 2, p. 89–122.

Knopoff, L., Drake, C. L., and Hart, P. J., eds., 1968, The crust and upper mantle of the Pacific area: Amer. Geophys. Union, Geophys. Monogr. 12, 522 p.

Kobayashi, T., 1941, The Sakawa orogenic cycle and its bearing on the origin of the Japanese islands: Tokyo, Univ., Fac. Sci., J., sect. 2, v. 5, p. 219–578.

Kobayashi, T., 1956, The insular arc of Japan, its hinter basin and its linking with the peri-Tunghai arc: Pacific Sci. Congr. Pacific Sci. Ass., 8th, 1953, Proc., v. 2A, p. 799–808.

Koning, L. P. G., 1942, On the determination of the fault planes in the hypocenter of the deep focus earthquakes of June 29, 1934 in the Netherlands East Indies: Ned. Akad. Wetensch., Proc., v. 45, p. 636–642.

Kono, M., 1971, Intensity of the earth's magnetic field during the Pliocene and Pleistocene in relation to the amplitude of mid-ocean ridge magnetic anomalies: Earth Planet. Sci. Lett., v. 11, p. 10–17.

Korea, Geological Survey, 1956, Geological map of Korea, 1:1,000,000: Seoul, Geological Survey of Korea and Geological Society of Korea.

Kraus, E., 1951, Die Baugeschichte der Alpen: Berlin, Akademie-Verlag, 2 v.

Kraus, E., 1960, Beobachtungen und Gedanken zur Geologie von Sardinien und Korsika: Deut. Geol. Gesell., Z., v. 112, p. 75–80.

Kraus, E., 1962, Le problème de l'espace en tectonique dans la region méditerranéenne: Livre à la mémoire du Prof. Paul Fallot: Soc. Géol. France, Bull., v. 1, p. 117–121.

Krause, D. C., 1964, Guinea fracture zone in the equatorial Atlantic: Science, v. 146, p. 57–59.

Krause, D. C., and Watkins, N. D., 1970, North Atlantic crustal genesis in the vicinity of the Azores: Roy. Astron. Soc., Geophys. J., v. 19, p. 261–283.

Krishnan, M. S., 1956, Geology of India and Burma: Madras, Higginbothams, 555 p.

Kuenen, P. H., 1946, Rate and mass of deep-sea sedimentation: Amer. J. Sci., v. 244, p. 563–572.

Kuhn, T. S., 1962, The structure of scientific revolutions: Chicago, Univ. of Chicago Press, 172 p.

Kuiper, G., 1954, On the origin of the lunar surface features: Nat. Acad. Sci., Proc., v. 40, p. 1096–1112.

Kuno, H., 1952, Cenozoic volcanic activity in Japan and surrounding areas: N. Y. Acad. Sci., Trans., ser. 2, v. 14, p. 225–231.

Kuno, H., 1959, Origin of Cenozoic petrographic provinces of Japan and surrounding areas: Bull. Volcanol., v. 20, p. 37–76.

Kuno, H., 1960, High-alumina basalt: J. Petrology, v. 1, p. 121–145.

Kuno, H., 1962, Frequency distribution of rock types in oceanic, orogenic and kratogenic volcanic associations, in Macdonald, G. A., and Kuno, H., eds., The crust of the Pacific basin: Amer. Geophys. Union, Geophys. Monogr. 6, p. 135–139.

Kuno, H., 1966, Lateral variation of basalt magma across continental margins and island arcs, in Poole, W. H., ed., Continental margins and island arcs: Canada, Geol Surv., Pap. 66–15, p. 317–336.

Kuno, H., 1967, Volcanological and petrological evidence regarding the nature of the upper mantle, in Gaskell, T. F., ed., The earth's mantle: New York, Academic Press, p. 89–110.

Kuno, H., 1968, Origin of andesite and its bearing on the island arc structure: Bull. Volcanol., v. 32, p. 141–176.

Kushiro, I., 1965, The system forsterite-nepheline-silica: Carnegie Inst. Wash., Yearbook, 1964–1965, v. 64, p. 106–109.

Kushiro, I., 1969, Effect of water on the compositions of magmas formed in the upper mantle (abs.): Amer. Geophys. Union, Trans., v. 50, p. 355.

Kushiro, I., and Kuno, H., 1963, Origin of primary basalt magmas and classification of basaltic rocks: J. Petrology, v. 4, p. 75–89.

Kushiro, I., Syono, Y., and Akimoto, S., 1968, Melting of a peridotite nodule at high pressures, and high water pressures: J. Geophys. Res., v. 73, p. 6023–6030.

Lachenbruch, A. H., 1968, Preliminary geothermal model of the Sierra Nevada: J. Geophys. Res., v. 73, p. 6977–6989.

Lacroix, A., 1924, Les caracteristiques lithologiques des Petites Antilles: Soc. Geol. Belg., Livre Jubilaire, v. 1, p. 385–405.

Lacroix, A., 1936, Le volcan actif de l'ile de la Réunion et ses produits: Paris, Gauthier Villars, 297 p.

Ladd, H. S., and Schlanger, S. O., 1960, Drilling operations on Eniwetok Atoll: U. S. Geol. Surv., Prof. Pap. 260-Y, p. 863–903.

Lamb, J. L., 1969, Planktonic forminiferal datums and late Neogene epoch boundaries in the Mediterranean, Caribbean, and Gulf of Mexico: Gulf Coast Ass. Geol. Soc., Trans., v. 19, p. 559–578.

Langseth, M. G., Le Pichon, X., and Ewing, M., 1966, Crustal structure of the mid-ocean ridges, 5, Heat flow through the Atlantic Ocean floor and convection currents: J. Geophys. Res., v. 71, p. 5321–5355.

Langseth, M. G., and Von Herzen, R. P., 1970, Heat flow through the floor of the world oceans, in Maxwell, A. E., ed., The sea, v. 4, pt. 1: New York, Wiley-Interscience, p. 299–352.

Larson, R. L., 1970, Near-bottom studies of the east Pacific rise crest and tectonics of the mouth of the Gulf of California: Ph.D thesis, University of California at San Diego, 164 p.

Larson, R. L., and Chase, C. G., 1970, Relative velocities of the Pacific, North America and Cocos plates in the middle America region: Earth Planet. Sci. Lett., v. 7, p. 425–428.

Larson, R. L., Menard, H. W., and Smith, S. M., 1968, Gulf of California, a result of ocean-floor spreading and transform faulting: Science, v. 161, p. 781–784.

Larson, R. L., and Spiess, F. N., 1969, East Pacific rise crest: a near-bottom geophysical profile: Science, v. 163, p. 68–71.

Laughlin, J. S., and Genna, S., 1966, Calorimetry, in, Attix, F. H., and Roesch, W. C., eds., Radiation dosimetry, v. 2: New York, Academic Press, p. 389–441.

Laughton, A. S., 1966, The Gulf of Aden: Roy. Soc. London, Phil. Trans., ser. A, v. 259, p. 150–171.

Laughton, A. S., 1971, South Labrador Sea and the evolution of the North Atlantic: Nature, v. 232, p. 612–617.

Lawley, E. A., 1970, The intensity of the geomagnetic field in Iceland during Neogene polarity transitions and systematic deviations: Earth Planet. Sci. Lett., v. 10, p. 145–149.

Lazareva, A. P., and Misharina, L. A., 1965, Stresses in earthquake foci in the arctic seismic belt: Acad. Sci. USSR, Bull., Geophys. Ser., no. 2, p. 84–87.

Leaton, B. R., and Malin, S. R. C., 1967, Recent changes in the magnetic dipole moment of the earth: Nature, v. 213, p. 1110.

Lee, W. H. K., ed., 1965, Terrestrial heat flow: Amer. Geophys. Union, Geophys. Monogr. 8, 276 p.

Lee, W. H. K., and Uyeda, S., 1965, Review of heat flow data, in Lee, W. H. K., ed., Terrestrial heat flow: Amer. Geophys. Union, Geophys. Monogr. 8, p. 87–190.

Lehmann, I., 1964a, On the travel times of P as determined from nuclear explosions: Seismol. Soc. Amer., Bull., v. 54, p. 123–139.

Lehmann, I., 1964b, On the velocity of P in the upper mantle: Seismol. Soc. Amer., Bull., v. 54, p. 1097–1103.

Lensen, G. J., 1960, Principal horizontal stress directions as an aid to the study of crustal deformation: Canada, Dominion Observ., Publ., v. 24, p. 389–397.

Le Pichon, X., 1968, Sea floor spreading and continental drift: J. Geophys. Res., v. 73, p. 3661–3697.

Le Pichon, X., 1969, Models and structure of the oceanic crust: Tectonophysics, v. 7, p. 385–401.

Le Pichon, X., Eittreim, S. L., and Ludwig, W. J., 1971a, Sediment transport and distribution in Argentine basin; 1, Antarctic bottom current passage through the Falkland fracture zone: Phys. Chem. Earth, v. 8, p. 1–28.

Le Pichon, X., and Fox, P. J., 1971, Marginal offsets, fracture zones, and the early opening of the North Atlantic: J. Geophys. Res., v. 76, p. 6294–6308.

Le Pichon, X., and Hayes, D. E., 1971, Marginal offsets, fracture zones, and the early opening of the South Atlantic: J. Geophys. Res., v. 76, p. 6283–6293.

Le Pichon, X., and Heirtzler, J. R., 1968, Magnetic anomalies in the Indian Ocean and sea floor spreading: J. Geophys. Res., v. 73, p. 2101–2117.

Le Pichon, X., Houtz, R. E., Drake, C. L., and Nafe, J. E., 1965, Crustal structure of the mid-ocean ridges: 1. Seismic refraction measurements: J. Geophys. Res., v. 70, p. 319–339.

Le Pichon, X., Hyndman, R. D., and Pautot, G., 1971b, Geophysical study of the opening of the Labrador Sea: J. Geophys. Res., v. 76, p. 4724–4743.

Le Pichon, X., and Langseth, M. G., 1969, Heat flow from the mid-ocean ridges and sea-floor spreading: Tectonophysics, v. 8, p. 319–344.

Le Pichon, X., Saito, T., and Ewing, J., 1968, Mesozoic and Cenozoic sediments from Rio Grande rise (abs.): Geol. Soc. Amer., Spec. Pap. 101, p. 121.

Lipman, P. W., Prostka, H. J., and Christiansen, R. L., 1970, Cenozoic volcanism and tectonism in the western United States and adjacent parts of the spreading ocean floor. Part I: Early and Middle Tertiary (abs.): Geol. Soc. Amer., Abstr., v. 2, p. 112–113.

Lipman, P. W., Prostka, H. J., and Christiansen, R. L., 1972, Cenozoic volcanism and plate tectonic evolution of the western United States, I. Early and Middle Cenozoic: Roy. Soc. London, Phil. Trans., ser. A., v. 271, p. 217–248.

Lipson, J., 1958, Potassium-argon dating of sedimentary rocks: Geol. Soc. Amer., Bull., v. 69, p. 137–150.

Lister, C. R. B., 1972, On the thermal balance of a mid-ocean ridge: Roy. Astron. Soc., Geophys. J., v. 26, p. 515–535.

Lliboutry, L., 1969, Sea-floor spreading, continental drift, and lithosphere sinking with an asthenosphere at melting point: J. Geophys. Res., v. 74, p. 6525–6540.

Loewinson-Lessing, F., 1923, Note sur les provinces petrographiques de la Russie: Soc. Géol. France, Bull., v. 23, p. 142–157.

Loncarevic, B. D., Mason, C. S., and Matthews, D. H., 1966, Mid-Atlantic ridge near 45° north, I. The median valley: Can. J. Earth Sci., v. 3, p. 327–349.

Loupekine, I. S., 1966, Uganda, the Toro earthquake of 20 March 1966: Paris, UNESCO, 38 p.

Lowenstam, H. A., 1964, Palaeotemperatures of the Permian and Cretaceous periods, in Nairn, A. E. M., ed., Problems in palaeoclimatology: London, Interscience, p. 227–252.

Ludwig, W. J., Ewing, J., Ewing, M., Murauchi, S., Den, N., Asano, S., Hotta, H., Hayakawa, M., Asanuma, T., Ichikawa, K., and Noguchi, I., 1966, Sediments and structure of the Japan trench: J. Geophys. Res., v. 71, p. 2121–2135.

Ludwig, W. J., Hayes, D. E., and Ewing, J., 1967, The Manila trench and West Luzon trough, 1. Bathymetry and sediment distribution: Deep-Sea Res., v. 14, p. 533–544.

Luyendyk, B. P., 1969, Origin of short-wavelength magnetic lineations observed near the ocean bottom: J. Geophys. Res., v. 74, p. 4869–4881.

Luyendyk, B. P., and Fisher, D. E., 1969, Fission track age of magnetic anomaly 10: A new point on the sea-floor spreading curve: Science, v. 164, p. 1516–1517.

McBirney, A. R., ed., 1969, Proceedings of Andesite Conference: Oreg., Dept. Geol. Mineral Ind., Bull., no. 65, 193 p.

McBirney, A. R., and Gass, I. G., 1967, Relations of oceanic volcanic rocks to mid-oceanic rises and heat flow: Earth Planet. Sci. Lett., v. 2, p. 265–276.

McCall, G. J. H., 1957, The Menengai caldera, Kenya Colony: Int. Geol. Congr., 20th, Contrib., v. 1, p. 55–69.

McCall, G. J. H., 1964, Kilombe caldera, Kenya: Geol. Ass. (London), Proc., v. 75, p. 563–572.

McConnell, R. B., 1970, The evolution of the rift system of eastern Africa: Int. Upper Mantle Comm., Sci. Rep., no. 27, p. 285–290.

McConnell, R. K., 1965, Isostatic adjustment in a layered earth: J. Geophys. Res., v. 70, p. 5171–5188.

McConnell, R. K., 1968, Viscosity of the mantle from relaxation time spectra of isostatic adjustment: J. Geophys. Res., v. 73, p. 7089–7105.

Macdonald, G. A., Davis, D. A., and Cox, D. C., 1960, Geology and ground-water resources of the island of Kauai, Hawaii: Hawaii, Div. Hydrography, Bull. 13, 212 p.

Macdonald, G. A., and Kuno, H., eds., 1962, The crust of the Pacific basin: Amer. Geophys. Union, Geophys. Monogr. 6, 195 p.

MacDonald, G. J. F., 1959, Calculations on the thermal history of the earth: J. Geophys. Res., v. 64, p. 1967–2000.

MacDonald, G. J. F., 1963, The deep structure of continents: Rev. Geophys., v. 1, p. 587–665.

MacDonald, G. J. F., 1964, The deep structure of continents: Science, v. 143, p. 921–929.

MacDonald, G. J. F., 1966, Mantle properties and continental drift, in Garland, G. D., ed., Continental drift: Toronto, Univ. of Toronto Press, p. 18–27.

McDonald, K. L., and Gunst, R. H., 1967, An analysis of the earth's magnetic field from 1835 to 1965: ESSA Tech. Rep. IER 46-IES 1, 86 p.

MacDougall, D., 1971, Deep-sea drilling: Age and composition of an Atlantic basaltic intrusion: Science, v. 171, p. 1244–1245.

McDougall, I., 1963, Potassium-argon ages from western Oahu, Hawaii: Nature, v. 197, p. 344–345.

McDougall, I., 1964, Potassium-argon ages from lavas of the Hawaiian Islands: Geol. Soc. Amer., Bull., v. 75, p. 107–128.

McDougall, I., 1966, Precision methods of potassium-argon isotopic age determination on young rocks, in Runcorn, S. K., ed., Methods and techniques in geophysics, v. 2: London, Interscience, p. 279–304.

McDougall, I., 1971, Volcanic island chains and sea floor spreading: Nature, Phys. Sci., v. 231, p. 141–144.

McDougall, I., Allsopp, H. L., and Chamalaun, F. H., 1966, Isotopic dating of the newer volcanics of Victoria, Australia, and geomagnetic polarity epochs: J. Geophys. Res., v. 71, p. 6107–6118.

McDougall, I., and Chamalaun, F. H., 1966, Geomagnetic polarity scale of time: Nature: v. 212, p. 1415–1418.

McDougall, I., and Chamalaun, F. H., 1969, Isotopic dating and geomagnetic polarity studies on volcanic rocks from Mauritius, Indian Ocean: Geol. Soc. Amer., Bull., v. 80, p. 1419–1442.

McDougall, I., and Compston, W., 1965, Strontium isotope composition and potassium-rubidium ratios in some rocks from Réunion and Rodriguez, Indian Ocean: Nature, v. 207, p. 252–253.

McDougall, I., and Tarling, D. H., 1963, Dating of polarity zones in the Hawaiian Islands: Nature, v. 200, p. 54–56.

McDougall, I., and Tarling, D. H., 1964, Dating geomagnetic polarity zones: Nature, v. 202, p. 171–172.

McDougall, I., and Wensink, H., 1966, Paleomagnetism and geochronology of the Pliocene-Pleistocene lavas in Iceland: Earth Planet. Sci. Lett., v. 1, p. 232–236.

McElhinny, M. W., 1967, The paleomagnetism of the southern continents – A survey and analysis: UNESCO-IUGS Symposium on Continental Drift, Montevideo, 1967. (Available on microfilm from American Geophysical Union.)

McElhinny, M. W., 1971, Geomagnetic reversals during the Phanerozoic: Science, v. 172, p. 157–159.

McElhinny, M. W., 1971, and Burek, P. J., 1971, Mesozoic palaeomagnetic stratigraphy: Nature, v. 232, p. 98–102.

McEvilly, T. V., 1966, Parkfield earthquakes of June 27–29, 1966, Monterey and San Luis Obispo counties, California – Preliminary report: Preliminary seismic data, June–July 1966: Seismol. Soc. Amer., Bull., v. 56, p. 967–971.

McEvilly, T. V., Bakun, W. H., and Casaday, K. B., 1967, The Parkfield, California, earthquakes of 1966: Seismol. Soc. Amer., Bull., v. 57, p. 1221–1244.

MacGregor, A. G., 1938, The volcanic history and petrology of Montserrat, with observations on Mt. Pelé, in Martinique: Roy. Soc. London, Phil. Trans., ser. B, v. 229, p. 1–90.

MacGregor, I. D., 1968, Mafic and ultramafic inclusions as indicators of the depth of origin of basaltic magmas: J. Geophys. Res., v. 73, p. 3737–3745.

McKee, E. H., Noble, D. C., and Silberman, M. L., 1970, Mid-Miocene hiatus in volcanic activity in the Great Basin area of the western United States: Earth Planet. Sci. Lett., v. 8, p. 93–96.

McKenzie, D. P., 1966, The viscosity of the lower mantle: J. Geophys. Res., v. 71, p. 3995–4010.

McKenzie, D. P., 1967, Some remarks on heat flow and gravity anomalies: J. Geophys. Res., v. 72, p. 6261–6273.

McKenzie, D. P., 1968a, The geophysical importance of high-temperature creep, in Phinney, R. A., ed., The history of the earth's crust: Princeton, N. J., Princeton Univ. Press, p. 28–44.

McKenzie, D. P., 1968b, The influence of the boundary conditions and rotation on convection in the earth's mantle: Roy. Astron. Soc., Geophys. J., v. 15, p. 457–500.

McKenzie, D. P., 1969a, The relation between fault plane solutions for earthquakes and the directions of the principle stresses: Seismol. Soc. Amer., Bull., v. 59, p. 591–601.

McKenzie, D. P., 1969b, Speculations on the consequences and causes of plate motions: Roy. Astron. Soc., Geophys. J., v. 18, p. 1–32.

McKenzie, D. P., 1970, Plate tectonics of the Mediterranean region: Nature, v. 226, p. 239–243.

McKenzie, D. P., and Julian, B., 1971, Puget Sound, Washington, earthquake and the mantle structure beneath the northwestern United States: Geol. Soc. Amer., Bull., v. 82, p. 3519–3524.

McKenzie, D. P., and Morgan, W. J., 1969, The evolution of triple junctions: Nature, v. 224, p. 125–133.

McKenzie, D. P., and Parker, D. L., 1967, The North Pacific: an example of tectonics on a sphere: Nature, v. 216, p. 1276–1280.

McKenzie, D. P., and Sclater, J. G., 1968, Heat flow inside the island arcs of the northwestern Pacific: J. Geophys. Res., v. 73, p. 3173–3179.

McKenzie, D. P., and Sclater, J. G., 1969, Heat flow in the eastern Pacific and sea floor spreading: Bull. Volcanol.,

v. 33, p. 101–118.

McMahon, B. E., and Strangway, D. W., 1967, Kiaman magnetic interval in the western United States: Science, v. 155, p. 1012–1013.

McManus, D. A., 1965, Blanco fracture zone, northeast Pacific Ocean: Mar. Geol., v. 3, p. 429–455.

McManus, D. A., 1967, Physiography of Cobb and Gorda rises, northeast Pacific Ocean: Geol. Soc. Amer., Bull., v. 78, p. 527–546.

McManus, D. A., and Burns, R. E., 1969, Scientific report on Deep-Sea Drilling Project, Leg V: Ocean Ind., v. 4, p. 40–42.

Malfait, B. T., and Dinkelman, M. G., 1972, Circum-Caribbean tectonic and igneous activity and the evolution of the Caribbean plate: Geol. Soc. Amer., Bull., v. 83, p. 251–272.

Malkus, W. V. R., 1968, Precession of the earth as the cause of geomagnetism: Science, v. 160, p. 259–264.

Malloy, R. J., 1964, Crustal uplift southwest of Montague Island, Alaska: Science, v. 146, p. 1048–1049.

Mark, H., and Fernbach, S., eds., 1969, Properties of matter under unusual conditions; in honor of Edward Teller's 60th birthday: New York, John Wiley, 389 p.

Markhinin, E. K., 1968, Volcanism as an agent of formation of the earth's crust, in Knopoff, L., Drake, C. L., and Pembroke, J. H., eds., The crust and upper mantle of the Pacific area: Amer. Geophys. Union, Geophys. Monogr. 12, p. 413–422.

Marshall, M., and Cox, A., 1971, Magnetism of pillow basalts and their petrology: Geol. Soc. Amer., Bull., v. 82, p. 537–552.

Mason, R. G., 1958, A magnetic survey off the west coast of the United States between latitudes 32° and 36° N, longitudes 121° and 128° W: Roy. Astron. Soc., Geophys. J., v. 1, p. 320–329.

Mason, R. G., and Raff, A. D., 1961, A magnetic survey off the west coast of North America 32° N to 42° N: Geol. Soc. Amer., Bull., v. 72, p. 1259–1265.

Massey, F. J., 1951, The Kolmogorov-Smirnov test for goodness of fit: Amer. Stat. Ass., J., v. 46, p. 68–78.

Mathews, J. H., and Gardner, W. K., 1963, Field reversals of "paleomagnetic" type in coupled disk dynamos: U. S. Naval Res. Lab., Rep. 5886, p. 1–11.

Matsuda, T., 1962, Crustal deformation and igneous activity in the south Fossa Magna, Japan, in Macdonald, G. A., and Kuno, H., eds., The crust of the Pacific basin: Amer. Geophys. Union, Geophys. Monogr. 6, p. 140–150.

Matsuda, T., 1964, Island arc features and the Japanese islands (in Japanese with English abstract): J. Geogr. (Tokyo), v. 73, p. 271–280.

Matsuda, T., and Kuriyagawa, S., 1965, Lower grade metamorphism in the eastern Akaishi Mtns. central Japan: Tokyo, Univ., Earthquake Res. Inst., Bull., v. 43, p. 209–235.

Matsuda, T., Nakamura, K., and Sugimura, A., 1967, Late Cenozoic orogeny in Japan: Tectonophysics, v. 4, p. 349–366.

Matsuda, T., and Uyeda, S., 1971, On the Pacific-type orogeny and its model – extension of the paired belts concept and possible origin of marginal seas: Tectonophysics, v. 11, p. 5–27.

Matsumoto, T., 1967, Fundamental problems in the circum-Pacific orogenesis: Tectonophysics, v. 4, p. 595–613.

Matsumoto, T., 1968, A hypothesis on the origin of the Late Mesozoic volcano-plutonic association in East Asia: Pac. Geol., v. 1, p. 77–83.

Matthews, D. H., 1963, A major fault scarp under the Arabian Sea displacing the Carlsberg ridge near Socotra: Nature, v. 198, p. 950–952.

Matthews, D. H., 1966, The Owen fracture zone and the northern end of the Carlsberg ridge: Roy. Soc. London, Phil. Trans., ser. A., v. 259, p. 227–239.

Matthews, D. H., and Bath, J., 1967, Formation of magnetic anomaly pattern of mid-Atlantic ridge: Roy. Astron. Soc., Geophys. J., v. 13, p. 349–357.

Matthews, D. H., Vine, F. J., and Cann, J. R., 1965, Geology of an area of the Carlsberg ridge, Indian Ocean: Geol. Soc. Amer., Bull., v. 76, p. 675–682.

Matuyama, M., 1929, On the direction of magnetization of basalt in Japan, Tyôsen, and Manchuria: Japan Acad., Proc., v. 5, p. 203–205.

Maxwell, A. E., 1969, Recent deep sea drilling results from the South Atlantic (abs.): Amer. Geophys. Union, Trans., v. 50, p. 113.

Maxwell, A. E., ed., 1970, The sea, v. 4: New concepts of sea floor evolution: New York, Wiley-Interscience, 2 v.

Maxwell, A. E., Von Herzen, R. P., Andrews, J. E., Boyce, R. E., Milow, E. D., Hsü, K. J., Percival, S. F., and Saito, T., 1970a, Initial reports of the Deep Sea Drilling Project, v. 3: Dakar, Senegal to Rio de Janeiro, Brazil, December 1968 to January 1969: Washington, D. C., Gov. Print. Off., 806 p.

Maxwell, A. E., Von Herzen, R. P., Hsü, K. J., Andrews, J. E., Saito, T., Percival, S. F., Milow, E. D., and Boyce, R. E., 1970b, Deep-sea drilling in the South Atlantic: Science, v. 168, p. 1047–1059.

Mayhew, M., Drake, C. L., and Nafe, J. E., 1968, Marine geophysical evidence for sea-floor spreading in the Labrador Sea (abs.): Amer. Geophys. Union, Trans., v. 49, p. 202.

Meade, B. K., and Small, J. B., 1966, Current and recent movement on the San Andreas fault, *in* Bailey, E. H., ed., Geology of northern California: Calif., Div. Mines Geol., Bull. 190, p. 385–391.

Meisser, O., ed., 1962, I. Internationales Symposium über rezente Erdkrustenbewegungen, Leipzig, 1962: Berlin, Akademie-Verlag, 508 p.

Melloni, M., 1853, Sur l'aimentation des roches volcaniques: Acad. Sci. Paris, C. R., v. 37, p. 229–231.

Melson, W. G., Thompson, G., and Van Andel, T. H., 1968, Volcanism and metamorphism in the mid-Atlantic ridge, 22° N lat.: J. Geophys. Res., v. 73, p. 5925–5941.

Melson, W. G., and Van Andel, T. H., 1968, Metamorphism in the mid-Atlantic: Mar. Geol., v. 4, p. 165–186.

Menard, H. W., 1955, Deformation of the northeastern Pacific basin and the west coast of North America: Geol. Soc. Amer., Bull., v. 66, p. 1149–1198.

Menard, H. W., 1958, Development of median elevations in the ocean basins: Geol. Soc. Amer., Bull., v. 69, p. 1179–1186.

Menard, H. W., 1959, Geology of the Pacific sea floor: Experientia, v. 15, p. 205–213.

Menard, H. W., 1964, Marine geology of the Pacific: New York, McGraw-Hill, 271 p.

Menard, H. W., 1965, Sea floor relief and mantle convection: Phys. Chem. Earth, v. 6, p. 315–364.

Menard, H. W., 1966, Fracture zones and offsets of the east Pacific rise: J. Geophys. Res., v. 71, p. 682–685.

Menard, H. W., 1967a, Extension of northeastern Pacific fracture zone: Science, v. 155, p. 72–74.

Menard, H. W., 1967b, Sea-floor spreading, topography, and the second layer: Science, v. 157, p. 923–924.

Menard, H. W., 1967c, Transitional types of crust under small ocean basins: J. Geophys. Res., v. 72, p. 3061–3073.

Menard, H. W., 1969a, Elevation and subsidence of oceanic crust: Earth Planet. Sci. Lett., v. 6, p. 275–284.

Menard, H. W., 1969b, Growth of drifting volcanoes: J. Geophys. Res., v. 74, p. 4827–4837.

Menard, H. W., and Atwater, T., 1968, Changes in the direction of sea floor spreading: Nature, v. 219, p. 463–467.

Menard, H. W., and Dietz, R. S., 1951, Submarine geology of the Gulf of Alaska: Geol. Soc. Amer., Bull., v. 62, p. 1263–1285.

Menard, H. W., and Dietz, R. S., 1952, Mendocino submarine escarpment: J. Geol., v. 60, p. 266–278.

Mercanton, P. L., 1910, Physique du globe: Acad. Sci. Paris, C. R., v. 151, p. 1092–1097.

Mercanton, P. L., 1926a, Inversion de l'inclinaison magnétique terrestre aux âges géologiques: J. Geophys. Res., v. 31, p. 187–190.

Mercanton, P. L., 1926b, Magnétisme terrestre – Aimentation de basaltes groenlandais: Acad. Sci. Paris., C. R., v. 182, p. 859–860.

Meyerhoff, A. A., 1968, Arthur Holmes: Originator of spreading ocean floor hypothesis: J. Geophys. Res., v. 73, p. 6563–6565.

Miller, D. J., 1953, Late Cenozoic marine glacial sediments and marine terraces of Middleton Island, Alaska: J. Geol., v. 61, p. 17–40.

Miller, D. J., Payne, T. C., and Gryc, G., 1959, Geology of possible petroleum provinces in Alaska: U.S. Geol. Surv., Bull. 1094, 131 p.

Miller, J., Shido, F., Banno, S., and Uyeda, S., 1961, New data on the age of orogeny and metamorphism in Japan: Jap. J. Geol. Geogr., v. 32, p. 145–151.

Miller, J. A., and Brown, P. E., 1965, Potassium-argon age studies in Scotland: Geol. Mag., v. 102, p. 106–134.

Minato, M., Gorai, M., and Hunahashi, M., 1965, the geologic development of the Japanese Islands: Tokyo, Tsukiji Shokan, 442 p.

Minear, J. W., and Toksöz, M. N., 1970, Thermal regime of a downgoing slab and new global tectonics: J. Geophys. Res., v. 75, p. 1397–1419.

Minikami, T., 1960, Fundamental research for predicting volcanic eruptions, 1, Earthquakes and crustal deformations originating from volcanic activities: Tokyo, Univ., Earthquake Res. Inst., Bull., v. 38, p. 497–544.

Misharina, L. A., 1964, On the stresses in the seismic foci of the Atlantic Ocean: Acad. Sci. USSR, Bull., Geophys. Ser., no. 10, p. 924–928.

Mitchell-Thome, R. C., 1970, Geology of the South Atlantic islands: Berlin, Gebrüder Borntraeger, 367 p.

Mitronovas, W., Seeber, L., and Isacks, B., 1968, Earthquake distribution and seismic wave propagation in the upper 200 km of the Tonga island arc (abs.): Amer. Geophys. Union, Trans., v. 49, p. 293.

Mitronovas, W., Seeber, L., and Isacks, B., 1971, Earthquake distribution and seismic wave propagation in the upper 200 km of the Tonga island arc: J. Geophys. Res., v. 76, p. 7154–7180.

Miyashiro, A., 1961, Evolution of metamorphic belts: J. Petrology, v. 2, p. 277–311.

Miyashiro, A., 1967, Orogeny, regional metamorphism and magmatism in the Japanese islands: Dansk Geol. Foren., Medd., v. 17, p. 390–446.

Mohr, P. A., 1970a, The Afar triple junction and sea-floor spreading: J. Geophys. Res., v. 75, p. 7340–7352.

Mohr, P. A., 1970b, Catalogue of chemical analyses of rocks from the African Gulf of Aden and Red Sea rift systems: Smithson. Contrib. Earth Sci., no. 2, 292 p.

Mohr, P. A., 1972, Regional significance of volcanic geochemistry in the Afar triple junction, Ethiopia: Geol. Soc. Amer., Bull., v. 83, no. 1, p. 213–222.

Molnar, P., and Oliver, J., 1969, Lateral variation of attenuation in the upper mantle and discontinuities in the lithosphere: J. Geophys. Res., v. 74, p. 2648–2682.

Molnar, P., and Sykes, L. R., 1969, Tectonics of the Caribbean and middle America regions from focal mechanisms and seismicity: Geol. Soc. Amer., Bull., v. 80, p. 1639–1684.

Moore, A. W., 1966, Pyrolytic graphite: Ned. Tijdsch. Natuurk., v. 32, p. 221–232.

Moore, G., 1962, Geology of Chirikof Island, Alaska: U.S. Geol. Surv., Tech. Lett: Aleutian-1, 10 p.

Moore, J. G., 1959, The quartz diorite boundary line in the western United States: J. Geol., v. 67, p. 198–210.

Moores, E. D., 1969, Petrology and structure of the Vourinos ophiolitic complex of northern Greece: Geol. Soc. Amer., Spec. Pap. 118, 74 p.

Moores, E. D., 1970, Ultramafics and orogeny, with models of the U.S. Cordillera and the Tethys: Nature, v. 228, p. 837–842.

Morgan, B. A., ed., 1970, Mineralogy and petrology of the upper mantle: Menasha, Wisc., Mineralogical Society of America, 319 p. (Miner. Soc. Amer., Spec. Pap. 3.)

Morgan, W. J., 1965, Gravity anomalies and convection currents: J. Geophys. Res., v. 70, p. 6189–6204.

Morgan, W. J., 1968a, Pliocene reconstruction of the Juan de Fuca ridge (abs.): Amer. Geophys. Union, Trans., v. 49, p. 327.

Morgan, W. J., 1968b, Rises, trenches, great faults, and curstal blocks: J. Geophys. Res., v. 73, p. 1959–1982.

Morgan, W. J., 1971, Convection plumes in the lower mantle: Nature, v. 230, p. 42–43.

Morgan, W. J., 1972, Plate motions and deep mantle convection: Geol. Soc. Amer, Mem. 132 (in press).

Morgan, W. J., 1972, Convection plumes and plate motions: Amer. Ass. Petrol. Geol., Bull., v. 56, p. 203–213.

Morgan, W. J., and Smith, M. L., 1969, Numerical solution for viscous flow in the upper mantle (abs.): Amer. Geophys. Union, Trans., v. 50, p. 316.

Morgan, W. J., Vogt, P. R., and Falls, D. F., 1969, Magnetic anomalies and sea-floor spreading on the Chile rise: Nature, v. 222, p. 137–142.

Morrison, R. B., and Wright, H. E., Jr., eds., 1968, Means of correlation of Quaternary successions: Int. Ass. Quatern. Res., 7th Congr. (1965), Proc., v. 8: Salt Lake City, Utah Univ. Press, 631 p.

Muir, I. D., 1965, Basalt types from the floor of the Atlantic Ocean: Geol. Soc. London, Proc., no. 1626, p. 141–142.

Muir, I. D., and Tilley, C. E., 1964, Basalts from the northern part of the rift zone of the mid-Atlantic ridge: J. Petrology, v. 5, p. 409–434.

Murauchi, S., 1966, On the origin of the Japan Sea: read at monthly meeting of Earthquake Research Institute, University of Tokyo (not verified).

Murphy, L. M., 1966, Worldwide seismic network, in ESSA Symposium on earthquake prediction; proceedings: Washington, D.C., Gov. Print. Off., p. 53–56.

Nagata, T., 1969, Length of geomagnetic polarity intervals (discussion of papers by A. Cox, 1968, 1969): J. Geomag. Geoelec., v. 21, p. 701–704.

Nagata, T., Akimoto, S., and Uyeda, S., 1951, Reverse thermo-remanent magnetism: Japan Acad., Proc., v. 27, p. 643–645.

Nagata, T., Akimoto, S., Uyeda, S., Momose, K., and Asami, E., 1954, Reverse magnetization of rocks and its connection with the geomagnetic field: J. Geomag. Geoelec., v. 6, p. 182–193.

Naidu, P. S., 1971, Statistical structure of geomagnetic field reversals: J. Geophys. Res., v. 76, p. 2649–2662.

Nairn, A. E. M., ed., 1964, Problems in palaeoclimatology: London, Interscience, 705 p.

Nakagawa, H., Niitsuma, N., and Hayasaka, I., 1969, Late Cenozoic geomagnetic chronology of the Boso Peninsula: Geol. Soc. Japan, J., v. 75, p. 267–280.

Nakamura, S., and Kikuchi, S., 1912, Permanent magnetism of volcanic bombs: Physico-Math. Soc. Japan, Proc., v. 6, p. 268–273.

Nakano, H., 1923, Notes on the nature of the forces which give rise to the earthquake motions: Japan, Central Meteorol. Observ., Seismol. Bull., v. 1, p. 92–120.

Néel, L., 1951, L'inversion de l'aimentation permanente des roches: Ann. Geophys., v. 7, p. 90–102.

Néel, L., 1955, Some theoretical aspects of rock-magnetism: Phil. Mag., Suppl., v. 4, p. 191–243.

Neuman Van Padang, M., 1951, Indonesia: Int. Volcanol. Ass., Catalogue of the active volcanoes of the world, pt. 1, 271 p.

Neumann, F., 1959, Crustal structure in the Puget Sound area: IUGG Ass. Seismol., Ser. A., Trav. Sci., v. 20, p. 153–167.

Nichols, G. D., 1965, Basalts from the deep ocean floor: Mineral. Mag., v. 34, p. 371–388.

Niggli, P., 1920–1926, Lehrbuch der Mineralogie: Berlin, Gebrüder Borntraeger, 2 v.

Ninkovich, D., Opdyke, N. D., Heezen, B. C., and Foster, J. H., 1966, Paleomagnetic stratigraphy, rates of deposition and tephrachronology in North Pacific deep-sea sediments: Earth Planet. Sci. Lett., v. 1, p. 476–492.

Noble, J. A., 1943, High-potash dikes in the Homestake mine, Lead, South Dakota: Geol. Soc. Amer., Bull., v. 59, p. 927–940.

Normark, W. R., Allison, E. C., and Curray, J. R., 1969, The geology of the Pacific continental margin of the peninsula of Baja California, Mexico: Geol. Soc. Amer., Abstr., v. 1, pt. 7, p. 163.

Nozawa, T., 1968, Radiometric ages of granitic rocks in outer zone of southwest Japan and its extension 1968 summary and north-shift hypothesis of igneous activity: Geol. Soc. Japan, J., v. 74, p. 485–489.

Oakeshott, G. B., 1966, San Andreas fault in the California Coast Ranges province, in Bailey, E. H., ed., Geology of northern California: Calif., Div. Mines Geol., Bull. 190, p. 357–373.

Obrochta, R. F., 1966, Geomagnetic measurements in the North Pacific Ocean aboard U.S.S. Rehoboth (AGS-50), 1961: U.S. Naval Hydrogr., Off., Informal Rep. H-3-66.

O'Connell, R. J., 1971, Pleistocene glaciation and the viscosity of the lower mantle: Roy, Astron. Soc., Geophys. J., v. 23, p. 299–327.

Odishaw, H., ed., 1964, Research in geophysics, v. 2: Solid earth and interface phenomena: Cambridge, Mass., M.I.T. Press, 595 p.

Officer, C. B., Ewing, J., Hennion, J. F., Harkrider, D. G., and Miller, D. E., 1959, Geophysical investigations in the eastern Caribbean: Summary of 1955 and 1956 cruises: Phys. Chem. Earth, v. 3, p. 17–109.

O'Hara, M. J., 1968, The bearing of phase equilibria studies in synthetic and natural systems on the origin and evolution of basic and ultrabasic rocks: Earth Sci. Rev., v. 4, p. 69–133.

O'Hara, M. J., and Yoder, H. S., 1967, Formation and fractionation of basic magmas at high pressures: Scot. J. Geol., v. 3, p. 67–117.

Oliver, J., and Isacks, B., 1967, Deep earthquake zones, anomalous structures in the upper mantle, and the lithosphere: J. Geophys. Res., v. 72, p. 4259–4275.

Oliver, J., and Isacks, B., 1968, Structure and mobility of the crust and mantle in the vicinity of island arcs: Can. J. Earth Sci., v. 5, p. 985–991.

Ono, C., and Isomi, H., 1967, Comparison of areas covered by different rocks in the Japanese islands: Japan, Geol. Surv., Bull., v. 18, p. 467–476.

Opdyke, N. D., 1972, Paleomagnetism of deep-sea cores: Rev. Geophys. Space Phys., v. 10, p. 213–249.

Opdyke, N. D., and Foster, J. H., 1970, The Paleomagnetism of cores from the North Pacific, in Hays, J. D., ed., Geological investigations of the North Pacific: Geol. Soc. Amer., Mem. 126, p. 83–120.

Opdyke, N. D., and Glass, B. P., 1969, The paleomagnetism of sediment cores from the Indian Ocean: Deep-Sea Res., v. 16, p. 249–261.

Opdyke, N. D., Glass, B. P., Hays, J. D., and Foster, J. H., 1966, Paleomagnetic study of Antarctic deep-sea cores: Science, v. 154, p. 349–357.

Opdyke, N. D., Ninkovich, D., Lowrie, W., and Hays, J. D., 1972, The paleomagnetism of two Aegean deep-sea cores: Earth Planet. Sci. Lett., v. 14, p. 145–159.

Opdyke, N. D., and Runcorn, S. K., 1956, New evidence for reversal of the geomagnetic field near the Pliocene-Pleistocene boundary: Science, v. 123, p. 1126–1127.

Operation Deep Freeze, 1962, Operation Deep Freeze 1961 – Marine geophysical investigation: U.S. Naval Oceanogr. Off., Tech. Rep. TR105, 231 p.

Orowan, E., 1966, Age of the ocean floor: Science, v. 154, p. 413–416.

Ostenso, N. A., 1965, Aeromagnetic evidence for a trans-Arctic ocean extension of the mid-Atlantic ridge (abs.): Amer. Geophys. Union, Trans., v. 46, p. 107.

Ostenso, N. A., and Wold, R. J., 1967, Aeromagnetic survey of the Arctic basin (abs.): Int. Ass. Geomag. Aeron., Bull., v. 24, p. 67.

Oversby, V. M., and Gast, P. W., 1970, Isotopic composition of lead from oceanic islands: J. Geophys. Res., v. 75, p. 2097–2114.

Oxburgh, E. R., and Turcotte, D. L., 1968, Problem of high heat flow and volcanism associated with zones of descending mantle convective flow: Nature, v. 218, p. 1041–1043.

Oxburgh, E. R., and Turcotte, D. L., 1969, Increased estimate for heat flow at oceanic ridges: Nature, v. 223, p. 1354–1355.

Oxburgh, E. R., and Turcotte, D. L., 1970, Thermal structure of island arcs: Geol. Soc. Amer., Bull., v. 81, p. 1665–1688.

Page, B. M., 1972, Oceanic crust and mantle fragment in subduction complex near San Luis Obispo, California: Geol. Soc. Amer., Bull., v. 83, p. 957–972.

Page, R., ed., 1966, Proceedings of the Second United States–Japan Conference on Research Related to Earthquake Prediction Problems: National Science Foundation–Japan Society for Promotion of Science, 106 p.

Pakiser, L. C., and Steinhart, J. S., 1964, Explosion seismology in the western hemisphere, in Odishaw, H., ed., Research in geophysics, v. 2: Cambridge, Mass., M. I. T. Press, p. 123–147.

Pannekoek, A. J., 1969, Uplift and subsidence in and around the western Mediterranean since the Oligocene: A review: Ned. Geol. Mijnbouwk. Genoot., Verh., v. 26, p. 57–77.

Parker, E. N., 1969, The occasional reversal of the geomagnetic field: Astrophys. J., v. 158, p. 815–827.

Parkin, E. J., 1966, Alaskan surveys to determine crustal movement, Part 2, Horizontal displacement: Washington, D. C., U. S. Dept. Commerce, 15 p.

Pecherski, D. M., 1970, Paleomagnetism and paleomagnetic correlation of Mesozoic formations of north-east USSR (in Russian): Akad. Nauk SSSR, Sib. Otd., Sev.-Vost. Kompleks. Nauch.-Issled. Inst., Tr., v. 37, p. 58–114.

Pekeris, C. L., 1935, Thermal convection in the interior of the earth: Roy. Astron. Soc., Month. Not., Geophys. Suppl., v. 3, p. 346–367.

Pekeris, C. L., 1966, The internal constitution of the earth: Roy. Astron. Soc., Geophys. J., v. 11, p. 85–132.

Peter, G., 1966, Magnetic anomalies and fracture pattern in the northeast Pacific Ocean: J. Geophys. Res., v. 71, p. 5365–5374.

Peter, G., Erickson, B. H., and Grim, P. J., 1970, Magnetic structure of the Aleutian and northeast Pacific basin, in Maxwell, A. E., ed., The sea, v. 4, pt. 2: New York, Wiley-Interscience, p. 191–222.

Peter, G., and Stewart, H. B., 1965, Ocean surveys: The systematic approach: Nature, v. 206, p. 1017–1018.

Peterson, M. N. A., 1966, Calcite: Rates of dissolution in a vertical profile in the central Pacific: Science, v. 154, p. 1542–1544.

Peterson, M. N. A., Edgar, N. T., Cita, M., Gartner, S., Goll, R., Nigrini, C., and Broch, C. von der, 1970, Initial reports of the Deep Sea Drilling Project, v. 2: Hoboken, N. J., to Dakar, Senegal, October–November 1968: Washington, D. C., Gov. Print. Off., 501 p.

Petree, B., and Lamperti, P., 1966, A comparison of absorbed dose determinations in graphite by cavity ionization measurements and by calorimetry: U. S. Nat. Bur. Stan., J. Res., C, v. 71C, p. 19–27.

Pevzner, M. A., 1970, Paleomagnetic studies of Pliocene-Quaternary deposits of Pridniestrovie: Palaeogeogr. Paleoclimatol. Palaeoecol., v. 8, p. 215–219.

Phillips, J. D., 1967, Magnetic anomalies over the mid-Atlantic ridge near 27° N: Science, v. 157, p. 920–922.

Phillips, J. D., Berggren, W. A., Bertels, A., and Wall, D., 1968, Paleomagnetic stratigraphy and micropaleontology of three deep sea cores from the central North Atlantic Ocean: Earth Planet. Sci. Lett., v. 4, p. 118–130.

Phillips, J. D., and Forsyth, D., 1972, Plate tectonics, paleomagnetism, and the opening of the Atlantic: Geol. Soc. Amer., Bull., v. 83, p. 1579–1600.

Phillips, J. D., and Luyendyk, B. P., 1970, Central North Atlantic plate motions over the last 40 million years: Science, v. 170, p. 727–729.

Phillips, J. D., Thompson, G., Von Herzen, R. P., and Bowen, W. T., 1969, Mid-Atlantic ridge near 43° N: J. Geophys. Res., v. 74, p. 3069–3081.

Phinney, R. A., ed., 1968, The history of the earth's crust – a symposium: Princeton, N. J., Princeton Univ. Press, 244 p.

Picard, L., 1966, Thoughts on the graben system in the Levant, in Irvine, T. N., ed., The world rift system: Canada, Geol. Surv., Pap. 66–14, p. 22–32.

Picard, M. D., 1964, Paleomagnetic correlation of units within Chugwater (Triassic) formation, west-central Wyoming: Amer. Ass. Petrol. Geol., Bull., v. 48, p. 269–291.

Pitman, W. C., III, 1967, Magnetic anomalies in the Pacific and ocean floor spreading: Ph.D thesis, Columbia University.

Pitman, W. C., III, and Hayes, D. E., 1968, Sea-floor spreading in the Gulf of Alaska: J. Geophys. Res., v. 73, p. 6571–6580.

Pitman, W. C., III, and Heirtzler, J. R., 1966, Magnetic anomalies over the Pacific-Antarctic ridge: Science, v. 154, p. 1164–1171.

Pitman, W. C., III., Herron, E. M., and Heirtzler, J. R., 1968, Magnetic anomalies in the Pacific and sea floor spreading: J. Geophys. Res., v. 73, p. 2069–2085.

Pitman, W. C., III, and Talwani, M., 1972, Sea floor spreading in the North Atlantic: Geol. Soc. Amer., Bull., v. 83, p. 619–646.

Pitman, W. C., III, Talwani, M., and Heirtzler, J. R., 1971, Age of the North Atlantic from magnetic anomalies: Earth Planet. Sci. Lett., v. 11, p. 195–200.

Plafker, G., 1965, Tectonic deformation associated with the 1964 Alaska earthquake: Science, v. 148, p. 1675–1687.

Plafker, G., 1967, Possible evidence for downward-directed mantle convection beneath the eastern end of the Aleutian arc (abs.): Amer. Geophys. Union, Trans., v. 48, p. 218.

Plafker, G., 1972, Alaskan earthquake of 1964 and Chilean earthquake of 1960: implications for arc tectonics: J. Geophys. Res., v. 77, p. 901–925.

Plafker, G., and MacNeil, R. H., 1966, Stratigraphic significance of Tertiary fossils from the Orca group in the Prince William Sound region, Alaska: U. S. Geol. Surv., Prof. Pap. 550-B, p. B62–B68.

Plafker, G., and Rubin, M., 1967, Vertical tectonic displacements in south-central Alaska during and prior to the great 1964 earthquake: Osaka Univ., J. Geosci., v. 10, p. 53–66.

Poldervaart, A., ed., 1955, Crust of the earth—a symposium: Geol. Soc. Amer., Spec. Pap. 62, 762 p.

Polyak, B. G., and Smirnov, Ya. B., 1966, Heat flow in the continents: Acad. Sci. USSR, Dokl., Earth Sci. Sect., v. 168, p. 170–172.

Polyak, B. G., and Smirnov, Ya. B., 1968, Relationship between terrestrial heat flow and the tectonics of continents: Geotectonics, p. 205–213.

Poole, W. H., ed., 1966, Continental margins and island arcs: Canada, Geol. Surv., Pap. 66–15, 486 p.

Press, F., 1968, Earth models obtained by Monte Carlo inversion: J. Geophys. Res., v. 73, p. 5223–5234.

Press, F., 1969, The suboceanic mantle: Science, v. 165, p. 174–176.

Press, F., 1970, Earth models consistent with geophysical data: Phys. Earth Planet. Interiors, v. 3, p. 3–22.

Press, F., and Brace, W. F., 1966, Earthquake prediction: Science, v. 152, p. 1575–1584.

Press, F., and Jackson, D., 1965, Alaskan earthquake, 27 March 1964: Vertical extent of faulting and elastic energy release: Science, v. 147, p. 867–868.

Press, F., and Kanamori, H., 1970, How thick is the lithosphere (abs.): Amer. Geophys. Union, Trans., v. 51, p. 363.

Quesnell, A. M., 1958, The structural and geomorphic evolution of the Dead Sea rift: Geol. Soc. London, Quart. J., v. 114, p. 1–24.

Quon, S. H., and Ehlers, E. G., 1963, Rocks of the northern part of the mid-Atlantic ridge: Geol. Soc. Amer., Bull., v. 74, p. 1–8.

Raasch, G. O., ed., 1961, Geology of the Arctic; proceedings of the 1st International Symposium on Arctic Geology: Toronto, Univ. of Toronto Press, 2 v.

Raff, A. D., 1963, Magnetic anomaly over Mohole drill EM7: J. Geophys. Res., v. 68, p. 955–956.

Raff, A. D., 1966, Boundaries of an area of very long magnetic anomalies in the northeast Pacific: J. Geophys. Res., v. 71, p. 2631–2636.

Raff, A. D., 1968, Sea-floor spreading—Another rift: J. Geophys. Res., v. 73, p. 3699–3705.

Raff, A. D., and Mason, R. G., 1961, Magnetic survey off the west coast of North America, 40° N latitude to 50° N latitude: Geol. Soc. Amer., Bull., v. 72, p. 1267–1270.

Raitt, R. W., 1956, Seismic refraction studies of the Pacific Ocean basin: Geol. Soc. Amer., Bull., v. 67, p. 1623–1640.

Raitt, R. W., 1963, The crustal rocks, in Hill, M. N., ed., The sea, v. 3: New York: Wiley-Interscience, p. 85–102.

Raitt, R. W., Fisher, R. L., and Mason, R. G., 1955, Tonga trench, in Poldervaart, A., ed., Crust of the earth—A symposium: Geol. Soc. Amer., Spec. Pap. 62, p. 237–254.

Raleigh, C. B., 1967, Plastic deformation of upper mantle silicate materials: Roy. Astron. Soc., Geophys. J., v. 14, p. 45–49.

Raleigh, C. B., and Kirby, S. H., 1970, Strength of the upper mantle, in Morgan, B. A., ed., Mineralogy and petrology of the upper mantle: Menasha, Wisc., Mineralogical Society of America, p. 113–121. (Mineral. Soc. Amer., Spec. Pap. 3.)

Raleigh, C. B., and Paterson, M. S., 1965, Experimental deformation of serpentinite and its tectonic implications: J. Geophys. Res., v. 70, p. 3965–3985.

Ramsay, J. G., 1963, Stratigraphy, structure, and metamorphism in the western Alps: Geol. Ass. (London), Proc., v. 74, p. 357–391.

Rasmussen, J., and Noe-Nyggard, A., 1970, Geology of the Faeroe Islands: Dan., Geol. Unders., ser. 1, no. 25, 142 p.

Reed, F. R. C., 1949, Geology of the British Empire: London, E. Arnold, 764 p.

Reid, H. F., 1910, The mechanics of the earthquake, in The California earthquake of April 18, 1906; report of the State Earthquake Investigation Commission, v. 2: Carnegie Inst. Wash., Publ. 87, v. 2, 192 p.

Research Group for Late Mesozoic Igneous Activity of Southwest Japan, 1967, Late Mesozoic igneous activity and tectonic history in the inner zone of southwest Japan: Ass. Geol. Collaboration Japan, Monogr. 13, p. 1–50.

Reynolds, J., 1956, High sensitivity mass spectrometer for noble gas analysis: Rev. Sci. Inst., v. 27, p. 928–934.

Rice, A. R., 1971, Mechanism of dissipation in mantle convection: J. Geophys. Res., v. 76, p. 1450–1459.

Rice, A. R., 1972, Some Benard convection experiments: their relationship to viscous dissipation and possible periodicity in sea-floor spreading: J. Geophys. Res., v. 77, p. 2514–2525.

Rice, L., 1930, Peculiarities in the distribution of barnacles in communities and their probable causes: Wash. (State) Univ., Puget Sound Biol. Sta., Publ., v. 7, p. 249–257.

Richey, J. E., 1948, British regional geology, Scotland: The Tertiary volcanic district: Edinburgh, H. M. Stat. Off., 105 p.

Richter, C. F., 1958, Elementary seismology: San Francisco, W. H. Freeman, 768 p.

Ricketts, E. F., and Calvin, J., 1968, Between Pacific tides, ed. 4: Stanford, Calif., Stanford Univ. Press, 614 p.

Riedel, W. R., and Funnell, B. M., 1964, Tertiary sediment cores and microfossils from the Pacific Ocean floor: Geol. Soc. London, Quart. J., v. 120, p. 305–368.

Rigg, G. B., and Miller, R. C., 1949, Intertidal plant and animal zonation in the vicinity of Neah Bay, Washington: Calif. Acad. Sci., Proc., v. 26, p. 323–351.

Rikitake, T., 1958, Oscillations of a system of disk dynamos: Cambridge Phil. Soc., Proc., v. 54, p. 89–105.

Rikitake, T., Miyamura, S., Tsubokawa, I., Murauchi, S., Uyeda, S., Kuno, H., and Gorai, M., 1968, Geophysical and geological data in and around the Japan arc: Can. J. Earth Sci., v. 5, p. 1101–1118.

Ringwood, A. E., 1958, Constitution of the mantle, 3: Geochim. Cosmochim. Acta, v. 15, p. 195–212.

Ringwood, A. E., 1969, Composition and evolution of the upper mantle, in Hart, P. J., ed., The earth's crust and upper mantle: Amer. Geophys. Union, Geophys. Monogr. 13, p. 1–17.

Ringwood, A. E., and Green, D. H., 1966, An experimental investigation of the gabbroeclogite transformation and some geophysical implications: Tectonophysics, v. 3, p. 383–427.

Ritsema, A. R., 1964, Some reliable fault plane solutions: Pure Appl. Geophys., v. 59, p. 58–74.

Ritsema, A. R., 1965, The mechanism of some deep and intermediate earthquakes in the region of Japan: Tokyo, Univ., Earthquake Res. Inst., Bull., v. 43, p. 39–52.

Ritsema, A. R., 1966, The fault-plane solutions of earthquakes of the Hindu Kush centre: Tectonophysics, v. 3, p. 147–163.

Ritsema, A. R., 1969, Seismic data of the west Mediterranean and the problem of oceanization: Ned. Geol. Mijnbouwk. Genoot., Verh., v. 26, p. 105–120.

Rivals, P. M., 1950, Etude sur la vegetation naturelle de l'ile de la Réunion: thesis, University of Paris.

Robertson, E. C., ed., 1972, The nature of the solid earth: New York, McGraw-Hill, 677 p.

Robson, G. R., and Tomblin, J. F., 1966, West Indies: Int. Volcanol. Ass., Catalogue of the active volcanoes of the world, pt. 20, 56 p.

Roche, A., 1951, Sur les inversions de l'aimentation rémanente des roches volcaniques dans les monts d'Auvergne: Acad. Sci. Paris, C. R., v. 233, p. 1132–1134.

Roche, A., 1956, Sur la date de la dernière inversion du champ magnétique terrestre: Acad. Sci. Paris, C. R., v. 243, p. 812–814.

Romney, C., 1957, Seismic waves from the Dixie Valley-Fairview Peak earthquakes: Seismol. Soc. Amer., Bull., v. 47, p. 301–319.

Rona, P. A., Brakl, J., and Heirtzler, J. R., 1970, Magnetic anomalies in the northeast Atlantic between the Canary and Cape Verde Islands: J. Geophys. Res., v. 75, p. 7412–7420.

Ross, C. S., Foster, M. D., and Myers, A. T., 1954, Origin of dunites and olivine rich inclusions in basaltic rocks: Amer. Mineral., v. 39, p. 693–737.

Ross, D. A., and Shor, G. G., Jr., 1965, Reflection profiles across the middle America trench: J. Geophys. Res., v. 70, p. 5551–5572.

Roy, J. L., 1972, A pattern of rupture of the eastern North American-western European paleoblock: Earth Planet. Sci. Lett., v. 14, p. 103–114.

Roy, R. F., Blackwell, D. D., and Birch, F., 1968, Heat generation of plutonic rocks and continental heat flow provinces: Earth Planet. Sci. Lett., v. 5, p. 1–12.

Rubey, W. W., 1951, Geologic history of sea water: Geol. Soc. Amer., Bull., v. 62, p. 1111–1148.

Runcorn, S. K., 1956, Palaeomagnetic comparisons between Europe and North America: Geol. Ass. Canada, Proc., v. 8, p. 77–85.

Runcorn, S. K., 1959, Rock magnetism: Science, v. 129, p. 1002–1011.

Runcorn, S. K., ed., 1960–1966, Methods and techniques in geophysics: New York, Interscience, 2 v.

Runcorn, S. K., ed., 1962, Continental drift: New York, Academic Press, 338 p.

Runcorn, S. K., 1965, Changes in convection pattern in earth's mantle and continental drift, evidence for cold origin of earth, in Blackett, P. M. S., Bullard, E., and Runcorn, S. K., eds., Symposium on continental drift: Roy. Soc. London, Phil. Trans., ser. A, v. 258, p. 228–251.

Runcorn, S. K., ed., 1967, Mantles of the earth and terrestrial planets: New York, Wiley, 584 p.

Runcorn, S. K., ed., 1969, The application of modern physics to the earth and planetary interiors: New York, Wiley-Interscience, 692 p.

Rusnak, G. A., Fisher, R. L., and Shepard, F. P., 1964, Bathymetry and faults of Gulf of California, in Van Andel, T. H., and Shor, G. G., Jr., eds., Marine geology of the Gulf of California: Tulsa, American Association of Petroleum Geologists, p. 59–75. (Amer. Ass. Petrol. Geol., Mem. 3.)

Rutten, M. G., 1959, Paleomagnetic reconnaissance of mid-Italian volcanos: Geol. Mijnbouw, v. 38, p. 373–374.

Rutten, M. G., 1960, Paleomagnetic dating of younger volcanic series: Geol. Rundschau, Sonderdr., v. 49, p. 161–167.

Rutten, M. G., and Den Boer, J. C., 1954, Inversion de l'aimantation des les basaltes du Coiron (Ardeche): Soc. Geol. France, C. R. S., no. 5, p. 106–108.

Rutten, M. G., and Wensink, H., 1960, Paleomagnetic dating, glaciation and the chronology of the Plio-Pleistocene in Iceland: Int. Geol. Congr., 21st., Proc., pt. 4, p. 62–70.

Sagan, C., 1965, Is the early evolution of life related to the development of the earth's core?: Nature, v. 206, p. 448.

Saggerson, E. P., 1963, Geology of the Simba-Kibwezi area: Kenya, Geol. Surv., Rep., no. 58, 70 p.

St. Amand, P., 1957, Geological and geophysical synthesis of the tectonics of portions of British Columbia, the Yukon Territory, and Alaska: Geol. Soc. Amer., Bull., v. 68, p. 1343–1370.

St. Amand, P., 1961, Los terremotos de Mayo, Chile 1960: China Lake, Calif., Michelson Lab., U. S. Naval Ordinance Test Stations NOTS TP2701.

Saito, R., 1940, Geological map of Manchuria and adjacent areas, 1:3,000,000: Manchuria, Geol. Surv., Rep. 100, Appendix.

Saito, T., Ewing, M., and Burckle, L. H., 1966, Tertiary sediment from the mid-Atlantic ridge: Science, v. 151, p. 1075–1079.

Saltzman, B., ed., 1962, Selected papers on the theory of thermal convection: New York, Dover, 461 p.

Sapper, K., 1927, Vulkankunde: Stuttgart, Engelhorn, 358 p.

Sato, T., and Ono, K., 1964, The submarine geology off San'in district, southern Japan Sea: Geol. Soc. Japan, J., v. 70, p. 434–445.

Savage, J. C., 1965, The effect of rupture velocity upon seismic first motions: Seismol. Soc. Amer., Bull, v. 55, p. 263–275.

Savage, J. C., and Hastie, L. M., 1966, Surface deformation associated with dip-slip faulting: J. Geophys. Res., v. 71, p. 4897–4904.

Scheidegger, A. E., 1966, Tectonics of the Arctic seismic belt in the light of fault-plane solutions of earthquakes: Seismol. Soc. Amer., Bull., v. 56, p. 241–245.

Schilling, J. G., 1971, Sea floor evolution: Rare-earth evidence: Roy. Soc. London, Phil. Trans., ser. A, v. 268, p. 663–706.

Schneider, E. D., and Vogt, P. R., 1968, Discontinuities in the history of sea-floor spreading: Nature, v. 217, p. 1212–1222.

Scholl, D. W., Buffington, E. C., and Hopkins, D. M., 1968a, Geologic history of the continental margin of North America in the Bering Sea: Mar. Geol., v. 6, p. 297–330.

Scholl, D. W., and Christensen, M. N., Von Huene, R., and Marlow, M. S., 1970, Peru-Chile trench sediments and sea-floor spreading: Geol. Soc. Amer., Bull., v. 81, p. 1339–1360.

Scholl, D. W., Von Huene, R., and Ridlon, J. B., 1968b, Spreading of the ocean floor: Undeformed sediments in the Peru-Chile trench: Science, v. 159, p. 869–871.

Scholz, C. H., Barazangi, M., and Sbar, M. L., 1971, Late Cenozoic evolution of the Great Basin, western United States, as an ensialic inter-arc basin: Geol. Soc. Amer., Bull., v. 82, p. 2979–2990.

Schubert, G., and Turcotte, D. L., 1972, One-dimensional model of shallow-mantle convection: J. Geophys. Res., v. 77, p. 945–951.

Schubert, G., Turcotte, D. L., and Oxburgh, E. R., 1970, Phase change instability in the mantle: Science, v. 169, p. 1075–1077.

Scientific Staff, Leg VI, 1969, Deep Sea Drilling Project, Leg VI: Geotimes, v. 14, p. 13–17.

Sclater, J. G., Anderson, R. N., and Bell, M. L., 1971, The elevation of ridges and the evolution of the central eastern Pacific: J. Geophys. Res., v. 76, p. 7888–7915.

Sclater, J. G., and Corry, C. E., 1967, Heat flow, Hawaiian area: J. Geophys. Res., v. 72, p. 3711–3715.

Sclater, J. G., and Francheteau, J., 1970, The implications of terrestrial heat flow observations on current tectonic and geochemical models of the crust and upper mantle of the earth: Roy. Astron. Soc., Geophys. J., v. 20, p. 509–537.

Sclater, J. G., and Harrison, C. G. A., 1971, Elevation of mid-ocean ridges and the evolution of the south-west Indian ridge: Nature, v. 230, p. 175–177.

Sclater, J. G., Hawkins, J. W., Mammerickx, J., and Chase, C. G., 1972, Crustal extension between the Tonga and Lau ridges: Petrologic and geophysical evidence: Geol. Soc. Amer., Bull., v. 83, p. 505–518.

Sclater, J. G., Mudie, J. D., and Harrison, C. G. A., 1970, Detailed geophysical studies on the Hawaiian arch near 24° 25' N, 157° 40' W: A closely spaced suite of heat flow stations: J. Geophys. Res., v. 75, p. 333–348.

Sclater, J. G., Vacquier, V., Greenhouse, J. P., and Dixon, F. S., 1968, Melanesian subcontinent presentation and discussion of recent heat flow measurements (abs.): Amer. Geophys. Union., Trans., v. 49, p. 217.

Sears, M., ed., 1959, International Oceanographic Congress preprints: Washington, D. C., American Association for the Advancement of Science, 1022 p.

Sears, M., ed., 1961, Oceanography: Washington, D. C., American Association for the Advancement of Science, 654 p.

Shagam, R., ed., in prep., Hess memorial volume: Geol. Soc. Amer., Mem. (not verified).

Shand, S. J., 1949, Rocks of the mid-Atlantic ridge, J. Geol., v. 57, p. 89–92.

Shaw, E. W., 1963, Canadian Rockies – orientation in time and space, in Childs, O. E., and Beebe, B. W., eds., Backbone of the Americas: tectonic history from pole to pole: Tulsa, American Association of Petroleum Geologists, p. 231–242. (Amer. Ass. Petrol. Geol., Mem. 2.)

Sherby, O. D., 1962, Factors affecting the high temperature strength of polycrystalline solids: Acta Met., v. 10, p. 135–147.

Sheridan, R. E., 1969, Subsidence of continental margins: Tectonophysics, v. 7, p. 219–229.

Sheridan, R. E., Drake, C. L., Nafe, J. E., and Hennion, J., 1966, Seismic refraction study of continental margin east of Florida: Amer. Ass. Petrol. Geol., Bull., v. 50, p. 1972–1991.

Shor, G. G., Jr., 1962, Seismic refraction studies off the coast of Alaska, 1956–1957: Seismol. Soc. Amer., Bull., v. 52, p. 37–57.

Shor, G. G., Jr., 1964, Structure of the Bering Sea and the Aleutian ridge: Mar. Geol., v. 1, p. 213–219.

Shor, G. G., Jr., 1966, Continental margins and island arcs of western North America, in Poole, W. H., ed., Continental margins and island arcs: Canada, Geol. Surv., Pap. 66–15, p. 216–222.

Sigurdssen, H., 1968, Petrology of acid xenoliths from Surtsey: Geol. Mag., v. 105, p. 440–453.

Sigurgeirsson, Th., 1957, Direction of magnetization in Icelandic basalts: Phil. Mag., Suppl., v. 6, p. 240–246.

Silgado, E., 1951, The Ancash, Peru earthquake of November 10, 1946: Seismol. Soc. Amer., Bull., v. 41, p. 83–100.

Silver, E. A., 1969a, Late Cenozoic underthrusting of the continental margin off northernmost California: Science, v. 166, p. 1265–1266.

Silver, E. A., 1969b, Structure of the continental margin off northern California north of the Gorda escarpment: Ph.D thesis, University of California, San Diego, 123 p.

Silver, E. A., 1971a, Small plate tectonics in the northeastern Pacific: Geol. Soc. Amer., Bull., v. 82, p. 3491–3496.

Silver, E. A., 1971b, Tectonics of the Mendocino triple junction: Geol. Soc. Amer., Bull., v. 82, p. 2965–2978.

Silver, E. A., 1971c, Transitional tectonics and Late Cenozoic structure of the continental margin off

northernmost California: Geol. Soc. Amer., Bull., v. 82, p. 1–22.

Simmons, G., and Nur, A., 1968, Granites: Relation of properties in situ to laboratory measurements: Science, v. 162, p. 789–791.

Simpson, J. F., 1966, Evolutionary pulsations and geomagnetic polarity: Geol. Soc. Amer., Bull., v. 77, p. 197–203.

Skinner, B. J., 1966, Thermal expansion, in Clark, S. P., ed., Handbook of physical constants: Geol. Soc. Amer., Mem. 97, p. 75–96.

Sleep, N. H., 1969a, Heat flow, gravity, and sea-floor spreading: J. Geophys. Res., v. 74, p. 542–549.

Sleep, N. H., 1969b, Sensitivity of heat flow and gravity to mechanism of sea floor spreading: J. Geophys. Res., v. 74, p. 542–549.

Sleep, N. H., 1971, Thermal effects of the formation of Atlantic continental margins by continental break-up: Roy. Astron. Soc., Geophys. J., v. 24, p. 325–350.

Smirnov, Ya. B., 1968, The relationship between the thermal field and the structure and development of the earth's crust and upper mantle: Geotectonics, p. 343–352.

Smith, A. G., and Hallam, A., 1970, The fit of the southern continents: Nature, v. 225, p. 139–144.

Smith, J. D., and Foster, J. H., 1969, Geomagnetic reversal in Brunhes normal polarity epoch: Science, v. 163, p. 365–367.

Smith, P. J., 1967a, Ancient geomagnetic field intensities from igneous rocks: Earth Planet, Sci. Lett., v. 2, p. 329–330.

Smith, P. J., 1967b, The intensity of the ancient geomagnetic field: A review and analysis: Roy. Astron. Soc., Geophys. J., v. 12, p. 321–362.

Smith, P. J., 1967c, Ancient geomagnetic field intensities – I. Historic and archaeological data – Sets H1-H9: Roy. Astron. Soc., Geophys. J., v. 13, p. 417–419.

Smith, P. J., 1967d, The intensity of the Tertiary geomagnetic field: Roy. Astron. Soc., Geophys. J., v. 12, p. 239–258.

Smith, R. L., Bailey, R. A., and Ross, C. S., 1961, Structural evolution of the Valles caldera, New Mexico, and its bearing on the emplacement of ring dikes: U. S. Geol. Surv., Prof. Pap. 424-D, p. D145–D149.

Smith, S. W. J., Dee, A. A., and Mayneord, W. V., 1924, The magnetism of annealed carbon steels: Phys. Soc. London, Proc., v. 37, p. 1–14.

Spaeth, M. G., 1964, Address presented on 7 May 1964 at a special meeting of the American Society of Photogrammetry, Washington D. C. (not verified).

Spain, I. L., Ubbelohde, A. R., and Young, D. A., 1967, Electronic properties of well oriented graphite: Roy. Soc. London, Phil. Trans., ser. A, v. 262, p. 345–386.

Sproll, W. P., and Dietz, R. S., 1969, Morphological continental drift fit of Australia and Antarctica: Nature, v. 222, p. 345–348.

Stacey, F. D., 1967, Convecting mantle as a thermodynamic engine: Nature, v. 214, p. 476–477.

Stanley, D. J., 1969, Atlantic continental shelf and slope of the United States – Color of marine sediments: U. S. Geol. Surv., Prof. Pap. 592-D, 15 p.

Stanton, R. L., 1967, A numerical approach to the andesite problem: Ned. Akad. Wetensch., Proc., ser. B, v. 70, p. 176–216.

Staub, R., 1924, Der Bau der Alpen: Beitr. Geol. Karte Schweiz, n.s. 52, 272 p.

Stauder, W., 1960, The Alaska earthquake of July 10, 1958: Seismic studies: Seismol. Soc. Amer., Bull., v. 50, p. 293–322.

Stauder, W., 1962, S wave studies of earthquakes of the North Pacific, 1, Kamchatka: Seismol. Soc. Amer., Bull., v. 52, p. 527–550.

Stauder, W., 1964, A comparison of multiple solutions of focal mechanisms: Seismol. Soc. Amer., Bull., v. 54, p. 927–937.

Stauder, W., 1967, Earthquake mechanisms (abs.): Amer. Geophys. Union, Trans., v. 48, p. 395.

Stauder, W., 1968a, Mechanism of the Rat Island earthquake sequence of February 4, 1965, with relation to island arcs and sea-floor spreading: J. Geophys. Res., v. 73, p. 3847–3858.

Stauder, W., 1968b, Tensional character of earthquake foci beneath the Aleutian trench with relation to sea-floor spreading: J. Geophy. Res., v. 73, p. 7693–7701.

Stauder, W., and Bollinger, G. A., 1964a, The S-wave project for focal mechanism studies: earthquakes of 1962: Seismol. Soc. Amer., Bull., v. 54, p. 2198–2208.

Stauder, W., and Bollinger, G. A., 1964b, The S wave project for focal mechanism studies, earthquakes of 1962: Air Force Off. Sci. Res., Grant AF-AFOSR 62–458, Rep.

Stauder, W., and Bollinger, G. A., 1965, The S wave project for focal mechanism studies — earthquakes of 1963: Air Force Off. Sci. Res., Grant AF-AFOSR 62–458, Rep.

Stauder, W., and Bollinger, G. A., 1966a, The focal mechanism of the Alaska earthquake of March 28, 1964, and of its aftershock sequence: J. Geophys. Res., v. 71, p. 5283–5296.

Stauder, W., and Bollinger, G. A., 1966b, The S wave project for focal mechanism studies, earthquakes of 1963: Seismol. Soc. Amer., Bull., v. 56, p. 1363–1371.

Stauder, W., and Udias, A., 1963, S-wave studies of earthquakes of the North Pacific, part II: Aleutian Islands: Seismol. Soc. Amer., Bull., v. 53, p. 59–77.

Stearns, H. T., 1946, Geology of the Hawaiian Islands: Hawaii, Div. Hydrography, Bull. 8, 105 p.

Stefansson, R., 1966, Methods of focal mechanism studies with application to two Atlantic earthquakes: Tectonophysics, v. 3, p. 209–243.

Steiner, A., 1958, Petrogenetic implications of the 1954 Ngauruhoe lava and its xenoliths, N. Z. J. Geol. Geophys., v. 1, p. 325–363.

Steinmann, G., 1926, Die ophiolithischen Zonen in den Mediterranen Kettengebirgen: Int. Geol. Congr., 14th, C. R., pt. 2, p. 637–667.

Stephenson, T. A., and Stephenson, A., 1949, The universal features of zonation between tide-marks on rocky coasts: J. Ecol., v. 37, p. 289.

Steuerwald, B. A., Clark, D. L., and Andrews, J. A., 1968, Magnetic stratigraphy and faunal patterns in Arctic Ocean sediments: Earth Planet. Sci. Lett., v. 5, p. 79–85.

Stocklin, J., 1968, Structural history and tectonics of Iran: A review: Amer. Ass. Petrol. Geol., Bull., v. 52, p. 1229–1258.

Stoiber, R. E., and Carr, M. J., 1971, Lithospheric plates, Benioff zones, and volcanoes: Geol. Soc. Amer., Bull., v. 82, p. 515–522.

Stommel, H. M., 1966, The large-scale oceanic circulation, in Hurley, P. M., ed., Advances in earth science: Cambridge, Mass., M. I. T. Press, p. 175–184.

Stoneley, R., 1967, The structural development of the Gulf of Alaska sedimentary province in southern Alaska: Geol. Soc. London, Quart. J., v. 123, p. 25–57.

Strangway, D. W., McMahon, B. E., Walker, T. R., and Larson, E. E., 1971, Anomalous Pliocene paleomagnetic pole positions from Baja California: Earth Planet. Sci. Lett., v. 13, p. 161–166.

Suess, E., 1885, Das Antlitz der Erde, v. 1: Prague, F. Tempsky, 778 p.

Sugimura, A., 1960, Zonal arrangement of some geophysical and petrological features in Japan and its environs: Tokyo, Univ., Fac. Sci., J., sect. 2, v. 12, p. 133–153.

Sugimura, A., Matsuda, T., Chinzei, K., and Nakamura, K., 1963, Quantitative distribution of Late Cenozoic volcanic materials in Japan: Bull. Volcanol., v. 26, p. 125–140.

Sugimura, A., and Uyeda, S., 1967, A possible anisotropy of the upper mantle accounting for deep earthquake faulting: Tectonophysics, v. 5, p. 25–33.

Sugimura, A., and Uyeda, S., 1972, Island arcs: Japan and its environs: Amsterdam, Elsevier (in press).

Summerhayes, C. P., 1967, New Zealand region volcanism and structure: Nature, v. 215, p. 610–611.

Sumner, J. S., 1954, Consequences of a polymorphic transition at the Mohorovicic discontinuity (abs.): Amer. Geophys. Union, Trans., v. 35, p. 385.

Suppe, J., 1970a, Offset of Late Mesozoic basement terrains by the San Andreas fault system: Geol. Soc. Amer., Bull., v. 81, p. 3253–3258.

Suppe, J., 1970b, Times of high-pressure metamorphism in the Franciscan terrain of California and Baja California (abs.): Geol. Soc. Amer., Abstr., v. 2, p. 151.

Sutton, G. H., and Berg, E., 1958, Direction of faulting from first-motion studies: Seismol. Soc. Amer., Bull, v. 48, p. 117–128.

Sutton, G. H., and Berg, E., 1959, Seismological studies of the western rift valley of Africa: Amer. Geophys. Union, Trans., v. 39, p. 474–481.

Sykes, L. R., 1963, Seismicity of the South Pacific Ocean: J. Geophys. Res., v. 68, p. 5999–6006.

Sykes, L. R., 1965, The seismicity of the Artic: Seismol. Soc. Amer., Bull., v. 55, p. 501–518.

Sykes, L. R., 1966, The seismicity and deep structure of island arcs: J. Geophys. Res., v. 71, p. 2981–3006.

Sykes, L. R., 1967a, Mechanism of earthquakes and nature of faulting on the mid-oceanic ridges: J. Geophys. Res., v. 72, p. 2131–2153.

Sykes, L. R., 1967b, Seismicity and earth structure in Central America and Mexico (in preparation) (not verified).

Sykes, L. R., 1968, Seismological evidence for transform faults, sea-floor spreading, and continental drift, in Phinney, R. A., ed., The history of the earth's crust: Princeton, N. J., Princeton Univ. Press, p. 120–150.

Sykes, L. R., 1969, Seismicity of the mid-ocean ridge system, in Hart, P. J., ed., The earth's crust and upper mantle: Amer. Geophys. Union, Geophys. Monogr. 13, p. 148–153.

Sykes, L. R., 1970a, Earthquake swarms and sea-floor spreading: J. Geophys. Res., v. 75, p. 6598–6611.

Sykes, L. R., 1970b, A possible nascent island arc in the Indian Ocean (abs.): Amer. Geophys. Union, Trans., v. 51, p. 356–357.

Sykes, L. R., 1970c, Seismicity of the Indian Ocean and a possible nascent island arc between Ceylon and Australia: J. Geophys. Res., v. 75, p. 5041–5055.

Sykes, L. R., 1971, Afrershock zones of great earthquakes, seismicity gaps, and earthquake prediction for Alaska and the Aleutians: J. Geophys. Res., v. 76, p. 8021–8041.

Sykes, L. R., and Ewing, M., 1965, The seismicity of the Caribbean region: J. Geophys. Res., v. 70, p. 5065–5074.

Sykes, L. R., Isacks, B., and Oliver, J., 1969, Spatial distribution of deep and shallow earthquakes of small magnitudes in the Fiji-Tonga region: Seismol. Soc. Amer., Bull., v. 59, p. 1093–1113.

Sykes, L. R., and Landisman, M., 1964, The seismicity of East Africa, the Gulf of Aden, and the Arabian and Red Seas: Seismol. Soc. Amer., Bull., v. 54, p. 1927–1940.

Takai, F., Matsumoto, T., and Toriyama, R., eds., 1963, Geology of Japan: Tokyo, Univ. of Tokyo Press, 279 p.

Takeuchi, H., and Uyeda, S., 1965, A possibility of present-day regional metamorphism: Tectonophysics, v. 2, p.

59–68.

Talwani, M., 1961, A review of marine geophysics: Mar. Geol., v. 2, p. 29–80.

Talwani, M., and Hayes, D. E., 1967, Continuous gravity profiles over island arcs and deep sea trenches (abs.): Amer. Geophys. Union, Trans., v. 48, p. 217.

Talwani, M., Heezen, B. C., and Worzel, J. L., 1961a, Gravity anomalies, physiography and crustal structure of the mid-Atlantic ridge: Bur. Central Séismol. Int. Publ., sér. A, Trav. Sci., no. 2, p. 81–111.

Talwani, M., Le Pichon, X., and Ewing, M., 1965a, Crustal structure of the mid-oceanic ridges, 2, Computed model from gravity and seismic refraction data: J. Geophys. Res., v. 70, p. 341–352.

Talwani, M., Le Pichon, X., and Heirtzler, J. R., 1965b, East Pacific rise: the magnetic pattern and the fracture zones: Science, v. 150, p. 1109–1115.

Talwani, M., Sutton, G. H., and Worzel, J. L., 1959, A crustal section across the Puerto Rico trench: J. Geophys. Res., v. 64, p. 1545–1555.

Talwani, M., Windisch, C. C., and Langseth, M. G., 1971, Reykjanes ridge crest: a detailed geophysical survey: J. Geophys. Res., v. 76, p. 473–517.

Talwani, M., Worzel, J. L., and Ewing, M., 1961b, Gravity anomalies and crustal section across the Tonga trench: J. Geophys. Res., v. 66, p. 1265–1278.

Tams, E., 1927, Die seismischen Verhältnisse des offenen Atlantischen Oceans: Z. Geophy., v. 3, p. 361–363.

Tarling, D. H., 1962, Tentative correlation of Samoan and Hawaiian Islands using "reversals" of magnetization: Nature, v. 196, p. 882–883.

Tarling, D. H., 1965, The palaeomagnetism of some of the Hawaiian Islands: Roy. Astron. Soc., Geophys. J., v. 10, p. 93–104.

Tarr, R. S., and Martin, L., 1912, The earthquakes at Yakutat Bay, Alaska, in September, 1899: U. S. Geol. Surv., Prof. Pap. 69, 135 p.

Tateiwa, I., 1925, Geological atlas of Chosen, no. 4: Kyokudo-Meisen-Shichihosan and Kotendo sheets: Seoul, Geological Survey of Korea.

Tatel, H. E., and Tuve, M. A., 1956, The earth's crust: Carnegie Inst. Wash., Yearbook, 1955–1956, v. 55, p. 81–85.

Teng, T., 1968, Attenuation of body waves and the Q structure of the mantle: J. Geophys. Res., v. 73, p. 2195–2208.

Terada, T., 1934, On bathymetrical features of the Japan Sea: Tokyo, Univ., Earthquake Res. Inst., Bull., v. 12, p. 650–656.

Thellier, E., and Rimbert, F., 1954, Sur l'analyse d'aimentation fossiles par action de champs magnétiques alternatifs: Acad. Sci. Paris, C. R., v. 239, p. 1399–1401.

Thellier, E., and Thellier, O., 1959, Sur l'intensité du champ magnétique terrestre dans le passé historique et géologique: Ann. Géophys., v. 15, p. 285–378.

Thompson, G. A., and Talwani, M., 1964, Crustal structure from Pacific basin to central Nevada: J. Geophys. Res., v. 69, p. 4813–4837.

Thorarinsson, S., 1967, Some problems of volcanism in Iceland: Geol. Rundschau, v. 57, p. 1–20.

Thorarinsson, S., Einarsson, T., and Kjartansson, G., 1959–1960, On the geology and geomorphology of Iceland: Geog. Ann., v. 41, p. 135–169; Int. Geogr. Congr., 19th, Excursion Guidebook E.I.1, 1960.

Tilley, C. E., 1947, The dunite-mylonites of St. Paul's Rocks (Atlantic): Amer. J. Sci., v. 245, p. 483–491.

Tobin, D. G., and Sykes, L. R., 1966, Relationship of hypocenters of earthquakes to the geology of Alaska: J. Geophys. Res., v. 71, p. 1659–1667.

Tobin, D. G., and Sykes, L. R., 1968, Seismicity and tectonics of the northeast Pacific Ocean: J. Geophys. Res., v. 73, p. 3821–3845.

Tocher, D., 1960, The Alaska earthquake of July 10, 1958: Movement on the Fairweather fault and field investigation of southern epicentral region: Seismol. Soc. Amer., Bull., v. 50, p. 267–292.

Toksöz, M. N., and Anderson, D. L., 1965, Velocities of long-period surface waves and structure of the upper mantle (abs.): Amer. Geophys. Union, Trans., v. 46, p. 157.

Toksöz, M. N., and Anderson, D. L., 1966, Phase velocities of long-period surface waves and structure of the upper mantle, 1, Great-circle Love and Rayleigh wave data: J. Geophys. Res., v. 71, p. 1649–1658.

Toksöz, M. N., Minear, J. W., and Julian, B. R., 1971, Temperature field and geophysical effects of a downgoing slab: J. Geophys. Res., v. 76, p. 1113–1138.

Tokuda, S., 1926, On the echelon structure of the Japanese archipelagoes: Jap. J. Geol. Geogr., v. 5, p. 41–76.

Tolstoy, I., 1951, Submarine topography in the North Atlantic: Geol. Soc. Amer., Bull., v. 62, p. 441–450.

Tolstoy, I., and Ewing, M., 1949, North Atlantic hydrography and the mid-Atlantic ridge: Geol. Soc. Amer., Bull., v. 60, p. 1527–1540.

Townes, C. H., ed., 1965, Symposium [on] earth sciences: Cambridge, Mass., M. I. T. Press (not verified).

Tozer, D. C., 1967, Towards a theory of thermal convection in the earth's mantle, in Gaskell, T. F., ed., The earth's mantle: New York, Academic Press, p. 325–353.

Trümpy, R., 1960, Paleotectonic evolution of the central and western Alps: Geol. Soc. Amer., Bull., v. 71, p. 843–908.

Turcotte, D. L., and Oxburgh, E. R., 1967, Finite amplitude convection cells and continental drift: J. Fluid Mech., v. 28, p. 29–42.

Turcotte, D. L., and Oxburgh, E. R., 1968, A fluid theory for the deep structure of dip-slip fault zones: Phys. Earth Planet. Interiors, v. 1, p. 381–386.

Turcotte, D. L., and Oxburgh, E. R., 1969, Convection in a mantle with variable physical properties: J. Geophys. Res., v. 74, p. 1458–1474.

Turcotte, D. L., and Schubert, G., 1971, Structure of the olivine-spinel phase boundary in the descending lithosphere: J. Geophys. Res., v. 76, p. 7980–7987.

Turner, D. L., Curtis, G. H. Berry F. A. F., and Jack, R. N., 1970, Age relationship between the Pinnacles and Parkfield felsites and felsite clasts in the southern Temblor range, California; implications for San Andreas fault displacement: Geol. Soc. Amer., Abstr., v. 2, p. 154–155.

Turner, F. J., 1968, Metamorphic petrology: New York, McGraw-Hill, 403 p.

Tuve, M. A., 1951, [Report of] Department of Terrestrial Magnetism: Carnegie Inst. Wash., Yearbook, 1950–1951, v. 50, p. 65–93.

Twenhofel, W. S., 1952, Recent shore-line changes along the Pacific coast of Alaska: Amer. J. Sci., v. 250, p. 523–548.

Udias, A., and Stauder, W., 1964, Application of numerical method for S-wave focal mechanism determinations to earthquakes of Kamchatka-Kurile Islands region: Seismol. Soc. Amer., Bull., v. 54, p. 2049–2065.

Udintsev, G. B., 1964, Bathymetry of the Pacific Ocean: Moscow, Akademiia Nauk, Institut Okeanologii (not verified).

Uffen, R. J., 1963, Influence of the earth's core on the origin and evolution of life: Nature, v. 198, p. 143–144.

Umbgrove, J. H. F., 1947, The pulse of the earth: The Hague, Martinus Nijhoff, 358 p.

U.S.S.R., Ministry of Geology, 1965, Geological map of the U.S.S.R., 1:2,500,000: Moscow, Ministry of Geology.

U.S. Coast and Geodetic Survey, 1935–1965, United States earthquakes (annual publication): Rockville, Md., Environmental Science Services Administration.

U.S. Coast and Geodetic Survey, 1964, Tidal datum plane changes: U. S. Coast and Geod. Surv., Oceanogr. Prelim. Rep. (Sept. 1964) (not verified).

U.S. Coast and Geodetic Survey, 1966, The Parkfield, California earthquake of June 27, 1966: Washington, D. C., Gov. Print. Off., 65 p.

U.S. Naval Oceanographic Office, 1965, Microfilm reel of Project Magnet total intensity data, reel 6 (not verified).

Uotila, U. A., 1962, Gravity anomalies for a model earth: Ohio State Univ., Dept. Geodetic Sci., Tech. Rep. 37, 15 p.

Upton, B. G. J., and Wadsworth, W. J., 1965, Geology of Réunion Island, Indian Ocean: Nature, v. 207, p. 151–154.

Urey, H. C., 1953, Comments on planetary convection as applied to the earth: Phil. Mag., ser. 7, v. 44, p. 227–230.

Urey, H. C., 1957, Boundary conditions for theories of the origin of the solar system: Phys. Chem. Earth, v. 2, p. 46–76.

Ustiyev, Ye. K., 1963, Problems of volcanism and plutonism: Volcano-plutonic formations: Int. Geol. Rev., v. 7, p. 1994–2016.

Utsu, T., 1967, Anomalies in seismic wave velocity and attenuation associated with a deep earthquake zone 1: Hokkaido Univ., Fac. Sci., J., ser. 7, Geophys., v. 3, p. 1–25.

Uyeda, S., 1958, Thermo-remanent magnetism as a medium of paleomagnetism, with special reference to reverse thermo-remanent magnetism: Jap. J. Geophys., v. 2, p. 1–123.

Uyeda, S., and Horai, K., 1964, Terrestrial heat flow in Japan: J. Geophys. Res., v. 69, p. 2121–2141.

Uyeda, S., and Vacquier, V., 1968, Geothermal and geomagnetic data in and around the island arc of Japan, in Knopoff, L., Drake, C. L., and Hart, P. J., eds., The crust and upper mantle of the Pacific area: Amer. Geophys. Union, Geophys. Monogr. 12, p. 349–366.

Uyeda, S., Vacquier, V., Yasui, M., Sclater, J., Sato, T., Lawson, J., Watanabe, T., Dixon, F., Silver, E., Fukao, Y., Sudo, K., Nishikawa, M., and Tanayka, T., 1967, Results of geomagnetic survey during the cruise of R/V Argo in western Pacific 1966 and the compilation of magnetic charts of the same area: Tokyo, Univ., Earthquakes Res. Inst., Bull., v. 45, p. 799–814.

Vacquier, V., 1959, Measurement of horizontal displacement along faults in the ocean floor: Nature, v. 183, p. 452–453.

Vacquier, V., 1962, Magnetic evidence for horizontal displacements in the floor of the Pacific Ocean, in Runcorn, S. K., ed., Continental drift: New York, Academic Press, p. 135–144.

Vacquier, V., 1965, Transcurrent faulting in the ocean floor, in Blackett, P. M. S., Bullard, E., and Runcorn, S. K., eds., Symposium on continental drift: Roy. Soc. London, Phil. Trans., ser. A., v. 258, p. 77–81.

Vacquier, V., Raff, A. D., and Warren, R. E., 1961, Horizontal displacements in the floor of the northeastern Pacific Ocean: Geol. Soc. Amer., Bull., v. 72, p. 1251–1258.

Vacquier, V., Sclater, J. G., and Correy, C. E., 1967, Studies of the thermal state of the earth. The 21st paper: Heat flow, eastern Pacific: Tokyo, Univ., Earthquake Res. Inst., Bull., v. 45, p. 375–393.

Vacquier, V., Uyeda, S., Yasui, M., Sclater, J., Corry, C., and Watanabe, T., 1966, Studies of the thermal state of the earth. The 19th paper: Heat-flow measurments in the northwestern Pacific: Tokyo, Univ., Earthquake Res. Inst., Bull., v. 44, p. 1519–1535.

Vacquier, V., and Von Herzen, R. P., 1964, Evidence for connection between heatflow and the mid-Atlantic ridge magnetic anomaly: J. Geophys. Res., v. 69, p. 1093–1101.

Van Andel, S. I., and Hospers, J., 1968, Palaeomagnetism and the hypothesis of an expanding earth—A new calculation method and its results: Tectonophysics, v. 5, p. 273–285.

Van Andel, T. H., and Bowin, C. P., 1968, Mid-Atlantic ridge between 22° and 23° north latitude and the tectonics of mid-ocean rises: J. Geophys. Res., v. 73, p. 1279–1298.

Van Andel, T. H., Corliss, J. B., and Bowen, V. T., 1967, The intersection between the mid-Atlantic ridge and the Vema fracture zone (abs.): Amer. Geophys. Union, Trans., v. 48, p. 133.

Van Andel, T. H., Phillips, J. D., and Von Herzen, R. P., 1969, Rifting origin for the Vema fracture in the North Atlantic: Earth Planet. Sci. Lett., v. 5, p. 296–300.

Van Andel, T. H., and Shor, G. G., Jr., eds., 1964, Marine geology of the Gulf of California: Tulsa, American Association of Petroleum Geologists, 408 p. (Amer. Ass. Petrol. Geol., Mem. 3.)

Van Bemmelen, R. W., 1964, The evolution of the Atlantic mega-undation: Tectonophysics, v. 1, p. 385–430.

Van Bemmelen, R. W., 1966, On mega-undations: A new model for the earth's evolution: Tectonophysics, v. 3, p. 83–127.

Van der Linden, W. J. M., 1969, Extinct mid-ocean ridges in the Tasman Sea and in the western Pacific: Earth Planet. Sci. Lett., v. 6, p. 483–490.

Van Dorn, W. G., 1964, Source mechanism of the tsunami of March 28, 1964 in Alaska: Ninth Conf. Coastal Eng., Amer. Soc. Civil Eng., Proc., p. 166–190.

Van Hilten, D., and Zijderveld, J. D. A., 1966, The magnetism of the Permian porphyries near Lugano (northern Italy, Switzerland): Tectonophysics, v. 3, p. 429–446.

Van Montfrans, H. M., 1971, Paleomagnetic dating in the North Sea basin: Earth Planet. Sci. Lett., v. 11, p. 226–235.

Van Montfrans, H. M., and Hospers, J., 1969, A preliminary report on the stratigraphical position of the Matuyama-Brunhes geomagnetic field reversal in the Quaternary sediments of the Netherlands: Geol. Mijnbouw, v. 48, p. 565–572.

Vening Meinesz, F. A., 1930, Maritime gravity surveys in the Netherlands East Indies, tentative interpretation of the results: Ned. Akad. Wetensch, Proc., ser. B, v. 33, p. 566–577.

Vening Meinesz, F. A., 1952, The origin of continents and oceans: Geol. Mijnbouw, v. 31, p. 373–384.

Vening Meinesz, F. A., 1954, Indonesian archipelago: A geophysical study: Geol. Sco. Amer., Bull., v. 65, p. 143–164.

Vening Meinesz, F. A., 1959, The results of development of the earth's topography in spherical harmonics up to the 31st order, provisional conclusions: Ned. Akad. Wetensch., Proc., ser. B, v. 115–136.

Vening Meinesz, F. A., 1962, Thermal convection in the earth's mantle, in Runcorn, S. K., ed., Continental drift: New York, Academic Press, p. 145–176.

Verhoogen, J., 1954, Petrological evidence on temperature distribution in the mantle of the earth: Amer. Geophys. Union, Trans., v. 35, p. 50–59.

Verhoogen, J., 1956a, Ionic ordering and self-reversal of magnetization in impure magnetites: J. Geophys. Res., v. 61, p. 201–209.

Verhoogen, J., 1956b, Temperatures within the earth: Phys. Chem. Earth, v. 1, p. 17–43.

Verhoogen, J., 1962, Oxidation of iron-titanium oxides in igneous rocks: J. Geol., v. 70, p. 168–181.

Vestine, E. H., 1953, On variations of the geomagnetic field fluid motions, and the rate of the earth's rotation: J. Geophys. Res., v. 58, p. 127–145.

Vine, F. J., 1966, Spreading of the ocean floor: new evidence: Science, v. 154, p. 1405–1415.

Vine, F. J., 1968, Magnetic anomalies associated with mid-ocean ridges, *in* Phinney, R. A., ed., The history of the earth's crust: Princeton, N. J., Princeton Univ. Press, p. 73–89.

Vine, F. J., and Hess, H. H., 1970, Sea floor spreading, *in* Maxwell, A. E., ed., The sea, v. 4, pt. 2: New York, Wiley-Interscience, p. 587–622.

Vine, F. J., and Matthews, D. H., 1963, Magnetic anomalies over oceanic ridges: Nature, v. 199, p. 947–949.

Vine, F. J., and Morgan, W. J., 1968, Simulation of mid-ocean ridge magnetic anomalies using a random injection model (abs.): Geol. Soc. Amer., Spec. Pap. 115, p. 228.

Vine, F. J., and Wilson, J. T., 1965, Magnetic anomalies over a young oceanic ridge off Vancouver Island: Science, v. 150, p. 485–489.

Vinogradov, A. P., 1961, The origin of the material of the earth's crust—communication 1: Geochemistry, no. 1, p. 1–32.

Vinogradov, A. P., Lavrukhina, A. K., and Revina, L. D., 1961, Yadernyye reaktsii v zheleznykh meteoritakh (with English abstract): Geokhim., no. 11, p. 955–966.

Vogt, P. R., 1971, Asthenosphere motion recorded by the ocean floor south of Iceland: Earth Planet. Sci. Lett., v. 13, p. 153–160.

Vogt, P. R., Anderson, C. N., and Bracey, D. R., 1971, Mesozoic magnetic anomalies, sea-floor spreading, and geomagnetic reversals in the southwestern North Atlantic: J. Geophys. Res., v. 76, p. 4796–4823.

Vogt, P. R., Anderson, C. N., Bracey, D. R., and Schneider, E. D., 1970a, North Atlantic magnetic smooth zones: J. Geophys. Res., v. 75, p. 3955–3967.

Vogt, P. R., Avery, O. E., Schneider, E. G., Anderson, C. N., and Bracey, D. R., 1969, Discontinuities in sea floor spreading: Tectonophysics, v. 8, p. 285–317.

Vogt, P. R., and Higgs, R. H., 1969, An aeromagnetic survey of the eastern Mediterranean and its interpretation: Earth Planet. Sci. Lett., v. 5, p. 439–448.

Vogt, P. R., and Johnson, G. L., 1971, Cretaceous sea floor spreading in the western North Atlantic: Nature, v. 234, p. 22–25.

Vogt, P. R., Johnson, G. L., Holcombe, T. L., Gilg, J. G., and Avery, O. E., 1972, Episodes of sea-floor spreading recorded by the North Atlantic basement: Tectonophysics, v. 12, p. 211–234.

Vogt, P. R., and Ostenso, N. A., 1970, Magnetic and gravity profiles across the Alpha Cordillera and their relation to Arctic sea floor spreading: J. Geophys. Res., v. 75, p. 4925–4937.

Vogt, P. R., Ostenso, N. A., and Johnson, G. L., 1970b, Bathymetric and magnetic data bearing on sea-floor spreading north of Iceland: J. Geophys. Res., v. 75, p. 903–920.

Von Huene, R., 1971, A possible relationship between the Transverse ranges of California and the Murray fracture (abs.): Geol. Soc. Amer., Abstr., v. 3, p. 213.

Wadati, K., Hirono, T., and Yumura, T., 1967, The absorption of transverse waves in the upper mantle under Japan (abs.): IUGG Gen. Assembly, Abstr. Pap. v. 2, p. 154.

Wadati, K., Hirono, T., and Yumura, T., 1969, On the attenuation of S-waves and the structure of the upper mantle in the region of Japanese islands: Pap. Meteorol. Geophys., v. 20, p. 49–78.

Waddington, C. J., 1967, Paleomagnetic field reversals and cosmic radiation: Science, v. 158, p. 913–915.

Wager, L. R., 1937, The Arun river drainage pattern and the rise of the Himalaya: Geogr. J., v. 89, p. 239–250.

Wager, L. R., and Deer, W. A., 1938, A dike swarm and crustal flexure in East Greenland: Geol. Mag., v. 75, p. 39–46.

Wakita, H., Nagasawa, H., Uyeda, S., and Kuno, H., 1967, Uranium and thorium contents in ultrabasic rocks:

Earth Planet. Sci. Lett., v. 2, p. 377–381.

Walker, G. P. L., 1963, The Breiddalur central volcano, eastern Iceland: Geol. Soc. London, Quart. J., v. 119, p. 29–64.

Walker, G. P. L., 1966, Acid volcanic rocks in Iceland: Bull. Volcanol., v. 29, p. 375–406.

Washington, H. S., 1906, The Roman comagmatic region: Carnegie Inst. Wash., Publ. 57, 199 p.

Washington, H. S., 1923, Comagmatic regions and the Wegener hypothesis: Wash. Acad. Sci. J., v. 13, p. 339–347.

Washington, H. S., 1924, The lavas of Barren Island and Narcondam: Amer. J. Sci., v. 207, p. 441–456.

Washington, H. S., 1930a, The origin of the mid-Atlantic ridge: Maryland Acad. Sci., J., v. 1, p. 20–29.

Washington, H. S., 1930b, The petrology of St. Paul's Rocks (Atlantic): Brit. Mus. (Nat. Hist.), Rep. Geol. Collections "Quest" Exped. 1921–22, p. 126–144.

Watkins, N. D., 1968a, Comments on the interpretation of linear magnetic anomalies: Pure Appl. Geophys., v. 69, p. 170–192.

Watkins, N. D., 1968b, Short period geomagnetic polarity events in deep-sea sedimentary cores: Earth Planet. Sci. Lett., v. 4, p. 341–349.

Watkins, N. D., 1969, Non-dipole behaviour during an Upper Miocene geomagnetic polarity transition in Oregon: Roy. Astron. Sco., Geophys. J., v. 17, p. 121–149.

Watkins, N. D., 1972, Review of the development of the geomagnetic polarity time scale and discussion of prospects for its finer definition: Geol. Soc. Amer., Bull., v. 83, p. 551–574.

Watkins, N. D., and Abdel-Monem, A., 1971, Detection of the Gilsa geomagnetic polarity event on the island of Madeira: Geol. Soc. Amer., Bull., v. 82, p. 191–198.

Watkins, N. D., and Goddell, H. G., 1967a, Confirmation of the reality of the Gilsa geomagnetic polarity event: Earth Planet. Sci. Lett., v. 2, p. 123–129.

Watkins, N. D., and Goodell, H. G., 1967b, Geomagnetic polarity changes and faunal extinction in the Southern Ocean: Science, v. 156, p. 1083–1089.

Watkins, N. D., Gunn, B. M., Baksi, A. K., York, D., and Ade-Hall, J., 1971, Paleomagnetism, geochemistry, and potassium-argon ages of the Rio Grande de Santiago volcanics, central Mexico: Geol. Soc. Amer., Bull., v. 82, p. 1955–1968.

Watkins, N. D., and Paster, T., 1971, Magnetic properties of submarine igneous rocks: Roy. Soc. London, Phil. Trans., ser. A, v. 268, p. 507–550.

Watkins, N. D., and Richardson, A., 1971, Intrusives, extrusives, and linear magnetic anomalies: Roy. Astron. Soc., Geophys. J., v. 23, p. 1–13.

Weed, W. H., and Pirsson, L. V., 1895, Highwood Mountains of Montana: Geol. Soc. Amer., Bull., v. 6, p. 389–422.

Weed, W. H., and Pirsson, L. V., 1896, The Bearpaw Mountains, Montana: Amer. J. Sci., v. 151, p. 283–301.

Weertman, J., 1970, The creep strength of the earth's mantle: Rev. Geophys. Space Phys., v. 8, p. 145–168.

Wegener, A., 1912, Die Entstehung der Kontinente: Geol. Rundschau, v. 3, p. 276–292.

Wegener, A., 1966, The origin of continents and oceans, ed. 4: New York, Dover, 246 p.

Wegmann, C. E., 1948, Geological tests of the hypothesis of continental drift in the Arctic regions, scientific planning: Med. Grønland, v. 144, no. 7, 48 p.

Weissel, J. K., and Hayes, D. E., 1971, Asymmetric spreading south of Australia: Nature, v. 231, p. 518–521.

Wellman, H. W., 1955, New Zealand Quaternary tectonics: Geol. Rundschau, v. 43, p. 248–257.

Wellman, H. W., 1971, Reference lines, fault classification, transform systems, and ocean-floor spreading: Tectonophysics, v. 12, p. 199–210.

Wells, A. J., 1969, The crush zone of the Iranian Zagros Mountains, and its implications: Geol. Mag., v. 106, p. 385–514.

Wentworth, C. M., 1968, Upper Cretaceous and Lower Tertiary strata near Gualala, California, and inferred large right slip on the San Andreas fault, *in* Dickinson, W. R., and Grantz, A., eds., Proceedings of the conference on geologic problems of San Andreas fault system: Standord Univ., Publ. Geol. Sci., v. 11, p. 130–143.

Weyl, R., 1966, Geologie der Antilles: Berlin, Gebrüder Borntraeger, 410 p.

White, D. A., Roeder, D. H., Nelson, T. H., and Crowell, J. C., 1970, Subduction: Geol. Soc. Amer., Bull., v. 81, p. 3431–3432.

Whittard, W. F., and Bradshaw, R., eds., 1965, Submarine geology and geophysics: London, Butterworth, 465 p.

Whitten, C. A., 1948, Horizontal earth movement, vicinity of San Francisco, California: Amer. Geophys. Union, Trans., v. 29, p. 318–323.

Whitten, C. A., 1955, Measurements of earth movements in California: Calif., Div. Mines Geol., Bull. 171, p. 75–80.

Whitten, C. A., 1956, Crustal movement in California and Nevada: Amer. Geophys. Union, Trans., v. 37, p. 393–398.

Williams, C. A., and McKenzie, D., 1971, The evolution of the North-East Atlantic: Nature, v. 232, p. 168–173.

Williams, L. A. J., 1970, The volcanics of the Kenya rift valley: Geol. Soc. London, Proc., no. 1663, p. 151.

Wilson, J. T., 1936, Foreshocks and aftershocks of the Nevada earthquake of December 20, 1932, and the Parkfield earthquake of June 7, 1934: Seismol. Soc. Amer., Bull., v. 26, p. 189–194.

Wilson, J. T., 1963a, Evidence from islands on the spreading of the ocean floor: Nature, v. 197, p. 536–538.

Wilson, J. T., 1963b, Hypothesis of earth's behavior: Nature, v. 198, p. 925–929.

Wilson, J. T., 1963c, A possible origin of the Hawaiian Islands: Can. J. Phys., v. 41, p. 863–870.

Wilson, J. T., 1965a, Evidence from ocean islands suggesting movement in the earth, *in* Blackett, P. M. S., Bullard, E., and Runcorn, S. K., eds., A symposium on continental drift: Roy. Soc. London, Phil. Trans., ser. A, v. 258, p. 145–167.

Wilson, J. T., 1965b, Transform faults, oceanic ridges and magnetic anomalies southwest of Vancouver Island: Science, v. 150, p. 482–485.

Wilson, J. T., 1965c, A new class of faults and their bearing on continental drift: Nature, v. 207, p. 343–347.

Wilson, J. T., 1965d, Submarine fracture zones, aseismic ridges and the International Council of Scientific Unions lines: Proposed western margin of the east Pacific ridge: Nature, v. 207, p. 907–910.

Wilson, J. T., 1966, Did the Atlantic close and then re-open?: Nature, v. 211, p. 676–681.

Wilson, R. L., 1962, The palaeomagnetism of baked contact rocks and reversals of the earth's magnetic field: Roy. Astron. Soc., Geophys. J., v. 7, p. 194–202.

Wise, D. U., 1963, An outrageous hypothesis for the tectonic pattern of the North American Cordillera: Geol. Soc. Amer., Bull, v. 74, p. 357–362.

Wiseman, J. D. H., 1965, Petrography, mineralogy, chemistry and mode of origin of St. Paul Rocks: Geol. Soc. London, Proc., no. 1626, p. 146–147.

Wiseman, J. D. H., 1966, St. Paul Rocks and the problem of the upper mantle: Roy. Astron. Soc., Geophys. J., v. 11, p. 519–525.

Wohlenberg, J., 1966, Remarks on the Uganda earthquake of March 20, 1966: Inst. Rech. Sci. Afrique Centrale, Chron., v. 1, p. 7–12.

Wollin, G., Ericson, D. B., and Ryan, W. B. F., 1971, Variations in magnetic intensity and climatic changes: Nature, v. 232, p. 549–551.

Worzel, J. L., 1965, Deep structure of coastal margins and mid-oceanic ridges: Colston Pap., v. 17, p. 335–361.

Worzel, J. L., 1966, Structure of continental margins and the development of ocean trenches, *in* Poole, W. H., ed., Continental margins and island arcs: Canada, Geol. Surv., Pap. 66-15, p. 357–375.

Wright, H. E., Jr., and Frey, D. G., eds., 1965, The Quaternary of the United States: Princeton, N. J., Princeton Univ. Press, 922 p.

Wright, J. B., 1966, Convection and continental drift in the southwest Pacific: Tectonophysics, v. 3, p. 69–81.

Wyss, M., and Brune, J. N., 1969, Shear stress associated with earthquakes between 0 and 700km depth (abs.): Amer. Geophys. Union, Trans., v. 50, p. 237.

Yamada, T., Kawachi, Y., Watanabe, T., Yokota, Y., and Kwanke, E., 1969, The Shimanto belt of the northern Akaishi mountainlands, central Japan: Geol. Soc. Japan, Mem., v. 4, p. 117–122.

Yamanari, F., 1925, Geological atlas of Chosen, no. 3: Kayoho-Kisshū-shiho and Rinmei sheets: Seoul, Geological Survey of Korea.

Yasui, M., Hashimoto, Y., and Uyeda, S., 1967, Geomagnetic studies of the Japan Sea, 1. Anomaly pattern in the Japan Sea: Oceanogr. Mag., v. 19, p. 221–231.

Yasui, M., Kishii, T., Watanabe, T., and Uyeda, S., 1968, Heat flow in the Sea of Japan, *in* Knopoff, L., Drake, C. L., and Hart, P. J., eds., The crust and mantle of the Pacific area: Amer. Geophys. Union, Geophys. Monogr. 12, p. 3–16.

Yerkes, R. F., McCulloh, T. H., Schoelhamer, J. E., and Vedder, J. G., 1965, Geology of the Los Angeles basin, California—an introduction: U. S. Geol. Surv., Prof. Pap. 420-A, 57 p.

Yoder, H. S., and Tilley, C. E., 1962, Origin of basalt magmas: an experimental study of natural and synthetic rock systems: J. Petrology, v. 3, p. 342–532.

Yoshida, K., ed., 1964, Studies in oceanography: Tokyo, Univ. of Tokyo Press, 560 p.

Zen, E., White, W. S., Hadley, J. B., and Thompson, J. B., eds., 1968, Studies of Appalachian geology: northern and maritime: New York, Wiley-Interscience, 475 p.

Indexes

Author Index

Subject Index